AF767775

2021 Feature Papers by Diversity's Editorial Board Members—Volume I

2021 Feature Papers by Diversity's Editorial Board Members—Volume I

Editor

Michael Wink

MDPI

Basel • Beijing • Wuhan • Barcelona • Belgrade • Novi Sad • Cluj • Manchester

Editor
Michael Wink
Heidelberg University
Heidelberg, Germany

Editorial Office
MDPI
St. Alban-Anlage 66
4052 Basel, Switzerland

This is a reprint of articles from the Special Issue published online in the open access journal *Diversity* (ISSN 1424-2818) (available at: https://www.mdpi.com/journal/diversity/special_issues/emb_diversity).

For citation purposes, cite each article independently as indicated on the article page online and as indicated below:

Lastname, A.A.; Lastname, B.B. Article Title. *Journal Name* **Year**, *Volume Number*, Page Range.

Volume I
ISBN 978-3-0365-9955-7 (Hbk)
ISBN 978-3-0365-9956-4 (PDF)
doi.org/10.3390/books978-3-0365-9956-4

Set
ISBN 978-3-0365-9953-3 (Hbk)
ISBN 978-3-0365-9954-0 (PDF)

Cover image courtesy of Michael Wink

© 2024 by the authors. Articles in this book are Open Access and distributed under the Creative Commons Attribution (CC BY) license. The book as a whole is distributed by MDPI under the terms and conditions of the Creative Commons Attribution-NonCommercial-NoDerivs (CC BY-NC-ND) license.

Contents

Preface

This is a Special Issue of high-quality papers in open access form by the Editorial Board Members of Diversity, or those recommended and invited by the Editorial Board Members and the Editor-in-Chief. The Special Issue comprises publications on all aspects of ecology and evolution of plants, animals, marine organisms and microbes, their biodiversity, phylogeny and evolution, and conservation.

Michael Wink
Editor

diversity

MDPI

Article

Ecology and Conservation of the Laotian langur *Trachypithecus laotum* in a Protected Area of Laos (Southeast Asia)

Johnny Souwideth [1], Phaivanh Phiapalath [2], Hai Dong Thanh [1], Peter Brakels [3], Thong Pham Van [4] and Luca Luiselli [5,6,7,*]

1 Vietnam National University of Forestry, Street 21, Xuan Mai Town, Chuong My District, Hanoi 100000, Vietnam; johnnyblack1991@gmail.com (J.S.); donghaifuv@gmail.com (H.D.T.)
2 Freelance Consultant, Vientiane, Laos; p.phiapalath@gmail.com
3 IUCN Vientiane Office, 391/24 Bourichane Road, Ban Naxay, Sayshetta District, Vientiane, Laos; peter.brakels@iucn.org
4 Save Vietnam's Wildlife, Cuc Phuong National Park, Cuc Phuong Commune, Nho Quan 430000, Vietnam; thong@svw.vn
5 Institute for Development, Ecology, Conservation and Cooperation, Via G. Tomasi di Lampedusa 33, 00144 Rome, Italy
6 Department of Applied and Environmental Biology, Rivers State University of Science and Technology, P.M.B. 5080 Port Harcourt, Nigeria
7 Département de Zoologie et Biologie Animale, Faculté des Sciences, Université de Lomé, B.P. 1515 Lomé, Togo
* Correspondence: l.luiselli@ideccngo.org

Abstract: Terrestrial species from the Lao People's Democratic Republic (PDR) are under high threat due to deforestation and overhunting. Previous studies have even defined these forests as subjected to an "empty forest syndrome", a condition in which forests that are apparently well preserved are instead almost deprived of vertebrate faunas due to extreme exploitation by local communities. Forest specialists, including several primates, are among the most threatened species in the country. The Laotian langur (*Trachypithecus laotum*) is endemic to Lao PDR, is listed as Endangered by the IUCN Red List, and it is one of the least studied species in the region. A survey on the local distribution, life history and conservation status of the Laotian langur was carried out in Phou Hin Poun National Protected Area, Khammouane Province of Lao PDR. The survey consisted of an initial phase with interviews to select key informants on the Laotian langur and the other primate species of the area. Then, a phase of field surveys along forest transects, totaling 64.1 km of 21 transects, yielded a record of 35 individuals in 9 groups. The highest encounter/detection rate of the Laotian langur was 1 group per km at one sector of the park. In contrast, it was much lower (0.18–0.34 groups/km) in the rest of the protected area. The group sizes were much lower than those observed in the same area between 1994 and 2010, thus suggesting a decline in the population size of langurs. This decline may be linked to habitat loss (timber extraction and mining). Still, also overhunting, as signs of poaching were observed during our field surveys. This was also supported by the reports of our interviewees. Laotian langurs were observed to be sympatric and interact while foraging with the Assamese macaque (*Macaca assamensis*). In the cases of sympatric occurrence between the two species, we observed that subtle mechanisms of niche partitioning may occur to reduce interspecific competition for food. Further research on the population and ecology of this endangered langur should be conducted to understand the species and aid its conservation.

Keywords: primates; Laotian langur; Assamese macaque; Phou Hin Poun National Protected Area; surveys; ecology; conservation; interspecific interactions

Citation: Souwideth, J.; Phiapalath, P.; Thanh, H.D.; Brakels, P.; Van, T.P.; Luiselli, L. Ecology and Conservation of the Laotian langur *Trachypithecus laotum* in a Protected Area of Laos (Southeast Asia). *Diversity* **2021**, *13*, 231. https://doi.org/10.3390/d13060231

Academic Editor: Michael Wink

Received: 27 April 2021
Accepted: 19 May 2021
Published: 25 May 2021

Publisher's Note: MDPI stays neutral with regard to jurisdictional claims in published maps and institutional affiliations.

Copyright: © 2021 by the authors. Licensee MDPI, Basel, Switzerland. This article is an open access article distributed under the terms and conditions of the Creative Commons Attribution (CC BY) license (https://creativecommons.org/licenses/by/4.0/).

1. Introduction

Extensive forest loss and overhunting are among the main threats affecting biodiversity worldwide, particularly in the megadiverse tropical countries [1–4]. Deforestation and logging rates have dramatically accelerated in recent years in many tropical countries, with a growing network of highways facilitating the entering of settlers, hunters/poachers and loggers into the heart of the mature forests. Therefore, the fragmentation of forests is becoming increasingly widespread with substantial changes in forest dynamics, structure, composition and microclimate [1,5–7]. These alterations also negatively affect a wide variety of animal species.

The forests of the Lao People's Democratic Republic (Laos) are threatened by rapid anthropic development, a changing economy and a growing population [8–11]. This country lost 8.4% of its primary forest between 2001 and 2018, with a primary forest/total tree cover of 48% (https://rainforests.mongabay.com/deforestation/archive/Laos.htm; last accessed on 24 April 2021). Due to unsustainable deforestation [12] and bushmeat consumption, the forests of Lao PDR are becoming increasingly empty [13–15], and many forest species are threatened [16–18]. Indeed, several studies have defined the Lao forests as affected by an "empty forest syndrome", a condition in which forests that are apparently well preserved but almost devoid of vertebrate fauna due to extreme exploitation by local communities [16–18].

Among the forest taxa targeted by hunting and that are affected by deforestation are those belonging to the primate genus *Trachypithecus*, which includes 14 Asian species of herbivorous arboreal species [19]. The Laotian langur *Trachypithecus laotum* (Figure 1) is endemic to Lao PDR [20]. This species is mainly restricted to karst forest habitats in northwestern Khammouane province and in southwestern Bolikhamxai province in central Laos [21]. Recently, *T. laotum* was categorized as Endangered by IUCN [22] and is listed on CITES Appendix II as well as under protection by the Lao wildlife and aquatic law No 07/NA dated 24 December 2007; it is categorized under the category "prohibited" on the threatened species list under the name *T. francoisi* sensu lato [23].

Given the threatened status of the Laotian langur and the scarce knowledge available on its ecology and population status, we (1) conducted several field surveys in one of the few presence areas for this primate species (Phou Hin Poun National Protected Area, Khammouane), and (2) compare our results to previous surveys that were conducted in the same territory during previous decades. Our objectives are to (1) provide information about the species distribution, ecology and conservation status at the study area, and (2) to collect data on the local knowledge of the ecology of the species as gathered from interviews with local people.

More specifically, we ask the following research questions:

(i) What is the density of the Laotian langur groups in the study area? Although scientists primarily use presence and absence data for conservation planning of species even in large landscapes, we can obtain a much better and accurate conservation planning if we include data on individual species abundance or density in the favored habitats [24].

(ii) What is the current group size of these monkeys, and has the average group size changed in the last twenty years in the study area? Although all studies on this species at Phou Hin Poun National Protected Area have been short-term, they have spread since 1994. Hence comparisons of the various collected datasets may show population size and conservation status trends. Given the heavy rates of deforestation and overhunting in Lao PDR [8–11], we predict that the group size of langur may be smaller nowadays than 20+ years ago, thus revealing an overall declining population.

(iii) Are Laotian langur groups sympatric with other primates, and if so, what are their interspecific interactions? Threatened species may suffer from interspecific competition with closely related species, representing a further threat to their conservation. In addition, Laotian forests are inhabited by a rich diversity of primate species [25]. Thus it is likely that the Laotian langur groups should share their habitat with other primate species.

Based on the answers that we obtained from the three questions mentioned above, we herein address the conservation perspectives for the target species in the study area are. It is expected that this contribution may serve to implement an action plan for *T. laotum* in Lao PDR.

Figure 1. Individuals of the Laotian langur *Trachypithecus laotum* at the study area.

2. Materials and Methods

2.1. Study Area

We studied the Laotian langur in Phou Hin Poun National Protected Area (hereby PHP-NPA). This area, situated in northwestern Khammouane province (17°26′–18°05′ N and 104°25′–105°10′ E), was formerly known as Khammouan Limestone National Biodiversity Conservation Area (Khammouan Limestone is called Phou Hin Poun in Lao) and declared in 1993. It covers a total actual area of 225,000 ha, mainly of spectacular karst formations situated at 180–850 m elevation, with pockets of tall mature forest in the valleys and depressions within the rocks [8]. The total area of the park is 150,000 ha (Decree 164). This area contains the largest habitat of the Laotian langur population in the country. The topography of depressions is flat, which contrasts with the surrounding. Depressions often contain primary and secondary mixed deciduous/semi-evergreen forest and less commonly deciduous dipterocarp forest [21]. However, the remaining langur habitat in PHP-NPA has become highly fragmented due to human destruction because many settlements are located on the adjacent plains and river valleys. In addition, agricultural fields often extend up to the base of the karst rocks, and timber and non-timber forest products are collected within the forests of the karst area. We surveyed the central part of the PHP-NPA both inside and edge of the total protection zone (TPZ). TPZ has the best quality forest in this protected area and covers approximately 351 km^2 [26], which accounts for about 15.6% of the actual park area. We selected this area for our surveys not only because of its pristine forest habitat but also because it has flat terrain and good accessibility. In the study area, the wet season was from June to October and the dry season from November to May. The rainfall peaks were in August (mean = 476.3 mm) and October (436 mm), whereas the lowest rainfall occurred in December (16.3 mm) and February (19.7 mm), with the monthly average precipitation being 149.8 mm.

2.2. Interviews and Historical Records

From 27 July to 1 August 2020, we interviewed 31 people (2 women, 29 men). These were 14 people belonging to the staff of six District Agriculture and Forestry Offices (DAFO, i.e., Hinboun, Khounkham, Nakai, Yommalath, Mahaxay and Thakhek), and 17 villagers, including three from Ban Natan (Nakai district), six from Ban Konglor (Khounkam district), three from Ban Buamlou, one from Ban Bo Neng and four from Ban Kuankacha (Hinboun district). We conducted standardized interviews with the above-mentioned people to determine the presence of primates in their monitored areas. To facilitate the process, we used photos of the various primate species potentially occurring in the area, uploaded on a smartphone. We showed photos of the following species: Laotian langur (*T. laotum*), black langur (*Trachypithecus ebenus*), southern white-cheeked gibbon (*Nomascus siki*), stump-tailed (*Macaca arctoides*) and Assamese (*Macaca assamensis*) macaques. During the interviews, we used the local names for each species: Laotian langur = khoung, black langur = khong, gibbon = tha nee, Assamese macaque = ling kang, stump-tailed macaque = ling nar daeng. Each interviewee was asked ten questions (see online Supplementary Table S1).

We spent about two hours interviewing the various members of the DAFO staff in each district and the villagers at survey camps. Interviews were authorized by the Department of Forestry of the Lao Government (protocol no. 3111, approved on 26 June 2020). We did not ask the age of the respondents. All eight DAFO interviewees in three districts (Khounkham, Nakai and Hinboun) correctly identified the Laotian langur. Still, two DAFO people in Yommalath correctly identified another species (*T. ebenus*) but thought that *T. laotum* and *T. ebenus* belonged to the same species. Two DAFO staff in Thakhek incorrectly identified the langurs but correctly identified the gibbon. Two other staff in Mahaxay did not know the Laotian langur at all. All villagers correctly identified the Laotian langur, but they also knew and correctly described its vocalization sound and its sleep sites (defined by direct observations) and identified some plant species eaten by the langur.

Based on the information we gathered from the interviewees, we defined some areas of the potential presence of *T. laotum* in the PHP-NPA. We established a field survey team to identify the areas to be carefully surveyed in each district.

We also compiled a list of all the historical species' observation sources inside the protected area [21,27].

2.3. Field Surveys

We focused our surveys in the central part (TPZ) of the PHP-NPA and divided the survey area into four sectors: (1) Konglor-Natan, (2) Konglor, (3) Buamlou and (4) Kuankacha sites (Figure 2) to observe the study species and record their locations, group sizes and threats. The length of each transect is summarized in the online Supplemental Table S2. Field surveys were carried out during the period from August to December 2020. Our team of 18 survey participants, including one main researcher, one person from the Provincial Agriculture and Forestry Office (PAFO), 4 from DAFO offices and 12 villagers, spent 37 days in the field.

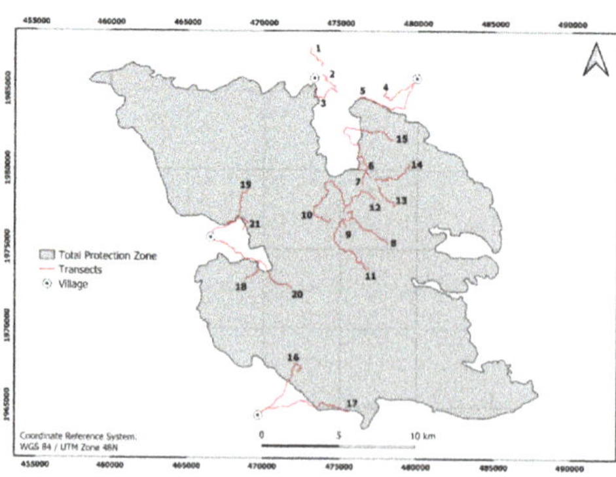

Figure 2. Map of the study area, including the 21 transects that were walked at the four sectors (Konglor-Natan, Konlgor, Buamlou and Kuankacha) of the Phou Hin Poun National Protected Area, Laos.

During the field surveys, we tried as much as possible to use a line transect methodology [28]. However, walking along a straight transect line was impossible in the limestone area in PHP-NPA. So we were forced to use often existing local trails as transects throughout the surveys. Overall, we walked 64.1 km along 21 transects (Figure 2). Our team walked silently along transects from about 7:00 to 17:00 P.M. For each sector, we conducted from 2 to 10 surveys along each transect. Since PHP-NPA has practically no sources of water for the survey team to drink in the dry season, we were forced to make the fieldwork only in the wet season (August to October 2020), but, due to the adverse weather (heavy rainfall making very dangerous to walk on karst areas), we had to postpone our working schedule and continued the field surveys to December 2020. The field surveys were carried out from 13 August to 17 December 2020.

The visibility was grossly similar in the four sectors of the study area, with a mean detection distance being about 150–200 m for direct sightings and 300–400 m for calls. For each observed group, we recorded the number, sex, and age of the individuals and the GPS coordinates, the altitude of the sighting spots (m a.s.l.), the time of sightings (Vientiane standard time), and the estimated distance from the observers. We classified the age classes of the observed langur individuals' age based on (1) body size, (2) fur color, and (3) sexual organs. Adult males are much larger in body size, with visible penis and with the breast being fully covered by fur. Adult females are clearly smaller in body size and with two breasts that can be easily seen at their chest and with no penis. Subadults of the two sexes are similar to adults of their respective sex, but with medium body size. Juveniles have a variable fur color, from yellowish or orange to black and white; the body and the tail are yellowish to black, and the head has three colors, yellowish, white and black. In contrast, the adults are black in the body and black and white in the head (online Supplemental Figure S2). In addition, the juveniles are normally spending their time nearby the adult females. Infants have orange or gold fur color and are carried by their mothers. We used the GPS device to mark the coordinate with altitude from a standing point and then used a laser rangefinder to measure the height from the standing point to the spotted animal above the transect. Then, the elevation of observer + elevation from observer to animal = estimated elevation.

2.4. Statistical Analyses

Correlations between group size and (i) elevation (m a.s.l.) of the sighting spots or (ii) year of observation were assessed by Spearman's rank correlation coefficient after having verified by Kolmogorov–Smirnov test that the two sets of variables did not attain normality and homoscedasticity. The effect of the yearly period of surveys (three groups: 1990s, 2010s, and 2020) on the group size was analyzed by Friedman ANOVA, followed by Mann–Whitney *U*-test for pairwise comparisons. To evaluate the diel activity patterns of the study species, we divided the daily time into 2-h-long sections. We then compared the frequencies of sightings across time sections with a contingency table χ^2 test. In the text, means are presented $\pm$ 1 standard deviation, with alpha set at 5%. All statistical analyses were made using SPSS 11.0 software version.

3. Results

3.1. Interviews and Historical Records

Interviewees in Hinboun, Khounkham and Nakai confirmed the presence of the Laotian langur in their districts. In contrast, interviewees in Yommalath and Thakhek were unsure about this species' presence in their districts. Both Laotian langur and black langur are suspected to occur in these two latter districts without any firm evidence. DAFO teams confirmed that no Laotian langur can be found in the Mahaxay district. Table 1 summarizes the interview results by village, including the identified threats and the feeling that interviewees had on the species' population status.

Table 2 summarizes the historical observations of the study species within the study area. Overall, and after excluding sightings with no number of observed individuals available, 38 sighting events of the Laotian langur in the study area between 1994 and 2010 (Table 2). Statistical comparisons of historical records with our current data are provided below.

Table 1. Species occurrence and threats as emerged from interview results. For more details, see the text.

District	Village	Occurrence	Threats	Population	Interviewees
Hinboun	Ban Pheepaeng	*T. laotum*, *N. siki*, Macaque indet.	Hunting/poaching	Decreased	DAFO staffs
	Ban Huana	*T. laotum*, *N. siki*, Macaque indet.	Hunting/poaching	Unknown	DAFO staffs
	Ban Buamlou	*T. laotum*, *N. siki*, Macaque indet.	Hunting/poaching	Unknown	DAFO staffs; villagers
	Ban Kuankacha	*T. laotum*, *N. siki*, Macaque indet.	Hunting/poaching/mining	Decreased	Villagers
	Ban Bo Neng	*T. laotum*, Macaque indet.	Hunting/poaching	Decreased	Villagers
Khounkham	Ban Konglor	*T. laotum*, *N. siki*, Macaque indet.	Hunting/poaching/logging	Decreased	DAFO staffs; villagers
	Ban Or	*T. laotum*	Hunting	Decreased	DAFO staffs
	Ban Kateup	*T. laotum*, *N. siki*, Macaque indet., serow sp.	Hunting	Unknown	Villagers
Nakai	Ban Natan	*T. laotum*. Macaque indet.	Poaching	Decreased	DAFO staffs; villagers
	Ban Kuam Sam	*T. laotum*	Unknown	Unknown	DAFO staffs
	Ban Vanghin	*T. laotum*	Unknown	Unknown	DAFO staffs
Yommalath	Ban Kuanphan	*T. ebenus*, Macaque indet., maybe *T. laotum*	Hunting	Unknown	DAFO staffs
Thakhek	Ban Doi	*T. laotum*, *N. siki*, maybe *T. ebenus*	Hunting	Decreased	DAFO staffs
	Ban Phalem	*T. laotum*, *N. siki*, maybe *T. ebenus*	Hunting	Decreased	DAFO staffs

Ban = Village; DAFO = District Agriculture and Forestry Office; Number of interviewees (31 people): 14 DAFO staffs + 17 villagers.

Table 2. Synopsis of the historical records of the Laotian langur in Phou Hin Poun National Protected Area, Lao PDR. Vis = Visual observations; Voc = vocalizations; Lat = latitude; Long = longitude. SEF = semi-evergreen forest; n/k = not known; MDF = mixed deciduous forest. "+" indicates the minimum number of individuals observed in each survey. For instance, 1+ would indicate that at least one individual was observed in that given survey.

Location	Lat and Long	Date	Habitat	Group Size	Type
Un-named	17°58′104°49′	21 December 1994	Open karst vegetation	1+	Vis
Un-named	18°02′104°47′	21 December 1994	Open karst vegetation	1+	Vis
Near Ban Vang dao	18°05′104°32′	16 May 1995	Open karst vegetation	8+	Vis
Near Ban Vang dao	18°03′104°31′	17 May 1995	n/k	n/k	Voc
North of Khua Din	17°51′104°51	6 February 1996	SEF	4+	Vis
Khua Din	17°50′104°50′	6 February 1996	n/k	probably several groups	Voc
South of Khua Din	17°49′104°50′	7 February 1996	SEF	8+	Vis
South of Ban Ak	17°52′104°52′	8 February 1996	Karst ridge	n/k	Vis
South of Ban Ak	17°52′104°53′	9 February 1996	n/k	n/k (probably several groups)	Voc
Near Nam Hinboun	18°01′104°26	31 January 1998	Karst above degraded SEF	9	Vis
Tham Pai	17°52′104°52′	15 March 1998	MDF	15–25	Vis
Tham Pai	17°51′104°52	16 March1998	MDF near cliff	n/k	Voc
Tham Pai	17°52′104°52′	21 March 1998	SEF on ridge	n/k	Voc
Khua Din	17°50′104°50′	18 March 1998	SEF near cliff	15–20	Vis
Khua Din	17°50′104°50′	20 March 1998	SEF not near cliff	15–20	Vis
Khua Din	17°50′104°50′	19 March 1998	SEF	Probably 2 group	Vis
Khouan Huy	17°42′104°48′	2 April 1998	Boulders at cave mouth	2+ (heads not seen)	Vis
Khouan Huy	17°41′104°49′	1 April 1998	Cliff face vegetation	n/k	Voc
Zone 1_Camp 1		2010		5 calls; 2 groups (8 individuals)	Vis
Zone 1_Camp 2		2010		2 calls, 1 group (4 individuals)	Vis + Voc
Zone 1_Camp 3		2010		4 calls, 1 group (5 individuals)	Vis + Voc
Zone 2_Camp 4		2010		2 calls, 1 group (3 individuals)	Vis + Voc
Zone 2_Camp 5		2010		1 call	Voc
Zone 2_Camp 6		2010		1 call	Voc
Zone 3_Camp 7		2010		2 calls	Voc
Zone 3_Camp 8		2010		2 calls	Voc
Zone 3_Camp 9		2010		1 call, 1 group (4 individuals)	Vis + Voc
Zone 3_Camp 10		2010		2 calls	Voc
Zone 3_Camp 11		2010		1 call, 1 group (6 individuals)	Vis + Voc
Zone 3_Camp 12		2010		1 call	Voc
Zone 3_Camp 13		2010		2 calls, 1 group (6 individuals)	Vis + Voc
Zone 4_Camp 14		2010		1 call	Voc
Zone 4_ Camp 15		2010		2 calls, 1 group (8 individuals)	Vis + Voc
Zone 4_ Camp 16		2010		2 calls, 2 groups (12 individuals)	Vis + Voc
Zone 4_Camp 17		2010		1 call, 1 group (6 individuals)	Vis + Voc
Zone 4_Camp 18		2010		1 call, 1 group (5 individuals)	Vis + Voc
Zone 5_Camp 19		2010		3 call, 3 groups (21 individuals)	Vis + Voc
Zone 5_Camp 20		2010		3 call, 3 groups (15 individuals)	Vis + Voc

3.2. Field Surveys

Details of walked transects and locations of the Laotian langur groups are given in the online supplemental Figure S1. Figure 3 shows the total sightings and vocalization locations. Statistics of sightings we made of the Laotian langur in each sector of the study area are compiled in Table 3. Sightings and calls were recorded at a mean distance from observers = 209.5 ± 101.6 m (n =19). 40% of the sightings were done between 06:00 and 08:00, 20% between 08:01–10:00, 15% between 10:01–12:00, and 25% between 14:01–16:00, with no sightings after 16:00. There were no significant differences in the frequency of sightings across 2-h-long daily phases (χ^2 = 5.58, df = 4, p = 0.232).

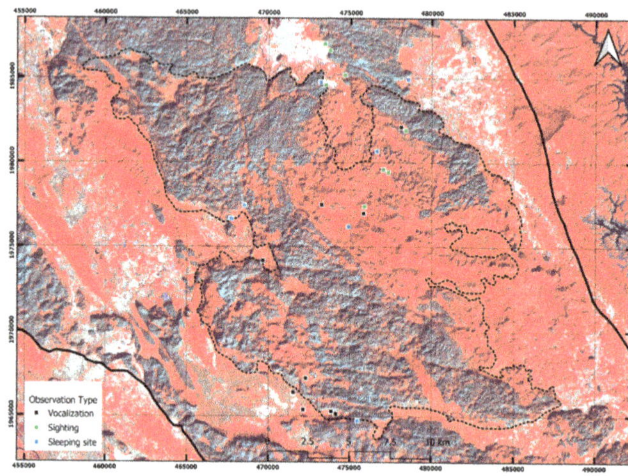

Figure 3. Sighting locations of the Laotian langur using Landsat 8 images. These images were taken from 22 January 2021 with bands 5–4–3 combination.

Table 3. Laotian langur encountered on trails walked in each sector. In this table, we include both the directly observed groups and those recorded by vocalizations during our surveys. For these latter groups, it was impossible to determine the size, sex or age of their members.

Sectors (Total Effort)	Frequency of Encountered Groups	No. of Individuals	Encounter Rates (Groups Per km Walked)
Konglor-Natan (14.9 km)	5	17	0.34
Konglor (32.8 km)	7	22	0.21
Buamlou (5 km)	5	6	1.00
Kuankacha (11.4 km)	2	2	0.18

The mean encounter rate of the langur groups, including those directly observed and those that were just heard, was 0.432 ± 0.38 km^{-1} (Table 3). Encounter rates were much higher at Buamlou than in the other sectors (Table 3). However, our samples are too small for any statistical evaluation. In total, we counted nine Laotian langur groups (35 individuals) that were directly observed. In these directly observed groups, we recorded 24 adults of both sexes, 10 subadults of both sexes and one juvenile. Group size ranged from 2 to 6 individuals (Table 4). However, larger group sizes (up to 20–25 individuals) were observed in the study area in 1998 (Table 2). Although overall group size did not vary significantly among the three yearly periods (Friedman ANOVA, χ^2 = 5.1, p = 0.079), there was a significant decrease of group size between 1994 and 2020 (Mann–Whitney pairwise comparison: p = 0.010), whereas there was no difference between the 2010s and 2020 (Mann–Whitney pairwise comparison: p = 0.250). The various individuals/groups were observed at a mean elevation = 313.8 ± 147.6 m a.s.l. (n = 19), and there was no correlation between size of the various groups and elevation (Spearman $r_s = 0.056$, $n = 19$, $p = 0.816$).

Table 4. Group size of the Laotian langur directly observed during this study.

Group No.	Location	Coordinate (GPS) WGS 84/UTM Zone 48 N		Adults	Subadults	Juveniles	Infants	Group Size
		Northing	Easting					
1	Tham Kuay	473,753	1,986,651	2	0	0	0	2
2	Tham Huator	473,457	1,987,041	3	2	0	0	5
3	Near Konglor cave	474,646	1,985,180	3	1	0	0	4
4	Poung Ta Tid Pha	473,512	1,984,602	3	0	0	0	3
5	Tham Huator	473,753	1,986,651	2	1	0	0	3
6	Pha Soung	476,995	1,979,610	2	2	0	0	4
7	Ang Nam Ta Ngon	475,844	1,977,413	4	2	0	0	6
8	Ang Khee Ther	477,304	1,979,456	2	1	0	0	3
9	Kuan Dik	478,413	1,981,923	3	1	1	0	5
Total				24	10	1	0	35

3.3. *Observations on Sympatry with Other Primates*

During the field surveys, we observed closely sympatric populations of the Laotian langur and Assamese macaque. Individuals of these two species used the same forest karst and even the same tree for feeding (Figure 4). At Tham Huator, we observed a group of the Laotian langur with three individuals eating the young leaves on trees. A group of five individuals of Assamese macaques came to the feeding tree site with a loud noise. It forced the langur group to abandon the tree within a few minutes of the interaction. We observed that the Laotian langur fed by early morning and finished feeding before the arrival of the macaque groups for feeding. Langurs were observed to feed mainly upon young tree leaves. In contrast, macaques preferred foraging upon leaves, but, including flowers of trees and flowers from vines, and they did not forage upon the same young tree leaves as the langur did at this site. Thus, it is possible that a food niche partitioning may occur between these two species. On another occasion, we observed a group of six individuals of langurs and a group of five individuals of macaques on the same tree for feeding in the morning. They did not fight each other, but we noticed that most langurs sat on branches in the middle of the canopy, eating young leaves. In contrast, most macaques occupied branches at the lower level of the tree canopy.

Figure 4. Laotian langur (on the left) and an Assamese macaque (on the right) interacting in the same tree. The two species are sympatric in several sites of the protected area where surveys were carried out.

In another case, we encountered a group of langurs at a site along transect 10; on the next day, at the same site, there were no langurs but a group of Assamese macaques that were feeding at the same site. We also heard macaque's vocalization and sighted the langurs at the same site along transect 15.

We observed langur individuals eating leaves at seven different and independent times during the field survey. On two of these occasions, the langur and the macaques were eating in the same spot.

3.4. Observations on Threats

Although not quantified, we observed logging in all four sectors of the study area. In particular, we recorded large numbers of valuable tree logging (*Diospyros embryopteris*) and other hardwood species from Konglor-Natan and Konglor sites. Ongoing logging of *Afzelia xylocarpa* and *Dalbergia cultrata* was mainly found at the Konglor site, where it occurred even far from the village. Many trails, camps and trash (energy drink cans, cigarettes, etc.) were also distributed inside the total protection zone. In addition, mining companies are operating in the Hinboun district. These companies have been implementing mining extraction activities close to Ban Kuankacha and built a new road for trucks to access the mining site. The mining site is located very close to the boundaries of the total protection zone and, therefore, potentially represents a source of disturbance for langur groups. Roads and trails also fragment the potential forest habitat of langurs.

We did not directly observe poaching during our survey. Still, it was reported by local villagers and confirmed by photos provided by hunters (Figure 5). In Ban Natan, it was reported that they hunted a group of 12 individuals nearby the local primary school in 2018. In Ban Konglor, a villager reported that, between the years 2000 and 2010, he and his friend spent many months in the protected area searching for agarwood tree and other valuable trees. Meanwhile, they killed at least 500 individuals of various primate species, including langurs, macaques, gibbon and loris from the NPA. He also reported that another group of poachers also killed above 200 additional primate individuals during the same time interval. In Ban Buamlou, villagers reported that the langur meat is consumed by families, and langur bones normally were sold for approximately 80,000 Kip/kg (approximately USD 9–10) to some Vietnamese shops in Ban Songhong. Other signals of poaching were that, for instance, in Ban Kuankacha, our survey team found two guns in the Kuankacha sector, and other guns were also found in other sectors of the surveyed territory. These guns were confiscated by the rangers of the protected area.

A B

Figure 5. Evidence of poaching within the protected area. Poaching is still one of the main threats to the survival of wildlife. In this figure: (**A**): two Laotian langur killed in Ban Natan; (**B**): serow killed in Hinboun. Photos obtained from anonymous hunters.

4. Discussion

Through the survey period from August to December 2020, we walked 64.1 km of 21 transects and recorded 35 langur individuals in 9 troops. The highest encounter/detection rate of the Laotian langur was at the Buamlou site (1 group per km). In contrast, it was much lower at Konglor-Natan (0.34 group/km), Konglor (0.21 group/km) and Kuankacha (0.18 group/km). No data are available for Laos in the past as the previously conducted survey aimed at understanding the species/subspecies distribution in the PHP-NPA [27] and not to estimate the relevant abundance of the various primates [27]. In 2010 Phiapalath [27] recorded 26 groups in 21 survey points. However, the current survey shows relatively similar encounter rates as those that were provided for conspecifics in the Phong Nha–Ke Bang National Park (0.19 to 0.48 group/km; see [29]) and for *Trachypithecus germaini* in Vietnam (range 0.21–1.15 group/km; [30]).

Compared to previous studies, group sizes observed at our study site in 2020 were much lower than those recorded in the Quang Binh Province, Vietnam (15 or more; see [31]), but similar to the size of langur groups observed by [32] (mean 7.3), by [31] (mean 8.2) and by [33]. The closely related *Trachypithecus delacouri* is territorial as probably also the Laotian langur is, and usually lives in single-male-multifemale groups of 5–30 individuals [34]. However, even subtle differences in survey technique may produce different results. Interestingly, comparing the historical dataset with the present survey results on the langur population from the study area, we found a significant decline in the number of individuals in each group (at least between 1994 and 2020). Although we do not have firm evidence of the reasons behind this pattern, it seems likely that this is a signal of overall population size decrease throughout the years. In addition, Table 4 shows that we observed just one juvenile and no infants. This demographic skew may indicate a declining population and perhaps unusual mortality pressures on younger aged langurs or possibly reflect a birth season not covered by our surveys [35]. Further surveys should confirm these population trends, but also circumstantial evidence provided by villagers may suggest an ongoing population decline of langur. For instance, (i) in Ban Natan, the langurs could be observed easily when they were feeding on the cliffs near the village to about 2010, but now it is hard to see them because villagers poached them, and (ii) there was a general feeling of our interviewees that the population of these monkeys is declining.

Based on the preliminary data provided in the present study, it is impossible to fully evaluate the population status of the study species in the study area. However, since langur groups were observed in a wide portion of the protected area (i.e., in all the four investigated sectors), it is likely that the species may still be relatively common at the local scale. In one sector (Buamlou), the encounter rates were similar to the highest rates recorded by [30], suggesting that Buamlou may be a crucial area to be protected if we want to guarantee the survival of the study species in the area. Buamlou sector is suitable for both the Laotian langur and the gibbon due to its inaccessibility. The rugged and steep terrain naturally protects these animals and makes it difficult for humans to access the site. Gibbon density in the Buamlou sector is higher than in other sectors of the NPA [36,37].

Interestingly, our study also showed that langurs can be sympatric and even use the same foraging and sleep sites as the Assamese macaques. A study of the coexistence mechanisms of these two species would be extremely interesting. Still, our preliminary observations may indicate that these two species may exhibit some kind of niche partitioning for both food types (eating preferentially different stages of leaves and flowers on the same tree) and foraging microhabitat (occupying different areas of the canopy when on the same tree). However, the Assamese macaque is essentially a young-leaves-eater in limestone habitats in China [38,39], but may also feed abundantly on fruits [40,41] and may opportunistically take a wide range of food types, including human food [41]. In contrast, the Laotian langur is typically folivorous (based on our small number of observations). Interestingly, François' langurs (*T. francoisi*) showed a more flexible diet composition than sympatric Assamese macaques. They increased dietary diversity and mature leaf consumption during periods of seasonal young leaves and fruit shortage. In contrast,

Assamese macaques relied heavily on young bamboo leaves (*Indocalamus calcicolus*) in most months [42]. Whereas interspecific interactions between sympatric Laotian langur and Assamese macaque were not observed previously, a previous survey [27] revealed that the Laotian langur and the black langur (*T. ebenus*) occurred sympatrically within the NPA. Based on Phiapalath's [27] observations, *T. laotum* and *T. ebenus* were found in the south and southwest (Ban Phalam) of the NPA. Our survey found that the Laotian langur and macaques also shared habitat in the central part of the NPA. In contrast, no *T. ebenus* was encountered during this survey. Moreover, black langur was observed to occasionally use the same sleep sites with Assamese and stump-tailed macaques in Hin Nam No NPA in 2008 (P. Phiapalath unpublished data). However, it should be borne in mind that our observations are just preliminary. It would be needed a specific methodology and hundreds of hours of observation to truly test niche partitioning between these sympatric species in the wild.

Threats and Conservation Considerations

Although we did not evaluate the threats quantitatively to langurs in the study area, we suggest that poaching and selective logging in its karst habitats may be the most critical issues to their survival within the PHP-NPA. This was also confirmed by the interviews that we conducted in the study area. Konglor site was the most accessible for local people and logging of valuable tree species. We frequently observed evidence of poaching on the transects inside the TPZ. This site is regularly visited by poachers/loggers because the terrain is easier to access. It has a constant water supply, and it contains large numbers of wild animals and valuable trees. Extensive logging has been throughout the country from 2013 to 2015, and often, it has been associated with hunting/poaching. Fortunately, the logging activity has been effectively banned in Laos since 2016 following the Order of Prime Minister no. 15/PM. Anyhow, TPZ should be prioritized for the conservation of key species in this reserve: indeed, our unpublished observations also showed that it is not only suitable habitat for the Laotian langur but also for southern white-cheeked gibbon (*Nomascus siki*), stump-tailed macaque (*Macaca arctoides*), Assamese macaque (*Macaca assamensis*), Asiatic black bear (*Ursus thibetanus*), sun bear (*Helarctos malayanus*), muntjac (*Muntiacus* sp.), hog badger (*Arctonyx collaris*), wild boar (*Sus scrofa*), Indochinese serow (*Capricornis sumatraensis maritimus*), silver pheasant (*Lophura nycthemera*), wreathed hornbills (*Rhyticeros undulatus*), bare-faced bulbul (*Nok hualon*) and probably black langur (*Trachypithecus ebenus*). PHP-NPA would hold important populations of other threatened species, including a recent clouded leopard *Neofelis nebulosa* record [43]. In addition, valuable and rare flora species such *Diospyros embryopteris*, agarwood *Aquilaris* sp., dragon's blood *Dracaena* sp., rosewood *Dalbergia cultrata* are also found in the TPZ [37].

To save the remaining Laotian langur population in PHP-NPA, there is still much work to be done by the protected area management, especially in Khammouane Province. Currently, IUCN provides technical support to protected area staff to conserve the protected area biodiversity and improve management plans for PHP-NPA and specifically for more effective TPZ management. Currently, IUCN and World Bank projects are supporting the management and biodiversity conservation in PHP. We believe that further research on the population and ecology of the endangered Laotian langur should be conducted, using these funds to better understand species living at TPZ and aid their long-term conservation.

Supplementary Materials: The following are available online at https://www.mdpi.com/article/10.3390/d13060231/s1, Figure S1: Field survey trails throughout the various sectors of the study area. Trails walked with locations of the *Laotian langur* encountered are presented, Figure S2: Coloration differences between adults and juvenile langurs at the study area, Table S1. List of the ten questions forming the standardized questionnaire for the various interviewees at the study area. For more details, see the methods, Table S2. Field effort (in terms of km walked) of each trail during this survey.

Author Contributions: Conceptualization, H.D.T. and J.S.; methodology, H.D.T., P.P. and J.S.; validation, P.B., formal analysis, L.L.; investigation, J.S., P.B., L.L., T.P.V. and J.S.; writing—original draft preparation, T.P.V., P.B., P.P. and H.D.T. All authors have read and agreed to the published version of the manuscript.

Funding: This research was funded by Rufford Foundation, grant number 29526-1.

Institutional Review Board Statement: Interviews and field surveys were authorized by the Department of Forestry of the Lao Government (protocol no. 3111, approved on 26 June 2020).

Informed Consent Statement: Informed consent was obtained from all subjects involved in the study.

Data Availability Statement: Data are presented in the paper and in the Online Supplemental Materials. The data are not publicly available due to conservation reasons, as the study species has a narrow distribution and maybe subjected to poaching.

Acknowledgments: We are thankful to the Rufford Foundation, which provided funding to undertake this study. We would like to thank the Department of Forestry to allow our team to conduct this study in Phou Hin Poun National Protected Area and also thank the protected area's staffs of the Provincial Agriculture and Forestry Office in Khammouane and District Agriculture and Forestry Offices as well as local people for their assisting the field survey. We also thank the Lao Wildlife Conservation Association for their help in managing the project fund. We are grateful to the staffs and the German International Cooperation (GIZ) adviser at Vietnam National University of Forestry and staff at Faculty of Forestry at the National University of Laos for their valuable advice and support and support, and to four anonymous referees for very helpful comments on the submitted draft.

Conflicts of Interest: The authors declare no conflict of interest.

References

1. Laurance, W.F.; Vasconcelos, H.L.; Lovejoy, T.E. Forest loss and fragmentation in the Amazon: Implications for wildlife conservation. *Oryx* **2000**, *34*, 39–45. [CrossRef]
2. Echeverria, C.; Coomes, D.A.; Hall, M.; Newton, A.C. Spatially explicit models to analyze forest loss and fragmentation between 1976 and 2020 in southern Chile. *Ecol. Model.* **2008**, *212*, 439–449. [CrossRef]
3. Liu, Y.; Feng, Y.; Zhao, Z.; Zhang, Q.; Su, S. Socioeconomic drivers of forest loss and fragmentation: A comparison between different land use planning schemes and policy implications. *Land Use Policy* **2016**, *54*, 58–68. [CrossRef]
4. Padalia, H.; Ghosh, S.; Reddy, C.S.; Nandy, S.; Singh, S.; Kumar, A.S. Assessment of historical forest cover loss and fragmentation in Asian elephant ranges in India. *Environ. Monit. Assess.* **2019**, *191*, 1–13. [CrossRef] [PubMed]
5. Bierregaard, R.O., Jr.; Lovejoy, T.E.; Kapos, V.; dos Santos, A.A.; Hutchings, R.W. The biological dynamics of tropical rainforest fragments. *Bioscience* **1992**, *42*, 859–866. [CrossRef]
6. Laurance, W.F. A crisis in the making: Responses of Amazonian forests to land use and climate change. *Trends Ecol. Evol.* **1998**, *13*, 411–415. [CrossRef]
7. Laurance, W.F. Reflections on the tropical deforestation crisis. *Biol. Conserv.* **1999**, *91*, 109–117. [CrossRef]
8. Timmins, R.J. *Notes on Wildlife and Habitats in Khammouan Limestone National Biodiversity Conservation Area, Khammouan Province, Lao PDR 1997*; Centre for Protected Areas and Watershed Management/The Wildlife Conservation Society Lao Program: Vientiane, Lao People's Democratic Republic, 1997.
9. Rigg, J.D. Forests, marketization, livelihoods and the poor in the Lao PDR. *Land Degrad. Dev.* **2006**, *17*, 123–133. [CrossRef]
10. McNamara, S.; Erskine, P.D.; Lamb, D.; Chantalangsy, L.; Boyle, S. Primary tree species diversity in secondary fallow forests of Laos. *For. Ecol. Manag.* **2012**, *281*, 93–99. [CrossRef]
11. Phompila, C.; Lewis, M.; Ostendorf, B.; Clarke, K. Forest cover changes in Lao tropical forests: Physical and socio-economic factors are the most important drivers. *Land* **2017**, *6*, 23. [CrossRef]
12. Lamb, D. Deforestation and its consequences in the asia-pacific region. In *Regreening the Bare Hills*; Springer: Dordrecht, Germany, 2011; pp. 1–39.
13. Long, B.; Gray, T.; Lynam, A.; Seng, T.; Laurance, W.; Scotson, L.; Ripple, W. Reversing 'empty forest syndrome' in Southeast Asia. *Natl. Geogr. Voices* **2017**, *8*, 1–6.
14. Pruvot, M.; Khammavong, K.; Milavong, P.; Philavong, C.; Reinharz, D.; Mayxay, M.; Theppangna, W. Toward a quantification of risks at the nexus of conservation and health: The case of bushmeat markets in Lao PDR. *Sci. Total Environm.* **2019**, *676*, 732–745. [CrossRef]
15. Zuklin, T.; Maury, N.; Sitthivong, S.; Pham, T.V.; Le Duc, O.; Bordes, C.; Leprince, B.; Ducotterd, C.; Oanh, L.V.; Vilay, P.; et al. The "empty forest syndrome" and the herpetofaunal communities in Laos (South-Eastern Asia). *Russ. J. Herpetol.* **2021**, in press.
16. Johnson, A.; Singh, S.; Duangdala, M.; Hedemark, M. The western black crested gibbon *Nomascus concolor* in Laos: New records and conservation status. *Oryx* **2005**, *39*, 311–317. [CrossRef]

17. Sacklokham, S.; Dufumier, M. Land-tenure policy, deforestation, and agricultural development in Lao PDR: The case of the Vientiane plain. *Moussons. Rech. En Sci. Hum. Sur L'asie Du Sud-Est* **2006**, *9*, 189–207. [CrossRef]
18. Suwannarong, K.; Chapman, R.S.; Lantican, C.; Michaelides, T.; Zimicki, S. Hunting, food preparation, and consumption of rodents in Lao PDR. *PLoS ONE* **2015**, *10*, e0133150. [CrossRef]
19. Groves, C. *Trachypithecus*. In *Mammal. Species of the World. A Taxonomic and Geographic Reference*, 3rd ed.; Wilson, D.E., Reeder, D.M., Eds.; Johns Hopkins University Press: Baltimore, MD, USA, 2005; pp. 175–178, ISBN 0-8018-8221-4.
20. Timmins, R.J.; Boonratana, R. *Trachypithecus laotum*. In *The IUCN Red List of Threatened Species*; The International Union for Conservation of Nature: Gland, Switzerland, 2008; E.T 22044A9350930. Available online: www.iucnredlist.org (accessed on 21 March 2021).
21. Steinmetz, R.; Timmins, R.J.; Duckworth, J.W. Distribution and conservation status of the Lao leaf monkey (*Trachypithecus (Fr.) Laotum*). *Int. J. Primatol.* **2011**, *32*, 587–604. [CrossRef]
22. IUCN. The IUCN Red List of Threatened Species, 2020–2021. IUCN Website. Available online: Iucnredlist.org (accessed on 27 April 2021).
23. Coudrat, C.N.Z.; Nadler, T.; Phiapalath, P.; Duckworth, J.W. *Trachypithecus laotum*. In *The IUCN Red List of Threatened Species*; The International Union for Conservation of Nature: Gland, Switzerland, 2020; E.T 22044A1795913. Available online: www.iucnredlist.org (accessed on 21 March 2021).
24. Veloz, S.; Salas, L.; Altman, B.; Alexander, J.; Jongsomjit, D.; Elliott, N.; Ballard, G. Improving effectiveness of systematic conservation planning with density data. *Conserv. Biol.* **2015**, *29*, 1217–1227. [CrossRef]
25. Hamada, Y.; Malaivijitnond, S.; Kingsada, P.; Bounnam, P. The distribution and present status of primates in the northern region of Lao PDR. *Trop. Nat. Hist.* **2007**, *7*, 161–191.
26. Brakels, P. *Summary Report of the Zoning Activities in Phou Hin Poun National Protected Area*; IUCN Lao PDR: Vientiane, Lao People's Democratic Republic, 2019.
27. Phiapalath, P. *Brief Report of Lao Langur Trachypithecus laotum Survey in Phou Hin Poun National Protected Area, Khammouane Province*; Lao Wildlife Conservation Association: Vientiane, Lao PDR, 2010.
28. Brockelman, W.Y.; Ali, R. Methods of surveying and sampling forest primate populations. In *Primate Conservation in the Tropical Forest*; Marsh, C.W., Mittermeier, R.A., Eds.; Alan R. Liss: New York, NY, USA, 1987; pp. 23–62.
29. Frankfurt Zoological Society in Vietnam. *Biodiversity Survey of the Macaque, Langur and Douq Monkey in and around the Phong Nha–Ke Bang National Park, Quang Binh, Vietnam*; Vietnam-Germany Development Cooperation: Hanoi, Vietnam, 2011.
30. Fiore, R.R. A Survey of Indochinese Silvered Langurs (*Trachypithecus germaini*) in Phu Quoc National Park, Vietnam. Ph.D. Thesis, University of Colorado at Boulder, Boulder, CO, USA, 2015.
31. Ha, N.M. Some observations on the Hatinh Langur, *Trachypithecus laotum hatinhensis* (Dao 1970), in north central Vietnam. *Primate Conserv.* **2006**, *21*, 149–154. [CrossRef]
32. Pham, N.; Do, T.; Truong, V.L. Preliminary survey for Hatinh langur in north central Vietnam. *Asian Primates* **1996**, *6*, 13–17.
33. Haus, T.; Vogt, M.; Forster, B. Observations on the Hatinh langur (*Trachypithecus hatinhensis*) during point and line transect sampling in the Phong Nha–Ke Bang National Park, Central Vietnam. *Vietnam. J. Primatol.* **2009**, *3*, 17–27.
34. Harding, L.E. *Trachypithecus delacouri* (Primates: Cercopithecidae). *Mamm. Species* **2011**, *43*, 118–128. [CrossRef]
35. Lhendup, S.; Tshering, U.; Tenzin, J. Population structure and habitat use of golden langur (*Trachypithecus geei*) in Royal Manas National Park, Bhutan. *Int. J. Fauna Biol. Stud.* **2018**, *5*, 97–101.
36. Phiapalath, P.; Bousa, A.; Paul Insua-Cao, P. *The Status and Distribution of Gibbons in Phou Hin Poun National Protected Area*; Fauna & Flora International/IUCN Lao PDR: Vientiane, Lao PDR, 2012.
37. Brakels, P. *The Status and Distribution of Gibbons in Phou Hin Poun National Protected Area*; Fauna & Flora International/IUCN Lao PDR: Vientiane, Lao PDR, 2019.
38. Zhou, Q.; Wei, H.; Huang, Z.; Huang, C. Diet of the Assamese macaque *Macaca assamensis* in limestone habitats of Nonggang, China. *Curr. Zool.* **2011**, *57*, 18–25. [CrossRef]
39. Huang, Z.; Huang, C.; Tang, C.; Huang, L.; Tang, H.; Ma, G.; Zhou, Q. Dietary adaptations of Assamese macaques (*Macaca assamensis*) in limestone forests in Southwest China. *Am. J. Primatol.* **2015**, *77*, 171–185. [CrossRef]
40. Schülke, O.; Pesek, D.; Whitman, B.J.; Ostner, J. Ecology of Assamese macaques (*Macaca assamensis*) at Phu Khieo Wildlife Sanctuary, Thailand. *J. Wildl. Thail.* **2011**, *18*, 1–15.
41. Koirala, S.; Chalise, M.K.; Katuwal, H.B.; Gaire, R.; Pandey, B.; Ogawa, H. Diet and activity of *Macaca assamensis* in wild and semi-provisioned groups in Shivapuri Nagarjun National Park, Nepal. *Folia Primatol.* **2017**, *88*, 57–74. [CrossRef]
42. Zhou, Q.H.; Huang, Z.H.; Wei, H.; Huang, C.M. Variations in diet composition of sympatric *Trachypithecus francoisi* and *Macaca assamensis* in the limestone habitats of Nonggang, China. *Zool. Res.* **2018**, *39*, 284–291. [CrossRef]
43. Brakels, P.; Somdachit, T. Record of cats from Phou Hin Poun National Protected Area. IUCN Lao PDR. *Cat News* **2020**, 71.

diversity

MDPI

Article

DNA Barcoding of Mullets (Family Mugilidae) from Pakistan Reveals Surprisingly High Number of Unknown Candidate Species

Ariba Hasan [1], Pirzada Jamal Ahmed Siddiqui [1], Shabir Ali Amir [2] and Jean-Dominique Durand [3,*]

1 Centre of Excellence in Marine Biology, University of Karachi, Karachi 75270, Pakistan; areebahasan52@gmail.com (A.H.); jamal.siidiqui@yahoo.com (P.J.A.S.)
2 Pakistan Museum of Natural History, Garden Avenue, Shakarparian, Islamabad 44000, Pakistan; shabiramir@gmail.com
3 MARBEC University Montpellier, IRD, Bat 24 cc093 Place Eugene Bataillon, 34095 Montpellier, France
* Correspondence: jean-dominique.durand@ird.fr

Abstract: The mullets are a widespread group of ecologically and economically important fishes of disputed taxonomy due to their uniform external morphology. Barcoding and phylogenetic studies from various locations around the world largely highlighted the species diversity underestimation using morphological criteria used to establish the taxonomy of the family. Here, we investigated the mullet species diversity from Pakistan, a biogeographic area where nearly no mullet species were genetically characterized. Morphological examination of 40 mullets reveals 6 known species (*Planiliza macrolepis*, *P. klunzingeri*, *P. subviridis*, *Crenimugil seheli*, *Ellochelon vaigiensis*, and *Mugil cephalus*). Using a references DNA barcode library, the DNA barcode-based species identification flagged eight molecular operational taxonomic units (MOTUs) belonging to five genera (*Crenimugil*, *Ellochelon*, *Mugil*, *Osteomugil*, and *Planiliza*). Among these MOTUs, only one was already present in Barcode of Life Data system, all other representing new Barcode Index Numbers (BIN). These results emphasize the importance of the recognition of cryptic species and the necessity to re-evaluate the overall diversity by the genetic characterization of different species of this family. DNA barcoding is an effective tool to reveal cryptic species that need to be considered in conservation and management measures of fisheries in Pakistan.

Keywords: Mugilidae; Cytochrome Oxidase I; Arabian Sea; cryptic species; sequence divergence

Citation: Hasan, A.; Siddiqui, P.J.A.; Amir, S.A.; Durand, J.-D. DNA Barcoding of Mullets (Family Mugilidae) from Pakistan Reveals Surprisingly High Number of Unknown Candidate Species. *Diversity* **2021**, *13*, 232. https://doi.org/10.3390/d13060232

Academic Editor: Michael Wink

Received: 26 April 2021
Accepted: 21 May 2021
Published: 26 May 2021

Publisher's Note: MDPI stays neutral with regard to jurisdictional claims in published maps and institutional affiliations.

Copyright: © 2021 by the authors. Licensee MDPI, Basel, Switzerland. This article is an open access article distributed under the terms and conditions of the Creative Commons Attribution (CC BY) license (https://creativecommons.org/licenses/by/4.0/).

1. Introduction

It is largely acknowledged that the current fish taxonomy based on the variation of morpho-anatomical characters greatly underestimated the species diversity [1,2]. Numerous phylogeographic, phylogenetic, and DNA barcoding studies flagged independent evolutionary lineages in widely geographically distributed species that are more and more recognized as cryptic species or at least as candidate species pending further taxonomic investigations [1,3–9]. The proper delineation of species is essential not only for better management and conservation of biodiversity [10,11] but also helps us to understand the causes of different evolutionary processes [12]. From a more pragmatic perspective, incorrect identification of commercially important species may lead to overexploitation and contribute to fish stock depletion [13].

In this context, the DNA barcoding method proved to be a useful and independent approach based on the variation of morphometric and meristic characters for species identification [14]. Instead of observable and definite morphological differences, mitochondrial cytochrome c oxidase subunit I (COI) gene has been used to resolve many taxonomic ambiguities [15,16].

The family Mugilidae currently consists of about 27 genera and 77 recognized species [17]. Mullets are important in marine fisheries and aquaculture in many temperate and tropical countries [18]. Owing to the significant morphological conservatism, the delimitation of mullet species is arduous, as a result, mullet species are often inadequately represented in field guides [19]. For this reason, several studies have been carried out to solve the phylogenetic relationship and identification of mullet species using different molecular methods [20–27]. Over the past decades, molecular phylogenetic studies have evidenced the presence of many species complexes within the Mugilidae family [2,19,28–32]. These species complex, often presented as complex of cryptic species due to the absence of evident diagnostic morphometric and meristic characters, are usually sibling species and thus with much more limited distribution range than described for the morpho-species [2]. Among those species *Mugil cephalus* Linnaeus, 1758, is a good example as the morpho species present in nearly all tropical, subtropical, and temperate waters of the world [33] consists of 14 molecular operational taxonomic units (MOTU's) with a distribution range generally limited to a biogeographic province [18]. In this context, it is important to better estimate the species diversity of the Mugilidae family and their distribution range to barcode most mullet species in the different biogeographic provinces. While some DNA barcoding or phylogenetic studies have been realized in various regions of the world such as Europe [21–23,34–42], Africa [43], South America [32,44], Asia [20,26,31], and India [45,46], no studies have been considered species diversity present in Pakistan or, more generally, in the Arabian sea.

Based on morphological variations, a variable number of mullet species have previously been reported from Pakistan: 6 species by Qureshi [47], 12 by Bianchi [48], 7 by Fahmida [49], 10 by Froese and Pauly [50], and 12 species by Psomadakis et al. [51]. Since DNA barcodes are not available for Pakistani mullet species, it is not possible to know if these morpho-species belong to already identified MOTU's (such as those listed in [29]) or represent cryptic diversity.

The present study was designed to evaluate the divergence threshold and barcoding gap for the accurate molecular delimitation of mullet species present in Pakistan and flag new MOTU's using sequences of the mitochondrial COI gene. The results will significantly contribute to BOLD systems and GenBank databases with new DNA barcodes and provide an overview of species diversity of mullets from Pakistan in comparison with species elsewhere.

2. Materials and Methods

2.1. DNA Barcode Reference Library and Taxonomical Nomenclature

The reference library used in this study originates from Durand et al. [28], Shen and Durand [29], and Delrieu-Trottin et al. [31]. This library consists of 76 DNA barcode records trimmed to 556 base pairs representing all the species and BIN diversity of genera *Planiliza*, *Ellochelon*, *Crenimugil*, *Osteomugil*, and *Mugil*. These DNA barcodes have been selected as reference for the DNA barcode-based species identification since most of specimens barcoded are stored in museum and have been identified by taxonomic experts of the Mugilidae family [2,19,28] (Table S1). They have been used in a number of phylogenetic studies dealing with the taxonomy of the Mugilidae family by several authors [2,19,29–32]. The nomenclature proposed by Durand et al. [52] and Xia et al. [30] was used for genera while cryptic or unidentified species followed the interim taxonomical nomenclatures established by Durand and Borsa [2]. However, in a state of clarity and traceability, we also mentioned the Barcode Index Numbers (BINs) that can also represent an interim taxonomical nomenclature when no clear species name can be assigned to a barcode. The two interim taxonomical nomenclatures are largely redundant since Durand et al. [19] demonstrated a large overlap of barcode gaps recovered with COI marker (used to establish the BIN by the BOLD system) or a longer marker composed of COI, 16S, and cytochrome b fragment (such as in Durand et al. [28] used by Durand et al. [2] for their interim taxonomical nomenclature). However, the advantage of BIN is to have using the BOLD system with a direct and dynamic vision of the distribution range of the putative species.

2.2. Sample Collection and Identification

Mullet fish samples were collected from the landing sites and fish markets of Pakistan located along with Sindh (Karachi Fish Harbor, Kakapir, and Keti Bunder) and Baluchistan coasts (Somniani, Pasni, Gwadar, and Jiwani) (Figure 1). All samples were morphologically identified using different identification keys [48,51] and other available literature [53,54]. Each specimen was photographed and fin clipped; then, all the samples were stored individually in an Eppendorf vial with 70% ethanol, and later, they were stored at −20 °C.

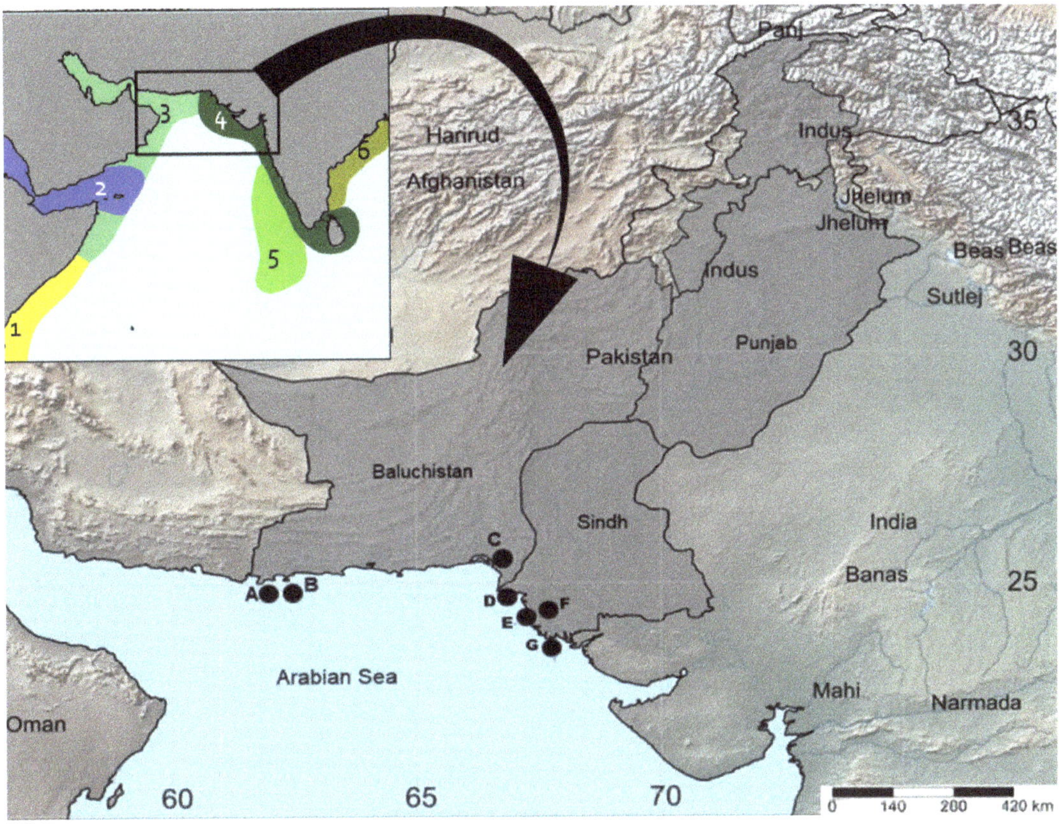

Figure 1. Biogeographic provinces present and surrounding the Pakistan marine region [55]: 1 = Western Indian Ocean, 2 = Red Sea and Gulf of Aden, 3 = Somali/Arabian, 4 = West and South Indian Shelf, 5 = Central Indian Ocean Islands, and 6 = Bay of Bengal. (B) Map showing the sampling locations of mullets analyzed in this study. A = Jiwani; B = Gwadar, C = Somniani, D = French Beach, E = Kakapir, F = Karachi Fish Harbor, G = Ibrahim Haideri and H = Keti Bunder.

2.3. DNA Amplification and Sequencing

Genomic DNA was isolated from fins using the G-Spin Total DNA extraction mini kit (iNtRON Biotechnology, Jungwon-gu, Gyeonggi, Korea) following the manufacturer's recommendations. Approximately, 652 base pairs (bp) of the cytochrome oxidase subunit I (COI) were amplified using primers FishF1+ FishF2/FishR1 [56]. Polymerase Chain Reaction (PCR) was conducted in the total volume of 40 μL containing 20 μL of MyTaq PCR Mastermix (Bioline, London, UK), 16 μL of ultrapure water, 0.8 μL of BSA (Euromedex, Souffelweyersheim, France), 0.6 μL of each primer (3 μM), and 2 μL of DNA template. The conditions used during PCR reaction were as follows: initial denaturation temperature at 92 °C for 5 min followed by 35 cycles of strand denaturation at 92 °C for 1 min, primer annealing at 52 °C for 45 s, primer extension at 72 °C for 1.5 min, and a final extension at 72

°C for 5 min. Sequencing was performed by Macrogen Inc. (Seoul, Korea). All nucleotide sequences were deposited in GenBank. The accession numbers are given in Table 1.

Table 1. List of mullet species (names inferred from morpho-anatomical keys), code numbers, locality information and GenBank's accession numbers.

Morpho Species	Code No.	Location	Co-Ordinates	Accession No.
Mugil cephalus	792	Kakapir, Karachi	24°50′42″ N 66°54′01″ E	MT943713
	794	Kakapir, Karachi	2450′42″ N 66°54′01″ E	MT943714
	PMNH-55212	Jiwani, Baluchistan	25°10′59″ N 61°46′24″ E	MN511974
	PMNH-55368	Gwadar, Baluchistan	24°06′53″ N 62°19′41″ E	MN511975
Planiliza macrolepis	PMNH-54728	Ibrahim Haideri, Karachi	24°47′39″ N 67°08′31″ E	MN512028
Planiliza subviridis	PAK Mu 851	Somniani, Baluchistan	25°09′25″ N 66°43′25″ E	MT943724
	806	Karachi Fish harbor	24°50′57″ N 66°58′35″ E	MT943723
	840	Keti Bunder, Sindh	24°07′49″ N 67°27′10″ E	MT943722
	PMNH-55121	Ibrahim Haideri, Karachi	24°47′39″ N 67°08′31″ E	MN511966
Planiliza klunzingeri	PAK Mu 884	Karachi Fish harbor	24°50′57″ N 66°58′35″ E	MT943743
	PAK Mu 804	Karachi Fish harbor	24°50′57″ N 66°58′35″ E	MT943734
	824	Karachi Fish harbor	24°50′57″ N 66°58′35″ E	MT943727
	816	Karachi Fish harbor	24°50′57″ N 66°58′35″ E	MT943737
	830	Karachi Fish harbor	24°50′57″ N 66°58′35″ E	MT943735
	828	Karachi Fish harbor	24°50′57″ N 66°58′35″ E	MT943736
	PaK Mu 3	Karachi Fish harbor	24°50′57″ N 66°58′35″ E	MT943740
	PAK Mu 872	Keti Bunder, Sindh	24°07′49″ N 67°27′10″ E	MT943730
	PAK Mu 870	Keti Bunder, Sindh	24°07′49″ N 67°27′10″ E	MT943731
	PAK Mu 881	Keti Bunder, Sindh	24°07′49″ N 67°27′10″ E	MT943729
	PAK Mu 882	Keti Bunder, Sindh	24°07′49″ N 67°27′10″ E	MT943728

Table 1. *Cont.*

Morpho Species	Code No.	Location	Co-Ordinates	Accession No.
	PAK Mu 883	Keti Bunder, Sindh	24°07′49″ N 67°27′10″ E	MT943744
	802	Keti Bunder, Sindh	24°07′49″ N 67°27′10″ E	MT943742
	827	Keti Bunder, Sindh	24°07′49″ N 67°27′10″ E	MT943726
	811	Keti Bunder, Sindh	24°07′49″ N 67°27′10″ E	MT943739
	812	Keti Bunder, Sindh	24°07′49″ N 67°27′10″ E	MT943738
	832	Gwadar, Baluchistan	25°06′53″ N 62°19′41″ E	MT943725
	PAK Mu 821	Gwadar, Baluchistan	25°06′53″ N 62°19′41″ E	MT943733
	823	Gwadar, Baluchistan	25°06′53″ N 62°19′41″ E	MT943732
	PMNH 55125	Gwadar, Baluchistan	25°06′53″ N 62°19′41″ E	MN12027
Crenimugil seheli	819	Gwadar, Baluchistan	25°06′53″ N 62°19′41″ E	MT943705
	PMNH-55231	Gwadar, Baluchistan	25°06′53″ N 62°19′41″ E	MN512029
Osteomugil sp.	829	Karachi Fish harbor	24°50′57″ N 66°58′35″ E	MT943716
	PAK Mu 866	Somniani, Baluchistan	25°09′25″ N 66°43′25″ E	MT943715
	PMNH-55122	Gwadar, Baluchistan	25°06′53″ N 62°19′41″ E	MN512031
	810	Karachi Fish harbor	24°50′57″ N 67°27′10″ E	MT943717
	PAK Mu 864	Karachi Fish harbor	24°50′57″ N 66°58′35″ E	MT943718
	PAK Mu 862	Karachi Fish harbor	24°50′57″ N 66°58′35″ E	MT943719
	PAK Mu 2	Karachi Fish harbor	24°50′57″ N 66°58′35″ E	MT943721
	PAK MU 885	Karachi Fish harbor	24°50′57″ N 66°58′35″ E	MT943720
Ellochelon vaigiensis	805	Keti Bunder, Sindh	24°07′49″ N 67°27′10″ E	MT943712
	PMNH-55070	French beach, Karachi	24°50′32″ N 66°48′53″ E	MN511887

2.4. DNA Barcode-Based Species Identification

Species identification based on specimen morphology was confronted to an independent species identification using DNA barcodes. All DNA barcodes generated in this study were uploaded on the BOLD system that assigned these barcodes to molecular operational taxonomic units, (MOTUs) called Barcode Index Numbers (BINs) using the RESL algorithm. [57]

This algorithm flag MOTU's boundaries by clustering DNA barcodes with high sequence similarity and connectivity using all DNA barcodes of the BOLD's library. BINs are used to confirm the concordance between species designations and barcode sequence clusters [57].

The composition and variations of nucleotides were analyzed by Mega V. 7.0 [58]. For the calculation of genetic distances between and within the species of mullets, Kimura-2-parameter (K2P) model was used [59]. A Neighbor-Joining tree was constructed with bootstrap analysis (500 replicates) to evaluate the reciprocal monophyly of species. To reveal and discriminate various species present in our sampling, we constructed phylogenetic tree using all COI barcodes generated in this study and secondarily with mullet reference barcodes. All trees were rooted using as outgroup a sequence of *Abudefduf vaigiensis* (Perciformes: Pomacentridae).

3. Results

A total of 41 specimens were successfully barcoded. All data relative to these specimens as well as their DNA barcodes were uploaded in BOLD's project PAKF. Among these specimens, 33 were morphologically identified at the species level. (Figure 2). These species are *Planiliza macrolepis, P. klunzingeri, P. subviridis, Crenimugil seheli, Ellochelon vaigiensis,* and *Mugil cephalus* (Table 1). The remaining eight specimens were identified at the genus level only: the *Osteomugil* genus. At exception of *Osteomugil* species, they were easily distinguishable using following criteria: length of the pectoral fin in regard to the birth of the first dorsal fin, presence of dot or blotch at the birth of the pectoral fin, presence and importance of adipose eyelid, color of the pectoral fin, position of the second dorsal in regard to the anal fin, form of the caudal fin, and scale margin (Figure 2).

All barcodes obtained in this study have been assigned in BOLD to 8 BINs (Figure 2). For specimens identified at the species taxonomic level, only one BIN has been recovered. The generated COI sequences were compared with the available COI sequences [2,19] and BOLD system revealed the presence of at least seven unknown candidate species.

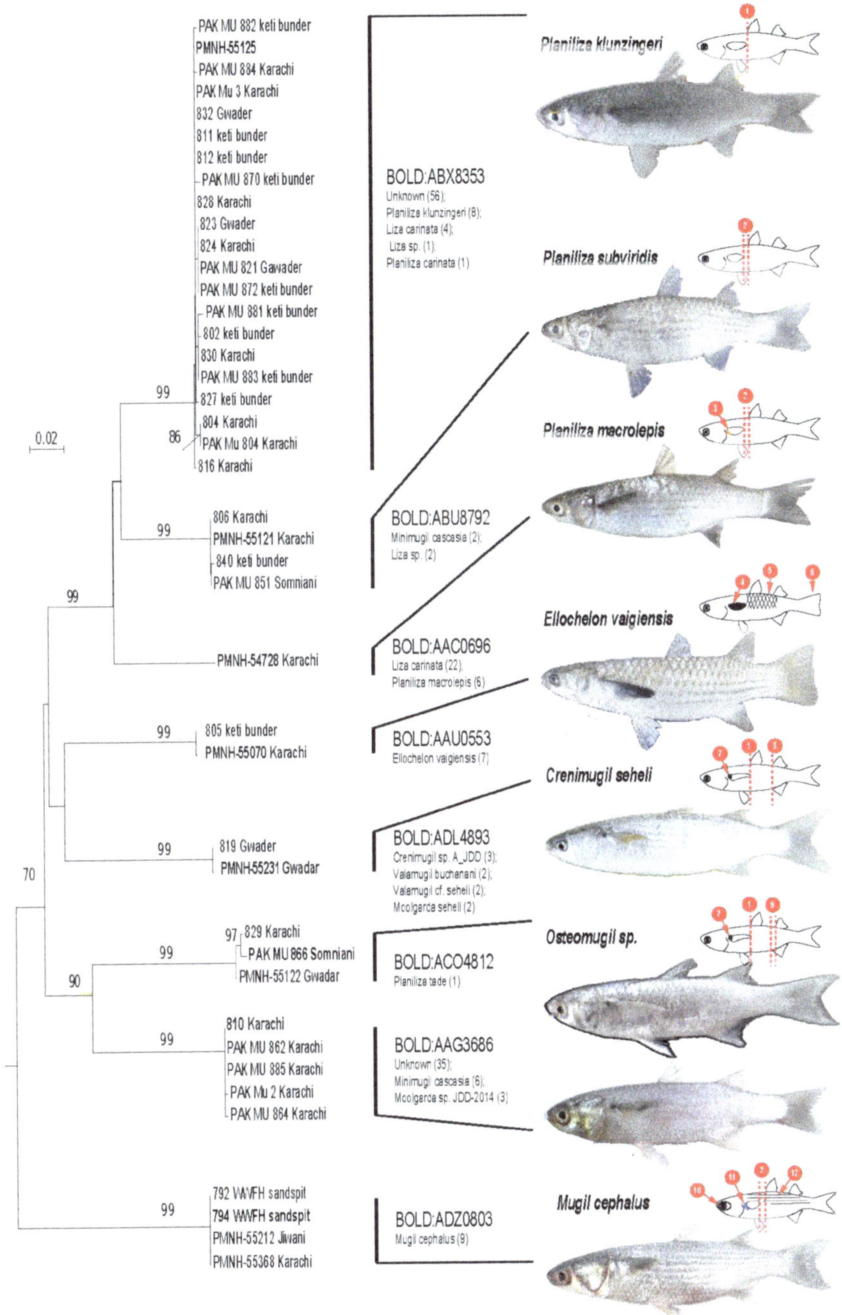

Figure 2. Phylogenetic relationships among Mugilidae specimens collected along Pakistan shores recovered using 622 bp of the COI and the Neighbor-Joining method. The percentage of replicate trees in which the associated taxa clustered together in the bootstrap test (500 replicates) are shown next to the branches [2]. The tree is drawn to scale, with branch lengths in the same units as those of the evolutionary distances used to infer the phylogenetic tree. The evolutionary distances were computed using the Kimura 2-parameter method [3] and are in the units of the number of base substitutions per site. Each

leaf of the tree corresponds to an individual sampled in Pakistani water, and leaves in bold correspond to the specimen's picture on the right. BINs provided from BOLD is mentioned for each clade as well as species name of specimen belonging to the BIN (in parentheses number of specimen). Species identified using morpho-anatomical criteria are indicated on the right of the figure. Fish draws highlight main morphometrical criteria that discriminate species collected in Pakistan. 1. Tip the pectoral fin at vertical of the birth of the first dorsal fin, 2. Tip the pectoral fin not reaching vertical of the birth of the first dorsal fin, 3. Presence of a golden blotch at the birth of the pectoral fin, 4. Pectoral fin black, 5. Scales with black margin, 6. Fin tail truncated, 7. Presence of a black dot at the birth of the pectoral fin, 8. Birth of the second dorsal fin at vertical of the birth of the birth of the anal fin, 9. Birth of the second dorsal fin not at vertical of the birth of the anal fin, 10. Large adipose eyelid, 11. Presence of a blue blotch at the birth of the pectoral fin, and 12. Black stripes on flank. Pictures provided by Ariba Hasan & Shabir Ali Amir (copyright).

4. Discussion

BOLD:AAC0696/*Planiliza macrolepis* (morphology)

Specimen from Pakistan identified morphologically as *Planiliza macrolepis* belongs to this BIN as well as Durand and Borsa's [2] reference sequences of *Planiliza macrolepis* (Figure 3A). Specimens identified morphologically as *P. macrolepis* belongs to two lineages with parapatric geographic distribution: one located in the East Indian Ocean (South Africa, Seychelles, and Oman) and one from the Central Indian Ocean to the Pacific Ocean (Maldives, Sri Lanka, Taiwan, Japan, New Caledonia, and Fiji) [28]. However, because the type locality of *P. macrolepis* Smith 1948 is in South Africa, Durand and Borsa [2] proposed to keep this name only for the NW Indian lineage, the second one being provisionally designated as *Planiliza* sp. H. Present data precise *P. macrolepis* (BOLD:AAC0696) distribution range eastward and, more importantly, its geographic limit with its sibling species *Planiliza* sp. H, to date, situated in India [29].

BOLD:ABX8353/*Planiliza klunzingeri* (morphology)

Specimens from Pakistan identified morphologically as *Planiliza klunzingeri* assigned to the BIN BOLD:ABX8353 that also included reference sequences of *Planiliza* sp. A of Durand and Borsa [2] (Figure 3A). There is no doubt that mullet specimens collected in the Persian Gulf and named "*Planiliza* sp. A" by Durand and Borsa [2] is actually *Planiliza klunzingeri* considering morphological characters records on our specimen as well as previous barcoding studies [29,60] The distribution range of *Planiliza klunzingeri* encompasses the Persian Gulf eastward to the coast of Karachi and Bombay [61]. Eastward distribution limits have been confirmed by [62], which provided DNA barcodes of *P. klunzingeri* (BOLD:ABX8353) collected in the Narmada River (NW India). Interestingly, *P. klunzingeri* (BOLD:ABX8353) and *P. macrolepis* (BOLD:AAC0696) probably shared the same eastward distribution limit, which suggests the presence of a biogeographic barrier in NW India and not along Pakistani shores as suggested by Spalding et al. [55] (Figure 1).

BOLD:ABU8792/*Planiliza subviridis* (morphology)

Pakistani specimens assigned to this BIN have been morphologically identified as *Planiliza subviridis* but does not correspond to any reference sequences nor *Planiliza subviridis* sensu Durand and Borsa [2] that are assigned in BOLD to 4 different BINs (Figure 3A). If specimens from Pakistan share a common ancestor with *P. subviridis* sensu Durand and Borsa [2], the divergence of Pakistani specimen with other *P. subviridis* specimen (6.9% K2P) largely exceeds divergence observed among *P. subviridis* sensu Durand and Borsa [2] (1.8% K2P).

In BOLD, the BIN BOLD: ABU8792 consists of only 4 specimens (ANGEN 113-115, DBFN284-12, DBFN295-12, and GBMINI126937-17) collected at two localities: Gujarat, India, and the Persian Gulf, Iran. These barcodes are labeled as *Liza* sp. or *Minimugil cascasia*, which indicate nomenclature mistake, i.e., *Liza* is no longer considered as valid [52], or misidentification, *Minimugil cascasia* is endemic to rivers of northern Bengal and present a very different phylogenetic position [63]. However, the barcode's geographic distribution origins describe a geographic distribution for this MOTU (BOLD: ABU8792) similar to *P. klunzingeri*, from the Persian Gulf to NW India. This distribution is fully parapatric to the distribution of *P. subviridis* sensu Durand and Borsa [2] as BIN assigned to this last species consists of specimen sampled from West to East in India (Maharashtra and Kochi to Puducherry, (BOLD:AAC0695); Indonesia and Malaysia (BOLD:ACC0823); Indonesia and Papua New Guinea (BOLD:ACV9440); Philippines, Taiwan, and China (BOLD:ABY5947). Considering the type locality of *P. subviridis*, Valenciennes 1836 Ganges River, Malabar, India, the name *subviridis* should be maintained only for the BIN BOLD: AAC0695, the other close candidate species being named "*Planiliza* cf. *subviridis*". In the case of the MOTU present in Pakistan, we assigned provisional species name *Planiliza* cf. *subviridis* (BOLD:ABU8792) pending further morpho-anatomical investigation to determine potential diagnostic feature.

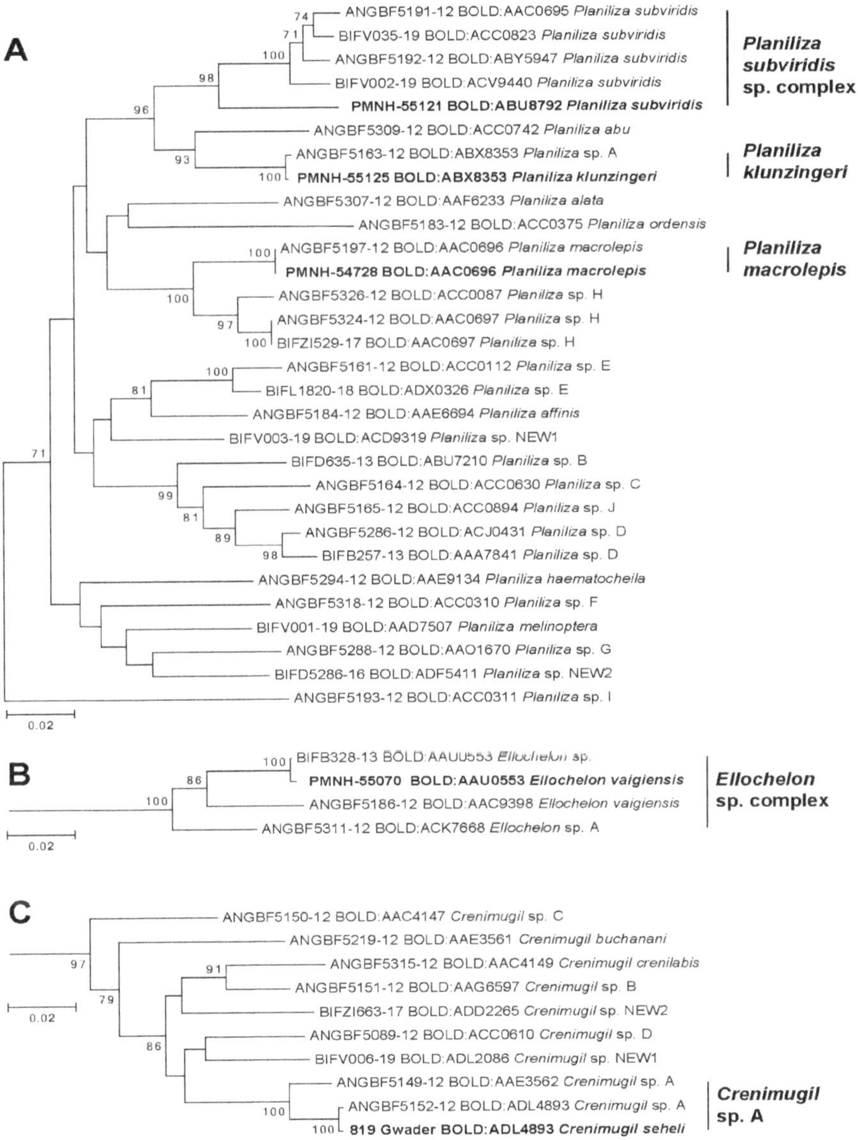

Figure 3. Phylogenetic tree of species in Genera *Planiliza* (**A**), *Ellochelon* (**B**), and *Crenimugil* (**C**). Leaves in bold represent a representative MOTU/BIN identified in Pakistan waters.

BOLD:AAU0553/*Ellochelon vaigiensis* (morphology)

Pakistani specimens assigned to the BIN BOLD:AAU0553 have been identified morphologically as *Ellochelon vaigiensis*. This BIN also includes the reference sequence from Indonesia labeled as *Ellochelon* sp. by Delrieu-Trottin et al. [31] but none of those depicted in Durand et al. [28] that identified two MOTUs in *Ellochelon vaigiensis* morpho species (Figure 3B). Later, based on the level of divergence that largely exceeds interspecific diversity, Durand et al. [2] proposed to provisionally name these MOTUs as *Ellochelon* sp. A for the lineage (BOLD:ACK7668) observed only in Australian specimens and maintained the name for the lineage (BOLD:AAC9398) present from Indonesia to French Polynesia. Following this logic, this third lineage corresponding to the BIN BOLD:AAU0553 with the divergence of 6.2% and 5.8% with *Ellochelon* sp. A and by *Ellochelon vaigiensis*, respectively, and is temporarily designated as *Ellochelon* cf. *vaigiensis* (BOLD:AAU0553). This MOTU is observed in specimens collected in Pakistan, as well as Iran, Malaysia, and Indonesia (BOLD, consultation 01/20/21). No significant phylogenetic relationship has been recovered in the COI phylogenetic tree; all MOTUs corresponding to *Ellochelon vaigiensis* sensu [2] descended from the same common ancestor (Figure 3B). A larger sampling scheme in the Indo-Pacific targeting *Ellochelon* spp. is necessary to better delineate the geographic structure of this species complex as well as its evolutionary history.

BOLD:ADL4893/*Crenimugil seheli* (morphology)

Specimen from Pakistan identified morphologically as *Crenimugil seheli* is assigned to the BIN BOLD:ADL4893 also included reference sequences of *Crenimugil* sp. A of Durand and Borsa [2] (Figure 3C). [28] Durand et al. [28] identified in *Crenimugil seheli* three lineages that occur sympatrically in the Indo-West Pacific; *Crenimugil* sp. A sensu Durand and Borsa [2] is one of this lineages. In BOLD, barcodes identified as *Crenimugil* sp. A by Durand and Borsa [2] are assigned to two BINs; BOLD:ADL4893 and BOLD:AAE3562 (Figure 3C). In BOLD, the BIN BOLD:ADL4893 is composed of barcodes observed in 9 specimens collected in the NW Indian Ocean (Iran, Saudi Arabia, Oman, and Seychelles), while BOLD:AAE3562 is composed of 50 specimens collected in the Indo-Pacific region (Reunion Island, Maldives, West Papua, China, Taiwan, Saipan, Australia, New Caledonia, and Fiji). Geographic distribution of these two MOTUs appears parapatric suggesting that these three are sibling species; the species present in Pakistan being assigned to provisional species name *Crenimugil* cf. *seheli* (BOLD:ADL4893).

BOLD: ACO4812/*Osteomugil* sp. (morphology)

Pakistani specimens assigned to this BIN have been morphologically identified as an *Osteomugil* species but their barcode does not correspond to any reference sequences. BIN BOLD:ACO4812 is associated with only one public specimen (LQDWL-TIS-31-12-2013-011) collected in India, Gujarat, close to Pakistan's border.

This specimen LGEN074-14 has been identified as *Planiliza tade* but the picture of specimen available in BOLD System (consultation 05/23/2021) indicates that it is an *Osteomugil* species: presence of a black dot at the birth of the pectoral fin, pectoral fin long reaching to the first dorsal fin vertically, and birth of second dorsal fin not to the birth of anal fin vertically. The phylogenetic position in the tree of this BIN also confirms that it is an *Osteomugil* species with a sister relationship with *Osteomugil* sp. C also collected in India. (Figure 4A). This MOTU is assigned to a provisional species name *Osteomugil* sp. (BOLD:ACO4812).

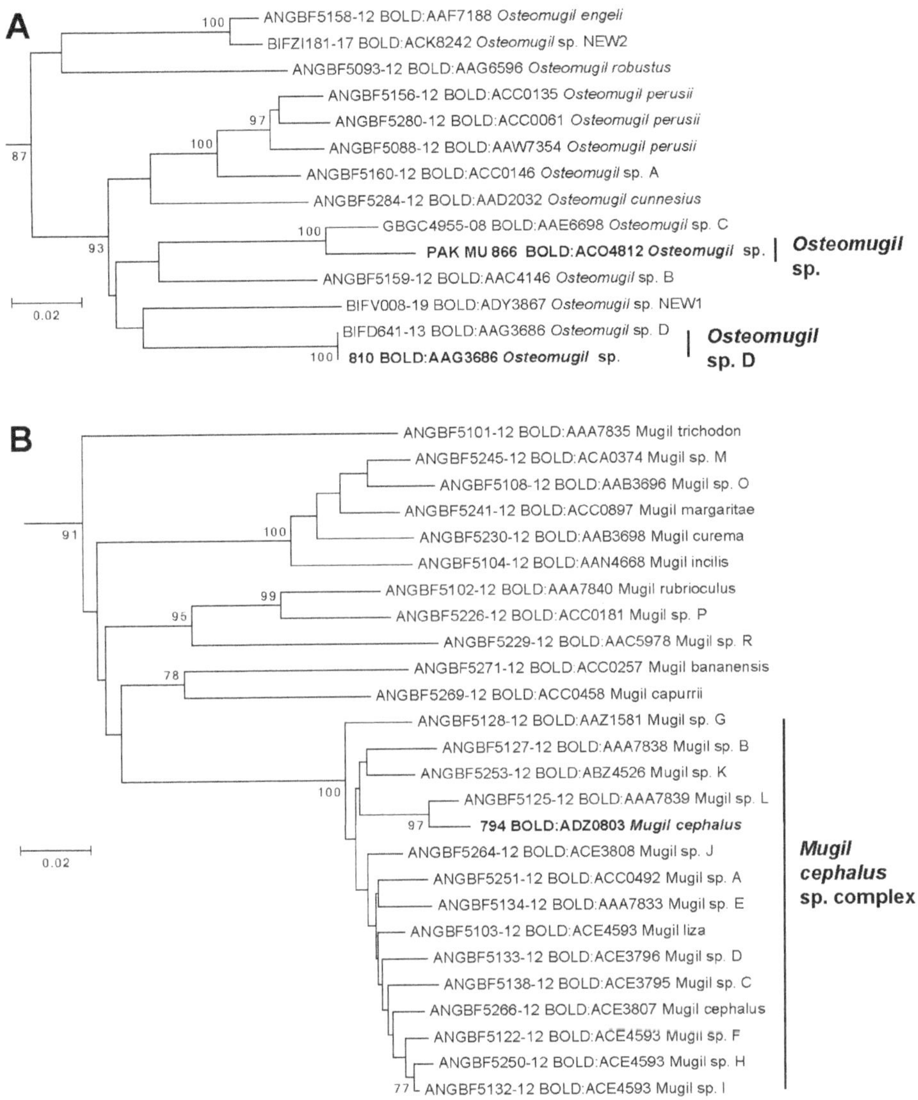

Figure 4. Phylogenetic tree of species in Genera *Osteomugil* (**A**) and *Mugil* (**B**). Leaves in bold represent a representative MOTU/BIN identified in Pakistan waters.

Pakistani specimen assigned to the BIN BOLD:AAG3686 present morphological characters of an *Osteomugil* species but the exact species was not identified. This BIN includes a reference sequence labeled as *Osteomugil* sp. D by Shen et al. [29] and Delrieu-Trottin et al. [31] but none depicted in Durand and Borsa [2] (Figure 4A). These reference sequences have been observed in a specimen collected in India and Indonesia. In BOLD, this BIN is associated with some additional barcodes obtained from India, Bangladesh, and Malaysia. This species provisionally named "*Osteomugil* sp. D" (BOLD: AAG3686) appear to be endemic to the Indian Ocean largely distributed from Pakistan to Indonesia. More taxonomical investigations are necessary to identify this species among all the *Osteomugil* species diversity described in the past.

BOLD: ADZ0803/*Mugil cephalus* (morphology)

Pakistani specimens assigned to this BIN BOLD: ADZ0803 have been identified morphologically as *Mugil cephalus*. No reference sequences have been observed in this BIN, while Durand et al. [28] and later Durand et al. [52] depicted in this morpho-species up to 13 and 14 MOTUs, respectively. In BOLD, this BIN is composed of 3 public specimens, collected in Bangladesh (GBMNB8388-20), India (GBGC9983-09), and one from unknown origin but the sequence produced from a laboratory in India (ANGBF54236-19). Within the *Mugil cephalus* species complex, these new MOTUs presents significant phylogenetic relationships with *Mugil* sp. L (BOLD:AAA7893) (Figure 4B) observed in the Pacific Ocean [2,18,64]. *Mugil cephalus* is considered as a species complex consisting of 15 candidate species including *Mugil liza* in the West Atlantic [2,18]. In some parts of the world, most of these species present parapatric distribution ranges such as the present new species provisionally named "*Mugil* cf. *cephalus*" (BOLD:ADZ0803) which is, to date, limited to the North Indian Ocean where no other *M. cephalus* MOTUs have been identified. When sympatry occurred, such as in the NW Pacific where three MOTU are present, reproductive isolation has been demonstrated, which confirms the validity of their species status [27,65]. No clear phylogenetic structure has been observed in the phylogenetic tree (Figure 4B) as well as a previous phylogenetic tree that included more molecular markers [28]. The diversification of *M. cephalus sensu lato* occurred during 5 million years (MYA) [32]. The divergence between the Indian *Mugil* cf. *cephalus* (BOLD:ADZ0803) and Pacific *Mugil* sp. L probably has rating of less than 2 MYA, considering its low level of divergence (1.7% K2P) by comparing to other lineages (mean distance 2.9% K2P).

More taxonomic and phylogenetic investigations are necessary to highlight the evolutionary history of this species complex present on a worldwide scale.

5. Conclusions

DNA barcoding appears to be the most efficient method for species identification and its advantage in the detection of cryptic species, an appealing application for many taxonomists [66]. The increasing number of new species detected through DNA barcoding suggests that the biodiversity level is greatly underestimated using solely the classical system of morphology-based identification.

In the present study, the COI gene was successfully used for species identification. Delimitation of MOTUs within the members of family Mugilidae found along the Pakistani waters was determined for the first time. Here, we morphologically characterized six species, although our specimens correspond genetically to eight MOTUs. The comparison of COI sequences generated in this study with the sequences available from different geographical regions [2,19] and BOLD system uncovered the existence of at least seven unknown candidate species from as much as a species complex. Analysis of the geographic distribution of *Planiliza* species present in Pakistan in light of the genetic diversity stressed the importance of Pakistan as a biogeographic border or transition between the NW Indian fauna and the rest of the Indo-Pacific region. This study calls for more taxonomic and phylogenetic investigations to describe Pakistani species and highlight the biogeographic component of Pakistan ichthyofauna in the Indo-Pacific area. This study will help in the development of DNA barcode reference data for the mullets of Pakistan which in turn would help in the management and conservation of fisheries. Furthermore, the novel sequences generated in this study and deposited in BOLD/GenBank will be available for future reference and research.

Supplementary Materials: The following are available online at https://www.mdpi.com/article/10.3390/d13060232/s1: Table S1: Mullet specimens list used as species reference from genera *Planiliza, Ellochelon, Crenimugil, Osteomugil* and *Mugil*.

Author Contributions: A.H., P.J.A.S. and S.A.A. collected the samples and performed morphological analysis. Molecular analysis was performed by J.-D.D. J.-D.D., A.H., and S.A.A. wrote the manuscript. All authors have read and agreed to the published version of the manuscript.

Funding: The molecular analysis was funded by IRD, France.

Institutional Review Board Statement: Not applicable.

Informed Consent Statement: Not applicable.

Data Availability Statement: The dataset generated/or analyzed during the current study are available from the corresponding author on reasonable request.

Conflicts of Interest: The authors declare that they have no conflict of interest.

References

1. Hubert, N.; Meyer, C.P.; Bruggemann, H.J.; Guerin, F.; Komeno, R.J.; Espiau, B.; Causse, R.; Williams, J.T.; Planes, S. Cryptic diversity in Indo-Pacific coral-reef fishes revealed by DNA-barcoding provides new support to the centre-of-overlap hypothesis. *PLoS ONE* **2012**, *7*, e28987. [CrossRef] [PubMed]
2. Durand, J.D.; Borsa, P. Mitochondrial phylogeny of grey mullets (Acanthopterygii: Mugilidae) suggest high proportion of cryptic species. *Comptes Rendus Biol.* **2015**, *338*, 226–277. [CrossRef] [PubMed]
3. Ward, R.D.; Holmes, B.H.; Yearsley, G.K. DNA barcoding reveals a likely second species of Asian sea bass (barramundi) (*Lates calcarifer*). *J. Fish Biol.* **2008**, *72*, 458–463. [CrossRef]
4. Ward, R.D.; Holmes, B.H.; White, W.T.; Last, P.R. DNA barcoding Australasian chondrichthyans: Results and potential uses in conservation. *Mar. Freshw. Res.* **2008**, *59*, 57–71. [CrossRef]
5. Lara, A.; PONCE de LEÓN, J.L.; Rodriguez, R.; Casane, D.; Cote, G.; Bernatchez, L.; Garcíamachado, E.R.I.K. DNA barcoding of Cuban freshwater fishes: Evidence for cryptic species and taxonomic conflicts. *Mol. Ecol. Resour.* **2010**, *10*, 421–430. [CrossRef]
6. Sriwattanarothai, N.; Steinke, D.; Ruenwongsa, P.; Hanner, R.; Panijpan, B. Molecular and morphological evidence supports the species status of the Mahachai fighter *Betta* sp. Mahachai and reveals new species of *Betta* from Thailand. *J. Fish Biol.* **2010**, *77*, 414–424. [CrossRef]
7. Smith, K.L.; Harmon, L.J.; Shoo, L.P.; Melville, J. Evidence of constrained phenotypic evolution in a cryptic species complex of agamid lizards. *Evolution* **2011**, *65*, 976–992. [CrossRef]
8. Puckridge, M.; Last, P.R.; White, W.T.; Andreakis, N. Phylogeography of the Indo West Pacific maskrays (Dasyatidae, *Neotrygon*): A complex example of chondrichthyan radiation in the Cenozoic. *Ecol. Evol.* **2013**, *3*, 217–232. [CrossRef]
9. Borsa, P.; Arlyza, I.S.; Hoareau, T.B.; Shen, K.N. Diagnostic description and geographic distribution of four new cryptic species of the blue-spotted maskray species complex (Myliobatoidei: Dasyatidae; *Neotrygon* spp.) based on DNA sequences. *J. Oceanol. Limnol.* **2018**, *36*, 827–841. [CrossRef]
10. Borsa, P.; Fauvelot, C.; Tiavouane, J.; Grulois, D.; Wabnitz, C.; Naguit, M.A. Andrefouet S Distribution of Noah's giant clam, *Tridacna noae*. *Mar. Biodivers.* **2015**, *45*, 339–344. [CrossRef]
11. Pante, E.; Puillandre, N.; Viricel, A.; Arnaud-Haond, S.; Aurelle, D.; Castelin, M.; Viard, F. Species are hypotheses: Avoid connectivity assessments based on pillars of sand. *Mol. Ecol.* **2015**, *24*, 525–544. [CrossRef] [PubMed]
12. Struck, T.H.; Cerca, J. Cryptic species and their evolutionary significance. *Encycl. Life Sci.* **2019**, 1–9. [CrossRef]
13. Fox, C.; Taylor, M.I.; Pereyra, R.; Rico, C. Mapping of the spawning grounds of Irish Sea gadoids using genetic identification of planktonic eggs. *Mol. Ecol.* **2005**, *14*, 879–884. [CrossRef]
14 Hubert, N.; Hanner, R.; Holm, E.; Mandrak, N.E.; Taylor, E.; Burridge, M.; Zhang, J. Identifying Canadian freshwater fishes through DNA barcodes. *PLoS ONE* **2008**, *3*, e2490. [CrossRef] [PubMed]
15. Bickford, D.; Lohman, D.J.; Sodhi, N.S.; Ng, P.K.L.; Meier, R.; Winker, K.; Das, I. Cryptic species as a window on diversity and conservation. *Trends Ecol. Evol.* **2007**, *22*, 148–155. [CrossRef] [PubMed]
16. Hyde, J.R.; Underkoffler, K.E.; Sundberg, M.A. DNA barcoding provides support for a cryptic species complex within the globally distributed and fishery important opah (*Lampris guttatus*). *Mol. Ecol. Resour.* **2014**, *14*, 1239–1247. [CrossRef] [PubMed]
17. Froese, R.; Pauly, D. FishBase. 2021. Available online: www.fishbase.org (accessed on 5 March 2021).
18. Whitfield, A.K.; Panfili, J.; Durand, J.D. A global review of the cosmopolitan flathead mullet *Mugil cephalus* Linnaeus 1758 (Teleostei: Mugilidae), with emphasis on the biology, genetics, ecology and fisheries aspects of this apparent species complex. *Rev. Fish Biol. Fish.* **2012**, *22*, 641–681. [CrossRef]
19. Durand, J.D.; Hubert, N.; Shen, K.N.; Borsa, P. DNA barcoding grey mullets. *Rev. Fish Biol. Fish.* **2017**, *27*, 233–243. [CrossRef]
20. Lee, S.C.; Chang, J.T.; Tsu, Y.Y. Genetic relationships of four Taiwan mullets (Pisces: Perciformes: Mugilidae). *J. Fish Biol.* **1995**, *46*, 159–162. [CrossRef]
21. Papasotiropoulos, V.; Klossa-Kilia, E.; Kilias, G.; Alahiotis, S. Genetic divergence and phylogenetic relationships in grey mullets (Teleostei: Mugilidae) using allozyme data. *Biochem. Genet.* **2001**, *39*, 155–168. [CrossRef]
22. Papasotiropoulos, V.; Klossa-Kilia, E.; Alahiotis, S.N.; Kilias, G. Molecular phylogeny of grey mullets (Teleostei: Mugilidae) in Greece: Evidence from sequence analysis of mtDNA segments. *Biochem. Genet.* **2007**, *45*, 623. [CrossRef]
23. Turan, C.; Caliskan, M.; Kucuktas, H. Phylogenetic relationships of nine mullet species (Mugilidae) in the Mediterranean Sea. *Hydrobiologia* **2005**, *532*, 45–51. [CrossRef]
24. Fraga, E.; Schneider, H.; Nirchio, M.; Santa-Brigida, E.; Rodrigues-Filho, L.F.; Sampaio, I. Molecular phylogenetic analyses of mullets (Mugilidae, Mugiliformes) based on two mitochondrial genes. *J. Appl. Ichthyol.* **2007**, *23*, 598–604. [CrossRef]

25. Erguden, D.; Gurlek, M.; Yaglioglu, D.; Turan, C. Genetic Identification and Taxonomic Relationship of Mediterranean Mugilid Species Based on Mitochondrial 16S rDNA Sequence Data. *J. Anim. Vet. Adv.* **2010**, *9*, 336–341.
26. Liu, J.Y.; Brown, C.L.; Yang, T.B. Phylogenetic relationships of mullets (Mugilidae) in China Seas based on partial sequences of two mitochondrial genes. *Biochem. Syst. Ecol.* **2010**, *38*, 647–655. [CrossRef]
27. Shen, K.N.; Jamandre, B.W.; Hsu, C.C.; Tzeng, W.N.; Durand, J.D. Plio Pleistocene sea level and temperature fluctuations in the northwestern Pacific promoted speciation in the globally distributed flathead mullet Mugil cephalus. *BMC Evol. Biol.* **2011**, *11*, 1–17.
28. Durand, J.D.; Shen, K.N.; Chen, W.J.; Jamandre, B.W.; Blel, H.; Diop, K.; Borsa, P. Systematics of the grey mullets (Teleostei: Mugiliformes: Mugilidae): Molecular phylogenetic evidence challenges two centuries of morphology-based taxonomy. *Mol. Phylogenet. Evol.* **2012**, *64*, 73–92. [CrossRef]
29. Shen, K.; Durand, J.D. The biogeography of Mugilidae in India, South-East and East Asia. In *Biology, Ecology and Culture of Grey Mullets (Mugilidae)*; Crosetti, D., Blaber, S., Eds.; CRC Press: Boca Racon, FL, USA, 2016; pp. 63–84. [CrossRef]
30. Xia, R.; Durand, J.D.; Fu, C. Multilocus resolution of Mugilidae phylogeny (Teleostei: Mugiliformes) implications for the family's taxonomy. *Mol. Phylogenet. Evol.* **2016**, *96*, 161–177. [CrossRef]
31. Delrieu-Trottin, E.; Durand, J.D.; Limmon, G.; Sukmono, T.; Sugeha, H.Y.; Chen, W.J.; Fitriana, Y. Biodiversity inventory of the grey mullets (Actinopterygii:Mugilidae) of the Indo Australian Archipelago through the iterative use of DNA based species delimitation and specimen assignment methods. *Evol. Appl.* **2020**, *13*, 1451–1467. [CrossRef]
32. Neves, J.M.; Almeida, J.P.; Sturaro, M.J.; FabrÉ, N.N.; Pereira, R.J.; Mott, T. Deep genetic divergence and paraphyly in cryptic species of *Mugil* fishes (Actinopterygii: Mugilidae). *Syst. Biodivers.* **2020**, *18*, 116–128. [CrossRef]
33. Briggs, J.C. Fishes of worldwide (circumtropical) distribution. *Copeia* **1960**, *3*, 171–180. [CrossRef]
34. Autem, M.; Bonhomme, F. Eléments de systématique biochimique chez les Mugilidés de Méditerranée. *Biochem. Syst. Ecol.* **1980**, *8*, 305–308. [CrossRef]
35. Caldara, F.; Bargelloni, L.; Ostellari, L.; Penzo, E.; Colombo, L.; Patarnello, T. Molecular phylogeny of grey mullets based on mitochondrial DNA sequence analysis: Evidence of a differential rate of evolution at the intrafamily level. *Mol. Phylogenet. Evol.* **1996**, *6*, 416–424. [CrossRef]
36. Rossi, A.R.; Capula, M.; Crosetti, D.; Campton, D.E.; Sola, L. Genetic divergence and phylogenetic inferences in five species of Mugilidae (Pisces: Perciformes). *Mar. Biol.* **1998**, *131*, 213–218. [CrossRef]
37. Rossi, A.R.; Ungaro, A.; De Innocentiis, S.; Crosetti, D.; Sola, L. Phylogenetic analysis of Mediterranean Mugilids by allozymes and 16S mt-rRNA genes investigation: Are the Mediterranean species of *Liza* monophyletic? *Biochem. Genet.* **2004**, *42*, 301–315. [CrossRef] [PubMed]
38. Murgia, R.; Tola, G.; Archer, S.N.; Vallerga, S.; Hirano, J. Genetic identification of grey mullet species (Mugilidae) by analysis of mitochondrial DNA sequence: Application to identify the origin of processed ovary products (bottarga). *Mar. Biotechnol.* **2002**, *4*, 119–126. [CrossRef]
39. Gornungm, E.; Colangelo, P.; Annesi, F. 5S ribosomal RNA genes in six species of Mediterranean grey mullets: Genomic organization and phylogenetic inference. *Genome* **2007**, *50*, 787–795. [CrossRef] [PubMed]
40. Imsiridou, A.; Minos, G.; Katsares, V.; Karaiskou, N.; Tsiora, A. Genetic identification and phylogenetic inferences in different Mugilidae species using 5S rDNA markers. *Aquac. Res.* **2007**, *38*, 1370–1379. [CrossRef]
41. Semina, A.V.; Polyakova, N.E.; Makhotkin, M.A.; Brykov, V.A. Mitochondrial DNA divergence and phylogenetic relationships in mullets (Pisces: Mugilidae) of the Sea of Japan and the Sea of Azov revealed by PCR-RFLP-analysis. *Russ. J. Mar. Biol.* **2007**, *33*, 187–192. [CrossRef]
42. Blel, H.; Chatti, N.; Besbes, R.; Farjallah, S.; Elouaer, A.; Guerbej, H.; Said, K. Phylogenetic relationships in grey mullets (Mugilidae) in a Tunisian lagoon. *Aquac. Res.* **2008**, *39*, 268–275. [CrossRef]
43. Durand, J.D.; Whitfield, A.K. Biogeography and Distribution of Mugilidae in the Western, Central and Southern Regions of Africa. In *Biology, Ecology and Culture of Grey Mullets (Mugilidae)*; Crosetti, D., Blaber, S., Eds.; CRC Press: Boca Racon, FL, USA, 2016; pp. 102–115. [CrossRef]
44. Menezes, N.A.; Nirchio, M.; Oliveira, C.D.; Siccharamirez, R. Taxonomic review of the species of *Mugil* (Teleostei: Perciformes: Mugilidae) from the Atlantic South Caribbean and South America, with integration of morphological, cytogenetic and molecular data. *Zootaxa* **2015**, *3918*, 1–38. [CrossRef]
45. Menezes, M.R.; Martins, M.; Naik, S. Interspecific genetic divergence in grey mullets from the Goa region. *Aquaculture* **1992**, *105*, 117–129. [CrossRef]
46. Rahman, M.A.U.; Khan, S.A.; Lyla, P.S.; Kumar, C.P. DNA barcoding resolves taxonomic ambiguity in mugilidae of Parangipettai waters (Southeast Coast of India). *Turk. J. Fish. Aquat. Sci.* **2013**, *13*, 321–330.
47. Qureshi, M.R. *Marine Fishes of Karachi and the Coast of Sindh and Makran*; Government of Pakistan Ministry of Food Agriculture; Government of Pakistan Press: Karachi, Pakistan, 1955; p. 80.
48. Bianchi, G. *Field Guide to the Commercial Marine and Brackish-Water Species of Pakistan*; FAO species identification sheets for fishery purposes; prepared with the support of PAK/77/033 and FAO (FIRM) Regular Programme; FAO: Italy, Rome, 1985; p. 200.
49. Fahmida, I. Mullets of Korangi Creek Karachi. *Zool. Surv. Rec. Pak.* **2002**, *14*, 11–18.
50. Froese, R.; Pauly, D. Fish Base. World Wide Web Electronic Publication. 2010. Available online: www.fishbase.org (accessed on 10 September 2010).

51. Psomadakis, P.N.; Osmany, H.B.; Moazzam, M. *Field Identification Guide to the Living Marine Resources of Pakistan*; FAO Species Identification Guide for Fishery Purposes; FAO: Italy, Rome, 2015; 386p.
52. Durand, J.D.; Chen, W.J.; Shen, K.N.; Borsa, P. Genus-level taxonomic changes imposed by the mitochondrial phylogene.y of grey mullets (Teleostei: Mugilidae). *Comptes Rendus Biol.* **2012**, *335*, 687–697. [CrossRef]
53. Thomson, J.M. The Mugilidae of the world. *Mem. Qld. Mus.* **1997**, *41*, 457–562.
54. Ghasemzadeh, J. Phylogeny and Systematics of Indo-Pacific Mullets (Teleostei: Mugilidae) with Special Reference to the Mullets of Australia. Ph.D. Thesis, Macquarie University, Sydney, Australia, 1998.
55. Spalding, M.D.; Fox, H.E.; Allen, G.R.; Davidson, N.; Ferdaña, Z.A.; Finlayson, M.A.X.; Robertson, J. Marine eco regions of the world: A bio regionalization of coastal and shelf areas. *BioScience* **2007**, *57*, 573–583. [CrossRef]
56. Ward, R.D.; Zemlak, T.S.; Innes, B.H.; Last, P.R.; Hebert, P.D. DNA barcoding of Australia's fish species. *Philos. Trans. R. Soc. Lond. B Biol. Sci.* **2005**, *360*, 1847–1857. [CrossRef]
57. Ratnasingham, S.; Hebert, P.D. A DNA-based registry for all animal species: The Barcode Index Number (BIN) system. *PLoS ONE* **2013**, *8*, e66213. [CrossRef] [PubMed]
58. Kumar, S.; Stecher, G.; Tamura, K. MEGA7: Molecular evolutionary genetics analysis version 7.0 for bigger datasets. *Mol. Biol. Evol.* **2016**, *33*, 1870–1874. [CrossRef]
59. Kimura, M. A simple method for estimating evolutionary rates of base substitutions through comparative studies of nucleotide sequences. *J. Mol. Evol.* **1980**, *16*, 111–120. [CrossRef] [PubMed]
60. Senou, H.; Yoshino, T.; Okiyama, M. A review of the mullets with a keel on the back, *Liza carinata* complex (Pisces: Mugilidae). *Publ. Seto Mar. Biol. Lab.* **1987**, *32*, 303–321. [CrossRef]
61. Liu, L.; Panhwar, S.K.; Gao, T.; Han, Z.; Li, C.; Sun, D.; Song, N. New genetic evidence from three keel-backed liza species based on DNA Barcoding confirms morphology-based identification. *Pak. J. Zool.* **2017**, *49*, 1901–1907.
62. Khedkar, G.D.; Jamdade, R.; Naik, S.; David, L.; Haymer, D. DNA barcodes for the fishes of the Narmada, one of India's longest rivers. *PLoS ONE* **2014**, *9*, e101460. [CrossRef] [PubMed]
63. Durand, J.D. Implications of Molecular Phylogeny for the Taxonomy of Mugilidae. In *Biology, Ecology and Culture of Grey Mullets (Mugilidae)*; Crosetti, D., Blaber, S., Eds.; CRC Press: Boca Racon, FL, USA, 2016; pp. 22–41. [CrossRef]
64. Viet Tran, T.T.; Ke Phan, L.; Durand, J.D. Diversity and distribution of cryptic species within the *Mugil cephalus* species complex in Vietnam. *Mitochondrial DNA Part A* **2017**, *28*, 493–501. [CrossRef] [PubMed]
65. Shen, K.N.; Chang, C.W.; Durand, J.D. Spawning segregation and philopatry are major prezygotic barriers in sympatric cryptic *Mugil cephalus* species. *Comptes Rendus Biol.* **2015**, *338*, 803–811. [CrossRef] [PubMed]
66. Knebelsberger, T.; Landi, M.; Neumann, H.; Kloppmann, M.; Sell, A.F.; Campbell, P.D.; Laakmann, S.; Raupach, M.J.; Carvalho, G.R.; Costa, F.O. A reliable DNA barcode reference library for the identification of the North European shelf fish fauna. *Mol. Ecol. Resour.* **2014**, *14*, 1060–1107. [CrossRef]

diversity

MDPI

Article

Polychaete Diversity Related to Different Mesophotic Bioconstructions along the Southeastern Italian Coast

Maria Flavia Gravina [1,2,*], Cataldo Pierri [2,3], Maria Mercurio [2,3], Carlotta Nonnis Marzano [2,3] and Adriana Giangrande [2,4]

1 Department of Biology, University of Rome "Tor Vergata", 00133 Rome, Italy
2 CoNISMa, Consorzio Interuniversitario per le Scienze del Mare, 00196 Rome, Italy
3 Department of Biology, University of Bari Aldo Moro, 70125 Bari, Italy
4 Department of Biological and Environmental Sciences and Technologies, University of Salento, 73100 Lecce, Italy
* Correspondence: maria.flavia.gravina@uniroma2.it

Abstract: In the different mesophotic bioconstructions recently found along the Southeastern Italian coast, polychaetes have been proved to show high species richness and diversity, hitherto never investigated. In the present study, the species composition and functional role of polychaete assemblages were analysed; the updated key to identification of the Mediterranean species of genus *Eunice* was presented and some taxonomic issues were also discussed. On the total of 70 species Serpulidae and Eunicida were the dominant polychaetes. Facing similar levels of α-diversity, the polychaete assemblages showed a high turnover of species along the north-south gradient, clearly according to the current circulation pattern, as well as to the different bioconstructors as biological determinants. Indeed, Serpulidae were dominant on the mesophotic bioconstructions primarily formed by the deep-sea oyster *Neopycnodonte cochlear*, while the Eunicida prevailed on the mesophotic bioconstructions mainly built by scleractinians. Lastly, the record of *Eunice dubitata* was the first for the Mediterranean and Italian fauna and proved this species to be characteristic of mesophotic bioconstructions.

Keywords: Polychaete Eunicida; Polychaete Serpulidae; marine bioconstructions; polychaete diversity; mesophotic bioconstructions; Mediterranean Sea; Southeastern Italian coast; Italian fauna

Citation: Gravina, M.F.; Pierri, C.; Mercurio, M.; Nonnis Marzano, C.; Giangrande, A. Polychaete Diversity Related to Different Mesophotic Bioconstructions along the Southeastern Italian Coast. *Diversity* **2021**, *13*, 239. https://doi.org/10.3390/d13060239

Academic Editors: Michael Wink and Bert W. Hoeksema

Received: 30 April 2021
Accepted: 27 May 2021
Published: 31 May 2021

Publisher's Note: MDPI stays neutral with regard to jurisdictional claims in published maps and institutional affiliations.

Copyright: © 2021 by the authors. Licensee MDPI, Basel, Switzerland. This article is an open access article distributed under the terms and conditions of the Creative Commons Attribution (CC BY) license (https://creativecommons.org/licenses/by/4.0/).

1. Introduction

Mediterranean polychaetes are proved to be good bioindicators of environmental conditions and ecological status both on sedimentary and rocky bottoms [1–3]. Such results are achieved following investigations especially in shallow habitats, as well as in circalittoral habitats where coralligenous formations occur [4–6].

The coralligenous is a characteristic Mediterranean biocoenosis which is an object of detailed studies, due to its role in shaping the seascape, formed by perennial algae and animal organisms with consistent calcareous concretions in sciaphilic environments, from 20 to 120 m depth [7–11]. Moreover, particular attention has recently been focused on more deep-sea habitats, such as sea mountains, non-symbiotic coral reefs, and submarine canyons, where peculiar invertebrates, mainly scleractinians, gorgonians, and antipatharians, act as the main habitat formers [12–18]. Also, the habitats located in the twilight zone, so-called "mesophotic zone" ranging from 30–40 to 150 m depth, are currently under study. Unfortunately, such habitats are still poorly investigated, and some possible confusion exists in the definition of the mesophotic zone (see Cerrano et al. [19]). Few recent studies have been conducted in the Ligurian Sea and Tyrrhenian Sea [20,21], as well as in the Southern Adriatic off the Italian coast [22–24].

The peculiar mesophotic communities recently found along the Apulian coast were described by Corriero et al. [22] and Cardone et al. [24], paying particular attention to their characterization and underlining their crucial ecological role in supporting high habitat complexity and enhancing local biodiversity. Only secondary to the primary bioconstructors of such bioconstructions, polychaetes have been proved to be a very rich and diverse component of such mesophotic communities. This group is very interesting because includes both vagile and sessile species, which may play diverse roles in forming the bioconstruction architecture and contribute to its functioning. Among the vagile forms the components of the family Eunicidae are known to be particularly relevant, some species of which are considered symbiotic with corals [25,26].

In the present study the polychaete assemblages associated with the mesophotic bioconstructions recently discovered along the Apulian coast are considered, with special focus on the following objectives: (i) to analyse their diversity patterns in the different bioconstructions both in terms of species composition and functional role; (ii) to analyse the taxonomic issues within the family Eunicidae in order to clarify the statement of the species referred to the genus *Eunice*; (iii) to update the checklist of the Mediterranean and Italian fauna of the polychaetes referred to the genus *Eunice*.

2. Material and Methods

2.1. Study Areas

The three study areas are located along the Southeastern Italian coast (Adriatic and Ionian coast of Apulia) (Figure 1), and harbour three different bioconstructions recently discovered and described [22,24].

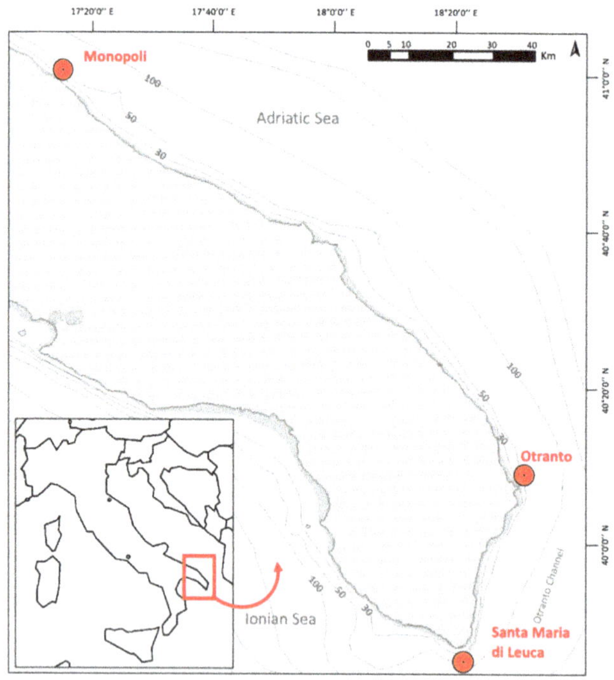

Figure 1. Location of the three areas under study off the Southeastern Italian coast (Apulia).

Monopoli site (MON) is the northernmost and located approximately 1.5 nautical miles off the coast in the Southern Adriatic Sea. At this site the bioconstruction is up to 2 m in thickness and occurs in a depth range between 30 and 55 m, along a fault line with NW-SE orientation. The bioconstruction is characterized by two species of non-symbiotic scleractinians *Phyllangia americana mouchezii* (Lacaze–Duthiers, 1897) and *Polycyathus muellerae* (Abel, 1959) as primary bioconstructors. The site of Otranto (OTR) is located at about 1 nautical mile off the coast in the Otranto Channel, which marks a passage from the Adriatic to the Ionian Sea, within the bathymetric range 45–64 m; the site of Santa Maria di Leuca (SML) is placed approximately at 1.5 nautical miles off the coast in the Ionian Sea within a 45–70 m depth range. In both these latter sites the bioconstructions are mainly built by the bivalve *Neopycnodonte cochlear* (Poli, 1795) and give rise to different structures like pinnacles and globose formations. The structural complexity of the bioconstruction at all the sites strongly supports a high level of biodiversity [22,24].

2.2. Sampling Methods and Taxonomic Analysis

Samplings were performed during the period August–September, in 2017 at MON and in 2018 at OTR and SML. The seafloor was dominated by a fault perpendicular to the coastline up to 55 m depth at MON and by a steep slope connecting the upper zone to the deeper areas, 64 m and 70 m, respectively, at OTR and SML. The coastal dynamics were characterized by a wave-dominated microtidal setting and the wave conditions produced prevailingly coastal longshore currents with NW–SE direction. For the characterization of the benthic assemblages, three samples (approximately three liters of total volume) were scraped from the bioconstruction at each study area by technical divers at 45–50 m depth, according to the sampling methods described by Corriero et al. [22] and Cardone et al. [24]. The collected biological material was sorted in the laboratory, and all the specimens were preserved in an alcohol solution and then identified at the lowest possible taxonomical level, using Zeiss stereo and optical microscopes. In the present study, the polychaete component of the benthos has been considered.

For the polychaete's faunal analysis, the vagile specimens were extracted from the samples and their position within the bioconstruction was observed and recorded. Similarly, the aggregation distribution, as well as the growth mode and superimposition of the calcareous tubes of the sessile worms were noticed and reported, with the aim to understand the role played by the polychaetes with respect to the bioconstruction organization and functioning. The polychaete specimens preserved in the authors' collections and in the collection of the Museum of Porto Cesareo Zoological Laboratory (PCZL), University of Salento, Italy, were also analysed and compared with the material of the present samples, in order to disentangle some criticisms about the identification of the Eunicidae. The following complete revisions and past and current significant literature regarding this family were also examined [27–38], with the purpose of producing a revised and updated dichotomous key for the identification of the Mediterranean species of the genus *Eunice sensu lato*, within the family Eunicidae.

2.3. Data Analysis

Polychaete diversity was measured in terms of species richness (α-diversity) and species turnover along the local geographical North-South gradient (β-diversity). This latter was computed using the Whittaker Index $\beta_W = (S/\bar{a}) - 1$, where S is the total number of species that results from merging the number of species of each pairwise of sites considered and $\bar{a}$ is the average number of species per each pairwise of a sample [39,40]. The species similarity between polychaete assemblages of the investigated sites was measured by means of the Sörensen Index.

3. Results

3.1. Species Composition and Diversity

A total of 70 species were found, most of them belonging to the family Serpulidae (54.3% with 38 species) and to the order Eunicida (23% with 15 species), this latter being mostly represented by species of the family Eunicidae accounting for 10 species (Table 1). Among them, *Eunice dubitata* constitutes a new record for the Italian fauna and amends the checklist [41].

Table 1. List of polychaete species recorded at three sites along the Apulian coasts.

Site		Monopoli	Otranto	Santa Maria di Leuca
Depth Main Structuring species		43 m Scleractinians	52 m *N. cochlear*	60 m *N. cochlear*
Family	**Species**			
Sabellidae Latreille, 1825	*Hypsicomus stichophthalmos* (Grube, 1863)	x		
Serpulidae Rafinesque, 1815	*Hydroides pseudouncinata* (Zibrowius, 1968)	x	x	
	Serpula vermicularis Linnaeus, 1767	x	x	x
	Serpula cavernicola Fassari & Mollica, 1991	x	x	x
	Serpula concharum Langerhans, 1880	x	x	x
	Serpula lobiancoi Rioja, 1917		x	x
	Serpula israelitica Amoureux, 1977			x
	Spiraserpula massiliensis (Zibrowius, 1968)	x	x	x
	Vermiliopsis infundibulum (Philippi, 1844)	x	x	x
	Vermiliopsis striaticeps (Grube, 1862)	x	x	x
	Vermiliopsis monodiscus Zibrowius, 1968	x		x
	Vermiliopsis labiata (O. G. Costa, 1861)	x	x	x
	Bathyvermilia eliasoni (Zibrowius, 1970)			x
	Metavermilia multicristata (Philippi, 1844)	x	x	x
	Semivermilia agglutinata (Marenzeller, 1893)		x	x
	Semivermilia crenata (O. G. Costa, 1861)	x	x	x
	Semivermilia cribrata (O. G. Costa, 1861)		x	x
	Semivermilia pomatostegoides (Zibrowius, 1969)		x	x
	Filogranula gracilis Langerhans, 1884	x		x
	Filogranula calyculata (O. G. Costa, 1861)		x	x
	Filogranula annulata (O. G. Costa, 1861)		x	x
	Janita fimbriata (Delle Chiaje, 1822)	x		x
	Spirobranchus lima (Grube, 1862)			x
	Spirobranchus polytrema (Philippi, 1844)			x
	Spirobranchus triqueter (Linnaeus, 1758)	x	x	x
	Placostegus tridentatus (Fabricius, 1779)		x	x
	Verminia cristallina (Philippi, 1844)	x	x	x
	Josephella marenzelleri Caullery & Mesnil, 1896		x	x
	Filograna implexa Berkeley, 1835	x	x	x
	Protula tubularia (Montagu, 1803)		x	x
	Spirorbis cuneatus Gee, 1964			x
	Spirorbis marioni Caullery & Mesnil, 1897		x	x
	Protolaeospira striata (Quiévreux, 1963)		x	x
	Janua heterostropha (Montagu, 1803)		x	x
	Neodexiospira pseudocorrugata (Bush, 1905)		x	x
	Pileolaria militaris Claparède, 1870		x	x
	Pileolaria heteropoma (Zibrowius, 1968)			x
	Vinearia koehleri (Caullery & Mesnil, 1897)			x
	Nidificaria clavus (Harris, 1968)		x	
Euphrosinidae Williams, 1852	*Euphrosine foliosa* Audouin & H Milne Edwards, 1833	x	x	x
Eunicidae Berthold, 1827	*Eunice dubitata* Fauchald, 1974	x	x	
	Eunice schizobranchia Claparède, 1870	x		
	Eunice pennata (Müller, 1776)			x
	Eunice floridana (Pourtalès, 1867)			x
	Eunice torquata (Quatrefages, 1866)	x	x	
	Palola siciliensis (Grube, 1840)	x	x	
	Palola valida (Gravier, 1900)		x	
	Paucibranchia fallax (Marion & Bobretzky, 1875)	x	x	
	Lysidice collaris Grube, 1870	x	x	x
	Lysidice ninetta Audouin & H Milne Edwards, 1833	x	x	
Lumbrineridae Schmarda, 1861	*Lumbrineris coccinea* (Renier, 1804)	x	x	
	Scoletoma laurentiana (Grube, 1863)	x		
Oenonidae Kinberg, 1865	*Arabella geniculata* (Claparède, 1868)	x	x	
	Arabella iricolor (Montagu, 1804)	x		
	Drilonereis filum (Claparède, 1868)	x	x	
Glyceridae Grube, 1850	*Glycera tesselata* Grube, 1863	x		
	Glycera unicornis Lamarck, 1818	x		
Goniadidae Kinberg, 1866	*Glycinde nordmanni* (Malmgren, 1866)	x		
	Goniada emerita Audouin & H Milne Edwards, 1833	x		
	Goniada maculata Örsted, 1843	x	x	
Nereididae Blainville, 1818	*Ceratonereis costae* (Grube, 1840)	x	x	
Polynoidae Kinberg, 1856	*Harmothoe antilopes* McIntosh, 1876	x		
	Harmothoe pagenstecheri Michaelsen, 1896		x	x
	Lepidasthenia elegans (Grube, 1840)	x		
Syllidae Grube, 1850	*Haplosyllis spongicola* (Grube, 1855)		x	
	Syllis alternata Moore, 1908	x		
	Syllis gracilis Grube, 1840	x	x	
	Syllis ferrani Alós & San Martín, 1987		x	
	Syllis variegata Grube, 1860	x		x
	Sphaerosyllis hystrix Claparède, 1863		x	

Species richness was highest at OTR, where 46 species were identified; whilst at SML and MON 42 and 40 were respectively found. However, when compared, sites showed a different species composition. Serpulids were the most responsible for the SML species richness, whilst eunicids for the richness of MON and OTR.

Among serpulids, at MON *Hydroides pseudouncinata*, a shallow and sciaphilic and coralligenous species, and *Serpula massiliensis*, a species characteristic of caves, were the most common species. Here the bioconstruction, mostly built by corals, was made in its basal layer by a dense intertwining of dead corallites and tubes of the largest serpulids, particularly *Serpula cavernicola* and *S. massiliensis*, which are typical inhabitants of submerged caves. Moreover, the exposed surface of the reef was extensively colonized by a high number of species: together with *S. cavernicola*, two main gregarious species, *S. massiliensis* and *Filograna implexa*, contributed significantly to the formation of the reef especially colonizing outlines and overhangs of the concretion: the first species was found with many intertwined tubes and the second one with assemblages which, even made by fragile and friable tubes, showed very large extension. Many serpulids were observed with their tubes in epibiosis on calcareous skeletons and valves of other sessile organisms, such as *Hydroides pseudouncinata, Spirobranchus triqueter, Serpula vermicularis, Serpula concharum* which are largely occurring also on shallow-shelf habitats; *Vermiliopsis monodiscus, Janita fimbriata, Metavermilia multicristata* which are deep-water and bathyal species; *Semivermilia crenata, Placostegus crystallinus, Filogranula gracilis, Vermiliopsis labiate*, which are characteristic of coralligenous and cave habitats.

At OTR and SML few highly dominant species of serpulids, including *Vermiliopsis infundibulum, S. crenata, Filogranula annulata* and *Semivermilia pomatostegoides*, occurred, and they were found together with some particularly abundant small species of spirorbids, such as *Protolaeospira (Protolaeospira) striata* and *Pileolaria militaris*. Different groups of species occurred in different microhabitats of the bioconstruction: species of shallow shelf (*Spirobranchus polytrema, Janua pagenstecheri, Pomatoceros triqueter*) and detritic bottoms (*Semivermilia cribrata* and *Spirorbis (Spirorbis) cuneatus*) mostly occurred on the outer edges of the reef, whilst species characteristic of deep cryptic and cave habitats (*Vermiliopsis monodiscus, Serpula israelitica, F. gracilis S. cavernicola, F. annulata*) were principally found within the *Neopycnodonte* valves and the reef interstices. Spirorbids showed particular adaptation to cryptic and deep crevices of the bioconstruction, as a result of their small dimensions and often-wrapped tubes; *P. (P.) striata, P. militaris* and *Vinearia koehleri* were also observed on the bare surfaces, such as the external edge and the smooth inner parts of the *Neopycnodonte* valves.

Among the vagile fauna, eunicids were the most abundant polychaetes at all the investigated sites: particularly *Eunice dubitata* and *Eunice torquata* were abundant especially at MON and OTR, whilst *Eunice pennata* and *Eunice floridana* were abundant at SML. In addition, *Lysidice collaris* was found abundant at all the sites, and the other abundant species *Palola siciliensis* and *Paucibranchia fallax* only occurred at MON and OTR.

Most of the specimens of *E. dubitata, E. torquata, E. floridana* and *P. siciliensis* were extracted from sinuous and twisted galleries entirely surrounded by corallites of scleractinians and *Neopycnodonte* valves; most of the complete specimens were large in dimensions, *E. dubitata* reaching a total length of 220 mm with 200 segments, *E. torquata* 180 mm with 210 segments, *E. floridana* 90 mm with 125 segments, respectively. Among the species of the genus *Eunice, E. torquata* and *E. pennata* are known from sciaphilic detritic and coralligenous habitats, while *E. dubitata* and *E. floridana* are reported from the deep to bathyal zone often associated to corals; contrarily, *P. siciliensis* and *P. fallax* were widely distributed both on infra-circalittoral detritic and shelf environments [5,42].

Lumbrineridae were also abundant even if represented by only two species. We paid particular attention to the identification of *Lumbrineris coccinea* which was found abundant at MON and OTR. This species, in fact, is often misidentified along the European coasts [32]. Our specimens were characterized by maxillary apparatus bearing five pairs of maxillae,

MI as long as MII, MIII clearly bidentate and by composite multidentate hooded hooks with short blade (Figure S1).

The polychaete α-diversity was similar at the three examined sites and ranged from 46 (OTR) to 40 species (MON). Conversely, the β-diversity varied from the highest values, 0.41 and 0.55, computed between MON and OTR and MON and SML respectively, to the lowest value, 0.33, found between OTR and SML. These results are in agreement with the measures of similarity which varied from a minimum of 0.43 between MON and SML and a maximum of 0.64 between OTR and SML (Table 2).

Table 2. Number of species (α), Sörensen Index (SI) and Whittaker Index (β) for the polychaete assemblages of the mesophotic bioconstructions along the Apulian coast.

Sites	α	SI	β
MON (α)/MON-OTR (SI, β)	40	0.59	0.41
OTR (α)/MON-SML (SI, β)	46	0.43	0.55
SML (α)/OTR-SML (SI, β)	42	0.64	0.33

3.2. Taxonomic Accounts

Polychaete members belonging to the genus *Eunice sensu lato* were particularly abundant in the mesophotic bioconstructions of the investigated Southeastern Italian coast, Apulian, and the collected material was also used for choosing the best diagnostic characters for the identification of the Mediterranean deep-water species. In this study the species found associated with the Mediterranean mesophotic bioconstructions were *E. torquata, E. dubitata, E. floridiana, E. pennata,* while *E. norvegica* was found associated with deep white corals [16,17]. Shape and articulation of the antennae was one of the best characters to separate such species (Figure 2).

Figure 2. Anterior end in dorsal view of the species of the genus *Eunice sensu lato* found in the mesophotic bioconstructions investigated in the Southeastern Italian coast: upper line *E. dubitata* (left), *E. floridana* (right); middle line *E. schizobranchia* (left), *E. torquata* (right); bottom line *E. pennata* (left) *E. norvegica* (right).

The specimens of the five eunicid species collected in the present work were compared with the material of the Mediterranean species preserved in the PCZL collection (Porto Cesareo Zoological Laboratory, University of Salento, Italy) and in the personal collections of two authors. Based on the comparison and on the dated and current literature on the subject, a new updated key to identification of *Eunice* species of the Mediterranean is here proposed. The following diagnostic characters were considered for the key: shape and length of the antennae, peristomial and dorsal cirri, shape and disposition of the branchiae, chaetal morphology and arrangement, colour pattern. For the identification procedure all the species have been ascribed to the genus *Eunice sensu lato*, because we refer the systematic issue on the distinction between the genera *Eunice* and *Leodice* to the discussion.

Key to identification of Mediterranean *Eunice sensu lato* species:

1. Antennae smooth, short and terminating at the same height. Branchiae starting far from the prostomium from 60°–70° chaetigers; peristomial cirri short not reaching the middle of the anterior peristomium ring . . . . *Eunice schizobranchia*
   Branchiae starting within the first chaetigers 2
2. Acicula yellow 3
   Acicula black 6
3. Acicular hooks bidentate 4
   Acicular hooks tridentate 5
4. Antennae indistinctly annulated or smooth; branchiae pectinate with up to 10–12 filaments, starting at the 3° chaetiger and lacking in the posterior part of the body . . . . *Eunice pennata*
   Antennae clearly annulated long, with long cylindrical articles; branchiae pectinate with up to 10–16 filaments; first branchiae at 3°–4° chaetiger and to near posterior end . . . . *Eunice harassii*
5. Antennae deeply annulated, moniliform with short segments. Branchiae pectinate with maximally 12 filaments, starting from chaetiger 4°–7° and present in the posterior segments . . . . *Eunice antennata*
   Antennae indistinctly articulated. Branchiae pectinate with maximally 12–14 filaments. starting from chaetiger 3 and lacking in the posterior segments . . . . *Eunice vittata*
6. Antennae smooth or indistinctly articulated 7
   Antennae regularly articulated or moniliformis 8
7. Antennae smooth and short, similar in length reaching the 2°–4° chaetiger, branchiae starting at 6°–10°, often 8°–9°, chaetiger, pectinate with numerous up to 15–40 filaments longer than the notopodial cirri, very large species . . . . *Eunice roussaei*
   Antennae indistinctly articulated and long, first branchiae on chaetiger 6°–10° often 7°, pectinate with 3–12 filaments, about as long as the notopodial cirri . . . . *Eunice norvegica*
8. Antennae articulated with cylindrical articulations; branchiae starting from 3°–4° chaetiger, with 2–3 rarely 5–6 filaments . . . . *Eunice oerstedi*
   Antennae clearly moniliform 9
9. First branchiae on chaetiger 7°–10° often 9°, branchiae pectinate with 4–8 filaments longer than the notopodial cirri; antennae articulated with moniliform increasingly distally articulations . . . . *Eunice floridana*
   First branchiae starting before 7° chaetiger 10
10. Branchiae poor developed, palmate with few 1–2 maximally 3 filaments, starting from 3° chatiger . . . . *Eunice dubitata*
    Branchiae well developed 11
11. Branchiae starting at 2°–3° chaetiger, pectinate with up to 8–10 filaments; body colour pattern uniform bright orange without spots nor whitish ring on anterior segment; subacicular hooks always in single arrangement . . . . *Eunice laurillardi*
    Branchiae starting from 3° chaetiger, pectinate with several up to 10–14, often 7, filaments,

body colour pattern red brown with a white collar at the 4° chaetiger; subacicular hooks in double arrangement in medial and posterior segments . . . . *Eunice torquata*

4. Discussion

Polychaete assemblages associated with the different mesophotic bioconstructions investigated off the Southeastern Italian coast, Apulia, proved to be very rich and diverse in species number and composition, with 70 species recorded overall, most of which are included both in the family Serpulidae and in the order Eunicida. In fact, polychaetes were dominant in terms of number of species compared with the other taxonomic groups associated with the bioconstructions, such as Bryozoans, with 50 species recorded throughout all sites [43], and Porifera with 59 and 65 species collected respectively from the coral reef of MON and from the *Neopycnodonte* bioconstructions of OTR and SML [22,24]. Facing the similar levels of α-diversity, ranging from 40 to 46 species, the polychaete assemblages of the different mesophotic bioconstructions showed a high turnover rate along the geographical North–South gradient as showed by β-diversity and by the similarity Sörensen index, with the highest affinity between southern sites: e.g., *Lepidastenia elegans* and *Harmothoe antiplopes* from MON were substituted by *Harmothoe pagenstecheri* at OTR and SML; *Eunice dubitata* and *E. schizobranchia* characterized MON and OTR instead of *E. pennata* and *E. floridana* which characterized SML; *Syllis alternata* and *S. variegata* from MON were replaced by *Syllis ferrani* and *Haplosyllis spongicola* at OTR; *Filogranula calyculata* and *Filogranula annulata* from OTR and SML took the place of *Filogranula gracilis* from MON. In addition, *Hydroides pseudouncinata*, exclusive to northern sites, was replaced by various species exclusive to southern sites i.e., *Serpula lobiancoi, Plagosteus tridentatus, Josephella marenzelleri, Protula tubularia*. On the basis of these results, it should be hypothesized that the observed increase of β-diversity is driven primarily by the typical hydrological features of the Apulian Adriatic coast [44–48]. This is supported by the circulation pattern of the area, where the surface current gyre flows southeastward, being responsible for moving larvae and propagules in such a direction and, as a consequence, for the different degree of connectivity between the sites (pre-settlement factors). However, this is not the only factor driving benthic community dynamics and population connectivity, because also competitive and facilitating factors are well known to be very influential (see Giangrande et al. [49]). Among the biological determinants, we highlight the relevant role of the bioconstructors, which are primarily responsible for the different bioconstructions, being mainly scleractinians at MON and deep oysters (*Neopycnodonte cochlear*) at OTR and SML [22,24]. In agreement with the above explanation, other studies come to similar conclusions, concerning the bryozoan assemblages from the same area [43] and other taxa from different habitats, e.g., molluscs [50], sponges [51] and brackish waters communities [52,53].

A further result of our study concerns the role of polychaetes in affecting the mesophotic bioconstruction structure. In fact, within the polychaete assemblages studied, we found a relevant functional guild diversity, among which the roles of secondary constructors, binders, dwellers, destroyers-borers may be detected. Serpulids are the main builder worms, which were subordinate in terms of carbonate production to scleractinians and deep oysters, but notwithstanding they contributed to increase the concretion structure, fixing their calcareous tubes in epibiosis on skeletons and valves of the primary constructors. So, they consistently enhanced surface heterogeneity, particularly by the species exhibiting gregarious habit, e.g., *Serpula massiliensis* and *Filograna implexa*, and created interstices and crevices which are suitable for collecting sediment particles close to hard surfaces of the substratum. In this way, these polychaetes acted as facilitators for the colonization of invertebrates with diverse substratum affinities. Some serpulids, e.g., *Serpula* spp., *Hydroides pseudouncinata, Sprirobranchus* spp., with their particular large-sized tubes coated other calcareous structures forming bridges and so they played the role of binders. Other serpulids, with small-sized tubes, e.g., most spirorbids *Protolaeospira (Protolaeospira) striata, Pileolaria militaris, Vinearia koehleri*, were observed on the bare surfaces, thus showing their pioneer role in the colonization pattern.

The growth of serpulids was noted to be essentially linked to the interaction with the other organisms forming the bioconstructions. In fact, they compete for space mainly with other sessile taxa such as bryozoans, which could extensively overgrow the calcareous tubes of the worms. Notwithstanding, the result of such interactions does not always lead to the death of serpulids, which continue to grow below the encrusting bryozoan colonies, like the observation that the tube openings remain free and not covered by the colonies supported. Our study also exhibits that the twisted calcareous tubes of large serpulids, particularly *Serpula cavernicola* and *S. massiliensis*, appeared to be lacking in erosion scars, in contrast to the evidence of large crevices on numerous corallites reported for coralligenous concretions, due to the excavating action of borer organisms such as clionid sponges [54]. These diverse patterns can be explained by differences in substratum microtexture, as the microcrystalline structure exhibited by serpulid tubes [55], as well as in terms of specificity of the boring sponge action [56]. Our observations advanced crucial implications on the role of bioerosion in balancing the growing and destroying phases of the mesophotic bioconstructions that deserve further investigations. A special mention is deserved, within the vagile fauna, to the eunicids, e.g., *Eunice dubitata, E. torquata, E. floridana*, the numerous and large specimens of which were recorded in association with corals and oysters, being directly living among the scleractinian corallites and the valve of *Neopycnodonte*. Unfortunately, this association is not yet clarified and causes some concerns on the nature of opportunistic "nestler" or true "bioeroder" for such species.

In short, in the southern bioconstructions, which are dominated by *N. cochlear*, the contribution of serpulids was more relevant, accounting for 36 species in SML, as compared with the 16 species collected at the northern site of MON, where the bioconstruction is dominated by scleractinians. An opposite situation was exhibited by the vagile fauna particularly with the Eunicida: even only considering the species richness, in fact, the species of this order are quite absent in the southern bioconstructions, with only 3 species at SML, against the 11 species of the northern site of MON, where *E. dubitata* was the dominant species.

5. Taxonomic Considerations

Our study revealed some relevant novelty for the polychaetological fauna of Mediterranean Sea concerning the genus *Eunice*. The analysis, based on comparisons of the sampled material with preserved scientific collections and with the relative scientific literature, made it possible to reach an updated baseline for the identification of the species, which have been recorded from the Mediterranean. However, confusion still exists on the synonymy and distribution of some species of the genus *Eunice sensu lato*. The genus is still deserving a clear definition and for this reason we referred all the species in the dichotomous key to *Eunice sensu lato*. In their recent phylogenetic revision Zanol et al. [36] stated the genera *Leodice* and *Nicidion* as resurrected to name monophyletic groups and including species previously ascribed to *Eunice* and *Marphysa*. The authors considered the articulation of prostomial appendages, other prostomial features and the regionalization of the body as characters supporting the monophyly of the family and genus level clades. According to Zanol et al. [36], some Mediterranean species such as *E. antennata, E. harassi, E. torquata, E. laurillardi, E. dubitata, E. floridana*, and *E. vittata* should be ascribed to the genus *Leodice*. By contrast, *Eunice norvegica* and *E. roussaei*, phylogenetically distant from the members of the genus *Leodice*, should be included in the genus *Eunice sensu strictu*. However, some other species, such as *E. oerstedi, E. schizobranchia* and *E. pennata*, need further analyses to clarify their systematic position. About other concerns, below we synthesize the issues on the species mostly subjected to the taxonomic debate. *Eunice roussaei* Quadrefages, 1866 is a very large species originally described on specimens reported from the Antilles Islands. It has been reviewed by Fauvel [57] and more recently by Fauchald [30]. This species has been separated from the similar *E. aphroditois* (Pallas, 1788) on the basis of the paired subacicular hooks disposition in most segments, the shape of pectinate chaetes, the branchiae morphology and the longer antennae reaching the fourth setiger; *E. aphroditois* has been considered synonym of *E. roussaei* in the checklist of the Italian fauna [41]. Zanol and Bettoso [33] as well proved that the specimens collected in the Adriatic Sea should be referred to *E. roussaei* and not to *E. aphroditois*. In the present study we analysed the specimens of the Giangrande's personal collection coming from Southern Adriatic Sea off the Apulian coast and agreed with the aforementioned Authors, thus we considered *E. roussaei* a valid species of the Mediterranean and Italian fauna and included it in the key to identification.

Some taxonomic confusion also concerns the species *Eunice purpurea* Grube, 1866, whose original description was based on specimens from the lagoon of Lesina, Southern Adriatic Sea. Afterwards, Fauvel [27] considered this species as a juvenile form of *E. roussaei*. Recently, Salazar et al. [35] noted the specimens from Adriatic to be similar to *E. purpurea*, redescribed by Fauchald [30] as a valid species. *E. purpurea* is also reported as synonym of *E. roussaei* in the checklist of the Italian fauna [41] and we agreed with this latter position regarding the synonymy of the two species and with the statement of Zanol and Bettoso [33] concerning the taxonomic debate on the identity of *E. roussaei*.

Regarding other novelty of our study, it should be mentioned the record of *E. dubitata* that was the first for the Mediterranean and the Italian coast and proved this species to be characteristic of the mesophotic bioconstructions, especially those which are primarily built by the non-symbiotic scleractinians, *Phyllangia americana mouchezii* and *Polycyathus muellerae* and by the deep-sea oyster *Neopycnodonte cochlear*.

This study also expands the list of Italian polychaete fauna with the new record of *E. dubitata*, so increasing to 11 the Mediterranean species of the genus *Eunice*.

Lastly, the checklist is also amended as far as the distribution of *E. norvegica*, a species living in association with deep-water white corals, as the recent records from Southern Adriatic and Ionian [16,17,58,59] and Tyrrhenian Sea [60] evidenced. By contrast, the record cited in the checklist from the coralligenous of the Marine Protected Area of Porto Cesareo [61] has to be exactly referred to *E. torquata*.

Supplementary Materials: The following are available online at https://www.mdpi.com/article/10.3390/d13060239/s1, Figure S1: Photo of the mandibles of *Lumbrineris coccinea*.

Author Contributions: M.F.G., Conceptualization, Data curation, Investigation, Taxononomical and Formal analysis, Writing—original draft, Writing—review & editing, Supervision, C.P., Data curation, Field methodology, Writing—review & editing. M.M., Data curation, Laboratory methodology, Writing—review & editing. C.N.M., Data curation, Investigation, Laboratory metodology, Writing—review & editing. A.G., Conceptualization, Data curation, Investigation, Taxonomical and Formal analysis, Writing—original draft, Writing—review & editing, Supervision. All authors have read and agreed to the published version of the manuscript.

Funding: This research received no external funding.

Institutional Review Board Statement: Not applicable. Ethical review and approval were waived for this study, due to nor humans or protected animals were involved.

Informed Consent Statement: Not applicable.

Data Availability Statement: All data generated or analysed during this study are included in this published article.

Conflicts of Interest: The authors declare no conflict of interest.

References

1. Giangrande, A.; Licciano, M.; Musco, L. Polychaetes as environmental indicators revisited. *Mar. Pollut. Bull.* **2005**, *50*, 1153–1162. [CrossRef] [PubMed]
2. Musco, L. Ecology and diversity of Mediterranean hard-bottom Syllidae (Annelida): A community level approach. *Mar. Ecol. Prog. Ser.* **2012**, *461*, 107–119. [CrossRef]
3. Giangrande, A.; Gravina, M.F. Brackish-water polychaetes, good descriptors of environmental changes in space and time. *Transit. Waters Bull.* **2015**, *9*, 42–55. [CrossRef]

4. Sardá, R. Polychaete communities related to plant covering in the midlittoral and infralittoral zones of the Balearic Islands (Western Mediterranean). *Mar. Ecol.* **1991**, *12*, 341–360. [CrossRef]
5. Giangrande, A.; Delos, A.L.; Musco, L.; Licciano, M.; Pierri, C. Polychaete assemblages of rocky shore along the South Adriatic coast (Mediterranean Sea). *Cah. Biol. Mar.* **2004**, *45*, 85–95.
6. Mikac, B.; Licciano, M.; Jaklin, A.; Iveša, L.; Giangrande, A.; Musco, L. Diversity and distribution patterns of hard bottom polychaete assemblages in the North Adriatic Sea (Mediterranean). *Diversity* **2020**, *12*, 408. [CrossRef]
7. Ballesteros, E. Mediterranean coralligenous assemblages: A synthesis of present knowledge. *Oceanogr. Mar. Biol. Annu. Rev.* **2006**, *44*, 123–195.
8. Bedini, R.; Bonechi, L.; Piazzi, L. Spatial and temporal variability of mobile macro-invertebrate assemblages associated to coralligenous habitat. *Medit. Mar. Sci.* **2014**, *15*, 302–312. [CrossRef]
9. Gatti, G.; Bianchi, C.N.; Morri, C.; Montefalcone, M.; Sartoretto, S. Coralligenous reefs state along anthropized coasts: Application and validation of the COARSE index, based on a rapid visual assessment (RVA) approach. *Ecol. Indic.* **2015**, *52*, 567–576. [CrossRef]
10. Chimienti, G.; Stithou, M.; Dalle Mura, I.; Mastrototaro, F.; D'Onghia, G.; Tursi, A.; Izzi, C.; Fraschetti, S. An Explorative Assessment of the Importance of Mediterranean Coralligenous Habitat to Local Economy: The Case of Recreational Diving. *J. Environ. Account. Manag.* **2017**, *5*, 315–325. [CrossRef]
11. Ingrosso, G.; Abbiati, M.; Badalamenti, F.; Bavestrello, G.; Belmonte, G.; Cannas, R.; Benedetti-Cecchi, L.; Bertolino, M.; Bevilacqua, S.; Bianchi, C.N.; et al. Mediterranean bioconstructions along the Italian coast. *Adv. Mar. Biol.* **2018**, *79*, 61–136.
12. Aguilar, R.; Garcia, S.; Perry, A.L.; Alvarez, H.; Blanco, J.; Chimienti, G.; Montesanto, F.; Mastrototaro, F. Deep-sea habitats and communities in the Aeolian Islands (North Sicily). In *Proceedings of the 2nd Mediterranean Symposium on the Conservation of Dark Habitats, Antalya, Turkey, 14–15 January 2019*; Langar, H., Ouerghi, A., Eds.; SPA/RAC: Tunis, Tunisia, 2019; pp. 27–33.
13. Bo, M.; Bavestrello, G.; Canese, S.; Giusti, M.; Angiolillo, M.; Cerrano, C.; Salvati, E.; Greco, S. Coral assemblage off the Calabrian Coast (South Italy) with new observations on living colonies of *Antipathes dichotoma*. *Ital. J. Zool.* **2011**, *78*, 231–242. [CrossRef]
14. Bo, M.; Bavestrello, G.; Angiolillo, M.; Calcagnile, L.; Canese, S.; Cannas, R.; Cau, A.; D'Elia, M.; Filippo D'Oriano, F.; Follesa, M.C.; et al. Persistence of pristine deep-sea coral gardens in the Mediterranean Sea (SW Sardinia). *PLoS ONE* **2015**, *10*, e0119393. [CrossRef] [PubMed]
15. Chimienti, G.; Bo, M.; Taviani, M.; Mastrototaro, F. Occurrence and biogeography of mediterranean cold-water corals. In *Mediterranean Cold-Water Corals: Past, Present and Future*; Orejas, C., Jiménez, C., Eds.; Springer International Publishing: Berlin/Heidelberg, Germany, 2019; pp. 213–243. [CrossRef]
16. D'Onghia, G.; Capezzuto, F.; Cardone, F.; Carlucci, R.; Carluccio, A.; Chimienti, G.; Corriero, G.; Longo, C.; Maiorano, P.; Mastrototaro, F.; et al. Macro- and megafauna recorded in the submarine Bari Canyon (Southern Adriatic, Mediterranean Sea) using different tools. *Mediterr. Mar. Sci.* **2015**, *16*, 180–196. [CrossRef]
17. Mastrototaro, F.; D'Onghia, G.; Corriero, G.; Matarrese, A.; Maiorano, P.; Panetta, P.; Gherardi, M.; Longo, C.; Rosso, A.; Sciuto, F.; et al. Biodiversity of the white coral bank off cape Santa Maria di Leuca (Mediterranean Sea): An update. *Deep-Sea Res. Part II* **2010**, *57*, 412–430. [CrossRef]
18. Taviani, M.; Angeletti, L.; Cardone, F.; Montagna, P.; Danovaro, R. A unique and threatened deep water coral-bivalve biotope new to the Mediterranean Sea offshore the Naples megalopolis. *Sci. Rep.* **2019**, *9*, 3411. [CrossRef] [PubMed]
19. Cerrano, C.; Bastari, A.; Calcinai, B.; Di Camillo, C.; Pica, D.; Puce, S.; Valisano, L.; Torsani, F. Temperate mesophotic ecosystems: Gaps and perspectives of an emerging conservation challenge for the Mediterranean Sea. *Eur. Zool. J.* **2019**, *86*, 370–388. [CrossRef]
20. Bo, M.; Bavestrello, G.; Canese, S.; Giusti, M.; Salvati, E.; Angiolillo, M.; Greco, S. Characteristics of a black coral meadow in the twilight zone of the central Mediterranean Sea. *Mar. Ecol. Prog. Ser.* **2009**, *397*, 53–61. [CrossRef]
21. Cerrano, C.; Danovaro, R.; Gambi, C.; Pusceddu, A.; Riva, A.; Schiaparelli, S. Gold coral (*Savalia savaglia*) and gorgonian forests enhance benthic biodiversity and ecosystem functioning in the mesophotic zone. *Biodivers. Conserv.* **2010**, *19*, 153–167. [CrossRef]
22. Corriero, G.; Pierri, C.; Mercurio, M.; Nonnis Marzano, C.; Tarantini, S.O.; Gravina, M.F.; Lisco, S.; Moretti, M.; De Giosa, F.; Valenzano, E.; et al. A Mediterranean mesophotic coral reef built by non-symbiotic scleractinians. *Sci. Rep.* **2019**, *9*, 3601. [CrossRef]
23. Angeletti, L.; Taviani, M. Offshore *Neopycnodonte* Oyster Reefs in the Mediterranean Sea. *Diversity* **2020**, *12*, 92. [CrossRef]
24. Cardone, F.; Corriero, G.; Longo, C.; Mercurio, M.; Tarantini, S.O.; Gravina, M.F.; Lisco, S.; Moretti, M.; De Giosa, F.; Giangrande, A.; et al. Massive bioconstructions built by *Neopycnodonte cochlear* (Mollusca, Bivalvia) in a mesophotic environment in the central Mediterranean Sea. *Sci. Rep.* **2020**, *10*, 6337. [CrossRef]
25. Roberts, J.M. Reef-aggregating behaviour by symbiotic eunicid polychaetes from cold-water corals: Do worms assemble reefs? *J. Mar. Biol. Assoc. U. K.* **2005**, *85*, 813–819. [CrossRef]
26. Mueller, C.E.; Lundälv, T.; Middelburg, J.J.; van Oevelen, D. The Symbiosis between *Lophelia pertusa* and *Eunice norvegica* stimulates coral calcification and worm assimilation. *PLoS ONE* **2013**, *8*, e58660. [CrossRef] [PubMed]
27. Fauvel, P. Polychètes Errantes. *Faune Fr.* **1923**, *5*, 1–488.
28. Day, J.H. *A Monograph on the Polychaeta of Southern Africa. Part I. Errantia*; Trustees of the British Museum (Natural History): London, UK, 1967; p. 458.
29. George, J.D.; Hartmann-Schröder, G. Polychaetes: British Amphinomida, Spintherida and Eunicida. Keys and notes for the identification of the species. *Synop. Br. Fauna* **1985**, *32*, 1–221.

30. Fauchald, K. A review of the genus *Eunice* (Eunicidae: Polychaeta) based upon type material. *Smithson. Contrib. Zool.* **1992**, *523*, 1–422. [CrossRef]
31. Cantone, G. Censimento dei policheti dei mari italiani: Eunicidae Berthold, 1827. *Atti Soc. Toscana Sci. Nat. Resid. Pisa Mem. Ser. B* **1993**, *100*, 229–243.
32. D'Alessandro, M.; Romeo, T.; Castriota, L.; Cosentino, A.; Perzia, P.; Martins, R. New records of Lumbrineridae (Annelida:.Polychaeta) in the Mediterranean biogeographic province, with an updated taxonomic key. *Ital. J. Zool.* **2016**, *83*, 233–243. [CrossRef]
33. Zanol, J.; Bettoso, N. Identity of *Eunice roussaei* (Eunicidae: Polychaeta: Annelida) from the Adriatic and Mediterranean Seas. *J. Mar. Biol. Assoc. U. K.* **2006**, *86*, 1017–1024. [CrossRef]
34. Zanol, J.; Fauchald, K.; Paiva, P.C. A phylogenetic analysis of the genus Eunice (Eunicidae, Polychaete, Annelida). *Zool. J. Linn. Soc.* **2007**, *150*, 413–434. [CrossRef]
35. Salazar-Vallejo, S.I.; Carrera-Parra, L.F.; de León-González, J.A. Giant Eunicid Polychaetes (Annelida) in shallow tropical and temperate seas. *Rev. Biol. Trop.* **2011**, *59*, 1463–1474. [CrossRef]
36. Zanol, J.; Halanych, K.M.; Fauchald, K. Reconciling taxonomy and phylogeny in the bristleworm family Eunicidae (Polychaete, Annelida). *Zool. Scr.* **2014**, *43*, 79–100. [CrossRef]
37. Arias, A.; Fernández-Álvarez, F.Á.; Martins, R.; Anadón, N. Rediscovery and redescription of *Leodice laurillardi* (Quatrefages, 1866) comb. nov. (Annelida: Eunicidae)-A rare European polychaete or just an overlooked species? *Zootaxa* **2015**, *3964*, 475–481. [CrossRef]
38. Zanol, J.; Carrera-Parra, L.F.; Steiner, T.M.; Amaral, A.C.Z.; Wiklund, H.; Ravara, A.; Budaeva, N. The Current State of Eunicida (Annelida) Systematics and Biodiversity. *Diversity* **2021**, *13*, 74. [CrossRef]
39. Whittaker, R.H. Evolution and measurement of species diversity. *Taxon* **1972**, *21*, 213–251. [CrossRef]
40. Koleff, P.; Gaston, K.J.; Lennon, J.J. Measuring beta diversity for presence-absence data. *J. Anim. Ecol.* **2003**, *72*, 367–382. [CrossRef]
41. Castelli, A.; Bianchi, C.N.; Cantone, G.; Çinar, M.E.; Gambi, M.C.; Giangrande, A.; Iraci Sareri, D.; Lanera, P.; Licciano, M.; Musco, L.; et al. Annelida Polychaeta. *Biol. Mar. Mediterr.* **2008**, *15*, 323–373.
42. Casellato, S.; Stefanon, A. Coralligenous habitat in the northern Adriatic Sea: An overview. *Mar. Ecol.* **2008**, *29*, 321–341. [CrossRef]
43. Giampaoletti, J.; Cardone, F.; Corriero, G.; Gravina, M.F.; Nicoletti, L. Sharing and Distinction in Biodiversity and Ecological Role of Bryozoans in Mediterranean Mesophotic Bioconstructions. *Front. Mar. Sci.* **2020**, *7*, 581292. [CrossRef]
44. Damiani, V.; Bianchi, C.N.; Ferretti, O.; Bedulli, D.; Morri, C.; Viel, M.; Zurlini, G. Risultati di una ricerca ecologica sul sistema marino pugliese. *Thalass. Salentina* **1988**, *18*, 153–169.
45. Russo, A.; Artegiani, A. Adriatic Sea hydrography. *Sci. Mar.* **1996**, *60*, 33–43.
46. Poulain, P.M. Adriatic Sea surface circulation as derived from drifter data between 1990 and 1999. *J. Mar. Syst.* **2001**, *29*, 3–32. [CrossRef]
47. Zavatarelli, M.; Pinardi, N.; Kourafalou, V.H.; Maggiore, A. Diagnostic and prognostic model studies of the Adriatic Sea general circulation: Seasonal variability. *J. Geophys. Res. Ocean.* **2002**, *107*, 2-1–2-20. [CrossRef]
48. Zavatarelli, M.; Pinardi, N. The Adriatic Sea modelling system: A nested approach. *Ann. Geophys.* **2003**, *21*, 345–364. [CrossRef]
49. Giangrande, A.; Gambi, M.C.; Gravina, M.F. Paradigm shifts in community ecology: Open versus closed units, challenges and limits of connectivity studies. *Mar. Ecol.* **2017**, *38*, e12480. [CrossRef]
50. Donnarumma, L.; Appolloni, L.; Chianese, E.; Bruno, R.; Baldrighi, E.; Guglielmo, R.; Russo, G.; Zeppilli, D.; Sandulli, R. Environmental and benthic community patterns of the shallow hydrothermal area of Secca delle Fumose (Baia, Naples, Italy). *Front. Mar. Sci.* **2019**, *6*, 685. [CrossRef]
51. Longo, C.; Cardone, F.; Pierri, C.; Mercurio, M.; Mucciolo, S.; Nonnis Marzano, C.; Corriero, G. Sponges associated with coralligenous formations along the Apulian coasts. *Mar. Biodivers.* **2018**, *48*, 2151–2163. [CrossRef]
52. Cardone, F.; Corriero, G.; Fianchini, A.; Gravina, M.F.; Nonnis Marzano, C. Biodiversity of transitional waters: Species composition and comparative analysis of hard bottom communities from the south-eastern Italian coast. *J. Mar. Biol. Assoc.* **2014**, *94*, 25–34. [CrossRef]
53. Gravina, M.F.; Cabiddu, S.; Como, S.; Floris, A.; Padedda, B.M.; Pusceddu, A.; Magni, P. Disentangling heterogeneity and commonalities in nanotidal Mediterranean lagoons through environmental features and macrozoobenthic assemblages. *Estuar. Coast. Shelf Sci.* **2020**, *237*, 106688. [CrossRef]
54. Cerrano, C.; Bavestrello, G.; Bianchi, C.N.; Calcinai, B.; Cattaneo-Vietti, R.; Morri, C.; Sarà, M. The Role of Sponge Bioerosion in Mediterranean Coralligenous Accretion. In *Mediterranean Ecosystems*; Faranda, F.M., Guglielmo, L., Spezie, G., Eds.; Springer: Milano, Italy, 2001; pp. 235–240. [CrossRef]
55. Sanfilippo, R.; Mòllica, E. *Serpula cavernicola* Fassari & Mòllica, 1991 (Annelida Polychaeta); diagnostic features of the tube and new Mediterranean records. *Mar. Life* **2000**, *10*, 27–32.
56. Calcinai, B.; Arillo, A.; Cerrano, C.; Bavastrello, G. Taxonomy-related differences in the excavating micro-patterns of boring sponges. *J. Mar. Biol. Assoc. U. K.* **2003**, *83*, 37–39. [CrossRef]
57. Fauvel, P. Annelides Polychetes de L'Australie meridionale. *Arch. Zool. Exp. Gen. (Paris)* **1917**, *56*, 159–277, plates 4–8.
58. Tursi, A.; Mastrototaro, F.; Matarrese, A.; Maiorano, P.; D'Onghia, G. Biodiversity of the white coral reefs in the Ionian Sea (Central Mediterranean). *Chem. Ecol.* **2004**, *20*, S107–S116. [CrossRef]

59. Vertino, A.; Savini, A.; Rosso, A.; Di Geronimo, I. Benthic habitat characterization and distribution from two representative sites of the deep-water SML Coral Province (Mediterranean). *Deep Sea Res. Part II Top. Stud. Oceanogr.* **2010**, *57*, 380–396. [CrossRef]
60. Taviani, M.; Angeletti, L.; Canese, S.; Cannas, R.; Cardone, F.; Cau, A.; Cau, A.B.; Follesa, M.C.; Marchese, F.; Montagna, P.; et al. The "Sardinian cold-water coral province" in the context of the Mediterranean coral ecosystems. *Deep Sea Res. Part II Top. Stud. Oceanogr.* **2015**, *145*, 61–78. [CrossRef]
61. Corriero, G.; Gherardi, M.; Giangrande, A.; Longo, C.; Mercurio, M.; Musco, L.; Nonnis Marzano, C. Inventory and distribution of hard bottom fauna from the marine protected area of Porto Cesareo (Ionian Sea): Porifera and Polychaeta. *It. J. Zool.* **2004**, *71*, 237–245. [CrossRef]

diversity

MDPI

Article

Quantifying Threats to Biodiversity and Prioritizing Responses: An Example from Papua New Guinea

Thomas H. White, Jr. [1,*], Patricia Bickley [2], Cory Brown [3], Dave E. Busch [4], Guy Dutson [5], Holly Freifeld [6], Douglas Krofta [7], Sean Lawlor [2], Dan Polhemus [8] and Rachel Rounds [8]

1 Puerto Rican Parrot Recovery Program, United States Fish and Wildlife Service, Rio Grande, PR 00745, USA
2 United States Department of Interior—International Technical Assistance Program, Washington, DC 20240, USA; Patricia_Bickley@ios.doi.gov (P.B.); lawlorsm@state.gov (S.L.)
3 Division of International Affairs, United States Fish and Wildlife Service, Washington, DC 20011, USA; cory_brown@fws.gov
4 College of Agricultural, Human and Natural Resources Sciences, Washington State University, Vancouver, WA 99164, USA; buschd27@gmail.com
5 Biodiversity Solutions—Australia, Ocean Grove, VIC 3221, Australia; guy@biodiversitysolutions.com.au
6 United States Fish and Wildlife Service-Pacific Region, Portland, OR 97232, USA; holly_freifeld@fws.gov
7 Branch of Listing Policy & Support, United States Fish and Wildlife Service (Retired), Washington, DC 22041, USA; baja_kayaker@hotmail.com
8 United States Fish and Wildlife Service-Pacific Islands Office, Honolulu, HI 96822, USA; dan_polhemus@fws.gov (D.P.); rachel_rounds@fws.gov (R.R.)
* Correspondence: thomas_white@fws.gov

Citation: White, T.H., Jr.; Bickley, P.; Brown, C.; Busch, D.E.; Dutson, G.; Freifeld, H.; Krofta, D.; Lawlor, S.; Polhemus, D.; Rounds, R. Quantifying Threats to Biodiversity and Prioritizing Responses: An Example from Papua New Guinea. *Diversity* **2021**, *13*, 248. https://doi.org/10.3390/d13060248

Academic Editors: Michael Wink and Karl Cottenie

Received: 23 April 2021
Accepted: 28 May 2021
Published: 4 June 2021

Publisher's Note: MDPI stays neutral with regard to jurisdictional claims in published maps and institutional affiliations.

Copyright: © 2021 by the authors. Licensee MDPI, Basel, Switzerland. This article is an open access article distributed under the terms and conditions of the Creative Commons Attribution (CC BY) license (https://creativecommons.org/licenses/by/4.0/).

Abstract: Accurately identifying threats to global biodiversity is the first step towards effectively countering or ameliorating them. However, such threats are usually only qualitatively categorized, without any comparative quantitative assessment of threat levels either within or across ecosystems. As part of recent efforts in Papua New Guinea to develop a long-term strategic plan for reducing threats to biodiversity at the national level, we developed a novel and quantitative method for not only assessing relative effects of specific biodiversity threats across multiple ecosystems, but also identifying and prioritizing conservation actions best suited for countering identified threats. To do so, we used an abbreviated quantitative SWOT (Strengths, Weaknesses, Opportunities, Threats) analysis and multivariate cluster analysis to identify the most significant threats to biodiversity in Papua New Guinea. Of 27 specific threats identified, there were nine major threats (each >5% of total) which accounted for approximately 72% of the total quantified biodiversity threat in Papua New Guinea. We then used the information to identify underlying crosscutting threat drivers and specific conservation actions that would have the greatest probability of reducing biodiversity threats across multiple ecosystem realms. We categorized recommended actions within three strategic categories; with actions within each category targeting two different spatial scales. Our integrated quantitative approach to identifying and addressing biodiversity threats is intuitive, comprehensive, repeatable and computationally simple. Analyses of this nature can be invaluable for avoiding not only wasted resources, but also ineffective measures for conserving biodiversity.

Keywords: Papua New Guinea; biodiversity; threats; conservation; SWOT analysis; ecosystem; planning; resources; prioritization

1. Introduction

The island of New Guinea (comprised of West Papua and Papua New Guinea) is the largest tropical island in the world, with a total land area of approximately 785,750 km^2. Due mainly to its great elevation range (sea-level to 4500 m ASL), combined with a tropical climate and diverse topography and geologic origins, Papua New Guinea (hereafter, PNG) harbors diverse and unique life zones and forested habitats not found elsewhere in the Pacific Islands [1–4] (Figures 1 and 2). After the Amazon and Congo basins, the forests of PNG comprise the third largest expanse of tropical forest in the world and are internationally recognized for their high biodiversity and ecological importance, both locally and globally [5–7]. Indeed, some areas of PNG (i.e., the islands of the Bismarck Archipelago) are currently classified as part of a global "biodiversity hotspot" ([8,9]. Host to one of the richest assemblages of vertebrates on Earth, the forests of PNG harbor at least 1786 species of birds, mammals, reptiles, and amphibians [7]—over 5 percent of the world's total—with many yet undiscovered and undescribed. Covering nearly 282,000 km^2 and 80 percent of the land mass of PNG [7], these forests have in recent decades come under increasing pressure from logging activities, agroforestry-related land clearing, mining activities, oil and gas development, and increasing subsistence agriculture arising from a rapidly increasing human population [4–11].

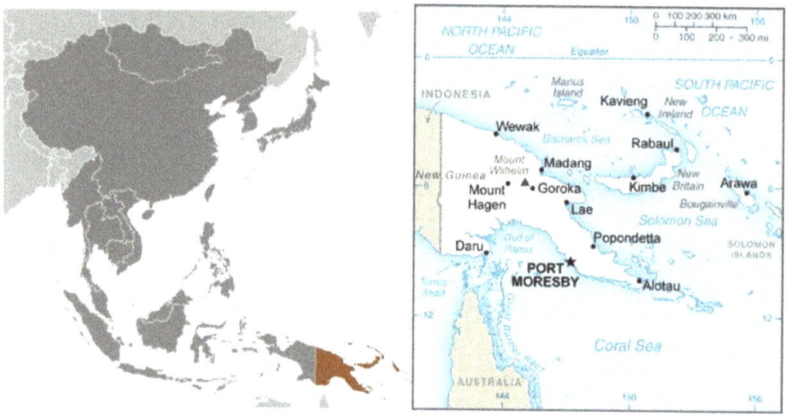

Figure 1. Regional map and major landmarks of PNG. From CIA World Factbook [https://www.cia.gov/library/publications/the-world-factbook/geos/pp.html (accessed on 3 June 2021)].

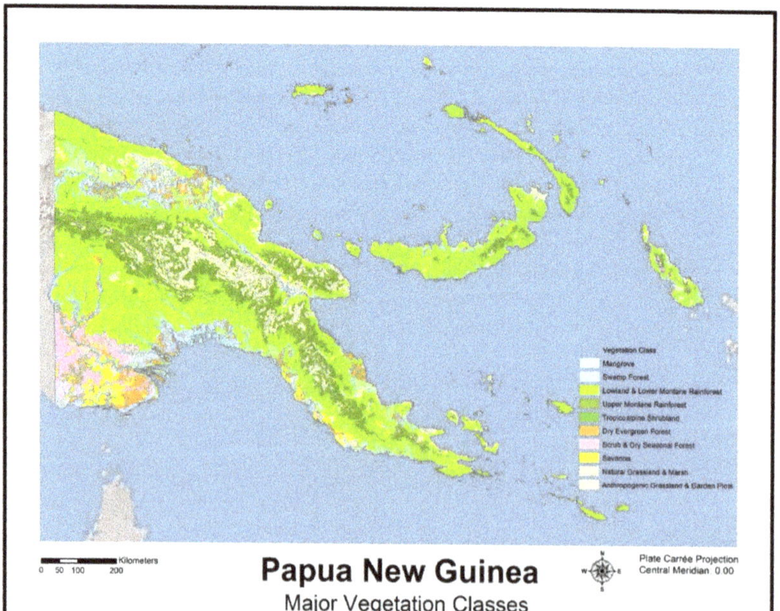

Figure 2. Extent and distribution of the major vegetation classes of Papua New Guinea. Data adapted from the Forest Inventory Mapping (FIM) System, sensu McAlpine and Quilley [12].

Additionally, because of PNG's unique land tenure system and widespread dependence on subsistence farming and harvesting, PNG's biodiversity continues to be fundamental to the health, economy, and culture of the nation's largely rural population [8]. However, PNG's rich and unique biodiversity is also among the most threatened, leading PNG's government to recognize major threats to its environment from deforestation and forest degradation; illegal, unreported and unregulated fishing; pollution; industrial activities; and climate change [13–16]. Recent remote sensing images have revealed significant ongoing habitat conversion in PNG, especially of old-growth and closed-canopy forest [7]. Moreover, a broad array of ecological studies demonstrate that threats such as logging, over-hunting, pollution, and invasive species reduce the intactness of ecosystems and cause significantly reduced numbers and local extirpation of many species [17].

Recognizing PNG's remarkable biodiversity and the threats to it, the Government of PNG and United States Agency for International Development (USAID) identified PNG as a priority country for assistance in biodiversity conservation, with funds apportioned to conduct an analysis of drivers and threats to biodiversity and a corresponding joint conservation strategy with the Government of PNG [17–20]. Consequently, the USAID requested the United States Department of the Interior's International Technical Assistance Program (DOI-ITAP) to provide technical support for a national biodiversity threat assessment of PNG.

The assessment was designed to provide a well-documented, objective analysis that (1) identifies and quantifies specific threats to PNG's biodiversity in marine, inland water, and terrestrial ecosystem realms, and (2) identifies the actions necessary to most effectively address these threats and benefit biodiversity conservation broadly in the country [21]. As such, the assessment builds upon, amplifies and increases precision of previous efforts at biodiversity threat identifications and assessments e.g., [22–29]. For this study, we define "terrestrial ecosystem" as all land-based environments and habitats, "inland water ecosystem" as all freshwater lentic and lotic systems, including all deltaic and brackish coastal and estuarine (mesohaline) zones, while "marine ecosystem" includes all euhaline offshore territorial waters and coral reefs. Threats to biodiversity are defined as those human activities that reduce the viability of an ecosystem, species or other types of biodiversity. This study uses threat nomenclature consistent with the IUCN lexicon developed by Salafsky et al., [30] but customized to PNG. The overarching goal was to provide a strong rationale for future biodiversity projects that USAID and its partners in PNG could develop as part of a long-term investment in PNG's unique and globally important biodiversity.

2. Materials and Methods

2.1. Threats Assessment

2.1.1. Literature Review

The DOI-ITAP assessment team (n = eight persons) first conducted an exhaustive and comprehensive review of the available literature—published and non-published—relative to biodiversity, resource conservation and associated issues in PNG. During this review, over 300 documents were obtained and reviewed, a complete bibliography of which is found in Brown et al. [21]. All documents so obtained were made available to all team members via internet using a shared Google Drive® storage and inventory system.

2.1.2. In-Country Site Visits and Interviews

In addition to literature-sourced information, the team also conducted numerous in-country interviews with PNG government and non-government entities. These interviews were conducted during two separate visits by team members to PNG; the first in November 2016, and the second during February–March 2017. During these visits, the team met with more than 110 individuals representing more than 30 organizational stakeholders having direct involvement or interests in PNG biodiversity see [21]. The assessment team also conducted site visits to several representative ecoregions of PNG to obtain additional first-hand information on current biodiversity threats and status thereof.

2.1.3. Quantitative Threat Assessment and Ranking

To provide an objective, quantitative and data-driven method for assessing relative threat levels, we used a modified version of quantitative SWOT (Strengths, Weaknesses, Opportunities, Threats) analysis [31]. Quantitative SWOT analysis has several desirable characteristics as applied to environmental assessments. The method is simple, versatile, repeatable and intuitive. Importantly, the ability to effectively combine mixed data (cardinal, categorical, and continuous) together with multidisciplinary expert knowledge [27,31] into a single structured model provides a comprehensive, practical and uniform metric for comparing multiple threats and prioritizing biodiversity conservation strategies [29,31]. Because our specific objective was to assess current threats to biodiversity, we therefore abbreviated the SWOT approach to focus only on the category of "Threats".

Specific threats were identified based on the literature review, and the collective knowledge and experience of team members, several of which (G.D., D.K., D.P.) also had extensive prior experience working in PNG. These threat tables were also periodically updated and refined during the analytical process as additional information became available. Once a complete listing of threats was compiled, a quantitative SWOT template [31] was used to assign a "weight" ranging from 100–1000 for each threat in terms of its relative impact (i.e., 100 = very minor impact; 1000 = extreme impact) on biodiversity in other similar settings. That is, we considered how each threat was known to affect biodiversity in other tropical, insular environments (e.g., Madagascar, Hawaii, Hispaniola, Indonesia, Mauritius, Jamaica). Threat weightings were based on the expert-based team consensus, the members of which all had prior or ongoing experience working in tropical, insular environments. The value of expert-based assessments in ecology and conservation has become increasingly recognized, particularly in cases of varying degrees of epistemic uncertainty, as in this assessment e.g., [27,29,31]. Once all threat weightings were determined, an importance coefficient (IC) was computed for each threat that reflected its relevance to overall threat levels. The individual ICs were calculated as follows: $IC_i = W_i/\Sigma W_{i\text{-}n}$; where W_i is the respective threat weight, $\Sigma W_{i\text{-}n}$ is the sum of all individual threat weights, and thus $\Sigma IC_{i\text{-}n} = 1$. Because there were ecosystem-specific differences in both the number and impact of individual threats, the individual threat weights and ICs also varied accordingly among the three ecosystem categories.

Threat levels were also affected by three additional coefficients: spatial, temporal and probability of occurrence. The spatial coefficient (SC) reflects the geographic extent of a given threat across PNG. If the threat applied equally over the entire country, the SC was 1. However, if the threat was spatially disproportional, then the SC was allocated accordingly. Spatial coefficients for each threat were determined based on a combination of published reports, PNG government documents and reports, and geospatial evaluation using overlaid data themes. Similarly, the temporal coefficient (TC) applies to threats that affect the country only periodically or cyclically during any given year (e.g., anthropogenic fires). Finally, the probability of occurrence (POC) coefficient indicates the likelihood that a given threat will be present or occur in any given year. A POC of 1 indicates that the factor is always present or occurs each year. Thus, these coefficients attenuate the relative weight (i.e., IC) of individual threats according to PNG-specific conditions. For example, two different threats may have the same general impact (i.e., weight) on biodiversity, but if one only affects 10 percent of the country, while the other affects 95 percent, then the relative impact (i.e., ICA) of the latter would be higher. All coefficients were based on existing data and information extracted from the available published literature and PNG Government reports and policy documents. Any threats not subject to measurable spatial, temporal or probability of occurrence variations were assigned default coefficients of 1, respectively.

Finally, for each threat the team assigned on a scale of 0–10, two numerical values for each threat level: optimal and actual. For any given threat, optimal level was assigned a default value of 10, as it represents the ideal scenario in terms of biodiversity conservation. Most critical however, were the values for actual level, as they represented the team consensus on the actual state of the threats for PNG. Actual values were based on how, or to what extent, the threats currently exist or operate in PNG relative to their impact on biodiversity. In this case, a value of 10 represents either absence of the threat, or the complete mitigation or abatement of it. Conversely, a value of zero would represent an imminent, existential threat to those aspects of biodiversity thus affected (i.e., "worst case scenario"). Importantly, estimates of actual values also considered relative impacts on overall biodiversity by the specific threats. For example, two different threats could each have the same weight and coefficients, but if one affects areas of greater species diversity and/or endemism (e.g., montane rainforests), then the actual value for this threat would be scored lower (i.e., greater severity) than that of a threat affecting primarily areas of lesser biodiversity (e.g., coastal scrub forests). The team used all available information—published, non-published and geospatial—as well as information obtained from in-country interviews and discussions, in order to assign actual values. For this estimate, the initial nonbiased starting point was a value of five. The team then decided whether the impact of a given threat on biodiversity in PNG was less severe (>5) or more severe (<5), and to what degree.

To estimate and quantitatively compare individual threats using the SWOT template, three indices were also calculated: optimal quality (OQ), actual quality (AQ), and the quality deviation index (QD). The OQ is the highest potential "quality" for the threat, while the AQ is the observed or actual quality based on the PNG-specific team assessments. For each threat, these indices were obtained by multiplying the associated optimal and actual quality values by all four corresponding coefficients. The final quality indices (optimal, actual) were then computed as the sum of the previous values as follows: $Q_{o,a} = \Sigma_{i\text{-}n} (V_{o,a} * I_i * S_i * T_i * P_i)$; where:

$Q_{o,a}$ = Threat level (optimal, actual).
$V_{o,a}$ = Optimal and actual values, respectively, of each threat.
I_i = Importance coefficient of each threat.
S_i = Spatial coefficient of each threat.
T_i = Temporal coefficient of each threat.
P_i = Probability of occurrence coefficient of each threat.

And, where the quality deviation index (QD) = $Q_o - Q_a$.

The QD provides detailed and valuable insights for threat evaluation. Analysis of individual partial QDs can reveal whether certain threats to biodiversity could feasibly be mitigated or improved by conservation actions. For example, some threats may be impractical or impossible to improve over the time frame of interest. Other factors however, may be more amenable to control or reduction via conservation actions. Thus, the relative contributions of individual threats to the total QD can be used to identify where specific conservation efforts (i.e., resource allocations) may have the greatest positive impact on biodiversity conservation. For instance, if threats "x" and "y" contribute 23 percent to the overall threat QD, then effective mitigation or reduction of those threats may reduce biodiversity threat level by a similar amount commensurate with conservation success see [31]. Examination of the component (i.e., ecosystem) subtotals for threat percentages can also indicate which ecosystem(s) face greater overall threat levels.

Finally, once overall percentages were calculated for all threats, we then assigned each threat to one of three practical and intuitive categories: High, Medium, and Low. We defined "High" threats as those factors currently contributing to substantial (>5% of total threat) ongoing biodiversity loss, "Medium" threats as those which have the potential to become more serious (i.e., High) threats to biodiversity in the immediately foreseeable future, and "Low" threats as those factors which are not considered to become significant causes of biodiversity loss in PNG in the immediately foreseeable future. For this, we defined "immediately foreseeable future" as a 5-year time horizon from date of the assessment.

Because the categorization of threats was based on their quantitative contributions to the overall biodiversity threat in PNG, we used multivariate cluster analysis as a quantitative threat classification method [32]. Specifically, we used a 3-centroid Euclidian distance model to assign individual threat percentages to one of three threat categories [33]. This allowed us to base our threat classifications on the actual data structure, rather than on any arbitrary or subjective considerations [21]. Threat classifications are presented graphically as a dendrogram generated from the cluster analysis. All statistical calculations were performed using the MINITAB® Version 13 Statistical Package for Windows®.

2.2. Situation Modeling

We used Situation Model (SM) guidance provided by USAID [34] to evaluate the biodiversity threat context, utilizing our literature review, in-country observations and discussions, as well as our SWOT and geospatial analyses [21]. Situation Modeling has been used for over two decades in the fields of public health and international development, and also provides a concise and effective way of depicting relationships among factors which affect biodiversity [34]. We relied partially upon the generic tropical forest and Philippines-oriented terrestrial ecosystem SM, as well as guidelines for the level of detail about drivers and threats to include [34,35]. Detailed descriptions and methods for constructing SMs are found in USAID [34].

Drivers, threats, and stressors identified were consistent with those identified in the USAID Tropical Forestry and Biodiversity Conservation Assessment (i.e., climate change, overexploitation and unsustainable harvest, habitat loss, invasive species and pollution) [36]. Our process enabled us to develop a comprehensive SM for biodiversity including terrestrial, inland water, and marine systems, as well as social, economic and governance drivers.

Within each strategic approach, we created recommendations at two different scales: community and national. Community level, for the purposes of this assessment, is defined to include Provincial, District and local-level governments along with local clan and village governance. The community-level recommendations focus on engaging communities in biodiversity conservation through education, training, and land use planning. The national-level recommendations focus on supporting government and educational institutions and other civil society. Government agencies can provide policies and regulation to guide conservation, law enforcement, and science-based technical assistance to communities and NGOs working in conservation. Higher education institutions can provide training and education across a wide-range of topics to support both the central government and local communities to develop a cohort of PNG citizens who are trained and empowered in conservation science and practice.

2.3. Results Chains and Response Prioritizations

To develop specific recommendations for biodiversity threat reduction we also looked at each of the "High" threats identified in the SWOT analysis and then identified those strategic approaches that could alter the linkages between drivers and threats or reduce the effect of the threats to biodiversity [36,37]. We then evaluated all potential strategic approaches relative to how many of the "High" threats each approach addressed. We distilled this into three primary strategic approaches: (1) Strengthening capacity for environmental decision-making, (2) Legal and policy education and training, and (3) Integrated land use planning. Although priority was given to addressing the top threats, the recommended strategic approaches also simultaneously address a large number of lesser (Medium, Low) threats. We used Results Chains to clearly illustrate the interactive linkages between recommended actions and desired results across all three strategic approaches at both the community and national levels [21,37]. Detailed descriptions and methods for constructing Results Chains are found in USAID [37].

3. Results

3.1. SWOT Threat Assessment and Rankings

Of the 27 individual threats identified across all ecosystem realms, nine (9) were ranked as "High", six (6) were ranked as "Medium", and twelve (12) were ranked as "Low" threats to biodiversity in PNG during the immediately foreseeable future (Figure 3 and Table 1). The complete composite SWOT analysis template is presented in Appendix B. These threat assessments took into consideration relative differences in overall impacts of specific threats on PNG biodiversity. For example, "undocumented harvests" of timber was classified as currently a "Low" threat. This is because most such harvests in PNG are conducted in areas previously degraded by selective logging (as opposed to primary forests), and by small independent operators (e.g., "walkabout" sawmills). Accordingly, the overall biodiversity impact of such harvests was considered a lesser threat (over next five years) than that of other broader threats such as the clear cutting or the selective commercial logging of more intact primary forests. Similarly, our analyses categorized "rates of harvest" of timber as a "Medium" threat over the next five years. This was because this threat, by definition, affects areas previously degraded by selective logging (Appendix A). Thus, the overall impact on biodiversity is primarily—but not limited to—that of delaying or interrupting secondary succession and the attendant recovery or restoration of biodiversity elements previously lost to other factors. Nevertheless, because this threat occupied the top position in the "Medium" category (Table 1), it has the potential to become a more serious threat over perhaps a longer time horizon than that considered by this study.

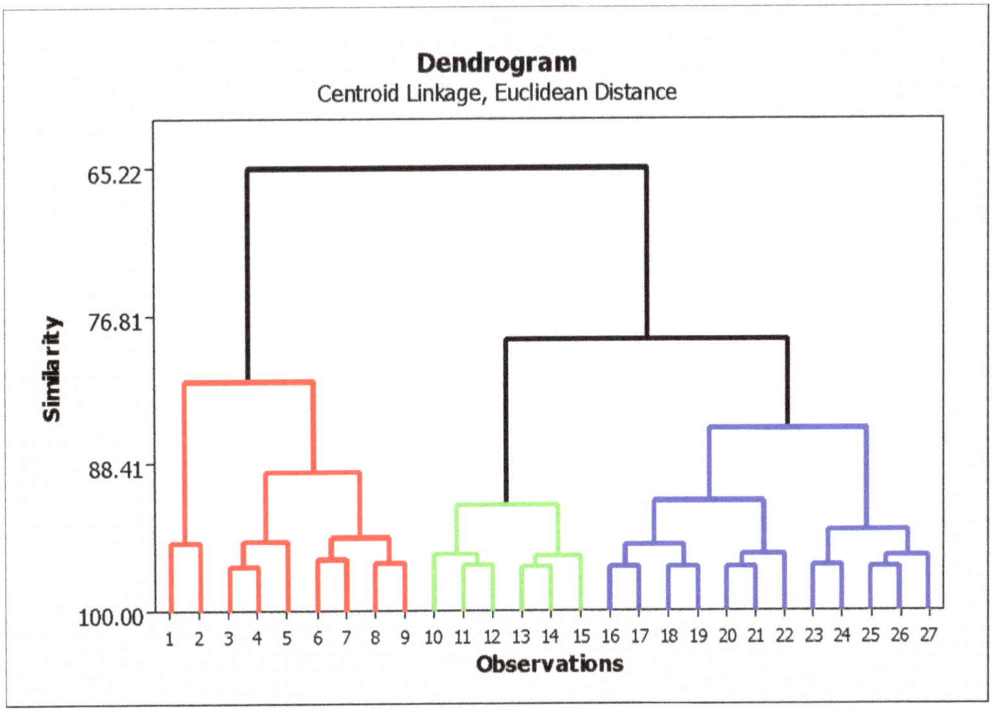

Figure 3. Dendrogram depicting the ranking and categorization of 27 specific threats to biodiversity in Papua New Guinea based on quantitative SWOT and multivariate cluster analyses. Red (1–9) denotes "High" threats, Green (10–15) denotes "Medium" threats, and Blue (16–27) denotes "Low" threats.

Table 1. Hierarchical listing of specific biodiversity threats in Papua New Guinea based on quantitative SWOT and multivariate cluster analyses. Heading colors and numbers (e.g., HIGH 1–9) correspond to those of the cluster analysis dendrogram (Figure 3). Threat nomenclatures and associated definitions are in Appendix A.

HIGH (1–9)	MEDIUM (10–15)	LOW (16–27)
Clear-cutting (all forms)	Rates of harvest	Anthropogenic fires
Commercial agroforestry	Subsistence agriculture	Development/village expansion
IUU marine fishing	Oil and gas extraction	Dams/hydroelectric projects
Selective cutting	Subsistence hunting	Unregulated plant/wildlife trade
Invasive species	Fuel wood harvesting	Regulated marine fishing (artisanal)
Pollution	Infrastructure developments	Traditional forest orchards
Regulated commercial marine fishing		Aquaculture
Harvest volumes		Renewable energy projects
Mineral extraction		Hunting for cultural items
		Seabed mining
		Undocumented harvests
		Regulated plant/wildlife trade

By ecosystem category, the SWOT results indicated that the greatest proportion (approximately 40%) of the total threat to PNG biodiversity directly affects the terrestrial ecosystem (Appendix B), with the inland water and marine ecosystems subjected to 33 percent and 27 percent of the overall threat, respectively. Moreover, of the nine threats ranked as "High", seven were threats common to two or more ecosystems, while two (i.e., IUU marine fishing, regulated commercial marine fishing) were specific to the marine ecosystem. In aggregate, the nine major threats accounted for approximately 72% (average 8% each) of the total quantified biodiversity threat in PNG (Appendix B). Clearly, in terms of overall biodiversity impact, these highest ranked threats are quantitatively noteworthy as reflected by their distinct "lineage" identified via the cluster analysis (Figure 3), compared to "Medium" and "Low" ranked threats which accounted for a remaining 16 percent (average 2.7%) and 12 percent (average 1%), respectively. In terms of overall impact on PNG biodiversity, "High" threats were on average three times more severe than "Medium" threats, which were in turn about three times more severe on average than "Low" ranked threats. Importantly for resource allocation purposes, the seven "High" threats that are shared by multiple ecosystems also accounted for 57 percent of the total biodiversity threat, and were also primarily terrestrial in origin (with IUU fishing and commercial marine fishing being the exceptions). Indeed, of the 27 total identified biodiversity threats, closer examination of only those shared by all three ecosystem realms reveals that these cross-realm threats (n = 8), and which originate within the terrestrial component also account—either directly or indirectly—for approximately 45 percent of the overall threat to biodiversity across all ecosystems in PNG (Appendix B). This suggests that conservation efforts focused on countering or mitigating the broader suite of these and other cross-realm terrestrial based threats would have the greatest positive impact in reducing current overall biodiversity loss in PNG, via a "multiplier effect" across multiple ecosystems.

3.2. Situation Model

We identified 10 drivers, 16 threats, and 7 stressors that impinge on biodiversity in PNG (Figure 4). Driver and threat categories are mostly synoptic and/or multi-faceted and apply broadly across PNG, as opposed to being highly-specific, unidimensional, or only applicable to tightly-constrained geographic areas or ecosystems. To simplify the visual presentation in the situation model (SM), we grouped some threats that were analyzed separately in the SWOT analysis. For example, the "timber harvest" threat was broken out into five subcategories that were each scored separately for SWOT analysis (see Table 1, Appendix B). Our SM follows the guidance from USAID in which drivers represent ultimate factors, usually social, economic, political, institutional or cultural, that enable or amplify the occurrence of one or more threats, which are the proximate human activities or processes causing degradation or loss of biodiversity [34,36]. Our process enabled us to develop a comprehensive SM for biodiversity including terrestrial, inland water and marine ecosystems, as well as social, economic and governance drivers (Figure 4).

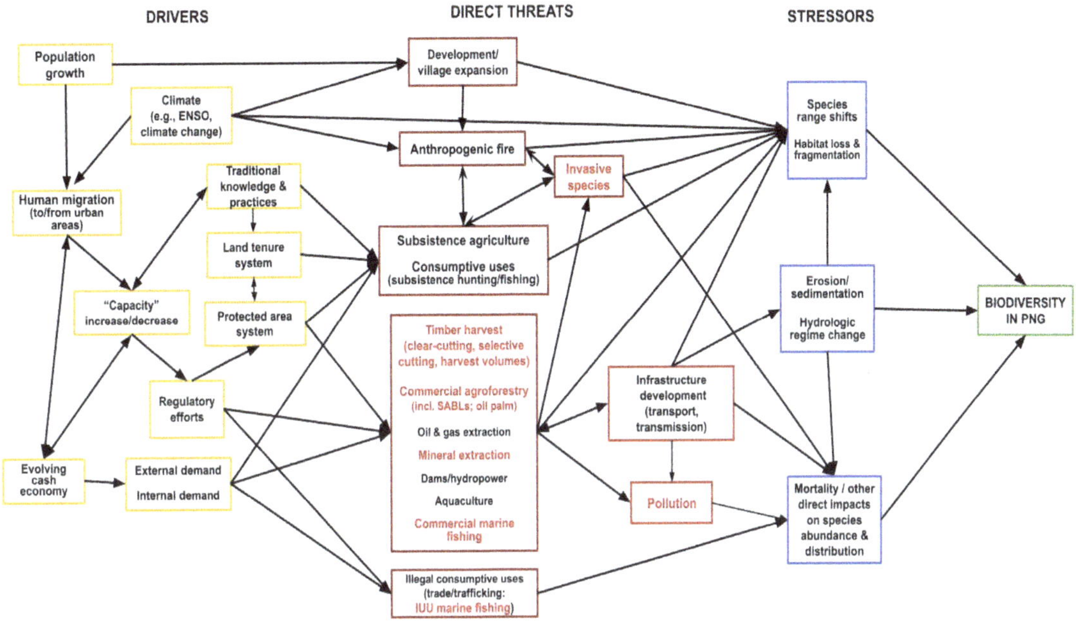

Figure 4. Situation Model based on quantitative assessment of biodiversity threats and associated drivers in Papua New Guinea. Bidirectional vectors represent interactions between component factors.

3.3. Results Chains and Response Prioritizations

Our recommendations resulted from extensive in-country discussions with stakeholders, a broad literature review, results of the quantitative SWOT analysis, and consensus of the assessment team. For each Strategic Approach we developed Results Chains [37] to link our recommendations to the Situation Model and SWOT analysis (Figures 5–7). The Results Chain linkages also provide a useful graphic-based guide for prioritizing recommended actions at both the national and local community levels. For example, those actions that influence or promote the greatest number of results could be considered as greater priority relative to those of lesser functional linkages.

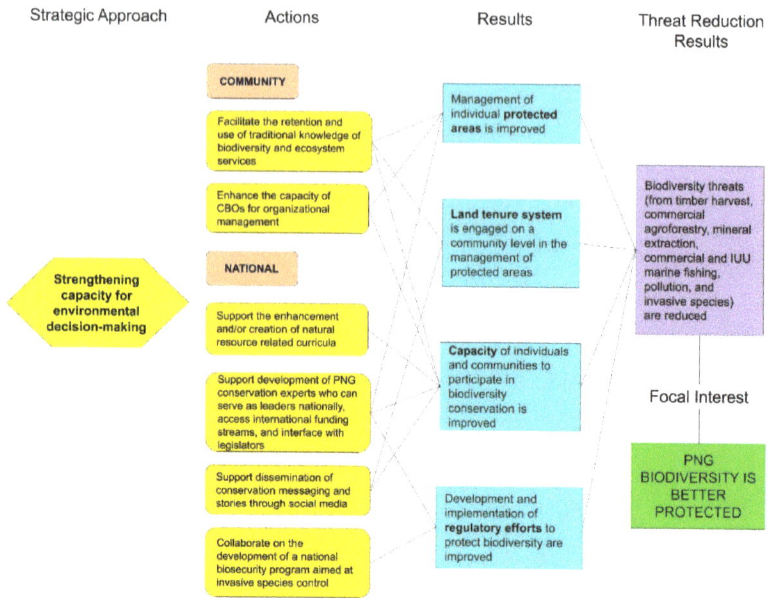

Figure 5. Results Chain for strategic goal of Strengthening Capacity for Environmental Decision-making in Papua New Guinea.

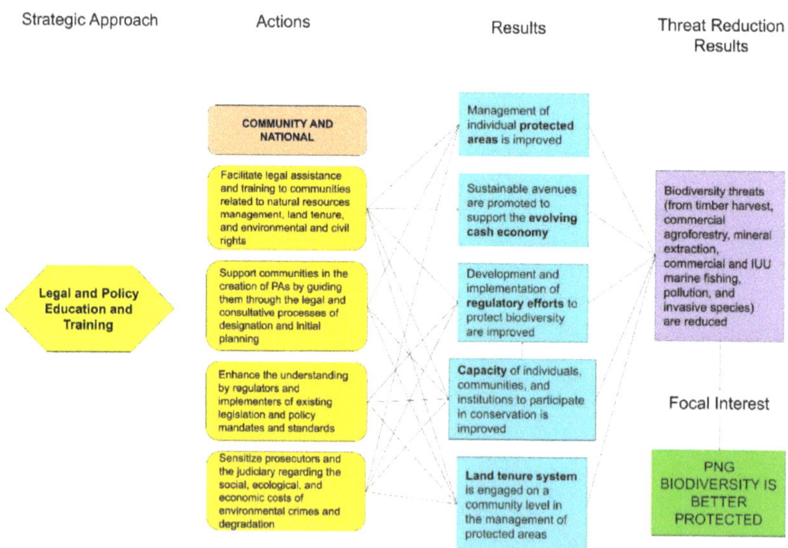

Figure 6. Results Chain for strategic goal of Legal and Policy Education and Training in Papua New Guinea.

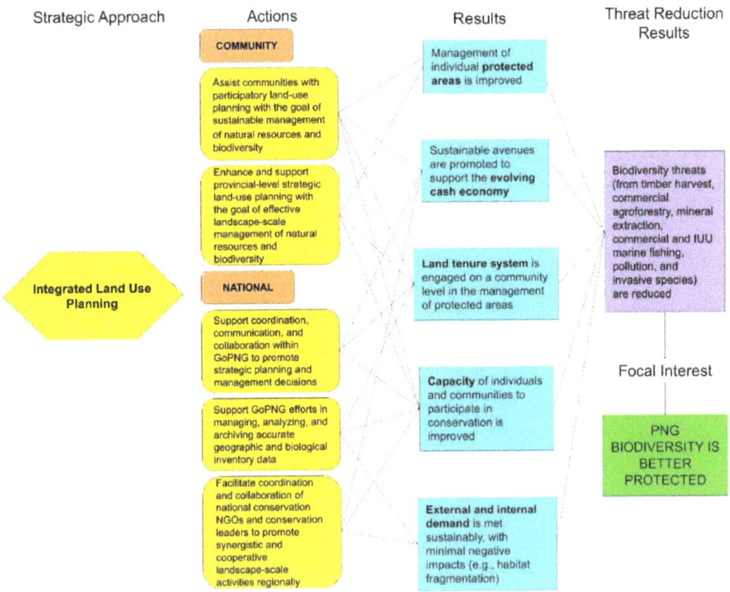

Figure 7. Results Chain for strategic goal of Integrated Land Use Planning in Papua New Guinea.

4. Discussion

The primary threats to biodiversity (those in the "High" category) identified in the assessment and combined for PNG's terrestrial, inland water, and marine ecosystems were: clear-cutting of forests; commercial agroforestry; illegal, unreported, and unregulated marine fishing; selective cutting (timber harvest); invasive species; pollution; regulated commercial fishing; timber harvest volumes; and mineral extraction. Through situation modeling, the underlying drivers were linked to these threats; these drivers are numerous and interconnected. Population growth, human migration, and climate change are ultimate drivers from which flow other, more proximate drivers, and resulting threats to biodiversity. The chief proximate drivers (those we identified as the best opportunities for intervention to address threats to biodiversity) included regulatory efforts, capacity enhancement, land tenure and protected area systems, internal and external demand for resources, and the evolving cash economy. Through the consultative process and results-chains exercises, and in light of these key drivers, we developed a set of three strategic approaches, each containing recommended programmatic actions at community and national levels to guide investments to help conserve PNG's biodiversity. Thus, our assessment and recommendations are consistent with the concept of "cross-realm decision-making", as recommended by Tulloch et al. [38].

A common thread throughout our in-country stakeholder interviews was the need to enhance the capacity of PNG citizens and institutions regarding environmental and biodiversity topics and best practices. The need for capacity enhancement stretches from individuals in rural communities to employees of government agencies in the capital of Port Moresby. Developing a cadre of people, across PNG, familiar with and educated in environmental and biodiversity topics, as well as other supporting disciplines such as finance and project management, will help to reduce threats to biodiversity across the landscape, provide sustainable opportunities for PNG communities, and create long-term conditions for successful conservation in PNG.

Because of the unique land tenure system of PNG, whereby up to 97 percent of land is legally owned and managed by resident communities and clans [39], it is also essential that PNG citizens have an adequate knowledge of their legal rights over communal lands to prevent inequitable and abusive land acquisitions and unsustainable resource exploitation by outside interests. For example, there have been numerous cases of rural communities and clans effectively losing permanent control over their legally-owned lands to unscrupulous developers and foreign resource extraction companies, e.g., [40–43]. Assisting local communities with the often complex legal and ecological issues associated with protecting and sustainably managing their lands can yield substantial dividends in terms of biodiversity conservation, including in development of protected areas. Probably nowhere in the world presents a better setting for an effective "bottom up" approach to natural resource management and conservation, e.g., [43]. Interestingly, because of both the weak central government and the community-based land tenure system of PNG, traditional environmental "law enforcement" (i.e., "reactive model") is largely supplanted by local community vigilance and control. This is why increasing capacity for effective environmental decision-making and environmental protection (i.e., "proactive model") at the local, community level is likely a more effective approach to long-term biodiversity conservation and associated enforcement issues in PNG. Nevertheless, in those inevitable cases in which the legal system must ultimately intervene, having a judiciary that is fully cognizant and understanding of the ecological, social and economic impacts of illegal appropriations and attendant natural resource abuse may more likely result in equitable judgments that protect and perpetuate not only citizen rights, but also the biodiversity of PNG. Indeed, we consistently heard from in-country stakeholders and regulators about the overwhelming need to provide education and training to legislators, regulators and policy implementers at all levels of government, from communities up to the national level, concerning existing laws, their specific mandates related to natural resource conservation, how to consistently coordinate across government, and how to implement and enforce existing laws and policies. Additionally, by providing education and training concerning existing laws, the respective governmental levels may learn what regulatory tools may be deficient or absent, and subsequently develop or amend them.

Land use planning, at both the community and national levels, was identified as a strategic approach that could help reduce threats from logging, agroforestry, mineral extraction, and overfishing as well as provide options for long-term sustainable management of PNG lands. For this study, land use planning includes all terrestrial and aquatic resources, including freshwater and marine, and includes components that contribute to sustainable livelihoods, economic opportunities, or other direct benefits to the community, in a relatively short timeframe. Land use planning can help communities identify lands or waters suitable for agriculture, fishing or hunting, village sites, and protected areas. At a national level, coordinated use of biological and geospatial data can identify suitable lands for biodiversity conservation and lands better suited for resource extraction. Land use planning helps prepare for population growth and climate change adaptation, top drivers of biodiversity loss in PNG, by identifying zones of differential land use and areas to be set aside for biodiversity conservation.

5. Conclusions

Attempts to address threats to biodiversity through traditional conservation and development paradigms that have worked in other parts of the world likely will fail in PNG without adapting activities and expectations to unique local cultures and contexts see, e.g., [44–47]. PNG is singular in the world with regard to its communal land tenancy structure and clan-based culture, which effectively divides the island into over 840 societies [21]. To be successful, conservation interventions in PNG must be deliberative, iterative, carefully considered, led by PNG nationals, supported over many years or decades, and yield both immediate and long-term benefits to affected communities. In addition, operational costs in PNG are extraordinarily high, thus conservation intervention costs are likewise much higher than in other parts of the world, and in-country conservation efforts therefore must be well and consistently funded. Biodiversity conservation in PNG is exceedingly challenging, slow, and costly, but the alternative—the irrevocable loss of the unique biological and cultural richness of this island nation—is unimaginable. The combined actions recommended in this study could deliver a strategic, cost-effective, and lasting legacy for PNG's biodiversity and people.

Author Contributions: Conceptualization, P.B., T.H.W.J., D.P., H.F., D.K.; methodology, T.H.W.J., H.F., C.B., D.E.B., G.D., D.K., D.P., R.R.; formal analysis, D.K., D.P., R.R., T.H.W.J.; investigation, C.B., D.E.B., G.D., H.F., D.K., S.L., D.P., R.R., T.H.W.J.; resources, P.B., S.L.; data curation, C.B., D.E.B., D.P., G.D., D.K., T.H.W.J.; writing—original draft preparation, T.H.W.J.; writing—review and editing, T.H.W.J., C.B.; supervision, P.B.; project administration, P.B., S.L. All authors have read and agreed to the published version of the manuscript.

Funding: This research was funded by the United States Agency for International Development and the United States Department of Interior.

Institutional Review Board Statement: Not applicable.

Informed Consent Statement: Not applicable.

Acknowledgments: We are grateful to the United States Department of Interior, United States Agency for International Development and the Government of Papua New Guinea for financial and logistical support of this study. A special thank you to the residents of Kenangi village, Ikanofi village, Pari village, Komunive Mudmen village and Gahavisuka National Park for hosting us and candidly sharing their insights on biodiversity, climate change, and development. We also thank Kenn Mondiai, Melinda Kera, Cecil Senive, Noel Gisawo, and all the staff of Partners with Melanesians who facilitated our in-country visits and logistics, arranged meetings and interviews with key stakeholders, and kept us in good spirits and good health. Michael Bongro and other colleagues from the PNG Conservation and Environmental Protection Authority (CEPA) also provided insights, support, and expertise that greatly assisted the research. Benedicte Veizaga provided technical assistance with graphics of Figure 4. The findings and conclusions in this article are those of the authors and do not necessarily represent the views of the U.S. Fish and Wildlife Service. Use of trade names in this article does not imply endorsement by the United States government.

Conflicts of Interest: The authors declare no conflict of interest.

Appendix A

Working definitions of the 27 distinct threats to biodiversity in Papua New Guinea identified for quantitative SWOT analysis.

1. CLEAR-CUTTING: The complete removal and elimination of forests and forested land cover, regardless of forest type/location.
2. SELECTIVE-CUTTING: The harvest and removal of trees from forests based on such criteria as size, form, or species and which leaves other non-target trees in place.
3. HARVEST VOLUMES: Total amount of timber removed from a given forested area(s) during timber harvests, usually measured in terms of stand basal area or cubic meters of merchantable wood.
4. RATES OF HARVEST: The time period (rotation) between successive loggings of a given forested tract or area.
5. UNDOCUMENTED HARVESTS: Removal of timber from a forest without official government and/or landowner sanction, nor with any archival records of such. Includes non-commercial small-scale reentry logging of previously logged areas.
6. COMMERCIAL AGROFORESTRY: Large-scale conversion of forests to production of agricultural products, typically for export (e.g., oil palm, tea, coconuts). This includes land clearing associated with SABLs (Special Agricultural and Business Leases).
7. SUBSISTENCE AGRICULTURE: Small-scale agricultural activities primarily for local consumption by individual families or villages. Includes small-scale production of agricultural products for sale in local markets or villages.
8. MINERAL EXTRACTION: All forms of mining (e.g., gold, copper, chromium, nickel) and associated activities. Does not include "seabed mining" which is considered a distinct and separate threat in this assessment.

9. OIL AND GAS EXTRACTION: All activities associated with the exploration, acquisition, and distribution of naturally occurring raw petroleum products (i.e., crude oil, natural gas).
10. DEVELOPMENT/VILLAGE EXPANSION: The increase in geographic extent ("footprint") of a given village. This includes the establishment of new villages due to population increases in existing villages.
11. ANTHROPOGENIC FIRES: Fires started in, or spread to, forested or other areas by human activities, such as from land-clearing activities, hunting activities, cooking fires, burning of garbage, or arson.
12. INFRASTRUCTURE DEVELOPMENTS: Increases in the physical structure(s) associated with human settlements and activities. Includes roads, bridges, communication towers, shopping centers, etc.
13. INVASIVE SPECIES: Any non-native organism (including plants, insects, vertebrates, fungi, diseases) that become – or may become – established in a given area,and which may eliminate, replace, infect, parasitize, or otherwise out-compete native species.
14. SUBSISTENCE HUNTING/FISHING/HARVESTING: The harvesting of wild animals (e.g., birds, mammals, reptiles, fish, mollusks) for non-commercial personal consumption by individuals, families, or villages.
15. HUNTING FOR CULTURAL ITEMS: The harvesting of wild animals (e.g., birds, mammals, reptiles, mollusks) to obtain parts of said animals for use in traditional or ceremonial objects or rituals.
16. PLANT AND WILDLIFE TRADE (REGULATED): The commercial exploitation of native plants and animals according to established laws and regulations.
17. PLANT AND WILDLIFE TRADE (UNREGULATED): The commercial exploitation of native plants and animals without established laws and regulations or other such guidelines or controls. Includes all forms of "illegal" trade and trafficking.
18. TRADITIONAL FOREST ORCHARDS: The long-standing Melanesian practice of the small-scale selective planting and inter-cropping of various tree species in a given area in order to provide a varied and long-term source of fruits, nuts, wood, fiber and other useful products for personal use and consumption.
19. FUEL WOOD HARVESTING: Cutting and/or collection of wood from the forest for use in household cooking or heating.
20. DAMS/HYDROELECTRIC PROJECTS: The use of artificial structures for impeding or altering stream flows for production of electricity, flood control, or to provide sources of irrigation water.
21. RENEWABLE ENERGY PROJECTS: Production of electricity via solar or wind power.
22. POLLUTION: Any environmental contaminants resulting from human activities. Includes household sewage, solid waste, industrial effluents, and agricultural chemicals.
23. REGULATED MARINE FISHING (ARTISANAL): Small-scale harvesting of fish or other aquatic organisms, for commercial or non-commercial purposes, via use of traditional gear and techniques and according to established laws and regulations.
24. AQUACULTURE: The production – commercial or otherwise – of aquatic organisms in controlled structures or environments. Includes activities such as "fish farming".
25. IUU MARINE FISHING: The "Illegal, Unregulated, Unreported" harvesting of fish or other marine species for commercial purposes. In PNG, includes (but not limited to) harvesting of tuna stocks and bêche de mer (sea cucumber) by foreign fishing fleets.
26. REGULATED MARINE FISHING (COMMERCIAL): The large-scale commercial harvest of marine fisheries resources according to established laws and regulations.
27. SEABED MINING: The extraction of submerged minerals and resources from the sea floor, either by dredging sand and/or sediments, or lifting benthic material in any other manner.

Appendix B

Table A1. Papua New Guinea biodiversity threat analysis complete template based on quantitative SWOT methods. Note that TOTAL % represents the total contribution of individual threats to overall biodiversity threat across all three ecosystem realms. ICA represents the product of (IC * SC * TC * POC).

No	THREATS	Weight	IC	SC	TC	POC	ICA	Optimal	Actual	OptimalQ	Actual Q	QD	%TQD	TOTAL%
	TERRESTRIAL													
1.1	Timber harvest													
	1.1.1. Clear cutting	1000	0.042	0.39	1	1	0.016	10	6	0.1625	0.0975	0.065	3.89	12.71
	1.1.2. Selective cutting	600	0.025	0.39	1	1	0.01	10	4	0.0975	0.039	0.059	3.5	8.66
	1.1.3. Harvest volumes	400	0.017	0.39	1	1	0.007	10	5	0.065	0.0325	0.033	1.94	5.52
	1.1.4. Rates of harvest (rotations)	600	0.025	0.39	1	1	0.01	10	4	0.0975	0.039	0.059	3.5	3.5
	1.1.5. Undocumented harvests	100	0.004	0.39	1	1	0.002	10	5	0.0163	0.0081	0.008	0.49	0.49
1.2	Commercial agroforestry (including SABLs)	1000	0.042	0.39	1	1	0.016	10	3	0.1625	0.0488	0.114	6.8	11.59
1.3	Subsistence agriculture	600	0.025	0.39	1	1	0.01	10	6	0.0975	0.0585	0.039	2.33	3.33
1.4	Mineral extraction	400	0.017	0.66	1	1	0.011	10	7	0.11	0.077	0.033	1.97	5.1
1.5	Oil and gas extraction	500	0.021	0.61	1	1	0.013	10	8	0.1271	0.1017	0.025	1.52	3.19
1.6	Development/village expansion	800	0.033	0.25	1	1	0.008	10	7	0.0833	0.0583	0.025	1.5	1.5
1.7	Anthropogenic fires	750	0.031	0.45	0.6	1	0.008	10	7	0.0844	0.0591	0.025	1.51	1.51
1.8	Infrastructure developments (roads, bridges, urbanization)	950	0.04	0.3	1	1	0.012	10	8	0.1188	0.095	0.024	1.42	2.02
1.9	Invasive species	500	0.021	0.85	1	1	0.018	10	8	0.1771	0.1417	0.035	2.12	7.7
1.1	Consumptive use of natural resources													
	1.10.1. Subsistence hunting	600	0.025	0.55	1	1	0.014	10	8	0.1375	0.11	0.028	1.64	2.08
	1.10.2. Hunting for cultural items	200	0.008	0.55	1	1	0.005	10	8	0.0458	0.0367	0.009	0.55	0.55
	1.10.3. Plant and wildlife trade (regulated)	200	0.008	0.55	1	1	0.005	10	9	0.0458	0.0413	0.005	0.27	0.42
	1.10.4. Plant and wildlife trade (unregulated)	300	0.013	0.55	1	1	0.007	10	7	0.0688	0.0481	0.021	1.23	1.38
	1.10.5. Traditional (subsistence) forest orchards	400	0.017	0.55	1	1	0.009	10	8	0.0917	0.0733	0.018	1.1	1.1
	1.10.6. Fuel wood harvesting	500	0.021	0.55	1	1	0.011	10	7	0.1146	0.0802	0.034	2.06	2.06
1.11	Dams/hydroelectric projects	700	0.029	0.13	1	1	0.004	10	8	0.0379	0.0303	0.008	0.45	1.43
1.12	Renewable energy projects (wind, solar)	200	0.008	0.4	1	1	0.003	10	8	0.0333	0.0267	0.007	0.4	0.7
1.13	Pollution (residential and industrial effluents, solid waste, agricultural chemicals)	100	0.004	0.32	1	1	0.001	10	8	0.0133	0.0107	0.003	0.16	6.56
	Subtotals	**11,400**	**0.475**										**40.36**	
	INLAND WATER													

Table A1. *Cont.*

No	THREATS	Weight	IC	SC	TC	POC	ICA	Optimal	Actual	OptimalQ	Actual Q	QD	%TQD	TOTAL%
	TERRESTRIAL													
2.1	Timber harvest													
	2.1.1. Clear cutting	1000	0.042	0.39	1	1	0.016	10	4	0.1625	0.065	0.098	5.83	
	2.1.2. Selective cutting	600	0.025	0.39	1	1	0.01	10	5	0.0975	0.0488	0.049	2.92	
	2.1.3. Harvest volumes	400	0.017	0.39	1	1	0.007	10	4	0.065	0.026	0.039	2.33	
2.2	Commercial agroforestry (including SABLs)	1000	0.042	0.32	1	1	0.013	10	4	0.1333	0.0533	0.08	4.79	
2.3	Subsistence agriculture	500	0.021	0.2	1	1	0.004	10	6	0.0417	0.025	0.017	1	
2.4	Consumptive use of natural resources													
	2.4.1. Subsistence fishing/harvesting	250	0.01	0.33	1	1	0.003	10	8	0.0344	0.0275	0.007	0.41	
	2.4.2. Regulated marine fishing (artisanal)	300	0.013	0.1	1	1	0.001	10	8	0.0125	0.01	0.003	0.15	1.15
	2.4.3. Plant/wildlife trade (regulated)	200	0.008	0.1	1	1	0.001	10	7	0.0083	0.0058	0.003	0.15	
	2.4.4. Plant/wildlife trade (unregulated)	300	0.013	0.1	1	1	0.001	10	7	0.0125	0.0088	0.004	0.22	
2.5	Aquaculture	300	0.013	0.33	1	1	0.004	10	6	0.0413	0.0248	0.017	0.99	0.99
2.6	Renewable energy projects (wind, solar)	200	0.008	0.59	1	1	0.005	10	9	0.0492	0.0443	0.005	0.29	
2.7	Pollution (residential and industrial effluents, solid waste, agricultural chemicals)	600	0.025	0.76	1	1	0.019	10	7	0.19	0.133	0.057	3.41	
2.8	Oil and gas extraction	500	0.021	0.25	1	1	0.005	10	8	0.0521	0.0417	0.01	0.62	
2.9	Mineral extraction	400	0.017	0.66	1	1	0.011	10	6	0.11	0.066	0.044	2.63	
2.1	Dams/hydroelectric projects	1000	0.042	0.13	1	1	0.005	10	7	0.0542	0.0379	0.016	0.97	
2.11	Invasive species	700	0.029	0.8	1	1	0.023	10	6	0.2333	0.14	0.093	5.58	
2.12	Infrastructure developments (roads, bridges, urbanization)	400	0.017	0.3	1	1	0.005	10	8	0.05	0.04	0.01	0.6	
	Subtotals	**8650**	**0.36**										**32.9**	
	MARINE													

Table A1. *Cont.*

No	THREATS	Weight	IC	SC	TC	POC	ICA	Optimal	Actual	OptimalQ	Actual Q	QD	%TQD	TOTAL%
	TERRESTRIAL													
3.1	Timber harvest (considered due to sediment transport, etc.)													
	3.1.1. Clear cutting	300	0.013	1	1	1	0.013	10	6	0.125	0.075	0.05	2.99	
	3.1.2. Selective cutting	150	0.006	1	1	1	0.006	10	4	0.0625	0.025	0.0375	2.24	
	3.1.3. Harvest volumes	100	0.004	1	1	1	0.004	10	5	0.0417	0.0208	0.0208	1.25	
3.2	Consumptive use of natural resources													
	3.2.1. Subsistence fishing/harvesting	200	0.008	0.01	1	1	0	10	6	0.0008	0.0005	0.0003	0.02	
	3.2.2. IUU marine fishing	700	0.029	1	1	1	0.029	10	5	0.2917	0.1458	0.1458	8.72	8.72
	3.2.3. Regulated marine fishing (commercial)	600	0.025	1	1	1	0.025	10	6	0.25	0.15	0.1	5.98	5.98
	3.2.4. Regulated marine fishing (artisanal)	200	0.008	1	1	1	0.008	10	8	0.0833	0.0667	0.0167	1	
3.3	Aquaculture	100	0.004	0.01	1	1	0	10	9	0.0004	0.0004	0	0	
3.4	Renewable energy projects (wind, solar)	100	0.004	0.01	1	1	0	10	9	0.0004	0.0004	0	0	
3.5	Pollution (residential and industrial effluents, solid waste, agricultural chemicals)	300	0.013	1	1	1	0.013	10	6	0.125	0.075	0.05	2.99	
3.6	Oil and gas extraction	700	0.029	0.2	1	1	0.006	10	7	0.0583	0.0408	0.0175	1.05	
3.7	Seabed mining	500	0.021	0.2	1	1	0.004	10	8	0.0417	0.0333	0.0083	0.5	0.5
	Subtotals	**3950**	**0.165**							**4.417**	**2.745**	**1.672**	**26.74**	
	COMPOSITE TOTAL	**24,000**	**1**							**4.783**	**3.026**	**1.758**	**100**	

References

1. Polhemus, D.A.; Englund, R.A.; Allen, G.R. Freshwater biotas of New Guinea and nearby islands: Analysis of endemism, richness, and threats. *Bishop Mus. Tech. Rep.* **2004**, *31*, 1–62.
2. Hope, G. The sensitivity of the high mountain ecosystems of New Guinea to climatic change and anthropogenic impact. *Arctic Antarct. Alp. Res.* **2014**, *46*, 777–786. [CrossRef]
3. Zeriga-Alone, T.; Whitmore, N.; Sinclair, J.R. (Eds.) *The Hindenburg Wall: A Review of Existing Knowledge*; Wildlife Conservation Society Papua New Guinea Program: Goroka, Papua New Guinea, 2012.
4. Shearman, P.L.; Ash, J.; Mackey, B.; Bryan, J.E.; Lokes, B. Forest conversion and degradation in Papua New Guinea 1972–2000. *Biotropica* **2009**, *41*, 379–390. [CrossRef]
5. Shearman, P.L.; Bryan, J.E. A bioregional analysis of the distribution of rainforest cover, deforestation and degradation in Papua New Guinea. *Aust. Ecol.* **2011**, *36*, 9–24. [CrossRef]
6. Gideon, O. The flora of New Guinea: Its origins, affinities, and patterns of diversity and endemism. In *The State of the Forests of Papua New Guinea 2014. Measuring Change over Period 2002–2014*; Bryan, J.E., Shearman, P.L., Eds.; University of Papua New Guinea: Port Moresby, Papua New Guinea, 2015; pp. 115–135.
7. Bryan, J.E.; Shearman, P.L.; Aoro, G.; Wavine, F.; Zerry, J. (Eds.) *The Current State of PNG's Forests and Changes Between 2002 & 2014*; University of Papua New Guinea: Port Moresby, Papua New Guinea, 2015.
8. Mittermeier, R.A.; Gil, P.R.; Mittermeier, C.G. *Megadiversity—Earth's Biologically Wealthiest Nations*; Agrupación Sierra Madre S.C.: D.F. Mexico, Mexico, 1997.
9. Aalsbersberg, B.; Lokani, P.; Tuiwawa, M.; Waqa-Sakiti, H.; Tordoff, A.W. *Ecosystem Profile: East Melanesian Islands Biodiversity Hotspot*; Critical Ecosystem Partnership Fund: Arlington, VA, USA, 2012.
10. Shearman, P.L.; Bryan, J.E.; Ash, J.; Hunnam, P.; Mackey, B.; Lokes, B. *The State of the Forests of Papua New Guinea. Mapping the Extent and Condition of Forest Cover and Measuring the Drivers of Forest Change in the Period 1972–2002*; University of Papua New Guinea: Port Moresby, Papua New Guinea, 2008.
11. Beehler, B.M.; Sine, R. *A Rapid Biodiversity Assessment of the Kaijende Highlands, Enga Province, Papua New: Birds of the Kaijende Highlands, Enga Province, Papua New Guinea*; Conservation International: Washington, DC, USA, 2007.
12. McAlpine, J.; Quigley, J. *Forest Resources of Papua New Guinea: Summary Statistics from the Forest Inventory Mapping (FIM) System*; Australian Agency for International Development: Canberra, ACT, Australian, 1998.
13. Osborne, P.L. Biological and cultural diversity in Papua New Guinea: Conservation, conflicts, constraints and compromise. *Ambio* **1995**, *24*, 231–237.
14. Alcorn, J.B. (Ed.) *Papua New Guinea Conservation Needs Assessment, Vol. 1*; The Biodiversity Support Program: Boroko, Papua New Guinea, 1993.
15. Filer, C. The nature of human threat to Papua New Guinea's biodiversity. In *Papua New Guinea Country Study on Biodiversity*; Sekhran, N., Miller, S.E., Eds.; PNG Department of Environment and Conservation Conservation Resource Centre: Boroko, Papua New Guinea, 1994; pp. 187–199.
16. Sekran, N.; Miller, S. (Eds.) *Papua New Guinea Country Study on Biological Diversity*; Papua New Guinea Department of Environment and Conservation: Port Moresby, Papua New Guinea, 1996.
17. United Nations Environment Programme and Global Environmental Facility (UNEP-GEF). *Papua New Guinea's Fourth National Report to the Convention on Biological Diversity*; United Nations Environment Programme: Port Moresby, Papua New Guinea; Global Environmental Facility: Washington, DC, USA, 2010.
18. Secretariat of the Convention on Biological Diversity (SEC CBD). *Global Biodiversity Outlook 3*; Secretariat of the Convention on Biological Diversity: Montréal, QC, Canada, 2010.
19. US Agency for International Development (USAID). *USAID's Biodiversity Conservation and Forestry Programs, 2015 Report: FY 2014 Results and Funding*; US Agency for International Development: Washington, DC, USA, 2015.
20. US Department of State (USDOS). *Project Agreement Between the United States of America and The Independent State of Papua New Guinea for Environmental Programming*; U.S. Department of State: Port Moresby, Papua New Guinea, 2016.
21. Brown, C.; Busch, D.; Dutson, G.; Freifeld, H.; Krofta, D.; Polhemus, D.; Rounds, R.; White, T. *Papua New Guinea Biodiversity Assessment: Final Report*; United States Department of the Interior—International Technical Assistance Program: Washington, DC, USA, 2017.
22. Salafsky, N.; Margoluis, R. Threat reduction assessment: A practical and cost-effective approach to evaluating conservation and development projects. *Conserv. Biol.* **1999**, *13*, 830–841. [CrossRef]
23. Faith, D.P.; Nix, H.A.; Margules, C.R.; Hutchinson, M.F.; Walker, P.A.; West, J.; Stein, J.L.; Kesteven, J.L.; Allison, A.; Natera, G. The BioRap biodiversity assessment and planning study for Papua New Guinea. *Pac. Conserv. Biol.* **2001**, *6*, 279–288. [CrossRef]
24. Faith, D.P.; Margules, C.R.; Walker, P.A.; Stein, J.; Natera, G. Practical application of biodiversity surrogates and percentage targets for conservation in Papua New Guinea. *Pac. Conserv. Biol.* **2001**, *6*, 289–303. [CrossRef]
25. Faith, D.P.; Margules, C.R.; Walker, P.A. A biodiversity of conservation plan for Papua New Guinea based on biodiversity trade-offs analysis. *Pac. Conserv. Biol.* **2001**, *6*, 304–324. [CrossRef]
26. Faith, D.P.; Walker, P.A.; Margules, C.R. Some future prospects for systematic biodiversity planning in Papua New Guinea—and for biodiversity planning in general. *Pac. Conserv. Biol.* **2001**, *6*, 325–343. [CrossRef]

27. MacMillan, D.C.; Marshall, K. The Delphi process–an expert-based approach to ecological modelling in data-poor environments. *Anim. Conserv.* **2006**, *9*, 11–19. [CrossRef]
28. Battisti, C.; Luiselli, L.; Pantano, D.; Teofili, C. On threats analysis approach applied to a Mediterranean wetland: Is the assessment of human-induced threats related to different level of expertise of respondents? *Biodivers. Conserv.* **2008**, *16*, 1529–1542. [CrossRef]
29. Battisti, C.; Fanelli, G.; Marini, F.; Amori, G.; Luiselli, L. Assessing the nature reserve management effort using an expert-based threat analysis approach. *Diversity* **2020**, *12*, 145. [CrossRef]
30. Salafsky, N.; Salzer, D.; Stattersfield, A.J.; Hilton-Taylor, C.; Neugarten, R.; Butchart, S.H.M.; Collen, B.; Cox, N.; Master, L.L.; O'Connor, S.; et al. A standard lexicon for biodiversity conservation: Unified classifications of threats and actions. *Conserv. Biol.* **2008**, *22*, 97–911. [CrossRef]
31. White, T.H.; Barros, Y.d.M.; Develey, P.F.; Llerandi-Roman, I.C.; Monsegur-Rivera, O.; Trujillo-Pinto, A. Improving reintroduction planning and implementation through quantitative SWOT analysis. *J. Nat. Conserv.* **2015**, *28*, 149–159. [CrossRef]
32. Johnson, R.; Wichern, D. *Applied Multivariate Statistical Methods*, 3rd ed.; Prentice Hall: Hoboken, NJ, USA, 1992.
33. Lance, G.N.; Williams, W.T. A general theory of classification sorting strategies, I. Hierarchical systems. *Comp. J.* **1967**, *9*, 373–380. [CrossRef]
34. US Agency for International Development (USAID). *Developing Situation Models in USAID Biodiversity Programming*; US Agency for International Development: Washington, DC, USA, 2016. Available online: https://biodiversitylinks.org/projects/completed-projects/measuring-impact/how-to-guides-for-usaid-biodiversity-programming/how-to-guides-for-usaid-biodiversity-programming/ (accessed on 22 May 2021).
35. US Agency for International Development (USAID). *Pacific Islands Tropical Forestry and Biodiversity Conservation Assessment*; USAID/Asia: Manila, PH, USA, 2012.
36. US Agency for International Development (USAID). *Defining Outcomes & Indicators for Monitoring, Evaluation, and Learning in USAID Biodiversity Programming*; US Agency for International Development: Washington, DC, USA, 2016. Available online: https://biodiversitylinks.org/projects/completed-projects/measuring-impact/how-to-guides-for-usaid-biodiversity-programming/how-to-guides-for-usaid-biodiversity-programming/ (accessed on 22 May 2021).
37. US Agency for International Development (USAID). *Using Results Chains to Depict Theories of Change in USAID Biodiversity Programming*; US Agency for International Development: Washington, DC, USA, 2016. Available online: https://biodiversitylinks.org/projects/completed-projects/measuring-impact/how-to-guides-for-usaid-biodiversity-programming/how-to-guides-for-usaid-biodiversity-programming/ (accessed on 22 May 2021).
38. Tulloch, V.J.D.; Atkinson, S.; Possingham, H.P.; Peterson, N.; Linke, S.; Allan, J.R.; Kaiye, A.; Keako, M.; Sabi, J.; Suruman, B.; et al. Minimizing cross-realm threats from land-use change: A national-scale conservation framework connecting land, freshwater and marine systems. *Biol. Conserv.* **2021**, *254*, 1–12. [CrossRef]
39. Australian Agency for International Development (AUSAID). *Making land work: Reconciling customary land and development in the Pacific—Case studies on customary land and development in the Pacific*; Commonwealth of Australia: Canberra, ACT, Australia, 2008.
40. Laurance, W.F.; Kakul, T.; Keenan, R.J.; Sayer, J.; Passingan, S.; Clements, G.R.; Villegas, F.; Sodhi, N.S. Predatory corporations, failing governance, and the fate of forests in Papua New Guinea. *Conserv. Lett.* **2011**, *4*, 95–100. [CrossRef]
41. Numapo, J. *Commission of Inquiry into the Special Agriculture and Business Lease (SABL): Final Report*; GoPNG Commission of Inquiry: Port Moresby, Papua New Guinea, 2013.
42. Gabriel, J.; Wood, M. The Rimbunan Hijau Group in the forests of Papua New Guinea. *J. Pac. Hist.* **2015**, *50*, 322–343. [CrossRef]
43. Mosseau, F.; Lau, P. *The Great Timber Heist: The Logging Industry in Papua New Guinea*; The Oakland Institute: Oakland, CA, USA, 2015.
44. Sakata, H.; Prideaux, B. An alternative approach to community based ecotourism: A bottom-up locally initiated non-monetised project in Papua New Guinea. *J. Sustain. Tour.* **2013**, *21*, 880–899. [CrossRef]
45. Filer, C. Why green grabs don't work in Papua New Guinea. *J. Peas. Stud.* **2012**, *39*, 599–617. [CrossRef]
46. Melick, D.R.; Kinch, J.P.; Govan, H. How global biodiversity targets risk becoming counterproductive: The case of Papua New Guinea. *Conserv. Soc.* **2012**, *10*, 344–353. [CrossRef]
47. Balboa, C.A. How successful transnational non-governmental organizations set themselves up for failure on the ground. *World Develop.* **2014**, *54*, 273–287. [CrossRef]

diversity

MDPI

Article

A Preliminary Survey on the Planktonic Biota in a Hypersaline Pond of Messolonghi Saltworks (W. Greece)

George N. Hotos

Plankton Culture Laboratory, Department of Animal Production, Fisheries & Aquaculture, University of Patras, 30200 Messolonghi, Greece; ghotos@upatras.gr

Abstract: During a survey in 2015, an impressive assemblage of organisms was found in a hypersaline pond of the Messolonghi saltworks. The salinity ranged between 50 and 180 ppt, and the organisms that were found fell into the categories of Cyanobacteria (17 species), Chlorophytes (4 species), Diatoms (23 species), Dinoflagellates (1 species), Protozoa (40 species), Rotifers (8 species), Copepods (1 species), *Artemia* sp., one nematode and *Alternaria* sp. (Fungi). *Fabrea salina* was the most prominent protist among all samples and salinities. This ciliate has the potential to be a live food candidate for marine fish larvae. *Asteromonas gracilis* proved to be a sturdy microalga, performing well in a broad spectrum of culture salinities. Most of the specimens were identified to the genus level only. Based on their morphology, as there are no relevant records in Greece, there is a possibility for some to be either new species or strikingly different strains of certain species recorded elsewhere.

Keywords: protists; cyanobacteria; rotifers; crustacea; hypersaline conditions; Messolonghi saltworks

Citation: Hotos, G.N. A Preliminary Survey on the Planktonic Biota in a Hypersaline Pond of Messolonghi Saltworks (W. Greece). *Diversity* **2021**, *13*, 270. https://doi.org/10.3390/d13060270

Academic Editor: Michael Wink

Received: 24 May 2021
Accepted: 12 June 2021
Published: 15 June 2021

Publisher's Note: MDPI stays neutral with regard to jurisdictional claims in published maps and institutional affiliations.

cc BY

Copyright: © 2021 by the author. Licensee MDPI, Basel, Switzerland. This article is an open access article distributed under the terms and conditions of the Creative Commons Attribution (CC BY) license (https://creativecommons.org/licenses/by/4.0/).

1. Introduction

It is well known that saltwork waters support high algal densities due to the abundance of nutrients concentrated by evaporation [1–3]. Apart from the fact that such ecosystems are of paramount ecological value, they are also a potential source for tolerant biota that can be exploited for aquaculture [4] or other uses [5,6]. Generally, in hypersaline systems, the microbial life in the prokaryotic level (halophilic archaea and bacteria) has been extensively described, (e.g., [7,8]) emphasizing their role (in addition to viruses) as highly essential in the biogeochemical processes. The eukaryotic invertebrate biota in hypersalinity lagged considerably in terms of diversity and interaction with all elements of this environment, resulting in a poor understanding of its role in the dynamics of food webs. In most works concerning protists or crustacea, the halotolerant green alga *Dunaliella* spp. (e.g., [9,10]) occupies the bulk of studies for algae, and the anostacan *Artemia* (e.g., [11,12]) for planktonic invertebrates.

Considering the scarcity of adequate information on organisms other than bacteria from hypersaline environments in Greece [13], a preliminary survey in the salterns of Messolonghi (W. Greece) was made throughout the spring and summer of 2015. The aim was to identify all organisms visible by optical microscopy to the genus level in order to gain an understanding of their presence and abundance as a guide for future, detailed studies in this biotope. A further aim involved testing the potential for maintenance and culture of all possible organisms in laboratory conditions for their use as live food for aquaculture and other general use. The situation is perplexing, as the topic of cyanobacteria and protists (algae and protozoa) from hypersalinity is highly varied in the literature. As images are essential for identifying species, pictures and live videos were taken by microscopy, and material is presented here.

2. Materials and Methods

The water samples were taken from a particular pond of the Messolonghi saltworks, lying between the coordinates 38°23′47″ to 38°23′31″ N and 21°24′17″ to 21°24′33″ E.

Samples were taken during April–September 2015 on a monthly basis, thus following the salinity range (50–180 ppt) of the changing water conditions. The pond (~12 ha) is located at the periphery of the saltworks complex and is connected to the Messolonghi lagoon by a narrow channel, the water of which is manageably diverted to fill the evaporation ponds of the saltworks. In contrast to the main evaporation ponds that produce salt, this pond is filled with water throughout the whole year. Although evaporation gradually increases the pond's salinity from April onwards, it never becomes dry, its water level remaining between 0.4 and 1.7 m at its deepest central area. The samples were only water with no benthos included. A 2 L plastic beaker was used, with only the surface of the bottom touching, in order for the water to be disturbed just enough for the top layer to be included in the sample (~0.7 m from the shore). Salinity and temperature were recorded, and the samples were immediately transported to the laboratory where the temperature was kept at 22 °C. The samples were kept in 2 L Erlenmeyer conical flasks, lit with ambient light of about 20 μmol m^{-2} s^{-1}, and aerated using pumped air via a pipette. There were 2 sub-samples of 50 mL taken and centrifuged mildly (SIGMA 3K10, Sigma Laborzentrifugen GmbH, Osterode, Germany) at 3000 rpm for 3 min after which the sediment with 2 mL of water was kept for microscopical examination. The decanted supernatant was free of organisms, as they were sedimented and then kept in the resuspended 2 mL of the vial. There were no adverse effects of the centrifugation on the motility and viability of the organisms. The procedure for the 50 mL sub-samples was repeated for 3 successive days in order to strengthen the detection. The 2 mL concentrated samples were apportioned to 0.1 mL sub-samples as droplets in shallow, glass, Petri dishes and examined microscopically (Leica Labovert FS inverted microscope and Leika Leitz DMRB) in order to count organisms larger than 30 μm. This included all protozoa, rotifers, copepods, *Artemia* and nematodes. After a thorough counting assessment of the live specimens, a drop of Lugol was added, and the immobilized organisms were counted once more. The other 2 mL sample was kept intact (without Lugol) in order to be examined microscopically for microalgae using the Leica DM-RB microscope at 400× magnification.

Additionally, from the live samples, organisms were targeted and removed by micropipette suction and placed in 6 well multi-chamber plates (SARSTEDT) with 4 mL of Walne's nutrient fertilized water (enriched with silica in case of diatoms) of similar salinity to that of their most abundance. This was carried out in order to test their potential for growth in culture conditions as a preliminary trial for the feasibility to be mass cultured. The culture plates were left to mature in the laboratory (at 22–23 °C, 12:12 h light:dark, illumination at 50–60 μmol $m^{-2}s^{-1}$ without aeration). The plates were examined after 10 days for population increase. It was noted that during this 10-day period, due to the initial low density of organisms in each chamber, there were no adverse effects of metabolic waste on them. When the cultures were algae, the density was calculated using a Fuchs–Rosenthal hematocytometer. If the cultures were protozoa, rotifers, copepods or *Artemia*, then to start, 0.2 mL of dense phytoplankton that was previously cultured and comprised of *Asteromonas gracilis* and *Dunaliella* was added in order to supply adequate food. These two microalgae were selected as food on the grounds that they dominated the chlorophyte microflora of the pond and presumably served as food for the above heterotrophs in their natural habitat. After 10 days, any increase in the population was recorded in a dissecting microscope (Nikon SMZ-U). The abundance of the various organisms (less than 30 μm in size) in the sub-sample mixtures taken from the various salinity samples was calculated as the counted individuals of each species in a 1 mm^2 area of the microscopy vision field. Counts were used for comparisons among salinities for a rough estimation of abundance. All photos presented here were taken (Jenoptik Progres Gryphax Arktur digital microscopy camera) from live specimens after immobilization in the freezer for about 1 h (Figures 1–7). For identification of the organisms, various studies were used as guides [14–19].

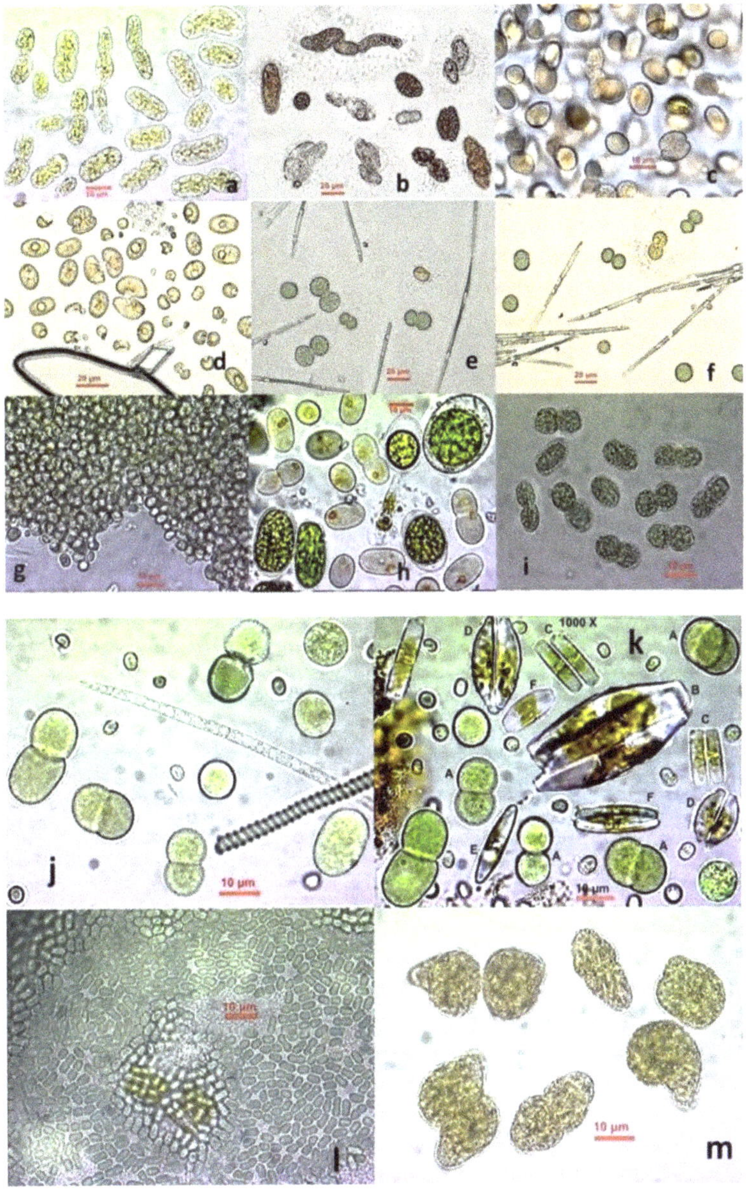

Figure 1. Cocciform Cyanobacteria from hypersalinity at Messolonghi saltworks. (**a**) Peculiar involuted cells of an unidentified species; (**b**) variously shaped cells, sole, involuted and dividing, probably of genus *Synechococcus* and in some of them with a mucilage layer around cells; (**c**) totally unknown species; (**d**) kidney-shaped cells of an unknown species; (**e**,**f**) various cells of genus *Cyanothece* in division state; (**g**) *Microcystis* sp. colony; (**h**) *Synechococcus*-like cells among normal and palmelloid cells of the chlorophyte *Tetraselmis marina*; (**i**) probably *Synechococcus* sp.; (**j**) *Cyanothece* sp. cells at various stages of division along with an *Arthrospira* sp. filament; (**k**) *Aphanothece* sp. and *Cyanothece* sp. cells along with pennate diatoms; (**l**) dense colony of small greenish cells of the *Synechococcus* type; (**m**) peculiar involuted cells of probably *Synechococcus* sp.

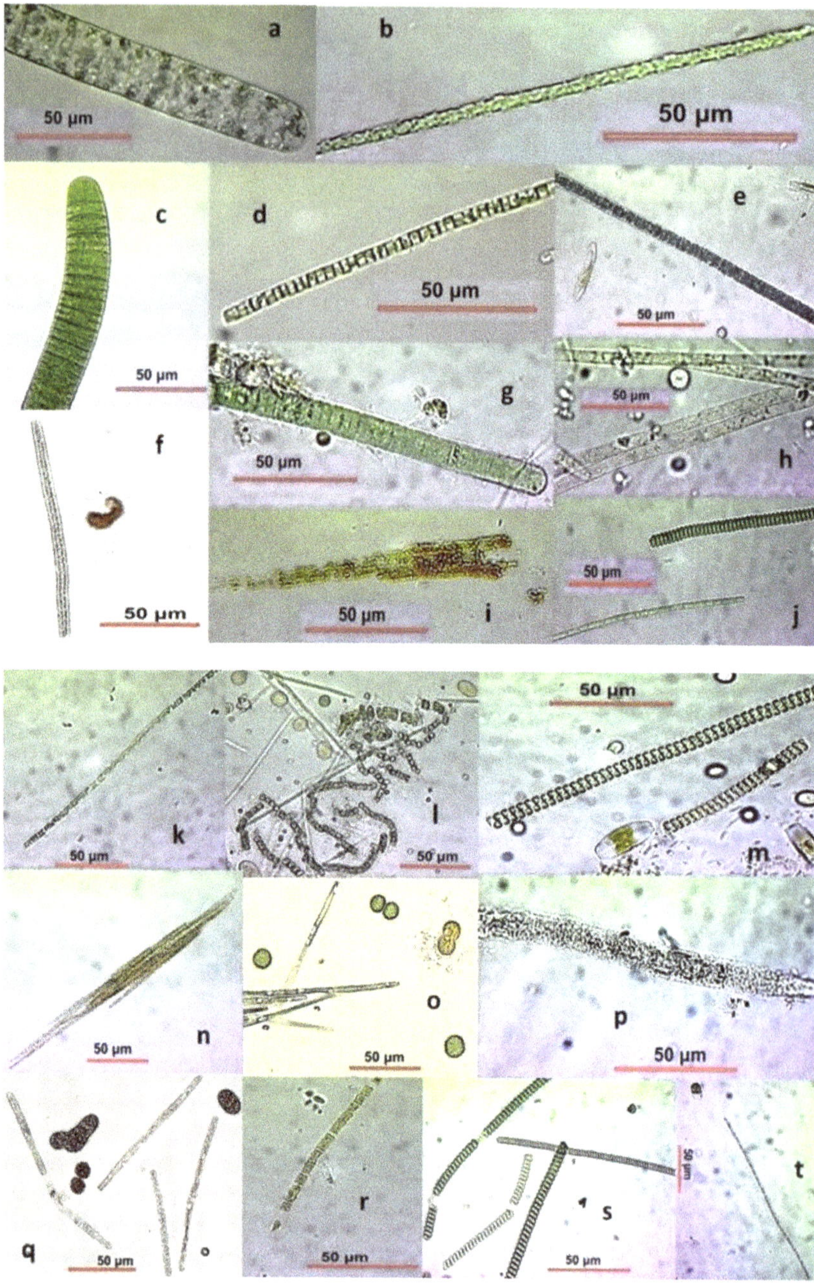

Figure 2. Filamentous Cyanobacteria from hypersalinity ponds. (**a**) *Oscillatoria* sp.; (**b**) unidentified trichome; (**c**) *Oscillatoria* sp.; (**d**) unidentified trichome; (**e**) *Beggiatoa* sp.?; (**f**) unidentified; (**g**) *Lyngbya* sp.; (**h**) *Tychonema* sp.; (**i**) *Aphanizomenon* sp.; (**j**) *Pseudoanabaena* sp.? and *Arthrospira* sp.; (**k**) *Prochlorothrix* sp.; (**l**) *Anabaena* sp.; (**m**) *Arthrospira* sp. thick and thin filaments; (**n**) *Aphanizomenon* sp.; (**o**) *Prochlorothrix* sp.?; (**p**) *Cylindrospermopsis* sp.?; (**q**) *Beggiatoa* sp. among *Synechococcus*; (**r**) *Cylindrospermopsis* sp.; (**s**) fragmented *Arthrospira* filaments of various thickness; (**t**) *Cylindrospermopsis* sp.?

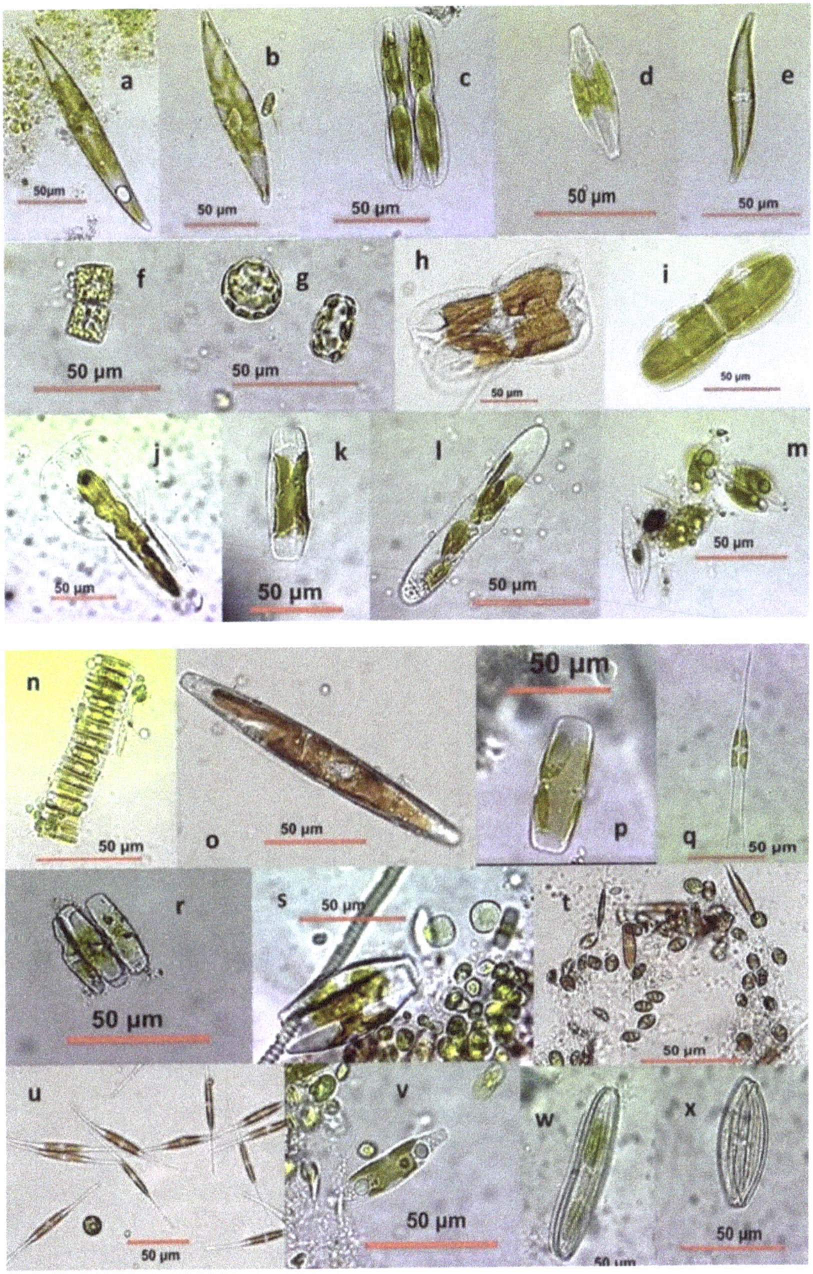

Figure 3. Diatoms from hypersalinity. (**a**) *Pleurosigma* sp., lateral view; (**b**) *Pleurosigma* sp., girdle view; (**c**) *Entomoneis* sp.; (**d**) *Navicula* sp.; (**e**) *Gyrosigma* sp.; (**f**) *Cyclotella* sp. dividing; (**g**) *Cyclotella* sp. round and elongated form; (**h**) *Entomoneis* sp.; (**i**) *Amphiprora* sp.?; (**j**) *Gomphonema* sp.?; (**k**) unidentified cymbelloid species; (**l**) *Pinnularia* sp.?; (**m**) *Cymbella* sp.; (**n**) *Eunotia* sp.?; (**o**) *Nitzschia* sp.; (**p**) unidentified diatom; (**q**) *Nitzschia* dividing; (**r**) *Eunotia* sp.?; (**s**) *Cymbella* sp.; (**t**) *Cocconeis* sp.; (**u**) *Cylindrotheca* sp.; (**v**) *Craticula* sp.; (**w**) *Epithemia* sp.?; (**x**) *Diatoma* sp.(same scale bar as that in w).

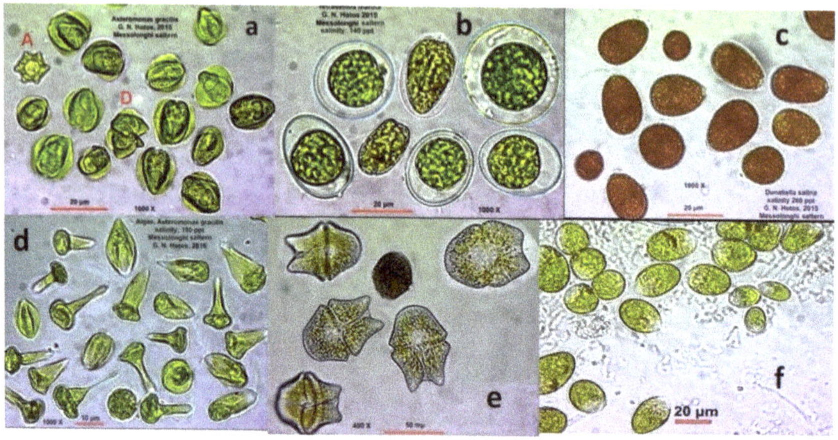

Figure 4. The dominant chlorophytes and dinophytes in hypersalinity. (**a**) *Asteromonas gracilis*; (**b**) *Tetraselmis marina*, normal and palmelloid cells; (**c**) *Dunaliella* sp., reddish cells full of carotenoids at 180 ppt salinity; (**d**) *Asteromonas gracilis* in peculiar cell shapes; (**e**) the dinoflagellate *Gymnodinium* sp.; (**f**) *Dunaliella* sp., green cells at 100 ppt salinity.

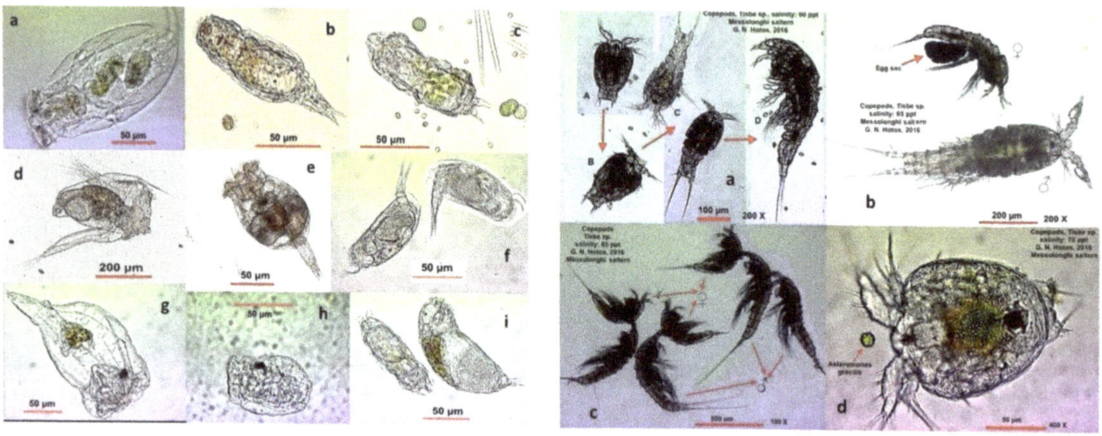

Figure 5. Metazoa from hypersalinity. Left plate: rotifers, (**a**) *Testudinella* sp.; (**b**) *Pleurotrocha* sp.; (**c**) *Lindia* sp.; (**d**) *Hexarthra* sp.; (**e**) *Brachionus plicatilis*; (**f**) *Colurella* sp.; (**g**) *Epiphanes* sp.; (**h**) unidentified marine rotifer; (**i**) *Encentrum* sp. Right plate: copepod *Tisbe* sp., (**a**) various ontogenic stages, A: early nauplius, B: late nauplius, C: copepodites, D: adult; (**b**) male and female individuals; (**c**) copulation captured photo; (**d**) *Tisbe* nauplius fed *Asteromonas* cells.

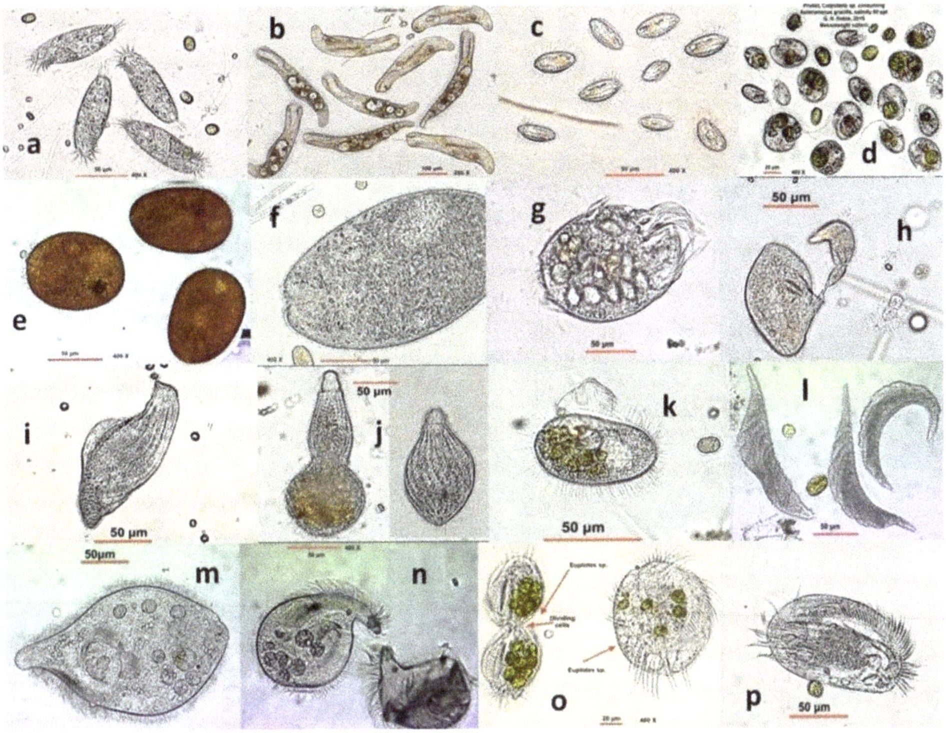

Figure 6. Representative ciliate Protozoa from a hypersaline pond of Messolonghi saltworks. (**a**) *Litonotus* sp.?; (**b**) *Condylostoma* sp.; (**c**) *Cyclidium* sp.; (**d**) *Colpodella* sp.; (**e**) *Frontonia* sp.; (**f**) *Climacostomum* sp.; (**g**) *Uronychia* sp.; (**h**) *Holophrya* sp. in budding reproduction; (**i**) *Loxodes* sp.; (**j**) *Phialina* sp.; (**k**) *Pleuronema* sp.; (**l**) *Amphileptus* sp.; (**m**) *Fabrea salina* at 90 ppt; (**n**) *Fabrea salina* at 170 ppt; (**o**) *Euplotes* sp. in division, full of *Asteromonas* cells (left); (**p**) *Euplotes* sp.

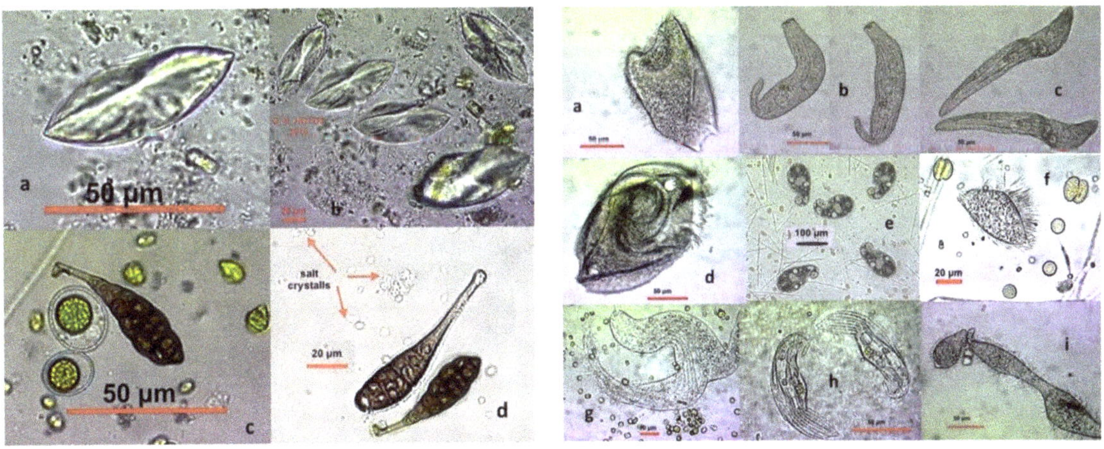

Figure 7. Unidentified organisms and fungi spores from the hypersalinity ponds of Messolonghi saltwork. Left plate: (**a**,**b**), unidentified microbes among peculiar salt crystals at 175 ppt, (**c**,**d**), *Alternaria* sp. (fungi) spores at 90 ppt. Right plate: (**a**–**i**), ciliates not resembling anything known from Protozoan atlases.

3. Results and Discussion

The organisms found (Table 1) can be categorized as Cyanobacteria, Protozoa, eucaryotic microalgae, rotifers, copepods, *Artemia*, a nematode and fungi. The salinity range clearly demarcated the presence of some organisms from others. In particular, at salinities over 160 ppt, only *Artemia* sp., *Dunaliella* sp., *Asteromonas gracilis, Fabrea salina* and some cocciform cyanobacteria were detected and able to stay alive and grow at similar (with their occurrence) salinities in laboratory conditions. An unusual finding was that while some genera of cyanobacteria were detected in mass at elevated salinities, their subsequent culture at similar salinities in the laboratory yielded poor results. It seems that a combination of elusive parameters in their specific natural habitat fulfills their needs. In the salinity range of 110–160 ppt, many more organisms (included those previously mentioned) were recorded with representatives from all categories except rotifers and copepods. At salinities of 60–110 ppt, cyanobacteria, rotifers and protozoa were more abundant than in higher salinities. *Fabrea salina*, a well-documented ciliate in saline water bodies (>35 ppt) all over the world (e.g., [20,21]), dominated in all salinities. It was easily mass cultured at almost every salinity in the range of 35–150 ppt, thus being a candidate live food for larval marine fish. At salinities higher than 160 ppt, *F. salina* encysts can remain viable for a long time [22], reviving again after lowering the salinity below 50 ppt (unpublished data). *F. salina* plays a crucial role within the food web in hypersaline waters through being a consumer of *Dunaliella* sp. [13,23] and in the quality of its salt production [24]. However, the statement in [13] that *F. salina* produces slime must be rejected, as this is rather the result of mucus excretion of several cocciform cyanobacteria (personal observations, unpublished) or glycerol overproduction of *Dunaliella*, a genus notorious for this process in high salinities [25]. The copepod *Tisbe* sp. also exhibited remarkable viability in a wide range of salinities (35–90 ppt) and was easily cultured with a high reproduction rate, feeding heavily on a wide spectrum of microalgae. Its culture can remain viable even in water with a heavy organic load without added food; therefore, it is considered to be a hardy species for larval aquaculture. The green Chlorophytes (*A. gracilis, Tetraselmis marina* and *Dunaliella* sp.) were easily mass cultured, showing better growth at salinities over 100 ppt. *T. marina* was the most sensitive of the three, as for unknown reasons, its cells often lose all four flagella and are transformed to palmelloid cells [26]. Nevertheless, these three halotolerant microalgae proved to be an excellent food source for the rotifer *Brachionus plicatilis*, copepods, *Artemia* and the ciliate *F. salina*. Considering the variety of information in the literature on the presence of all the above categories of organisms in hypersalinity (e.g., [20,21,27]), a wide field awaits to be studied in detail. In particular, the spectrum of the actual number of species of cyanobacteria and protists may be much broader than presented here. Endemicity may also be much more intense than conservatively thought. The species in Greece may be different from saltworks in adjacent countries, noting that there are few natural hypersaline lakes in Europe. This remains true, especially when more remote areas on Earth are considered. Saltworks are not naturally formed and evolved biotopes but rather reflect the conditions in which the extreme edge of acclimation and adaptation of the marine organisms that are constantly transported from the sea to the saltpans can be observed. Consequently, the endemicity theory refers to the sea habitat. In that sense, Foissner's (2008) [28] moderate endemicity distribution model in protists as opposed to the ubiquity distribution model seems to explain the findings of the present study, even in this case whereby recognition was based on morphology and confined to the genus and not to the species level of the encountered organisms. It seems that apart from protists, this hypothesis also applies to hypersaline cyanobacteria; thus, an entire unexplored eco-habitat awaits a multidisciplinary approach. The present study should be considered as a preliminary attempt to outline the wealth of micro-biota in a specific, local hypersaline environment. With the aim of igniting interest for further elaboration in future studies, the organisms presented in Figures 1–7 are representatives of the whole collection.

Table 1. The organisms recorded in hypersalinity at Messolonghi saltworks identified to the genus level. "+" stands for the least presence, "++++" for maximum and "-" for absence in relation to the counts sum of each particular organism across salinities (absolute abundance) and in combination with a rather rough estimation of their relative abundance among all other organisms in each particular sample examined. Concerning their response to the culture trials, notations in the column "Culture Response" mean: "0" = no change or decrease in the initial number of organisms, "1×–2× " = increase of 1–2 times in the initial number of organisms, "2×–3×", = increase of 2–3 times, "3×–4×" = increase of 3–4 times, ">4×" = increase of over 4 times. Detailed records in the Supplementary Material of Table S1.

Salinity Range (ppt)	50–80	81–110	111–130	131–160	>160	Culture Response
CYANOBACTERIA						
Synechococcus	+++	++++	++++	+	+	1×–2×
Aphanothece	++	+++	++++	+	-	0
Microcystis	++++	+++	++	-	-	0
Cyanothece	+	++	++++	+++	+	3×–4×
Oscillatoria	++++	+++	++	-	-	1×–2×
Lyngbya	++++	++	+	-	-	0
Aphanizomenon	+++	++++	++	-	-	
Cylindrospermopsis	++	+++	+	+	-	
Anabaena	+++	+	-	-	-	1×–2×
Arthrospira	+++	++++	++++	++	-	1×–2×
Beggiatoa	++	+	-	-	-	
Scytonema	++	+	-	-	-	
Prochlorothrix	+	-	-	-	-	
Microcoleus	+	-	-	-	-	
Tychonema	+	-	-	-	-	
Pseudoanabaena	++	+	-	-	-	
Phormidium	++++	+	-	-	-	>4×
PROTOZOA						
Euplotes	++++	++++	++	+	-	>4×
Uronychia	++++	+	-	-	-	1×–2×
Diophrys	++++	+	-	-	-	
Frontonia	++++	++	+	-	-	0
Dysteria	+					
Aspidisca	++++	++++	++	-	-	
Paramecium	++++	++	-	-	-	1×–2×
Euglena	++	-	-	-	-	1×–2×
Paraurostyla	+++	++	+	-	-	
Colpoda	++++	+++	++	-	-	
Coleps	++	-	-	-	-	1×–2×
Amphileptus	+++	+	+	-	-	
Condylostoma	++++	+++	++	+	-	2×–3×
Amoeba	++++	++++	++	+	-	2×–3×
Holophrya	++++	++	+	+	-	

Table 1. *Cont.*

Salinity Range (ppt)	50–80	81–110	111–130	131–160	>160	Culture Response
Halteria	++	+	-	-	-	0
Pleuronema	++++	++	++	+	-	1×–2×
Cyclidium	++++	++++	+++	++	-	2×–3×
Loxodes	++	++	+	-	-	
Litonotus	++	+	+	-	-	1×–2×
Chaetospira	+++	+	+	+	-	
Stichotria	+++	+	+	-	-	
Bursaridium	++	+++	-	-	-	
Climacostomum	++++	+++	++	+	-	
Blepharisma	++++	+++	++	-	-	
Holosticha	++++	++	+	-	-	
Vorticella	++++	+++	++	+	-	2×–3×
Remanella	++++	++	+	+	-	
Lembandion	++	-	-	-	-	
Strobidium	++	+	-	-	-	
Uronema	++++	++++	++	+	-	
Bursaria	++	-	-	-	-	
Tracheloraphis	++	-	-	-	-	
Lacrymaria	+	-	-	-	-	
Hemiophrys	++	+	-	-	-	
Fabrea salina	++++	++++	++++	++++	++	>4×
Dileptus	++++	+	-	-	-	
Colpodella	++++	+++	++	-	-	2×–3×
Phialina	+++	++	+	-	-	
Choanoflagellates	++	+	-	-	-	
MICROALGAE (Chlorophytes)						
Asteromonas gracilis	++	++++	++++	++++	++++	>4×
Dunaliella	++++	++++	++++	++++	++++	>4×
Tetraselmis marina	++	++++	+++	+	-	2×–3×
Hymenomonas	++++	++	-	-	-	0
MICROALGAE (Diatoms)						
Cymbella	++++	+++	+++	+	-	1×–2×
Caloneis	++	+	-	-	-	
Cyclotella	++++	+	-	-	-	3×–4×
Craticula	++	+	-	-	-	
Navicula	++++	++++	+++	++	-	
Nitzschia	++++	++++	++++	+++	-	>4×
Pleurosigma	++++	+++	++	-	-	
Entomoneis	+++	+	-	-	-	

Table 1. *Cont.*

Salinity Range (ppt)	50–80	81–110	111–130	131–160	>160	Culture Response
Encyonema	++	+	-	-	-	
Ulnaria	+	-	-	-	-	
Pinnularia	++	+	-	-	-	
Surinella	+	+	-	-	-	
Neidium	++	-	-	-	-	
Synendra	++++	++	+	+	-	
Stauroneis	+	+	-	-	-	
Gyrosigma	++++	++	+	-	-	2×–3×
Amphiprora	+	-	-	-	-	
Eunotia	++	-	-	-	-	
Epithemia	+	-	-	-	-	
Diatoma	+	-	-	-	-	
Cymatopleura	++	-	-	-	-	
Cocconeis	++++	+	+	-	-	2×–3×
Cylindrotheca	++	++	+	+	-	1×–2×
DINOFLAGELLATES						
Gymnodinium	++++	++	-	-	-	
ROTIFERS						
Hexarthra	++	-	-	-	-	0
Pleurotrocha	++++	+	-	-	-	
Epiphanes	++	-	-	-	-	0
Encentrum	+++	-	-	-	-	
Lindia	++++	+++	-	-	-	0
Colurella	+++	++	-	-	-	
Testudinella	++	+	-	-	-	1×–2×
Brachionus plicatilis	++	-	-	-	-	>4×
COPEPODS						
Tisbe	++++	+++	-	-		>4×
ANOSTRACA						
Artemia	++++	++++	++++	++++	++++	>4×
NEMATODE						
Mesacanthoides	++++	++++	+	+	-	>4×
FUNGI						
Alternaria	+	+	-	-	-	

Supplementary Materials: The following are available online at https://www.mdpi.com/article/10.3390/d13060270/s1, Table S1: Records of organisms found in the monthly samples at various salinities and of their response in the culture trials.

Funding: This research received no external funding.

Institutional Review Board Statement: Not applicable.

Informed Consent Statement: Not applicable.

Data Availability Statement: Not applicable.

Conflicts of Interest: The author declare no conflict of interest.

References

1. Javor, B.J. Planktonic standing crop and nutrients in a saltern ecosystem 1. *Limnol. Oceanogr.* **1983**, *28*, 153–159. [CrossRef]
2. Wen, Z.; Zhi-Hui, H. Biological and ecological features of inland saline waters in North Hebei, China. *Int. J. Salt Lake Res.* **1999**, *8*, 267–285. [CrossRef]
3. Davis, S.J. Biological and physical management information for commercial solar saltworks. In Proceedings of the 1st Intern Conf. Ecological Importance of Solar Saltworks (CEISSA 06), Santorini, Greece, 20–22 October 2006; pp. 5–14.
4. Borowitzka, M.A. Microalgae for aquaculture: Opportunities and constraints. *J. Appl. Phycol.* **1997**, *9*, 393–401. [CrossRef]
5. DasSarma, S.; Arora, P. Halophiles. In *Encyclopedia of Life Sciences*; Nature Publ. Group: London, UK, 2002; Volume 8, pp. 458–466.
6. Oren, A. Diversity of halophilic microorganisms: Environments, phylogeny, physiology, and applications. *J. Ind. Microbiol. Biotechnol.* **2002**, *28*, 56–63. [CrossRef] [PubMed]
7. Harris, J.; Caporaso, J.G.; Walker, J.J.; Spear, J.; Gold, N.J.; Robertson, C.; Hugenholtz, P.; Goodrich, J.; McDonald, D.; Knights, D.; et al. Phylogenetic stratigraphy in the Guerrero Negro hypersaline microbial mat. *ISME J.* **2012**, *7*, 50–60. [CrossRef] [PubMed]
8. Podell, S.; Ugalde, J.A.; Narasingarao, P.; Banfield, J.F.; Heidelberg, K.; Allen, E.E. Assembly-Driven Community Genomics of a Hypersaline Microbial Ecosystem. *PLoS ONE* **2013**, *8*, e61692. [CrossRef] [PubMed]
9. Oren, A. A hundred years of *Dunaliella* research: 1905–2005. *Saline Syst.* **2005**, *1*, 2. [CrossRef] [PubMed]
10. Ramos, A.A.; Polle, J.; Tran, D.; Cushman, J.C.; Jin, E.-S.; Varela, J.C. The unicellular green alga *Dunaliella salina* Teod. as a model for abiotic stress tolerance: Genetic advances and future perspectives. *Algae* **2011**, *26*, 3–20. [CrossRef]
11. Camara, M.R. Dispersal of *Artemia franciscana* Kellogg (Crustacea, Anostraca) Populations in the Coastal Saltworks of Rio Grande do Norte, Northeastern Brazil. *Hydrobiologia* **2001**, *466*, 145–148.
12. Saygi, Y. Characterization of the parthenogenic Artemia populations from Camalti (Izmir, Turkey) and Kalloni (Lesvos, Greece). Survival, growth, maturation, biometrics, fatty acid profiles and hatching characteristics. *Hydrobiologia* **2004**, *52*, 227–239. [CrossRef]
13. Dolapsakis, P.N.; Tafas, T.; Abatzopoulos, J.T.; Ziller, S.; Economou-Amilli, A. Abundance and growth response of micro-algae at Megalon Embolon solar saltworks in northern Greece: An aquaculture prospect. *J. Appl. Phycol.* **2005**, *17*, 39–49. [CrossRef]
14. Sournia, A. *Atlas du Phytoplankton Marin. Cyanophycées, Dictyochophycées, Dinophycées, Raphidophycées*; CNRS: Paris, France, 1986; p. 219.
15. Ricard, M. *Atlas du Phytoplankton Marin. Diatomophycées*; CNRS: Paris, France, 1987; p. 297.
16. Foissner, W.; Berger, H. A user-friendly guide to the ciliates (Protozoa, Ciliophora) commonly used by hydrobiologists as bioindicators in rivers, lakes, and waste waters, with notes on their ecology. *Freshw. Biol.* **1996**, *35*, 375–482. [CrossRef]
17. Tomas, C.R. *Identifying marine Diatoms and Dinoflagellates*; Academic Press: San Diego, CA, USA, 1996; 565p.
18. Lee, J.; Leedale, G.; Bradbury, P. *An Illustrated Guide to the Protozoa, Volumes I & II*; Society of the Protozoologists: Law-Rence, KS, USA, 2000; pp. 1–1432.
19. Fontaneto, D.; De Smet, H.W.; Melone, G. Identification key to the genera of marine rotifers worldwide. In *Meiofauna Marina*; Verlag, Dr. Friedrich Pfeil: München, Germany, 2008; Volume 16, pp. 75–99.
20. Basuri, C.K.; Pazhaniyappan, E.; Munnooru, K.; Chandrasekaran, M.; Vinjamuri, R.R.; Karri, R.; Mallavarapu, R.V. Composition and distribution of planktonic ciliates with indications to water quality in a shallow hypersaline lagoon (Pulicat Lake, India). *Environ. Sci. Pollut. Res.* **2020**, *27*, 18303–18316. [CrossRef] [PubMed]
21. Shadrin, N.V.; Anufriieva, E.V. Structure and Trophic Relations in Hypersaline Environments. *Biol. Bull. Rev.* **2020**, *10*, 48–56. [CrossRef]
22. Esteban, F.G.; Finlay, J.B. Marine ciliates (Protozoa) in central Spain. *Ophelia* **2004**, *58*, 13–22. [CrossRef]
23. Hotos, G. Feeding with various microalgae the salt "loving" ciliate *Fabrea salina* in normal salinity 35 ppt. *J. Sci. Food Agric.* **2019**, 3, 150–152. [CrossRef]
24. Korovesis, A.K.; Hotos, G.; Zalidis, G. The role of the ciliate protozoan *Fabrea salina* in solar salt production. In Proceedings of the 10th World Salt Symposium, Park City, UT, USA, 19–21 June 2018.
25. Avron, M.; Ben-Amotz, A. (Eds.) *Dunaliella: Physiology, Biochemistry, and Biotechnology*; CRC Press: Boca Raton, FL, USA, 1992; 256p.
26. Hotos, G.N. A Short Review on the Halotolerant Green Microalga *Asteromonas gracilis* Artari with Emphasis on Its Uses. *Asian J. Fish. Aquat. Res.* **2019**, *4*, 1–8. [CrossRef]
27. Anufriieva, E.; Shadrin, N. The long-term changes in plankton composition: Is Bay Sivash transforming back into one of the world's largest habitats of *Artemia* sp. (Crustacea, Anostraca)? *Aquac. Res.* **2020**, *51*, 341–350. [CrossRef]
28. Foissner, W. Protist diversity and distribution: Some basic considerations. *Biodivers. Conserv.* **2007**, *17*, 235–242. [CrossRef]

diversity

MDPI

Article

Target Species and Other Residents—An Experiment with Nest Boxes for Red Squirrels in Central Poland

Jakub Gryz [1], Tomasz Jaworski [2] and Dagny Krauze-Gryz [3,]*

[1] Department of Forest Ecology, Forest Research Institute, Braci Leśnej 3, Sękocin Stary, 05-090 Raszyn, Poland; j.gryz@ibles.waw.pl
[2] Department of Forest Protection, Forest Research Institute, Braci Leśnej 3, Sękocin Stary, 05-090 Raszyn, Poland; t.jaworski@ibles.waw.pl
[3] Department of Forest Zoology and Wildlife Management, Institute of Forest Sciences, Warsaw University of Life Sciences WULS-SGGW, Nowoursynowska 159, 02-776 Warsaw, Poland
* Correspondence: dagny_krauze_gryz@sggw.edu.pl

Abstract: The red squirrel typically nests in dreys and tree hollows, but also (when given an opportunity) in large nest boxes. We assessed the occupancy rate of nest boxes by red squirrel and non-target species (120 boxes in the continuous forest, habitat mosaic and urban park, checked annually for eight years). Habitat type explained the variability in the occupancy of nest boxes by different species/taxa. Red squirrels used nest boxes in all habitats but occupancy rates were highest in the urban park (>50% of the boxes at maximum) and lowest in the forest. This could be explained by high population density, competition for shelters and willingness to explore alternative sheltering opportunities by urban squirrels. The yellow-necked mouse inhabited nest boxes infrequently and mostly in habitat mosaic. Tits mostly occurred in the forest and least often in the park, which suggests limited availability of natural cavities in managed forest. Nest box occupancy by starlings increased with an anthropopression level, which reflects high densities of urban and rural populations of the species. Hymenoptera (mainly wasps) were present only in rural areas, which may be due to their persecution by humans or use of anti-mosquito pesticides in urban parks. Additionally, 24 insect species were found to inhabit squirrel dreys.

Keywords: *Sciurus vulgaris*; yellow-necked mouse; tits; common starling; Hymenoptera; nest-dwelling invertebrates; anthropogenic transformation

Citation: Gryz, J.; Jaworski, T.; Krauze-Gryz, D. Target Species and Other Residents—An Experiment with Nest Boxes for Red Squirrels in Central Poland *Diversity* **2021**, *13*, 277. https://doi.org/10.3390/d13060277

Academic Editor: Michael Wink

Received: 19 May 2021
Accepted: 15 June 2021
Published: 21 June 2021

Publisher's Note: MDPI stays neutral with regard to jurisdictional claims in published maps and institutional affiliations.

Copyright: © 2021 by the authors. Licensee MDPI, Basel, Switzerland. This article is an open access article distributed under the terms and conditions of the Creative Commons Attribution (CC BY) license (https://creativecommons.org/licenses/by/4.0/).

1. Introduction

The red squirrel *Sciurus vulgaris* is the only *Sciurus* species in Poland, widely distributed in the whole country [1]. The species primarily inhabits forests [2] but can easily adapt to mild environmental changes. That is why the red squirrel can be found in natural areas [3], but also in small woods in agricultural lands [4], as well as in the city parks [5–8]. These last habitats are, according to some studies, potentially suitable refuges for red squirrels [9]. In the cities, red squirrel abundance is approximately twice as high in urban habitats than in forests [10]. However, both in rural and urban areas, red squirrels need well-connected patches of habitat, as the species is sensitive to fragmentation [11].

Red squirrels typically nest in dreys and tree hollows, but also in large nest boxes (review in [12]). In an urban park in Warsaw, the capital city of Poland, red squirrels were also found to have their resting sites in the buildings and in other anthropogenic structures (i.e., wooden animal figures, street lamps, etc.) [13].

Anthropogenic habitat changes negatively affect the availability of natural cavities, i.e., in managed forests tree cavities are usually less abundant than in less disturbed ones [14,15] or in urban and rural areas (e.g., [16]). This negatively influences many vertebrate species that use tree cavities for breeding, hibernation or food storage [17–19], but also numerous invertebrates that develop in decaying wood and other organic material that accumulates in these microhabitats [20–22]. To compensate for the loss of cavity trees, man-made nest boxes are widely used in forests and non-forest areas [23].

In Poland, about 25 types of nest boxes for birds and small mammals are used, of which more than 400,000 are currently installed only in the wooded areas managed by the Polish State Forests [24]. Apart from state forests, nest boxes are widely used in other wooded areas (including protected ones), urban parks and private gardens, where they are provided by municipal authorities, NGOs, scientific institutions and private land owners. Most of these nest boxes are dedicated to small birds from the Paridae family and the starling (*Sturnus vulgaris*). Nest boxes for mammals are hung less often and are dedicated mainly to bats, less often to mammals from the Gliridae family or the red squirrel.

Nest boxes can be inhabited by invertebrates, e.g., social wasps (Hymenoptera) [25], who build their nests inside boxes. Additionally, organic debris (e.g., plant parts, feces, hair) that accumulate in vertebrate nests can be exploited by numerous arthropods, primarily insects [26–28].

In Great Britain and Italy, the red squirrel populations have been in a decline due to disease-mediated competition from the introduced grey squirrel (*Sciurus carolinensis*) [29]. Replacement of red by grey squirrels occurs primarily due to exploitation competition [30] (and references therein). In Great Britain and Ireland the replacement is indeed accelerated by SQPV, but the virus is absent in Italy and replacement still occurs, albeit more slowly [31,32]. The remaining populations (and squirrel shelters) are protected [33]. Therefore, the use of nest boxes is considered a conservation tool for the red squirrel [12] allowing also the mitigation of the impact of forest operations on this threatened species [33]. Nest boxes for the red squirrel were also found to be a valuable research tool for monitoring breeding performance [34], interactions between the red squirrel and birds inhabiting nest boxes [35] or the response to predation [36].

As the red squirrel population in Poland is stable and is not threatened by alien and invasive squirrel species, it is not targeted by conservation programs. Nevertheless, one of the first nest boxes for the red squirrel was exposed in 2007 in central Poland, and was quickly inhabited by this species with breeding success (four juveniles, Figure A1, [37]). Additionally, our experience from other studies showed that bird nest boxes were inhabited by the red squirrel. For example, in 15 nest boxes for stock pigeon *Columba oenas* checked yearly (2014–2020), four dreys of the red squirrel were found [38]. In turn, in 47 nest boxes dedicated to the tawny owl *Strix aluco* checked between 2012–2018 [39], we found four red squirrel dreys [38].

These observations were a trigger to start a scientific project in order to assess the occupancy rate of artificial shelters (i.e., nest boxes) by red squirrels in Poland as a potential conservation and research tool. We assumed that, especially in urban parks, red squirrels from an abundant population would inhabit nest boxes readily. The nest box design (i.e., entrance hole placement) was supposed to limit occupancy by non-target species (i.e., birds) so we also focused on birds building their nests in the boxes. As previous studies showed, faunistic exploration of nest boxes for the tawny owl led to new records of some moths [40,41]. This is why we also focused on invertebrates building their nests in the boxes, and other invertebrates inhabiting dreys built by the red squirrel, assuming that drey material may provide sheltering opportunities for nest-inhabiting insects.

2. Materials and Methods

2.1. Study Areas

The study was conducted in central Poland, a region that is under the influence of the mild oceanic climate of Western Europe and the harsh and dry continental climate of Eastern Europe and Asia. The duration of the growing season is 210 days; the total precipitation is 600 mm per year; and the mean ambient temperature ranges from −4 °C in January to +18 °C in July.

We conducted the research in study areas representing three habitat types of various levels of anthropogenic pressure, two of which were located in rural areas (1. continuous forest, 2. habitat mosaic), and one in an urban area (3. urban park). The first two study areas were managed by the Experimental Forest Station of Warsaw University of Life Sciences in Rogów. Here, forests accounted for 25% of the area (approx. 2400 ha), formed seven complexes (70–1000 ha) and were surrounded by a mosaic of different crops, pastures, fallow land and forest stands. The remaining area included arable lands (59%), orchards (5%), grasslands (5%) and scattered buildings [42–44]. The first study area was Głuchów forest (51°45′11.8″ N, 20°06′24.1″ E), the biggest unfragmented forest complex of 1000 ha, surrounded by open fields, woodlots and the Rawka River valley (which is a nature reserve). The main forest-forming species was Scots pine (*Pinus sylvestris*), with an admixture of oaks (*Quercus* spp). Part of the area was covered by wet black alder (*Alnus glutinosa*) forest. The second study area (habitat mosaic) was located near Rogów village (51°49′17.98″ N, 19°53′54.15″ E) in smaller and much more fragmented forests covered by Scots pine and oaks with an admixture of larches (*Larix* spp.), with a busy, national road and railway cutting through the forested area. Part of the forest complex was also covered by a forest nursery. The study area also encompassed the neighboring Rogów arboretum and WULS campus. The arboretum was founded in 1925. Fragments of the forest which existed before the arboretum was organized included 150-year-old pines. The third study area was located in Łazienki Park, a 76 ha urban park located in the central district of Warsaw (the capital city of Poland, approximately 2 million inhabitants). It is a representative area with numerous old trees that provide a rich and natural food base [5]. In addition, as a result of being founded in the 17th century, it also contains historical buildings. The park has 92 species of trees and shrubs, comprising both native and exotic species. The tree cover in the park is approximately 70% and roughly one in every five trees is more than 140 years old [5]. The park is very popular among both tourists and the local inhabitants, and is visited by more than 2 million people every year [45]. Squirrels are often fed nuts by visitors [46,47], and their population here reaches very high density (i.e., more 2 ind./ha, [13]). Squirrels occupy small home ranges and human-delivered food plays an important role in the diet of some individuals [7,8].

2.2. Data Collection and Laboratory Analyses

In autumn 2012, 40 wooden boxes dedicated to the red squirrel were hung in the second (landscape mosaic) and the third (urban park) study area. Next, 40 nest boxes were hung in autumn 2013 in the first study area (continuous forest). These were modified nest boxes for dormice, i.e., they had an entrance hole at the back (facing the tree trunk); inner bottom measurements were 16 × 16 cm, maximum height was 49 cm and the board was 2 cm thick. The entrance hole diameter was 5.5 cm (Figures A1 and A2). The nest boxes were attached to a tree trunk with nails at 5 m height. They were set at least 100 m apart. All nest boxes were checked and cleaned every year in winter (usually December/January), old nests were removed and nest boxes were repaired if necessary. During control, the species (or higher taxa) were determined on the basis of the nest appearance built in the nest box. The nest boxes in the landscape mosaic and the urban park (study areas no. 2 and 3) were first checked in winter 2013/14 (which delivered data on their occupancy in 2013), and nest boxes in the forest (study area no. 1) were first checked in winter 2014/15 (data on their occupancy in 2014). The last nest box check was carried out in January 2021 in all three study areas (data on their occupancy in 2020).

In addition, to monitor vertebrate species visiting the nest boxes, six trail cameras were placed next to selected nest boxes (Figure A3) in the second study area (habitat mosaic) shortly after their installation (in November 2012). Trail cameras were exposed for one year (until 8 December 2013), they were checked every six weeks (to replace memory cards and batteries) and were then moved to another nest box. Altogether, 21 of 40 nest boxes were monitored by trail cameras. Three different camera trap models were used: one RECONYX PC900 HyperFire Professional IR, one RECONYX PC90 and four Ecotone SGN-5220. All cameras were set to make a series of three photos at 1 s intervals. Each registered animal was considered as a single observation if a minimum of 15 min elapsed between subsequent photos or series of photos of the animal. The exception to this rule was only made in circumstances clearly indicating that the animal in the photo was different from the previously registered one (different gender, age, color pattern, etc.).

To study insects inhabiting nests of red squirrels, dreys found in the nest boxes were taken to a laboratory and placed separately in plastic containers with ventilated lids. The containers were left in a non-heated room for a few weeks to allow the insects to complete their winter diapause. They were then brought to the heated laboratory and checked every 2–3 days. Adult insects were then collected and preserved as dry specimens or in ethyl alcohol. The insects were identified on the basis of their morphology and in certain cases (moths) were also identified by the morphology of their genitalia. Species identification was done by: Radosław Plewa (Forest Research Intitute, Sękocin Stary, Poland)—Coleoptera, Seweryn Grobelny (Poznań, Poland)—Dermaptera, Grzegorz Hebda (University of Opole, Opole, Poland)—Hemiptera, Waldemar Żyła (Upper Silesian Museum, Bytom, Poland)—Hymenoptera, Tomasz Jaworski (author)—Lepidoptera and Roland Dobosz (Upper Silesian Museum, Bytom, Poland)—Neuroptera and Raphidioptera. Insect larvae were not identified due to difficulties in their identification. Additionally, some insect groups (e.g., flies—Diptera) were left out due to lack of specialists being able to do the identification.

2.3. Statistical Analysis

As nest boxes in the first study area (continuous forest) were hung one year later than in the second and third ones, to compare occupancy of nest boxes by various species/taxa between study areas, only data from the years 2014–2020 were taken for comparisons. Data on the recorded species were pooled for single nest boxes. To investigate differences in the occupancy of a nest box by main species/taxa (i.e., red squirrel, yellow-necked mouse (*Apodemus flavicollis*), tits, common starling, Hymenoptera) between the three studied areas, a redundancy analysis (RDA) was implemented in CANOCO software. The pooled number of records of a given species was calculated for each nest box, which gave 40 samples for each study area. The occupancy rates by each species for a given nest box were defined as the response variables, whereas the study area served as the explanatory variable. A Monte Carlo test (499 permutations) was used to test the statistical significance of the differences between the occupancy rates by different taxa in three habitats. To compare the occupancy rates separately of each species/taxa (red squirrel, yellow-necked mouse, tits, starling, Hymenoptera) between study areas we used the Kruskal–Wallis test (data did not follow a normal distribution—Shapiro–Wilk W test), with a Mann–Whitney pairwise test for post-hoc analysis (Bonferroni-corrected p values).

To compare the proportion of nest boxes used by vertebrates/all animals each year between study areas we used the Kruskal–Wallis test.

The analyses were done in Canoco 4.5 [48] and Past 4.05 [49] software.

3. Results

RDA revealed that the landscape type (continuous forest, habitat mosaic, urban park) significantly explained the variability (the amount of variation explained by landscape type was 38.5%) in occupancy of nest boxes by different species/taxa (pseudo-F = 71.272, $p < 0.005$) (Figure 1). Red squirrel and starling were mostly connected to the urban park, yellow-necked mouse and Hymenoptera to the habitat mosaic and tits to the forest.

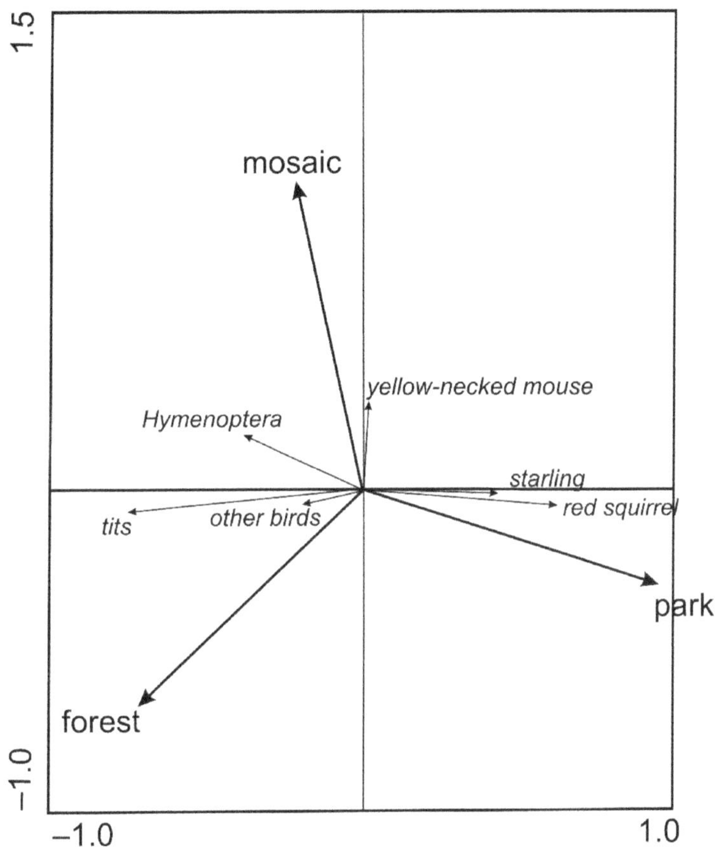

Figure 1. Biplot from redundancy analysis (RDA) showing the influence of the habitat (continuous forest, habitat mosaic, urban park) on the variability in species of animals building their nests in the boxes dedicated to the red squirrel in central Poland in the years 2014–2020.

The red squirrel most often inhabited nest boxes in the urban park (i.e., 2.3 records per one nest box in the years 2014–2020), and the lowest value was recorded for the continuous forest (0.18) (Figure 2a). The occupancy rate differed between habitats (Kruskal-Wallis test, H = 42.96, $p < 0.001$) and in post-hoc analysis differences were found between all pairs of habitats. In total (data from all seasons, 2013–2020), in the urban park red squirrels were recorded 95 times, in the habitat mosaic 27 times and just 7 times in the forest (Table A1). In the urban park, the highest number of red squirrel dreys were recorded in the second year of exposition (2014) when they were found in more than half (n = 21) of the boxes, and the lowest was in the last but one year, 2019 (n = 9). In the habitat mosaic, 7 boxes were occupied by the red squirrels at maximum (in 2017), while in the forest only 0–2 boxes were inhabited each year.

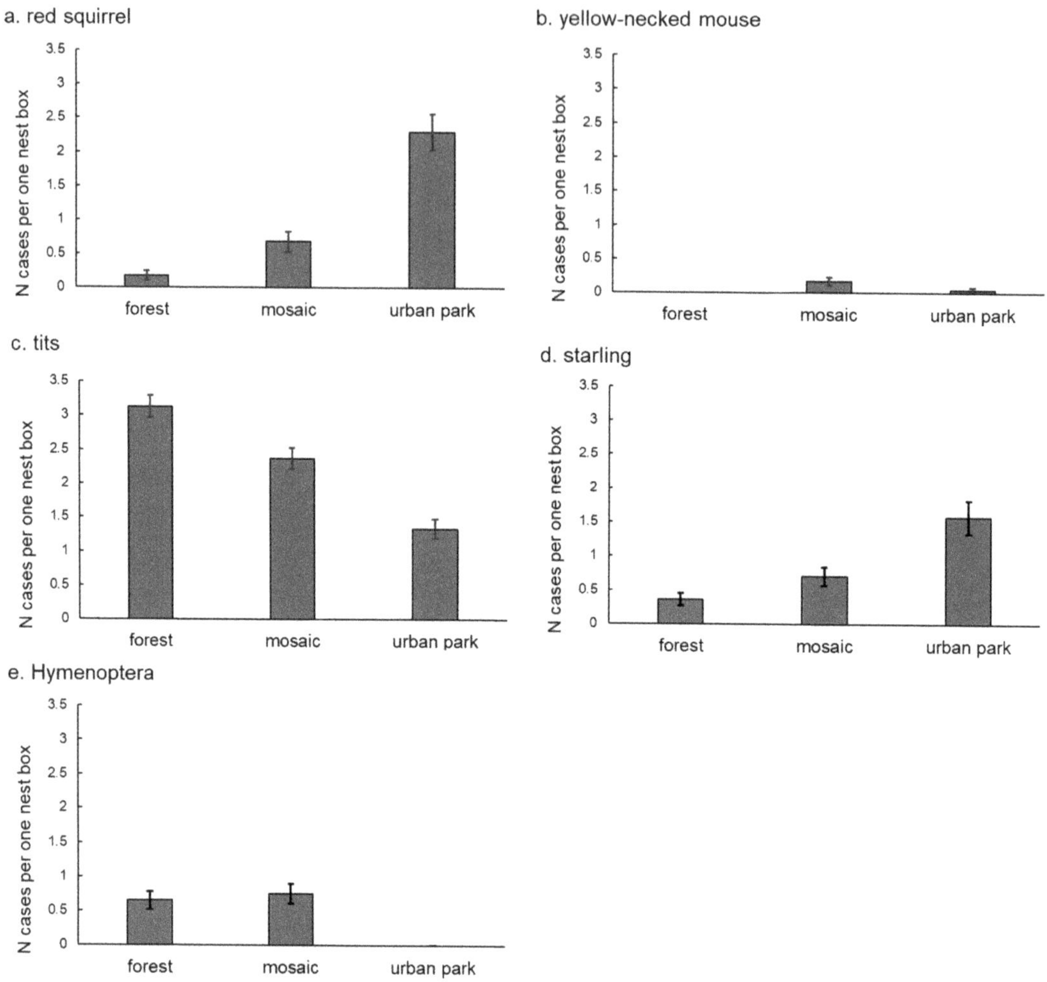

Figure 2. The occupancy of the nest boxes dedicated to the red squirrel (mean ± SE occupancy per one nest box in the 2014–2020 research period) by a target (i.e. red squirrel—figure (**a**)) and non-target species/taxa ((**b**). yellow-necked mouse, (**c**). tits, (**d**). European starling, (**e**). Hymenoptear) in the three habitats of various anthropopression in central Poland. Each nest box was checked once a year (in winter), and data on records (presence of the nest inside a box) of species/taxa for each nest box from the whole research period were pooled together.

The other mammal species present in the nest boxes was the yellow-necked mouse. Yet, its occupancy rate was much lower: mice were never found in the forest, in the mosaic the occupancy rate was 0.18 records per one nest box and in the urban park it was 0.05 (Figure 2b) and the occupancy rate did not differ between the two habitats ($H = 0.92$, $p > 0.05$).

Among birds, tits inhabited the nest boxes most often (they were recorded 240 times in total and were the most frequently recorded from all species, Table A1). Additionally, in this case a clear difference between study areas was found ($H = 25.9$, $p < 0.001$, post-hoc differences between all pairs). The highest occupancy rate was recorded for the forest complex (3.13 records/nest box), where nests were found 125 times (Table A1), while the lowest was for the urban park (1.33) (Figure 2c). The other bird whose nests were often recorded in the boxes was the starling (Figure 2d). In contrast to tits, starlings were most often recorded in the urban park (1.58 records per one nest box), and least often in the forest (0.35); this difference was statistically significant (H = 15.03, $p < 0.001$, post-hoc differences found to be significant for all pairs). Other birds that nested occasionally in the boxes were flycatchers (Muscicapidae) and the Eurasian nuthatch (*Sitta europea*).

Among invertebrates, nests of Hymenoptera (mainly wasps, Vespidae) were often found (68 records) in the boxes dedicated to red squirrels, but only in rural areas (H = 19.09, $p < 0.001$, Figure 2e).

On average, signs of animals' presence in the nest boxes were found during 62.7 (min = 35.0, max = 92.5) and 54.2% (min = 27.5, max = 32.5) of nest box checks (n = 920) for all species and vertebrates, respectively (data from years 2013–2020). These values did not differ between study areas in the subsequent years, neither for all species (Kruskal–Wallis test, H = 1.25, $p > 0.05$), nor for vertebrates (H = 1.03, $p > 0.05$). Out of 40 nest boxes installed in each study area, 7 were used by the red squirrel in the forest, 18 in the habitat mosaic and 34 in the urban park. All nest boxes were used by tits in the forest (min number of records per one nest box = 1, max = 5), 36 in the habitat mosaic and 14 in the urban park. Overall, all nest boxes were used at least once by a vertebrate inhabitant. Wasps were recorded in 18 nest boxes in the forest and 26 nest boxes in the habitat mosaic.

On the basis of data collected from trail cameras installed next to nest boxes in the habitat mosaic, the red squirrel was the most often recorded species (43.2% of all records). Additionally, the great spotted woodpecker (*Dendrocopos major*) and great tit (*Parus major*) were observed frequently (17.3 and 15.5% of records, respectively). The other species were birds: Eurasian nuthatch, tawny owl, common chaffinch (*Fringilla coelebs*), Eurasian jay (*Garrulus glandarius*), European robin (*Erithacus rubecula*), European green woodpecker (*Picus viridis*) and common buzzard (*Buteo buteo*); and mammals: yellow-necked mouse, stone marten (*Martes foina*) and domestic cat (*Felis catus*). The animals were seen to enter the nest box, examine the entrance hole or perch on the nest box.

Our study revealed 24 insect species inhabiting dreys of red squirrels built in nest boxes. Beetles (Coleoptera) were the most diverse, with 16 species, followed by moths (Lepidoptera; 5 species), earwigs (Dermaptera; 1 species), true bugs (Hemiptera; 1 species) and net-winged insects (Neuroptera; 1 species) (Table A2).

4. Discussion

Nest-site availability limits cavity-using populations in many managed forests (review in [50]) and numerous birds (review in [23]) and mammals (review in [12]) are known to use nest boxes as natural cavity replacements.

In our study, red squirrels used nest boxes in all habitats but occupancy rates were highest in the urban park. This can be explained by the very high density of the red squirrel population in this area, driven by supplementary feeding [7]. The occupancy rates recorded in the urban park in our study (from a few to around 50% of the nest boxes available) were very similar to what was found in Great Britain (2 to maximum 53%) in a very abundant red squirrel population [12]. Local density may inversely affect the number of nests used by individual squirrels [51] and high density of population may result in competition for shelters and willingness to explore new resting possibilities. Squirrels in the urban park were also observed to build their dreys inside various anthropogenic shelters, such as buildings, street lamps or wooden ornamental figures of animals [13], which may be because of the high behavioral plasticity of urban populations of this species. In most cases the nest boxes in the urban park did not seem to be used by red squirrels for breeding. Nevertheless, squirrels can use several nests in sequence, and for various purposes (such as daytime or nighttime rest, breeding, etc.) [51]. On one hand, nest boxes offer a more permanent sheltered nest site than bolus nests, which are more vulnerable to unfavorable weather conditions (such as strong wind, rain, heavy snowfall). On the other hand, they are more permanent, which may place occupants at a greater risk of predation from mammalian predators [34]. In the urban park under study (as in other green spaces of Warsaw) the stone marten was abundant and this species was also often recorded to visit and look into the squirrel nest boxes. Both marten species (*M. martes* and *M. foina*) were also present in our rural study areas [39,52]. Martens are also known to use nest (or den) boxes readily for breeding [53]. In the case of our earlier studies, both pine and stone martens were recorded in nest boxes dedicated to the tawny owl [52]. Yet, nest boxes provided for the red squirrels in our study were much smaller than the tawny owl boxes and (as assumed at the construction stage) entrance holes were not large enough for martens to enter their heads. Other mammalian predators that were recorded by camera traps to look into nest boxes were domestic cats. Domestic cats are abundant in our rural study areas [52,54] and both in rural and urban sites are one of the most important mammalian predators [55,56]. Additionally, in rural areas, the northern goshawk (*Accipiter genitilis*), who is known to be an important red squirrel predator [36], was present [43], and the predation risk might have affected nest box use by the red squirrel [36].

In this study, nest box occupancy by the red squirrel was dependent on the habitat, i.e., boxes placed in rural areas were occupied less often than in urban areas. Especially in the first study area (continuous forest), nest box occupancy was low. Additionally, in Finland, red squirrel occurrence in nest boxes was positively associated with the amount of farmland in the landscape surrounding the nest box (which may be explained by the fact that farmland is usually established on rich soils and forest edges are dominated by deciduous trees, a source of high-quality seeds), and also with built-up areas (which may be related to availability of supplemental food in bird feeders). In turn, mature birch–spruce forest had a negative impact on the red squirrel occurrence [36]. This is in contrast to British studies where, in a coniferous habitat, nest boxes were used by adults, subadults, lactating females and their young. It was concluded that the box occupancy may partly depend upon the density of squirrels (and the density of boxes). Indeed, in the British study site, population density was very high (3.5–4 ind./ha) thanks to supplemental feeding [34]. We may expect that densities in rural areas in our study were far lower [52,57]. Nest boxes for red squirrels may be especially important in heavily managed forests or younger stands. In another study performed in Great Britain, thinning in a commercial forest led to squirrels occupying more nest boxes, which were assessed as useful to mitigate the impacts of forest operations. However, no evidence of breeding in nest boxes was found in that case; the boxes could be important as shelters during forest operations and potentially for juveniles during natal dispersal [33]. In the case of another squirrel species, the grey squirrel, a population living in a young forest reacted with an abundance increase after artificial dens were erected. This was attributed to better survival of both young and adult individuals, thanks to providing structures that could afford secure nest sites [58]. Yet, when assessing the significance of nest boxes for the red squirrel, it must be taken into account that squirrels use various shelters (review in [59]) so they may switch from bolus nests to nest boxes and vice versa. Indeed, in the mature mixed forests of British Columbia, the American red squirrel and northern flying squirrel (*Glaucomys sabrinus*) used many exposed nest boxes (and this nest box addition presumably led to an increase in squirrel abundance) but this could have been attributed to squirrels moving from bolus nests to nest boxes, not solely to population increase [50]. Overall, our results suggest that nest box use by red squirrels can (at least to some extent) depend on the density of the population. It may also be assumed that urban red squirrels, being accustomed to various anthropogenic changes, will more readily accept alternative shelters in the form of nest boxes. In our case, no significant differences were found between nest box occupancy in the subsequent years. Yet, in the case of the urban park, the most (n = 21) nest boxes were used by red squirrels in the second year of their deployment. This resembles what was shown in the other study in the forest in Great Britain [33], and suggests that squirrels habituate to nest boxes over time. In our case, as in Finland, nest box occupancy decreased over time [36], which may be explained by the nest box natural deterioration over time.

The other rodents whose nests were found in the nest boxes were yellow-necked mice. This species is known to use nest boxes, e.g., dedicated to a dormouse (*Muscardinus avellanarius*) [60]. Yet, the occupancy rates were rather low and mice were present mostly in rural areas.

Squirrels and birds may interact when using the same cavities. First of all, squirrels will compete for cavities with birds. In a forest in Great Britain, the presence of great tit clutches and broods did not prevent red squirrels from occupying boxes (only 28% of all recorded clutches were found to have produced fledglings, squirrels were found to eat some of the chicks and eggs) [35]. There was a strong negative interaction between spring great tit box occupancy and the use of boxes by red squirrels [12]. On the other hand, when the jackdaw *Corvus monedula* occupied nest boxes, the red squirrel started to avoid this shelter [35]. On the other hand, the dry material of abandoned squirrel nests can be explored and used by songbirds [61]. In this study, nest boxes in the forest were mostly occupied by tits. This may suggest that natural cavities are limited in this area (but might have also been driven by differences in the population density; this could not be tested as no density estimates were available). Indeed, in the urban park there are numerous deciduous trees which provide tree holes. In the habitat mosaic there were old trees (in the arboretum, in the form of experimental stands and in a nature reserve) which were absent in the first study area (continuous forest). It was shown that there was a high difference in the cavity resources between a Białowieża national park (12.5/ha) and managed stands (3.0/ha), as most cavities were in dead (mostly pine) trees (over 74%) [14]. Similar results were obtained in northern Swedish boreal forests, i.e., unmanaged forests had a significantly higher density of cavities (2.4/ha) than managed old forests (1.1/ha) [15].

In Great Britain, the occupancy of nest boxes for red squirrels by great tits varied between 10 and 43%, and the occupancy rate of great tits increased over the years [12]. Similarly, in our case nest box occupancy by tits seemed to rise in the first few years but then dropped, probably due to natural wood deterioration and unfavorable conditions inside the box. It was shown that massive nest box supplementation boosted fecundity, survival and immigration in a recovering hoopoe (*Upupa epops*) population [62]. Additionally, density of mountain chickadee (*Poecile gambeli*, Paridae) increased after nest box addition and declined following box removal in mature mixed conifer forest in British Columbia, suggesting that populations were limited by cavity availability [50]. On the other hand, hole-nesting birds seem to be very flexible and often locate their nests in anthropogenic breeding sites (in the suburbs of Warsaw this included even vertical metal pipes), and in the case of great tits this type of shelter was dominant [63].

The other bird species that occupied nest boxes was the common starling, which in our case was mostly connected to the urban park but also to the habitat mosaic. In Poland, this species is most common in rural and urban areas, nesting also in the ecotone but avoiding continuous forests [64]. This species is known to breed in various habitats, although urbanization can negatively affect the reproductive parameters of starlings [65]. Starlings often occupy nest boxes, but they prefer deeper ones and those without old nests [66], which may be explained by the cost of old nest material removal [67]. Starlings are able to compete for nesting holes with many primary and secondary cavity nesters. In an experiment conducted in Poland, in an urban forest, starlings took some of the holes chosen by nuthatches for breeding. Yet, it is claimed that where starling is a native species, it is not able to influence the cavity-nesting community to a great extent [68]. The boxes exposed in our study were very similar in dimensions to nest boxes dedicated to starlings, so it was not surprising these birds occupied a high proportion of the boxes. Yet, we may expect that the red squirrel, being heavier, would be able to successfully compete with starlings for these shelters.

The last group of animals that built their nests in the nest boxes in our study were Hymenoptera (mainly wasps, *Vespa* sp., and less often hornets, *Vespa crabro*). Wasps are known to build their nests in various spaces, including nests boxes for birds (e.g., [69,70]). Other species from this group (bumblebees, *Bombus* spp., and honey bees, *Apis* spp.) were found in the nest boxes in Great Britain [12,35]. Yet, it was shown that occupation of nest boxes by these insects was generally low (and, interestingly, these insects were never found in tree cavities) [25]. It is hard to explain absence of Hymenoptera in the urban park in our study as this insect group is widespread and common. We can only hypothesize than in the cities, wasps and hornets are persecuted due to their harmfulness for humans. Additionally, some insecticides are used in the park to limit the number of mosquitoes (or other pests), which may also affect Hymenoptera negatively (e.g., [71], and the literature cited therein). Moreover, urban habitats are characterized by a relatively low abundance of insects [72], so the food base for predatory insects, including wasps and hornets, may be limited.

Not only nest boxes themselves provided animals with sheltering opportunities, but the nests of red squirrels built in nest boxes were also inhabited by numerous species of nidicolous, i.e., nest-associated, insects. Most of them were saprophagous, i.e., feeding on plant or animal detritus. In the first group, there were representatives of a few families of beetles (Histeridae, Ptinidae, Latridiidae, Tenebrionidae), which, both as larvae and adults, feed on the various remains building the nest lining. The second group included beetles, whose larvae feed on keratin and animal fat (fur, animals carcasses) accumulating in the nests. This mainly included beetles from the Dermestidae family [73] and moths from the Tineidae [74–77] and Oecophoridae [78] families.

Seasonally unfavorable conditions, mainly low winter temperatures, force most insects to stop their activity and search for refuges (e.g., [79]). Usually, natural cavities are used, e.g., tree holes, spaces under tree bark. Yet, some species have adapted to inhabiting shelters of an anthropogenic origin, which includes nest boxes for birds [40,41,80]. As we have shown in this study, nest boxes for the red squirrel can also be used by numerous insects as overwintering sites. Insects that inhabited dreys in nest boxes were some beetles (e.g., *Harmonia axyridis* in this study), Hemiptera, including Lygaeidae recorded in this study, Neuroptera, including green lacewings (*Chrysoperla* spp.), and also some moths (Bucculatricidae, Elachistidae) [41,81]. Further, wasp nests inside nest boxes for vertebrates can serve as habitats for other insects, i.e., the bee moth *Aphomia sociella*, whose caterpillars feed on wax and honey and are even predators of wasp larvae (e.g., [81]). The group of nidicolous insects include predatory species, such as earwigs (Dermaptera) (e.g., [82]).

Nest boxes can be a valuable conservation and research tool, yet variables such as their dimensions, placement height and the way they are constructed and maintained may affect their use, thus producing potential bias [23]. As shown by other studies, both positive and negative influences of nest box provision on the reproductive output of target species have been documented and this may depend on the habitat. It may be strongly positive in one habitat and smaller or even negative in another habitat within the same study area. Furthermore, nest boxes may create an ecological trap by causing a supra-optimal breeding density (review in [83]).

To conclude, as we have shown in our study, the occupancy of nest boxes by the red squirrel depended on the habitat type and was presumably driven by the red squirrel population density, higher competition for shelters in urban habitats and willingness to explore alternative sheltering opportunities by urban populations. This shows that nest boxes can help limit the competition for shelters. Nest boxes dedicated to and designed for red squirrels were readily occupied by two bird species/groups (tits and starlings) and Hymenoptera. Additionally, in this case, habitat type (and the anthropopression level) affected occupancy rate. Nevertheless, none of these taxa are supposed to effectively limit the occupancy of nest boxes by the red squirrel. High occupancy of nest boxes by tits in the forest suggests limited availability of natural cavities, while high occupancy of nest boxes by starlings in the urban park may result from the high densities of urban populations of the species.

Author Contributions: Conceptualization, J.G., T.J. and D.K.-G.; methodology, J.G., T.J. and D.K.-G.; investigation, J.G., T.J. and D.K.-G.; data curation, J.G., T.J. and D.K.-G.; writing—original draft preparation, J.G. and D.K.-G., writing—review and editing, J.G., T.J. and D.K.-G.; visualization, D.K.-G.; project administration, T.J. and D.K.-G.; funding acquisition, T.J. and D.K.-G. All authors have read and agreed to the published version of the manuscript.

Funding: The study was financed by the Polish Ministry of Science and Higher Education with funds from the Institute of Forest Sciences, Warsaw University of Life Sciences (WULS), for the purpose of scientific research and by the Forest Research Institute within the framework of the research topic entitled 'Insects inhabiting nests and pellets of selected forest bird species'.

Institutional Review Board Statement: Not applicable.

Informed Consent Statement: Not applicable.

Data Availability Statement: The data presented in this study are available on request from the corresponding author.

Acknowledgments: We would like to thank Adam Tarłowski from the USSURI Company for preparing and erecting the nest boxes, Agata Beliniak for her help in nest box checks and Kinga Mazur for her help in nest box checks and collecting data from the trail cameras. We would to express our gratitude to Barbara Werner from the Royal Łazienki Museum and to Piotr Banaszczak from the Rogów Arboretum for site access.

Conflicts of Interest: The authors declare no conflict of interest.

Appendix A

Figure A1. Young squirrel *Sciurus vulgaris* in one of the first nest boxes dedicated to the red squirrels in Poland. The nest box was placed in 2007 in central Poland and the next year a reproductive success was recorded in it. The box has an entrance hole at the back, i.e., facing the tree trunk. Such nest boxes were used in this study in the years 2013–2020.

Figure A2. A check of the nest box dedicated to the red squirrel; the front panel is removed. The box was occupied by the target species (photo by Adam Tarłowski).

Figure A3. A red squirrel on nest box dedicated to this species (a camera-trap-taken photo). The nest box was placed autumn 2013 in a habitat mosaic of central Poland (study area no. 2) and monitored by photo traps.

Appendix B

The nest boxes (40 nest boxes in each study area) were checked once a year (in winter). The total number of records of a given species/group of animals in all nest boxes was given for each year

Table A1. The number of records of presence of a species/higher taxa (red squirrel, yellow-necked mouse, tits, European starling, Hymenoptera) in the nest boxes dedicated to the red squirrels in the three habitats of different level of anthropopression in central Poland.

Years/Study Area	Squirrel	Mouse	Tits	Starling	Hymenoptera
		n of records			
		1. Continous forest			
2014	1		23		6
2015			29	5	
2016	1		24	4	2
2017	2		13	3	3
2018	1		12		5
2019	1		16	1	7
2020	1		8	1	3
Total	7		125	14	26
		2. Habitat mosaic			
2013		1	14	3	12
2014	4	2	16	4	5
2015	5	2	14	5	1
2016	1	2	15	6	2
2017	7		11	3	6
2018	4		10	1	5
2019	3		9	5	9
2020	3	1	8	4	2
Total	27	8	97	31	42
		3. Urban park			
2013	3		2	10	
2014	21		2	13	
2015	13		4	10	
2016	15		1	10	
2017	13		1	10	
2018	10		5	4	
2019	9	2	1	7	
2020	11		2	9	
Total	95	2	18	73	
All study areas	129	10	240	118	68

Appendix C

Table A2. Species of insects found in nesting material accumulated in the nest boxes occupied by the red squirrel in central Poland in the three habitats (continuous forest, habitat mosaic and urban park) in the years 2013–2020.

Order	Family	Family/Species	Total Abundance
Coleoptera	Coccinellidae	*Harmonia axyridis* (Pallas, 1773)	1
	Cryptophagidae	*Cryptophagus* sp.	1
	Curculionidae	*Phyllobius arborator* (Herbst, 1797)	1
	Dermestidae	*Anthrenus museorum* (Linnaeus, 1761)	2
		Anthrenus pimpinellae Fabricius, 1775	3
		Attagenus pellio (Linnaeus, 1758)	1
		Ctesias serra (Fabricius, 1792)	1
		Dermestes lardarius Linnaeus, 1758	2
		Globicornis corticalis (Eichhoff, 1863)	1
		Megatoma undata (Linnaeus, 1758)	12
	Latridiidae	*Latridius minutus* (Linnaeus, 1767)	1
	Malachiidae	*Malachius bipustulatus* (Linnaeus, 1758)	3
	Ptinidae	*Ptinus fur* (Linnaeus, 1758)	1
		Ptinus raptor Sturm, 1837	1
		Stegobium paniceum (Linnaeus, 1758)	1
	Tenebrionidae	*Tenebrio molitor* Linnaeus, 1758	2
Dermaptera	Forficulidae	*Forficula auricularia* Linnaeus, 1758	9
Hemiptera	Lygaeidae	*Arocatus melanocephalus* (Fabricius, 1798)	1
Lepidoptera	Bucculatricidae	*Bucculatrix thoracella* (Thunberg, 1794)	1
	Elachistidae	*Agonopterix* sp.	1
	Oecophoriadae	*Borkhausenia minutella* (Linnaeus, 1758)	1
	Pyralidae	*Aphomia sociella* (Linnaeus, 1758)	4
	Tineidae	*Tinea pellionella* Linnaeus, 1758	1
Neuroptera	Chrysopidae	*Chrysopa* sp.	1

References

1. Atlas of Polish Mammals 2021. Wiewiórka Pospolita *Sciurus vulgaris* Linnaeus. 1758. Available online: https://www.iop.krakow.pl/Ssaki/gatunek/57 (accessed on 18 May 2021).
2. Shar, S.; Lkhagvasuren, D.; Bertolino, S.; Henttonen, H.; Kryštufek, B.; Meinig, H. Available online: www.iucnredlist.org (accessed on 18 May 2021).
3. Stachura, K.; Niedziałkowska, M.; Bartoń, K. Biodiversity of forest mammals. In *Essays on Mammals of Białowieża Forest*; Jędrzejewska, B., Wójcik, M., Eds.; Mammal Research Institute, Polish Academy of Sciences: Białowieża, Poland, 2004; pp. 13–24.
4. Wauters, L.A. The ecology of red squirrels in fragmented habitats: A review. In *The Conservation of Red Squirrels, Sciurus vulgaris L.*; Gurnell, J., Lurz, P., Eds.; People's Trust for Endangered Species: London, UK, 1997; pp. 5–12.
5. Babińska-Werka, J.; Żółw, M. Urban Populations of the Red Squirrel (*Sciurus vulgaris*) in Warsaw. *Ann. Zool. Fenn.* **2008**, *45*, 270–276. [CrossRef]
6. Reher, S.; Dausmann, K.H.; Warnecke, L.; Turner, J.M. Food availability affects habitat use of Eurasian red squirrels (*Sciurus vulgaris*) in a semi-urban environment. *J. Mammal.* **2016**, *97*, 1543–1554. [CrossRef]
7. Krauze-Gryz, D.; Gryz, J.; Brach, M. Spatial organisation, behaviour, and feeding habits of red squirrels: Differences between an urban park and an urban forest. *J. Zool.* **2021**, *314*. [CrossRef]
8. Krauze-Gryz, D.; Gryz, J.; Klich, D.; Brach, M. Same yet different—individual red squirrels (*Sciurus vulgaris*) react differently to human presence in an urban park. *Hystrix Ital. J. Mammal.* **2021**, *32*. [CrossRef]
9. Rézouki, C.; Dozières, A.; Coeur, C.; Le Thibault, S.; Pisanu, B.; Chapuis, J.L.; Baudry, E. A viable population of the European red squirrel in an urban park. *PLoS ONE* **2014**, *9*, 1–7. [CrossRef] [PubMed]
10. Jokimäki, J.; Selonen, V.; Lehikoinen, A.; Kaisanlahti-Jokimäki, M.-L.L. The role of urban habitats in the abundance of red squirrels (*Sciurus vulgaris*, L.) in Finland. *Urban For. Urban Green.* **2017**, *27*, 100–108. [CrossRef]
11. Koprowski, J. The response of tree squirrels to fragmentation: A review and synthesis. *Anim. Conserv.* **2005**, *8*, 369–376. [CrossRef]
12. Shuttleworth, C.M.; Schuchert, P. Are nest boxes a useful tool in regional red squirrel conservation programs? *Hystrix Ital. J. Mammal.* **2014**, *25*, 91–94. [CrossRef]
13. Krauze-Gryz, D.; (Warsaw University of Life Sciences, Warsaw, Poland). Personal communication, 2021.

14. Walankiewicz, W.; Czeszczewik, C.; Stański, S.; Sahel, M.; Ruczyński, I. Tree cavity resources in spruce-pine managed and protected stands of the Białowieża Forest, Poland. *Nat. Areas J.* **2014**, *34*, 423–428. [CrossRef]
15. Andersson, J.; Gómez, E.D.; Michon, S.; Roberge, J.-M. Tree cavity densities and characteristics in managed and unmanaged Swedish boreal forest. *Scand. J. For. Res.* **2018**, *33*, 233–244. [CrossRef]
16. Le Roux, D.S.; Ikin, K.; Lindenmayer, D.B.; Manning, A.D.; Gibbons, P. The future of large old trees in urban landscapes. *PLoS ONE* **2014**, *9*, e99403. [CrossRef] [PubMed]
17. Lacki, M.J.; Hayes, J.P.; Kurta, A. *Bats in Forests: Conservation and Management*; Johns Hopkins University Press: Baltimore, MD, USA, 2007.
18. Cockle, K.L.; Martin, K.; Wesołowski, T. Woodpeckers, decay and the future of cavity-nesting vertebrate communities worldwide. *Front. Ecol. Environ.* **2011**, *9*, 377–382. [CrossRef]
19. van der Hoek, Y.; Gaona, G.V.; Martin, K. The diversity, distribution and conservation status of the tree-cavity-nesting birds of the world. *Divers. Distrib.* **2017**, *23*, 1120–1131. [CrossRef]
20. Siitonen, J. Dead wood in agricultural and urban habitats. In *Biodiversity in Dead Wood. Ecology, Biodiversity and Conservation*; Stokland, J.N., Siitonen, J., Jonsson, B.G., Eds.; Cambridge University Press: New York, NY, USA, 2012; pp. 380–401.
21. Müller, J.; Jarzabek-Müller, A.; Bussler, H.; Gossner, M.M. Hollow beech trees identified as keystone structures for saproxylic beetles by analyses of functional and phylogenetic diversity. *Anim. Conserv.* **2013**, *17*, 154–162. [CrossRef]
22. Micó, E. Saproxylic Insects in Tree Hollows. Diversity, Ecology and Conservation. In *Saproxylic Insects*; Zoological Monographs; Ulyshen, M., Ed.; Springer: Cham, Switzerland, 2018; Volume 1, pp. 693–727.
23. Lambrechts, M.; Adriaensen, F.; Ardia, D.R.; Artemyev, A.V.; Atienzar, F.; Banbura, J.; Barba, E.; Bouvier, J.-C.; Camprodon, J.; Cooper, C.B.; et al. The design of artificial nestboxes for the study of secondary hole-nesting birds: A review of methodological inconsistencies and potential biases. *Acta Ornithol.* **2010**, *45*, 1–26. [CrossRef]
24. General Directorate of State Forests. unpublished data. 2021.
25. Broughton, R.K.; Hebda, G.; Maziarz, M.; Smith, K.W.; Smith, L.; Hinsley, S.A. Nest-site competition between bumblebees (Bombidae), social wasps (Vespidae) and cavity-nesting birds in Britain and the Western Palearctic. *Bird Study* **2015**, *62*, 427–437. [CrossRef]
26. Hicks, E.A. *Check-List and Bibliography on the Occurrence of Insects in Birds' Nests*; Iowa State College Press: Ames, IA, USA, 1959.
27. Petersen, G. Tineiden als Bestandteil der Nidicolenfauna. *Beiträge Zur Entomol.* **1963**, *13*, 411–427.
28. Robinson, G.S. Moth and bird interactions: Guano, feathers, and detritophagous caterpillars (Lepidoptera, Tineidae). In *Insect and Bird Interactions*; van Emden, H.F., Rothschild, M., Eds.; Intercept: Andover, UK, 2004; pp. 271–285.
29. Rushton, S.P.; Lurz, P.W.W.; Gurnell, J.; Nettleton, P.; Bruemmer, C.; Shirley, M.D.F.; Sainsbury, A.W. Disease threats posed by alien species: The role of a poxvirus in the decline of the native red squirrel in Britain. *Epidemiol. Infect.* **2006**, *134*, 521–533. [CrossRef]
30. Gurnell, J.; Wauters, L.A.; Lurz, P.W.W.; Tosi, G. Alien species and interspecific competition: Effects of introduced eastern grey squirrels on red squirrel population dynamics. *J. Anim. Ecol.* **2004**, *73*, 26–35. [CrossRef]
31. Romeo, C.; McInnes, C.J.; Dale, T.D.; Shuttleworth, C.; Bertolino, S.; Wauters, L.A.; Ferrari, N. Disease, invasions and conservation: No evidence of squirrelpox virus in grey squirrels introduced to Italy. *Anim. Conserv.* **2019**, *22*, 14–23. [CrossRef]
32. Bertolino, S.; di Montezemolo, N.C.; Preatoni, D.G.; Wauters, L.A.; Martinoli, A. A grey future for Europe: Sciurus carolinensis is replacing native red squirrels in Italy. *Biol. Invasions* **2014**, *16*, 53–62. [CrossRef]
33. de Raad, A.L.; Balafa, F.; Heitkonig, I.; Lurz, P.W.W. Mitigating the impact of forest management for conservation of an endangered forest mammal species: Drey surveys and nest boxes for red squirrels (*Sciurus vulgaris*). *Hystrix Ital. J. Mammal.* **2021**, *32*. [CrossRef]
34. Shuttleworth, C.M. The use of nest boxes by the Red Squirrel *Sciurus vulgaris* in a coniferous habitat. *Mammal Rev.* **1999**, *29*, 63–68. [CrossRef]
35. Shuttleworth, C.M. Interactions between the red squirrel (*Sciurus vulgaris*), great tit (*Parus major*) and jackdaw (*Corvus monedula*) whilst using nest boxes. *J. Zool.* **2001**, *255*, 269–272. [CrossRef]
36. Turkia, T.; Korpimäki, E.; Villers, A.; Selonen, V. Predation risk landscape modifies flying and red squirrel nest site occupancy independently of habitat amount. *PLoS ONE* **2018**, *13*, e0194624. [CrossRef] [PubMed]
37. Krauze-Gryz, D.; (Warsaw University of Life Sciences, Warsaw, Poland); Gryz, J.; (Forest Research Institute, Sękocin Stary, Poland). Personal communications, 2007.
38. Gryz, J.; (Forest Research Institute, Sękocin Stary, Poland). Personal communication, 2021.
39. Gryz, J.; Chojnacka-Ożga, L.; Krauze-Gryz, D. Long-term stability of tawny owl (*Strix aluco*) population despite varying environmental conditions–a case study from central Poland. *Pol. J. Ecol.* **2019**, *67*, 75–83. [CrossRef]
40. Jaworski, T.; Gryz, J.; Buszko, J. *Monopis fenestratella* (Heyden, 1863) (Lepidoptera, Tineidae)–new records from Poland with notes on species biology. *Fragm. Faun.* **2011**, *54*, 149–151. [CrossRef]
41. Jaworski, T.; Gryz, J.; Krauze-Gryz, D. Skrzynki lęgowe puszczyków (*Strix aluco* L.) jako środowisko występowania niektórych gatunków motyli (Lepidoptera). *Wiadomości Entomol.* **2012**, *31*, 17–22.
42. Goszczyński, J.; Gryz, J.; Krauze, D. Fluctuations of a common buzzard *Buteo buteo* population in Central Poland. *Acta Ornithol.* **2005**, *40*, 75–78. [CrossRef]
43. Gryz, J.; Krauze-Gryz, D. Pigeon and poultry breeders, friends or enemies of northern goshawk *Accipiter gentilis*? A long term study of the population in central Poland. *Animals* **2019**, *9*, 141. [CrossRef]

44. Gryz, J.; Krauze-Gryz, D. Common buzzard *Buteo buteo* population in a changing environment, central Poland as a case study. *Diversity* **2019**, *11*, 35. [CrossRef]
45. Kruczek, Z. Analiza frekwencji w polskich atrakcjach turystycznych. *Turyzm* **2015**, *25*, 47–55. [CrossRef]
46. Krauze-Gryz, D.; Gryz, J. A review of the diet of the red squirrel (Sciurus vulgaris) in different types of habitats. In *Red Squirrels: Ecology, Conservation & Management in Europe*; Shuttleworth, C.M., Lurz, P.W., Hayward, M., Eds.; European Squirrel Initiative: Stoneleigh Park, Warwickshire CV8 2L, England, UK, 2015; pp. 39–50.
47. Kostrzewa, A.; Krauze-Gryz, D. The choice and handling of supplemental food by red squirrels in an urban park. *Behav. Process.* **2020**, *178*, 104153. [CrossRef] [PubMed]
48. ter Braak, C.J.F.; Smilauer, P. CANOCO Reference Manual and CanoDraw for Windows User's Guide: Software for Canonical Community Ordination (version 4.5). 2002. Available online: https://edepot.wur.nl/405659 (accessed on 15 May 2021).
49. Hammer, Ø.; Harper, D.A.T.; Ryan, P.D. PAST: Paleontological statistics software package for education and data analysis. *Palaeontol. Electron.* **2001**, *4*. Available online: http://palaeo-electronica.org/2001_1/past/issue1_01.htm (accessed on 15 May 2021).
50. Aitken, K.E.H.; Martin, K. Experimental test of nest-site limitation in mature mixed forests of central British Columbia, Canada. *J. Wildl. Manag.* **2012**, *76*, 557–565. [CrossRef]
51. Wauters, L.; Dhondt, A. Nest-use by red squirrels (*Sciurus vulgaris* Linnaeus, 1758). *Mammalia* **1990**, *54*, 377–390. [CrossRef]
52. Gryz, J.; Krauze-Gryz, D.; Lesiński, G. Mammals in the vicinity of Rogów (central Poland). *Fragm. Faun.* **2011**, *54*, 183–197. [CrossRef]
53. Croose, E.; Birks, J.D.S.; Martin, J. Den boxes as a tool for pine marten *Martes martes* conservation and population monitoring in a commercial forest in Scotland. *Conserv. Evid.* **2016**, *13*, 57–61.
54. Krauze-Gryz, D.; Gryz, J.; Goszczyński, J.; Chylarecki, P.; Żmihorski, M. The Good, the Bad and the Ugly: Space use and intraguild interactions among three opportunistic predators–cat *Felis catus*, dog *Canis familiaris* and fox *Vulpes vulpes* –under human pressure. *Can. J. Zool.* **2012**, *90*, 1402–1413. [CrossRef]
55. Krauze-Gryz, D.; Żmihorski, M.; Gryz, J. Annual variation in prey composition of domestic cats in rural and urban environment. *Urban Ecosyst.* **2017**, *20*, 945–952. [CrossRef]
56. Krauze-Gryz, D.; Gryz, J.; Żmihorski, M. Cats kill millions of vertebrates in Polish farmland annually. *Glob. Ecol. Conserv.* **2019**, *17*, e00516. [CrossRef]
57. Krauze-Gryz, D.; Mazur, K.; Gryz, J. Density of red squirrels and their use of non-native tree species in the Rogów arboretum. *For. Res. Pap.* **2016**, *77*, 42–49. [CrossRef]
58. Burger, G. Response of gray squirrels to nest boxes at Remington Farms, Maryland. *J. Wildl. Manag.* **1969**, *33*, 796–801. [CrossRef]
59. Bosch, S.; Lurz, P. *The Eurasian Red Squirrel*; Westarp Wissenschaften: Hohenwarsleben, Germany, 2012.
60. Marsh, A.C.W.; Morris, P.A. The use of dormouse *Muscardinus avellanarius* nest boxes by two species of *Apodemus* in Britain. *Acta Theriol.* **2000**, *45*, 443–453. [CrossRef]
61. Bosch, S.; Lurz, P.W.W. The process of drey construction in red squirrels-nestbox observations based on a hidden camera. *Hystrix Ital. J. Mammal.* **2013**, *24*, 199–202. [CrossRef]
62. Berthier, K.; Leippert, F.; Fumagalli, L.; Arlettaz, R. Massive nest-box supplementation boosts fecundity, survival and even immigration without altering mating and reproductive behaviour in a rapidly recovered bird population. *PLoS ONE* **2012**, *7*, e36028. [CrossRef] [PubMed]
63. Lesiński, G. Location of bird nests in vertical metal pipes in suburban built-up area of Warsaw. *Acta Ornithol.* **2000**, *35*, 211–214. [CrossRef]
64. Chylarecki, P.; Chodkiewicz, T.; Neubauer, G.; Sikora, A.; Meissner, W.; Woźniak, B.; Wylegała, P.; Ławicki, Ł.; Marchowski, D.; Betleja, J.; et al. *Trendy Liczebności ptaków w Polsce*; GIOŚ: Warsaw, Poland, 2018.
65. Mennechez, G.; Clergeau, P. Effect of urbanisation on habitat generalists: Starlings not so flexible? *Acta Oecologica* **2006**, *30*, 182–191. [CrossRef]
66. Mazgajski, T. Nest site choice in relation to the presence of old nests and cavity depth in the starling *Sturnus vulgaris*. *Ethol. Ecol. Evol.* **2003**, *15*, 273–281. [CrossRef]
67. Mazgajski, T. Effect of old nest material in nestboxes on ectoparasites abundance and reproductive output in the European Starling (*Sturnus vulgaris* L.). *Pol. J. Ecol.* **2007**, *55*, 377–385.
68. Mazgajski, T. Competition for nest sites between the starling *Sturnus vulgaris* and other cavity nesters–study in forest park. *Acta Ornithol.* **2000**, *35*, 103–107.
69. Pawlikowski, T.; Osmański, M. Atrakcyjność środowisk miejskich dla os społecznych (Hymenoptera: Vespinae) na obszarze Torunia. *Wiadomości Entomol.* **1998**, *17*, 95–104.
70. Pawlikowski, K.; Pawlikowski, T. Nesting interactions of the social wasp Dolichovespula saxonica [F.] (Hymenoptera: Vespinae) in wooden nest boxes for birds in the forest reserve "Las Piwnicki" in the Chełmno Land (Northern Poland). *Ecol. Quest.* **2010**, *13*, 67–72.
71. Harriott, N. Mosquito control and pollinator health. *Pestic. You* **2016**, *36*, 9–17.
72. Seress, G.; Sándor, K.; Evans, K.L.; Liker, A. Food availability limits avian reproduction in the city: An experimental study on great tits *Parus major*. *J. Anim. Ecol.* **2020**, *89*, 1570–1580. [CrossRef] [PubMed]
73. Mroczkowski, M. Dermestidae. Skórnikowate. (Insecta: Coleoptera). *Fauna Pol. Warszawa* **1975**, *4*, 1–163.

74. Buszko, J. Mole (Tineidae, Lepidoptera) zasiedlające huby i gniazda ptaków w rezerwacie Las Piwnicki. *Acta Univ. Nicolai Copernic* **1996**, *96*, 49–55.
75. Buszko, J.; Pałka, K. Nowe dla fauny Polski gatunki Tineidae i Tortricidae (Lepidoptera). *Wiadomości Entomol.* **1992**, *11*, 105–111.
76. Buszko, J.; Śliwinski, Z. Nowe dla fauny Polski i rzadko spotykane gatunki motyli (Lepidoptera). *Wiadomości Entomol.* **1979**, *49*, 653–662.
77. Gaedike, R. *Tineidae II (Myrmecozelinae, Perissomasticinae, Tineinae, Hieroxestinae, Teichobiinae and Stathmopolitinae). Microlepidoptera of Europe*; Brill: Leiden, The Netherlands, 2019; Volume 9.
78. Tokár, Z.; Lvovsky, A.; Huemer, P. *Die Oecophoridae s. l. (Lepidoptera) Mitteleuropas. Bestimmung-Verbreitung-Habitat–Bionomie*; Slamka: Bratislava, Slovakia, 2005.
79. Lees, A.D. The physiology and biochemistry of diapause. *Annu. Rev. Entomol.* **1956**, *1*, 1–16. [CrossRef]
80. Buszko, J.; Pacuk, B. Uwagi o zimowaniu imago Gracillariidae (Lepidoptera) w budkach lęgowych w Polsce. *Wiadomości Entomol.* **2009**, *28*, 255.
81. Nadolski, J. Factors restricting the abundance of wasp colonies of the European hornet *Vespa crabro* and the Saxon wasp *Dolichovespula saxonica* (Hymenoptera: Vespidae) in an urban area in Poland. *Entomol. Fenn.* **2013**, *24*, 204–215. [CrossRef]
82. Bourne, A.; Fountain, M.T.; Wijnen, H.; Shaw, B. Potential of the European earwig (*Forficula auricularia*) as a biocontrol agent of the soft and stone fruit pest *Drosophila suzukii*. *Pest Manag. Sci.* **2019**, *75*, 3340–3345. [CrossRef] [PubMed]
83. Mänd, R.; Tilgar, V.; Lõhmus, A.; Leivits, A. Providing nest boxes for hole-nesting birds–Does habitat matter? *Biodivers. Conserv.* **2005**, *14*, 1823–1840. [CrossRef]

diversity

MDPI

Review

Rotifer Species Diversity in Mexico: An Updated Checklist

S. S. S. Sarma [1,*], Marco Antonio Jiménez-Santos [2] and S. Nandini [1]

[1] Laboratory of Aquatic Zoology, FES Iztacala, National Autonomous University of Mexico, Av. de Los Barrios No. 1, Tlalnepantla 54090, Mexico; nandini@unam.mx
[2] Posgrado en Ciencias del Mar y Limnología, Universidad Nacional Autónoma de México, Ciudad Universitaria, Mexico City 04510, Mexico; antoniojimenez@comunidad.unam.mx
* Correspondence: sarma@unam.mx; Tel.: +52-55-56231256

Abstract: A review of the Mexican rotifer species diversity is presented here. To date, 402 species of rotifers have been recorded from Mexico, besides a few infraspecific taxa such as subspecies and varieties. The rotifers from Mexico represent 27 families and 75 genera. Molecular analysis showed about 20 cryptic taxa from species complexes. The genera *Lecane*, *Trichocerca*, *Brachionus*, *Lepadella*, *Cephalodella*, *Keratella*, *Ptygura*, and *Notommata* accounted for more than 50% of all species recorded from the Mexican territory. The diversity of rotifers from the different states of Mexico was highly heterogeneous. Only five federal entities (the State of Mexico, Michoacán, Veracruz, Mexico City, Aguascalientes, and Quintana Roo) had more than 100 species. Extrapolation of rotifer species recorded from Mexico indicated the possible occurrence of more than 600 species in Mexican water bodies, hence more sampling effort is needed. In the current review, we also comment on the importance of seasonal sampling in enhancing the species richness and detecting exotic rotifer taxa in Mexico.

Keywords: rotifera; distribution; checklist; taxonomy

Citation: Sarma, S.S.S.; Jiménez-Santos, M.A.; Nandini, S. Rotifer Species Diversity in Mexico: An Updated Checklist *Diversity* **2021**, *13*, 291. https://doi.org/10.3390/d13070291

Academic Editor: Michael Wink

Received: 29 May 2021
Accepted: 25 June 2021
Published: 28 June 2021

Publisher's Note: MDPI stays neutral with regard to jurisdictional claims in published maps and institutional affiliations.

Copyright: © 2021 by the authors. Licensee MDPI, Basel, Switzerland. This article is an open access article distributed under the terms and conditions of the Creative Commons Attribution (CC BY) license (https://creativecommons.org/licenses/by/4.0/).

1. Introduction

Taxonomical studies involving species richness provide information on the global patterns of species distribution and are helpful to detect changes associated with climate or global trade. For example, in Mexico, the number of exotic and thus invasive species has been steadily increasing during the last two decades [1,2]. The existence of taxonomic checklists is helpful to confirm this.

Mexico is one of the megadiverse countries and accounts for about 10% of the world's biodiversity [3]. Despite well-classified geographical regions of Mexico [4], the description of the distribution of different groups of animal species is still fragmentary, especially with reference to invertebrates, including rotifers. Freshwater zooplanktonic groups are mainly composed of ciliates, rotifers, cladocerans, and copepods. Rotifers, being important trophic links in aquatic ecosystems, have been the focus of basic research, such as taxonomy and autecology, and applied aspects, such as ecotoxicology, aquaculture, and water quality indicators [5].

Studies on the rotifer species richness in Mexico have been steadily gaining importance during the last 25 years. Earlier studies were mainly sporadic and, at times, biased, with a limnological perspective [6]. Species checklists of rotifers from the Mexican territory are available only for selected regions. For example, information about the distribution of rotifers exists for the State of Mexico, Aguascalientes, Michoacán, Mexico City, and a few regions of the Yucatan Peninsula [7–13]. However, larger parts of the Mexican territory still lack such information. The first national checklist of rotifers from Mexico was produced during the late 1990s [14]. Since then, considerable progress has been made on the distribution of rotifers in different regions, although no attempts have been made to update the checklist.

Numerous models and computer programs are available to predict the possible number of species in a region or nation based on species accumulation and rarefaction curves, the presence or absence of given taxa, etc. For example, for understanding the state of biodiversity, models such as ICE, Chao 2, Jackknife, and Bootstrap are traditionally used to obtain species estimates for different groups of organisms [15]. Significant errors may still occur if the published reports of species are not corrected or weak data with large sampling gaps are used. In Mexico, the National Commission for the Knowledge and Use of Biodiversity (Comisión Nacional para el Conocimiento y Uso de la Biodiversidad, CONABIO) contains data on the biodiversity of different groups of organisms, yet information on patterns of distribution of groups such as rotifers within its territorial jurisdiction are limited.

This work aimed to provide a comprehensive list of rotifer species recorded and document their distributional patterns from different regions of Mexico.

2. Materials and Methods

A bibliographic review of rotifer diversity studies from Mexico was conducted using the standard databases in the Web of Science using the search words "rotifer*", "Mexic*" and "diversity" during the entire period available in each database (retrieved during May 2021). The records were then consulted in the full text, and we checked each work for the records of rotifer species. We also consulted works from other non-indexed journals but avoided contributions that contained only genus-level descriptions for rotifers. The data were sorted out into Excel files according to the geographical entities of Mexico. In addition, the documents available from CONABIO were also considered. For species nomenclature, we followed standard works on Rotifera [16,17]. The checklist provided here does not contain a listing of the infraspecific taxa. Therefore, only species were enumerated. However, infraspecific taxa were reported in the checklist without assigning an additional number.

Due to the increased accessibility of molecular tools in the study of systematics of rotifers, several cryptic taxa of commonly distributed species within genera such as *Brachionus*, *Keratella*, *Asplanchna*, and *Lecane* have been documented. However, cryptic species without formal description were not included in the checklist, although references to such studies are made in a separate table. When a known species was already reported from Mexico (e.g., *Philodina roseola*), the same taxon with *conferatur* status (e.g., *Philodina* cf. *roseola*) was not numbered. However, if a taxon was reported only with *conferatur*, it was considered for numbering (e.g., *Notholca* cf. *liepetterseni*). Further, taxa that have been identified as having potential species status but not described are not included here, for example, *Brachionus* sp. "Mexico" [18] and *Hexarthra* n. sp. [19]. In addition, as far as possible, we used published reports of species. When necessary, we checked the species identifications based on the illustrations provided in the articles with those from standard literature [20–23]. Yet, some taxa with species *inquirenda status* (e.g., *Polyarthra trigla*) were retained as such pending further studies. The species checklist was not arranged based on phylogeny of Rotifera. Rotifer families were arranged alphabetically, and within each family and genus, the species were all in alphabetic order. This facilitated reporting new records in future research.

A nonparametric analysis of species richness of Rotifera reported from Mexico was performed using the updated checklist. Models/computer simulations based on Chao 2, Jackknife 2, and Bootstrap were performed using EstimatesS 9 [24]. From the diversity estimators, we derived the efficiency percentage of each estimator with the following formula:

$$\frac{S_{observed}}{S_{estimated}} \times 10$$

3. Results

Mexico has 31 states and a capital, Mexico City. The total number of rotifer species reported from Mexico was 402, besides a few infraspecific taxa such as subspecies and varieties. The list of consulted works is available in Supplementary 1 with coordinates for each federal entity obtained from the Mexican National Institute of Statistics and Geography (INEGI). The database, created using published works from Mexican Rotifera, is presented in Supplementary 2. Rotifers from Mexico represented 27 families and 75 genera (Table 1). Only eight genera, viz., *Lecane*, *Trichocerca*, *Brachionus*, *Cephalodella*, *Lepadella*, *Keratella*, *Ptygura*, and *Notommata*, of rotifers had more than 50% of the total species recorded from the Mexican territory. Each of these genera had at least 10 species, while the remaining genera had less than 10 species each. Of the 15 species recorded with *conferatur* status, 11 were from Chihuahua and Quintana Roo. To date, molecular analysis has revealed the existence of 17 taxa as species complexes consisting of cryptic species (Table 2).

Table 1. Checklist of rotifer species recorded from Mexico.

Subclass: Bdelloidea Hudson, 1884	
Order: Adinetida Melone & Ricci, 1995	
Family: Adinetidae Hudson & Gosse, 1886	
1	*Adineta vaga* (Davis, 1873)
Order: Philodinida Melone & Ricci 1995	
Family: Philodinidae Ehrenberg, 1838	
2	*Dissotrocha aculeata* (Ehrenberg, 1830)
3	*Macrotrachela sonorensis* Örstan, 1995
4	*Philodina acuticornis* Murray, 1902
5	*Philodina megalotrocha* Ehrenberg, 1832
6	*Philodina roseola* Ehrenberg, 1832
	Philodina cf. *roseola* Ehrenberg, 1832
7	*Pleuretra africana* Murray, 1911
8	*Rotaria citrina* (Ehrenberg, 1838)
9	*Rotaria elongata* (Weber, 1888)
10	*Rotaria magnacalcarata* (Parsons, 1892)
11	*Rotaria neptunia* (Ehrenberg, 1830)
12	*Rotaria rotatoria* (Pallas, 1766)
Subclass: Monogononta Plate, 1889	
Order: Collothecacea Harring, 1913	
Family: Atrochidae Harring, 1913	
13	*Atrochus tentaculatus* Wierzejski, 1893
14	*Cupelopagis vorax* (Leidy, 1857)
Family: Collothecidae Harring, 1913	
15	*Collotheca ambigua* (Hudson, 1883)
16	*Collotheca campanulata* (Dobie, 1849)
17	*Collotheca coronetta* (Cubitt, 1869)
18	*Collotheca crateriformis* Offord, 1934
19	*Collotheca ornata* (Ehrenberg, 1832)
20	*Collotheca pelagica* (Rousselet, 1893)
21	*Collotheca riverai* Vilaclara & Sládeček, 1989
22	*Collotheca tenuilobata* (Anderson, 1889)
23	*Collotheca trilobata* (Collins, 1872)
24	*Stephanoceros millsii* (Kellicott, 1885)
Order: Flosculariacea Harring, 1913	
Family: Conochilidae Harring, 1913	
25	*Conochilus coenobasis* (Skorikov, 1914)
26	*Conochilus dossuarius* Hudson, 1885
27	*Conochilus hippocrepis* (Schrank, 1803)
28	*Conochilus natans* (Seligo, 1900)
29	*Conochilus unicornis* Rousselet, 1892
Family: Flosculariidae Ehrenberg, 1838	
30	*Beauchampia crucigere* (Dutrochet, 1812)
31	*Floscularia melicerta* (Ehrenberg, 1832)
32	*Limnias ceratophylli* Schrank, 1803
33	*Limnias melicerta* Weisse, 1848
34	*Octotrocha speciosa* Thorpe, 1893

Table 1. *Cont.*

	35	*Ptygura beauchampi* Edmondson, 1940
	36	*Ptygura brachiata* (Hudson, 1886)
	37	*Ptygura brevis* (Rousselet, 1893)
	38	*Ptygura crystallina* (Ehrenberg, 1834)
	39	*Ptygura furcillata* (Kellicott, 1889)
	40	*Ptygura libera* Myers, 1934
	41	*Ptygura* cf. *linguata* Edmondson, 1939
	42	*Ptygura longicornis* (Davis, 1867)
	43	*Ptygura melicerta* Ehrenberg, 1832
	44	*Ptygura pedunculata* Edmondson, 1939
	45	*Ptygura tacita* Edmondson, 1940
	46	*Ptygura tridorsicornis* Summerfiel-Wright, 1957
	47	*Ptygura velata* (Gosse, 1851)
	48	*Sinantherina ariprepes* Edmondson, 1939
	49	*Sinantherina semibullata* (Thorpe, 1893)
	50	*Sinantherina socialis* (Linnaeus, 1758)
Family: Hexarthridae Bartoš, 1959		
	51	*Hexarthra fennica* (Levander, 1892)
	52	*Hexarthra intermedia* (Wiszniewski, 1929)
		Hexarthra intermedia f. *braziliensis* Hauer, 1953
	53	*Hexarthra jenkinae* (de Beauchamp, 1932)
	54	*Hexarthra mira* (Hudson, 1871)
	55	*Hexarthra oxyuris* (Sernov, 1903)
	56	*Hexarthra polyodonta* (Hauer, 1957)
Family: Testudinellidae Harring, 1913		
	57	*Pompholyx complanata* Gosse, 1851
	58	*Pompholyx sulcata* Hudson, 1885
	59	*Testudinella caeca* (Parsons, 1892)
	60	*Testudinella emarginula* (Stenroos, 1898)
	61	*Testudinella incisa* (Ternetz, 1892)
	62	*Testudinella mucronata* (Gosse, 1886)
	63	*Testudinella parva* (Ternetz, 1892)
	64	*Testudinella patina* (Hermann, 1783)
	65	*Testudinella reflexa* (Gosse, 1887)
Family: Trochosphaeridae Harring, 1913		
	66	*Filinia brachiata* (Rousselet, 1901)
	67	*Filinia cornuta* (Weisse, 1847)
	68	*Filinia longiseta* (Ehrenberg, 1834)
	69	*Filinia opoliensis* (Zacharias, 1898)
	70	*Filinia pejleri* Hutchinson, 1964
	71	*Filinia saltator* (Gosse, 1886)
	72	*Filinia terminalis* (Plate, 1886)
	73	*Horaella thomassoni* Koste, 1973
	74	*Trochosphaera aequatorialis* Semper, 1872
Order: Ploima Hudson & Gosse, 1886		
Family: Asplanchnidae Eckstein, 1883		
	75	*Asplanchna brightwellii* Gosse, 1850
	76	*Asplanchna girodi* de Guerne, 1888
	77	*Asplanchna herrickii* de Guerne, 1888
	78	*Asplanchna intermedia* Hudson, 1886
	79	*Asplanchna priodonta* Gosse, 1850
	80	*Asplanchna sieboldii* (Leydig, 1854)
	81	*Asplanchna silvestrii* Daday, 1902
	82	*Asplanchnopus multiceps* (Schrank, 1793)
Family: Brachionidae Ehrenberg, 1838		
	83	*Anuraeopsis fissa* Gosse, 1851
	84	*Anuraeopsis quadriantennata* (Koste, 1974)
	85	*Brachionus ahlstromi* Lindeman, 1939
	86	*Brachionus angularis* Gosse, 1851

Table 1. *Cont.*

87	*Brachionus araceliae* Silva-Briano, Galván-De la Rosa, Pérez-Legaspi & Rico-Martínez, 2007
88	*Brachionus bidentatus* Anderson, 1889
89	*Brachionus budapestinensis* Daday, 1885
90	*Brachionus calyciflorus* Pallas, 1766
	Brachionus calyciflorus calyciflorus Pallas, 1766
91	*Brachionus caudatus* Barrois & Daday, 1894
92	*Brachionus dimidiatus* Bryce, 1931
93	*Brachionus dolabratus* Harring, 1914
94	*Brachionus durgae* Dhanapathi, 1974
95	*Brachionus falcatus* Zacharias, 1898
96	*Brachionus forficula* Wierzejski, 1891
97	*Brachionus havanaensis* Rousselet, 1911
98	*Brachionus josefinae* Silva-Briano & Segers, 1992
99	*Brachionus leydigii* Cohn, 1862
100	*Brachionus paranguensis* Guerrero-Jiménez, Vannucchi, Silva-Briano, Adabache-Ortiz, Rico-Martínez, Roberts, Neilson & Elías-Gutiérrez, 2019
101	*Brachionus plicatilis* Müller, 1786
	Brachionus plicatilis longicornis Fadeev, 1925
102	*Brachionus pterodinoides* Rousselet, 1913
103	*Brachionus quadridentatus* Hermann, 1783
	Brachionus quadridentatus quadridentatus Herman, 1783
104	*Brachionus rotundiformis* Tschugunoff, 1921
105	*Brachionus rubens* Ehrenberg, 1838
106	*Brachionus urceolaris* Müller, 1773
107	*Brachionus variabilis* Hempel, 1896
108	*Kellicottia bostoniensis* (Rousselet, 1908)
109	*Kellicottia longispina* (Kellicott, 1879)
110	*Keratella americana* Carlin, 1943
111	*Keratella cochlearis* (Gosse, 1851)
	Keratella cochlearis cochlearis (Gosse, 1851)
112	*Keratella hiemalis* Carlin, 1943
113	*Keratella irregularis* (Lauterborn, 1898)
114	*Keratella lenzi* Hauer, 1953
115	*Keratella mexicana* Kutikova & Silva-Briano, 1995
116	*Keratella morenoi* Modenutti, Diéguez & Segers, 1998
117	*Keratella procurva* (Thorpe, 1891)
	Keratella procurva robusta Koste & Shiel, 1980
118	*Keratella quadrata* (Müller, 1786)
119	*Keratella serrulata* (Ehrenberg, 1838)
120	*Keratella taurocephala* Myers, 1938
121	*Keratella tecta* (Gosse, 1851)
122	*Keratella ticinensis* (Callerio, 1921)
123	*Keratella tropica* (Apstein, 1907)
124	*Keratella valga* (Ehrenberg, 1834)
125	*Notholca acuminata* (Ehrenberg, 1832)
126	*Notholca bipalium* (Müller, 1786)
127	*Notholca foliacea* (Ehrenberg, 1838)
128	*Notholca* cf. *liepetterseni* Godske Björklund, 1972
129	*Notholca squamula* (Müller, 1786)
130	*Notholca striata* (Müller, 1786)
131	*Plationus patulus* (Daday, 1905)
	Plationus patulus macracanthus (Müller, 1786)
132	*Plationus polyacanthus* (Ehrenberg, 1834)
133	*Platyias leloupi* Gillard, 1967
134	*Platyias quadricornis* (Ehrenberg, 1832)
Family: Dicranophoridae Harring, 1913	
135	*Aspelta angusta* Harring & Myers, 1928
136	*Aspelta curvidactyla* Bērziņš, 1949
137	*Aspelta lestes* Harring & Myers, 1928

Table 1. *Cont.*

138	*Dicranophoroides caudatus* (Ehrenberg, 1834)
139	*Dicranophoroides claviger* (Hauer, 1965)
140	*Dicranophorus epicharis* Harring & Myers, 1928
141	*Dicranophorus forcipatus* (Müller, 1786)
142	*Dicranophorus grandis* (Ehrenberg, 1832)
143	*Dicranophorus prionacis* Harring & Myers, 1928
144	*Dicranophorus robustus* Harring & Myers, 1928
145	*Encentrum* cf. *cruentum* Harring & Myers, 1928
146	*Encentrum saundersiae* (Hudson, 1885)
147	*Encentrum uncinatum* (Milne, 1886)
148	*Paradicranophorus sordidus* Donner, 1968
Family: Epiphanidae Harring, 1913	
149	*Cyrtonia tuba* (Ehrenberg, 1834)
150	*Epiphanes brachionus* (Ehrenberg, 1837)
151	*Epiphanes clavulata* (Ehrenberg, 1832)
152	*Epiphanes macroura* (Barrois & Daday, 1894)
153	*Epiphanes senta* (Müller, 1773)
154	*Proalides subtilis* Rodewald, 1940
155	*Proalides tentaculatus* de Beauchamp, 1907
Family: Euchlanidae Ehrenberg, 1838	
156	*Beauchampiella eudactylota* (Gosse, 1886)
157	*Dipleuchlanis elegans* (Wierzejski, 1893)
158	*Dipleuchlanis propatula* (Gosse, 1886)
159	*Euchlanis calpidia* Myers, 1930
160	*Euchlanis deflexa* (Gosse, 1851)
161	*Euchlanis dilatata* Ehrenberg, 1832
	Euchlanis dilatata lucksiana Hauer, 1930
162	*Euchlanis incisa* Carlin, 1939
163	*Euchlanis lyra* Hudson, 1886
164	*Euchlanis* cf. *mikropous* Koch-Althaus, 1962
165	*Euchlanis oropha* Gosse, 1887
166	*Euchlanis pyriformis* Gosse, 1851
167	*Euchlanis triquetra* Ehrenberg, 1838
168	*Tripleuchlanis plicata* (Levander, 1894)
Family: Gastropodidae Harring, 1913	
169	*Ascomorpha ecaudis* Perty, 1850
170	*Ascomorpha ovalis* (Bergendal, 1892)
171	*Ascomorpha saltans* Bartsch, 1870
172	*Gastropus hyptopus* (Ehrenberg, 1838)
173	*Gastropus stylifer* (Imhof, 1891)
Family: Ituridae Sudzuki, 1964	
174	*Itura aurita* (Ehrenberg, 1830)
175	*Itura chamadis* Harring & Myers, 1928
176	*Itura myersi* Wulfert, 1935
Family: Lecanidae Remane, 1933	
177	*Lecane aculeata* (Jakubski, 1912)
178	*Lecane aeganea* Harring, 1914
179	*Lecane arcuata* (Bryce, 1891)
180	*Lecane arcula* Harring, 1914
181	*Lecane aspasia* Myers, 1917
182	*Lecane bifurca* (Bryce, 1892)
183	*Lecane bulla* (Gosse, 1851)
184	*Lecane candida* Harring & Myers, 1926
185	*Lecane clara* (Bryce, 1892)
186	*Lecane closterocerca* (Schmarda, 1859)
187	*Lecane cornuta* (Müller, 1786)
188	*Lecane crenata* (Harring, 1913)
189	*Lecane crepida* Harring, 1914
190	*Lecane curvicornis* (Murray, 1913)

Table 1. *Cont.*

	191	*Lecane decipiens* (Murray, 1913)
	192	*Lecane doryssa* Harring, 1914
	193	*Lecane elasma* Harring & Myers, 1926
	194	*Lecane elegans* Harring, 1914
	195	*Lecane elsa* Hauer, 1931
	196	*Lecane flexilis* (Gosse, 1886)
	197	*Lecane furcata* (Murray, 1913)
	198	*Lecane grandis* (Murray, 1913)
	199	*Lecane haliclysta* Harring & Myers, 1926
	200	*Lecane hamata* (Stokes, 1896)
	201	*Lecane hastata* (Murray, 1913)
		Lecane cf. *hastata* (Murray, 1913)
	202	*Lecane hornemanni* (Ehrenberg, 1834)
	203	*Lecane inermis* (Bryce, 1892)
	204	*Lecane inopinata* Harring & Myers, 1926
	205	*Lecane latissima* Yamamoto, 1955
	206	*Lecane leontina* (Turner, 1892)
	207	*Lecane levistyla* (Olofsson, 1917)
	208	*Lecane ludwigii* (Eckstein, 1883)
	209	*Lecane luna* (Müller, 1776)
	210	*Lecane lunaris* (Ehrenberg, 1832)
	211	*Lecane margarethae* Segers, 1991
	212	*Lecane monostyla* (Daday, 1897)
	213	*Lecane nana* (Murray, 1913)
	214	*Lecane nelsoni* Segers, 1994
	215	*Lecane obtusa* (Murray, 1913)
	216	*Lecane ohioensis* (Herrick, 1885)
	217	*Lecane papuana* (Murray, 1913)
	218	*Lecane perpusilla* (Hauer, 1929)
	219	*Lecane pertica* Harring & Myers, 1926
	220	*Lecane punctata* (Murray, 1913)
	221	*Lecane pyriformis* (Daday, 1905)
	222	*Lecane quadridentata* (Ehrenberg, 1830)
	223	*Lecane rhenana* Hauer, 1929
	224	*Lecane rhytida* Harring & Myers, 1926
	225	*Lecane rugosa* (Harring, 1914)
	226	*Lecane ruttneri* Hauer, 1938
	227	*Lecane satyrus* Harring & Myers, 1926
	228	*Lecane scutata* (Harring & Myers, 1926)
	229	*Lecane signifera* (Jennings, 1896)
	230	*Lecane sola* Hauer, 1936
	231	*Lecane spinulifera* Edmondson, 1935
	232	*Lecane stenroosi* (Meissner, 1908)
	233	*Lecane stichaea* Harring, 1913
	234	*Lecane stokesii* (Pell, 1890)
	235	*Lecane subtilis* Harring & Myers, 1926
	236	*Lecane subulata* (Harring & Myers, 1926)
	237	*Lecane tenuiseta* Harring, 1914
	238	*Lecane thalera* (Harring & Myers, 1926)
	239	*Lecane thienemanni* (Hauer, 1938)
	240	*Lecane uenoi* Yamamoto, 1951
	241	*Lecane undulata* Hauer, 1938
	242	*Lecane unguitata* (Fadeev, 1925)
	243	*Lecane ungulata* (Gosse, 1887)
	244	*Lecane venusta* Harring & Myers, 1926
	245	*Lecane yatseni* Wei & Xu, 2010
Family: Lepadellidae Harring, 1913		
	246	*Colurella adriatica* Ehrenberg, 1831
	247	*Colurella colurus* (Ehrenberg, 1830)

Table 1. *Cont.*

	248	*Colurella hindenburgi* Steinecke, 1917
	249	*Colurella oblonga* Donner, 1943
	250	*Colurella obtusa* (Gosse, 1886)
	251	*Colurella uncinata* (Müller, 1773)
		Colurella uncinata bicuspidata (Ehrenberg, 1832)
	252	*Lepadella acuminata* (Ehrenberg, 1834)
	253	*Lepadella apsida* Harring, 1916
	254	*Lepadella astacicola* Hauer, 1926
	255	*Lepadella benjamini* Harring, 1916
	256	*Lepadella biloba* Hauer, 1958
	257	*Lepadella* cf. *cornuta* Koste & Shiel, 1989
	258	*Lepadella cristata* (Rousselet, 1893)
	259	*Lepadella dactyliseta* (Stenroos, 1898)
	260	*Lepadella discoidea* Segers, 1993
	261	*Lepadella donneri* Koste, 1972
	262	*Lepadella ehrenbergii* (Perty, 1850)
	263	*Lepadella heterostyla* (Murray, 1913)
	264	*Lepadella latusinus* (Hilgendorf, 1899)
	265	*Lepadella ovalis* (Müller, 1786)
	266	*Lepadella patella* (Müller, 1773)
		Lepadella patella patella (Müller, 1786)
	267	*Lepadella punctata* Wulfert, 1939
	268	*Lepadella quadricarinata* (Stenroos, 1898)
	269	*Lepadella quinquecostata* (Lucks, 1912)
		Lepadella quinquecostata quinquecostata (Lucks, 1912)
	270	*Lepadella rhomboides* (Gosse, 1886)
	271	*Lepadella triba* Myers, 1934
	272	*Lepadella triptera* (Ehrenberg, 1832)
	273	*Squatinella lamellaris* (Müller, 1786)
Family: Lindiidae Harring & Myers, 1924		
	274	*Lindia ecela* Myers, 1933
	275	*Lindia tecusa* Harring & Myers, 1922
	276	*Lindia torulosa* Dujardin, 1841
	277	*Lindia truncata* (Jennings, 1894)
Family: Mytilinidae Harring, 1913		
	278	*Lophocharis oxysternon* (Gosse, 1851)
	279	*Lophocharis salpina* (Ehrenberg, 1834)
	280	*Mytilina acanthophora* Hauer, 1938
	281	*Mytilina bisulcata* (Lucks, 1912)
	282	*Mytilina mucronata* (Müller, 1773)
		Mytilina mucronata spinigera (Ehrenberg, 1830)
	283	*Mytilina ventralis* (Ehrenberg, 1830)
		Mytilina ventralis brevispina (Ehrenberg, 1830)
		Mytilina ventralis ventralis (Ehrenberg, 1830)
Family: Notommatidae Hudson & Gosse, 1886		
	284	*Cephalodella apocolea* Myers, 1924
	285	*Cephalodella calosa* Wulfert, 1956
	286	*Cephalodella catellina* Müller, 1786
	287	*Cephalodella exigua* (Gosse, 1886)
	288	*Cephalodella forficula* (Ehrenberg, 1830)
	289	*Cephalodella gibba* (Ehrenberg, 1830)
	290	*Cephalodella gigantea* Remane, 1933
	291	*Cephalodella globata* (Gosse, 1887)
	292	*Cephalodella gracilis* (Ehrenberg, 1830)
	293	*Cephalodella* cf. *graciosa* Wulfert, 1956
	294	*Cephalodella hoodii* (Gosse, 1886)
	295	*Cephalodella macrodactyla* (Stenroos, 1898)
	296	*Cephalodella* cf. *marina* Myers, 1924
	297	*Cephalodella megalocephala* (Glascott, 1893)
	298	*Cephalodella misgurnus* Wulfert, 1937

Table 1. *Cont.*

299	*Cephalodella panarista* Myers, 1924
300	*Cephalodella physalis* Myers, 1924
	Cephalodella cf. *physalis* Myers, 1924
301	*Cephalodella rotunda* Wulfert, 1937
302	*Cephalodella stenroosi* Wulfert, 1937
303	*Cephalodella sterea* (Gosse, 1887)
304	*Cephalodella tenuiseta* (Burn, 1890)
305	*Cephalodella ventripes* (Dixon-Nuttall, 1901)
306	*Enteroplea lacustris* Ehrenberg, 1830
307	*Eosphora anthadis* Harring & Myers, 1922
308	*Eosphora ehrenbergi* Weber & Montet, 1918
309	*Eosphora najas* Ehrenberg, 1830
310	*Eosphora thoa* Harring & Myers, 1830
311	*Eosphora thoides* Wulfert, 1935
312	*Eothinia carogaensis* Myers, 1937
313	*Eothinia elongata* (Ehrenberg, 1832)
314	*Monommata actices* Remane, 1933
315	*Monommata diaphora* Myers, 1930
316	*Notommata aurita* (Müller, 1786)
317	*Notommata cerberus* (Gosse, 1886)
318	*Notommata copeus* Ehrenberg, 1834
319	*Notommata cyrtopus* Gosse, 1886
320	*Notommata falcinella* Harring & Myers, 1922
321	*Notommata glyphura* Wulfert, 1935
322	*Notommata haueri* Wulfert, 1939
	Notommata cf. *haueri* Wulfert, 1939
323	*Notommata pachyura* (Gosse, 1886)
324	*Notommata saccigera* Ehrenberg, 1830
325	*Notommata tripus* Ehrenberg, 1838
326	*Pleurotrocha petromyzon* (Ehrenberg, 1830)
327	*Resticula gelida* (Harring & Myers, 1922)
328	*Resticula melandocus* (Gosse, 1887)
329	*Resticula nyssa* Harring & Myers, 1924
330	*Sphyrias lofauna* (Rousselet, 1910)
331	*Taphrocampa annulosa* Gosse, 1851
332	*Taphrocampa selenura* Gosse, 1887
Family: Proalidae Harring & Myers, 1924	
333	*Proales cognita* Myers, 1940
334	*Proales daphnicola* Thompson, 1892
335	*Proales decipiens* (Ehrenberg, 1832)
336	*Proales fallaciosa* Wulfert, 1937
337	*Proales globulifera* (Hauer, 1921)
338	*Proales sigmoidea* (Skorikov, 1896)
339	*Proales similis* de Beauchamp, 1907
340	*Proales sordida* Gosse, 1886
341	*Proales* cf. *wesenbergi* Wulfert, 1960
342	*Wulfertia ornata* Donner, 1943
Family: Scaridiidae Manfredi, 1927	
343	*Scaridium botsjani* Daems & Dumont, 1974
344	*Scaridium longicaudum* (Müller, 1786)
Family: Synchaetidae Hudson & Gosse, 1886	
345	*Ploesoma hudsoni* (Imhof, 1891)
346	*Polyarthra dolichoptera* Idelson, 1925
	Polyarthra cf. *dolichoptera* Idelson, 1925
347	*Polyarthra euryptera* Wierzejski, 1891
348	*Polyarthra longiremis* Carlin, 1943
349	*Polyarthra luminosa* Kutikova, 1962
350	*Polyarthra major* Burckhardt, 1900
351	*Polyarthra remata* Skorikov, 1896
352	*Polyarthra trigla* Ehrenberg, 1834 (species inquirenda)

Table 1. *Cont.*

353	*Polyarthra vulgaris* Carlin, 1943
354	*Synchaeta bicornis* Smith, 1904
355	*Synchaeta elsteri* Hauer, 1963
356	*Synchaeta hyperborea* Smirnov, 1932
357	*Synchaeta longipes* Gosse, 1887
358	*Synchaeta oblonga* Ehrenberg, 1832
359	*Synchaeta pectinata* Ehrenberg, 1832
360	*Synchaeta stylata* Wierzejski, 1893
361	*Synchaeta tremula* (Müller, 1786)
362	*Synchaeta tremuloida* Pourriot, 1965
Family: Tetrasiphonidae Harring & Myers, 1924	
363	*Tetrasiphon hydrocora* Ehrenberg, 1840
Family: Trichocercidae Harring, 1913	
364	*Ascomorphella volvocicola* (Plate, 1886)
365	*Trichocerca bicristata* (Gosse, 1887)
366	*Trichocerca bidens* (Lucks, 1912)
367	*Trichocerca brachyura* (Gosse, 1851)
368	*Trichocerca braziliensis* (Murray, 1913)
369	*Trichocerca capucina* (Wierzejski & Zacharias, 1893)
370	*Trichocerca collaris* (Rousselet, 1896)
371	*Trichocerca cylindrica* (Imhof, 1891)
372	*Trichocerca dixonnuttalli* (Jennings, 1903)
373	*Trichocerca elongata* (Gosse, 1886)
374	*Trichocerca iernis* (Gosse, 1887)
375	*Trichocerca insignis* (Herrick, 1885)
376	*Trichocerca insulana* (Hauer, 1937)
377	*Trichocerca* cf. *intermedia* (Stenroos, 1898)
378	*Trichocerca longiseta* (Schrank, 1802)
379	*Trichocerca marina* (Daday, 1890)
380	*Trichocerca mollis* Edmondson, 1936
381	*Trichocerca mucosa* (Stokes, 1896)
382	*Trichocerca multicrinis* (Kellicott, 1897)
383	*Trichocerca musculus* (Hauer, 1937)
384	*Trichocerca porcellus* (Gosse, 1851)
385	*Trichocerca pusilla* (Jennings, 1903)
386	*Trichocerca rattus* (Müller, 1776)
387	*Trichocerca rosea* (Stenroos, 1898)
388	*Trichocerca rousseleti* (Voigt, 1902)
389	*Trichocerca ruttneri* Donner, 1953
390	*Trichocerca similis* (Wierzejski, 1893)
391	*Trichocerca stylata* (Gosse, 1851)
392	*Trichocerca tenuior* (Gosse, 1886)
393	*Trichocerca tigris* (Müller, 1786)
394	*Trichocerca vernalis* (Hauer, 1936)
395	*Trichocerca weberi* (Jennings, 1903)
Family: Trichotriidae Harring, 1913	
396	*Macrochaetus collinsii* (Gosse, 1867)
397	*Macrochaetus longipes* Myers, 1934
398	*Macrochaetus sericus* (Thorpe, 1893)
399	*Macrochaetus subquadratus* (Perty, 1850)
400	*Trichotria pocillum* (Müller, 1776)
401	*Trichotria tetractis* (Ehrenberg, 1830)
402	*Wolga spinifera* (Western, 1894)

Table 2. Some species complexes and cryptic species of rotifers reported from Mexico.

Species Complex	Reference
Ascomorpha ovalis	[25]
Asplanchna brightwellii	[26]
Asplanchna girodi	[27]
Brachionus calyciflorus	[25]
Brachionus plicatilis	[18,28]
Brachionus quadridentatus	[25]
Euchlanis dilatata	[29]
Keratella cochlearis	[25]
Lecane bulla	[30]
Lecane cornuta	[25]
Lecane crepida	[25]
Lecane curvicornis	[25]
Lecane hastata	[25]
Lecane lunaris	[25]
Mytilina ventralis	[25]
Platyias quadricornis	[25]
Testudinella patina	[25]

The faunal diversity of rotifers from the different states of the country was highly heterogeneous. Only five federal entities (the State of Mexico, Michoacán, Veracruz, Mexico City, and Quintana Roo) had more than 100 species. The total number of genera per state followed the same trend of species richness (Table 3). Thus, seven federal entities (the State of Mexico, Michoacán, Veracruz, Mexico City, Quintana Roo, Aguascalientes, and Chihuahua) had more than 30 genera.

Table 3. Number of genera and species of rotifers reported from different States of Mexico. The states are represented by bold numbers. 1: Aguascalientes, 2: Campeche, 3: Chiapas, 4: Chihuahua, 5: Colima, 6: Guanajuato, 7: Guerrero, 8: Hidalgo, 9: Jalisco, 10: Mexico City, 11: Michoacan, 12: Morelos, 13: Nayarit, 14: Oaxaca, 15: Puebla, 16: Quintana Roo, 17: San Luis Potosi, 18: Sinaloa, 19: Sonora, 20: State of Mexico, 21: Tabasco, 22: Tlaxcala, 23: Veracruz, 24: Yucatán, 25: Zacatecas. Other states do not have published records of rotifers, and these were not included.

Species/States	1	2	3	4	5	6	7	8	9	10	11	12	13	14	15	16	17	18	19	20	21	22	23	24	25
Adineta	0	0	0	0	0	0	0	0	0	0	0	0	0	0	0	1	0	0	0	0	0	1	0	0	0
Anuraeopsis	1	0	0	1	0	0	1	0	1	1	1	1	0	0	1	1	1	0	0	2	1	0	1	0	1
Ascomorpha	3	0	0	1	0	2	0	0	0	1	1	0	0	0	0	2	0	1	0	3	1	0	3	0	0
Ascomorphella	1	0	0	0	0	0	0	0	0	0	0	0	0	0	0	0	0	0	0	1	0	0	1	0	0
Aspelta	0	0	0	0	0	0	0	0	0	0	0	1	0	0	0	0	0	0	0	2	0	0	1	0	0
Asplanchna	5	0	1	0	0	2	2	0	1	4	3	1	0	0	0	0	0	0	0	6	2	0	4	0	1
Asplanchnopus	0	0	0	0	0	0	0	0	0	0	0	0	0	0	0	1	0	0	0	1	0	0	0	0	0
Atrochus	0	0	0	0	0	0	0	0	0	1	0	0	0	0	0	0	0	0	0	0	0	0	0	0	0
Beauchampia	0	0	0	0	0	0	0	0	0	1	0	0	0	0	0	0	0	0	0	1	0	0	0	0	0
Beauchampiella	1	1	0	0	0	0	0	0	0	0	1	0	0	0	0	1	0	0	0	1	0	0	1	0	0
Brachionus	14	4	5	5	0	4	11	6	2	13	14	8	0	0	0	8	2	0	0	11	9	1	17	0	12
Cephalodella	2	0	0	12	0	0	2	0	0	4	8	3	0	0	0	3	0	0	0	14	0	0	9	0	0
Collotheca	1	0	0	2	0	0	0	0	0	5	2	0	0	0	0	1	1	0	0	2	0	0	0	0	0
Colurella	0	0	0	4	0	0	0	0	0	3	2	2	1	0	0	3	0	0	0	4	0	0	5	0	0
Conochilus	4	0	0	0	0	0	1	0	1	1	4	0	0	0	0	0	0	0	0	4	0	0	2	0	1
Cupelopagis	0	0	0	0	0	0	1	0	0	1	1	0	0	0	0	1	0	0	0	1	0	0	0	0	0
Cyrtonia	0	0	0	0	0	0	0	0	0	0	1	0	0	0	0	0	0	0	0	0	0	0	0	0	0
Dicranophoroides	0	0	0	0	0	0	1	0	0	1	1	1	0	0	0	1	0	0	0	1	0	0	2	0	0
Dicranophorus	0	0	0	2	0	0	1	0	0	2	1	2	1	0	0	3	0	0	0	4	0	1	3	0	0
Dipleuchlanis	0	0	0	0	0	0	1	0	0	1	1	0	0	0	0	1	0	0	0	2	0	0	1	0	0
Encentrum	0	0	0	2	0	0	0	0	0	0	0	1	0	0	0	0	0	0	0	2	0	0	0	0	0
Enteroplea	0	0	0	0	0	0	0	0	0	0	0	0	0	0	0	0	0	0	0	1	0	0	1	0	0
Eosphora	0	0	0	1	0	0	1	0	0	1	1	1	0	0	0	1	0	0	0	3	1	0	2	0	0
Eothinia	0	0	0	0	0	0	0	0	0	0	1	0	0	0	0	1	0	0	0	1	0	0	0	0	0
Epiphanes	1	0	0	0	0	0	1	0	0	2	2	0	0	0	0	1	0	0	0	2	1	0	2	0	0
Euchlanis	2	0	0	3	0	1	1	2	0	5	2	1	0	0	0	2	1	0	0	6	1	2	4	0	0
Filinia	4	0	2	3	0	2	3	0	1	4	3	1	0	0	0	1	0	0	0	5	1	1	2	0	3
Floscularia	0	0	0	0	0	0	0	0	0	1	0	0	0	0	0	0	0	0	0	0	0	0	0	0	0
Gastropus	1	0	0	0	0	0	0	1	0	0	1	0	0	0	0	0	1	0	0	1	0	1	0	0	0
Hexarthra	2	0	1	2	0	2	1	0	1	2	1	1	0	0	1	1	0	0	0	3	0	0	2	0	0
Horaëlla	1	0	0	0	0	0	1	0	0	0	1	0	0	0	0	0	0	0	0	1	0	0	1	0	0
Itura	0	0	0	0	0	0	0	1	0	0	2	0	0	0	0	1	0	0	0	3	0	0	1	0	0
Kellicottia	2	0	0	0	0	0	1	0	1	0	1	0	0	0	0	0	0	0	0	2	0	0	0	0	1
Keratella	10	0	3	3	0	4	3	5	2	5	11	3	0	0	0	3	0	0	0	10	3	4	5	1	5
Lecane	10	13	2	21	3	0	15	3	0	29	29	21	17	0	0	40	6	1	0	34	4	2	39	0	11
Lepadella	2	0	0	5	1	1	1	4	0	6	11	5	1	0	0	9	0	2	0	11	0	2	5	0	1
Limnias	0	0	0	1	0	0	0	0	0	2	0	0	0	0	0	0	0	0	0	1	0	0	0	0	0
Lindia	0	0	0	0	0	0	0	0	0	1	2	1	1	0	0	1	0	0	0	2	0	0	2	0	0
Lophocharis	0	0	0	0	0	0	0	0	0	2	1	0	0	0	0	1	0	0	0	2	0	0	1	0	0
Macrochaetus	0	1	0	0	0	0	0	0	0	0	1	0	0	0	0	3	0	0	0	1	0	0	1	0	0
Monommata	0	0	0	0	0	0	0	0	0	0	1	1	0	0	0	1	0	0	0	1	0	0	0	0	0
Mytilina	1	0	0	1	0	0	0	0	0	3	3	1	1	0	0	3	0	0	0	4	0	0	4	0	0
Notholca	0	0	0	1	0	0	0	0	0	0	1	0	0	0	0	0	0	0	0	2	0	0	2	0	0
Notommata	0	0	0	2	0	0	1	0	0	1	4	0	3	0	0	3	0	0	0	7	0	0	2	0	0
Octotrocha	0	0	0	0	0	0	0	0	0	0	0	0	0	0	0	0	0	0	0	1	0	0	0	0	0
Paradicranophorus	0	0	0	1	0	0	0	0	0	0	0	0	0	0	0	0	0	0	0	0	0	0	0	0	0
Philodina	1	0	0	2	0	0	0	0	0	0	0	0	0	0	0	1	0	0	1	0	0	0	0	0	0

Table 3. *Cont.*

Species/States	1	2	3	4	5	6	7	8	9	10	11	12	13	14	15	16	17	18	19	20	21	22	23	24	25
Plationus	2	1	1	1	0	1	0	1	0	1	1	0	0	0	0	1	0	0	0	2	0	0	1	0	1
Platyias	1	0	1	1	0	1	1	1	0	2	1	1	1	0	0	1	0	0	0	2	1	0	1	0	1
Pleuretra	0	0	0	0	0	0	0	0	0	0	0	0	0	1	0	0	0	0	0	0	0	0	0	0	0
Pleurotrocha	0	0	0	1	0	0	0	0	0	1	1	0	0	0	0	0	0	0	0	1	0	1	1	0	0
Ploesoma	0	0	0	0	0	0	1	0	0	0	0	0	0	0	0	0	0	0	0	0	1	0	0	0	0
Polyarthra	5	0	1	2	0	2	2	1	3	3	4	1	0	0	0	1	2	0	0	6	0	1	1	0	0
Pompholyx	1	0	0	0	1	0	1	0	0	0	2	0	0	0	0	0	0	0	0	2	0	0	0	0	0
Proales	1	0	0	5	0	0	0	0	0	2	1	1	0	0	0	2	0	0	0	3	0	0	1	0	0
Proalides	1	0	0	1	0	0	0	0	0	1	1	0	0	0	0	0	0	0	0	1	0	0	0	0	0
Ptygura	0	0	0	2	0	0	0	0	0	8	0	0	0	0	0	3	0	0	0	3	0	0	0	0	0
Resticula	0	0	0	1	0	0	0	0	0	0	2	0	0	0	0	1	0	0	0	1	0	0	1	0	0
Rotaria	1	0	0	0	0	0	0	0	0	0	0	0	0	0	0	0	0	0	0	3	0	0	1	0	0
Scaridium	0	0	0	1	0	0	0	0	0	0	1	0	0	0	0	2	0	0	0	1	0	0	1	0	0
Sinantherina	0	0	0	0	0	0	0	0	0	3	2	0	0	0	0	1	0	0	0	2	0	1	1	0	0
Sphyrias	0	0	0	0	0	0	0	0	0	0	1	0	0	0	0	0	0	0	0	0	0	0	0	0	0
Squatinella	0	0	0	0	0	0	0	1	0	1	1	0	0	0	0	1	0	0	0	1	0	0	1	0	0
Stephanoceros	0	0	0	0	0	0	0	0	0	1	0	0	0	0	0	0	0	0	0	0	0	0	0	0	0
Synchaeta	4	0	1	2	0	0	0	0	1	3	4	1	0	0	0	2	0	0	0	3	4	1	1	0	0
Taphrocampa	0	0	0	0	0	0	0	0	0	0	1	0	1	0	0	1	0	0	0	2	0	1	1	0	0
Testudinella	1	1	1	1	0	0	1	0	0	3	5	2	1	0	0	4	0	0	0	6	0	1	3	0	0
Tetrasiphon	0	0	0	0	0	0	0	0	0	0	1	0	0	0	0	0	0	0	0	0	0	0	0	0	0
Trichocerca	7	0	0	2	0	1	3	0	0	15	12	4	2	0	0	8	0	1	0	23	6	2	10	0	1
Trichotria	2	0	0	1	0	0	1	0	0	2	2	1	0	0	0	0	0	0	0	2	0	1	2	0	2
Tripleuchlanis	0	0	0	1	0	0	0	0	0	1	1	0	1	0	0	1	0	0	0	1	0	0	1	0	0
Trochosphaera	0	0	0	0	0	0	0	0	0	0	0	0	0	0	0	0	0	0	0	0	0	0	1	0	0
Wolga	0	0	0	1	0	0	1	0	0	0	0	0	0	0	0	0	0	0	0	1	0	0	0	0	0
Wulfertia	0	0	0	1	0	0	0	0	0	0	0	0	0	0	0	0	0	0	0	0	0	0	0	0	0

Seasonally collected samples offered a higher number of species than those collected sporadically. Data on the species richness of rotifers collected seasonally from selected water bodies are presented in Table 4.

Table 4. The number of rotifer species reported from selected waterbodies through seasonal sampling.

Waterbody	Total Species	Reference
Valle de Bravo Reservoir (State of Mexico)	50	[31]
Madín reservoir (State of Mexico)	28	[32]
Llano reservoir (State of Mexico)	84	[33]
Iturbide reservoir (State of Mexico)	55	[34]
Lake Zumpango (State of Mexico)	33	[35]
Chimaliapan wetland (State of Mexico)	75	[36]
Lake Xochimilco (Mexico City)	81	[37]
Lake Cantera Oriente (Mexico City)	68	[38]
Benito Juárez Reservoir (Mexico City)	80	[39]
River Antigua (Veracruz State)	125	[40]
Amacuzac River Basin (State of Morelos)	65	[41]
Valerio Trujano Reservoir (Guerrero State)	64	[42]

Biogeographic distribution of selected species recorded from Nearctic and Neotropical regions of Mexico showed some of them to be out of known range based on global patterns. More than 20 taxa distributed in Palearctic region were reported from Nearctic or Neotropical regions (Table 5).

Table 5. Out of known range distribution of Rotifera recorded from Mexico. The known range from different geographical regions was based on [16], and for the national biogeographic provinces, Ref. [4] was followed. Afr: Afrotropical region; Ant: Antarctic region; Aus: Australian region; Nea: Nearctic region; Neo: Neotropical region; Ori: Oriental region; Pac: Pacific region and Pal: Palearctic region.

Species and Distribution	Records from Mexico
Adineta vaga: Afr, Pal	Quintana Roo and Tlaxcala: Neo
Atrochus tentaculatus: Aus, Pal, Ori	Mexico City: Nea
Collotheca crateriformis: Pal	Chihuahua: Nea
Colurella colurus: Pal	State of Mexico and Chihuahua: Nea; Veracruz and Quintana Roo: Neo
Colurella oblonga: Pal	Veracruz: Neo
Dicranophorus forcipatus: Pal	State of Mexico, Michoacán, Mexico City, Chihuahua and Tlaxcala: Nea; Morelos, Veracruz, Quintana Roo, Guerrero and Nayarit: Neo
Epiphanes brachionus: Pal	Mexico City: Nea; Guerrero: Neo
Horaella thomassoni: Neo	State of Mexico, Michoacán and Aguascalientes: Nea
Keratella procurva robusta: Aus	State of Mexico, Michoacán and Aguascalientes: Nea and Tabasco: Neo
Lecane unguitata: Afr, Aus, Ori, Pal	State of Mexico, Michoacán and Mexico City: Nea, Quintana Roo and Veracruz: Neo
Lecane yatseni: Ori	Veracruz: Neo
Lepadella discoidea: Afr, Aus, Ori	State of Mexico: Nea
Lepadella punctata: Ori, Pal	State of Mexico: Nea
Mytilina mucronata spinigera: Pal	Aguascalientes: Nea
Mytilina ventralis: Afr, Pac, Pal	State of Mexico, Mexico City, Morelos, Michoacán: Nea; Veracruz, Quintana Roo and Nayarit: Neo
Notholca acuminata: Afr, Pal	Chihuahua: Nea
Notommata haueri: Pal	Chihuahua: Nea
Paradicranophorus sordidus: Ant, Pal	Chihuahua: Nea
Philodina acuticornis: Pal	Chihuahua: Nea
Plationus polyacanthus: Pal	State of Mexico and Aguascalientes: Nea
Proales globulifera: Pal	State of Mexico: Nea
Ptygura brevis: Aus, Pal	Chihuahua deserts: Nea
Ptygura tridorsicornis: Pal	State of Mexico: Nea
Sphyrias lofauna: Afr, Pac	Michoacán: Nea
Squatinella lamellaris: Pac, Pal	State of Mexico, Michoacán and Mexico City: Nea; Veracruz y Quintana Roo: Neo
Synchaeta elsteri: Pal	Michoacán: Nea
Synchaeta hyperborea: Pal	Tabasco: Neo
Synchaeta tremuloida: Pal	Jalisco: Nea

Different estimators of species diversity indicated the asymptote in all cases (Figure 1). The efficiency percentage of species estimates varied between 62% and 86% (Chao 2 and Bootstrap, respectively). In addition, these estimators indicated that the potential richness of rotifers from Mexico could be from 450 to 600 species.

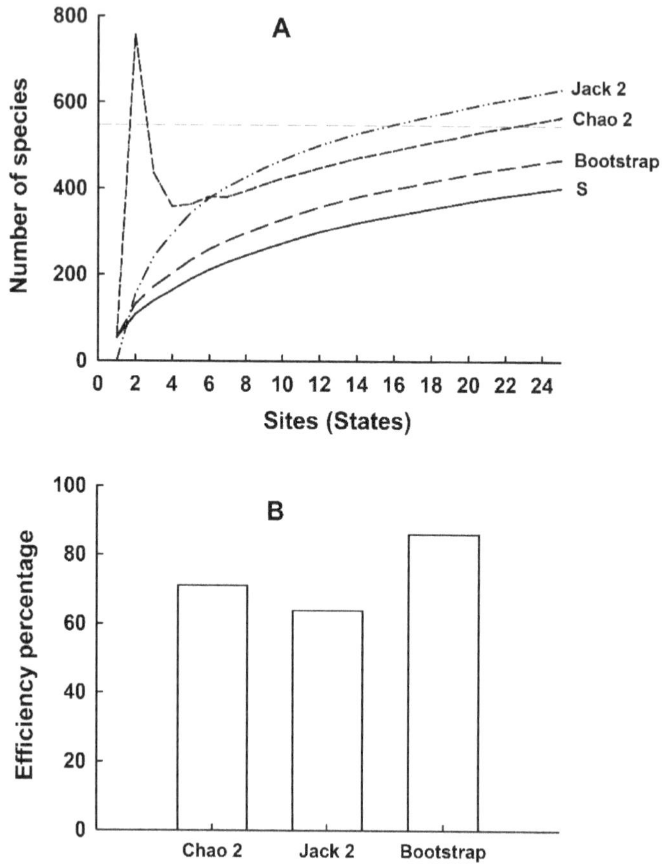

Figure 1. Diversity estimators (**A**) and efficiency percentage (**B**) using Jack 2, Chao 2, and Bootstrap methods.

4. Discussion

Taxonomical studies on Mexican rotifers date back more than 100 years. However, increased awareness of their role in limnological studies began only during the last 25 years. Figure 2 shows some of the interesting rotifer species from Mexico. Conventional limnological investigations in Mexico included rotifers as part of plankton [6], yet rarely quantified their abundances. One of the earliest studies on the seasonal variations of freshwater rotifers showed just seven rotifer taxa [43]. Thereafter, many studies on the seasonal variations of rotifers have been carried out from different water bodies such as ponds, lakes, reservoirs, and rivers. For certain freshwater ecosystems, zooplankton sampling was carried out for many years, for example, in the Valle de Bravo reservoir [44]. Long-term studies of riverine plankton are rare in Mexico, although the country has more than 200 rivers. Seasonal studies from River Antigua in the State of Veracruz have revealed 125 species REF. The importance of seasonal studies in understanding the rotifer species richness began receiving considerable attention after it became clear that certain exotic taxa appear only in certain months of the year. For example, *Notholca* cf. *liepetterseni* and *Lecane yatseni* have been recorded in River Antigua, Veracruz sporadically [40], although these species are native to the Scandinavian region and China, respectively.

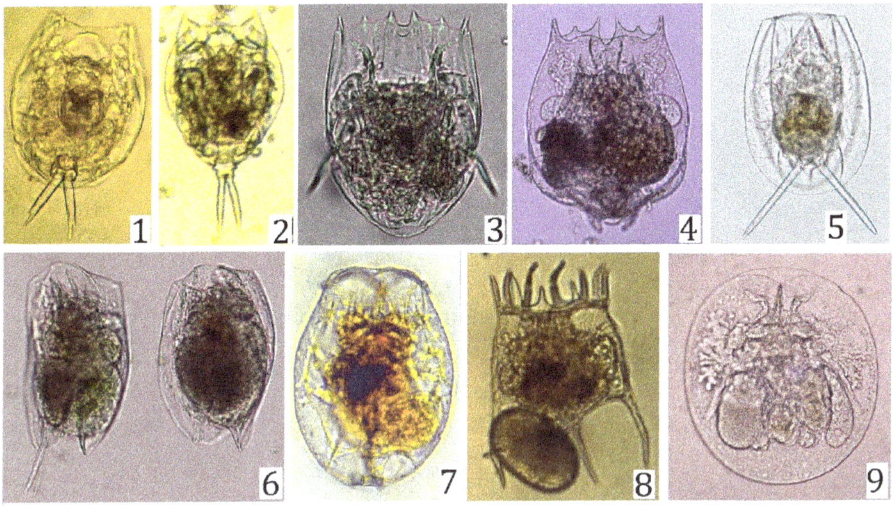

Figure 2. Some interesting rotifers from Mexico. (**1**) *Lecane yatseni*, (**2**) *Lecane rhytida*, (**3**) *Notholca* cf. *liepetterseni*, (**4**) *Brachionus bidentatus*, (**5**) *Dipleuchlanis propatula;* (**6**) *Euchlanis* cf. *mikropous*, (**7**) *Brachionus dimidiatus*, (**8**) *Plationus patulus macracanthus* and (**9**) *Testudinella patina*. All photos from authors' previous works.

The first comprehensive list of Mexican rotifers was documented about 3 decades ago and contained 283 species [14]. Since then, many studies on Mexican freshwaters have reported the presence of 120 additional rotifer taxa. This, however, does not include close to 20 cryptic taxa, which require formal description. From the mean of species estimators, it appears that there is a possibility of encountering more than 600 species in Mexico. This may be a sub-estimation of the actual reality, since it is based on the diversity of rotifers which have been well studied only in 5–7 of the 32 federal entities in the country. This number is not unreasonable if one considers the numerous habitats that exist in Mexico which confer it a megadiverse status (CONABIO), as well as the existence of cryptic taxa within Rotifera. For example, the *Brachionus plicatilis* complex has as many as 15 cryptic species [45]. Several species complexes have already been reported in Mexico [25,28–30]. The geographic location of Mexico (as a corridor between South and North America) [46] also supports the possible occurrence of diverse rotifer species in different federal entities. This is further evidenced by the poor sampling in certain regions, especially in states such as Baja California, Durango, and Coahuila. Mexico has 70 large lakes (area: 1000 to 10,000 hectares), 14,000 reservoirs (85% with <10 hectares), and >200 rivers [6]. The rotifer species list presented in this work was based on only a handful of waterbodies and many more are yet to be studied.

Desert temporary ponds, rivers, and marine ecosystems have great potential for enhancing the species richness of rotifers to the Mexican fauna. For example, ephemeral waterbodies from the desert states in Mexico have yielded more than 100 rotifer species [47]. Yet, many temporary water bodies in Mexico have not been sampled even once. Rotifer fauna in riverine habitats have been rarely studied, although the species richness in these aquatic systems is high [40]. Mexico is bestowed with 9330 km of coastline. Yet, knowledge on the marine rotifers from Mexico is more fragmentary than inland saline waters [48]. For example, seasonal sampling efforts from the brackish water ecosystem in the State of Tabasco showed the presence of more than 35 rotifer species [49]. Of the three classes of rotifers, Bdelloidea, Monogononta, and Seisonacea, the last is represented by two marine genera, *Seison* and *Paraseison*. *Seison* is epizoic on the crustacean genus *Nebalia* but has not been so far reported from marine waters of Mexico, although *Nebalia* occurs in these waters [50]. Therefore, further studies on marine rotifers may be oriented for identifying *Seison* from *Nebalia*.

An aspect often overlooked in taxonomic studies is the culture of rotifer species, which is important for many reasons. The first is that, when studying the molecular taxonomy of predatory taxa, prey in the stomach contents may interfere with the analysis [26]. The second reason is that culturing species may reveal the presence of different phenotypes from the same genotype as observed in the case of *Euchlanis* cf. *mikropous* [51]. Third, some descriptions are vague and incomplete. For example, culturing a rare taxon with appearance of *Collotheca monoceros* [52] resolved the issue, showing that it was a regeneration by *Stephanoceros millsii*. Fourth, cryptic species have different life histories which cannot be identified from fixed samples [53]. Finally, for certain analysis of taxonomic characters such as measurements of trophi on SEM, culturing is needed to obtain sufficient quantity for the description of size range [54].

The occurrence of some rotifer species known from the geographic regions such as the Palearctic, Afrotropical, and Oriental were reported from Nearctic region and Neotropical regions of Mexico. For example, *Lecane yatseni*, typical to the Oriental region, was recorded from Mexico. Similarly, *Sphyrias lofauna*, common to Afrotropical and Pacific regions, was documented from Nearctic region of Mexico [14]. This suggests not only extensive sampling, but also distributional aspects, including the possible roles that global climatic changes and trade involving aquatic species play a role in the dispersion of rotifers.

5. Conclusions

A taxonomic survey of rotifers so far has revealed the occurrence of about 400 species of rotifers from Mexico. Many Mexican states still do not have formal rotifer checklists. Only a few states in Mexico have some information on the diversity of rotifers. Yet, the species richness reported in this work is based on only a few selected water bodies. Species estimators have predicted the possible occurrence of about 600 rotifer species within the Mexican territory. Thus, further studies are still needed to understand rotifer diversity in Mexico.

Supplementary Materials: The following are available online at https://www.mdpi.com/article/10.3390/d13070291/s1, supplementary 1. List of consulted works for works on rotifer taxonomy and supplementary 2. Database compiled by the authors on the occurrence of different rotifer species from Mexico.

Author Contributions: Conceptualization, S.S.S.S.; formal analysis, M.A.J.-S.; interpretation, original draft preparation, S.N. All authors have prepared the final manuscript. All authors have read and agreed to the published version of the manuscript.

Funding: This research was funded by PAPIIT-IG 200820.

Institutional Review Board Statement: Not applicable.

Informed Consent Statement: Not applicable.

Data Availability Statement: All the data were taken from literature and are available from the publishers. Authors will provide data on request.

Acknowledgments: MAJC thanks Posgrado en Ciencias del Mar y Limnología (UNAM) and CONACyT (582568). Three anonymous reviewers have improved our presentation.

Conflicts of Interest: The authors declare no conflict of interest.

References

1. Weber, D.; Hintermann, U.; Zangger, A. Scale and trends in species richness: Considerations for monitoring biological diversity for political purposes. *Glob. Ecol. Biogeogr.* **2004**, *13*, 97–104. [CrossRef]
2. Llorente-Bousquets, J.; Ocegueda, S. Estado del conocimiento de la biota. In *Capital Natural de México. Conocimiento Actual de la Biodiversidad*; Conabio: México City, Mexico, 2008; Volume 1, pp. 283–322.
3. Ramírez-Albores, J.E.; Badano, E.I.; Flores-Flores, J.; Yáñez-Espinosa, L. Scientific literature on invasive alien species in a megadiverse country: Advances and challenges in Mexico. *NeoBiota* **2019**, *48*, 113–127. [CrossRef]
4. Morrone, J.J.; Escalante, T.; Rodríguez-Tapia, G. Mexican biogeographic provinces: Map and shapefiles. *Zootaxa* **2017**, *4277*, 277–279. [CrossRef]

5. Wallace, R.L.; Snell, T.W.; Ricci, C.; Nogrady, T. Rotifera. In *Biology, Ecology and Systematics*, 2nd ed.; Segers, H., Ed.; Guides to the Identification of the Microinvertebrates of the Continental Waters of the World; Kenobi Productions: Ghent, Belgium; Backhuys Publishers: Leiden, The Netherlands, 2006; Volume 23.
6. De la Lanza, E.G.; García, C.J.L. *Lagos y Presas de México*; AGT Editor S.A.: México City, Mexico, 2002; p. 680.
7. Rico-Martínez, R.; Silva-Briano, M. Contribution to the knowledge of the rotifera of Mexico. *Hydrobiologia* **1993**, *255/256*, 467–474. [CrossRef]
8. Rico-Martínez, R. Rotíferos. In *La Biodiversidad en Aguascalientes: Estudio de Estado*; Comisión Nacional para el Conocimiento y Uso de la Biodiversidad (CONABIO): Aguascalientes, Mexico; Instituto del Medio Ambiente del Estado de Aguascalientes (IMAE): Aguascalientes, Mexico; Universidad Autónoma de Aguascalientes (UAA): Aguascalientes, Mexico, 2008.
9. Sarma, S.S.S.; Serranía-Soto, C.; Nandini, S. Diversidad de Rotíferos. In *La Diversidad Biológica del Estado de México*; Ceballos, G.R., Ed.; Gobierno del Estado de México y Comisión Nacional para el Conocimiento y Uso de la Biodiversidad (CONABIO): México, Mexico, 2009.
10. Castellanos-Páez, M.E.; Zamora-García, M.; Benítez-Díaz-Mirón, M.I.; Mouriño, G.G.; Contreras-Tapia, R.A. Abundancia y biomasa de la comunidad de rotíferos y su relación con parámetros ambientales en tres estaciones del Canal Cuemanco, Xochimilco. *Soc. Rural. Prod. Medio Ambiente* **2014**, *14*, 10–21.
11. Arroyo-Castro, J.L.; Alvarado-Flores, J.; Uh-Moo, J.C.; Koh-Pasos, C.G. Monogonont rotifers species of the island Cozumel, Quintana Roo, México. *Biodivers. Data J.* **2019**, *7*, e34719. [CrossRef]
12. Ortega-Murillo, M.d.R.; Alvarado-Villanueva, R.; Sánchez-Heredia, J.D.; Muñóz-Gaytán, A.A.; Morales, R.H. El Plancton de Agua dulce. In *La Biodiversidad en Michoacán*; Estudio de Estado 2. CONABIO; Comisión Nacional Para el Conocimiento y Uso de la Biodiversidad (CONABIO): México, Mexico, 2019.
13. Delgado-Saucedo, J.J.; Silva-Briano, M.; Sigala-Rodríguez, J.J. Rotíferos. In *La Biodiversidad en Zacatecas*; Estudio de Estado. CONABIO; Comisión Nacional para el Conocimiento y Uso de la Biodiversidad (CONABIO): México, Mexico; Gobierno del Estado de Zacatecas: México, Mexico, 2020.
14. Sarma, S.S.S. Checklist of rotifers (Rotifera) from Mexico. *Environ. Ecol.* **1999**, *17*, 978–983.
15. Gwinn, D.C.; Allen, M.S.; Bonvechio, K.I.; Hoyer, M.V.; Beesley, L.S. Evaluating estimators of species richness: The importance of considering statistical error rates. *Methods Ecol. Evol.* **2016**, *7*, 294–302. [CrossRef]
16. Segers, H. Annotated checklist of the rotifers (Phylum Rotifera), with notes on nomenclature, taxonomy and distribution. *Zootaxa* **2007**, *1564*, 1–104. [CrossRef]
17. Jersabek, C.D.; Leitner, M.F. The Rotifer World Catalog. World Wide Web Electronic Publication. 2013. Available online: http://www.rotifera.hausdernatur.at/ (accessed on 20 May 2021).
18. Alcántara-Rodríguez, J.A.; Ciros-Pérez, J.; Ortega-Mayagoitia, E.; Serrania-Soto, C.R.; Piedra-Ibarra, E. Local adaptation in populations of a *Brachionus* group *plicatilis* cryptic species inhabiting three deep crater lakes in Central Mexico. *Freshw. Biol.* **2012**, *57*, 728–740. [CrossRef]
19. Brown, P.D.; Schröder, T.; Ríos-Arana, J.V.; Rico-Martínez, R.; Silva-Briano, M.; Wallace, R.L.; Walsh, E.J. Patterns of rotifer diversity in the Chihuahuan desert. *Diversity* **2020**, *12*, 393. [CrossRef]
20. Koste, W. Rotatoria. In *Die Rädertiere Mitteleuropas. Ein Bestimmungswerk Begründet von Max Voigt*; Bornträger: Stuttgart, Germany, 1978; Volume 1–2, pp. 234, 673.
21. Segers, H. *Rotifera 2: The Lecanidae Monogononta*; Guides to the Identification of Microinvertebrates of the Continental Waters of the World Series; Balogh Scientific Books: The Hague, The Netherlands, 1994; Volume 6.
22. Wallace, R.L.; Snell, T.W. Rotifera. Chapter 8. In *Ecology and Classifications of North American Freshwater Invertebrates*, 3rd ed.; Thorp, J., Covich, A., Eds.; Elsevier: Oxford, UK, 2010.
23. Wallace, R.L.; Snell, T.W.; Walsh, E.J.; Sarma, S.S.S.; Segers, H. Chapter 8. Phylum Rotifera. Keys to Palaearctic Fauna. In *Thorp and Covich's Freshwater Invertebrates*, 4th ed.; Academic Press: Cambridge, MA, USA; Elsevier: Amsterdam, The Netherlands, 2019; Volume 4, pp. 219–267. [CrossRef]
24. Colwell, R.K.; Mao, C.X.; Chang, J. Interpolating, extrapolating, and comparing incidence-based species accumulation curves. *Ecology* **2004**, *85*, 2717–2727. [CrossRef]
25. García-Morales, A.E.; Elías-Gutiérrez, M. DNA barcoding of freshwater rotifera in Mexico: Evidence of cryptic speciation in common rotifers. *Mol. Ecol. Resour.* **2013**, *13*, 1097–1107. [CrossRef] [PubMed]
26. Jiménez-Contreras, J.; Sarma, S.S.S.; Piedra-Ibarra, E.; Calderón-Torres, M.; Nandini, S. Morphological, morphometrical and molecular (CO1 and ITS) analysis of the rotifer *Asplanchna brightwellii* from selected freshwater bodies in Central Mexico (Mexico). *J. Environ. Biol.* **2013**, *34*, 1039–1046.
27. Jiménez-Contreras, J.; Sarma, S.S.S.; Piedra-Ibarra, E.; Nandini, S. Morphometric and molecular (COX 1) variations of *Asplanchna girodi* clones from Central Mexico. *J. Environ. Biol.* **2017**, *38*, 1229–1239. [CrossRef]
28. Nandini, S.; Peña-Aguado, F.; Arreguin-Rebolledo, U.; Sarma, S.S.S.; Murugan, G. Molecular identity and demographic responses to salinity of a freshwater strain of *Brachionus plicatilis* from the shallow Lake Pátzcuaro, Mexico. *Fundam. Appl. Limnol.* **2019**, *192*, 319–329. [CrossRef]
29. Kordbacheh, A.; Shapiro, A.N.; Walsh, E.J. Reproductive isolation, morphological and ecological differentiation among cryptic species of *Euchlanis dilatata*, with the description of four new species. *Hydrobiologia* **2019**, *844*, 221–242. [CrossRef]

30. Walsh, E.J.; Schröder, T.; Wallace, R.L.; Rico-Martínez, R. Cryptic speciation in *Lecane bulla* (Monogononta: Rotifera) in Chihuahuan Desert waters. *Int. Ver. Für Theor. Angew. Limnol. Verh.* **2009**, *30*, 1046–1050.
31. Nandini, S.; Merino-Ibarra, M.; Sarma, S.S.S. Seasonal changes in the zooplankton abundances of the reservoir Valle de Bravo (State of Mexico, Mexico). *Lake Reserv. Manag.* **2008**, *24*, 321–330. [CrossRef]
32. Moreno-Gutiérrez, R.M.; Sarma, S.S.S.; Sobrino-Figueroa, A.S.; Nandini, S. Population growth potential of rotifers from a high altitude eutrophic waterbody, Madín reservoir (State of Mexico, Mexico): The importance of seasonal sampling. *J. Limnol.* **2018**, *77*, 441–451. [CrossRef]
33. Muñoz-Colmenares, M.E.; Sarma, S.S.S.; Nandini, S. Seasonal variation of rotifers from the high altitude Llano reservoir (State of Mexico, Mexico). *J. Environ. Biol.* **2017**, *38*, 1171–1181. [CrossRef]
34. Sarma, S.S.S.; Osnaya-Espinosa, L.R.; Aguilar-Acosta, C.R.; Nandini, S. Seasonal variations in zooplankton abundances in the Iturbide reservoir (Isidro Fabela, State of Mexico, Mexico). *J. Environ. Biol.* **2011**, *32*, 473–480.
35. Figueroa-Sanchez, M.A.; Nandini, S.; Sarma, S.S.S. Zooplankton community structure in the presence of low levels of cyanotoxins: A case study in a high altitude tropical reservoir (Valley de Bravo, Mexico). *J. Limnol.* **2014**, *73*, 157–166. [CrossRef]
36. García-García, G.; Nandini, S.; Sarma, S.S.S.; Martínez-Jerónimo, F.; Jiménez-Contreras, J. Impact of chromium and aluminium pollution on the diversity of zooplankton: A case study in the Chimaliapan wetland (Ramsar site) (Lerma basin, Mexico). *J. Environ. Sci. Health Part. A* **2012**, *47*, 534–547. [CrossRef]
37. Jiménez-Contreras, J.; Nandini, S.; Sarma, S.S.S. Diversity of Rotifera (Monogononta) and egg ratio of selected taxa in the canals of Xochimilco (Mexico City). *Wetlands* **2018**, *38*, 1033–1044. [CrossRef]
38. Gutiérrez, S.G.; Sarma, S.S.S.; Nandini, S. Seasonal variations of rotifers from a high altitude urban shallow water body, La Cantera Oriente (Mexico City, Mexico). *Chin. J. Oceanol. Limnol.* **2017**, *35*, 1387–1397. [CrossRef]
39. Espinosa-Rodríguez, C.A.; Sarma, S.S.S.; Nandini, S. Zooplankton community changes in relation to different macrophyte species: Effects of *Egeria densa* removal. *Ecohydrol. Hydrobiol.* **2021**, *21*, 153–163. [CrossRef]
40. Nandini, S.; Sarma, S.S.S.; Gulati, R.D. A seasonal study reveals the occurrence of exotic rotifers in the river Antigua, Veracruz, close to the Gulf of Mexico. *River Res. Appl.* **2017**, *33*, 970–982. [CrossRef]
41. Nandini, S.; Ramírez-García, P.; Sarma, S.S.S.; Gutiérrez-Ochoa, R.A. Planktonic indicators of water quality: A case study in the Amacuzac River Basin (State of Morelos, Mexico). *River Res. Appl.* **2019**, *35*, 268–279. [CrossRef]
42. Vázquez-Sánchez, A.; Reyes-Vanegas, G.; Nandini, S.; Sarma, S.S.S. Diversity and abundance of rotifers (Rotifera) during an annual cycle in the reservoir Valerio Trujano (Tepecoacuilco, Mexico). *Inland Waters* **2004**, *4*, 293–302. [CrossRef]
43. Suárez-Morales, E.; Vázquez-Mazy, A.; Solís, M.E. Preliminary investigations on the zooplankton community of a Mexican eutrophic reservoir, a seasonal survey. *Hidrobiológica* **1993**, *3*, 71–80.
44. Ramírez-García, P.; Nandini, S.; Sarma, S.S.S.; Robles-Valderrama, E.; Cuesta, I.; Hurtado-Maria, D. Seasonal variations of zooplankton abundance in the freshwater reservoir Valle de Bravo (Mexico). *Hydrobiologia* **2002**, *467*, 99–108. [CrossRef]
45. Mills, S.J.; Alcántara-Rodríguez, A.; Ciros-Pérez, J.; Gómez, A.; Hagiwara, A.; Galindo, K.H.; Jersabek, C.D.; Malekzadeh-Viayeh, R.; Leasi, F.; Lee, J.-S.; et al. Fifteen species in one: Deciphering the *Brachionus plicatilis* species complex (Rotifera, Monogononta) through DNA taxonomy. *Hydrobiologia* **2017**, *796*, 39–58. [CrossRef]
46. Elías-Gutiérrez, M.; Suárez-Morales, E.; Sarma, S.S.S. Diversity of freshwater zooplankton in the neotropics: The case of Mexico. *Verh. Internat. Verein. Limnol.* **2001**, *27*, 4027–4031. [CrossRef]
47. Ríos-Arana, J.V.; del Carmen Agüero-Reyes, L.; Wallace, R.L.; Walsh, E.J. Limnological characteristics and rotifer community composition of Northern Mexico Chihuahuan desert springs. *J. Arid. Environ.* **2019**, *160*, 32–41. [CrossRef]
48. Walsh, E.J.; Schröder, T.; Wallace, R.L.; Ríos-Arana, J.V.; Rico-Martínez, R. Rotifers from selected inland saline waters in the Chihuahuan Desert of México. *Saline Syst.* **2008**, *4*, 7. [CrossRef]
49. Sarma, S.S.S.; Nandini, S.; Ramírez-García, P.; Cortés-Muñoz, J.E. New records of brackish water Rotifera and Cladocera from Mexico. *Hidrobiológica* **2000**, *10*, 121–124.
50. Escobar-Briones, E.; Villalobos-Hiriart, J.L. *Nebalia lagartensis* (Leptostraca) a new species from the Yucatán Peninsula, Mexico. *Crustaceana* **1995**, *68*, 1–11.
51. Nandini, S.; Sarma, S.S.S. Adaptive toe morphology of *Euchlanis* cf. *mikropous* Koch-Althaus, 1962 (Rotifera: Euchlanidae) exposed directly and indirectly to invertebrate predators. *Limnologica* **2019**, *78*, 125693. [CrossRef]
52. Sarma, S.S.S.; Jiménez-Santos, M.A.; Nandini, S.; Wallace, R.L. Review on the ecology and taxonomy of sessile rotifers (Rotifera) with special reference to Mexico. *J. Environ. Biol.* **2020**, *41*, 3–12. [CrossRef]
53. Gabaldón, C.; Fontaneto, D.; Carmona, M.J.; Montero-Pau, J.; Serra, M. Ecological differentiation in cryptic rotifer species: What we can learn from the *Brachionus plicatilis* complex. *Hydrobiologia* **2017**, *796*, 7–18. [CrossRef]
54. Kordbacheh, A.; Wallace, R.L.; Walsh, E.J. Evidence supporting cryptic species within two sessile microinvertebrates, *Limnias melicerta* and *L ceratophylli* (Rotifera, Gnesiotrocha). *PLoS ONE* **2018**, *13*, e0205203. [CrossRef]

diversity

MDPI

Article

Characterizing the Influence of Domestic Cats on Birds with Wildlife Rehabilitation Center Data

K. Grace Demezas and W. Douglas Robinson *

Department of Fisheries, Wildlife and Conservation Sciences, 104 Nash Hall, Oregon State University, Corvallis, OR 97331, USA; kgdemezas@gmail.com
* Correspondence: douglas.robinson@oregonstate.edu

Abstract: Depredation of birds by domestic cats is hypothesized to be one of many significant sources of mortality leading to global bird declines. Direct observations are relatively rarely documented compared with large numbers of birds hypothesized to be killed or wounded by cats. We analyzed data from two wildlife rehabilitation centers located in Salem and Grants Pass, Oregon USA, to understand which species were most likely to interact with a cat, and the species traits associated with cat interactions and habitats (urban vs. rural) of rescued birds. Interaction with a cat was the second-most commonly reported cause of admission, representing 12.3% of 6345 admissions. Half to two-thirds of birds were rescued from cats in urban settings and were usually species foraging on or near the ground. Most species were admitted to rehabilitation centers in direct proportion to their regional abundance. An exception was the absence of common species weighing less than 70 g, which we conclude is an effect of sampling bias. We conclude that cats most often interact with regionally common near-ground-dwelling bird species in both urban and rural habitats. Wildlife rehabilitation centers can provide valuable sources of data for cat-bird interactions but potential sources of uncertainty and bias in their data need to be considered carefully.

Keywords: avian mortality; cat-bird interactions; cat predation; citizen science; domestic cat; human-wildlife conflict; wildlife rehabilitation

Citation: Demezas, K.G.; Robinson, W.D. Characterizing the Influence of Domestic Cats on Birds with Wildlife Rehabilitation Center Data *Diversity* **2021**, *13*, 322. https://doi.org/10.3390/d13070322

Academic Editor: Michael Wink

Received: 18 June 2021
Accepted: 13 July 2021
Published: 15 July 2021

Publisher's Note: MDPI stays neutral with regard to jurisdictional claims in published maps and institutional affiliations.

Copyright: © 2021 by the authors. Licensee MDPI, Basel, Switzerland. This article is an open access article distributed under the terms and conditions of the Creative Commons Attribution (CC BY) license (https://creativecommons.org/licenses/by/4.0/).

1. Introduction

Recent reports of widespread declines of birds have elevated interest in sources of mortality, particularly anthropogenic sources [1,2]. Depredation by domestic cats (*Felis catus*), both owned and feral, and collisions with buildings, automobiles, wind turbines, and power lines and communication towers have been identified as the primary mortality sources [3]. Domestic cats kill an estimated 1.3 to 4.0 billion birds each year in the United States alone [4], a number much greater than the estimated 365 to 988 million birds killed from collisions with buildings [5], or 89 million to 340 million killed in collisions with automobiles [6].

Despite the apparent impacts of cats on bird populations, important information gaps continue to present challenges. For example, several basic aspects of cat–bird interactions remain poorly described in most locations, including which species are most often captured, the role of habitat differences in determining the identities of species attacked by cats, whether year-round residents or migrants are more susceptible, and which other traits (e.g., body size, propensity to forage on or near the ground) might influence the likelihood of birds being attacked by cats [7]. Furthermore, current estimates of cat-influenced mortality have been generated from a number of variables difficult to measure accurately. Statistics estimating proportions of households owning cats, rates at which owned cats are allowed outdoors and the rate at which free-roaming cats kill wildlife have all been estimated but exhibit high levels of uncertainty in most studies [4].

Estimating feral (non-owned) cat population sizes is also difficult [4]. Rates of cat ownership are often measured via survey [8], which may bias estimates if cat owners are more or less likely to respond to such surveys [9,10]. Cat owners' likelihood to allow their cats to roam outdoors is also influenced by a complex variety of factors, such as whether the cat was adopted or found as a stray or how long the cat has lived with that owner [11]. The rate at which free-roaming cats kill wildlife has also been difficult to estimate, as cats do not bring all prey items home, so it may be common to underestimate true impact [9]. Finally, quantification of effects of feral cats requires estimates of feral cat population size [12], and the relative rates at which feral versus owned cats kill wildlife. All of these differences are further complicated by the diversity of domestic cat ownership behavior, which ranges from owned cats not allowed outdoors to cats allowed to fully access the outdoors [13]. Some of these uncertainties are being addressed using rapidly improving technology. For instance, video cameras have been used to better quantify predation by free-roaming cats without the need to retrieve prey [14]. However, data gathered from existing, perhaps under-utilized, sources may provide an efficient and more cost-effective way to gain additional information on interactions between birds and domestic cats [15–18].

Wildlife rehabilitation centers provide a relatively untapped source of data on cat–bird interactions because people regularly bring birds injured by cats into centers [19]. Such data provide information on species (and therefore traits that may predict vulnerability to cats as well), location of the cat–bird interaction, and even potential for successful rehabilitation after treatment. Although the data have limitations [20,21], their availability in coordinated databases such as WILD-ONe (Wildlife Center of Virginia, Waynesboro, Virginia, USA) and the tens of thousands of animals admitted each year provide a potentially unique source of information on cat–bird interactions [22]. For example, concerns over bird losses sometimes focus on Neotropical migratory birds, a group known to be declining overall [1]. Most Neotropical migratory species are found in rural instead of urban areas [23], so if most cats occupy urban landscapes where their owners live, then cats may interact less with sensitive Neotropical migrants than they do with common species tolerant of human modification of landscapes.

Here, we analyzed data from two wildlife rehabilitation centers in western Oregon, USA. Our objectives were to: (1) understand what proportion of birds admitted were hypothesized or known to be admitted as a result of interacting with a cat; (2) enumerate the species most commonly delivered to rehabilitation centers after interactions with cats; (3) identify traits of bird species correlated with numbers of admissions to the centers; (4) evaluate the relationship between indices of species' abundances and the numbers of admissions; and (5) compare the sources of cat-influenced admissions across urban versus rural habitats. If admissions to wildlife rehabilitation centers reflect trends in interactions between birds and cats, the information gained from our analyses could help predict which species are most at risk from cat interactions and in what habitats, guiding conservation, management and outreach actions aimed at reducing influence of domestic cats on avian mortality.

2. Materials and Methods

2.1. Data Sources

We used data on cat–bird interactions archived in WILD-ONe, an online database for wildlife rehabilitators and researchers. We selected two Oregon wildlife rehabilitation centers included in the database for inclusion in our study. The first, Turtle Ridge Wildlife Center, is located in Salem, a city of approximately 174,000 people [24] located in Marion County. The second, Wildlife Images Rehabilitation and Education Center, is in Grants Pass, a city of approximately 37,000 people [25] in Josephine County. Both centers draw admissions of injured, presumed injured, ill and presumed orphaned animals from across urban and rural habitats generally characterized as patchworks of woodland, agricultural areas and grasslands. We used data from November, 2014, when the two centers began submitting data to WILD-ONe, through the end of 2018. We extracted data on species admitted, location of rescue, and cause of admission. We chose the two centers because of the large number of admissions, the sampling of both urban and rural habitats and our familiarity with the regional avifauna.

2.2. Determination of Cause of Admission

We focused on records of birds whose admission documentation indicated contact with a cat. In such cases, we included the record whether the contact had been directly observed or reported as suspected to have happened, the latter of which typically occurred when cats were observed in the vicinity of injured birds. Some records listed multiple potential reasons for admission. We therefore included records when another primary cause of admission was listed in documentation, but when interaction with a cat was also indicated. Finally, records indicating that the bird was rescued due to immediate danger from a cat, or suspicion thereof, were included. Additional details on admission procedures and definitions of cat interactions as a cause of admission are in Appendix A.

2.3. Identification of Locations and Habitats

We categorized the location of each cat–bird interaction as occurring in urban (including suburban and moderate to high density of dwellings and/or impermeable surface cover) or rural sites (low density of dwellings and/or impermeable surface, typically *Quercus* oak and *Pseudotsuga* fir woodlands or grasslands). All classifications were made using Google Maps (maps.google.com, accessed on 18 June 2021), with the address of the site of rescue located and placed in the center of a circle with a diameter of approximately 300 m. The dominant land cover type within each area was identified. As all classifications included only two options, whichever classification better fit greater than 50% of the visible area was used. To be conservative, we utilized such broad categories because no information to independently verify the exact locations at which injured birds were obtained was available in the wildlife rehabilitation center databases. We excluded any record for which the site of rescue was not provided or was too vague (i.e., reported at the city or county level only) to allow for classification.

2.4. Species Traits

2.4.1. Species and Species Groups

We assumed that bird species reported in the database were identified correctly, except for a few cases involving out-of-range rarities where we concluded the identification was likely incorrect. For example, a few reports of "blue jay", unlikely to refer to the eastern North American species Blue Jay *Cyanocitta cristata* (see Table S1 for all scientific names), were probably misreported California Scrub-Jay *Aphelocoma californica* or Steller's Jay *Cyanocitta stelleri*, both commonly found in western Oregon. In a few cases, we combined similarly appearing congeners into one "species group" (e.g., *Sphyrapicus* sapsuckers and *Spinus* goldfinches; Appendix A).

2.4.2. Body Mass

We included mean body mass of each species [26]. For species with sexual dimorphism or large geographic variation in mass, we selected data from the site nearest Oregon and used the smallest mean mass (e.g., male raptor masses instead of females). To focus on the subset of species likely to interact with cats as potential prey, we removed all species with an average adult mass >200 g [27,28], as well as aquatic species. A list of species and their characteristics is in Table S2.

2.4.3. Residency Status

We categorized species based on their presence in the study sites year-round (primarily non-migratory residents) or their absence in some months of the year (migratory). Some migratory species were absent during the winter (e.g., flycatchers), whereas others were absent during the summer (e.g., Varied Thrush *Ixoreus naevius*). We treated both as migrants. A few species could be found in very small numbers year-round (e.g., Wilson's Snipe

Gallinago delicata) but if their abundances changed dramatically owing to migration of most individuals out of the region, we treated them as migrants.

2.4.4. Aquatic or Non-Aquatic

We categorized species as primarily aquatic or non-aquatic and filtered out the aquatic species from our analyses because cats rarely attack aquatic species unless they are on land. We called ducks, geese, rails and Marsh Wren *Cistothorus palustris* aquatic species. Most ducks and geese are too large as adults to be attacked, but small offspring are sometimes admitted to rehabilitation centers, presumably having been attacked while on land. Because we did not know the mass of these immature individuals when admitted to rehabilitation centers, we excluded them from analyses.

2.4.5. Terrestrial or Not

Cats forage mostly on or near the ground, so we categorized each species as foraging mostly on or within 1 m of the ground versus foraging mostly well above the ground. Some individuals of nearly all species may occasionally be found on the ground, particularly fledglings.

2.4.6. Feeder Use

We also categorized species into one of two groups: those species expected to occasionally or commonly visit bird feeders versus rarely or never visiting bird feeders, based on our own experience.

2.5. *Indices of Bird Abundance*

We used eBird data [29] to create an index of avian abundance in the counties where birds were admitted to each rehabilitation center. Species were rank-ordered from most to least common based on the proportion of complete eBird checklists on which each species was included. Checklists were contributed by birders from 2011 through 12 November 2020 in Marion (N = 39,255 checklists; Turtle Ridge) and Josephine (N = 11,644; Wildlife Images) counties. Because of the large sample size, inclusion of multiple years, and alignment with our own experiences surveying birds in western Oregon [30], we assumed the ranks were positively correlated with abundance and we used the ranks as indices of relative abundance. Our analyses of habitat cover around recovery locations indicated birds were rarely delivered from neighboring counties. Even so, we inspected species ranked orders on eBird checklists in the counties adjacent to each center's home county and found them to be all highly correlated ($r > 0.9$). To evaluate influence of potential errors in the relationships between abundance and ranked order of species occurrences on checklists, we analyzed data with both the raw ranked information from eBird and with ranks categorized into ten intervals. All results associating variables with raw rank-ordered data and categorical ranks were qualitatively similar.

We quantified differences in ranks of species in eBird lists versus in the species involved in cat-influenced admissions to the two centers. Our objective was to identify species that were under- or over-represented in the center admissions and to discover potential correlates between discrepancies in the ranks and traits, such as mass (Figure 1).

2.6. *Data Analyses*

We used non-parametric statistics because of uncertainty in the distributional shapes of variables, driven by occasional lack of clarity in the reliability of reported values in the rehabilitation center data and to be cautious in our use of eBird data when generating indices of relative abundance. For example, current protocols at the rehabilitation centers provide no independent checks of most information, such as species identification, confirmation of locations where birds were obtained prior to transport to rehabilitation centers, or identification of admission causes. Although we screened the data for obvious or potential errors, without independent verification of the accuracy of such data, we elected to compare ranks to discover patterns in most of the data. When comparing proportions of species obtained in urban versus rural environments, we used Chi-square tests when sample sizes permitted, and Fisher's Exact tests when sample sizes were fewer than five individuals in each habitat type. When assessing associations with habitat types, we determined the expected values by randomly choosing locations (N = 55) from the dataset after excluding admissions noted to involve or potentially involve interaction with a cat. We assumed that if cat-interacted birds were brought to centers in proportion to admissions from all other causes, then the proportion of urban versus rural sites in the entire dataset would be the appropriate expected values. In Salem, those values were 63% urban and 37% rural, and they were similar in Grants Pass, at 68% urban and 32% rural. We assumed that each admission was an independent event. For comparisons of species ranks, we used Wilcoxon tests. All analyses were performed using JMP [31].

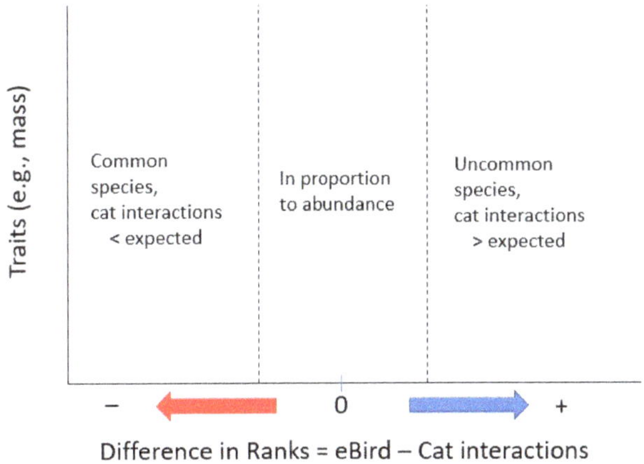

Figure 1. Schematic for evaluating differences among species admitted to the rehabilitation centers in their ranked abundances from the eBird data and the cat interaction data. When species ranked approximately the same in each list, the differences in ranks are near zero and those species are hypothesized to be admitted after cat interactions in approximately the same proportion to their abundance in the study areas. Species uncommonly reported in eBird (i.e., lower position in a ranked list, so higher numerical value for ranking) than in the rehabilitation center data are hypothesized to be species admitted to centers in higher numbers than expected based on their abundance alone. Species commonly included in eBird lists but much less often admitted to rehabilitation centers are hypothesized to be under-represented in cat interaction data relative to their abundance in the study areas.

3. Results

Of the 6345 animals admitted to the two rehabilitation centers, birds comprised one-quarter to one-third of admissions once use of the WILD-ONe database became consistent in 2016 (Table 1). A slightly higher proportion of birds admitted came from Turtle Ridge in Salem (51.9%; N = 3293) than from Wildlife Images in Grants Pass (48.1%, N = 3052).

The top ten most frequently admitted bird species comprised 53.0% and 42.7% of the total bird admissions at Turtle Ridge and Wildlife Images, respectively. Seven species (Mallard *Anas platyrynchos*, European Starling *Sturnus vulgaris*, American Crow *Corvus brachyrhynchos*, California Scrub-Jay, American Robin *Turdus migratorius*, Eurasian Collared-Dove *Streptopelia decaocto*, Mourning Dove *Zenaida macroura*, and Red-tailed Hawk *Buteo jamaicensis*) were in the top ten at both centers (Table 2). Turtle Ridge, in a larger city, received more Rock Pigeons *Columba livia* and Vaux's Swifts *Chaetura vauxi*, while Wildlife Images, in a smaller city, received more Western Screech-Owls *Megascops kennicottii* and Canada Geese *Branta canadensis*. Several of the bird species admitted were large-bodied, such as Red-tailed Hawk, frequently admitted after being struck by vehicles, and waterfowl. Although information was inconsistently recorded in the database, many waterfowl admitted were juveniles.

Table 1. Total annual admissions to the rehabilitation centers (N) and the percentages of those admissions that were birds.

Year	Percentage (N) Total Admissions		
	Turtle Ridge	Wildlife Images	Combined
2014	0.03 (1)	0.3 (10)	0.2 (11)
2015	2.3 (76)	26.3 (802)	13.8 (878)
2016	28.0 (922)	21.0 (642)	24.6 (1564)
2017	32.6 (1073)	24.4 (744)	28.6 (1817)
2018	37.1 (1221)	28.0 (854)	32.7 (2075)
Total	100 (3293)	100 (3052)	100 (6345)

Table 2. Ten most frequently admitted avian species at each wildlife center, regardless of presumed cause; Turtle Ridge, Salem, and Wildlife Images, Grants Pass, Oregon.

Turtle Ridge		Wildlife Images	
Species	Percentage (N) Total Admissions	Species	Percentage (N) Total Admissions
Mallard	11.0 (361)	European Starling	5.3 (162)
European Starling	8.5 (279)	Mourning Dove	4.8 (146)
American Crow	6.5 (214)	Western Screech-Owl	4.8 (146)
California Scrub-Jay	5.1 (169)	California Scrub-Jay	4.7 (144)
American Robin	4.9 (162)	American Robin	4.4 (135)
Rock Pigeon	4.3 (143)	Canada Goose	4.2 (128)
Eurasian Collared-Dove	3.6 (117)	American Crow	3.8 (116)
Mourning Dove	3.3 (109)	Red-tailed Hawk	3.8 (116)
Red-tailed Hawk	3.1 (103)	Mallard	3.7 (113)
Vaux's Swift	2.9 (97)	Eurasian Collared-Dove	3.6 (110)

Birds were admitted for a wide variety of reasons, approximately one-quarter of which were undetermined (Table 3). Cat interactions accounted for the second-highest fraction of reports (12.3%), behind 'orphaned' and 'behavioral stranding', which we interpret as being synonymous (Table 3). Nearly one-quarter of admissions had unreported causes.

Table 3. Percentages and numbers of total admissions for the ten most frequently reported causes of admission from both wildlife rehabilitation centers, as well as frequency of undetermined cause of admission.

Cause	Percentage (N) of Admissions		
	Turtle Ridge	Wildlife Images	Both Centers
Orphaned	34.0 (1120)	8.3 (252)	21.6 (1372)
Interaction with a cat	12.4 (409)	12.2 (373)	12.3 (782)
Collision with a car, truck, or motorcycle	7.0 (228)	10.9 (333)	8.8 (561)
Nest/habitat disturbance or destruction	6.7 (222)	7.0 (213)	6.9 (435)
Behavioral stranding	0.4 (12)	13.0 (396)	6.4 (408)
Abduction with intent of rescue	0.0 (0)	10.2 (311)	4.9 (311)
Collision with a wall or window	4.2 (141)	5.5 (169)	4.9 (310)
Interaction with a dog	1.7 (56)	2.7 (81)	2.2 (137)
Failure to thrive/maladaptation	0.4 (13)	3.3 (102)	1.8 (115)
Interaction with a non-domestic animal of another species	1.8 (60)	1.7 (52)	1.8 (112)
Undetermined	27.7 (911)	18.7 (570)	23.3 (1481)

Turtle Ridge admitted 55 bird species after cat interactions, with 36 species (64%) having fewer than ten individuals admitted during the study period. Wildlife Images admitted 61 species; 30 (49%) had fewer than ten individuals admitted (Table S1).

Birds admitted to a wildlife rehabilitation center after interacting with a cat were generally representative of birds admitted for any cause, although species with a larger adult body mass (> 200 g) that were in the top ten most admitted species for any cause were not found on the top ten species admitted to each center after interaction with a cat (Table 4). The exception to this was the American Crow, which would be the tenth most frequently admitted species at Turtle Ridge due to interaction with a cat.

Table 4. Top ten species at each center admitted after interacting with a cat. One species (American Crow) with an adult body mass > 200 g was excluded. Percentages are calculated from all cat-related admissions at each center.

Turtle Ridge		**Wildlife Images**	
Species	**Percentage (N) Admissions**	**Species**	**Percentage (N) Admissions**
California Scrub-Jay	13.0 (53)	American Robin	11.8 (44)
American Robin	11.2 (46)	California Scrub-Jay	11.0 (41)
European Starling	6.8 (28)	Mourning Dove	8.0 (30)
Mourning Dove	6.1 (25)	Eurasian Collared-Dove	6.2 (23)
Eurasian Collared-Dove	5.6 (23)	Spotted Towhee	3.2 (12)
Spotted Towhee	4.9 (20)	European Starling	3.2 (12)
Dark-eyed Junco	3.9 (16)	Northern Flicker	2.9 (11)
House Sparrow	3.7 (15)	Black-headed Grosbeak	2.9 (11)
Northern Flicker	3.2 (13)	Acorn Woodpecker	2.7 (10)
Varied Thrush	2.7 (11)	House Sparrow	2.7 (10)

3.1. Rescue Location

The proportions of urban versus rural rescue locations for all admission records were similar between Turtle Ridge and Wildlife Images, with approximately two-thirds of admissions being reported as originating from urban locations (Table 5). Birds admitted at Turtle Ridge due to interaction with a cat were more likely than the overall average to come from urban areas than from rural areas, while at Wildlife Images, these birds were nearly equally likely to come from a rural versus urban environment. However, especially at Wildlife Images, a sufficiently high fraction (16.9%) of admissions were from unknown locations to obscure potential differences in origination habitat.

Table 5. Percentage of bird admissions coming from urban or rural locations. Records involving interaction with a cat do not add up to 100%, as some records from this category had incomplete location information. Records from other causes of admission were not included in analysis of location information if this information was incomplete.

	Cat Interaction			**Other Admission Causes**		
Rescue Location	**Turtle Ridge**	**Wildlife Images**	**Combined**	**Turtle Ridge**	**Wildlife Images**	**Combined**
Percent (N) Urban	70.9 (290)	42.6 (159)	57.4 (449)	63.0 (17)	67.9 (19)	65.5 (36)
Percent (N) Rural	22.5 (92)	40.5 (151)	31.1 (243)	37.0 (10)	32.1 (9)	34.5 (19)
Percent (N) Unknown	6.6 (27)	16.9 (63)	11.5 (90)	-	-	-
Total	100 (409)	100 (373)	100 (782)	100.0 (27)	100.0 (28)	100.0 (55)

When data from both centers were combined, cat-influenced admissions appeared to be more common from urban locations (57%) than rural locations (31%) but 11% of reports did not include address information.

Proportions of urban versus rural rescue locations differed significantly from expected proportions for only a few species at each rehabilitation center (Table S3). At Turtle Ridge, California Scrub-Jays were more likely to have interactions with cats in urban areas ($n = 53$, $\chi^2 = 10.91$, $p < 0.001$), as were European Starlings ($n = 27$, $\chi^2 = 10.14$, $p = 0.0014$). American Crows tended to be more likely to come from urban areas ($N = 12$, Fisher's Exact $p = 0.093$). At Wildlife Images, Spotted Towhees *Pipilo maculatus* were more likely to interact with a cat in rural areas ($N = 9$, Fisher's Exact $p = 0.009$).

3.2. *Species Traits*

3.2.1. Indices of Abundance

The rank orderings of species in the bird communities based on eBird data in Marion (Turtle Ridge) and Josephine (Wildlife Images) Counties were significantly correlated ($R^2 = 0.424$, $p < 0.0001$, Table S1). The rank orderings of species by frequency of cat-interacted admissions to wildlife rehabilitation centers were positively correlated to the eBird indices of abundances in both Marion ($R^2 = 0.303$, $p = 0.0406$) and Josephine Counties ($R^2 = 0.272$, $p < 0.0001$). Thus, species interacting with cats were admitted to rehabilitation centers directly in proportion to their indices of abundance in the counties served by each center. An important exception was the absence of several common species with small (<70 g) body mass.

3.2.2. Body Mass

We found no difference in masses among species that were admitted after interaction with a cat ($N = 61$) and species that were not noted to have been involved in interactions with a cat ($N = 56$, $t = 0.77$, $df = 55$, $p < 0.22$). The relationships between species' mass and their rank-ordered abundances were best fit by quadratic functions (Turtle Ridge: $R^2 = 0.265$, $p < 0.0016$; Wildlife Images: $R^2 = 0.297$, $p < 0.0006$). At both centers, species with a higher mean adult mass were disproportionately represented among the top five species admitted after interaction with a cat (Table 4). At Turtle Ridge mass of the top five species admitted ranged from 70 to 140 g and at Wildlife Images it ranged from 70 to 150 g. The list of top ten species under 200 g admitted after interaction with a cat was also highly correlated between the centers ($R^2 = 0.279$, $p = 0.0027$). Common species below 70 g were under-represented from lists of cat-related admissions at both centers.

While there was not a significant relationship between body mass and our index of eBird abundance (that is, small species are common just as often as they are uncommon) in either the Salem or Grants Pass datasets, smaller birds (less than 70 g) were less likely to be brought into wildlife rehabilitation centers for any reason (Turtle Ridge $R^2 = 0.104$, $p = 0.0073$, Wildlife Images $R^2 = 0.191$, $p = 0.0373$, Figure 2).

3.2.3. Feeder Use

Species regularly using feeders had a higher-ranking index of abundance in both Marion ($z = -4.81$, $p < 0.0001$) and Josephine ($z = -4.65$, $p < 0.0001$) counties and species regularly using feeders were delivered to rehabilitation centers in proportions expected from their indices of regional abundance (Figure 3). At Wildlife Images in Josephine County, there was a significant relationship between ranked index of abundance of species using feeders and admission due to interaction with a cat ($z = 3.35$, $p = 0.0008$). However, the relationship was much weaker for Turtle Ridge in Marion County ($z = 1.34$, $p = 0.179$).

3.2.4. Residency

Resident species outnumbered migrants at both rehabilitation centers. We found no significant differences from expected proportions of residents and migrants admitted to the rehabilitation centers, either as a whole or limited to species noted to have interacted with cats.

3.2.5. Terrestrial Species

Birds that forage on or near the ground were more likely to rank high on the list of species reported to have interacted with a cat in the Salem area ($z = -3.42$, $p = 0.0006$) but not in Grants Pass ($z = 1.55$, $p = 0.12$).

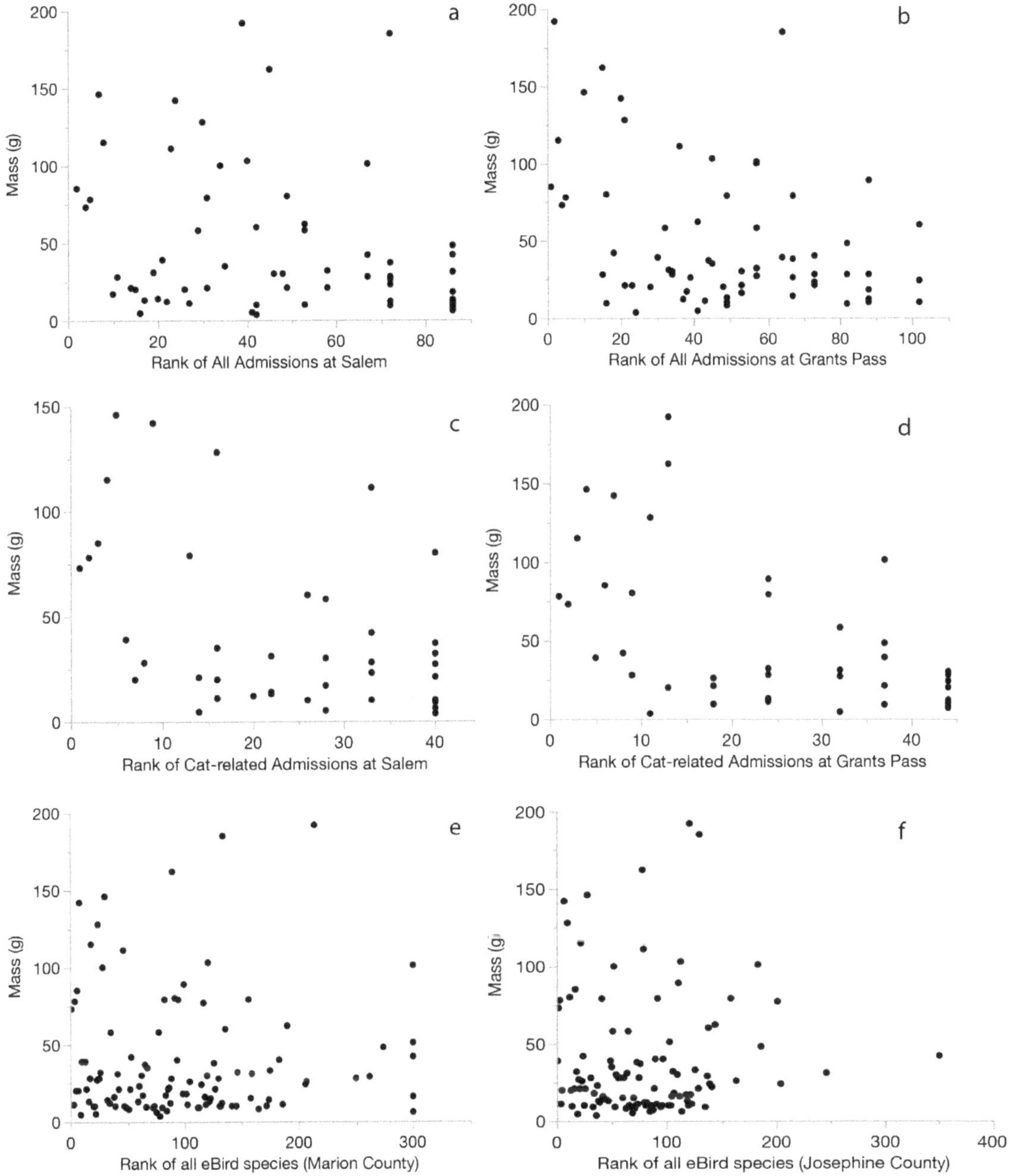

Figure 2. Species with a mean adult mass of less than 70 g were much less likely than larger species to be admitted for any reason to a wildlife rehabilitation center in Salem (**a**) and Grant Pass (**b**). This was also true for birds admitted after interacting with a cat in Salem (**c**) and Grants Pass (**d**). Local abundances, based on eBird checklists, did not show the same pattern in either Marion County (**e**) or Josephine County (**f**).

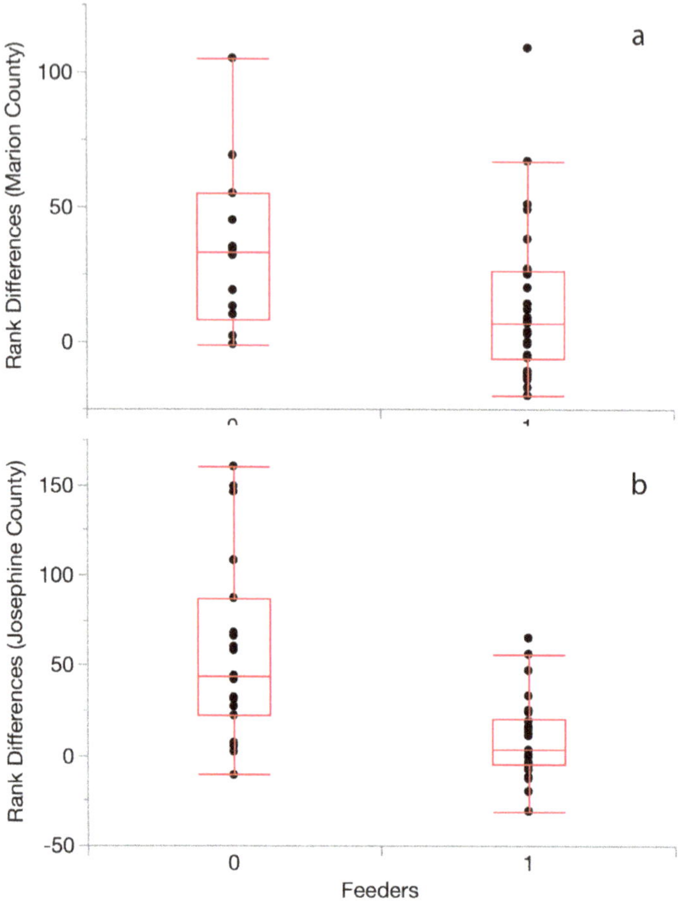

Figure 3. Rank differences between species using feeders (1) and species that do not use feeders (0) as related to their indices of abundance in Marion (**a**) and Josephine (**b**) Counties.

4. Discussion

Of reported causes of bird admissions to wildlife rehabilitation centers, cat-related interactions were second behind orphaned/behavioral stranding and accounted for 12.3% of all admissions. Admissions of bird species generally reflected local species abundances. We found strong positive correlations between rank orderings of species on eBird checklists and in lists of species admitted because of cat interactions. Residency versus migratory status did not influence likelihood of cat interactions. While species admissions were correlated with our indices of local abundance, species with an adult body mass of 70 g or less were distinctly under-represented in rehabilitation center data. The absence of such species was not limited to presumed cat interactions, meaning their absences were indicative of a general source of bias where small, injured birds were infrequently detected, rescued or delivered to rehabilitation centers relative to their regional abundance. Perhaps unsurprisingly, species more likely to have been reported as injured by a cat tended to be terrestrial species, but associations with use of bird feeders was equivocal with a significant positive association at one study site but not the other. Both total admissions and cat-related admissions came more from urban than rural areas. Overall, the sample included more terrestrial and near-terrestrial species than species foraging higher above ground, species larger than 70 g but less than the 200 g typical upper range of size cats regularly attack [27], and species with higher indices of relative abundance.

The proportion of cat-related causes of admission at Turtle Ridge (12.4%) and Wildlife Images (12.2%) aligned closely to most previous reports from centers in the United States, including 5% in Maine [17], 8% in Ohio [21], 9.8% in Wisconsin [19], 13.7% in Virginia [22] and 25.4% in Florida [18], but was lower than a Tennessee study [15] reporting 48.3% admissions owing to cat interactions. Cat interaction was the second-most frequently identified admission reason at each center, following "orphaned" (34%) at Turtle Ridge and "behavioral stranding" (13%) at Wildlife Images. Wildlife Images had much higher rates for "behavioral stranding" and "abduction with intent to rescue", which likely overlapped with Turtle Ridge's use of "orphaned," reasonably placing cat interaction as the second most common form of reported reason at both centers. Despite possible differences in use of admission categories at the two centers, we conclude that cat interactions were a common reason for admission at both centers. Some uncertainty is involved, however, for two reasons. First, rehabilitation center data did not provide independently verifiable information of cat interactions. They relied on reports of the people who rescued each animal. Second, the rate at which birds interacting with cats were delivered to rehabilitation centers also remains obscured because approximately 23% of admissions did not include a reason for admission in the WILD-ONe database. The authors in [21] found similar results with no cause of admission being reported in 20% of cases. Thus, our average of 12.3% may be a low estimate if we conclude the importance of undetermined causes of admission outnumber incorrect reports of cat-related interactions. At a minimum, hundreds of birds per year were brought to the two rehabilitation centers after rescuers determined involvement of a cat was likely.

Despite uncertainty in the rates of cat-related admissions, it is clear that the most common bird species, especially those foraging on or near the ground, were admitted most often. The two centers sampled from largely similar bird communities, sharing eight of the top ten most frequently admitted species. A few regional and habitat-related differences in species composition and abundances were apparent. At both sites, bird species tended to be admitted at rates reflecting their local abundances. That is, when rank-ordered from most to least common, common species were also the most commonly admitted after cat interactions. Common species also comprised the majority of birds admitted due to interaction with a cat at the Wildlife Center of Virginia [22] and throughout North America [21].

While admission rates appeared to reflect local abundance, species with an adult body mass of under 70 g were under-represented in rehabilitation center admissions for any cause, including interaction with a cat. These common species included Anna's Hummingbirds *Calypte anna*, Black-capped Chickadees *Poecile atricapillus*, Dark-Eyed Juncos *Junco hyemalis*, and Spotted Towhees in both Salem and Grants Pass, as well as Song Sparrows *Melospiza melodia* in Salem. We hypothesize that small-bodied birds may be under-represented because they are less likely to survive traumatic events, to be found if they do survive injuries, or to be transported successfully to rehabilitation centers. An additional explanation could involve identification challenges. Rehabilitation center volunteers might confuse species, such as sparrows, with other similar species and dilute the rate at which individual species are reported relative to sparrows or some other similarly appearing species as a whole. We do not think the low rates of admission for species under 70 g can be explained by higher rates of cat-caused mortality instead of injury because this class of small birds was absent from all admissions regardless of indicated cause of admission.

Birds were more likely to be rescued from urban areas than from rural areas, even as evaluated with our coarse categorization of habitat type. The trend for more admissions from urban areas was seen in all admissions, although it was much less pronounced for admissions due to interaction with a cat at Wildlife Images, where birds were nearly equally likely to come from an urban or rural area. This contrasts with admissions at Turtle

Ridge, where birds admitted for cat-related reasons were more likely to have come from urban areas than for all other admission reasons. Wildlife Images is located near a much smaller city than Turtle Ridge, which may affect where birds are found and rescued. Only a few species were more strongly associated with urban or rural rescue locations when cause of admission was interaction with a cat. In Salem, these species included California Scrub-Jays, European Starlings and, to a lesser extent, American Crows. These three species were found to be associated with urban areas. In Grants Pass, Spotted Towhees were more likely to have come from rural environments. Overall, significant differences were uncommon because samples sizes of most individual species were small. Our study region included two cities of moderate size, neither of which abutted protected natural areas. A South African study of cat–bird interactions in an urbanized landscape using video footage from cat-borne cameras found cats caught few non-native species in urbanized areas and suggested cats whose home ranges adjoin natural areas near urbanized areas could post greater risks for native species [32]. Thus, landscape context can influence the species most at risk of predation from free-ranging cats.

Our interpretation of the results requires consideration of assumptions associated with use of rehabilitation center data. We consider several caveats and offer recommendations for improving the scientific value of data collected at wildlife rehabilitation centers and archived in the WILD-ONe database. Several points of uncertainty in wildlife rehabilitation data stem from admission procedures and circumstances of rescue. Patient intake procedures vary between rehabilitation centers, and the quality of information collected at admission can vary at each center. Rehabilitation centers are often run by a combination of staff and volunteers and may experience high volunteer turnover leading to a low level of experience by those completing admissions. Information collected at the time of admission may also be incomplete, due to incorrect use of admission forms or because the person bringing in the animal may have incorrect or incomplete information about the circumstances of rescue. An undetermined cause of admission, for example, was noted in 23% of cases at the western Oregon centers. When thousands of patients are admitted each year, this represents a substantial number of cases for which an admission cause was not determined. Although we screened data for records indicating interactions with cats, the empirical evidence for such interactions is sparse, and not well documented on admission forms either at time of admission or after inspection by rehabilitation personnel. The rate at which cat interactions occurred could be under-estimated because of the large proportion of admissions owing to undetermined causes or even over-estimated if rescuers incorrectly attribute animal injuries to cats.

Some centers may prioritize recording of certain causes of admission over others. While it is possible to enter multiple causes of admission just as it is possible to list multiple injury details, employees or volunteers may need to prioritize the most apparent cause, or the cause that will require the most aggressive treatment. When birds are potentially injured by cats, for example, it is common practice to treat them immediately with antibiotics [33], a step not normally taken if injuries may have resulted from a collision. The accuracy of diagnoses, therefore, certainly influences the interpretation of proportion of admissions to centers as a function of presumed causes. Diagnoses vary between centers as well. For example, from the different distributions of admission causes at Turtle Ridge and Wildlife Images it appears that the rescue of fledglings was coded quite differently, orphaned at one center and as behavioral stranding or abduction with intent of rescue at the other.

Beyond uncertainty in cause of admission, other data collected from wildlife rehabilitation centers may be incorrect or incomplete. We found several instances of probable misidentification of species, as well as suspected misidentifications. When identifications are clearly wrong, they are normally easy to correct (for example, correcting a presumed erroneous identification or data entry mistake from Eastern to Western Bluebird). In other cases, the identifications of similarly appearing species do cause uncertainty in data. To reduce these effects, we combined similar species into species groups (e.g., *Sphyrapicus* sapsuckers and *Spinus* goldfinches), but other potential misidentifications are difficult to detect. A step that could reduce such uncertainty could be to require pictures of each patient be uploaded into the WILD-ONe database and available for independent verification. We recognize that this step may not always be convenient because of the degree of injuries, need for expediency of treatment, or other sensitive situations. Given the scientific utility of rehabilitation center data, however, getting the species identifications correct should be a top priority. It is probably unreasonable to expect rehabilitation center staffs, characterized by high turnover of volunteers, to be skilled at species levels identifications of all potential patients, particularly when many of those patients may be fledglings or juveniles in unfamiliar plumages. The difficulty of proper identification of species may also justify more flexibility on the part of data collection services. While WILD-ONe currently provides an option to categorize species as an "undetermined bird," more potential levels of identification, such as genera or species groups, may benefit both rehabilitation centers and researchers. Using the same species taxonomy as iNaturalist, for example, and submitting to iNaturalist pictures of admissions for which species identity is uncertain could provide an opportunity to connect with the artificial intelligence programs and supporting assembly of taxonomic experts that identify species for iNaturalist.

Overall, wildlife rehabilitation centers provide an important opportunity to gather scientific data of relevance to conservation biology, but do need improvements that could increase the scientific rigor of data collected [19,22]. Future efforts to address uncertainties associated with rehabilitation center data should target both the centers and the database management service collecting the data (Table 6). Wildlife rehabilitation centers vary widely in size and funding, which will cause variation in the ability of individual rehabilitators or centers to meet these recommendations, some of which require significant financial or professional resources. Training of center personnel to transfer more complete accounting of potential explanations for injuries onto admission forms could reduce the fraction of records with undetermined causes. Although widespread training may be impractical given the characteristics of most volunteer-based rehabilitation centers, training focused on centers most interested in collaborating with scientific research groups might be productively implemented. A network of well-funded centers with lower turnover of human resources might be identified and adjustments in intake and data collection procedures could be implemented to maximize the scientific reliability of information gathered from admitted patients.

Table 6. Recommendations for wildlife rehabilitators and database designers to improve accuracy and scientific utility of data collected by wildlife rehabilitation centers.

Wildlife Rehabilitators or Centers	Database Designers and Operators
Consistent training on admission procedures Designing admissions paperwork to reflect database requirements Access to individuals trained in species identification for all species groups admitted by the rehabilitator Consistent procedures with paperwork, including disposition information	Greater flexibility with species information Allowing less specific information pertaining to injury/illness Readily available definitions of all terms used

Determination of which bird species are more likely to have interactions with a cat matters because these birds often die, even after being admitted to a wildlife rehabilitation center. The authors of [20] found a 78% mortality rate for birds brought to a wildlife rehabilitation center after having been attacked by a cat. This number did not include birds that died immediately during the interaction, but included birds that died during transport, or died or were euthanized at the rehabilitation center. The authors of [15] found that 71.3% of birds admitted to a wildlife hospital in Tennessee for a cat-related reason either died or were euthanized. The authors of [21] noted that 68% of birds admitted to wildlife rehabilitation centers due to interaction with a cat died or were euthanized, and only 24% were released. These high rates indicate that even birds injured and escaping (or being rescued from) cats are likely to die, suggesting that estimates of cat-caused bird deaths extrapolated from data on rates at which cats deliver prey back to their homes are low. At the least, our data corroborate the concern that estimating the impacts of cats on bird populations is a complex problem.

5. Conclusions

Despite our concerns of possible biases in the rehabilitation center data, several conclusions from our analyses should be robust to such issues. First, cat-related interactions are an important source of injured birds being delivered to rehabilitation centers. The general proportions of such causes at the centers we studied align closely with such proportions at other centers. Second, hundreds of birds per year per center are injured by cats and rescued by the public. Third, cat interactions occurred most often with the commonest species in each region, being largely in direct proportion to each species' prevalence on checklists in the eBird database. Fourth, species foraging on or near the ground were most often admitted to the rehabilitation centers. Finally, small (<70 g) common birds were under-represented, probably resulting from sampling biases associated with discovering the birds when they were injured.

Wildlife rehabilitation centers provide important services to the public and to wildlife generally [17,19]. The data they collect, when shared through structured databases such as WILD-ONe, can provide useful information on sources of mortality, rehabilitation success, and locations of high-risk areas for wildlife [22]. Additional steps to improve the ability to verify data taken at time of admission will increase the value of the data even more.

Supplementary Materials: The following are available online at https://www.mdpi.com/article/10.3390/d13070322/s1, Table S1: Percentages and total numbers (N) of admissions for each species; Table S2: Names and characteristics of bird species included in this study; Table S3: Table of all species with at least one cat interaction record.

Author Contributions: Conceptualization, K.G.D. and W.D.R.; methodology, K.G.D. and W.D.R.; validation, K.G.D. and W.D.R.; formal analysis, K.G.D. and W.D.R.; writing—original draft preparation, K.G.D.; writing—review and editing, K.G.D. and W.D.R. All authors have read and agreed to the published version of the manuscript.

Funding: This research received no external funding.

Institutional Review Board Statement: Not applicable.

Data Availability Statement: The data in this study are available in Tables and Supplementary Tables.

Acknowledgments: We thank WILD-ONe and the Wildlife Center of Virginia for providing these data, as well as Turtle Ridge Wildlife Center, and Wildlife Images Rehabilitation and Education Center. B. Dugger and S. Dunham provided feedback on an earlier draft. Tara Kate Designs assisted with figures.

Conflicts of Interest: The authors declare no conflict of interest.

Appendix A

Appendix A.1. Defining Interaction with a Cat

We necessarily used a broad definition of interaction with a cat because admission data report observations made by rescuers and are not independently verifiable. The interactions range from the abduction or rescue of birds due to concern of a potential interaction with a cat (e.g., fledglings being rescued to protect them from cats) to direct interactions where the bird was rescued from the mouth of a cat. Our choice to use a broad definition results from the difficulty in confirming the level of interaction that has occurred when detailed information is not taken at the time of admission.

Center admission forms contain basic information (date, time of intake, contact name and address of the rescuer, address where the animal was rescued) requested from the rescuer. The forms from our centers also requested information on possible cause of injury, listing common causes that might be quickly noted by the intake personnel (e.g., car hit, window hit, orphaned, contact with a cat). The forms also include space to list intake procedures undertaken, typically without any prompts to direct the personnel toward particular information, and the disposition of the animal after its admission.

Appendix A.2. Handling of Potential Species Identification Issues

Staff and volunteers at wildlife rehabilitation centers are often not experts in species identification. In many cases, animals are not adults (e.g., fledgling birds), making identification even more difficult.

In two cases, we combined species into a single category.

Sapsuckers (*Sphyrapicus* spp.) include three species that hybridize and pose identification challenges even for experienced observers. Thus, we combined all sapsuckers into one group.

American Goldfinches (*Spinus tristis*) and Lesser Goldfinches (*Spinus psaltria*) were combined into a single category, Goldfinches. The two species of goldfinch were combined because of complexities of correctly identifying them, particularly in the cases of juveniles, females, and non-breeding males.

References

1. Rosenberg, K.V.; Dokter, A.M.; Blancher, P.J.; Sauer, J.R.; Smith, A.C.; Smith, P.A.; Stanton, J.C.; Panjabi, A.; Helft, L.; Parr, M.; et al. Decline of the North American Avifauna. *Science* **2019**, *366*, 120–124. [CrossRef]
2. Inger, R.; Gregory, R.; Duffy, J.P.; Stott, I.; Voříšek, P.; Gaston, K.J. Common European Birds Are Declining Rapidly While Less Abundant Species' Numbers Are Rising. *Ecol. Lett.* **2015**, *18*, 28–36. [CrossRef]
3. Loss, S.R.; Will, T.; Marra, P.P. Direct Mortality of Birds from Anthropogenic Causes. *Annu. Rev. Ecol. Evol. Syst.* **2015**, *46*, 99–120. [CrossRef]
4. Loss, S.R.; Will, T.; Marra, P.P. The Impact of Free-Ranging Domestic Cats on Wildlife of the United States. *Nature Communications* **2013**, *4*, 1396. [CrossRef]
5. Loss, S.R.; Marra, P.P. Bird-Building Collisions in the United States: Estimates of Annual Mortality and Species Vulnerability. *Condor* **2014**, *16*, 8–23. [CrossRef]
6. Loss, S.R.; Marra, P.P. Estimation of Annual Bird Mortality from Vehicle Collisions on U.S. Roads. *J. Wildl. Manag.* **2014**, *78*, 763–771. [CrossRef]
7. Woinarski, J.C.Z.; Murphy, B.P.; Legge, S.M.; Garnett, S.T.; Lawes, M.J.; Comer, S.; Dickman, C.R.; Doherty, T.S.; Edwards, G.; Nankivell, A.; et al. How Many Birds Are Killed by Cats in Australia? *Biol. Conserv.* **2017**, *214*, 76–87. [CrossRef]
8. Lepczyk, C.; Mertig, A.G.; Liu, J. Landowners and Cat Predation across Rural-to-Urban Landscapes. *Biol. Conserv.* **2003**, *115*, 191–201. [CrossRef]
9. Dauphine, N.I.C.O.; Cooper, R.J. Impacts of free-ranging domestic cats (Felis catus) on birds in the United States: A review of recent research with conservation and management recommendations. In *Proceedings of the Fourth International Partners in Flight Conference: Tundra to Tropics*; Partners in Flight: McAllen, Texas, USA, 2008; pp. 205–219.
10. Baker, P.J.; Bentley, A.J.; Ansell, R.J.; Harris, S. Impact of Predation by Domestic Cats Felis Catus in an Urban Area. *Mammal Review* **2005**, *35*, 302–312. [CrossRef]
11. Clancy, E.A.; Moore, A.S.; Bertone, E.R. Evaluation of Cat and Owner Characteristics and Their Relationships to Outdoor Access of Owned Cats. *J. Am. Vet. Med Assoc.* **2003**, *222*, 1541–1545. [CrossRef]
12. Lepczyk, C.A.; Duffy, D.C. Feral cats. In *Ecol. Manag. Terr. Invasive Species*; CRC Press: Boca Raton, FL, USA, 2018; pp. 267–285.
13. Crowley, S.L.; Cecchetti, M.; McDonald, R.A. Diverse Perspectives of Cat Owners Indicate Barriers to and Opportunities for Managing Cat Predation of Wildlife. *Front. Ecol. Environ.* **2020**, *18*, 544–549. [CrossRef]
14. Loyd, K.A.T.; Hernandez, S.M.; Carroll, J.P.; Abernathy, K.J.; Marshall, G.J. Quantifying Free-Roaming Domestic Cat Predation Using Animal-Borne Video Cameras. *Biol. Conserv.* **2013**, *160*, 183–189. [CrossRef]
15. Schenk, A.N.; Souza, M.J. Major Anthropogenic Causes for and Outcomes of Wild Animal Presentation to a Wildlife Clinic in East Tennessee, USA, 2000-2011. *PLoS ONE* **2014**, *9*, e93517. [CrossRef] [PubMed]
16. Molina-Lopez, R.A.; Mañosa, S.; Torres-Riera, A.; Pomarol, M.; Darwich, L. Morbidity, Outcomes and Cost-Benefit Analysis of Wildlife Rehabilitation in Catalonia (Spain). *PLoS ONE* **2017**, *12*, e0181331. [CrossRef] [PubMed]
17. Duffy, M.M. *Wildlife Rehabilitation Datasets as an Underutilized Resource to Understand Avian Threats, Mortality, and Mitigation Opportunities*; University of Maine: Orono, ME, USA, 2020.
18. Kratter, A.W.; Steadman, D.W. Mortality in Birds from Florida Wildlife Rehabilitation Clinics. *Fla. Field Nat.* **2020**, *48*, 147–166.
19. Long, R.B.; Krumlauf, K.; Young, A.M. Characterizing Trends in Human-Wildlife Conflicts in the American Midwest Using Wildlife Rehabilitation Records. *PLoS ONE* **2020**, *15*, e0238805. [CrossRef] [PubMed]

20. Baker, P.; Thompson, R.; Grogan, A. Survival Rates of Cat-Attacked Birds Admitted to RSPCA Wildlife Centres in the UK: Implications for Cat Owners and Wildlife Rehabilitators. *Anim. Welf.* **2018**, *27*, 305–318. [CrossRef]
21. Loyd, K.A.T.; Hernandez, S.M.; McRuer, D.L. The Role of Domestic Cats in the Admission of Injured Wildlife at Rehabilitation and Rescue Centers. *Wildl. Soc. Bull.* **2017**, *41*, 55–61. [CrossRef]
22. McRuer, D.L.; Gray, L.C.; Horne, L.-A.; Clark Jr., E. E. Free-Roaming Cat Interactions with Wildlife Admitted to a Wildlife Hospital. *J. Wildl. Manag.* **2017**, *81*, 163–173. [CrossRef]
23. Stratford, J.A.; Robinson, W.D. Distribution of Neotropical Migratory Bird Species across an Urbanizing Landscape. *Urban Ecosyst* **2005**, *8*, 59–77. [CrossRef]
24. U. S. Census Bureau. *ACS Total Popul. Salem City, Oregon. Table B01003*; United States Government Printing Office: Washington, DC, USA, 2019.
25. U. S. Census Bureau. *ACS Total Popul. Grants Pass City, Oregon. Table B01003*; United States Government Printing Office: Washington, DC, USA, 2018.
26. Dunning, J.B. *CRC Handbook of Avian Body Masses*; CRC Press: Boca Raton, FL, USA, 2008.
27. Dickman, C.R. *Overview of the Impacts of Feral Cats on Australian Native Fauna*; Australian Nature Conservation Agency: Canberra, Australia, 1996.
28. Fleming, P.A.; Crawford, H.M.; Auckland, C.H.; Calver, M.C. Body Size and Bite Force of Stray and Feral Cats—Are Bigger or Older Cats Taking the Largest or More Difficult-to-Handle Prey? *Animals* **2020**, *10*, 707. [CrossRef]
29. Sullivan, B.L.; Wood, C.L.; Iliff, M.J.; Bonney, R.E.; Fink, D.; Kelling, S. EBird: A Citizen-Based Bird Observation Network in the Biological Sciences. *Biol. Conserv.* **2009**, *142*, 2282–2292. [CrossRef]
30. Robinson, W.D.; Hallman, T.A.; Curtis, J.R. Benchmarking the Avian Diversity of Oregon in an Era of Rapid Change. *Northwestern Nat.* **2020**, *101*, 180–193. [CrossRef]
31. *JMP, Version 18*; SAS Institute, Inc.: Cary, NC, USA, 2018.
32. Seymour, C.L.; Simmons, R.E.; Morling, F.; George, S.T.; Peters, K.; O'Riain, M.J. Caught on Camera: The Impacts of Urban Domestic Cats on Wild Prey in an African City and Neighbouring Protected Areas. *Glob. Ecol. Conserv.* **2020**, *23*, e01198. [CrossRef]
33. Gage, L.J.; Duerr, R.S. *Hand-Rearing Birds*; Wiley-Blackwell Publishing: Hoboken, NJ, USA, 2007.

diversity

MDPI

Review

First Insights into the Microbiology of Three Antarctic Briny Systems of the Northern Victoria Land

Maria Papale [1,†], Carmen Rizzo [1,2,†], Gabriella Caruso [1], Rosabruna La Ferla [1], Giovanna Maimone [1], Angelina Lo Giudice [1,*], Maurizio Azzaro [1,‡] and Mauro Guglielmin [3,‡]

[1] Institute of Polar Sciences, National Research Council (CNR-ISP), Spianata San Raineri 86, 98122 Messina, Italy; maria.papale@isp.cnr.it (M.P.); carmen.rizzo@cnr.it (C.R.); gabriella.caruso@cnr.it (G.C.); rosabruna.laferla@cnr.it (R.L.F.); giovanna.maimone@cnr.it (G.M.); maurizio.azzaro@cnr.it (M.A.)
[2] Stazione Zoologica Anton Dohrn, Department BIOTECH, National Institute of Biology, Villa Pace, Contrada Porticatello 29, 98167 Messina, Italy
[3] Dipartimento di Scienze Teoriche e Applicate, University of Insubria, Via J.H. Dunant 3, 21100 Varese, Italy; mauro.guglielmin@uninsubria.it
* Correspondence: angelina.logiudice@cnr.it; Tel.: +39-090-6015-414
† Equal contribution as first author.
‡ Equal contribution as last author.

Abstract: Different polar environments (lakes and glaciers), also in Antarctica, encapsulate brine pools characterized by a unique combination of extreme conditions, mainly in terms of high salinity and low temperature. Since 2014, we have been focusing our attention on the microbiology of brine pockets from three lakes in the Northern Victoria Land (NVL), lying in the Tarn Flat (TF) and Boulder Clay (BC) areas. The microbial communities have been analyzed for community structure by next generation sequencing, extracellular enzyme activities, metabolic potentials, and microbial abundances. In this study, we aim at reconsidering all available data to analyze the influence exerted by environmental parameters on the community composition and activities. Additionally, the prediction of metabolic functions was attempted by the phylogenetic investigation of communities by reconstruction of unobserved states (PICRUSt2) tool, highlighting that prokaryotic communities were presumably involved in methane metabolism, aromatic compound biodegradation, and organic compound (proteins, polysaccharides, and phosphates) decomposition. The analyzed cryoenvironments were different in terms of prokaryotic diversity, abundance, and retrieved metabolic pathways. By the analysis of DNA sequences, common operational taxonomic units ranged from 2.2% to 22.0%. The bacterial community was dominated by Bacteroidetes. In both BC and TF brines, sequences of the most thermally tolerant and methanogenic Archaea were detected, some of them related to hyperthermophiles.

Keywords: cryoenvironments; prokaryotic diversity; prokaryotic abundance; microbial metabolic activities; predictive functional profiling; Antarctic lakes' brines

Citation: Papale, M.; Rizzo, C.; Caruso, G.; La Ferla, R.; Maimone, G.; Lo Giudice, A.; Azzaro, M.; Guglielmin, M. First Insights into the Microbiology of Three Antarctic Briny Systems of the Northern Victoria Land *Diversity* **2021**, *13*, 323. https://doi.org/10.3390/d13070323

Academic Editor: Michael Wink

Received: 22 June 2021
Accepted: 11 July 2021
Published: 15 July 2021

Publisher's Note: MDPI stays neutral with regard to jurisdictional claims in published maps and institutional affiliations.

Copyright: © 2021 by the authors. Licensee MDPI, Basel, Switzerland. This article is an open access article distributed under the terms and conditions of the Creative Commons Attribution (CC BY) license (https://creativecommons.org/licenses/by/4.0/).

1. Introduction

In continental Antarctica, lakes are often characterized by the presence of icing blisters, which develop annually on their surface. Differently from their Arctic counterparts, Antarctic icing blisters mostly derive from the generation of hydrostatic pressures by the progressive freezing of high salt-content water beneath the lake-ice cover during winter. In fact, even if the lake ice cover never melts, heating by direct insolation or enhanced thaw and seepage at the permafrost table probably allows free water, in the form of liquid and saline brine lenses, to accumulate beneath the lake-ice cover in warmer periods [1,2]. Studying briny ecosystems is multifaceted, and assumes special relevance for geological aspects, and also for unravelling the functional potential and roles in biogeochemical cycling of the psychrophilic lifeforms. Microorganisms inhabit extreme environments on

Earth [3–5] that are also intriguingly similar to other worlds within our Solar system [6]. Current knowledge about the drivers of prokaryotic diversity, distribution, and metabolism in Antarctic briny systems is patchy, and therefore deserves to be deepened. The TF and BC areas in the NVL host several small perennially ice-covered lakes [7], and some of them conceal brine lenses [8–10]. Recently, brines from three Tarn Flat and the Boulder Clay lakes have been analyzed for the prokaryotic component by culture-dependent and -independent approaches [4,5,11,12]. Overall, the prokaryotic community included methanogens, strictly anaerobes, halophiles, and (hyper)thermophiles [4,12]. Diversification in terms of abundance, metabolic potentials, and enzymatic activities of the prokaryotic assemblages among the examined brines has been highlighted [4,5]. In the Tarn Flat brines, previous analyses have found the presence of proteolytic activity, as well as a comparatively lower alkaline phosphatase activity than in Boulder Clay. Enzymes that could degrade polysaccharides were also detected, whose hydrolytic activity rates were quantitatively different between the studied samples [4]. Conversely, in Boulder Clay, the microbial community was mostly active in the decomposition of organic phosphates and lower proteolytic and glycolytic activity rates were recorded. Moreover, decreasing patterns of aminopeptidase and phosphatase activities were observed with increasing depth of the collection site [5]. These peculiar features were ascribed to a lot of factors including brines' historical origin, depth horizon, and time of segregation.

The prediction of microbial functions from 16S rRNA gene sequencing data has been proposed as a valuable alternative to the shotgun metagenomics approach. As an additional in-depth analysis, in this study, we used the phylogenetic investigation of communities by reconstruction of unobserved states (PICRUSt2), as a bioinformatics tool applied to 16S rRNA gene data, to infer the functional profile of prokaryotic communities in Antarctic lakes' brines. Recently, this approach provided useful insights into predictive metagenomics of microbial communities in the thalassohaline brine of Lake Tuz, [13]. To the best of our knowledge, predictions on the metabolic functions of prokaryotes in cryosystems, such as Antarctic briny systems, have never been performed. This approach may be particularly suitable in extreme environments to rapidly survey microbial metabolic mechanisms and their possible relationships with well-known environmental constrains, allowing one to explore unique adaptation strategies, as well as to give insights into the biotechnological potentialities of microbes inhabiting unusual habitats.

In this study, we aimed at coupling previous information on the microbiological and physicochemical features of five distinct briny systems of the Northern Victoria Land and an attempt to predict functional profiles of microbial communities. All available data were collected, comparatively analyzed, and further reprocessed to answer the following questions: (i) Which environmental parameters exert a main influence on the prokaryotic communities in terms of composition, structure, and activities? (ii) Does brine separation overwhelm environmental conditions in shaping prokaryotic diversity? (iii) To what extent do the prokaryotic assemblages show differences among the brines? (iv) How does the metabolic potential of prokaryotes vary among the brines? This research provides further insights into the diversity, biotechnological potential, and ecological role of prokaryotes in these under-investigated and peculiar Antarctic cryosystems.

2. Materials and Methods

2.1. Brine Sample Collection

Five brine samples were collected, in 2014, from three perennially frozen lakes lying in the Tarn Flat (TF) and Boulder Clay (BC) areas of the NVL, Antarctica (coordinates 75°4′ S, 162°30′ E and 74°44′ S, 164°01′ E, respectively). Samples TF4 and TF5 were taken at two different depths (3.78–3.98 and 4.10–4.94 m depth, respectively) of the same borehole drilled in the TF lake. These brine pockets were separated by a thick ice layer of 12 cm [10]. Samples BC1 and BC2 were collected from two different sampling points (at 2.5 and 0.9 m depth, respectively) at Lake 16 in BC, whilst sample BC3 (depth 2.0 m) was collected from the adjacent Lake L-2 [14]. Sampling was carried out using presterilized polycarbonate bottles, sterilized peristaltic pump, and tubing.

2.2. Available Dataset

The microbiological and physicochemical results previously obtained on BC and TF lake brines, and used in this study, are summarized in Supplementary Material Table S1. Briefly, trace elements (Table S1) were analyzed by inductively coupled plasma sector field mass spectrometry (ICP-SFMS) [3,11,14]. Anions (Table S1) were analyzed by ion chromatography [3,14]. Total organic and inorganic carbon contents were determined using a Shimadzu 5050 A TOC analyser [3,14] (Table S1). The salinities and pHs of the brines were measured using refractive index and a potentiometric method, respectively [10,14] (Table S1). The bacterial viable counts (BVC) were determined by spread plating on agar plates [11]. Prokaryotic cell abundances (PA), respiration (CTC), viability (L/D), biomass (PB) (Table S1), and morphometric features (Table S1) were estimated by epiflurescence microscopy, as reported in Papale et al. [4] and Azzaro et al. [5]. Microbial enzymatic activities on proteinaceous and glucidic organic matter as well as on organic phosphates (i.e., AP, beta-GLU, and LAP) were fluorometrically estimated [4,5] (Table S1). Physiological profiles (PP) were determined by the Biolog EcoPlate™ microplate assay and spectrophotometrically measured [4,5] (Table S1). The prokaryotic (Bacteria and Archaea) community diversity and composition (Table S1) were evaluated by Ion Torrent sequencing [4,12]. The total OTU number, retrieved in each sample, was used to estimate alpha diversity by considering the Shannon, evenness, and Chao1 indices.

2.3. Predictive Functional Profiling

PICRUSt2 (phylogenetic investigation of communities by reconstruction of unobserved states) was selected among the tools for the functional prediction of the prokaryotic communities. To date, PICRUSt2 translates 16S RNA sequences into the most accurate prediction method [15] and acts as a predictor of functional metagenomic content based on the frequency of detected 16S rRNA gene sequences corresponding to genomes in regularly updated, functionally annotated genome databases [16]. It is in fact based on a new algorithm that uses short reads and recent positioning tools that insert sequences into an existing phylogenetic tree. Briefly, 16S RNA sequences with the Phred quality score less than 20 per base, and more than four consecutive low-quality base calls, were filtered by Trimmomatic (SLIDINGWINDOW 4:20). Then, denoising was carried out with DADA2 with the denoise-single command (using –p-trim-left 20 –p-trunc-len 0 –i-demultiplexed-seqs). Sequences were taxonomically classified in QIIME2 (version 2019.4) by Silva reference files (Silva release 132 full-length sequences and taxonomy references) using classify-consensus-blast.

Metagenomes from 16S data were predicted with the PICRUSt2 tool (version 2.3.0). The Hidden state prediction method with the mp (maximum parsimony) approach was used. To specify how distantly a sequence needs to be placed in the reference phylogeny, before it is excluded, the command –p-max-nsti (cut-off 2) was applied. The accuracy of metagenome predictions was tested trough the nearest sequenced taxon index (NSTI). The accuracy prediction is related to the presence of closely representative bacterial genomes. The lower values reveal a closer mean relationship. The data obtained by PICRUSt2 were examined by Kyoto Encyclopaedia of Genes and Genomes (KEGG) [17].

2.4. Statistical Analyses

Not-metric multidimensional scaling (nMDS) was performed by using Bray–Curtis similarity matrices calculated on transformed data of the relative abundances at phylum level (Primer 7, Plymouth Marine Laboratory, Roborough, UK). A Venn diagram was constructed to evidence the taxonomic sharing level between brines by using a web-based tool for the analysis of dataset of genera distribution [18].

After normalization, the entire dataset (consisting of previously obtained data plus metabolic prediction results) was then processed by calculating similarity and performing the cluster analysis. The obtained matrix was used to perform the principal component analysis (PCA) by Primer 7 software, using the origin of each Lake as a factor and overlaying vectors of the most affecting factors. Additional nMDS was performed using the Bray–Curtis similarity matrices calculated on transformed data of all biological data and the BEST Spearman rank correlation with physicochemical data (Primer 7). To achieve the main objectives, first, brines samples were compared for their main microbiological and physicochemical data (previously individually reported for BC and TF) (Table S1), and predictive functional attributes (determined in this study). In a second step, the complete dataset was statistically analyzed.

3. Results

3.1. Main Features of Analyzed Antarctic Brines: A Comparison

3.1.1. Chemical Data

From a physicochemical point of view, the brines from BC were richer in some elements (e.g., Ca, Mg, and K) than TF samples (Table S1). Specifically, samples from BC1 and BC3 showed high concentrations of Ca (362.9 and 148.1 mg L^{-1}, respectively), K (70.6 and 18.7 mg L^{-1}, respectively), and Mg (167.4 and 21.8 mg L^{-1}, respectively). Conversely, BC2 had a composition that was similar to TF brines, with values ranging from 0.3 to 2.8 mg L^{-1}. Sulfate ions were detected at a concentration of 5–10 mg L^{-1} in TF and BC2 brines, while in BC1 and BC3 samples they accounted for more than 1000 mg L^{-1}. The remaining elements did not differ among all samples (Table S1). Salinity and TOC were higher in TF than in BC brines (Table S1).

3.1.2. Microbial Abundance

Prokaryotic cell abundance (PA) was in the order of 10^8–10^9 cells L^{-1} (Table S1). The highest values were observed in TF5 (8.1×10^9 cells L^{-1}), followed by BC1 (6.1×10^9 cells L^{-1}) and TF4 (5.0×10^9 cells L^{-1}). Differently, BC2 and BC3 showed one order of magnitude lower abundance, i.e., 0.38 and 0.41×10^9 cells L^{-1}, respectively. Cell volume (VOL) exhibited remarkable differences among the investigated brines, with the smallest and largest cell sizes in TF4 (mean value 0.04 μm^3) and BC1 (0.242 μm^3), respectively. In BC2, VOL was on average 0.176 μm^3 while in BC3 and TF5 it was 0.139 and 0.105 μm^3, respectively. In addition, the prokaryotic biomass (PB) strongly differed among the brine samples, reaching surprisingly high values of 567 and 242 μg C L^{-1} in BC1 and TF5, respectively.

The highest percentage of live cells, accounting for 67%, was observed in TF4, while the lowest was observed in BC2 and BC3 (Figure 1a). Moreover, the percentages of live cells in BC1 and TF5 were comparable. Respiring cells (CTC) were higher in TF brines (18 and 30% of the total cells in TF4 and TF5, respectively) than in BC brines (range 1–8.5) (Table S1).

Concerning morphotype composition (Table S1), the prokaryotic cells were grouped into six classes: cocci, coccobacilli, vibrios, rods, curved rods, and filamentous forms (i.e., cells exceeding 4 μm in length). In this case, each morphotype also contributed differently to the composition of the microbial assemblage among the brines. Overall, the most abundant cell groups were represented by rods (range 27–74% of the total cells), coccobacilli (range 26–50%, except for BC1), and cocci (range 23–44%, except for BC3) (Table S1). Vibrios and curved rods were almost lacking or negligible. Finally, the filamentous forms with a relatively high percentage (27% of the total cells) were observed in BC1 only.

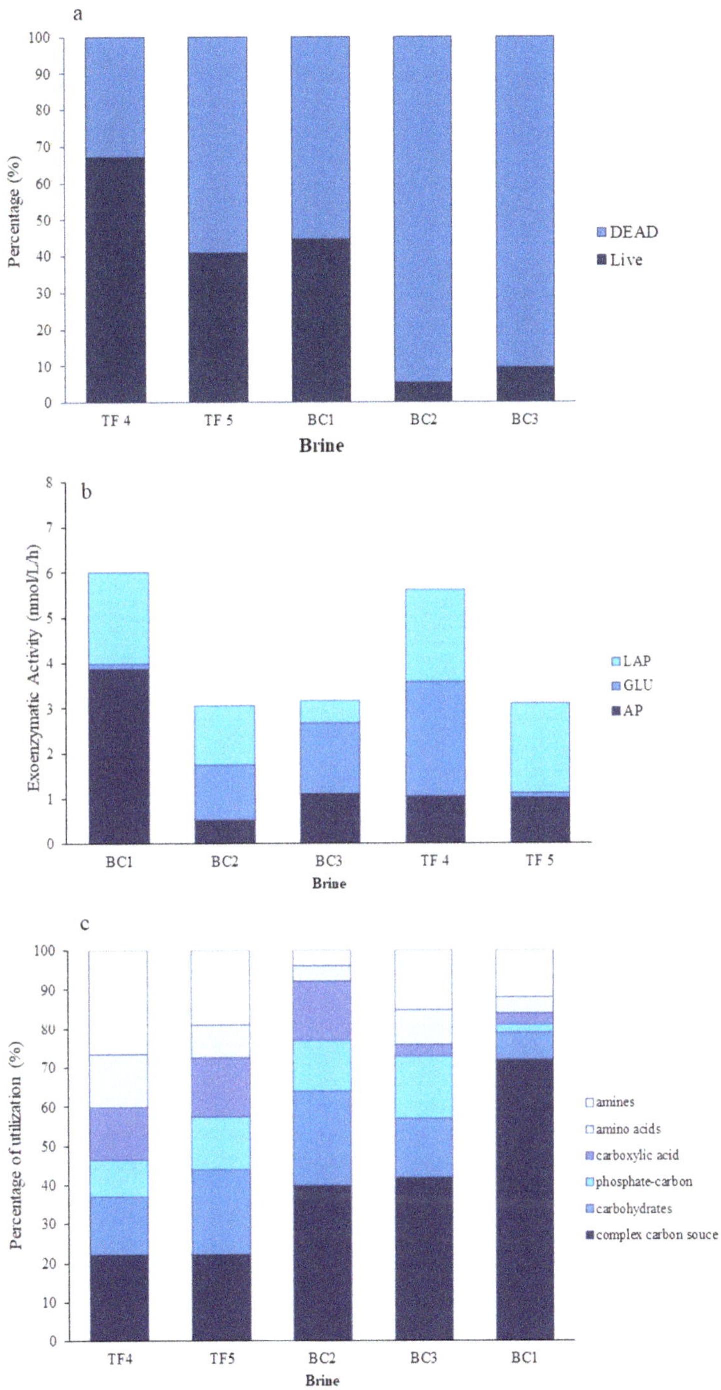

Figure 1. Microbial abundance and activities: (**a**) Viable (live/dead) cells in brine samples; (**b**) enzymatic activities measured in brine samples; (**c**) percentages of carbon source utilization obtained by Biolog Ecoplate.

3.1.3. Microbial Activities

AP activity showed very low rates in BC2 (0.6 nmol L^{-1} h^{-1}), higher in BC1 (3.9 nmol L^{-1} h^{-1}), and reciprocally similar values in BC3, TF4, and TF5 (1.1, 1.1, and 1 nmol L^{-1} h^{-1}, respectively) (Table S1). Differently, β-GLU activity was at low levels in BC1 and TF5 (0.1 and 0.1 nmol L^{-1} h^{-1}, respectively), while it showed high values in BC2, BC3, and TF4 brines (1.2, 1.6, and 2.5 nmol L^{-1} h^{-1}, respectively). Finally, LAP activity was measured with minimum value in BC3 (0.5 nmol L^{-1} h^{-1}); higher rates were found in the remaining samples (ranging from 1.3 nmol L^{-1} h^{-1} in BC2 to 2.0 nmol L^{-1} h^{-1} in TF4) (Figure 1b).

Overall, the percentages of carbon source utilization obtained by Biolog Ecoplate showed that the complex carbon sources were well utilized polymers in each brine, and mainly in BC brines (Table S1). In particular, in BC3, they accounted for the 72% of the total utilized sources. Differently, amines were better used in TF than in BC. Carbohydrates were well utilized in BC1 and TF5 brines and to a lesser extent in BC2 and TF4 brines. BC1, TF5, and TF4 brines expressed discrete utilization patterns for carboxylic acids, while BC2, BC1, and TF5 brines expressed discrete utilization patterns for phosphate carbon sources. Finally, aminoacids were scantly utilized everywhere, except in TF4 (Figure 1c).

3.1.4. Prokaryotic Community Composition

Main data on total sequence reads, quality trimming, OTU information, and diversity indices obtained for brine samples are reported in Supplementary Material Table S2 [4,12]. Overall, a comparable number of bacterial OTUs were generated in BC and TF samples using the Silva database (Figure 2a). Conversely, Archaeal reads were numerically higher in BC than TF brines (Figure 2b). In analyzed brines, the calculated alpha diversity, in terms of Shannon and Chao1 indices, showed a positive trend in TF samples. In particular, all the Archaea samples had higher values than bacteria. In terms of OTU composition, all brines shared 40 bacterial OTUs, while 19 OTUs were shared only between TF brines (Figure 2a and Figure S1a). Similarly, all brines shared 35 archaeal OTUs, with 19 OTUs that were shared among BC brines and 59 between TF brines (Figure 2b and Figure S1b).

With respect to the bacterial community composition (Figure 3a,b and Table S1), TF and BC brines mainly differed for the predominant phyla (Gammaproteobacteria and Bacteroidetes, respectively), a higher abundance of Actinobacteria in TF than in BC, and the almost exclusive occurrence of Delta- and Epsilonproteobacteria sequences in TF brines. Alphaproteobacterial abundances were comparable in TF5 and BC1 samples. Betaproteobacteria were particularly abundant in BC1 and BC3 brine samples, whereas in BC2, they showed lower relative percentages, but similar to TF brines (Figure 3a and Table S1). As shown in Figure 3b, brines BC2 and BC3 grouped together in a cluster with 80% of similarity, mainly due to the highest and similar Bacteroidetes abundance. Together with BC1, they formed a bigger cluster with a similarity of 60%, while brines from TF clustered separately in a 60% similarity group.

The archaeal community composition (Figure 3c,d) was relatively similar in TF and BC brines. They harbored almost the same taxonomic groups, even if differences in abundance values occurred (Figure 3c and Table S1). In all the brine samples, the archaeal community was mainly represented by Euryarchaeota, with higher abundance in BC than TF brine samples. Crenarchaeota displayed a concentration almost twice in BC than that in TF brines. Two other archaeal groups, i.e., the Ancient Archaeal Group and Korarchaeota, were considerably less abundant in all brines, with the highest percentages (ca. 2%) in BC3 (Figure 3c and Table S1). According to the distribution of archaeal phyla, BC and TF brines clustered in two distinct groups (with a similarity of 80% each). A bigger cluster grouped TF and BC brines with a similarity of 60% (Figure 3d).

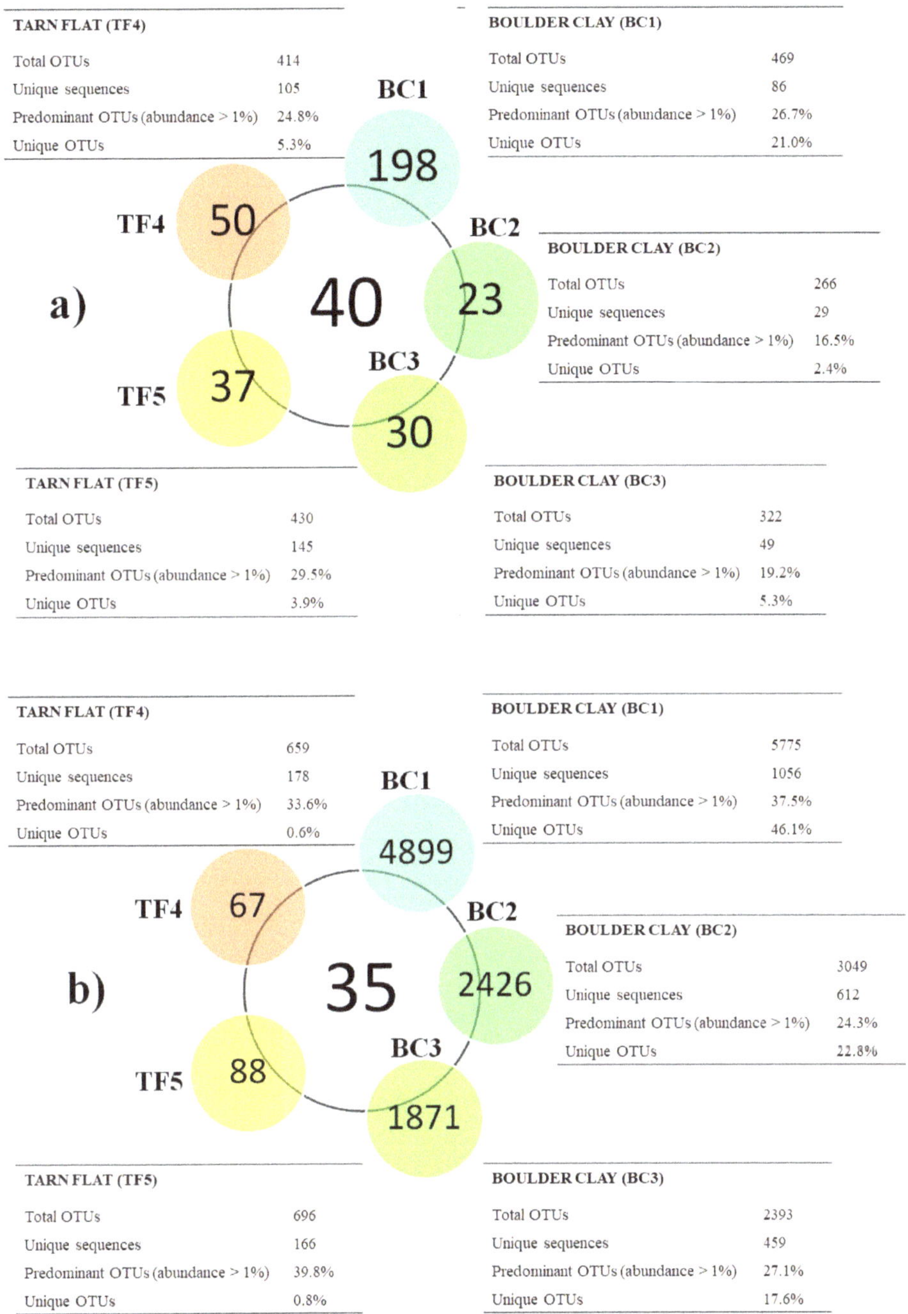

Figure 2. Diagram comparing bacterial (**a**) and archaeal (**b**) communities. Central circles show the number of OTUs shared among brines, while the lateral-colored circles report unique OTUs per brine. The number of total OTUs and unique sequences, as well as the percentages of the predominant and unique OTUs, for each brine, are shown in the tables, which are located close to the sample names.

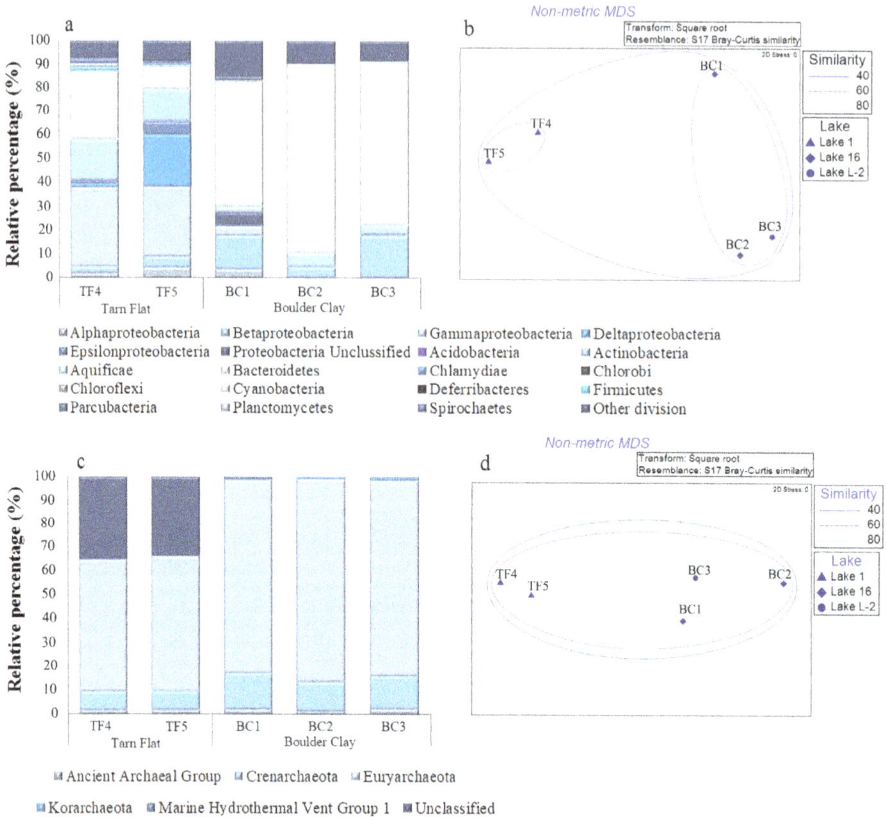

Figure 3. Bacterial (**a**) and archaeal (**c**) community composition at phylum level in TF and BC brines. Nonmetric multidimensional scaling analysis (nMDS) computed on transformed and clustered abundance data for bacterial (**b**) and archaeal (**d**) taxonomic groups at the phylum level.

At the genus level, TF and BC brines strongly differed (Figure 4 and Supplementary Material Figure S2). With respect to bacteria, *Flavobacterium*, *Psychroserpens*, and *Salmonella* were the relatively most abundant genera in BC brines (relative abundances ranging from ca. 2 to 44% of the total community), BC1 and BC3 shared with TF4 sequences related to *Flavobacterium*, whereas *Illumatobacter* was a common genus among BC1, BC3, and TF brines. The highest genus-sharing level was observed among BC3 and both TF brines (i.e., *Psychroserpens*, *Prevotella*, *Bosea*, *Bradyrhizobium*, *Belnapia*, *Arcobacter*, *Paraglaciecola* with TF4; *Hydrogenophilus*, *Pelomonas* and *Collimonas* with TF5; *Flavobacterium* and *Salmonella* were common genera to BC3 and both TF brines) (Figure 4a and Supplementary Material Figure S2a).

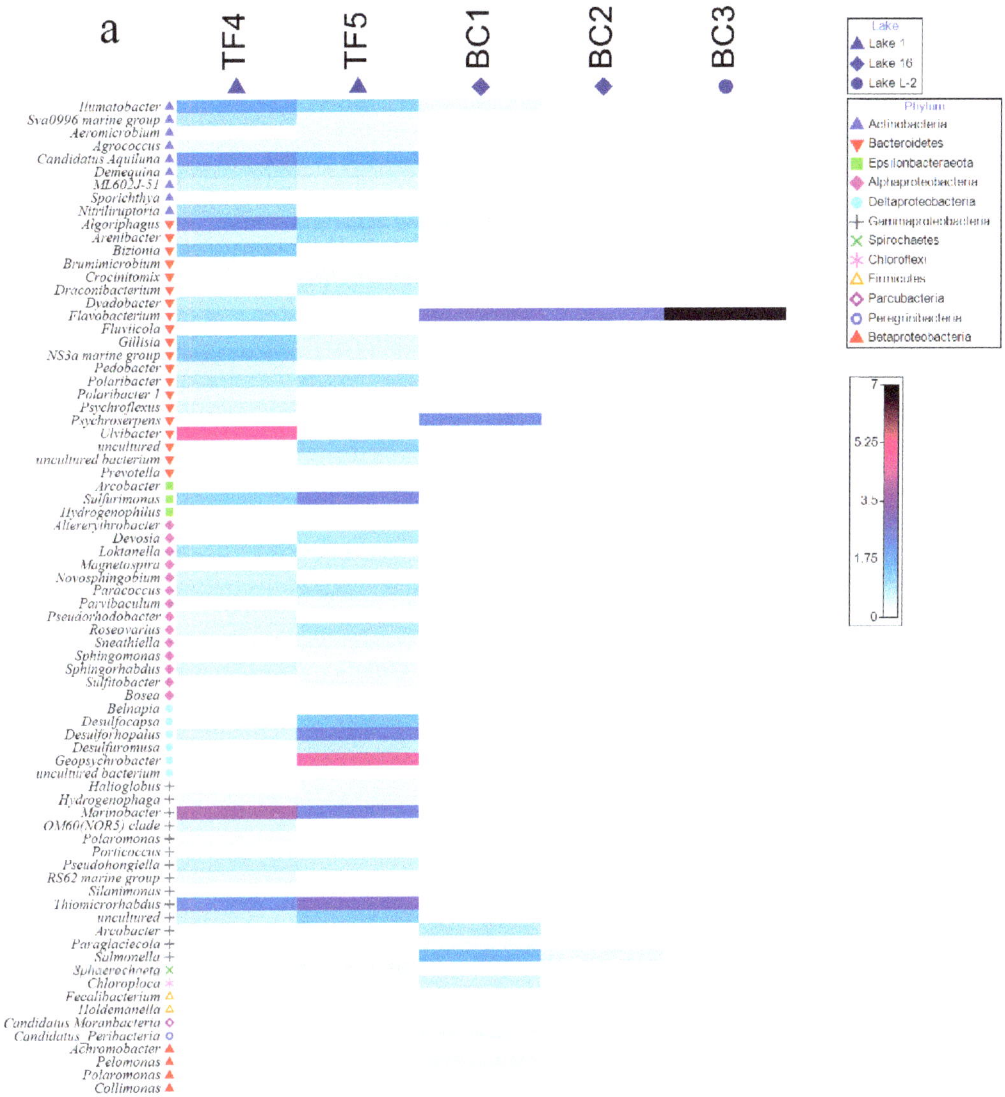

Figure 4. *Cont.*

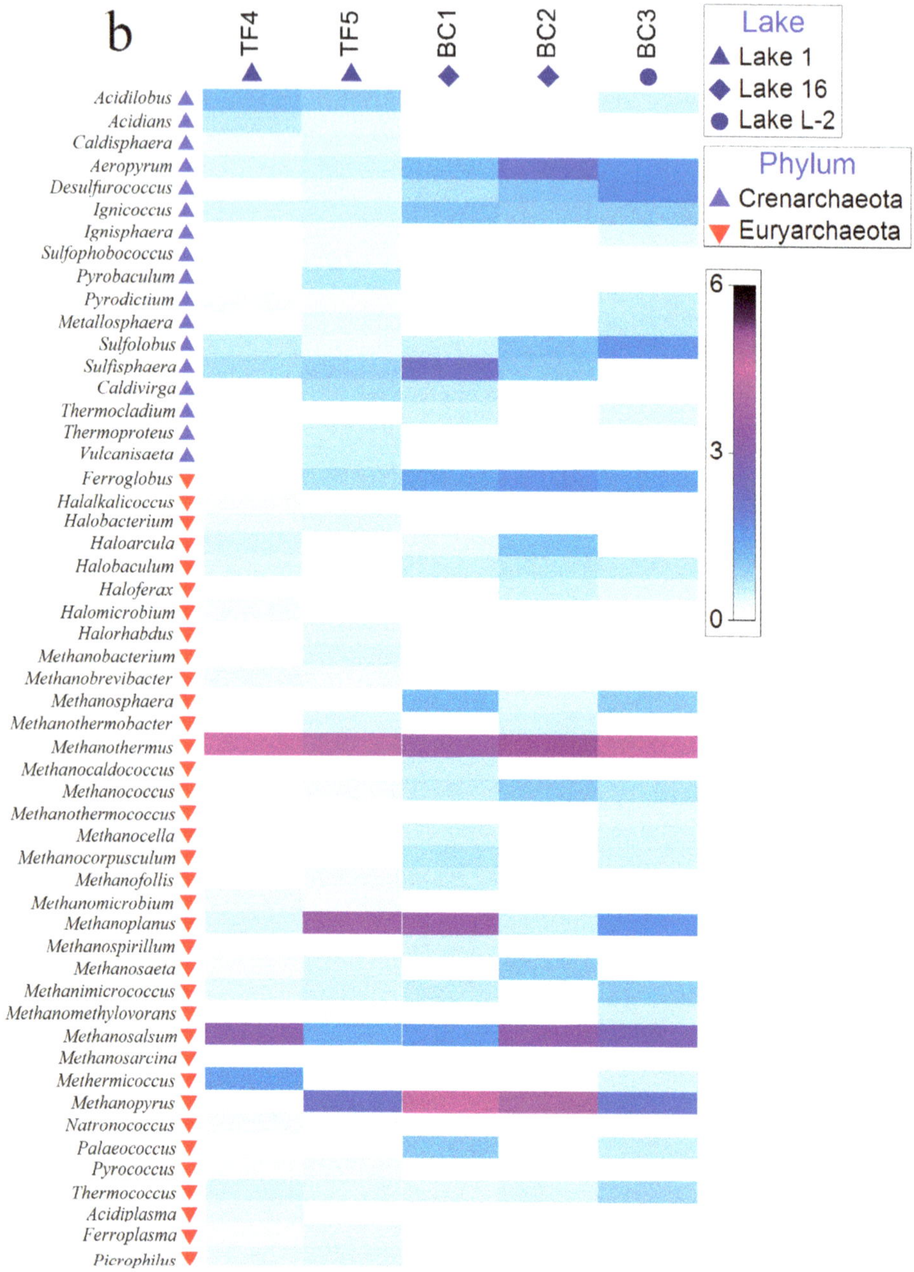

Figure 4. Shade matrix plots showing (**a**) the bacterial community structure at the genus level and (**b**) the archaeal community structure at the genus level. Shading intensity within the matrix indicates the square root transformed relative abundance of each genus.

Archaeal genera were distributed quite homogeneously among TF and BC brines. In particular, the genera *Aeropyrum*, *Desulfurococcus*, *Ignicoccus*, *Sulfolobus*, *Methanothermus*, *Methanoplanus*, *Methanosalsum*, and *Thermococcus* were common to all brines. Among them, *Methanothermus* was the most abundant genus in all brines (relative abundances ranging from ca. 13 to 25% of the total community for BC brines and from 17 to 19% of the total community for TF brines), whereas *Aeropyrum* (together with *Ferroglobus* and *Methanopyrus*) was more abundant in BC (Figure 4b and Supplementary Material Figure S2b).

3.1.5. Overall Comparison among Brine Samples with Respect to Previous Data

The PCA computed on the dataset showed the spatial separation of brine samples (Supplementary Material Figure S3). The two main components explained the 91.1% of the total variance, with PC1 and PC2 accounting for the 61.8 and 29.3% of the variance, respectively. The first component was mainly expressed by the Ca concentration and PB values (negative correlation), while the second component was mainly expressed by the S, TOC, and PB values (positive correlation). BC1 was distinct from other groups. A bigger cluster was composed of all brines (with the exception of BC1) and included two subclusters (i.e., BC2 plus BC3 and TF4 plus TF5).

3.2. Predicted Functional Genes

3.2.1. General Aspects

The KEGG pathway database accounted for a total of 1333 predicted KEGG orthologs (Kos, i.e., sets of homologous sequences) in the brine samples. Among the total KOs associated with metabolic processes, 642 (48.16%) had an abundance >0.1% in at least one sample. Overall, the predicted gene sequences annotated based on KEGG pathways resulted in 130 different biological processes, mostly related to functions such as "metabolic pathways" (range ≈ 18–19% of the total bacterial pathways in each sample), followed by "biosynthesis of secondary metabolites" (range ≈ 8–9% of the total bacterial pathways in each sample), "microbial metabolism in diverse environments" (≈5% of the total bacterial pathways in each sample), "biosynthesis of amino acids" (≈4% of the total bacterial pathways in each sample), "biosynthesis of cofactors" (≈4% of the total bacterial pathways in each sample), "ABC transporters" (range ≈ 1–2% of the total bacterial pathways in each sample), and "two components system" (range ≈ 1–2% of the total bacterial pathways in each sample).

The results for predicted biological processes in which unique KEGG showed relative abundance >0.01% for the whole briny prokaryotic community are reported in Figure 5.

Less represented pathways were related to degradation processes (≈1.5%), methane metabolism (0.6%), sulfur metabolism (≈0.5%), and quorum sensing (≈1.5%). The percentages of retrieved KOs, i.e., related to a single pathway, were calculated.

The occurrence of generic metabolic pathways, which included more specific processes, did not allow one to fully appreciate all the predicted functions within the analyzed communities. Therefore, all data were reclassified by grouping together pathways correlated to the same specific biological function (as reported in Supplementary Material Table S3). The relative abundance of predicted genes identified by the PICRUSt analysis within the prokaryotic communities is reported in Figure 6. The biological processes related to metabolic pathways were responsible for the higher number of KOs (approximately 19% in almost all samples) than the other detected biological processes. The second most represented category included all processes related to the "metabolism of amino acids", with a global value of 15% of predicted genes in all brines, followed by the "carbohydrate metabolism" (10% of predicted genes in all samples). "Biosynthesis of secondary metabolites", "metabolism of cofactors and vitamins", "energy metabolism" and "microbial metabolism in different environments" accounted for about 9, 7, 6, and 5% of total predicted genes, respectively. The most represented or interesting metabolic pathways are detailed below.

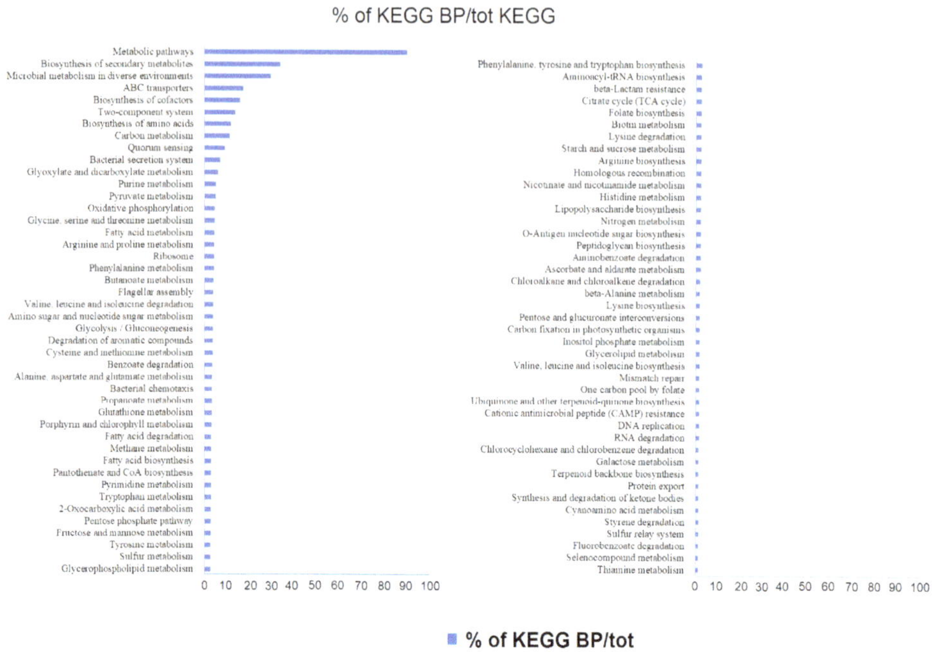

Figure 5. Percentages of unique KEGGs on the total retrieved KEGGs related to predicted biological processes (abundance >0.01%).

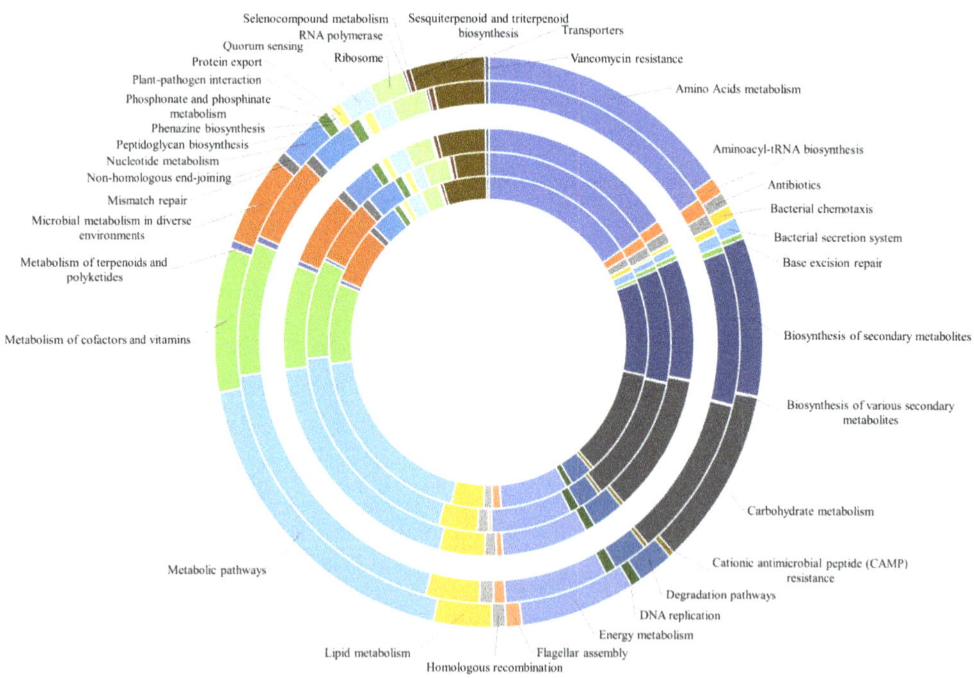

Figure 6. Relative abundance of predicted genes of the most abundant pathways identified in the bacterial and archaeal populations by the PICRUSt analysis. The pathways are presented according to KEGGs. Brines are in the order BC1 (as the innermost sample)→BC2→BC3→TF4→TF5 (as the outermost sample).

3.2.2. Metabolism of Amino Acids

The generic pathway "biosynthesis of aminoacids" was retrieved at similar percentages (between 3.8% and 4.1% of the total predicted genes) in all brines. Overall, the most prominent predicted pathways were glycine, serine, and threonine metabolism > cysteine and methionine metabolism > phenylalanine, tyrosine, and tryptophan biosynthesis > alanine, aspartate, glutamate, arginine, lysine, valine, leucine, and isoleucine degradation. The molecular process identified with KEGG16370 (*pfk*B, 6-phosphofructokinase 2) was more abundant in BC than in TF brines, whereas the process L-serine/L-threonine ammonia lyase (SDS, KEGG17989) was particularly represented in BC1, absent in BC2 and BC3, and less abundant in TF brines (Figure S4a).

3.2.3. Metabolism of Carbohydrates

The carbohydrate metabolism was mainly mediated by predicted pathways such as pyruvate metabolism, glyoxylate and dicarboxylate metabolism, and glycolysis/gluconeog enesis, followed by the 2-oxocarboxylic acid metabolism, citrate cycle (TCA cycle), butanoate, and propanoate metabolism (equally represented in all brine samples). Some predicted molecular processes were better represented in TF5, as in the case of glycolate oxidase iron-sulfur subunit (*glc*F, KEGG11473), glycolate oxidase (*glc*D, KEGG00104), and formate dehydrogenase major subunit (*fdo*G, KEGG00123) (Figure S4b).

3.2.4. Metabolism of Cofactors and Vitamins

The biosynthesis of cofactors was mainly supported by the activity of a number of enzymes, as follows: 3-oxoacyl-[acyl-carrier protein] reductase (*fab*G, KEGG00059) and 3-oxoacyl-[acyl-carrier-protein] synthase II (*fab*F, KEGG09458), both involved in the synthesis of fatty acids; aldehyde dehydrogenase (NAD+) (ALDH, KEGG00128); dihydrolipoamide dehydrogenase (DLD, KEGG00382); branched-chain amino acid aminotransferase (KEGG00826); dihydroorotase (URA4, KEGG01465); oxygen-independent coproporphyrinogen III oxidase (*hem*N, KEGG02495). The molecular processes were similarly represented in all samples. In addition to the biosynthesis of cofactors, other well represented predicted biological processes included folate biosynthesis, nicotinate, and nicotinamide metabolism, one carbon pool by folate and porphyrin and chlorophyll metabolism, with percentages ranging from 0.4% to 0.8% on the total of KOs retrieved in all samples (Figure S4c).

3.2.5. Energy Metabolism

The energy metabolism category was mainly represented by a high number of KOs involved in the "carbon metabolism" and "oxidative phosphorylation" (3 and 1% of the retrieved KOs, respectively) (Figure S4d). In addition, KEGGs also involved in the "methane and sulfur metabolism" were detected. Molecular functions related to methane metabolism were found in all samples. Molecular functions related to sulfur metabolism were detected in all samples. Among them, the most represented were serine O-acetyltransferase (*cys*E, KEGG00640), thiosulfate/3-mercaptopyruvate sulfurtransferase (TST, KEGG01011), and cysteine synthase (*cys*K, KEGG01738) involved in the cysteine and sulfur metabolism (in the case of the enzyme form *cys*K use thiosulfate instead of sulfide, to produce cysteine). Moreover, the molecular process sulfide:quinone oxidoreductase (*sqr*, KEGG17218), involved in the sulfide oxidation pathways responsible for the sulfide-dependent reduction of quinones, was detected in all samples.

3.2.6. Transporters

The category of transporters was represented by three predicted pathways such as ABC transporters, two-components system, and, to a lesser extent, by the phosphotransferase system (PTS) (Figure S4e). The pathway of ABC transporters, responsible for the transport of substrates across the cell membrane, was similarly represented in all brine samples, with predominant for molecular processes involved in the amino acids and phospholipid transport. In detail, two KEGGs related to amino acid transport via ATP-binding protein (*liv*G, branched-chain amino acid transport system ATP-binding protein, KEGG01995 and *liv*F, branched-chain amino acid transport system, ATP-binding protein, KEGG01996), two KEGGs related to amino acid transport via permease protein (*liv*H, branched-chain amino acid transport system permease protein, KEGG01997 and *liv*M, branched-chain amino acid transport system permease protein, KEGG01998), and one KEGG related to amino acid transport via substrate-binding protein (*liv*K, branched-chain amino acid transport system substrate-binding protein, KEGG01999) were found in all samples. In total, three KEGGs related to phospholipid transport were detected in the brine samples, namely phospholipid/cholesterol/gamma-HCH transport system ATP-binding protein (*mla*F, KEGG02065), phospholipid/cholesterol/gamma-HCH transport system permease protein (*mla*E, KEGG02066), and phospholipid/cholesterol/gamma-HCH transport system substrate-binding protein (*mla*D, KEGG02067).

Some molecular processes were more represented in BC1 brine samples (*cco*N, cytochrome c oxidase cbb3-type subunit I, KEGG00404; *cco*O, cytochrome c oxidase cbb3-type subunit II, KEGG00405; *dct*B, two-component system, NtrC family, C4-dicarboxylate transport sensor histidine kinase DctB, KEGG10125; *dct*D, two-components system, NtrC family, C4-dicarboxylate transport response regulator DctD, KEGG10126), while cytochrome c (CYC, KEGG08738) was highly represented in BC samples as compared with in TF brines.

3.2.7. Degradation Pathways

The predictive analysis also showed the presence of several pathways involved in different degradation processes. The most abundant calculated on the total of all KOs retrieved for degradation pathways were the benzoate degradation and RNA degradation pathways. The most represented molecular processes detected within the RNA degradation are linked to the action of two helicases involved in the maintenance of the genome and ribosome assembly, namely ATP-dependent DNA helicase RecQ (*rec*Q, KEGG03654) and ATP-dependent RNA helicase RhlE (*rhl*E, KEGG11927). Finally, although to a lower extent, a series of genes involved in the degradation of aliphatic hydrocarbons (chloroalkanes), terpenes (i.e., limonene and pinene), aromatic compounds (i.e., chlorobenzene, styrene, toluene, xylene, and naphthalene) were predicted with similar percentages in all samples (Figure S4f).

3.2.8. Biosynthesis of Secondary Metabolites

The most represented molecular functions associated with the "biosynthesis of secondary metabolites" pathway were the 3-oxoacyl-[acyl-carrier protein] reductase (KEGG00059); aldehyde dehydrogenase (NAD+) (ALDH, KEGG00128); the glyceraldehyde 3-phosphate dehydrogenase (GAPDH, KEGG00134); acetyl-CoA C-acetyltransferase (ACAT, KEGG00626); and acyl carrier protein (*acp*P, KEGG02078). Some of them were common to other predicted metabolic pathways, and therefore involved in carbohydrate and amino acid metabolism. The molecular function beta-glucosidase (*bgl*X, KEGG05349) was mostly represented in the BC2 brine sample.

3.2.9. Microbial Metabolism in Diverse Environments

The "microbial metabolism in diverse environments" pathway involves all processes related to different kinds of metabolism, i.e., carbohydrate metabolism, energy metabolism, amino acid metabolism, and xenobiotic degradation. In the brine samples, the most represented molecular functions were related to carbohydrate metabolism, such as aldehyde dehydrogenase (NAD+) (ALDH, KEGG00128); GAPDH, glyceraldehyde 3-phosphate dehydrogenase (KEGG00134); succinate-semialdehyde dehydrogenase/glutarate-semialdehyde dehydrogenase (*gab*D, KEGG00135); dihydrolipoamide dehydrogenase (DLD, KEGG00382); acetyl-CoA C-acetyltransferase (ACAT, KEGG00626); 4-hydroxy-tetrahydrodipicolinate synthase (*dap*A, KEGG01714); and glutamine synthetase (*gln*A, KEGG1915).

3.3. Statistical Analyses on the Entire Dataset

The PCA, shown in Figure 7, was computed on the entire dataset, including results from enzymatic activities, physicochemical parameters, microbiological data, archaeal and bacterial community composition, and predictive analysis of metabolic pathways. The following spatial clustering of brines was obtained: a group including TF4 and TF5 brines; a group including BC2 and BC3 brines; BC1, with a strong positive correlation with some chemical elements (i.e., Ca, Mg, Ni, and K), and some specific substrates (i.e., carbohydrates and carboxylic acids). The two principal components explained 77.9% of the total variance, with PC1 accounting for 48.8% of the variance and the PC2 accounting for 29.1% of the variance. The overlapping of the vectors related to the predicted metabolic pathways showed a higher positive correlation among the BC brines.

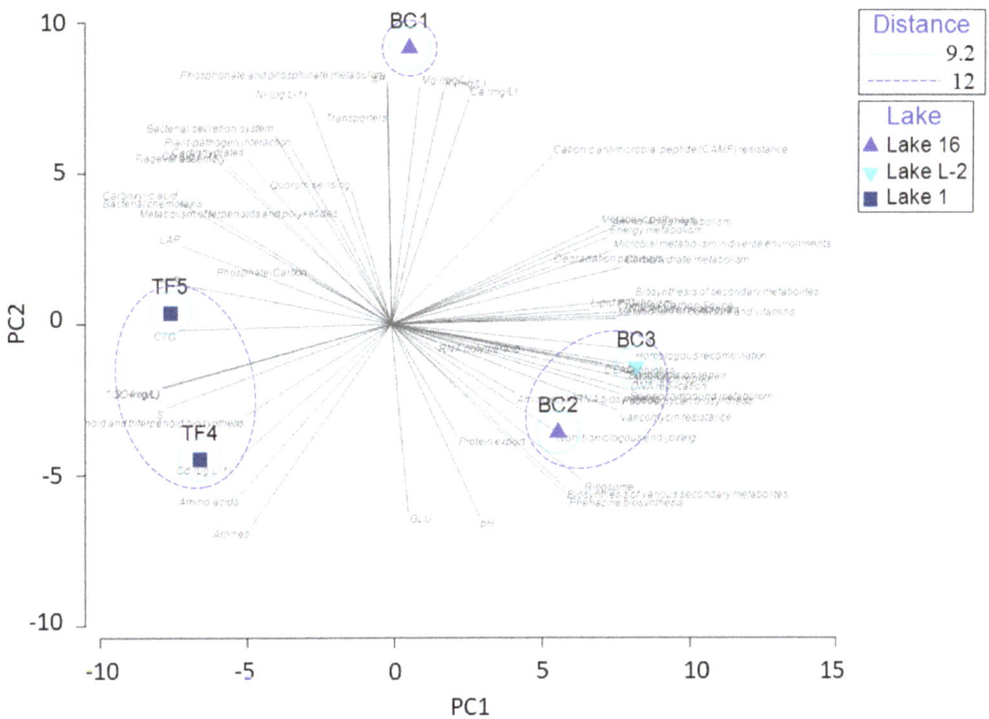

Figure 7. Principal component analysis computed on the entire dataset, including results from enzymatic activities, physicochemical parameters, microbiological data, archaeal and bacterial community composition, and predictive analysis of the metabolic pathways of brine samples.

The nMDS, shown in Figure 8, was performed by using all normalized biological data (enzymatic activities, microbiological data, archaeal and bacterial community composition, and predictive analysis), by superimposing the cluster analysis performed on physicochemical results. It evidenced a slight difference in brine sample separation, with BC2 that was included in a bigger cluster together with TF4 and TF5 brines, and BC1 and BC3 samples grouping together in another cluster. The bubble plots show the parameters having the most influence (Co, Cd, Ni, pH, and S) on the multidimensional scaling, as determined by the BEST Spearman rank correlation test.

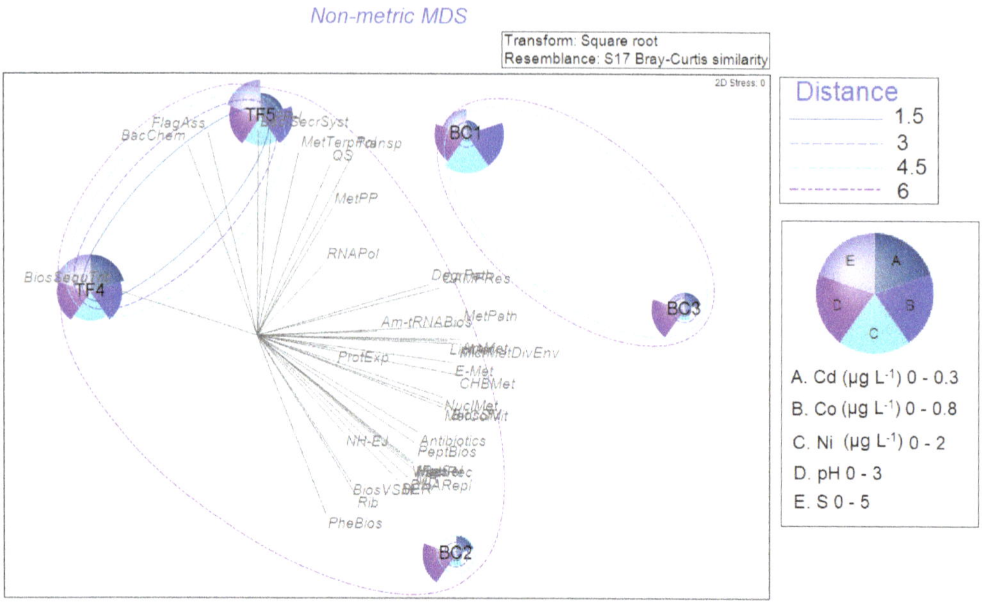

Figure 8. Non-metric multidimensional scaling analysis computed on normalized biological data (enzymatic activities, microbiological data, archaeal and bacterial community composition, and predictive analysis), by superimposing the cluster analysis performed on the physicochemical results. The bubble plots show the parameters having the most influence on the multidimensional scaling, as determined by the BEST Spearman rank correlation test.

4. Discussion

This study was devised with the intention of contributing to the poor existing knowledge on the microbiology of Antarctic briny systems. Complex datasets, which were previously obtained for three perennially ice-covered Antarctic lakes, were merged and compared. Then, available information was integrated with the analysis of predictive functional metabolic profiles of prokaryotes inhabiting such underexplored cryoenvironments. Physicochemically, three subsystems were distinguished: (1) TF4 plus TF3, (2) BC2, and (3) BC1 plus BC3. This latter was more positively affected by the concentration of certain chemicals (i.e., Ca, Mg, K, and Ni), while TF brines were more affected by salinity and TOC than BC. The contextual analysis of physicochemical and biological data confirms the separation between the two systems (BC and TF), with slight differences. Remarkable differences were observed among the analyzed brines in relation to their microbiological parameters, as described below.

4.1. Microbial Abundance, Biomass, and Morphometric Traits

Microbial abundance was comparable in both TF brines and BC1, where it was higher than BC2 and BC3. The observed discrepancy could depend on several interacting factors such as the possible competition among organisms inhabiting the same environment (i.e., grazers) or the availability of trophic resources. For instance, the specific occurrence of filamentous forms in BC1 and the high PA in TF5 contributed to modulate the microbial biomass in these brines. In terms of VOL, diverse community structures were observed, with larger sizes in BC than in TF brines. These features were presumably related to stress, starvation, or dormant cells in extremely cold and salty environments such as permafrost below the active layer [19,20].

In terms of morphotypes, cocci and rods were differently relevant in TF and BC brines, respectively. Specifically, these aforementioned shapes are connected with inputs of organic substances less or more difficult to be decomposed [21]. In fact, prokaryotic cells modify their size and shape to meet different environmental conditions [22,23]. In this context, studies of the genomic diversity could help to understand the intriguing properties of these cryo ecosystems. For instance, Lo Giudice et al. [12] observed in BC1 brine the massive presence of *Flavobacterium* spp., a bacterial strain with a mean size of 0.4 μm^3.

By comparing the brines for their prokaryotic viability, the viable and respiring prokaryotic communities were more abundant in TF4, TF5, and BC1 samples than in BC2 and BC3, where high proportions of dead cells were present. These clear differences among the brines make us assume that the more limited the resources, the more complex the community is. Overall, the respiring cells were always lower than the LIVE ones by L/D, with the exception of BC3 in whih their percentages were reciprocally similar. This disagreement between L/D and CTC results could be due to the different approaches used to recognize the living cells based on the cell membrane integrity (L/D) and the active respiring capacity (CTC), respectively. Cellular respiratory activity and cell membrane integrity are among the most suitable parameters for evaluating the viability of cells in a microbial assemblage [24]. According to Posch et al. [25], the low percentages of CTC+ cells could be attributed to alive but dormant cells or with respiration rates below the detection limit of the method.

4.2. Microbial Activities

The occurrence of viable and respiring prokaryotic cells implies the capability to metabolize in situ thermogenic or biogenic sources, as reported by Murray et al. [8]. In fact, the organic carbon resources contained in the brines could enable microbial populations to survive for millennia, as suggested in Lake Vida. In our study, the microbial assemblages of each brine were found to be able to potentially metabolize the carbon sources. The main characteristic was the good utilization of the complex carbon sources in all the brines, probably owing to the easy degradability and antifreeze characteristics of these compounds [26]. In cryosystems, they influenced the adaptability of the prokaryotic community to freezing conditions, avoiding the intracellular ice crystal growth, and the consequent cell damage [27,28]. Carbohydrates were also efficiently used in almost all brines, and mostly in BC1 and TF5. They are functional biomolecules that allow energy storage, particularly important to survive in extreme environments [29]. Differently, the capability to utilize amines, mainly in TF4, suggested the occurrence of nitrogen compounds, nitrifying bacteria, and nitrogen fixation [30]. Finally, the discrete utilization of carboxylic acids, and among them the pyruvic acid (particularly in BC1), reinforced the results on the occurrence of methanogens, as it was determined by molecular analyses (see below).

Substrate utilization is strictly dependent on the production of enzymes. Under extremely cold conditions, such as those in icy brines, enzymes work at freezing temperatures [31] or under saturating salt conditions [32]. Nevertheless, their activity under both salinity and temperature extremes has been the subject of limited investigations [33]. Bacteria isolated from cold environments have been shown to produce several extracellular enzymes, such as leucine aminopeptidases or proteases [34,35], amylase [36,37], and esterase [38]. The peculiar characteristics of sea brines such as low temperature and high salt increase the viscosity of such matrices, limiting the diffusion processes [39], as well as reducing water activity within them, which is critical for enzyme activities [33]. The LAP, AP, and ß-GLU activity rates detected in our studies [4,5] suggested that the microbial communities inhabiting the brines of TF and BC lakes had the ability to produce a relatively wide spectrum of enzymes involved in the degradation of organic polymers.

Extracellular enzyme activity is the first step for the decomposition of high molecular weight organic matter into monomers or small molecules that microbial communities can uptake [40]. The low molecular weight compounds released by the active hydrolytic process could be beneficial to the brine communities and also to the surrounding communities, enabling them to exploit the available organic matter. Within sea-ice brines, enhanced enzyme production may help balance this resource need, where characteristic extremes of temperature and salinity already act to reduce growth rates [41]. Bacterial use of extracellular enzymes to degrade high molecular weight material under extreme conditions, such as in the brines, also involves chemotaxis, halotaxis, and chemohalotaxis that bacteria may adopt to position themselves in response to environmental gradients encountered within the sea-ice brine network; the study of organic matter decomposition in this context can give insights into the potential bacterial activity in an extraterrestrial ice layer [41]. Active proteolytic activity suggests that brine pockets are enriched in proteinaceous material, such as observed in sea-ice brines [42].

4.3. Prokaryotic Community Compositions

The analyzed brine systems hosted peculiar communities of microorganisms, sulfate-reducing bacteria, and methanogenic Archaea, as has been previously discussed [4,12]. However, diversity analyses on the bacterial and archaeal communities highlighted a clear separation between the Tarn Flat and Boulder Clay brine systems, thus, reflecting the above discussed data on microbial abundances and activities. For Bacteria, main differences relied on the predominance of Bacteroidetes in BC brines, and the exclusive occurrence of Delta- and Epsilonproteobacteria in TF brines (particularly in TF5). More markedly, BC and TF brines differed at the genus level. The highest genus-sharing (mainly dependent on Bacteroidetes affiliates) was observed among BC3 and both TF brines. The separation of the BC3 brines from both BC1 and BC2 samples (deriving from an adjacent lake) suggests the occurrence of lake-specific bacterial communities. For Archaea, even if the same taxonomic groups were detected, they occurred at a different extent, with Euryarchaeota that were more abundant in BC than in TF brines.

4.4. Prediction of the Metabolic Profiles

To the best of our knowledge, the prediction of metabolic profiles has never been reported for microbial communities inhabiting brines of perennially ice-covered Antarctic lakes. The PICRUSt software compares the identified 16S rRNA gene sequences to those of known genome-sequenced species, thereby, estimating the possible gene contents of the uncultured microbial communities [43]. Although this tool has been designated and validated for humane microbiomes [44], its accuracy to detect the phylogenetic proximity of the reference genomes to the environmental strains has been evaluated. The authors assessed the prediction efficiency of the bioinformatics tool to compare potential functions of samples driving from a wider variety of habitats [16]. According to the authors, the method is valid for application on 16S datasets providing valuable accuracy measure, for example, the nearest sequenced taxon index (NSTI, in our study it ranged between 0.08 and 0.1).

In our study, a very high rate of putative biological processes derived from the predictive analysis, revealing a microbial metabolism mainly based on the processes of synthesis and biodegradation of organic molecules. BC and TF brines did not show a sharp separation in terms of predicted metabolic profile, suggesting that, despite the differences encountered at both the taxonomic and environmental levels, the communities globally envisaged similar metabolic capacities. However, some small point variations delineate the possible presence of specific molecular processes.

Among the best represented metabolic pathways, the biosynthesis of amino acids was identified as one of the most predicted in the microbial communities. As part of processes related to amino acid metabolism, some pathways were strongly predicted in BC brines, such as those for 6-phosphofructokinase 2 and L-serine/L-threonine ammonia-lyase. This latter protein is involved in the first step of the sub-pathway synthesizing 2-oxobutanoate from L-threonine, and in the L-isoleucine biosynthesis (also involved in the amino acid biosynthesis).

The molecular function beta-glucosidase (*bglX*, KEGG05349), detected within the metabolic pathway of carbohydrate metabolism, was mostly represented in the BC2 brine. This process is responsible for the hydrolysis of the glycosidic bonds in various glycosides and oligosaccharides with the release of glucose [44]. Differently, other molecular processes were better represented in TF brines, glycolate oxidase iron-sulfur subunit (involved in the oxidation of glycolate to glyoxylate, which generally requires using glycolate as a sole carbon source) and the formate dehydrogenase major subunit. This enzyme is also involved in the energy metabolism pathway and belongs to a set of enzymes able to catalyse the formate oxidation by using a second substrate as electron acceptor, for example, NAD^+. Generally, this kind of dehydrogenases are involved in the metabolic processes of methylotrophic microorganisms and are crucial in the catabolism of C1 compounds, such as methanol [45].

The most prominent energy-generating metabolic pathways were predicted to be carbon metabolism, methane metabolism, and oxidative phosphorylation, indicating that ATP was generated by electron transfer to terminal acceptor, i.e., oxygen, nitrate, or sulfate. In particular, the prediction of genes encoding enzymes involved in methane metabolism confirmed the considerations by Stibal et al. [46], who highlighted the occurrence of methanogens in polar subglacial systems. The results reinforce our observation on microbial activities and are in line with the abundance of sequences related to Methanopyrales in TF5 brine [4] and strictly anaerobic methanogens in the BC active community [12], respectively.

Most of the studies related to the N-cycling functional markers in Antarctica have focused on the core genes involved in N-fixation, nitrification, and denitrification processes [47]. Nitrate is, here, predicted to be reduced through two dissimilatory membrane-bound respiratory (Nar) nitrate reductases, which generate a transmembrane proton motive force allowing ATP synthesis. Membrane-bound nitrate reductases are generally associated with denitrification and anaerobic nitrate respiration processes. Ammonia is then probably used as raw material for L-glutamate synthesis, as suggested by the predicted *gln*A gene, which catalyses the ATP-dependent biosynthesis of glutamine from glutamate and ammonia.

Sulfur metabolism was predicted as a common pathway for energy in all brines. This seemed to be mainly based on assimilatory sulfate reduction, a process entirely occurring in plants and microorganisms, necessary for the formation of sulfur-containing amino acids. The processes are mediated through sulfite reduction by reductases encoded from genes of *cys*JIH operon, almost equally predicted in all brine samples. A crucial step in the assimilatory sulfate reduction metabolic pathway is the formation of 3′-phosphoadenosine-5′-phosphosulfate [48]. The molecular process is catalysed by the enzyme phosphoadenosine phosphosulfate reductase, detected in this study through the prediction of the encoding *cys*H gene. The occurrence of such molecular processes mediating sulfur metabolism is also supported by the prediction of the gene *cys*K, encoding for the cysteine synthase, and by the occurrence of genes encoding for enzymes involved in sulfate transportation, such as *cys*P, *cys*U, *cys*W, and *cys*A. Interestingly, the sulfide quinone reductase (SQR) activity, here, detected at similar extent in all samples, was found to be widely distributed among prokaryotes, and the protein sequence comparison leads to the conclusion that SQR is a phylogenetically very old enzyme that was acquired early in evolution [49]. These findings are in line with previous evidence of the presence of sulfur cycling microorganisms in Antarctic subglacial environments [50–55], able to support sulfur transformations providing energy for growth.

Overall, the predictive analysis alone did not separate the briny microbial communities at a metabolic potential level. The number of samples was exiguous, due to the low accessibility and the difficulty to collect from such remote and peculiar ecosystems. However, in this study, we try to use some statistical elaboration to highlight possible distinguishing features among brines, which should be considered cautiously. TF and BC brines appeared distinguished, probably by differences intrinsic to individual systems. Interestingly, the spatial distribution obtained by analysing the entire dataset reflected that obtained from the elaboration of the sole physicochemical data (Figure 7 and Supplementary Material Figure S3) but suggested that biological parameters tend to separate more markedly BC2 and BC3 brines from TF4 and TF5 that, instead, grouped all together in a bigger cluster in the physicochemical level analysis. The BEST Spearman rank correlation test detected five parameters, namely Cd, Co, Ni concentrations, and pH and salinity values, as those that most affected the brine systems. Except for pH, all these parameters affected mostly TF4, TF5, and BC1 brines.

5. Conclusions

The brine systems differed in geochemical conditions and phylogenetic diversity of the hosted microorganisms, but they shared a similar metabolic potential even at conditions commonly considered prohibitive for life. The environmental parameters, mainly salinity and metal concentrations, affected these habitats by shaping the taxonomic composition of the microbial communities. The observation of microbial communities that can grow and survive in icy systems on Earth is an important tool for expanding our awareness of life subsistence and persistence under challenging conditions. The collective ability of brine microbial communities to catabolize several carbon sources, and to play a pivotal role in nutrient cycling, strongly suggests a nutritional versatility which could provide a fitness advantage. Bioinformatics tools have given a new edge to the research approach. Exploratory analyses can be very useful to focus more in-depth studies appropriately and are of greater value especially for remote sites that are difficult to access. Here, the predictive analysis gained more insights into the ecological role of microbial communities in perennially ice-covered Antarctic lakes and prospected environmental parameters shaping their structure.

Supplementary Materials: The following are available online at https://www.mdpi.com/article/10.3390/d13070323/s1, Figure S1: Venn diagram for (**a**) Bacteria and (**b**) Archaea, Figure S2: Venn diagram showing the sharing level of bacterial (**a**) and archaeal (**b**) genera between BC and TF brine samples, Figure S3: Principal component analysis computed on the previous dataset including physicochemical data related to BC and TF brine samples, Figure S4: Circos diagrams showing the connection among brine samples and metabolic pathways. The ribbon size represents the cell value corresponding to a brine/metabolic pathway segment pair, while the stacked bars show ribbon contribution for each segment, Table S1: Microbiological and physicochemical data previously reported for brines, Table S2: The sequence reads, good quality reads, observed number of OTUs, and Shannon diversity, Evenness, and Chao 1 indices per sample of the bacterial and archaeal 16S rRNA gene datasets. Data are from Papale et al. [4] and Lo Giudice et al. [12], Table S3: Relative abundance (% on the total predicted pathways) of predicted metabolic pathways identified in prokaryotic populations produced by the PICRUSt analysis. The icons are referred to percentages within the same subgroup (in bold).

Author Contributions: Conceptualization, A.L.G.; methodology, all authors; software, C.R. and M.P.; formal analysis, M.P., C.R., G.C. and G.M.; investigation, all authors; data curation, all authors; writing—original draft preparation, A.L.G., C.R. and M.P.; writing—review and editing, all authors; supervision, A.L.G.; funding acquisition, M.A. and M.G. All authors have read and agreed to the published version of the manuscript.

Funding: This research was funded by the Programma Nazionale di Ricerche in Antartide (PNRA): PNRA16_00194—A1 "Climate Change and Permafrost Ecosystems in Continental Antarctica" and PNRA18_00186-E "Interactions between permafrost and ecosystems in Continental Antarctica".

Institutional Review Board Statement: Not applicable.

Informed Consent Statement: Not applicable.

Data Availability Statement: Data on bacterial abundances and activities are available upon request. Ion Torrent sequence data obtained from this study have been registered as NCBI Bioproject PRJNA435985.

Conflicts of Interest: The authors declare no conflict of interest.

References

1. French, H.M.; Guglielmin, M. Frozen ground phenomena in the vicinity of Terra Nova Bay, Northern Victoria Land, Antarctica: A preliminary report. *Geogr. Ann.* **2000**, *82*, 513–526. [CrossRef]

2. Guglielmin, M.; Lewkowicz, A.G.; French, H.M.; Strini, A. Lake-ice blisters, Terra Nova Bay area, Northern Victoria Land, Antarctica. *Geogr. Ann. Ser. A Phys. Geogr.* **1999**, *91*, 99–111. [CrossRef]
3. Borruso, L.; Sannino, C.; Selbmann, L.; Battistel, D.; Zucconi, L.; Azzaro, M.; Turchetti, B.; Buzzini, P.; Guglielmin, M. A thin ice layer segregates two distinct fungal communities in Antarctic brines from Tarn Flat (Northern Victoria Land). *Sci. Rep.* **2018**, *8*, 6582. [CrossRef]
4. Papale, M.; Lo Giudice, A.; Conte, A.; Rizzo, C.; Rappazzo, C.; Maimone, G.; Caruso, G.; La Ferla, R.; Azzaro, M.; Gugliandolo, C.; et al. Microbial assemblages in pressurized Antarctic brine pockets (Tarn Flat, Northern Victoria Land): A hotspot of biodiversity and activity. *Microorganisms* **2019**, *7*, 333. [CrossRef]
5. Azzaro, M.; Maimone, G.; La Ferla, R.; Cosenza, A.; Rappazzo, A.C.; Caruso, G.; Paranhos, R.; Cabral, A.S.; Forte, E.; Guglielmin, M. The prokaryotic community in an extreme Antarctic environment: The brines of Boulder Clay lakes (Northern Victoria Land). *Hydrobiologia* **2021**, *848*, 1837–1857. [CrossRef]
6. Preston, L.; Dartnell, L. Planetary habitability: Lessons learned from terrestrial analogues. *Int. J. Astrobiol.* **2014**, *13*, 81–98. [CrossRef]
7. Porcino, N.; Cosenza, A.; Azzaro, M. A review on the geochemistry of lakes in Victoria Land (Antarctica). *Chemosphere* **2020**, *251*, 126229. [CrossRef] [PubMed]
8. Murray, A.E.; Kenig, F.; Fritsen, C.H.; McKay, C.P.; Cawley, K.M.; Edwards, R.; Kuhn, E.; McKnight, D.M.; Ostrom, N.E.; Peng, V.; et al. Microbial life at −13 °C in the brine of an ice-sealed Antarctic lake. *Proc. Natl. Acad. Sci. USA* **2012**, *109*, 20626–20631. [CrossRef] [PubMed]
9. Dugan, H.A.; Doran, P.T.; Wagner, B.; Kenig, F.; Fritsen, C.H.; Arcone, S.A.; Kuhn, E.; Ostrom, N.E.; Warnock, J.P.; Murray, A.E. Stratigraphy of Lake Vida, Antarctica: Hydrologic implications of 27 m of ice. *Cryosphere* **2015**, *9*, 439–450. [CrossRef]
10. Forte, E.; Dalle Fratte, M.; Azzaro, M.; Guglielmin, M. Pressurized brines in continental Antarctica as a possible analogue of Mars. *Sci. Rep.* **2016**, *6*, 33–58. [CrossRef] [PubMed]
11. Rizzo, C.; Conte, A.; Azzaro, M.; Papale, M.; Rappazzo, A.C.; Battistel, D.; Roman, M.; Lo Giudice, A.; Guglielmin, M. Cultivable bacterial communities in brines from perennially ice-covered and pristine Antarctic lakes: Ecological and biotechnological implications. *Microorganisms* **2020**, *8*, 819. [CrossRef]
12. Lo Giudice, A.; Conte, A.; Papale, M.; Rizzo, C.; Azzaro, M.; Guglielmin, M. Prokaryotic diversity and metabolically active communities in brines from two perennially ice-covered Antarctic lakes. *Astrobiology* **2021**, *21*, 551–565. [CrossRef] [PubMed]
13. Oyewusi, H.A.; Wahab, R.A.; Edbeib, M.F.; Mohamad, M.A.N.; Hamid, A.A.A.; Kaya, Y.; Huyop, F. Functional profiling of bacterial communities in Lake Tuz using 16S rRNA gene sequences. *Biotechnol. Biotechnol. Equip.* **2021**, *35*, 1–10. [CrossRef]
14. Sannino, C.; Borruso, L.; Mezzasoma, A.; Battistel, D.; Zucconi, L.; Selbmann, L.; Azzaro, M.; Onofri, S.; Turchetti, B.; Buzzini, P.; et al. Intra- and inter-cores fungal diversity suggests interconnection of different habitats in an Antarctic frozen lake (Boulder Clay, Northern Victoria Land). *Environ. Microbiol.* **2020**, *22*, 3463–3477. [CrossRef] [PubMed]
15. Douglas, G.M.; Maffei, V.J.; Zaneveld, J.; Yurgel, S.N.; Brown, J.N.; Taylor, C.M.; Huttenhower, C.; Langille, M.G.I. PICRUSt2: An improved and extensible approach for metagenome inference. *bioRxiv* **2019**, 672295. [CrossRef]
16. Langille, M.G.I.; Zaneveld, J.; Caporaso, J.G.; McDonald, D.; Knights, D.; Reyes, J.A.; Clemente, J.C.; Burkepile, D.E.; Thurber, R.L.V.; Knight, R.; et al. Predictive functional profiling of microbial communities using 16S rRNA marker gene sequences. *Nat. Biotechnol.* **2013**, *31*, 814–821. [CrossRef] [PubMed]
17. Kanehisa, M.; Furumichi, M.; Tanabe, M.; Sato, Y.; Morishima, K. KEGG: New perspectives on genomes, pathways, diseases and drugs. *Nucleic Acids Res.* **2017**, *45*, 353–361. [CrossRef]
18. Heberle, H.; Meirelles, G.V.; da Silva, F.R.; Telles, G.P.; Minghim, R. InteractiVenn: A web-based tool for the analysis of sets through Venn diagrams. *BMC Bioinform.* **2015**, *16*, 169. [CrossRef]
19. La Ferla, R.; Azzaro, M.; Michaud, L.; Caruso, G.; Lo Giudice, A.; Paranhos, R.; Cabral, A.S.; Conte, A.; Cosenza, A.; Maimone, G.; et al. Prokaryotic abundance and activity in permafrost of the Northern Victoria Land and Upper Victoria Valley (Antarctica): A study case. *Microb. Ecol.* **2017**, *74*, 402–415. [CrossRef]
20. Papale, M.; Conte, A.; Mikkonen, A.; Michaud, L.; La Ferla, R.; Azzaro, M.; Caruso, G.; Paranhos, R.; Cabral, A.S.; Maimone, G.; et al. Prokaryotic assemblages within permafrost active layer at Edmonson Point (Northern Victoria Land, Antarctica). *Soil Biol. Biochem.* **2018**, *123*, 165–179. [CrossRef]
21. Kalcheva, H.; Beshkova, M.; Pehlivanov, L.; Kalcev, R. Bacterioplankton dynamics and the influence of environmental factors on it in the Srebarna Lake. In Proceedings of the The Third International Scientific Conference BALWOIS, Ohrid, Republic of Macedonia, 27–31 May 2008.
22. Young, K.D. The selective value of bacterial shape. *Microbiol. Mol. Biol. Rev.* **2006**, *70*, 660–703. [CrossRef]
23. Gentile, G.; Maimone, G.; La Ferla, R.; Azzaro, M.; Catalfamo, M.; Genovese, M.; Santisi, S.; Maldani, M.; Macrì, A.; Cappello, S. Phenotypic variations of *Oleispira antarctica* RB8T in different growth conditions. *Curr. Microbiol.* **2020**, *77*, 3414–3421. [CrossRef]
24. Rezaeinejad, S.; Ivanov, V. Heterogeneity of *Escherichia coli* population by respiratory activity and membrane potential of cells during growth and long-term starvation. *Microbiol. Res.* **2011**, *166*, 129–135. [CrossRef]
25. Posch, T.; Pernthaler, J.; Alfreider, A.; Psenner, R. Cell-specific respiratory activity of aquatic bacteria studied with the tetrazolium reduction method, cyto-clear slides, and image analysis. *Appl. Environ. Microbiol.* **1997**, *63*, 867–873. [CrossRef]
26. Kenarova, A.; Encheva, M.; Chipeva, V. Physiological diversity of bacterial communities from different soil locations on Livingston Island, South Shetland archipelago, Antarctica. *Pol. Biol.* **2013**, *36*, 223–233. [CrossRef]

27. Loferer-Krössbacher, M.; Klima, J.; Psenner, R. Determination of bacterial cell dry mass by transmission electron microscopy and densitometric image analysis. *Appl. Environ. Microbiol.* **1998**, *64*, 688–694. [CrossRef]
28. Deming, J.W. Psychrophiles and polar regions. *Curr. Opin. Microbiol.* **2002**, *5*, 301–309. [CrossRef]
29. Wagner, D.; Kobabe, S.; Liebner, S. Bacterial community structure and carbon turnover in permafrost-affected soils of the Lena Delta, North Eastern Siberia. *Can. J. Microbiol.* **2009**, *55*, 73–83. [CrossRef]
30. Laybourn-Parry, J.; Pearce, D.A. The biodiversity and ecology of Antarctic lakes: Models for evolution. *Philos. Trans. R. Soc. B Biol. Sci.* **2007**, *362*, 2273–2289. [CrossRef] [PubMed]
31. Huston, A.L.; Methe, B.; Deming, J.W. Purification, characterization, and sequencing of an extracellular cold-active aminopeptidase produced by marine psychrophile *Colwellia psychrerythraea* strain 34H. *Appl. Environ. Microbiol.* **2004**, *70*, 3321–3328. [CrossRef] [PubMed]
32. Ortega, G.; Lain, A.; Tadeo, X.; Lopez-Mendez, B.; Castano, D.; Millet, O. Halophilic enzyme activation induced by salts. *Sci. Rep.* **2011**, *1*, 6. [CrossRef]
33. Karan, R.; Capes, M.D.; DasSarma, S. Function and biotechnology of extremophilic enzymes. *Biochemistry* **2011**, *76*, 686–693.
34. Lei, F.; Zhao, Q.; Sun-Waterhouse, D.; Zhao, M. Characterization of a salt-tolerant aminopeptidase from marine *Bacillus licheniformis* SWJS33 that improves hydrolysis and debittering efficiency for soy protein isolate. *Food Chem.* **2016**, *214*, 347–353. [CrossRef]
35. Salwan, R.; Sharma, V.; Chand Kasan, R.; Gulati, A. Bioprospecting psychrotrophic bacteria for serine-type proteases from the cold areas of the Western Himalayas. *Curr. Microbiol.* **2020**, *77*, 795–806. [CrossRef]
36. Qin, Y.; Huang, Z.; Liu, Z. A novel cold-active and salt-tolerant α-amylase from marine bacterium *Zunongwangia profunda*: Molecular cloning, heterologous expression and biochemical characterization. *Extremophiles* **2014**, *18*, 271–281. [CrossRef]
37. Wang, Q.F.; Miao, J.L.; Hou, Y.H.; Ding, Y.; Wang, G.D.; Li, G.Y. Purification and characterization of an extracellular cold-active serine protease from the psychrophilic bacterium *Colwellia* sp. NJ341. *Biotechnol. Lett.* **2005**, *27*, 1195–1198. [CrossRef]
38. Tchigvintsev, A.; Tran, H.; Popovic, A.; Kovacic, F.; Brown, G.; Flick, R.; Hajighasemi, M.; Egorova, O.; Somody, J.C.; Tchigvintsev, D.; et al. The environment shapes microbial enzymes: Five cold-active and salt-resistant carboxylesterases from marine metagenomes. *Appl. Microbiol. Biotechnol.* **2015**, *99*, 2165–2178. [CrossRef] [PubMed]
39. Cox, G.F.N.; Weeks, W.F. CRREL Report 82-30, Equations for determining the gas and brine volumes in sea ice samples. *J. Glaciol.* **1983**, *29*, 306–316. [CrossRef]
40. Hoppe, H.G.; Arnosti, C.; Herndl, G.F. Ecological significance of bacterial enzymes in marine environment. In *Microbial Enzymes in the Environment Activity, Ecology and Applications*; Burns, R.C., Dick, R.P., Eds.; Marcel Dekker Inc.: New York, NY, USA, 2002; pp. 73–107.
41. Maxwell, S.G. Acquisition, Degradation, and Cycling of Organic Matter within Sea-Ice Brines by Bacteria and Their Viruses. Ph.D. Thesis, University of Washington, Washington, DC, USA, 2020; pp. 1–168.
42. Stedmon, C.A.; Thomas, D.N.; Papadimitriou, S.; Granskog, M.A.; Dieckmann, G.S. Using fluorescence to characterize dissolved organic matter in Antarctic sea ice brines. *J. Geophys. Res. Biogeosci.* **2011**, *116*, G03027. [CrossRef]
43. Cox, M.; Lehninger, A.L.; Nelson, D.R. *Lehninger Principles of Biochemistry*; Worth Publishers: New York, NY, USA, 2000; pp. 306–308.
44. Sun, S.; Jones, R.B.; Fodor, A.A. Inference-based accuracy of metagenome prediction tools varies across sample types and functional categories. *Microbiome* **2020**, *8*, 46. [CrossRef]
45. Chistoserdova, L.; Laukel, M.; Portais, J.C.; Vorholt, J.A.; Lidstrom, M.E. Multiple formate dehydrogenase enzymes in the facultative methylotroph *Methylobacterium extorquens* AM1 are dispensable for growth on methanol. *J. Bacteriol.* **2004**, *186*, 22–28. [CrossRef]
46. Stibal, M.; Wadham, J.L.; Lis, G.P.; Telling, J.; Pancost, R.D.; Dubnick, A.; Sharp, M.J.; Lawson, E.C.; Butler, C.E.H.; Hasan, F.; et al. Methanogenic potential of Arctic and Antarctic subglacial environments with contrasting organic carbon sources. *Glob. Chang. Biol.* **2012**, *18*, 3332–3345. [CrossRef]
47. Ortiz, M.; Bosch, J.; Coclet, C.; Johnson, J.; Lebre, P.; Salawu-Rotimi, A.; Vikram, S.; Makhalanyane, T.; Cowan, D. Microbial nitrogen cycling in Antarctic soils. *Microorganisms* **2020**, *8*, 1442. [CrossRef] [PubMed]
48. Santos, A.A.; Venceslau, S.S.; Grein, F.; Leavitt, W.D.; Dahl, C.; Johnston, D.T.; Pereira, I.A. A protein trisulfide couples dissimilatory sulfate reduction to energy conservation. *Science* **2015**, *350*, 1541–1545. [CrossRef] [PubMed]
49. Griesbeck, C.; Hauska, G.; Schütz, M. Biological Sulfide-Oxidation: Sulfide-Quinone Reductase (SQR), the Primary Reaction. In *Recent Research Developments in Microbiology*; Pandalai, S.G., Ed.; Research Signpost: Trivadrum, India, 2000; Volume 4, pp. 129–203.
50. Karr, E.A.; Sattley, W.M.; Jung, D.O.; Madigan, M.T.; Achenbach, L.A. Remarkable diversity of phototrophic purple bacteria in a permanently frozen Antarctic lake. *Appl. Environ. Microbiol.* **2003**, *69*, 4910–4914. [CrossRef]
51. Karr, E.A.; Sattley, W.M.; Rice, M.R.; Jung, D.O.; Madigan, M.T.; Achenbach, L.A. Diversity and distribution of sulfate-reducing bacteria in permanently frozen Lake Fryxell, McMurdo Dry Valleys, Antarctica. *Appl. Environ. Microbiol.* **2005**, *71*, 6353–6359. [CrossRef]
52. Christner, B.C.; Priscu, J.C.; Achberger, A.M.; Barbante, C.; Carter, S.P.; Christianson, K.; Michaud, A.B.; Mikucki, J.A.; Mitchell, A.C.; Skidmore, M.L.; et al. WISSARD Science Team. Subglacial Lake Whillans: A microbial ecosysytem beneath the West Antarctic Ice Sheet. *Nature* **2014**, *512*, 310–313. [CrossRef]

53. Mikucki, J.; Pearson, A.; Johnston, D.T.; Turchyn, A.V.; Farquhar, J.; Schrag, D.P.; Anbar, A.D.; Priscu, J.C.; Lee, P.A. A contemporary microbially maintained subglacial ferrous ocean. *Science* **2009**, *324*, 397–400. [CrossRef]
54. Sattley, W.M.; Madigan, M.T. Isolation, characterization, and ecology of cold-active, chemolithotrophic, sulfur-oxidizing bacteria from perennially ice-covered Lake Fryxell, Antarctica. *Appl. Environ. Microbiol.* **2006**, *72*, 5562–5568. [CrossRef]
55. Sattley, W.M.; Madigan, M.T. Temperature and nutrient induced responses of Lake Fryxell sulfate-reducing prokaryotes and description of *Desulfovibrio lacusfryxellense* sp. nov., a pervasive, cold-active, sulfate-reducing bacterium from Lake Fryxell, Antarctica. *Extremophiles* **2010**, *14*, 357–366. [CrossRef]

diversity

MDPI

Article

What Makes a Hot-Spring Habitat "Hot" for the Hot-Spring Snake: Distributional Data and Niche Modelling for the Genus *Thermophis* (Serpentes, Colubridae)

Sylvia Hofmann [1,*], Peter Fritzsche [2], Tsering Dorge [3], Georg Miehe [4] and Michael Nothnagel [5,6]

1 Centre of Taxonomy and Evolutionary Research, Zoological Research Museum Alexander Koenig, Adenauerallee 160, 53113 Bonn, Germany
2 Institute of Zoology, Martin-Luther University, Domplatz 4, 06108 Halle, Germany; fritzsche@zoologie.uni-halle.de
3 Department of Biology, Faculty of Science, Tibet University, Lhasa 850000, China; tseringdorge@protonmail.com
4 Department of Geography, Philipps-Universität Marburg, Deutschhausstrasse 12, 35032 Marburg, Germany; miehe@geo.uni-marburg.de
5 Statistical Genetics and Bioinformatics, Cologne Center for Genomics, University of Cologne, Weyertal 115b, 50931 Cologne, Germany; michael.nothnagel@uni-koeln.de
6 Faculty of Medicine, The Cologne University Hospital, Kerpener Str. 62, 50937 Cologne, Germany
* Correspondence: s.hofmann@leibniz-zfmk.de

Abstract: Knowledge about species' distributions is central to diverse applications in ecology, biogeography, and conservation science. Hot-spring snakes of the genus *Thermophis* share a distribution restricted to geothermal sites at the Tibetan Plateau (*T. baileyi*) and in the Hengduan Mountains (*T. zhaoermii*, *T. shangrila*). Although the suture zones of these regions are widely covered with hot springs, *Thermophis* populations are restricted to only a few of these habitats. Here, we use bioclimatic, topographic, and land cover data to model the potential distribution of the genus. Moreover, using logistic regression on field survey data of *T. zhaoermii*, we test whether hot-spring water parameters and landscape features correlate with the species' presence or absence. Hot springs with temperatures between 45 and 100 °C and winter precipitation showed the most predictive power. At small scale, our data support the relevance of the hot-spring temperature on the species' occurrence and indicate that also the along-valley distance from the hot-spring site to the major river might influence the distribution of *Thermophis* species. Our findings contribute to better understand factors shaping the current distribution of the genus and will aid in setting priorities in applied conservation biology for the hot-spring snakes.

Keywords: hot-spring keel-back; distribution; habitat suitability; Hengduan Mountains; Qinghai-Tibet-Plateau

Citation: Hofmann, S.; Fritzsche, P.; Dorge, T.; Miehe, G.; Nothnagel, M. What Makes a Hot-Spring Habitat "Hot" for the Hot-Spring Snake: Distributional Data and Niche Modelling for the Genus *Thermophis* (Serpentes, Colubridae) *Diversity* **2021**, *13*, 325. https://doi.org/10.3390/d13070325

Academic Editor: Luca Luiselli

Received: 5 July 2021
Accepted: 14 July 2021
Published: 16 July 2021

Publisher's Note: MDPI stays neutral with regard to jurisdictional claims in published maps and institutional affiliations.

Copyright: © 2021 by the authors. Licensee MDPI, Basel, Switzerland. This article is an open access article distributed under the terms and conditions of the Creative Commons Attribution (CC BY) license (https://creativecommons.org/licenses/by/4.0/).

1. Introduction

Understanding the factors underlying the distribution patterns of species is one of the major tasks of ecology and biogeography [1,2]. It is, moreover, of prime concern for biodiversity conservation.

The Tibetan Plateau and the Hengduan Mountains pose as unique biogeographic zones since they are among the world's ecologically most diverse areas [3,4]. The Hengduan Mountains comprise more than 500,000 km^2 of temperate and alpine ecosystems that run along a north–south orientation at the eastern margin of the Himalayan range, in the southeastern corner of the Tibetan Plateau. Its southeastern quarter is one of the world's most important biodiversity hotspots characterized by high levels of species richness and endemism [5]. Similarly, the Tibetan Plateau supports high species diversity. With an area of 2.5 million km^2 and an average elevation of 4500 m above sea level (a.s.l.), the highlands comprise 95% of the Earth's surface above 4400 m [6]. It encompasses the world's largest subnival and alpine ecosystems, and a multitude of temperate and subtropical mountain forest ecosystems [7,8].

Their biodiversity faces major threats from biotope destruction, habitat loss, hydropower projects and pollution due to urbanization and economic development, illegal wildlife trade, and from the consequences of global climate warming [9–11]. Among the most critical conservation actions needed for the flora and fauna of both regions are the identification, establishment, and protection of areas that are important for species populations to mitigate a loss of biodiversity [12–14]. Information on species' distribution patterns and ecological niches is crucial for these tasks.

The rare hot-spring snakes *Thermophis zhaoermii*, GUO, LIU, FENG, and HE, 2008, and *T. shangrila*, PENG, LU, HUANG, GUO, and ZHANG, 2014, are geographically separated from their sister species, namely the Tibetan hot-spring snake *T. baileyi*, WALL, 1907. All species share a strong association with geothermal sites in high elevation areas and, thus, also a severe threat from habitat destruction due to the construction of geothermal power stations or other commercial establishments at the hot springs [15,16]. A major deficiency in knowledge about these species is the extent of their geographic range, especially with respect to *T. zhaoermii* and *T. shangrila*. While a patchy distribution along the main part of the Yarlung suture zone in southern central Tibet has been suggested for *T. baileyi* [17], *T. zhaoermii* has only been recorded from its type locality in Litang County, Sichuan Province [18,19], while *T. shangrila* is known from two localities in Shangri-La, Yunnan province (Figure 1; [20,21]).

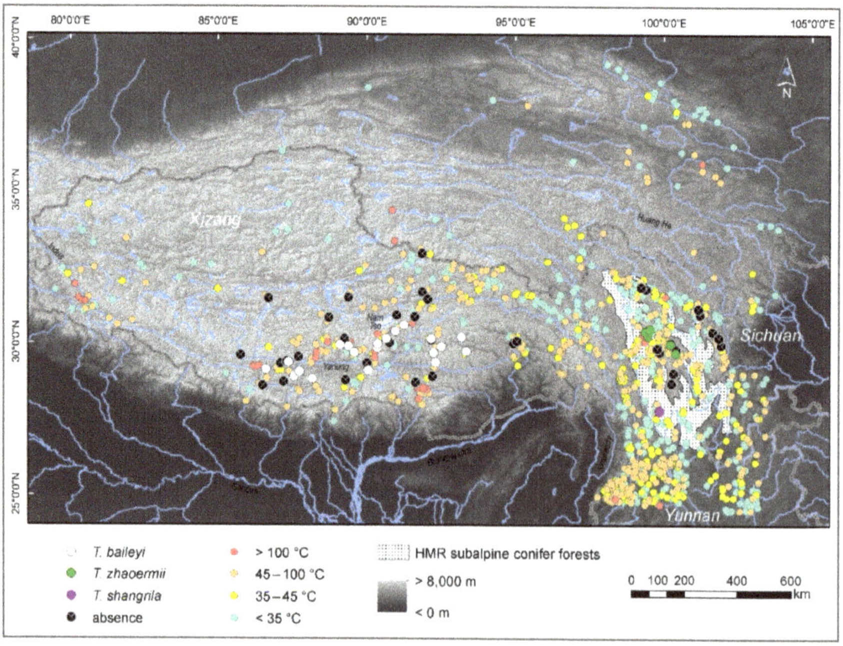

Figure 1. Map of the currently known distribution of *Thermophis baileyi*, *T. zhaoermii*, and *T. shangrila*. Records are according to [17,18,21] and based on our survey data. Locations where we did not find *Thermophis* are included in the map (absence). Hot-spring sites are drawn according to the "Atlas of Tibet" [22]. Type localities of the three species are imprecise in the original descriptions and cannot be mapped exactly–*T. baileyi*: "Thibet"; *T. shangrila*: "Shangri-La, Northern Yunnan, China" but coordinates match Yulong, Lijiang, Yunnan Province, China; *T. zaoermii*: Litang County, Sichuan, China (probably locality no. 12 according to [20]). The extension of the Hengduan Mountains region (HMR) is roughly reflected by its typical temperate coniferous forests ecoregion [23].

Compared with the Tibetan Plateau where hot springs are much more sparsely distributed, the Hengduan Mountains are densely covered with hot springs [22] (Figure 1). Surprisingly, the distribution of *T. zhaoermii* and *T. shangrila* is virtually unknown for the Hengduan Mountains and adjacent mountain ranges, despite multiple herpetological field surveys that have been carried out in these regions (e.g., [24–28]). We suppose that this apparent contradiction between the large number of potential habitats and lack of hot-spring snake records might be explained by specific environmental characteristics that impact the suitability of a hot-spring locality for species of *Thermophis* and, thus, determine the presence or absence of this taxon in apparently similar habitats. Therefore, in the present study we aimed to model the habitat suitability and potential distribution of the genus to determine the environmental factors related to its occurrence. We further sought to identify chemical, physical, and spatial hot-spring water characteristics that may impact the distribution of the hot-spring snakes by collecting a comprehensive data set on water quality of surveyed hot springs. Our data represent the first evaluation of environmental features and occurrences of this enigmatic genus. The results encourage discussion on our understanding of the distribution of *Thermophis* and can have important implications for the management and conservation of these rare high-elevation snakes.

2. Materials and Methods

2.1. Occurrence Data and Ecological Niche Analyses

A total of 20 records of *T. baileyi*, 6 records of *T. zhaoermii*, and 1 record of *T. shangrila* were available to us (Figure 1, Supplementary Materials Figure S1). All of them, except the locality of *T. shangrila*, were obtained from field surveys during 2003–2014 [17,29] (Supplementary Text S1 and Table S1). Since many hot springs are located in remote, inaccessible areas, a systematic survey of such sites is difficult, if not impossible due to logistic and financial challenges. Therefore, locations were selected based on hot-spring literature references [22] and information provided by residents. Sites were visited between May and August, and were surveyed intensively for up to four days, depending on weather conditions (for details on site visits see Supplementary Text S1). We are aware that the absence of secretive species like snakes from a given site can only be determined with a certain probability and strongly depends, among other factors, on population density and number of visits [30]. However, hot-spring snakes are not difficult to find if looked for under suitable weather conditions during their active season. The availability of snake distribution data and the assistance of local residents were essential components of our successful searches. Geographical coordinates and altitude (ALT) were recorded with a GPS receiver using the WGS84 datum and imported into ArcGIS 10.8 (ESRI, Redlands, CA, USA). Overall, we surveyed 66 habitats for presence or absence of *T. baileyi* (n = 44; [17]) and *T. zhaoermii* (n = 22; Supplementary Table S1). The third species, *T. shangrila*, is known only from the holotype location and a further site in Shangri-La (Dêqên). However, since the coordinates given in the original description do not match the correspondingly specified locality "Shangri-La" [20] but Yulong, Lijiang, Yunnan Province, China, we excluded this record due to its spatial discrepancy. Due to the low number of reported localities for the three species and the general strong preference of the genus for hot-spring habitats, we combined all *Thermophis* records into a single data set.

Grids of 19 standard bioclimatic variables for the current climate and elevation were downloaded from the WorldClim database [31] (http://www.worldclim.org (accessed on 7 June 2021)). Hot springs were digitized from, and according to "The Atlas of Tibet" [22] using ArcGIS 10.8, subdividing them into the following groups (layers): 5–35 °C, >35–45 °C, >45–100 °C, and >100 °C. All layers were projected to WGS84 at a resolution of 30 arc-seconds (~1 km grid cells at the equator) and clipped to the extent of Tibet s.l. (incl. the northern part of the Indian state Arunachal Pradesh), and Sichuan and Yunnan Provinces, which cover all known *Thermophis* records, and which together correspond roughly to the main area where hot springs exist (Figure 1). Prior to final model construction, we explored all climate variables and the elevation data for autocorrelation by calculating Pearson's correlation coefficients (r) using the python script SDMtoolbox v.2.4 [32] available for ArcGIS 10.8 and removed highly correlated variables (r > 0.8). The final set of environmental predictors comprised six variables: BIO1 = annual mean temperature, BIO2 = mean diurnal temperature range, BIO3 = isothermality, BIO4 = temperature seasonality, BIO15 = precipitation seasonality, and BIO19 = precipitation of coldest quarter. Principal components analysis (PCA) was carried out for these variables with the SDMtoolbox, reducing them to three orthogonal principal components describing the majority (>99%) of the variability in climate. We then used these three components to assess the climatic heterogeneity across the area of interest. To eliminate spatial clusters of localities we spatially filtered our presence data by Euclidian distances (min 2 km, max 20 km; 5 distance classes) according to climate heterogeneity using the rarefying module in SDMtoolbox. This resulted in the exclusion of one site (no. 15, see Supplementary Figure S1, Table S1). We also included land cover (LC) information in the model using GlobCover Land Cover V2.2 provided by the European Spatial Agency (http://due.esrin.esa.int/page_globcover.php (accessed on 7 June 2021)), as well as the river network (RN) using vmap0 data obtained from the National Geospatial-Intelligence Agency (https://www.nga.mil (accessed on 7 June 2021)). The original land cover map was rescaled to match the 30 arc-seconds resolution of the other variables. Vector data were converted to a raster using the same cell size as all other layers.

The final models were generated with the Maxent algorithm [33,34] which is extensively used for analyzing presence-only data and based on the principle of maximum entropy. Maxent is known for its high predictive accuracy compared with other modelling methods [35], and robustness to small sample sizes [36]. In the occurrence dataset, we implemented the filtered 25 georeferenced localities corresponding to the surveyed populations of *Thermophis* (Figure 1, Supplementary Table S1; [17]). These occurrences were randomly split into training data (75%) and test data (25%), and 100 subsampled (model 1) or bootstrapped (model 2) replicates were run for model evaluation using the following parameters: regularization = 1.5; max. iterations = 50,000; min. threshold = 0.00001; output = logistic. Both of the two models comprised BIO1-4, 15, 19, LC, RN, and the hot-spring layers. Model performance and the importance of the environmental variables to the model were assessed using the mean area under the curve (AUC) of the receiver operating characteristics (ROC; [37,38]), and jack-knife testing. Jack-knife testing obtains models by first omitting each variable and then using only one variable to determine the importance of an environmental variable to the predictive distribution of a species [33]. The AUC ranges from 0 to 1 and can be interpreted as the probability that a randomly chosen presence site will be ranked above a randomly chosen absence or pseudo-absence (background) site [39]. An AUC value close to 1 indicates higher model fit, a value of 0.5 implies random prediction, and values <0.5 indicate performance worse than random; models with values above 0.75 are considered potentially informative [35], good between 0.8 and 0.9, and excellent for AUC between 0.9 and 1 [40]. Although, shortcomings of AUC have been made evident [41,42], we considered that criterion a reliable measurement to compare models generated for the same genus, within the same area and under the same settings.

2.2. Hot-Spring Water Characteristics

To include microhabitat variables in our assessment, in particular water characteristics of the hot springs, we additionally obtained several water chemistry measurements from a total of 16 hot-spring sites that were surveyed for *T. zhaoermii* in the Hengduan Mountains. All the surveyed sites, except for no. 1, were located in Sichuan Province, China (Supplementary Figure S1; Table S1) and presented the following characteristics, which were common to all habitats where the species has been recorded [17]: sites were in river valleys with rocky slopes and vegetated shorelines, within a short distance to a river (0–720 m), and could be considered as low-sulfur sites (i.e., without characteristic sulfur smell).

Since hot-spring snakes are suggested to spread mainly along river valleys [17], we assume increasing distance of a hot-spring habitat from a (main) riverine corridor to cause higher costs for the migratory pathway and, thus, to lower the probability that a

Thermophis population has been established itself in that habitat. Therefore, we determined the length of the riverine corridor from the respective hot spring (located in the river valley) to the point where this river merges with the superordinate river and, further, with a major river that runs through the Hengduan Mountains (Dajin/Dadu, Jinsha, Litang or Yalong River) to explore the impact of the corridor length on species presence or absence. Data on distances of a hot-spring site to (i) the nearest small (seasonal) river (SRD), (ii) the next merging point with the superordinate river (tributary flow path of SRD; TRIB), and (iii) to the junction with the major river (MRD) were extracted as pathways along the river valleys based on Google Earth layer using ArcGIS 10.8. Measurements of hot-spring water chemistry were performed under field conditions next to the respective hot spring. We recorded the following parameters with a portable digital data logger PCE-PHD 1 (PCE Instruments, Meschede, Germany): temperature (TEMP; 0.1 $\pm$ 0.8 °C), conductivity (COND; 0.1 µS $\pm$ 2%), total dissolved solids (TDS; 1 ppm $\pm$ 2%), pH (0.01 $\pm$ 0.02 pH), oxidation reduction potential (ORP; 1 mV $\pm$ 0.5%), salinity (SAL; 0.01 $\pm$ 0.05%), and dissolved oxygen (OX; 0.1 $\pm$ 0.4 µg/L). Measurements of COND, pH, and OX were automatically temperature compensated. OX was moreover manually compensated for altitude. In order to determine several ionic concentrations, we used semi-quantitative (colorimetric) test kits (NO_3^-, S^{2-}, PO_4^{3-}, NH_3/NH_4, carbonate hardness: JBL GmbH, Neuhofen, Germany; S^{2-}: HC393108 and HC41474, Merck, KGaA, Darmstadt, Germany; SO_4^2: Hanna Instruments, Vöhringen, Germany). Since the Maxent models showed the importance of BIO19 for the distribution of *Thermophis*, we extracted these data for the surveyed sites from WorldClim.

2.3. Logistic Regression Models

We used binomial regression models to test whether certain hot-spring water characteristics, climatic, and/or landscape features were favorable for the species' presence or absence (modelled as dependent variable and coded as 1 and 0, respectively). Given the very small sample size of our dataset, we were left to avoid model over-fitting of the data by applying exclusively an exploratory approach. In a first step, we calculated Spearman correlation coefficients for all pairs of water characteristics and removed variables with coefficients >0.7 in order to avoid multicollinearity (Supplementary Table S2). Next, we performed a forward model selection based on Akaike's information criterion (AIC; [43]), adding the remaining variables stepwise to the null model, to explore which variables could best discriminate between presence and absence of the species. All statistical analyses were carried out using the glm, step, and anova functions in R v.4.1.0 [44].

3. Results

3.1. Climate Heterogeneity, Maxent Models

For the set of six analyzed climate variables, 99.7% of the variance was explained by the first ordination axis (Supplementary Figure S2A). Overall, climate across the modelled area is spatially heterogeneous. Areas in the Yarlung River catchment at the Tibetan Plateau and in the central Hengduan Mountains, where *Thermophis* occur, share similar climatic conditions (indicated by similar, blue-reddish color, see Supplementary Figure S2A), while regions between and around these areas are different. Particularly southeastern Tibet, with a topography dropping below 3000 m a.s.l. in many valleys, shows varying and higher mean annual temperatures and precipitation rates. Similarly, northern areas of Tibet and Sichuan, central and southern Yunnan, and the Sichuan Basin are climatically different from those areas where *Thermophis* has been recorded. Regions of higher climate heterogeneity can be found in the southeastern part of Tibet, the southern Hengduan Mountains, and along the margin of the Sichuan basin; the known distribution of *Thermophis* matches climatically more homogeneous areas (Supplementary Figure S2B).

The two Maxent models produced consistent results and had high predictive power (AUC_{m1} 0.98 and AUC_{m2} 0.99; Figure 2, Supplementary Figure S3). The test omission rate of m1 (subsample approach) is close to the predicted omission, while the omission rate of m2 (bootstrap approach) slightly deviates from the predictive values for cumulative thresholds <0.5 (Figure 2, Supplementary Figure S3).

Hot springs with a temperature between 45 °C and 100 °C and the amount of precipitation in the coldest quarter of the year (i.e., winter snowfall; BIO19) had the highest percentage contribution to the models (58%, 15% in m1, and 56%, 14% in m2, respectively; Figure 2, Supplementary Figure S3, Table S2). Both maps of the two models show very similar patchy distribution patterns of habitat suitability with high values (yellow to red) corresponding to pixels that overlap with several hot-spring sites in the central Hengduan Mountains and in the tributary valleys of the Yarlung River in southern central Tibet (Figure 2, Supplementary Figure S3). These suitable habitats are located in areas with lower winter precipitation than surrounding regions (Figure 3; Supplementary Figure S4). Hot springs outside these ranges are associated with low values of habitat suitability (green), and areas without hot springs are unsuitable (blue).

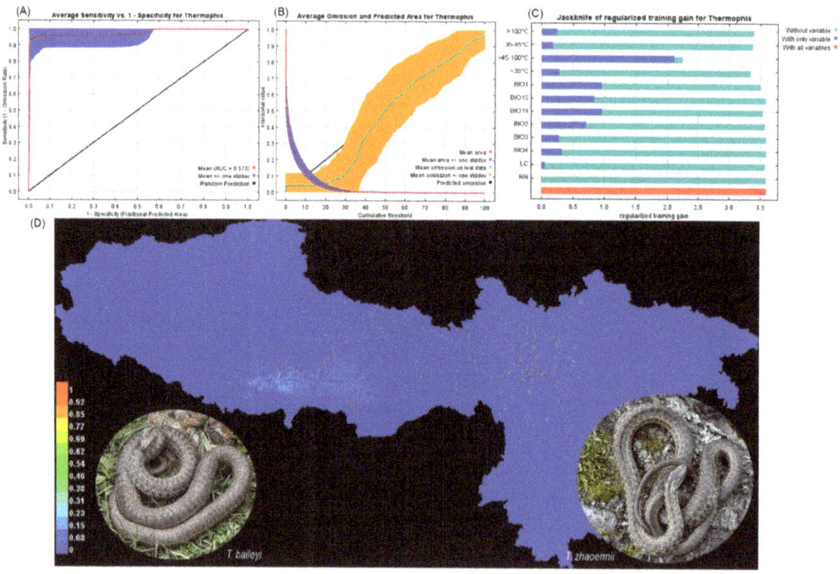

Figure 2. Diagnostic plots and resulting map for the Maxent model of *Thermophis* based on subsample algorithm and 100 replicates. Panel (**A**) depicts the mean area under the curve (AUC) of the Maxent receiver operator characteristic curve, (**B**) shows the omission rate and predicted area as a function of the cumulative threshold, averaged over the replicate runs, (**C**) represents the results of the jackknife test of variable importance, and (**D**) shows the Maxent predictions of suitable habitat for *Thermophis* (point-wise mean of the 100 output grids). Photo credits on *Thermophis* images: Sylvia Hofmann.

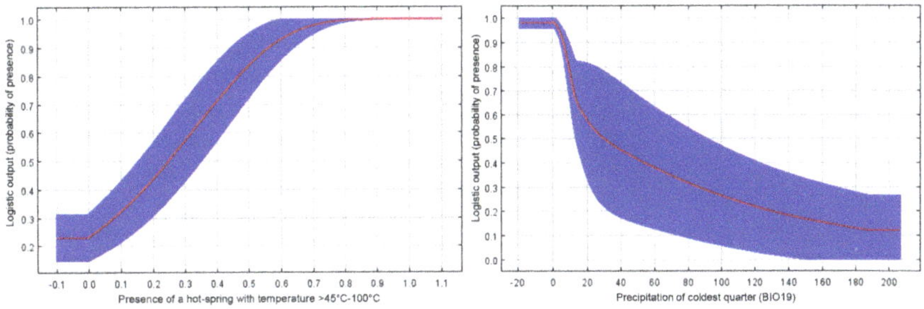

Figure 3. Response curves of the hot-spring layer (>45–100 °C) and the precipitation of the coldest quarter in the Maxent model. The curves show how the predicted probability of presence changes as the respective environmental variable is varied, keeping all other environmental variables at their average sample value.

3.2. Habitat Characteristics

We detected *T. zhaoermii* in 6 out of the 22 surveyed sites, including the type locality in Litang county (Supplementary Figure S1; Table S1). All of these few "positive" sites were located west of the Yangtze River at elevations between 3600 and 4090 m (a.s.l.). Three further sites were considered as "uncertain" with respect to the presence of the species and were excluded from subsequent analysis. At these sites, we observed snakes only from long distances in difficult mountainous terrain and could not verify them morphologically.

Spearman correlation coefficients >0.5 for all pairs of variables measured at the hot-spring sites ranged between 0.52 and 1.00. Variables TDS, SAL, and NH_3/NH_4 were excluded from subsequent analysis due to high pairwise correlation (>0.7) with other parameters (Supplementary Table S3); S^{2-} was removed due to invariability. The final dataset contained 14 variables. Forward model selection among these variables did not reveal an important contribution of any hot-spring water characteristics, the distance of a hot spring to the nearest small river or to the superordinate river. The final model comprised only two variables: MRD and TEMP (Supplementary Table S4). The former was nominally insignificant in a single-variable likelihood-ratio test (LRT), while TEMP turned out to be highly significant ($p = 8.73 \times 10^{-05}$; Supplementary Table S4), suggesting an impact of the hot-spring temperature on the presence of *T. zhaoermii*. Mean hot-spring water temperature at sites where hot-spring snakes occur was about 56 °C (41–70 °C), i.e., ~10° K warmer than hot-springs at which the snakes did not occur; the mean distance from these "presence" sites to a major river was ca. 70 km, in contrast to distances >120 km from sites without snakes (Supplementary Tables S5 and S6).

4. Discussion

Determining how species are distributed in space is a key issue in ecology and conservation biology and this knowledge is particularly important in endangered species. Our study provides first information on the potential geographic distribution range of the rare species *T. zhaoermii* (and *T. shangrila*) and largely confirms previous data on the occurrence area of *T. baileyi*. The most important predictors in our niche models were hot springs with temperatures ranging between 45 °C and 100 °C, and winter snowfall. Moreover, results of the fine-scale modelling support the relevance of the hot-spring temperature for the presence or absence of *Thermophis*. They also indicate an impact of the distance from a hot-spring site to a major river on the distribution pattern of the snakes. Although based on a relatively small data set, our study presents valuable insights into the potential distribution of *Thermophis*, which can assist in setting conservation priorities for those snakes and should encourage future ecological studies in the genus.

Species of the genus *Thermophis* are the only snakes with a vertical distribution up to 4900 m a.s.l. [29]. The only snake taxon to come close in the Himalaya is *Gloydius himalayanus* which has been recorded at 4772 m [45]. Living in such extreme habitats requires specific physical and or behavioral adaptations to cope with the high, cold, and arid environments. Besides molecular adaptive responses to oxidative stress and UV radiation that have been shown in *Thermophis* [21], the strong association of these snakes to hot-spring habitats has been interpreted as an ecological adaptation that evolved during the geological uplift of the Tibetan Plateau [19,29]. Irrespective of the variable taxonomic placement of *Thermophis* (Pseudoxenodontinae vs. Dipsadinae; [46] and references therein) and wide range of its estimated divergence time (MRCA *Thermophis* and xenodontine snakes ~10–28 Mya; [47]), ancestral lineages of this genus might have been present in the area of Paleo-Tibet long before the final uplift of the Plateau. With the continuously rising Himalaya-Tibet orogen and the associated cooling of the environment, these ancestors have probably been forced to retreat into geothermal active areas, or, otherwise, would have gone extinct. At least some of the hot-spring sites have served as refuges for the snakes during the Pleistocene [19]. It can be assumed that in the alpine zone hot springs may have become essential for *Thermophis* to be present to survive during cold periods and to achieve enough activity time that determines the species' fitness [48]. However, not every hot-spring site represents a suitable habitat for *Thermophis*. The high predictive power of springs with water temperatures between 45 °C and 100 °C in our models suggests a clear preference of the snakes for moderately but not too warm places. We suspect that the geothermal sites are particularly important to the snakes during and soon after hibernation in wintertime, and less relevant throughout the highest activity peak in summer. This may also explain why *Thermophis* can occasionally be found more distant from a hot-spring site during the middle of the year ([49]; pers. obs.), especially throughout the mating season [50]. Since hibernation sites at high elevations must protect snakes from colder conditions for longer periods of time, we assume that *Thermophis* hibernates underground near the geothermal field in rodent burrows, old root systems, or rocky outcrops. Here, they probably prefer thermal sites that are warm enough to protect them from the cold, but not hot enough to achieve their body temperatures necessary for appetite and digestion. Geothermal fields with very hot water temperatures may heat up deep soil layers that then become too warm and unsuitable for hibernating snakes.

The broad-scale influencing climatic factor for the distribution of *Thermophis* taxa is linked to precipitation during the coldest quarter of the year, with an increase in winter snowfall decreasing the probability of the snake's presence in a hot-spring habitat. Therefore, regions suitable for *Thermophis* experience significantly lower precipitation during the winter than do areas predicted to be unsuitable for the genus. A similar finding has been reported for *Eirenis persicus* in western Iran and Turkey [51]. In *Thermophis*, an intuitive explanation for this pattern is that the advantage of the snowpack by insulating the belowground hibernation site from ambient temperatures [52,53] might be negated in geothermal active locations. Moreover, due to the generally long hibernation season, high-altitude reptiles depend on the relatively short active season to maximize growth, reproduction, and energy storage. A higher winter precipitation, however, can result in the accumulation of a large seasonal snowpack, prolonged cold and, thus, an even shorter activity period for the snakes.

At a fine scale, our results support the hypothesis that the hot-spring temperature, and probably also the access to the hot spring over short distances via a major river, influence the distribution of *Thermophis*, although these findings have to be treated with caution due to the low number of observations. If true, long riverine pathways may have prevented a suitable hot-spring site being colonized by *Thermophis* in the past. It is well-known that river-mediated migration can be a key factor for the distribution of species, for example in the European pool frog *Rana lessonae* [54], and the common wall lizard, *Podarcis muralis* [55]. In a previous molecular study, we could show that populations of *T. baileyi* are connected through riverine corridors by the Yarlung River and its tributary system [17]. We suggest that, for hot-spring snakes, the suitability of a geothermal habitat may depend on the trade-off between species dispersal capacity and the prospects for successful settling that could be decreasing with increasing along-valley distance to the hot spring.

5. Conclusions

Today, the main areas of occurrence for *Thermophis* cover the southern central region of the Tibetan Plateau along the Yarlung suture zone, and the central range of the Hengduan Mountains (Chola Shan, Shaluli Shan). The fact that only these relatively narrow areas are predicted to be suitable, and host recorded presence localities suggests that the potential distribution of the species might not be much wider than our current records indicate [17]. The distribution map of the genus presented here provides details on hot-spring localities with a high probability of the species' presence, which could guide future surveys and conservation activities. It also indicates that the current distribution of these snakes depends on the amount of winter precipitation, while other climatic predictors seem to be less informative with respect to the distributional patterns of the genus. According to the IUCN Red List *T. baileyi* is considered as "near threatened", while *T. zhaoermii* as "endangered"; *T. shangrila* is not even registered in the IUCN red list. Both *T. baileyi* and *T. zhaoermii* are flagged with a decreasing population trend and their presence in any protected areas is unknown [15]. Given the geographic distribution maps in the most recent status assessment for the IUCN Red List, our predicted potential distribution of *Thermophis* differs substantially from these data. However, despite their exceptional value for biodiversity, no specific conservation action plans exist for the hot-spring snakes. Our results contribute to the knowledge about the distribution of these species and may guide future management of the genus, e.g., by surveying and protecting suitable habitats for the hot-spring snakes.

Supplementary Materials: The following are available online at https://www.mdpi.com/article/10.3390/d13070325/s1, Text S1: Details on site visits. Table S1: Hot-spring localities surveyed for *Thermophis zhaoermii*. Water parameters were measured at hot-spring sites 1–16 (see Figure S1 for geographic reference, below); localities 17–22 were surveyed for snakes only. Table S2: Estimates of relative contributions of the environmental variables to the Maxent model based on subsample (left) and bootstrap algorithm (right). Table S3: Spearman correlation coefficients (SCE) of microhabitat characteristics. The following variables showed pairwise correlation > |0.5|. Table S4: Analysis of deviance table (generalized linear model). Table S5: Summary statistics of variables in the model for "absence" localities of *Thermophis zhaoermii*. Table S6: Summary statistics of variables for "presence" localities of *Thermophis zhaoermii*. Figure S1: Known records of *Thermophis zhaoermii* and *T. shangrila* and sites surveyed for *Thermophis*. Details to the locality numbers are listed in Table S1. Figure S2: Climate heterogeneity in the distribution area of *Thermophis* species. Figure S3: Diagnostic plots for the Maxent model of *Thermophis* based on bootstrap algorithm and 100 replicates. Figure S4: Precipitation of the coldest quarter of the year (BIO19) in the modelled area.

Author Contributions: Conceptualization, S.H., P.F., T.D. and G.M.; methodology, S.H., P.F. and M.N.; analysis, S.H. and M.N.; investigation, S.H., P.F. and T.D.; resources, S.H., P.F., T.D., G.M. and M.N.; data curation, S.H. and M.N.; writing—original draft preparation, S.H.; writing—review and editing, all authors; visualization, S.H. and M.N.; funding acquisition, G.M. and S.H. All authors have read and agreed to the published version of the manuscript.

Funding: This research was funded by the German Research Foundation (DFG), grant number MI 271/21 to GM and SH.

Institutional Review Board Statement: The study was conducted according to the regulations for the protection of terrestrial wild animals.

Informed Consent Statement: Not applicable.

Data Availability Statement: All data are presented in the Supplementary Materials.

Acknowledgments: We are grateful to Markus Sandner for valuable help in the field and Shane Lindsey for linguistic improvement. We thank the Regional Computing Center of the University of Cologne (RRZK) for providing computing time on the DFG-funded High Performance Computing (HPC) system CHEOPS as well as technical support. We especially thank the Editor and anonymous reviewers for constructive input which improved the manuscript.

Conflicts of Interest: The authors declare no conflict of interest. The funders had no role in the design of the study; in the collection, analyses, or interpretation of data; in the writing of the manuscript, or in the decision to publish the results.

References

1. Krebs, C.J. *Ecology: The Experimental Analysis of Distribution and Abundance*, 4th ed.; Harper Collins College Publishers: New York, NY, USA, 1994.
2. Lomolino, M.V.; Riddle, B.R.; Whittaker, R.J.; Brown, J.H. *Biogeography*; Sinauer Associates: Sunderland, UK, 2010.
3. Boufford, D.E.; Van Dyck, P.P. *South-Central China in Hotspots: Earth's Biologically Richest and Most Endangered Terrestrial Ecoregions*; Mittermeier, R.A., Myers, N., Mittermeier, C.G., Eds.; CEMEX: Mexico City, Mexico, 1999; pp. 338–351.
4. Tang, Z.; Wang, Z.; Zheng, C.; Fang, J. Biodiversity in China's mountains. *Front. Ecol. Environ.* **2006**, *4*, 347–352. [CrossRef]
5. Mittermeier, R.A.; Robles Gil, P.; Hoffman, M.; Pilgrim, J.; Brooks, T.; Mittermeier, C.G.; Lamoreux, J.; da Fonseca, G.A.B. *Hotspots Revisited*; CEMEX: Mexico City, Mexico, 2004.
6. Zhang, Y.L.; Li, B.Y.; Zheng, D. A discussion on the boundary and area of the Tibetan Plateau in China. *Geogr. Res.* **2002**, *21*, 1–8.
7. Miehe, G.; Schleuss, P.M.; Seeber, E.; Babel, W.; Biermann, T.; Braendle, M.; Chen, F.; Coners, H.; Foken, T.; Gerken, T.; et al. The *Kobresia pygmaea* ecosystem of the Tibetan highlands—Origin, functioning and degradation of the world's largest pastoral alpine ecosystem: *Kobresia* pastures of Tibet. *Sci. Total Environ.* **2019**, *648*, 754–771. [CrossRef] [PubMed]
8. Xu, B.; Li, Z.M.; Sun, H. *Seed Plants of the Alpine Subnival Belt from the Hengduan Mountains*; SW China; China Science Publishing & Media Ltd. Group: Beijing, China, 2013.
9. Yang, Y.; Tian, K.; Hao, J.; Pei, S.; Yang, Y. Biodiversity and biodiversity conservation in Yunnan, China. *Biodivers. Conserv.* **2004**, *13*, 813–826. [CrossRef]
10. Li, Z.; He, Y.; An, W.; Song, L.; Zhang, W.; Catto, N.; Wang, Y.; Wang, S.; Liu, H.Q.; Cao, W.; et al. Climate and glacier change in southwestern China during the past several decades. *Environ. Res. Lett.* **2011**, *6*, 1–24. [CrossRef]
11. Qiu, J. Double threat for Tibet—Climate change and human development are jeopardizing the plateau's fragile environment. *Nature* **2014**, *512*, 240–241. [CrossRef]
12. Sherman, R.; Mullen, R.; Haomin, L.; Zhendong, F.; Yi, W. Alpine ecosystems of northwest Yunnan, China: An initial assessment for conservation. *J. Mt. Sci.* **2007**, *4*, 181–192. [CrossRef]
13. Sherman, R.; Mullen, R.; Haomin, L.; Zhendong, F.; Yi, W. Spatial patterns of plant diversity and communities in Alpine ecosystems of the Hengduan Mountains, northwest Yunnan, China. *J. Plant Ecol.* **2008**, *1*, 117–136. [CrossRef]
14. Watson, J.E.M.; Rao, M.; Ai-Li, K.; Yan, X. Climate Change Adaptation Planning for Biodiversity Conservation: A Review. *Adv. Clim. Chang. Res.* **2012**, *3*, 1–11. [CrossRef]
15. The IUCN Red List of Threatened Species. IUCN. Version 3.1. 2014. Available online: www.iucnredlist.org (accessed on 7 June 2021).
16. Huang, S.; Peng, L. Tibetan hot-spring snakes under threat. *Science* **2019**, *366*, 193–194. [CrossRef]
17. Hofmann, S.; Kraus, S.; Dorge, T.; Nothnagel, M.; Fritzsche, P.; Miehe, G. Effects of Pleistocene climatic fluctuations on the phylogeography, demography and population structure of a high-elevation snake species, *Thermophis baileyi*, on the Tibetan Plateau. *J. Biogeogr.* **2014**, *41*, 2162–2172. [CrossRef]
18. Guo, P.; Liu, S.Y.; Feng, J.C.; Miao, H.E. The description of a new species of *Thermophis* (Serpentes: Colubridae). *Sichuan J. Zool.* **2008**, *27*, 321.
19. Hofmann, S. Population genetic structure and geographic differentiation in the hot spring snake *Thermophis baileyi* (Serpentes, Colubridae): Indications for glacial refuges in southern-central Tibet. *Mol. Phylogenet. Evol.* **2012**, *63*, 396–406. [CrossRef] [PubMed]
20. Peng, L.; Lu, C.; Huang, S.; Guo, P.; Zhang, Y. A new species of the genus *Thermophis* (Serpentes: Colubridae) from Shangri-La, Northern Yunnan, China, with a proposal for an eclectic rule for species delimitation. *Asian Herpetol. Res.* **2014**, *5*, 228–239.
21. Li, J.-T.; Gao, Y.-D.; Xie, L.; Deng, C.; Shi, P.; Guan, M.-L.; Huang, S.; Ren, J.-L.; Wu, D.-D.; Ding, L.; et al. Comparative genomic investigation of high-elevation adaptation in ectothermic snakes. *Proc. Natl. Acad. Sci. USA* **2018**, *115*, 8406–8411. [CrossRef]
22. CAS. *Atlas of Tibet Plateau*; CAS, Institute of Geography: Bejing, China, 1990.
23. Olson, D.M.; Dinerstein, E.; Wikramanayake, E.D.; Burgess, N.D.; Powell, G.V.N.; Underwood, E.C.; D'Amico, J.A.; Itoua, I.; Strand, H.E.; Morrison, J.C.; et al. Terrestrial ecoregions of the world: A new map of life on Earth. *Bioscience* **2001**, *51*, 933–938. [CrossRef]
24. Kou, Z. Preliminary reports on the herpetofauna of Shuitang and Zhelong districts of the eastern slope of Mt. Ailao, with description of a new Species. *Acta Herpetol. Sin. Chengdu* **1984**, *3*, 39–45.
25. Chen, X.; Huang, Y. A herpetological survey of Muli county, Sichuan. *Acta Herpetol. Sin. Guiyang* **1995**, *4–5*, 117–123.
26. Chen, L.M.; Gao, Z.F.; Ou, W.F.; Chen, W.L.; Ma, W.-H. Surveys of amphibians and reptiles in Tangjiahe Nature Reserve, Sichuan Province. *Sichuan J. Zool.* **1999**, *18*, 132–134.
27. Pan, X.-F.; Zhou, W.; Zhou, Y.-W.; Wu, F.; Zhang, Q. Amphibian and Reptile in Zhongdian Area of Northwest Yunnan. *Sichuan J. Zool.* **2002**, *21*, 88–91.
28. Wang, D.; Song, Z.; Yue, B.; Ye, H.; Liu, S. A survey of herpetological resources in National Dafengding Nature Reserve, Meigu Co., Sichuan, China. *Sichuan J. Zool.* **2004**, *23*, 238–242.

29. Dorge, T.; Hofmann, S.; Wangdwei, M.; Duoje, L.; Solhøy, T.; Miehe, G. The ecological specialist, *Thermophis baileyi* (Wall, 1907)—New records, distribution and biogeographic conclusions. *Herpetol. Bull.* **2007**, *101*, 8–12.
30. Kéry, M. Inferring the Absence of a Species: A Case Study of Snakes. *J. Wildl. Manag.* **2002**, *66*, 330. [CrossRef]
31. Hijmans, R.J.; Cameron, S.E.; Parra, J.L.; Jones, P.G.; Jarvis, A. Very high resolution interpolated climate surfaces for global land areas. *Int. J. Climatol.* **2005**, *25*, 1965–1978. [CrossRef]
32. Brown, J.L. SDMtoolbox: A python-based GIS toolkit for landscape genetic, biogeographic and species distribution model analyses. *Methods Ecol. Evol.* **2014**, *5*, 694–700. [CrossRef]
33. Phillips, S.J.; Anderson, R.P.; Schapire, R.E. Maximum entropy modeling of species geographic distributions. *Ecol. Model.* **2006**, *190*, 231–259. [CrossRef]
34. Phillips, S.J.; Dudık, M.; Schapire, R.E. A maximum entropy approach to species distribution modeling. In Proceedings of the Twenty-First International Conference on Machine Learning, New York, NY, USA, 4–8 July 2004; pp. 655–662.
35. Elith, J.; Graham, C.H.; Anderson, R.P.; Dudík, M.; Ferrier, S.; Antoine, G.; Hijmans, R.J.; Huettmann, F.; Leathwick, J.R.; Lehmann, A.; et al. Novel methods improve prediction of species' distributions from occurrence data. *Ecography* **2006**, *29*, 129–151. [CrossRef]
36. Pearson, R.G.; Raxworthy, C.J.; Nakamura, M.; Peterson, A.T. Original article: Predicting species distributions from small numbers of occurrence records: A test case using cryptic geckos in Madagascar. *J. Biogeogr.* **2006**, *34*, 102–117. [CrossRef]
37. Hanley, J.A.; McNeil, B.J. The meaning and use of the area under a receiver operating characteristic (ROC) curve. *Radiology* **1982**, *143*, 29–36. [CrossRef]
38. Robin, X.; Turck, N.; Hainard, A.; Tiberti, N.; Lisacek, F.; Sanchez, J.-C.; Muller, M.J. pROC: An open-source package for R and S+ to analyze and compare ROC curves. *BMC Bioinform.* **2011**, *12*, 77. [CrossRef]
39. Phillips, S.J.; Dudik, M. Modeling of species distributions with Maxent: New extensions and a comprehensive evaluation. *Ecography* **2008**, *31*, 161–175. [CrossRef]
40. Préau, C.; Trochet, A.; Bertrand, R.; Isselin-Nondedeu, F. Modeling potential distributions of three European amphibian species comparing ENFA and MaxEnt. *Herpetol. Conserv. Biol.* **2018**, *13*, 91–104.
41. Lobo, J.; Jiménez-Valverde, A.; Real, R. AUC: A misleading measure of the performance of predictive distribution models. *Glob. Ecol. Biogeogr.* **2008**, *17*, 145–151. [CrossRef]
42. Hanczar, B.; Hua, J.; Sima, C.; Weinstein, J.; Bittner, M.; Dougherty, E.R. Small-sample precision of ROC-related estimates. *Bioinformatics* **2010**, *26*, 822–830. [CrossRef]
43. Akaike, H. Autoregressive model fitting for control. *Ann. Inst. Stat. Math.* **1971**, *23*, 163–180. [CrossRef]
44. R Core Team. *R: A Language and Environment for Statistical Computing*; R Foundation for Statistical Computing: Vienna, Austria, 2021; Available online: https://www.R-project.org/ (accessed on 20 June 2021).
45. Gloyd, H.K.; Conant, R. *Snakes of the Agkistrodon Complex. A Monographic Review. Contributions to Herpetology, No. 6*; Society for the Study of Amphibians and Reptile: Oxford, OH, USA, 1990; Volume 614, p. 52.
46. Pyron, R.A.; Burbrink, F.T.; Wiens, J.J. A phylogeny and revised classification of Squamata, including 4161 species of lizards and snakes. *BMC Evol. Biol.* **2013**, *13*, 1–54. [CrossRef] [PubMed]
47. He, M.; Feng, J.C.; Liu, S.Y.; Guo, P.; Zhao, E.M. The phylogenetic position of *Thermophis* (Serpentes: Colubridae), an endemic snake from the Qinghai-Xizang Plateau, China. *J. Nat. Hist.* **2009**, *43*, 479–488. [CrossRef]
48. Sinervo, B.; Méndez-de-la-Cruz, F.; Miles, D.B.; Heulin, B.; Bastiaans, E.; Villagrán-Santa Cruz, M.; Lara-Resendiz, R.; Martínez-Méndez, N.; Calderón-Espinosa, M.L.; Meza-Lázaro, R.N.; et al. Erosion of lizard diversity by climate change and altered thermal niche. *Science* **2010**, *328*, 894–899. [CrossRef]
49. Rao, D.Q. Complimentary survey of the herpetofauna of Xizang Autonomous Region (Tibet) with discussion of their distribution and current status. *Sichuan J. Zool.* **2000**, *19*, 107–112.
50. Hofmann, S.; Fritzsche, P.; Solhøy, T.; Dorge, T.; Miehe, G. Evidence of sex-biased dispersal in *Thermophis baileyi* inferred from microsatellite markers. *Herpetologica* **2012**, *68*, 514–522. [CrossRef]
51. Rajabizadeh, M.; Nagy, Z.T.; Adriaens, D.; Avcı, A.; Masroor, R.; Schmidtler, J.; Nazarov, R.; Esmaeili, H.R.; Christiaens, J. Alpine-Himalayan orogeny drove correlated morphological, molecular, and ecological diversification in the Persian dwarf snake (Squamata: Serpentes: *Eirenis persicus*). *Zool. J. Linn. Soc. Lond.* **2015**, *176*, 878–913. [CrossRef]
52. O'Connor, J.H.; Rittenhouse, T.A.G. Snow cover and late fall movement influence wood frog survival during an unusually cold winter. *Oecologia* **2015**, *181*, 635–644. [CrossRef] [PubMed]
53. Simmons, A.M.; Goulet, J.M.; Bellehumeur, K.F. The effect of snow depth on overwinter survival in *Lobelia inflata*. *Oikos* **2010**, *119*, 1685–1689. [CrossRef]
54. Zeisset, I.; Bebee, T.J.C. Determination of biogeographical range: An application of molecular phylogeography to the European pool frog *Rana lessonae*. *Proc. R. Soc. B Biol. Sci.* **2001**, *268*, 933–938. [CrossRef] [PubMed]
55. Gassert, F.; Schulte, U.; Husemann, M.; Ulrich, W.; Rödder, D.; Hochkirch, A.; Engel, E.; Meyer, J.; Habel, J.C. From southern refugia to the northern range margin: Genetic population structure of the common wall lizard, *Podarcis muralis*. *J. Biogeogr.* **2013**, *40*, 1475–1489. [CrossRef]

diversity MDPI

Review

Sunscreens' UV Filters Risk for Coastal Marine Environment Biodiversity: A Review

Samuele Caloni [1], Tiziana Durazzano [1], Giada Franci [1] and Letizia Marsili [1,2,*]

[1] Department of Physical, Earth and Environmental Sciences, University of Siena, Via Mattioli, 4, 53100 Siena, Italy; samuele.caloni@student.unisi.it (S.C.); tiziana.durazzano@student.unisi.it (T.D.); giada.franci@student.unisi.it (G.F.)

[2] Department of Physical, Earth and Environmental Sciences, Inter-University Center of Cetacean Research (CIRCE), University of Siena, Via Mattioli, 4, 53100 Siena, Italy

* Correspondence: letizia.marsili@unisi.it; Tel.: +39-0577-232083

Abstract: Considering the rapid growth of tourism in recent years and the acknowledgement that exposure to solar UV radiation may cause skin cancer, sunscreens have been widely used by beachgoers in recent decades. UV filters contained in sunscreens, however, were recently identified as emerging pollutants in coastal waters since they accumulate in the marine environment with different adverse effects. In fact, exposure to these components was proven to be toxic to most invertebrate and vertebrate marine species. Some UV filters are linked to the production of significant amounts of reactive oxygen species (ROS), such as hydrogen peroxide, and the release of inorganic micronutrients that may alter the status of coastal habitats. Bioaccumulation and biomagnification have not yet been fully addressed. This review highlights recent progress in research and provides a comprehensive overview of the toxicological and ecotoxicological effects of the most used UV filters both on the abiotic and biotic compartments in different types of coastal areas, to gain a better understanding of the impacts on coastal biodiversity.

Keywords: sunscreens; UV filters; nanoparticles; coastal areas; coral reef; ecotoxicology

Citation: Caloni, S.; Durazzano, T.; Franci, G.; Marsili, L. Sunscreens' UV Filters Risk for Coastal Marine Environment Biodiversity: A Review *Diversity* **2021**, *13*, 374. https://doi.org/10.3390/d13080374

Academic Editor: Bert W. Hoeksema

Received: 6 July 2021
Accepted: 10 August 2021
Published: 12 August 2021

Publisher's Note: MDPI stays neutral with regard to jurisdictional claims in published maps and institutional affiliations.

Copyright: © 2021 by the authors. Licensee MDPI, Basel, Switzerland. This article is an open access article distributed under the terms and conditions of the Creative Commons Attribution (CC BY) license (https://creativecommons.org/licenses/by/4.0/).

1. Introduction

During the last decade, tourism has seen massive growth and is among the economic sectors expected to experience constant development in the future. It was estimated that by 2035, the rate of global tourism will increase by 179%, and is set to generate substantial anthropic stress on natural environments [1]. In fact, in 2017, the Mediterranean sea alone attracted over 267 million international tourists [2]. Water environments are at high risk, and plenty of research has been devoted to studying them: fragile balances regulate these environments, particularly in the coastal areas, for they are very rich in biodiversity and the ecosystem services provided by these areas sustain half of the planet population [3]. Coastal tourism, and the related recreational activities, have led to a massive use of photoprotective personal care products (PCPs), which are highly and widely recommended to prevent skin damage from sun exposure [4–6], resulting in a direct input from swimming and bathing (non-point sources). These inputs, together with industrial wastewater discharges (point sources) [7–9], are capable of starting decay processes, irreversible at times [6]. In fact, coastal tourism is acknowledged as a source of impact on shallow-water marine habitats [1], as well as lakes and rivers [10,11]. Unfortunately, recent data about the global annual production of these PCPs are lacking. Last available data from a market study in 2005 estimated a 10,000 tons of sunscreens global production per year [12]. This means that nowadays there is a gap in judging the threat currently posed to the environment. Nonetheless, it was evaluated that, during in-water activities, at least 25% of sunscreens and PCPs applied to the skin get washed off [13]. A study carried out in France estimated that a sample of 3000 beachgoers applied, on average, 52.5 kg of sunscreen per day, releasing

15.7 kg of it into the water [14]. Since the widespread use of photoprotective PCPs, UV filters contained in sunscreens have become emerging contaminants in various environments. Only in recent years, the scientific community has started studying and investigating the causes and the effects of their accumulation in different ecosystems [4,15–17].

1.1. *Sunscreen Definition*

Sunscreen lotions are defined as PCPs containing UV filters, substances whose main function is to reflect, to refract, and to dissipate the wavelengths of sunlight considered harmful to human skin (UVA 320–400 nm and UV-B 280–320 nm). These lotions are designed for external application and the UV filters contained in the general PCP formula can be distinguished into organic and inorganic [18].

1.1.1. Organic Filters

Organic filters (or *chemical*) are synthesized substances, which include the derivatives of cinnamic acid such as ethylhexyl methoxycinnamate (EHMC), benzophenones (BPs) such as the commonly used benzophenone-3 (BP-3), salicylates such as ethylhexyl salicylate (OCS), benzoyl derivatives such as diethylamino hydroxybenzoyl hexyl benzoate (DHHB), and butyl methoxydibenzoylmethane (BMDBM). These compounds usually have single or multiple aromatic structures, sometimes conjugated with carbon-carbon double bonds and carbonyl moieties, able to attenuate the transmission of energetic solar photons that reach the surface of the Earth [6]. These molecules typically get to an excited state when hit by UV radiations and release the energy as fluorescence or heat, and, in this way, are able to dissipate a part of it and transform the rest into a non-harmful wavelength for the skin [4].

1.1.2. Inorganic Filters

Inorganic (also referred to as *physical* or *mineral*) filters provide filtering action against sunlight via two mechanisms: (1) the crystals refract and scatter a significant amount of the incoming radiation, and (2) the molecules themselves get to an excited state and then de-excite the same way as organic filters. These cycles of excitement and de-excitement entail a collateral photocatalytic activity, which is capable of producing reactive oxygen species (ROS), such as $O_2^{-\bullet}$, $HO^\bullet$, and H_2O_2 [19]. There are only two mineral filters widely approved and used around the world: titanium dioxide (TiO_2) and zinc oxide (ZnO), which can be used in both micrometric (TiO_2 and ZnO) and nanometric form (n-TiO_2 and n-ZnO). In the latter, the particles can be referred to as engineered nanoparticles (NPs or ENPs) and, if they are made of TiO_2, they are often coated with inert compounds to avoid undesired chemical reactions capable of skin damage [20]. The coating often has one or two layers: the innermost, which is made of an inert material, e.g., alumina (Al_2O_3), aluminum hydroxide ($Al(OH)_3$) or silica (SiO_2) [21], and the outer, e.g., silicone, which is optional and used to give hydrophobic properties to improve the blending capacities of TiO_2 [22].

1.1.3. Other Compounds

Apart from UV filters, sunscreen lotions contain other ingredients such as preservatives, emulsifiers, colorants, foams, and perfumes [5].

2. Abiotic Compartment

The most analyzed matrices to evaluate the behavior of UV filters are waters, sediments, and SML (surface microlayer). Water samples are used to evaluate the water solubility of the substances examined and the relative concentrations [23–27], while sediments and SML are used because they are more suitable for the identification of lipophilic compounds released into the environment [4,5,28].

The measurement of the production of ROS species can be carried out directly or indirectly: the direct way involves spectrometry, setting the reading of the sample at a specific wavelength characteristic of each chemical species; while the indirect way predicts the determination, spectrometric [22] or chromatographic [29], of the oxidized amount of a compound acting as a "trap" for ROS, i.e., 2,3-bis(2-methoxy-4-nitro-5-sulfophenyl)-2H-tetrazolium-5-car-boxanilide (XTT) [22] and furfuryl alcohol (FFA) [29].

2.1. *Mineral Filters Behavior in Water*

TiO_2 and ZnO are among the most employed particles in sunscreens. These two raw materials, depending on the size used (micrometric or nanostructured), show different behaviors in water. It was evaluated that micrometric mineral filters may be released into seawater by up to 49% of the quantity used, meaning they are extremely washable [14], due to their hydrophilicity. As per the nanostructured UV filters, they can be released in the environment in variable amount from 8% up to 72%, with the cosmetic formula having a large influence on the leaching rate (mean ± SD: 45 ± 33%) [30].

Once leached into the water, they undergo further modifications since the external silicone layer can be easily degraded in slightly acidic (pH = 5) or slightly alkaline (pH = 9) waters [22]. As time passes and the surface becomes more and more degraded, NPs can enter into suspension from 5% to over 30% of the total amount of sunscreen dispersed in water [21,31]. The presence of organic matter in water represents an important contribution for the stabilization of the particles of $nTiO_2$, which, once dispersed in water, may remain isolated or form aggregates together with macromolecules capable of forming complexes (e.g., humic acids) [32,33] that endure in the environment. Moreover, there is evidence that salinity and pH play a role in leading NPs to aggregate and to descend the water column until they reach the bottom, where they may lay and eventually sediment [21].

The main negative side of inorganic UV filters is their ability to transfer the absorbed energy to other surrounding molecules, causing ROS formation. These oxygen compounds, characterized by a high reactivity, cause oxidative stress in organisms exposed to higher concentrations. In particular, the photocatalytic activity of TiO_2 is also linked to the size of the particles used in the formulation: the microparticles have a moderate reactivity, which does not require countermeasures beyond the respect of a maximum percentage in the formulation; nanoparticles, on the other hand, are much more reactive and therefore require a coating [21,34,35].

2.2. *Organic UV Filters and Derivatives Behavior in Water*

Organic UV filters tend to be more concentrated on the SML and could, therefore, influence the availability of sunlight for photosynthetic organisms, a phenomenon which would be especially harmful in areas where barrier reefs are present [13,36]. This happens because some organic UV filters have photocatalytic activity, a feature that makes them co-responsible for the overproduction of ROS in aquatic environments [29,37]. The main responsible organic UV filters for ROS production in aquatic environment are octinoxate (EHMC), octocrylene (OCR), 4-aminobenzoic acid (PABA), and 2-ethylhexyl 4-(dimethylamino)benzoate (OD-PABA). In this context, benzophenones (particularly BP-3 and BP-8) and ethylhexyl salicylate (OCS) are more suitable because they seem to be incapable of forming singlet oxygen or other ROS when exposed to light [29]. In a well-lit environment, sunscreens can also undergo photodegradation, often generating less toxic compounds than the original UV filter: benzophenone derivatives showed, in laboratory studies, a modest genotoxic potential if present in concentrations of >250 ng/L, comparable to those that they are found in crowded parts of the coast or areas with low water exchange [38–40]. Other UV filters, such as OD-PABA, EHMC and iso-amylmethoxycinnamate (IAMC), are overall less toxic, especially if exposed to intense illumination due to their higher photolability, when compared to the previous case [40,41].

2.3. *Release of Inorganic Nutrients and Metals in the Aquatic Environment*

There is proof that the introduction of sunscreens into shallow-water environments lead to the release of heavy metals, purportedly present within the lotion as leftovers and production process debris, and micronutrients, such as NO_3^-, NH_4^+, and PO_4^{3-}, which are residuals derived from the degradation of some organic compounds and linked to events of eutrophication and anomalous algal growth [4,42]. The photoprotective PCPs formulations contain not only water and sunscreens, but also a vast set of substances that are dispersed into the environment. Elements that play important biological roles, for example, Fe, Cu, N, and P, or are highly toxic to most organisms, such as Pb and Cd, can also be released [31,38]. Conservative simulations carried out for a typical Mediterranean beach showed an increase, compared to background concentrations, of close to 20% for Ti and 5% for Al. All the other elements taken into consideration also had a small increase, mostly less than 0.1% [31].

3. Biotic Compartment

The toxicity of various UV filters contained in sunscreens, both organic and inorganic, on marine organisms varies considerably depending on the UV filter and organism physiology. Many studies emphasized biological and toxicological responses, which may affect survival, behavior, growth, development, and reproduction, that were observed at various trophic levels. Coral reefs will be discussed separately in Section 4 as they are unique environments highly exposed to both climate change and human activities.

Table 1 summarizes recent studies carried out on the exposure of various organisms to UV filters and the effects of these exposures.

Table 1. Effects of various UV filters from different studies.

UV Filter(s)	Organism(s)	Exposure Conditions	Effects	Reference
4-MBC BP-3 BP-4 EHMC	Mediterranean mussel (*Mytilus galloprovincialis*), sea urchin (*Paracentrotus lividus*)	EC_{50}	EHMC and 4-MBC toxicity assessed from 4–5 mg/L, followed by BP-3 and finally BP-4	[43]
n-TiO_2	Mediterranean mussel (*Mytilus galloprovincialis*)	From 0.05 to 5 mg/L for 24 h	Cellular damage NRR in hemocytes and digestive glands; stimulated glutathione-S-transferase (GST)	[44]
n-TiO_2	Mediterranean mussel (*Mytillus galloprovincialis*)	From 2.8 to 280 μg/L for 24 h	Adaptive response in gills at 28 μg/L; oxidative stress and neurotoxicity over 280 μg/L	[45]
n-TiO_2	Marine abalone (*Haliotis diversicolor supertexta*)	Acute toxicity stress: from 0.1 to 10 mg/L for 96 h	Oxidative stress: SOD increased (1 mg/L), GSH decreased (1 mg/L), LPO dose-dependent increase	[46]
n-TiO_2	Lungworm (*Arenicola marina*)	Sub-lethal OECD/ASTM 1990 acute toxicity test	Decrease in casting rate; increase in cellular damage (NRR); DNA damage in coelomocytes	[47]
n-ZnO	Sea urchin (*Paracentrotus lividus*)	21-day exposure via food to reach 10 mg Zn/kg food	Damages to immune cells (33% of damaged nucleus); transmissible effects to offspring (75.5% of malformed larvae)	[48]

Table 1. *Cont.*

UV Filter(s)	Organism(s)	Exposure Conditions	Effects	Reference
4-MBC	Senegalese sole (*Solea senegalensis*)	Mortality and growth assessment 96 h egg exposure from 0.235 to 0.935 mg/L; biochemical markers from 0.068 to 0.360 mg/L	Induced mortality and malformations in a dose-response manner; reduced growth with increasing concentrations; increased activity of AChE on larvae exposed to 0.085 mg/L; significantly lower LDH activity (p b 0.05); swimming behavior was affected by 4-MBC at low concentrations.	[49]
BP-1 BP-2 BP-3 BP-4 BP-7 BP-8	Marine bacterium (*Photobacterium phosphoreum*) and planktonic crustacean (*Daphnia magna*)	EC_{50} protocol and QSAR modelling	Toxicity evaluated for both species	[50]
PBSA	Rainbow trout (*Oncorhynchus mykiss*)	21 and 42 days; from 1 to 1000 µg/L	Increased activity of P450 cytochromes	[51]
4-MBC BP-3 BMDBM EHMC OCR HMS	Ciliate (*Tetrahymena thermophila*)	IC_{50}	4-MBC, BP-3 and BMDBM could significantly inhibit the activity of the MXR system, IC_{50} values of 4-MBC, BP-3, and BMDBM were 23.54, 40.59, and 26.37 lM	[52]
BP 2-HBP BP-3 BP-4	Bioluminescent bacterium (*Vibrio fischeri*) in vitro and zebrafish (*Danio rerio*) larvae in vitro	EC_{50}, SOS/umu assay and yeast estrogen screen assay (YES assay)	Luminescent bacteria toxicity, expressed as $\log EC_{50}$, increased with the lipophilicity ($\log K_{ow}$) of BP-derived UV filters; estrogenic activity in dose-effect relationship. *V. fischeri* toxicity order is BP-3 > 2-HBP > BP > BP-4	[53]
BP-1 BP-3	Green alga (*Chlamydomonas reinhardtii*)	Response surface methodologies (RSM)	Exposure to the combined BP-1 and BP-3 negatively affected cell growth and pigments production, with dose-dependent inhibition, affecting the photosynthesis process	[54]
BP-1 BP-2 3-BC Et-PABA	Fathead minnow (*Pimephales promelas*)	14-day BP-1 from 8.9 to 4919.4 µg/L; BP-2 from 10.3 to 8782.9 µg/L; 3BC from 8.7 to 952.5 µg/L e Et-PABA from 6.9 to 4394 µg/L	Induction of vitellogenin: 3-BC from 3 µg/L and BP-2 from 1.2 mg/L caused feminization in male fish, alteration of gonads in male and female fish, and decrease in fertility and reproduction	[55]

Table 1. *Cont.*

UV Filter(s)	Organism(s)	Exposure Conditions	Effects	Reference
BP-3	Zebrafish (*Danio rerio*)	Fish and embryos were exposed for 14 days and 120 h post-fertilization, respectively, to 2.4–312 µg/L and 8.2–438 µg/L BP-3.	BP-3 was partly transformed to BP-1 and both compounds were accumulated in adult fish; BP-3 exposure led to similar alterations of gene expression in both adult fish and eleuthero embryos with antiandrogenic activity	[56]
BP-3	Japanese medaka (*Oryzias latipes*)	14 days from 0 to 90 µg/L. First generation eggs (F1) reproduced were counted and further exposed up to 30 µg/L of BP-3	After 14 days, plasma concentrations of testosterone (T) significantly increased in male fish. The 17-β-estradiol (E2) to T (E2/T) ratio showed significant decreases in both male and female fish during 28 day exposure; daily average egg reproduction per female was significantly reduced at 26 µg/L of BP-3; hatchability of F1 eggs was not affected	[57]
BP-3 EHMC IAMC OD-PABA OCR 4-MBC	Green alga (*Scenedesmus vacuolatus*)	EC_{50}	BP-3 showed 43-fold higher toxicity than theoretically predicted. BP-3 and IAMC seem to have a more specific mode of action on algal cells	[40]
BMDBM EHMC OCR	Non-biting midge (*Chironomus riparius*), oligochaete (*Lumbriculus variegatus*), and snails (*Melanoides tuberculata* and *Potamopyrgus antipodarum*).	56 days (*L. variegatus*) or 28 days (*Chironomus riparius*, *M. tuberculata*, *P. antipodarum*) sediment test	EHMC caused a toxic effect on reproduction in both snails with lowest observed effect concentrations (LOEC) of 0.4 mg/kg (*Potamopyrgus antipodarum*) and 10 mg/kg (*Melanoides tuberculata*). BDMDM and OCR showed no effects on any of the tested organisms	[58]
EHMC OCR BMDBM	Planktonic crustacean (*Daphnia magna*)	EC_{10}, EC_{25}, and EC_{50} EHMC up to 80.0 µg/mL; OCR and BMDBM up to 640.0 µg/mL;	EHMC, OCR, and BMDBM highly toxic at low concentration (>1 µg/mL) and resulted in immobilization higher than 25%; immobilization reached more than 90% at concentrations of 40 µg/mL; EC_{50} values for EHMC, OCR, and BMDBM were 2.73, 3.18, and 1.95 µg/mL, respectively, indicating that OCR had the lowest toxic effect on *Daphnia*; reduction of toxic effects in the mixtures of the three UV-filters, caused by antagonistic action of the components	[59]

Table 1. *Cont.*

UV Filter(s)	Organism(s)	Exposure Conditions	Effects	Reference
n-TiO_2	Cyanobacterium (*Anabaena variabilis*)	24 h to 6 days from 0.5 to 250 mg/L	Reduced N fixation activity, growth rate, toxicity time, and dose-dependency	[60]
n-TiO_2	Fathead minnow (*Pimephales promelas*)	Exposed to 2 ng/g and 10 mg/g body weight. Challenged with fish bacterial pathogens, *Aeromonas hydrophila* or *Edwardsiella ictaluri*	Fish mortality during bacterial challenge with *Aeromonas hydrophila* and *Edwardsiella ictalurid*; reduced neutrophil phagocytosis of *A. hydrophila*; significant histopathological alterations	[61]
n-TiO_2	European sea bass (*Dicentrarchus labrax*)	7 days, 1 mg/L	Chromosomal alteration	[62]
n-TiO_2	Marine scallop (*Chlamys farreri*)	14 days, 1 mg/L	Elevated superoxide dismutase (SOD), catalase (CAT) activities, and malondialdehyde (MDA) contents, increased acetylcholinesterase (AChE) activities; histopathological alterations in gills and digestive gland (dysplastic and necrosis)	[63]
n-TiO_2 n- ZnO	Diatoms (*Skeletonema marinoi, Thalassiosira pseudonana*), green alga (*Dunaniella tertiolecta*), and Haptophyta alga (*Isochrysis galbana*)	24 and 96 h from 0.10 to 1000 μg/L	n-TiO_2 did not affect the growing rate, n-ZnO depressed growth in all species	[64]
n-ZnO	Diatoms (*Thalassiosira pseudonana, Chaetocerus gracilis, Phaedacttylum tricornutum*)	72 h, from 10 to 80 mg/L	Growth stopped in *T. pseudonana* and *C. gracilis*; growth rate inversely proportional to NP concentration in *P. tricornutum*; Zn bioaccumulation killed *T. pseudonana*	[65]
n-ZnO	Diatoms (*Skeletonema costatum* and *Thalassiosia pseudonana*), crustaceans (*Tigriopus japonicus* and *Elasmopus rapax*), and medaka fish (*Oryzias melastigma*)	IC_{50}	n-ZnO toxic towards algae; ZnO toxic towards crustaceans; up-regulation of SOD and MT. Toxicity attributed mainly to dissolved Zn ions	[66]
n-ZnO	Green alga (*Dunaliella tertiolecta*), bioluminescent bacterium (*Vibrio fischeri*), brine shrimp (*Artemia salina*)	*V. fischeri* bioluminescence test for 5, to 30 min from 0.3 to 40 mg/L; *D. tertiolecta* algal growth test 24, 48 and 72 h from 0.1 to 10 mg/L; *A. salina* acute toxicity at 24–96 h from 10 to 100 mg/L, *A. salina* chronic exposure for 14 days from 0.03 to 0.5 mg/L	ZnO 14-day chronic exposure of *A. salina* significant inhibition of vitality and body length (EC_{50} 14d 0.02 mg Zn/L). ZnO NPs were more toxic towards algae (EC_{50} 2.2 mg Zn/L), but relatively less toxic towards bacteria (EC_{50} 17 mg Zn/L) and crustaceans (EC_{50} 96 h 58 mg Zn/L)	[67]

Table 1. *Cont.*

UV Filter(s)	Organism(s)	Exposure Conditions	Effects	Reference
OD-PABA OCR	Haptophyta alga (*Isochrysis galbana*), Mediterranean mussel (*Mytilus galloprovincialis*), and sea urchin (*Paracentrotus lividus*) in early stage	*I. galbana* 72 h to 2 and 90 ng/L, *M. galloprovincialis* and *P. lividus* 48 h EC_{50}	OCR was the more toxic compound for *P. lividus*; OD-PABA caused a severe negative effect on both *M. galloprovincialis* and *I. galbana*	[68]
n-TiO_2	Mediterranean mussel (*Mytilus galloprovincialis*)	96 h from 1 to 100 μg/L	Lysosomal and oxidative stress; decreased transcription of antioxidant and immune-related genes; decreased lysosomal membrane stability and phagocytosis; increased oxyradical production and transcription of antimicrobial peptides; pre-apoptotic processes	[69]
Sunscreen containing BP-3, sunscreen containing TiO_2	Clownfish (*Amphiprion ocellaris*)	97 h from 0 mg/L, 1 mg/L, 3 mg/L, 10 mg/L, 30 mg/L and 100 mg/L	Exposure level of 100 mg/L of BP-3 containing sunscreen led to 25% death and 100% disrupted swimming behavior by the end of the 97-h testing period. 100% of the animals failed to feed over the first 49 h of testing TiO_2 sunscreen at 100 mg/L had 6.7% mortality, swimming behavior was disrupted during the first 25 h of testing (26.7% abnormal movement), animals recovered well over the remainder of the testing period (out to 97 h)	[70]
4-MBC	Japanese clam (*Ruditapes philippinarum*)	0, 1, 10, 100 μg/L over a 7-day period followed by a 3-day depuration period (total 10 days)	Assessed mortality reached up to 100 % at concentration of 100 μg/L. LC50 value of 7.71 μg/L-was derived	[71]
4-MBC	Copepod (*Tigriopus japonicus*)	Exposed to three different salinity conditions (20, 30, and 40 ppt) prior to exposure to 0, 1, and 5 μg/L for multiple generations (F0–F3)	Environmentally relevant concentrations of 4-MBC had toxic effects on *T. japonicus*. Higher salinity levels increased the lethal, developmental, and reproductive toxicities of 4-MBC in *T. japonicus*	[72]

Table 1. *Cont.*

UV Filter(s)	Organism(s)	Exposure Conditions	Effects	Reference
BP-3 BEMT BMDBM MBBT OCS DHHB DBT EHT HMS OCR	Brine shrimp (*Artemia salina*) and green algae (*Tetraselmis* spp.)	*A. salina* 48 h exposure at 0, 0.02, 0.2, 2, 20, 200, and 2000 µg/L; *Tetraselmis* spp. 7-day exposure at 10, 100, and 1000 µg/L	HMS and OCR were the most toxic, followed by BMDBM, on *A. salina* at high concentrations (1 mg/L). OCS, BP3 and DHHB affected metabolic activity of green algae at 100 µg/L. BEMT, DBT, EHT, and MBBT had no effects, even at high concentrations (2 mg/L).	[73]

Legend: benzophenone (BP) and its derivatives (2-HBP, BP-1, BP-2, BP-3, BP-4, BP-7, and BP-8); 3-benzylidene camphor (3-BC); octyl methoxycinnamate or ethylhexyl methoxycinnamate (EHMC); octocrylene (OCR); butyl methoxydibenzoylmethane or avobenzone (BMDBM); homosalate (HMS); iso-amylmethoxy-cinnamate (IAMC); 4-methylbenzylidene camphor (4-MBC); ethyl-4-aminobenzoate (Et-PABA); 2-ethylhexyl 4-dimethylaminobenzoate (OD-PABA); 2-phenylbenzimidazole-5-sulfonic acid (PBSA); bis-ethylhexyloxyphenol methoxyphenyl triazine (BEMT); methylene bis-benzotriazolyl tetramethylbutylphenol (MBBT); 2-ethylhexyl salicylate (OCS); diethylaminohydroxybenzoyl hexyl benzoate (DHHB); diethylhexyl butamido triazone (DBT); ethylhexyl triazone (EHT); nanostructured titanium dioxide (n-TiO_2); nanostructured zinc oxide (n-ZnO).

It should be noted that most of these experiments were performed in laboratory settings and some of the UV filters were tested in isolation. Moreover, the concentrations used as stressor are usually higher than those observed in the environment.

3.1. Toxicity of Organic UV Filters

Samplings of wild *Mytilus edulis* and *Mytilus galloprovincialis* in ten sites along the French Atlantic and Mediterranean coasts from June to November 2008 showed accumulation of EHMC, OCR, and OD-PABA, highlighting how these concentrations significantly increased with the rising air temperature in summer and recreational pressure, although they also depended on the geomorphological structure of the sampling sites [74]. Studies carried out in the Hong Kong coastal area showed that the occurrence of these compounds was linked to the level of anthropogenic activities [75,76]. To validate patterns and the occurrence of PCPs in coastal sites impacted by recreational activities, diurnal variations (mirroring variations in recreational activities) as well as the tourist season [77] must be taken into consideration when writing monitoring protocols. In mussels, diurnal variations in OCR were observed, with the lowest concentrations recorded in the morning and then increasing throughout the day [26]. An alarming fact about organic UV filters is their diffusion in the planet's waters, wherein some of these compounds can be indicated as ubiquitous contaminants in the oceans: in a study conducted on marine water between the Pacific and the Atlantic Ocean and the Arctic Sea noted the presence, in each sample, of four UV filters (BP-3, OCR, BMDBM, and EHMC). The least polluted samples of the 12 organic UV filters tested were those of Shantou and Chaozhou (5 OUVs each), two cities in southern China near the mouth of the Han river, while the most polluted ones came from Hong Kong, in whose waters all 12 of the compounds analyzed were found [23]. Organic UV filters were reported as present in Arctic waters, far away from anthropogenic sources, and it's been hypothesized that these molecules were transported there by major oceanic currents from the conveyor belt [23].

The benthic community seems to be the most impacted by the presence of PCPs, since hydrophobic UV filters accumulate in the sediment phase [24], but the presence of UV filters may also enhance the spread of viral infection on both benthic and pelagic organisms [13]. At present, studies performed on the general formula or with a combination of UV filters are scarce both for the human body [78] and the environment [16]. Moreover, some organic UV filters seem to have estrogenic effects, but their activity and interactions in mixtures are largely unknown [55,79]. In particular, laboratory studies seemed to show that BP-3 showed anti-androgenic activities in zebrafish (*Danio rerio*) and Japanese medaka (*Oryzias latipes*) [56,57].

The analysis of biological tissues is used to identify bioaccumulation or biomagnification of organic UV filters along the food chain. Organic UV filters seem to accumulate with patterns similar to PCBs, highly persistent pollutants, [80] with the potential to reach marine mammals [81]. In a laboratory experiment performed on swamp crayfish (*Procambarus clarkii*), five organic UV filters (BP-3, 4-MBC, OCR, EHMC, and HMS) were tested for bioaccumulation and both 4-MBC and OCR showed accumulation in fecal matter, while EHMC and HMS showed the highest bioaccumulation factors [82]. In a natural environment, the presence of organic UV filters was ubiquitous in Lebranche mullet (*Mugil liza)* samples taken in the highly urbanized Guanabara Bay (Rio de Janeiro, Brazil) and data suggested an estimated daily intake in humans, via diet, from 0.3 to 15.2 ng of UV filters (kg/body weight). Therefore, UV filters might pose a hazard to human health as well [83]. To date, few data are available regarding the bioaccumulation and biomagnification processes, even if bioaccumulation has been detected [26,84]. This suggests that further evaluation must be undertaken to gain knowledge on the fate of these compounds along the trophic chain.

3.2. *Toxicity of Inorganic UV Filters*

As concerns the biotic field, particularly important is the tendency of inorganic UV filters to move vertically within the water column, starting from superficial layers, depending on surface charge, particle shape and size, and the pH and ionic strength of the water. The main problem, from a biological point of view, is that the suspended NPs can be captured by filtering organisms directly in the water column and, if not, will otherwise settle on the bottom and be taken up by detritivore organisms [85]. Studies carried out on the bivalve species *Mytilus galloprovincialis* showed the ability of the nano-TiO_2 to generate a moderate oxidative stress at concentrations of 0.2 mg/L. The stress was measured as the destabilization of the lysosomal membranes of hemocytes and digestive gland cells, and as an increase in the activity of the GST (glutathione-S-transferase) and catalase enzymes [44]. NPs, especially n-TiO_2, are strongly suspected of being bio-available and potentially gatherable by living organisms, unveiling a biomagnification phenomenon along the trophic chain [86,87].

In highly contaminated areas, their interaction with other pollutants may also be taken into consideration. A study performed in artificial seawater linked an antagonistic immune response towards 2,3,7,8-TCDD to the presence of n-TiO_2 in European sea bass (*Dicentrarchus labrax*) after 7 days in vivo exposure, suggesting that n-TiO_2 negatively influenced immune response induced by 2,3,7,8-TCDD in the spleen [88].

Zinc oxide, on the other hand, can, once released into the environment, cause very serious damage to ecosystems because it is highly toxic to bacteria and to marine invertebrates [89]. Studies on populations of the planktonic crustacean (*Daphnia magna)* fed with microalgae (*Pseudokirchneriella subcapitata*) exposed to different concentrations of ZnO showed an important reduction in the reproductive rate of the *D. magna* population [90]. These data are particularly alarming since the presence of ZnO may lead to the shrinking of planktonic organisms at the lowest levels of the trophic web, potentially causing a "cascade effect" within the whole ecosystem [64].

4. Toxicity on Coral Reef

Barrier reefs are unique ecosystems that, in recent years, have been threatened by increasingly frequent bleaching events. A bleaching event refers to the loss of symbiotic zooxanthellae hosted within scleractinian corals, often causing the death of the whole coral and therefore a loss of biodiversity in the ecosystem. It is thought that up to 10% of all coral reefs on the planet are menaced by these events [13]. Latent infections are common in symbiotic zooxanthellans [91], but a link was established between the weakening of coral due to exposition to sunscreen and the occurrence of viral infections, suggesting that the presence of PCPs, especially BP-3 and BP-8, could be a joint cause [13,91,92]. For example, BP-3 exceeded the threshold values by over 20% in hard corals *(Acropora* sp. and *A. pulchra*) in Hong Kong beaches located near snorkeling spots. It should be noted that these two compounds were detected widely and frequently at high concentrations in most of the sampled locations, causing larval deformity and mortality [93]. BP-3 is so far a ubiquitous presence in coastal seawater, sediment, and coral tissue, as also determined from sampling at sites around Oahu, Hawaii [94]. Taking into consideration the official data of the UNWTO, it was evaluated that 10% of the total sunscreen used is used in barrier reef tropical areas, and these data raise consistent concerns for the conservation of these endangered environments. Even so, relatively few studies have been conducted to identify environmental concentrations and potential toxicity of organic and inorganic UV filters [13,23,36,94,95]. Overall, there is a strong need to improve our understanding of the in situ concentrations of UV filters and preservatives, as well as their individual and combined effects. The environmentally measured concentrations are generally significantly lower than the nominal concentrations used in the laboratory to assess toxicity, but co-effects with other parameters may be crucial to assess risks for these compounds. Recently, it was discovered that mostly organic filters, such as BP-3, showed exacerbated adverse effects in the light [96], confirming that the concentration itself may not be the only parameter to consider. The assessment of risk should include biotic parameters (e.g., sensitivities, life stages of coral, metabolic capacities focus on both the host and symbionts) as well as abiotic parameters (e.g., solar irradiation, presence of other pollutants, and water temperature). Furthermore, adult corals were proven to accumulate and metabolize BPs during exposure in laboratory [92], but these effects have not yet been fully evaluated.

Concerning inorganic UV filters, uncoated ZnO induced severe bleaching and stimulated a microbial enrichment in the seawater that surrounds the corals [97]. Moreover, the maximum photosynthetic efficiency (Fv/Fm) of symbiotic zooxanthellae in scleractinian coral (*Stylophora pistillata*) when exposed to 90 µg/L of ZnO for 35 days, was reduced by 38% as compared to the control [98]. This clearly shows that ZnO is not an environmentally friendly compound and that its impact should be carefully evaluated.

In contrast, TiO_2 coated with alumina and dimethicone and TiO_2 modified with manganese caused minimal alterations in symbiotic interactions and did not cause bleaching, thus making it more eco-friendly than ZnO [97]. Alongside the direct impact on corals, UV filters also seem to pose a significant threat to reef biota, suggesting population and colony decline, as well as behavioral changes, for some common inhabitants of the reefs [13,36].

The studies taken into consideration are synthesized in Table 2.

Table 2. Effects of various UV filters on corals and reef biota.

UV Filter(s)	Organism(s)	Exposure Conditions	Effects	Reference
ZnO	*Acropora* spp. coral nubbins	24 and 48 h, up to 6.3 mg/L	67% coral nubbins surface bleached	[97]
BMDBM 2% BP-3 6% EHMC 6% OCR 6% OCS 5% 4-MBC 3% Butylparaben 0.5% and commercial sunscreens	*Acropora* spp. coral nubbins, *Stylophora pistillata* and *Millepora complanata*	18, 48 and 96 h, final concentrations of 10, 33, 50, and 100 µL/L	Sunscreen even in very low quantities (i.e., 10 µL/L) resulted in the release of large amounts of coral mucus (composed of zoo-xanthellae and coral tissue) within 18–48 h and complete bleaching of hard corals within 96 h	[13]

Table 2. *Cont.*

UV Filter(s)	Organism(s)	Exposure Conditions	Effects	Reference
BP-3	*Stylophora pistillata* (larval form)	PB-3 EC_{50} and LC_{50}, with different light exposure (8 h in the light, 8 h in the dark, a full diurnal cycle of 24 h, beginning at 08:00 in daylight and darkness from 18:00 in the evening until 08:00 h the next day, and a full 24 h in darkness), at 0.00001, 0.0001, 0.001, 0.01, 0.1 and 1 mM	BP-3 transformed planulae from a motile state to a deformed and sessile condition, showing genotoxicant, skeletal, and endocrine disruptor activity. BP-3 effects exacerbated in the light	[96]
ZnO Ethylparaben Butylparaben TDSA DTS EHT BMDBM OCR	*Stylophora pistillata*	35 days: ZnO from 10 to 1000 µg/L, UV filters from 10 to 5000 µg/L, preservatives (Ethylparaben and Butylparaben) from 0.1 a 1000 µg/L	ZnO reduced photosynthetic efficiency Fv/Fm by 38%, no adverse effects on the other UV filters tested up to the concentration corresponding to their water solubility limit. Butylparaben decreased the Fv/Fm by 25% at the highest concentration of 100 µg/L	[98]
BP-1 BP-3 BP-4 BP-8	*Pocillopora damicornis, Seriatopora caliendrum*	7–12 days from 0.1 to 1000 µg/L. <1000 µg/L (S. *caliendrum* nubbins)	No bleaching was observed in the *P. damicornis* larval tests, while bleaching was observed in the *P. damicornis* nubbin tests. Overall, BP-1 and BP-8 were more toxic to the two tested species than BP-3 and BP-4, which matches the relative bioaccumulation potential of the four BPs (BP-8 > BP-1 ≈ BP-3 > BP-4)	[92]
HMS 13% BP-3 6% OCR 5% OCS 5% BMDBM 3%	Flatworm *(Convolutriloba macropyga);* pulse corals (*Xenia* sp.); glass anemones (*Aiptasia* spp.); Diatoms (*Nitzschia* spp.)	Flatworms: 72 h from 0.1 to 1 mL/L; pulse corals: 72 h, 1 mL in 3.8 L seawater; glass anemones: 7 days from 0.1 to 1 mL/L; diatoms: 72 h 1 mL on 3.8 L seawater	Flatworm populations exposed to sunscreen showed a highly reduced growing rate. Pulse corals showed effects on growing rate, with a drastic decrease during the first week of treatment and partially recovering in the following period, and polyp pulses per minute, slowed down after about 10 min of exposition. All anemones exposed to sunscreen were categorized as unhealthy since pedal disks were weakly or not attached to the container walls, tentacles or body columns were not extended, individuals did not clearly respond to touch and appeared dark brown to black. Diatoms were less green with the average green fluorescent content showing a decrease	[36]

Table 2. *Cont.*

UV Filter(s)	Organism(s)	Exposure Conditions	Effects	Reference
BP-3 HMS OCS OCR	Concentrations in water, sediment, and coral tissue (Ka'a'awa, Waikiki Beach, Kaneohe Bay in October 2017)		Total mass concentrations of all UV-filters detected in seawater were <750 ng/L, in sediment < 70 ng/g and in coral tissue < 995 ng/g dry weight (dw). UV-filter concentrations generally varied as follows: **Water:** HMS > OCS > BP-3 > OCR, concentrations in surface seawater highest at Waikiki beach; **Sediment:** HMS > OCS > OCR > BP-3; **Coral:** OCS ≈ HMS > OCR ≈ BP-3	[94]

Legend: benzophenone derivatives (BP-1, BP-3, BP-4, BP-8); octyl methoxycinnamate or ethylhexyl methoxycinnamate (EHMC); octocrylene (OCR); butyl methoxydibenzoylmethane or avobenzone (BMDBM); homosalate (HMS); 4-methylbenzyliden camphor (4-MBC); micrometric zinc oxide (ZnO); ethylhexyl triazone (EHT); terephthalylidene dicamphor sulfonic acid (TDSA); drometrizole trisiloxane (DTS); 2-ethylhexyl 2-cyano-3,3-diphenylacrylate (EHCDA); 2-ethylhexyl salycilate (OCS).

5. Conclusions

Although a significant development was reached by global research on the impact of sunscreens and other photoprotective PCPs in nature, much more needs to be understood through future and more in-depth studies. The fields to be explored are many, given the recent interest in this area of environmental toxicology: while studies on nanoparticles in the Mediterranean and on organic UV filters in tropical countries are relatively abundant, ecotoxicological investigations on the average toxicity thresholds are deficient. When assessing the effects on natural coastal environments and coastal biota, we need to take into consideration parameters such as variation in pH, salinity, solar irradiation, level of anthropogenic activities, and currents etc. For example, increasing salinity levels posed a significant risk for the marine copepod *Tigropus japonicus* in the presence of different concentrations of 4-MBC by exacerbating oxidative stress and the uptake of this chemical [72]. A special focus must be taken to monitor these compounds in natural environments and to evaluate their co-existence in shallow waters as the combination of UV filters and co-formulants may enhance or alter the toxic effects of each component. On this matter, a worldwide protocol should be created to make data easily comparable. Important gaps are also related to research on bioaccumulation and biomagnification, of both organic and inorganic UV filters, towards the trophic levels of marine ecological networks.

These new pieces of information will be necessary to improve and integrate the knowledge we have about the environmental effects of sunscreens and allow us to correct our actions and to start empowering institutions and the global population towards a greater respect for the environment. It should be added that, in recent years, we have also seen the first steps in this direction by some tropical countries that care about the fate of the coral reefs along their coasts. For example, the American State of Hawaii applied important restrictions to the ingredients of sunscreen products that can be marketed within their territory to counteract the phenomena of coral bleaching. Moreover, in this case, correct information must be made available to dissuade people from using sunscreens with banned chemicals purchased outside of the State [99] and to reduce misunderstandings on the correct use of sunscreen [100]. Furthermore, special attention needs to be given on Marine Protected Areas [77].

New conservation strategies are needed to drastically reduce the impact on ecosystems [101], possibly developed according to the most vulnerable habitats (e.g., tropical atolls, coral reefs, the Mediterranean coral reef, and other biodiversity hotspots).

Environmental issues are becoming more recognized due to the increasing media coverage provided in this regard, but comprehensive knowledge is lacking. Future legislation for a "coral safe" labelling might be addressed to help people make informed purchases [99]. By pushing this, initiatives could be promoted to decrease individual impacts on the environment with small gestures that can make a big difference when adopted by many people. For example, reducing the surface of application and the use of opaque garments, such as one-piece swimsuits instead of two-piece swimsuits. The research on new photoprotective compounds, extracted directly from plants, algae and animals, should be encouraged to identify sustainable molecules, easily degradable by organisms. This could be a promising development sector for research institutions and industries working towards a more sustainable future.

Author Contributions: Conceptualization, S.C., T.D., G.F. and L.M.; resources, L.M.; writing—original draft preparation, S.C., T.D. and G.F.; writing—review and editing, S.C., T.D., G.F. and L.M.; supervision, L.M. All authors have read and agreed to the published version of the manuscript.

Funding: This research received no external funding.

Conflicts of Interest: The authors declare no conflict of interest.

References

1. UNEP. Sustainable Coastal Tourism—An Integrated Planning and Management Approach. Available online: https://wedocs.unep.org/bitstream/handle/20.500.11822/7819/-Sustainable%20Coastal%20Tourism%E2%80%94An%20integrated%20planning%20and%20management%20approach-2009913.pdf?sequence=3&isAllowed=y (accessed on 30 April 2020).
2. Tovar-Sánchez, A.; Sánchez-Quiles, D.; Rodríguez-Romero, A. The mediterranean sea. In *Handbook of Environmental Chemistry*; Springer: Berlin/Heidelberg, Germany, 2020; Volume 94, pp. 131–161.
3. UNDP. *Annual Report*; UNDP: New York, NY, USA, 2017.
4. Sánchez-Quiles, D.; Tovar-Sánchez, A. Are sunscreens a new environmental risk associated with coastal tourism? *Environ. Int.* **2015**, *83*, 158–170. [CrossRef]
5. Tovar-Sánchez, A.; Sánchez-Quiles, D.; Basterretxea, G.; Benedé, J.L.; Chisvert, A.; Salvador, A.; Moreno-Garrido, I.; Blasco, J. Sunscreen Products as Emerging Pollutants to Coastal Waters. *PLoS ONE* **2013**, *8*, e65451. [CrossRef]
6. Giokas, D.L.; Salvador, A.; Chisvert, A. UV filters: From sunscreens to human body and the environment. *TrAC Trends Anal. Chem.* **2007**, *26*, 360–374. [CrossRef]
7. Balmer, M.E.; Buser, H.R.; Müller, M.D.; Poiger, T. Occurrence of some organic UV filters in wastewater, in surface waters, and in fish from Swiss lakes. *Environ. Sci. Technol.* **2005**, *39*, 953–962. [CrossRef] [PubMed]
8. Amine, H.; Gomez, E.; Halwani, J.; Casellas, C.; Fenet, H. UV filters, ethylhexyl methoxycinnamate, octocrylene and ethylhexyl dimethyl PABA from untreated wastewater in sediment from eastern Mediterranean river transition and coastal zones. *Mar. Pollut. Bull.* **2012**, *64*, 2435–2442. [CrossRef] [PubMed]
9. Schneider, S.L.; Lim, H.W. Review of environmental effects of oxybenzone and other sunscreen active ingredients. *J. Am. Acad. Dermatol.* **2019**, *80*, 266–271. [CrossRef]
10. Buser, H.R.; Balmer, M.E.; Schmid, P.; Kohler, M. Occurrence of UV filters 4-methylbenzylidene camphor and octocrylene in fish from various Swiss rivers with inputs from wastewater treatment plants. *Environ. Sci. Technol.* **2006**, *40*, 1427–1431. [CrossRef] [PubMed]
11. Ramos, S.; Homem, V.; Alves, A.; Santos, L. Advances in analytical methods and occurrence of organic UV-filters in the environment—A review. *Sci. Total Environ.* **2015**, *526*, 278–311. [CrossRef] [PubMed]
12. Gago-Ferrero, P.; Díaz-Cruz, M.S.; Barceló, D. An overview of UV-absorbing compounds (organic UV filters) in aquatic biota. *Anal. Bioanal. Chem.* **2012**, *404*, 2597–2610. [CrossRef] [PubMed]
13. Danovaro, R.; Bongiorni, L.; Corinaldesi, C.; Giovannelli, D.; Damiani, E.; Astolfi, P.; Greci, L.; Pusceddu, A. Sunscreens cause coral bleaching by promoting viral infections. *Environ. Health Perspect.* **2008**, *116*, 441–447. [CrossRef] [PubMed]
14. Labille, J.; Slomberg, D.; Catalano, R.; Robert, S.; Apers-Tremelo, M.L.; Boudenne, J.L.; Manasfi, T.; Radakovitch, O. Assessing UV filter inputs into beach waters during recreational activity: A field study of three French Mediterranean beaches from consumer survey to water analysis. *Sci. Total Environ.* **2020**, *706*. [CrossRef]
15. la Farré, M.; Pérez, S.; Kantiani, L.; Barceló, D. Fate and toxicity of emerging pollutants, their metabolites and transformation products in the aquatic environment. *TrAC Trends Anal. Chem.* **2008**, *27*, 991–1007. [CrossRef]
16. Sendra, M.; Sánchez-Quiles, D.; Blasco, J.; Moreno-Garrido, I.; Lubián, L.M.; Pérez-García, S.; Tovar-Sánchez, A. Effects of TiO2nanoparticles and sunscreens on coastal marine microalgae: Ultraviolet radiation is key variable for toxicity assessment. *Environ. Int.* **2017**, *98*, 62–68. [CrossRef] [PubMed]
17. Matranga, V.; Corsi, I. Toxic effects of engineered nanoparticles in the marine environment: Model organisms and molecular approaches. *Mar. Environ. Res.* **2012**, *76*, 32–40. [CrossRef] [PubMed]
18. European Council Regulation (EC). No 1223/2009 of the European Parliament and of the Council of 30 November 2009 on Cosmetic Products 1–21. *Off. J. Eur. Union* **2009**, *342*, 59–209.

19. Sánchez-Quiles, D.; Tovar-Sánchez, A. Sunscreens as a source of hydrogen peroxide production in coastal waters. *Environ. Sci. Technol.* **2014**, *48*, 9037–9042. [CrossRef]
20. Baek, S.; Joo, S.H.; Blackwelder, P.; Toborek, M. Effects of coating materials on antibacterial properties of industrial and sunscreen-derived titanium-dioxide nanoparticles on Escherichia coli. *Chemosphere* **2018**, *208*, 196–206. [CrossRef]
21. Botta, C.; Labille, J.; Auffan, M.; Borschneck, D.; Miche, H.; Cabié, M.; Masion, A.; Rose, J.; Bottero, J.Y. TiO_2-based nanoparticles released in water from commercialized sunscreens in a life-cycle perspective: Structures and quantities. *Environ. Pollut.* **2011**, *159*, 1543–1550. [CrossRef]
22. Auffan, M.; Pedeutour, M.; Rose, J.; Masion, A.; Ziarelli, F.; Borschneck, D.; Chaneac, C.; Botta, C.; Chaurand, P.; Labille, J.; et al. Structural degradation at the surface of a TiO_2-based nanomaterial used in cosmetics. *Environ. Sci. Technol.* **2010**, *44*, 2689–2694. [CrossRef]
23. Tsui, M.M.P.; Leung, H.W.; Wai, T.C.; Yamashita, N.; Taniyasu, S.; Liu, W.; Lam, P.K.S.; Murphy, M.B. Occurrence, distribution and ecological risk assessment of multiple classes of UV filters in surface waters from different countries. *Water Res.* **2014**, *67*, 55–65. [CrossRef]
24. Fagervold, S.K.; Rodrigues, A.S.; Rohée, C.; Roe, R.; Bourrain, M.; Stien, D.; Lebaron, P. Occurrence and Environmental Distribution of 5 UV Filters During the Summer Season in Different Water Bodies. *Water Air Soil Pollut.* **2019**, *230*, 1–13. [CrossRef]
25. Ma, B.; Lu, G.; Liu, F.; Nie, Y.; Zhang, Z.; Li, Y. Organic UV Filters in the Surface Water of Nanjing, China: Occurrence, Distribution and Ecological Risk Assessment. *Bull. Environ. Contam. Toxicol.* **2016**, *96*, 530–535. [CrossRef]
26. Picot-Groz, M.; Fenet, H.; Martinez Bueno, M.J.; Rosain, D.; Gomez, E. Diurnal variations in personal care products in seawater and mussels at three Mediterranean coastal sites. *Environ. Sci. Pollut. Res.* **2018**, *25*, 9051–9059. [CrossRef]
27. Tovar-Sanchez, A.; Sparaventi, E.; Gaudron, A.; Rodriguez-Romero, A. A new approach for the determination of sunscreen levels in seawater by ultraviolet absorption spectrophotometry. *PLoS ONE* **2020**, *15*, e0243591. [CrossRef] [PubMed]
28. Engel, A.; Bange, H.W.; Cunliffe, M.; Burrows, S.M.; Friedrichs, G.; Galgani, L.; Herrmann, H.; Hertkorn, N.; Johnson, M.; Liss, P.S.; et al. The ocean's vital skin: Toward an integrated understanding of the sea surface microlayer. *Front. Mar. Sci.* **2017**, *4*, 1–14. [CrossRef]
29. Allen, J.M.; Gossett, C.J. Photochemical formation of singlet molecular oxygen in illuminated aqueous solutions of several commercially available sunscreen active ingredients. *Chem. Res. Toxicol.* **1996**, *9*, 605–609. [CrossRef]
30. Wong, S.W.Y.; Zhou, G.J.; Leung, P.T.Y.; Han, J.; Lee, J.S.; Kwok, K.W.H.; Leung, K.M.Y. Sunscreens containing zinc oxide nanoparticles can trigger oxidative stress and toxicity to the marine copepod Tigriopus japonicus. *Mar. Pollut. Bull.* **2020**, *154*, 111078. [CrossRef]
31. Rodríguez-Romero, A.; Ruiz-Gutiérrez, G.; Viguri, J.R.; Tovar-Sánchez, A. Sunscreens as a New Source of Metals and Nutrients to Coastal Waters. *Environ. Sci. Technol.* **2019**, *53*, 10177–10187. [CrossRef]
32. Keller, A.A.; Wang, H.; Zhou, D.; Lenihan, H.S.; Cherr, G.; Cardinale, B.J.; Miller, R.; Zhaoxia, J.I. Stability and aggregation of metal oxide nanoparticles in natural aqueous matrices. *Environ. Sci. Technol.* **2010**, *44*, 1962–1967. [CrossRef]
33. Thio, B.J.R.; Zhou, D.; Keller, A.A. Influence of natural organic matter on the aggregation and deposition of titanium dioxide nanoparticles. *J. Hazard. Mater.* **2011**, *189*, 556–563. [CrossRef] [PubMed]
34. Lewicka, Z.A.; Yu, W.W.; Oliva, B.L.; Contreras, E.Q.; Colvin, V.L. Photochemical behavior of nanoscale TiO_2 and ZnO sunscreen ingredients. *J. Photochem. Photobiol. A Chem.* **2013**, *263*, 24–33. [CrossRef]
35. Li, Y.; Yang, D.; Lu, S.; Qiu, X.; Qian, Y.; Li, P. Encapsulating TiO_2 in Lignin-Based Colloidal Spheres for High Sunscreen Performance and Weak Photocatalytic Activity. *ACS Sustain. Chem. Eng.* **2019**, *7*, 6234–6242. [CrossRef]
36. McCoshum, S.M.; Schlaib, A.M.; Baum, K.A. Direct and indirect effects of sunscreen exposure for reef biota. *Hydrobiologia* **2016**, *776*, 139–146. [CrossRef]
37. Inbaraj, J.J.; Bilski, P.; Chignell, C.F. Photophysical and Photochemical Studies of 2-Phenylbenzimidazole and UVB Sunscreen 2-Phenylbenzimidazole-5-sulfonic Acid. *Photochem. Photobiol.* **2002**, *75*, 107. [CrossRef]
38. Morel, F.M.M.; Price, N.M. The biogeochemical cycles of trace metals in the oceans. *Science* **2003**, *300*, 944–947. [CrossRef] [PubMed]
39. Kotnik, K.; Kosjek, T.; Žegura, B.; Filipič, M.; Heath, E. Photolytic fate and genotoxicity of benzophenone-derived compounds and their photodegradation mixtures in the aqueous environment. *Chemosphere* **2016**, *147*, 114–123. [CrossRef] [PubMed]
40. Rodil, R.; Moeder, M.; Altenburger, R.; Schmitt-Jansen, M. Photostability and phytotoxicity of selected sunscreen agents and their degradation mixtures in water. *Anal. Bioanal. Chem.* **2009**, *395*, 1513–1524. [CrossRef]
41. MacManus-Spencer, L.A.; Tse, M.L.; Klein, J.L.; Kracunas, A.E. Aqueous photolysis of the organic ultraviolet filter chemical octyl methoxycinnamate. *Environ. Sci. Technol.* **2011**, *45*, 3931–3937. [CrossRef] [PubMed]
42. Tovar-Sánchez, A.; Sánchez-Quiles, D.; Rodríguez-Romero, A. Massive coastal tourism influx to the Mediterranean Sea: The environmental risk of sunscreens. *Sci. Total Environ.* **2019**, *656*, 316–321. [CrossRef]
43. Paredes, E.; Perez, S.; Rodil, R.; Quintana, J.B.; Beiras, R. Ecotoxicological evaluation of four UV filters using marine organisms from different trophic levels *Isochrysis galbana*, *Mytilus galloprovincialis*, *Paracentrotus lividus*, and *Siriella armata*. *Chemosphere* **2014**, *104*, 44–50. [CrossRef]
44. Canesi, L.; Fabbri, R.; Gallo, G.; Vallotto, D.; Marcomini, A.; Pojana, G. Biomarkers in Mytilus galloprovincialis exposed to suspensions of selected nanoparticles (Nano carbon black, C60 fullerene, Nano-TiO_2, Nano-SiO_2). *Aquat. Toxicol.* **2010**, *100*, 168–177. [CrossRef]

45. Sureda, A.; Capó, X.; Busquets-Cortés, C.; Tejada, S. Acute exposure to sunscreen containing titanium induces an adaptive response and oxidative stress in Mytillus galloprovincialis. *Ecotoxicol. Environ. Saf.* **2018**, *149*, 58–63. [CrossRef] [PubMed]
46. Zhu, X.; Zhou, J.; Cai, Z. The toxicity and oxidative stress of TiO_2 nanoparticles in marine abalone (Haliotis diversicolor supertexta). *Mar. Pollut. Bull.* **2011**, *63*, 334–338. [CrossRef] [PubMed]
47. Galloway, T.; Lewis, C.; Dolciotti, I.; Johnston, B.D.; Moger, J.; Regoli, F. Sublethal toxicity of nano-titanium dioxide and carbon nanotubes in a sediment dwelling marine polychaete. *Environ. Pollut.* **2010**, *158*, 1748–1755. [CrossRef] [PubMed]
48. Manzo, S.; Schiavo, S.; Oliviero, M.; Toscano, A.; Ciaravolo, M.; Cirino, P. Immune and reproductive system impairment in adult sea urchin exposed to nanosized ZnO via food. *Sci. Total Environ.* **2017**, *599–600*, 9–13. [CrossRef] [PubMed]
49. Araújo, M.J.; Rocha, R.J.M.; Soares, A.M.V.M.; Benedé, J.L.; Chisvert, A.; Monteiro, M.S. Effects of UV filter 4-methylbenzylidene camphor during early development of Solea senegalensis Kaup, 1858. *Sci. Total Environ.* **2018**, *628–629*, 1395–1404. [CrossRef]
50. Liu, H.; Sun, P.; Liu, H.; Yang, S.; Wang, L.; Wang, Z. Acute toxicity of benzophenone-type UV filters for Photobacterium phosphoreum and Daphnia magna: QSAR analysis, interspecies relationship and integrated assessment. *Chemosphere* **2015**, *135*, 182–188. [CrossRef]
51. Grabicova, K.; Fedorova, G.; Burkina, V.; Steinbach, C.; Schmidt-Posthaus, H.; Zlabek, V.; Kocour Kroupova, H.; Grabic, R.; Randak, T. Presence of UV filters in surface water and the effects of phenylbenzimidazole sulfonic acid on rainbow trout (Oncorhynchus mykiss) following a chronic toxicity test. *Ecotoxicol. Environ. Saf.* **2013**, *96*, 41–47. [CrossRef]
52. Gao, L.; Yuan, T.; Cheng, P.; Zhou, C.; Ao, J.; Wang, W.; Zhang, H. Organic UV filters inhibit multixenobiotic resistance (MXR) activity in Tetrahymena thermophila: Investigations by the Rhodamine 123 accumulation assay and molecular docking. *Ecotoxicology* **2016**, *25*, 1318–1326. [CrossRef]
53. Zhang, Q.; Ma, X.; Dzakpasu, M.; Wang, X.C. Evaluation of ecotoxicological effects of benzophenone UV filters: Luminescent bacteria toxicity, genotoxicity and hormonal activity. *Ecotoxicol. Environ. Saf.* **2017**, *142*, 338–347. [CrossRef]
54. Mao, F.; He, Y.; Gin, K.Y.H. Evaluating the joint toxicity of two benzophenone-type UV filters on the green alga Chlamydomonas reinhardtii with response surface methodology. *Toxics* **2018**, *6*, 8. [CrossRef]
55. Fent, K.; Kunz, P.Y.; Gomez, E. UV filters in the aquatic environment induce hormonal effects and affect fertility and reproduction in fish. *Chimia* **2008**, *62*, 368–375. [CrossRef]
56. Blüthgen, N.; Zucchi, S.; Fent, K. Effects of the UV filter benzophenone-3 (oxybenzone) at low concentrations in zebrafish (Danio rerio). *Toxicol. Appl. Pharmacol.* **2012**, *263*, 184–194. [CrossRef]
57. Kim, S.; Jung, D.; Kho, Y.; Choi, K. Effects of benzophenone-3 exposure on endocrine disruption and reproduction of japanese medaka (Oryzias latipes)-A two generation exposure study. *Aquat. Toxicol.* **2014**, *155*, 244–252. [CrossRef]
58. Kaiser, D.; Sieratowicz, A.; Zielke, H.; Oetken, M.; Hollert, H.; Oehlmann, J. Ecotoxicological effect characterisation of widely used organic UV filters. *Environ. Pollut.* **2012**, *163*, 84–90. [CrossRef] [PubMed]
59. Park, C.B.; Jang, J.; Kim, S.; Kim, Y.J. Single- and mixture toxicity of three organic UV-filters, ethylhexyl methoxycinnamate, octocrylene, and avobenzone on Daphnia magna. *Ecotoxicol. Environ. Saf.* **2017**, *137*, 57–63. [CrossRef]
60. Cherchi, C.; Gu, A.Z. Impact of titanium dioxide nanomaterials on nitrogen fixation rate and intracellular nitrogen storage in Anabaena variabilis. *Environ. Sci. Technol.* **2010**, *44*, 8302–8307. [CrossRef] [PubMed]
61. Jovanović, B.; Whitley, E.M.; Kimura, K.; Crumpton, A.; Palić, D. Titanium dioxide nanoparticles enhance mortality of fish exposed to bacterial pathogens. *Environ. Pollut.* **2015**, *203*, 153–164. [CrossRef] [PubMed]
62. Nigro, M.; Bernardeschi, M.; Costagliola, D.; Della Torre, C.; Frenzilli, G.; Guidi, P.; Lucchesi, P.; Mottola, F.; Santonastaso, M.; Scarcelli, V.; et al. n-TiO_2 and $CdCl_2$ co-exposure to titanium dioxide nanoparticles and cadmium: Genomic, DNA and chromosomal damage evaluation in the marine fish European sea bass (*Dicentrarchus labrax*). *Aquat. Toxicol.* **2015**, *168*, 72–77. [CrossRef]
63. Xia, B.; Zhu, L.; Han, Q.; Sun, X.; Chen, B.; Qu, K. Effects of TiO_2 nanoparticles at predicted environmental relevant concentration on the marine scallop Chlamys farreri: An integrated biomarker approach. *Environ. Toxicol. Pharmacol.* **2017**, *50*, 128–135. [CrossRef]
64. Miller, R.J.; Lenihan, H.S.; Muller, E.B.; Tseng, N.; Hanna, S.K.; Keller, A.A. Impacts of metal oxide nanoparticles on marine phytoplankton. *Environ. Sci. Technol.* **2010**, *44*, 7329–7334. [CrossRef] [PubMed]
65. Peng, X.; Palma, S.; Fisher, N.S.; Wong, S.S. Effect of morphology of ZnO nanostructures on their toxicity to marine algae. *Aquat. Toxicol.* **2011**, *102*, 186–196. [CrossRef]
66. Wong, S.W.Y.; Leung, P.T.Y.; Djurišić, A.B.; Leung, K.M.Y. Toxicities of nano zinc oxide to five marine organisms: Influences of aggregate size and ion solubility. *Anal. Bioanal. Chem.* **2010**, *396*, 609–618. [CrossRef] [PubMed]
67. Schiavo, S.; Oliviero, M.; Li, J.; Manzo, S. Testing ZnO nanoparticle ecotoxicity: Linking time variable exposure to effects on different marine model organisms. *Environ. Sci. Pollut. Res.* **2018**, *25*, 4871–4880. [CrossRef] [PubMed]
68. Giraldo, A.; Montes, R.; Rodil, R.; Quintana, J.B.; Vidal-Liñán, L.; Beiras, R. Ecotoxicological Evaluation of the UV Filters Ethylhexyl Dimethyl p-Aminobenzoic Acid and Octocrylene Using Marine Organisms Isochrysis galbana, Mytilus galloprovincialis and Paracentrotus lividus. *Arch. Environ. Contam. Toxicol.* **2017**, *72*, 606–611. [CrossRef]
69. Barmo, C.; Ciacci, C.; Canonico, B.; Fabbri, R.; Cortese, K.; Balbi, T.; Marcomini, A.; Pojana, G.; Gallo, G.; Canesi, L. In vivo effects of n-TiO_2 on digestive gland and immune function of the marine bivalve Mytilus galloprovincialis. *Aquat. Toxicol.* **2013**, *132–133*, 9–18. [CrossRef]

70. Barone, A.N.; Hayes, C.E.; Kerr, J.J.; Lee, R.C.; Flaherty, D.B. Acute toxicity testing of TiO_2-based vs. oxybenzone-based sunscreens on clownfish (*Amphiprion ocellaris*). *Environ. Sci. Pollut. Res.* **2019**, 14513–14520. [CrossRef] [PubMed]
71. Santonocito, M.; Salerno, B.; Trombini, C.; Tonini, F.; Pintado-Herrera, M.G.; Martínez-Rodríguez, G.; Blasco, J.; Lara-Martín, P.A.; Hampel, M. Stress under the sun: Effects of exposure to low concentrations of UV-filter 4-methylbenzylidene camphor (4-MBC) in a marine bivalve filter feeder, the Manila clam Ruditapes philippinarum. *Aquat. Toxicol.* **2020**, *221*. [CrossRef]
72. Hong, H.; Wang, J.; Shi, D. Effects of salinity on the chronic toxicity of 4-methylbenzylidene camphor (4-MBC) in the marine copepod Tigriopus japonicus. *Aquat. Toxicol.* **2021**, *232*, 105742. [CrossRef] [PubMed]
73. Thorel, E.; Clergeaud, F.; Jaugeon, L.; Rodrigues, A.M.S.; Lucas, J.; Stien, D.; Lebaron, P. Effect of 10 UV filters on the brine shrimp Artemia salina and themarinemicroalga *Tetraselmis* sp. *Toxics* **2020**, *8*, 29. [CrossRef]
74. Bachelot, M.; Li, Z.; Munaron, D.; Le Gall, P.; Casellas, C.; Fenet, H.; Gomez, E. Organic UV filter concentrations in marine mussels from French coastal regions. *Sci. Total Environ.* **2012**, *420*, 273–279. [CrossRef]
75. Sang, Z.; Leung, K.S.Y. Environmental occurrence and ecological risk assessment of organic UV filters in marine organisms from Hong Kong coastal waters. *Sci. Total Environ.* **2016**, *566–567*, 489–498. [CrossRef]
76. Amine, H.; Gomez, E.; Halwani, J.; Casellas, C.; Fenet, H.; Araújo, M.J.; Rocha, R.J.M.; Soares, A.M.V.M.; Benedé, J.L.; Chisvert, A.; et al. Photostability and phytotoxicity of selected sunscreen agents and their degradation mixtures in water. *Sci. Total Environ.* **2018**, *25*, 368–375. [CrossRef]
77. Moschino, V.; Schintu, M.; Marrucci, A.; Marras, B.; Nesto, N.; Da Ros, L. An ecotoxicological approach to evaluate the effects of tourism impacts in the Marine Protected Area of La Maddalena (Sardinia, Italy). *Mar. Pollut. Bull.* **2017**, *122*, 306–315. [CrossRef] [PubMed]
78. Kunz, P.Y.; Fent, K. Estrogenic activity of UV filter mixtures. *Toxicol. Appl. Pharmacol.* **2006**, *217*, 86–99. [CrossRef] [PubMed]
79. Lorigo, M.; Mariana, M.; Cairrao, E. Photoprotection of ultraviolet-B filters: Updated review of endocrine disrupting properties. *Steroids* **2018**, *131*, 46–58. [CrossRef] [PubMed]
80. Silvia Díaz-Cruz, M.; Llorca, M.; Barceló, D. Organic UV filters and their photodegradates, metabolites and disinfection by-products in the aquatic environment. *TrAC Trends Anal. Chem.* **2008**, *27*, 873–887. [CrossRef]
81. Gago-Ferrero, P.; Alonso, M.B.; Bertozzi, C.P.; Marigo, J.; Barbosa, L.; Cremer, M.; Secchi, E.R.; Azevedo, A.; Lailson-Brito, J.; Torres, J.P.M.; et al. First determination of UV filters in marine mammals. octocrylene levels in Franciscana dolphins. *Environ. Sci. Technol.* **2013**, *47*, 5619–5625. [CrossRef]
82. He, K.; Hain, E.; Timm, A.; Blaney, L. Bioaccumulation of estrogenic hormones and UV-filters in red swamp crayfish (*Procambarus clarkii*). *Sci. Total Environ.* **2021**, *764*, 142871. [CrossRef]
83. Molins-Delgado, D.; Muñoz, R.; Nogueira, S.; Alonso, M.B.; Torres, J.P.; Malm, O.; Ziolli, R.L.; Hauser-Davis, R.A.; Eljarrat, E.; Barceló, D.; et al. Occurrence of organic UV filters and metabolites in lebranche mullet (*Mugil liza*) from Brazil. *Sci. Total Environ.* **2018**, *618*, 451–459. [CrossRef]
84. Zenker, A.; Schmutz, H.; Fent, K. Simultaneous trace determination of nine organic UV-absorbing compounds (UV filters) in environmental samples. *J. Chromatogr. A* **2008**, *1202*, 64–74. [CrossRef]
85. Baker, T.J.; Tyler, C.R.; Galloway, T.S. Impacts of metal and metal oxide nanoparticles on marine organisms. *Environ. Pollut.* **2014**, *186*, 257–271. [CrossRef]
86. Chen, J.; Li, H.; Han, X.; Wei, X. Transmission and Accumulation of Nano-TiO_2 in a 2-Step Food Chain (*Scenedesmus obliquus* to *Daphnia magna*). *Bull. Environ. Contam. Toxicol.* **2015**, *95*, 145–149. [CrossRef] [PubMed]
87. Wang, Z.; Yin, L.; Zhao, J.; Xing, B. Trophic transfer and accumulation of TiO_2 nanoparticles from clamworm (Perinereis aibuhitensis) to juvenile turbot (*Scophthalmus maximus*) along a marine benthic food chain. *Water Res.* **2016**, *95*, 250–259. [CrossRef]
88. Della Torre, C.; Buonocore, F.; Frenzilli, G.; Corsolini, S.; Brunelli, A.; Guidi, P.; Kocan, A.; Mariottini, M.; Mottola, F.; Nigro, M.; et al. Influence of titanium dioxide nanoparticles on 2,3,7,8-tetrachlorodibenzo-p-dioxin bioconcentration and toxicity in the marine fish European sea bass (*Dicentrarchus labrax*). *Environ. Pollut.* **2015**, *196*, 185–193. [CrossRef] [PubMed]
89. Hou, J.; Wu, Y.; Li, X.; Wei, B.; Li, S.; Wang, X. Toxic effects of different types of zinc oxide nanoparticles on algae, plants, invertebrates, vertebrates and microorganisms. *Chemosphere* **2018**, *193*, 852–860. [CrossRef]
90. De Schamphelaere, K.A.C.; Canli, M.; Van Lierde, V.; Forrez, I.; Vanhaecke, F.; Janssen, C.R. Reproductive toxicity of dietary zinc to Daphnia magna. *Aquat. Toxicol.* **2004**, *70*, 233–244. [CrossRef]
91. Danovaro, R.; Armeni, M.; Corinaldesi, C.; Mei, M.L. Viruses and marine pollution. *Mar. Pollut. Bull.* **2003**, *46*, 301–304. [CrossRef]
92. He, T.; Tsui, M.M.P.; Tan, C.J.; Ng, K.Y.; Guo, F.W.; Wang, L.H.; Chen, T.H.; Fan, T.Y.; Lam, P.K.S.; Murphy, M.B. Comparative toxicities of four benzophenone ultraviolet filters to two life stages of two coral species. *Sci. Total Environ.* **2019**, *651*, 2391–2399. [CrossRef]
93. Tsui, M.M.P.; Lam, J.C.W.; Ng, T.Y.; Ang, P.O.; Murphy, M.B.; Lam, P.K.S. Occurrence, Distribution, and Fate of Organic UV Filters in Coral Communities. *Environ. Sci. Technol.* **2017**, *51*, 4182–4190. [CrossRef]
94. Mitchelmore, C.L.; He, K.; Gonsior, M.; Hain, E.; Heyes, A.; Clark, C.; Younger, R.; Schmitt-Kopplin, P.; Feerick, A.; Conway, A.; et al. Occurrence and distribution of UV-filters and other anthropogenic contaminants in coastal surface water, sediment, and coral tissue from Hawaii. *Sci. Total Environ.* **2019**, *670*, 398–410. [CrossRef]
95. Brausch, J.M.; Rand, G.M. A review of personal care products in the aquatic environment: Environmental concentrations and toxicity. *Chemosphere* **2011**, *82*, 1518–1532. [CrossRef]

96. Downs, C.A.; Kramarsky-Winter, E.; Segal, R.; Fauth, J.; Knutson, S.; Bronstein, O.; Ciner, F.R.; Jeger, R.; Lichtenfeld, Y.; Woodley, C.M.; et al. Toxicopathological Effects of the Sunscreen UV Filter, Oxybenzone (Benzophenone-3), on Coral Planulae and Cultured Primary Cells and Its Environmental Contamination in Hawaii and the U.S. Virgin Islands. *Arch. Environ. Contam. Toxicol.* **2016**, *70*, 265–288. [CrossRef]
97. Corinaldesi, C.; Marcellini, F.; Nepote, E.; Damiani, E.; Danovaro, R. Impact of inorganic UV filters contained in sunscreen products on tropical stony corals (*Acropora* spp.). *Sci. Total Environ.* **2018**, *637–638*, 1279–1285. [CrossRef]
98. Fel, J.P.; Lacherez, C.; Bensetra, A.; Mezzache, S.; Béraud, E.; Léonard, M.; Allemand, D.; Ferrier-Pagès, C. Photochemical response of the scleractinian coral Stylophora pistillata to some sunscreen ingredients. *Coral Reefs* **2019**, *38*, 109–122. [CrossRef]
99. Levine, A. Sunscreen use and awareness of chemical toxicity among beach goers in Hawaii prior to a ban on the sale of sunscreens containing ingredients found to be toxic to coral reef ecosystems. *Mar. Policy* **2020**, 103875. [CrossRef]
100. Sirois, J. Examine all available evidence before making decisions on sunscreen ingredient bans. *Sci. Total Environ.* **2019**, *674*, 211–212. [CrossRef] [PubMed]
101. Corinaldesi, C.; Damiani, E.; Marcellini, F.; Falugi, C.; Tiano, L.; Brugè, F.; Danovaro, R. Sunscreen products impair the early developmental stages of the sea urchin *Paracentrotus lividus*. *Sci. Rep.* **2017**, *7*, 7815. [CrossRef] [PubMed]

diversity

MDPI

Review

Dittrichia viscosa: Native-Non Native Invader

Barbara Sladonja [1], Danijela Poljuha [1,*], Marin Krapac [1], Mirela Uzelac [1] and Maja Mikulic-Petkovsek [2]

[1] Institute of Agriculture and Tourism, Karla Huguesa 8, 52440 Poreč, Croatia; barbara@iptpo.hr (B.S.); marin@iptpo.hr (M.K.); mirela@iptpo.hr (M.U.)
[2] Department of Agronomy, Biotechnical Faculty, University of Ljubljana, Jamnikarjeva 101, SI-1000 Ljubljana, Slovenia; maja.mikulic-petkovsek@bf.uni-lj.si
* Correspondence: danijela@iptpo.hr; Tel.: +385-52-408-336

Abstract: *Dittrichia viscosa* (L.) Greuter is a shrub native to the Mediterranean, however, declared as a very invasive species in Australia and North America. Environmental (climatic) and socio-economic (land abandonment) changes can trigger different adaptive mechanisms and cause changes in species behavior, influencing invasion dynamics. Motivated by the recently noticed change of *D. viscosa* behavior in its native Mediterranean habitat, we discuss the invasion properties, its behavior in the native habitat and new areas, and its management options. We review the species' adverse effects and its positive ecosystem services in the Millennium Ecosystem Assessment framework. In this review, we provide information on the phytochemical properties of *D. viscosa* and highlight its potential use in ecological agriculture, phytopharmacy, and medicine. The presented data is useful for developing effective management of this contentious species, with emphasis on mitigating environmental and economic damages, especially in agriculture. The final aim is to achieve a balanced ecosystem, providing a high level of possible services (provisioning, regulating, cultural and supporting).

Keywords: contentious invasive species; invasive properties; ecosystem services; phytochemical properties; weed

Citation: Sladonja, B.; Poljuha, D.; Krapac, M.; Uzelac, M.; Mikulic-Petkovsek, M. *Dittrichia viscosa*: Native-Non Native Invader *Diversity* **2021**, *13*, 380. https://doi.org/10.3390/d13080380

Academic Editor: Michael Wink

Received: 14 July 2021
Accepted: 11 August 2021
Published: 15 August 2021

Publisher's Note: MDPI stays neutral with regard to jurisdictional claims in published maps and institutional affiliations.

Copyright: © 2021 by the authors. Licensee MDPI, Basel, Switzerland. This article is an open access article distributed under the terms and conditions of the Creative Commons Attribution (CC BY) license (https://creativecommons.org/licenses/by/4.0/).

1. Introduction

Dittrichia viscosa (L.) Greuter, Stinkwort or False Yellowhead, is a perennial ruderal plant belonging to the Compositae family, native to the Mediterranean and Western Asia [1–3]. It is declared and listed on the Alert List of Environmental Weeds in Australia and North America [1]. It is abundant in anthropically altered areas [4]. In the Mediterranean, *D. viscosa* belongs to the vegetation of abandoned Mediterranean cultures, class of *INULETEA VISCOSAE* Trinajstić (1965) 1978; characteristic of the association *Helichryso italici-Dittrichietum viscosae* Trinajstić ex Di Pietro, Germani & Fortini 2017 [5], a pasture of immortelle and false yellowhead [3,6]. Owing to its high seed production and spreading, good adaptability, and resistance to adverse conditions, it is considered an important environmental weed [1,6]. Recently an unusual spread of *D. viscosa* was recorded in coastal Croatia (personal records), causing functional and economic problems (Figure 1). Although there are no published quantitative data showing the spread of *D. viscosa* in Croatia, according to GBIF database (which included six different datasets) [7] from 2009 to 2020 number of observations have increased supporting our observations (Figure 2).

In neighboring Slovenia, it is considered indigenous and widespread [8]. Habitat changes in the Mediterranean basin, as well as in Croatia, are commonly linked to environmental and socio-economic alterations, land abandonment, and loss of traditional agricultural areas [9]. Climate change is one of the most significant environmental threats to biodiversity and has a number of potential effects on weeds in both agricultural and native vegetation landscapes [10,11]. It is known that climate variations can influence the spreading of invasive plant species [12–14]. Hence, established non-native species could become invasive if climate change increases their competitive ability or spreading rate [15].

Figure 1. High abundance of *Dittrichia viscosa* on abandoned agriculture land in coastal Croatia.

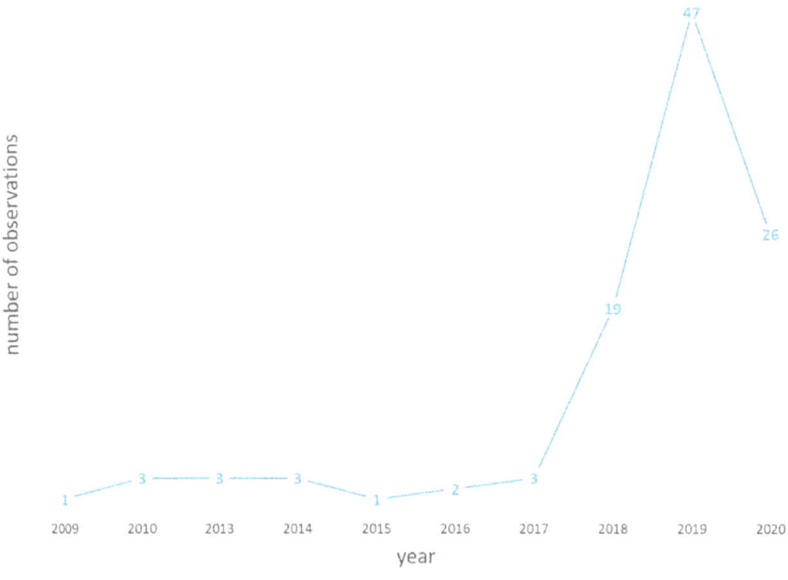

Figure 2. Number of observations of *D. viscosa* in the Republic of Croatia from 2009 to 2020.

However, to date, the characteristics that make species a successful invader or the ecosystem features that predispose it to invasion are still poorly understood [16,17]. Moreover, it is unclear which species or community attributes enhance invader success or explain spread dynamics [18]. Nevertheless, it has been shown that invasion success is dependent on unique interactions between the invader and the environment [19–22]. There are many hypotheses about why invasive species are successful. Some hypothesized mechanisms are reproductive ability, vegetative growth, predation, adaptation, and allelopathy [18]. These competitive mechanisms can provide substantial socio-economic, landscape, and ecological services, both in native and non-native species [23].

Generally, ecosystem services are defined as the output of natural systems from which humans can derive benefits. They have been categorized into four categories by the

Millennium Ecosystem Assessment [24]: cultural, provisioning, supporting, and regulating. This assessment has been created to qualitatively test the effects of singular species and give useful inputs for effective environmental management. *D. viscosa* is known to offer a large variety of ecosystem functions in the Mediterranean basin. It is an important species that can be used for phytoremediation, as a bio accumulator or bioindicator, as well as an additional tool in ecological agriculture [25–27]. For example, *D. viscosa* is a host plant for natural enemies (the parasitic complex of *Myopites stylata* and *Eupelmus urozonus*) of plant pests like olive fruit fly (*Bactrocera oleae* Gmel.) [28].

Provisioning of potentially useful compounds to humans is an important ecosystem service. Plant species can be a consolidated source of bioactive components, and many higher plants possess allopathic potential [29,30], which can be investigated and used for the development of commercial natural herbicides. Plant-derived natural products have also served as an important resource for medicinal compounds. *D. viscosa* contains several biologically active compounds: flavonoids, sesquiterpene lactones, sesquiterpene acids, and triterpenes [31].

The aims of this review are to: (i) review and summarize the current state of knowledge regarding *D. viscosa* invasiveness potential; (ii) provide an overview of *D. viscosa* ecosystem services; (iii) provide information for the effective management of the species in changing habitats and asses its contribution to the ecosystem.

2. *Dittrichia viscosa* Invasive Properties

It is difficult to identify factors associated with the degree of invasion by alien plants. It is certain that invasive species are more plastic in a variety of traits. Still, it remains extremely difficult to define a set of traits responsible for particular species invasiveness [32]. *D. viscosa* has the potential to be a serious environmental weed both in native and introduced areas.

2.1. *D. viscosa in Native Areas*

It shows a great pioneer character and, in recent years, has largely expanded its range in many Mediterranean countries, possibly due to increased human disturbances [33]. According to Wacquant [34] the species areal has been expanding in the Mediterranean for the previous 25 years. He described the capability of *D. viscosa* to colonize new habitats and threaten biodiversity, mostly due to its phenotypic plasticity [35]. It has been proven that climatic changes can drive the spread of thermophilic species, such as *D. viscosa*. Vesperinas et al. [12] correlated the expansion of these species with mean temperature increases.

According to different authors *D. viscosa* has a number of characteristics that makes it highly competitive in its native range:

2.1.1. Biology

Morphological and anatomical characteristics that make *D. viscosa* a good invader are: substantial roots even in small plants, dense canopy, presence of glandular hairs on leaves and stems, strong odor, the glandular hairs secrete lipids, polysaccharides, and proteins.

2.1.2. Reproduction, Regeneration, and Dispersal

D. viscosa is a prolific seed producer, but the longevity and viability of *D. viscosa* seeds are unknown [36]. *D. viscosa* tolerates salty soil with small amounts of available water and germination is favored by ground disturbance and fire. It can withstand soils with salinity of sea water level, approx. 30 g NaCl/L [37]. *Dittrichia viscosa* can be regenerated in several ways. One is by seeds which can be collected at the end of October, the second is through cuttings, and the third is the transplantation of whole plants. Although in nature germinates the following year without interventions, the germination rates in the pots are poor [36]. The impact of salinity on the germination rate of *D. viscosa* in Slovenia was studied by Grašič et al. [38]. The authors concluded that even though *D. viscosa* seed is highly resistant to salinity, elevated salinity levels cannot be considered the main factor in determining its occurrence.

2.1.3. Allelopathy

D. viscosa is rich in secondary compounds such as flavonoids, sesquiterpenes and essential oils [36]. The allelopathic potential of the leaf exudates increases during the dry, hot, and sunny summer. Seasonal and geographic variations have also been found, but all inside native growing areas [39]. Dor and Hershenhorn [40] also revealed that *D. viscosa* produces sesquiterpene lactone, tayunin, which inhibits the growth of other nearby plants. Moreover, *D. viscosa* extracts can delay germination of other plants by reducing the frequency of dividing cells in the root [4,41].

2.1.4. Tolerance of a Broad Range of Environmental Conditions

Different authors [8,33,42] have described its high tolerance to salinity and unfavorable conditions. Al Hassan et al. [33] have evaluated the potential risk that *D. viscosa* represents for Mediterranean salt marsh vegetation, and it depended on its salt tolerance. They proved that *D. viscosa* prefers anthropically influenced or degraded communities and that the lower salinity makes it more competitive. *D. viscosa* can compromise some less tolerant species and general biodiversity. They also explain mechanisms responsible for stress tolerance in this species. *D. viscosa* is a ruderal plant, grows in abandoned fields, roadsides, walking trails, urban areas, and all modified and altered areas by anthropic activities [36,43,44].

2.1.5. Lack of Predators

2.2. D. viscosa in New Areas

De Laurentis et al. [39] recorded that the chemical composition and concentration of volatile constituents of *D. viscosa* from different areas in Italy were different. This indicates the possibility that, also in the introduced areas, the chemical composition could be different. Still, site-specific studies are needed to prove this assumption.

In the USA, California pest rating was performed according to plant's invasive characteristics and observed behavior on the site [45]. *D. viscosa* was rated as a species with high risk (score 3/3) to establish a widespread distribution, high risk for a host range, high risk for reproduction and dispersal, medium economic impact, and high environmental impact.

In Australia, where *D. viscosa* is considered a serious environmental weed marked for eradication flowering occurs between December and April. As in native areas, *D. viscosa* spreads by seed. Seed can also be spread during construction works or when attached to machinery. Germination is generally enhanced by fire or mechanical disturbance which creates bare ground. Under laboratory conditions, *D. viscosa* seeds undergo a deep dormancy, which is broken by a lack of light [46]. Generally, it is indicated as a drought-resistant and salt tolerant species [37,47], but it is occasionally found in Australia in swamps and along waterways [45]. There are no specific investigations on the invasive properties of *D. viscosa* in introduced areas and available information on which characteristics are climate or site-specific.

3. *Dittrichia viscosa* Ecosystem Services

Ecosystem services are the outputs of natural systems from which humans derive benefits [23]. Invasiveness can induce losses in ecosystem services. However, novel services are resulting from the characteristics of invasive species as well. The review of *D. viscosa* ecosystem services shows that it successfully meets many contemporary environmental and social needs. In its native range, *D. viscosa* provides several ecosystem services and functions, presented in Table 1.

These useful services have been arranged and grouped within the Millennium Ecosystem Assessment [24]. In the Table, all ecosystem services provided by *D. viscosa* have been investigated and summarized within four main groups. In general, provisional ecosystem services include food, fiber, genetic resources, pharmaceuticals and fresh water. Within this group, *D. viscosa* provides several services. The most studied services of *D. viscosa* (Table 1) are medicinal services and weed and pest control. For example, leaf extracts of *D. viscosa* contain antifungal agents which inhibit the growth of dermatophytes and *Candida albicans*, a group of fungi that cause a skin disease in animals and humans [31]. Additionally, both leaf extracts and dry leaves proved to be effective herbicides for use in organic agriculture. The extracts decreased seed germination of several weed species: *Sinapis arvensis* L., *Amaranthus palmeri* S. Wats., and *Solanum nigrum* L. [40]. Leaf extracts can also be used for the preparation of natural nematicides. The root-knot nematode *Meloidogyne javanica* causes severe damage in vegetable crops in Israel, especially in organic vegetable production systems. The use of *D. viscosa* extracts reduced nematode infection on tomato plants [58]. Regulating ecosystem services such as pollination, climate regulation, air quality regulation, and supporting services are important but global and not specific to *D. viscosa*. Cultural ecosystem services are expressed through aesthetic value, ecotourism, preservation of traditional practices. *D viscosa* is a plant known in Mediterranean traditional medicine and agricultural practices (Table 1).

Table 1. Ecosystem services, as defined by the Millennium Ecosystem Assessment [24], provided by *Dittrichia viscosa*.

Category	Example of Service Provided by *D. viscosa*	References
Provisioning	Medicinal	[31,48–53]
	Weed management	[4,8,54,55]
	Pest control	[40,56–60]
	Antifungal control	[31,61–64]
	Biomass production	[65]
Regulating	Bioindicator	[36,66,67]
	Bioaccumulator	[36,68–71]
	Phytoremediation	[36,43,55,72,73]
	Desertification control	[33,42,74]
Cultural	Traditional use	[34]
	Biogeographical component	[51,75,76]
Supporting	Primary production	[77,78]
	Nutrient cycling	[41,79]
	Soil formation	[80]

Invasive plant species represent a natural resource that can be freely collected without anthropogenic pressure on local resources in their natural habitats. The use of this resource can be a control method of invasive species and contributes to the ecological balance. Moreover, the collection of these species can produce economic and social benefits, especially in rural areas. Additional potential remains for *D. viscosa* in provisioning of useful phytochemicals. Certainly, in order to realize this potential, site-specific researches are needed.

4. Plant Chemistry as a Competitive Advantage and Potential Ecosystem Service

It is often assumed that alien plants can become invasive when they possess novel secondary metabolites compared to the native plants in the introduced range. High chemical diversity and phytochemical uniqueness in alien species could indicate biological invasion potential [30]. Moreover, data on chemical properties can provide a background for possible approaches to restrict and control invasive populations and are of considerable taxonomic interest. A study by Brahmi-Chendouh et al. [81] revealed the chemical composition of *D. viscosa* extracts, highlighting its diversity in polyphenolic constituents and the abundance of bioactive nutraceutical phytochemicals. Trimech et al. [82] described the metabolic profile of *D. viscosa* and the structures of the major polyphenolic constituents tentatively assigned based on their MS and UV–VIS spectra. Caffeic acid, also present in *D. viscosa* derivatives, plays an important role in the plants' defence against pathogens and insects [83,84]. They also participate in processes that promote herbicidal properties [85].

The search for new natural phytotoxins that can be applied in agriculture as target herbicides recently increased [86]. The allelopathic potential of *D. viscosa* can be found in its leaf leachate, which could be used in irrigation for sustainable weed management [55]. The high diversity of bioactive compounds in *D. viscosa* extracts highlights its significant potential for use in agriculture and pharmaceutics.

5. Invasive Native-Non Native *D. viscosa*

Despite the broad scientific interest on the topic, the "invasive-native-non-native" terminology is still unresolved [33]. Some scientists identify native species with invasive characteristics as "expansive" or "super-dominant" [87]. Others distinguish invasive species from transformers, not harmful species, and weeds, which can be non-native species, but not necessarily [88]. In our case, *D. viscosa* is classified as a weed, a native species that often grows in the Mediterranean area in sites where it is not wanted and has detectable economic or environmental impacts [89]. It is also classified as a non-native invasive outside its range of origin. Invasiveness is a dynamic process influenced by changing environmental parameters [15,90]. Many studies relate invasiveness to biotic and abiotic conditions, but the underlying mechanisms are still unclear, i.e., which variables best correlate to the observed patterns of richness and abundance [91]. Human activities such as pollution and land-use change often result in rapid environmental shifts [92,93]. Moreover, climate change is considered one of the major causes of disturbances to ecological conditions and poses an additional challenge to our ability to manage invasiveness. Due to their adaptability to climate change disturbance, such as increased temperatures and CO_2 concentrations, invasive species have increased opportunities for a spread in a broad range of geographic conditions [94]. Under changing local conditions, any species, regardless of origin (native or non-native), can respond with novel ecological behaviors and assume invasive characteristics. There are several examples where native species are considered invasive; for instance, the fern *Pteridium aquilinum* (L.) Kuhn. and the shrub *Ulex europaeus* L. [95].

In this review, we are discussing an example of an expressed change in the species distribution pattern, noticed on a local scale in Croatia. Local changes provoked native *D. viscosa* behavior change, expressing its competitive characteristics and becoming an abundant weed in Croatian coastal areas (Figure 1). As in many places in the Mediterranean, complex socio-economic and environmental factors resulted in such changes. Rural depopulation and land abandonment are common in Croatia, but little is known about their impact on biodiversity and population dynamics. Although there are numerous case studies on the impacts of land abandonment in the Mediterranean basin [96], there are limited studies on these processes in Croatia [97]. Among the most significant processes of global environmental change [98], the termination of crop cultivation and livestock grazing, seen in rural areas, could be some of the main reasons behind recent increases in abundance of *D. viscosa* and other similar weeds in coastal Croatia. The invasive properties overviewed in this study are the background of evidence for spreading, which will develop under specific conditions. However, the data on its invasion dynamics remains scarce [36].

When given the opportunity, native species can move into new areas where they have a competitive advantage over indigenous ones [99]. They can exhibit aggressive characteristics—such as demographic explosions, biomass accumulation, high reproductive output, phenotypic plasticity, and novel reproductive strategies—becoming weeds [100,101]. Some weeds are particularly concerning and have been listed as a priority for agricultural management or legislation. *D. viscosa* is listed in the Alert List of Environmental Weeds of

Australia [2]. The list comprises 28 non-native plants that threaten biodiversity and cause other environmental and economic damage. These weeds have the potential to degrade Australia's ecosystems seriously. It is not clear how *D. viscosa* was introduced to Australia, but it may have been introduced by horticulturalists. Throughout Australia, weeds are spreading uncontrollably fast, and their management is consuming an enormous amount of resources. Particularly, considering their potential for negative effect, such as reducing agricultural production, damaging infrastructure, and can negatively influence human health or well-being [102].

6. Species Management

When *D. viscosa* is treated as weed it needs proper management measures, but it can also be used as a management mean for other weed removals. Effective management strategies require detailed assessments of both the positive aspects of species in an ecosystem and the negative impacts of their invasiveness [103]. The goal is to manage invasive species in a way that establishes balanced and controlled ecosystems, providing a high level of possible ecosystem services. This is especially valid for native habitats in which there are natural mechanisms of species control. In new habitats, though, these species deserve special attention in recognition of their costs and management models. In the case study of buffelgrass in Australia, Grice and the authors created broadscale strategic solutions for the management of this species. Many invasive species, *D. viscosa* including, are contentious species and management approaches could follow the same direction as Grice et al. [104] proposed. The results and observations presented in this review highlight the fragility of environmental balance, as well as the unexplored potential of *D. viscosa*. Systematic research on the invasion dynamics and ecosystem services provisioning of *D. viscosa*, as well as complex large scale strategies and actions, are needed to establish sustainable management plans.

Different authors [40,105] have proved that *D. viscosa* chemical properties can be useful in weed germination inhibition, and therefore used as a biological herbicide. They identified that allelopathic activity changed during the vegetation; extracts from leaves collected in spring were more active than those collected in autumn. *D. viscosa* could also be used as a secondary plant in biological control [1]. Moreover, this species has shown a significant potential in soil management, in particular in use for phytoremediation in mining-affected semiarid soils, since it is an efficient bioaccumulator of trace metals [33].

7. Conclusions

Motivated by observed changes in *Dittrichia viscosa* behavior, with a particularly noticeable negative impact on agriculture, in this review, we add and summarize valuable information to the existing knowledge on this species. Ecosystem disturbances and constant changes in environmental conditions can easily compromise the ecological balance. Climate change poses risks that vary greatly geographically. Phytochemicals present in the plant organs of *D. viscosa* make it a promising resource for use in organic agriculture and phytomedicine. The positive aspects of species in any ecosystem have to be weighed against the loss of other ecosystem services. Habitat-specific studies are essential to adopt goal-specific measures in the management of *D. viscosa*. Still, it is more likely to use the plant in the native environment as an ecosystem provider and try to control its spread and eradicate when possible in the new habitats. Additional studies on the possible uses of extracts from *D. viscosa* for the purpose of controlling weeds in agriculture and other uses of this species in soil management would also be beneficial.

Author Contributions: All authors have contributed significantly to the present work. In particular, B.S. and D.P. to conceptualization; B.S., D.P., M.U. and M.M.-P. to methodology development, analysis performance, and investigation; M.K. and M.U. to field sampling and data curation; M.U. to original draft preparation, and M.K. to vizualisation. All authors have contributed to writing, editing and proof checking. B.S. and M.M.-P. moreover to project administration and funding acquisition. All authors have read and agreed to the published version of the manuscript.

Funding: This work was supported by the Croatian-Slovenian bilateral project funded by the Croatian Ministry of Science and Technology and Horticulture program no. P4-0013-0481 funded by the Slovenian Research Agency (ARRS).

Acknowledgments: We especially thank Slavko Brana and Istrian Botanical Society for help in the fieldwork and availability.

Conflicts of Interest: The authors declare no conflict of interest. The funders had no role in the design of the study; in the collection, analyses, or interpretation of data; in the writing of the manuscript, or in the decision to publish the results.

References

1. Parolin, P.; Scotta, I.; Bresch, C. Biology of *Dittrichia viscosa*, a Mediterranean ruderal plant: A review. *Phyton* **2014**, *83*, 251–262.
2. Ounoughi, A.; Ramdani, M.; Lograda, T.; Chalard, P.; Figuérédo, G. Chemotypes and antibacterial activities of *Inula viscosa* essential oils from Algeria. *Biodiversitas J. Biol. Divers.* **2020**, *21*, 1504–1517. [CrossRef]
3. Glanznig, A.; Kessal, O. *Invasive Plants of National Importance and Their Legal Status by State and Territory*; WWF-Australia: Sydney, Australia, 2004; pp. 19–21.
4. Araniti, F.; Lupini, A.; Sunseri, F.; Abenavoli, M.R. Allelopatic Potential of *Dittrichia viscosa* (L.) W. Greuter Mediated by VOCs: A Physiological and Metabolomic Approach. *PLoS ONE* **2017**, *12*, e0170161. [CrossRef]
5. Šugar, I. *Vegetacijska Karta SR Hrvatske List 77*; Botanički zavod PMF: Pula, Hrvatska, 1978.
6. GBIF Secretariat. GBIF Backbone Taxonomy. 2021. Checklist Dataset. Available online: https://www.gbif.org/species/3101184 (accessed on 2 August 2021).
7. Di Pietro, R.; Germani, D.; Fortini, P. A phytosociological investigation on the mixed hemycryptophitic and therophitic grasslands of the Cornicolani mountains (Lazio Region-central Italy). *Plant Sociol.* **2017**, *54*, 107–128. [CrossRef]
8. Hulme, P. Weed risk assessment: A way forward or a waste of time? *J. Appl. Ecol.* **2011**, *49*, 10–19. [CrossRef]
9. Grašič, M.; Anžlovar, S.; Strgulc Krajšek, S. Germination rate of stinkwort (*Dittrichia graveolens* (L.) Greuter) and false yellowhead (*D. viscosa* (L.) Greuter) in relation to salinity and the impact of their extracts on germination of selected plant species. In Proceedings of the 6th Slovenian Symposium on Plant Biology, Hoče, Maribor, Slovenia, 12 September 2014.
10. Cramer, V.; Hobbs, J.R.; Standish, R. What's new about old fields? Land abandonment and ecosystem assembly. *Ecol. Evol.* **2008**, *23*, 104–112. [CrossRef]
11. Seabloom, E.; Williams, J.; Slayback, D.; Stoms, D.; Viers, H.J.; Dobson, A.P. Human impacts, plant invasion, and imperiled plant species in California. *Ecol. Appl.* **2006**, *16*, 1338–1350. [CrossRef]
12. Brook, B.; Sodhi, S.N.; Bradshaw, C. Synergies among extinction drivers under global change. *Ecol. Evol.* **2008**, *23*, 453–460. [CrossRef]
13. Vesperinas, E.S.; Moreno, A.G.; Elorza, M.S.; Sánchez, E.D.; Mata, D.S.; Gavilán, R. The Expansion of Thermophilic Plants in the Iberian Peninsula as a Sign of Climatic Change. In *"Fingerprints" of Climate Change*; Walther, G.R., Burga, C.A., Edwards, P.J., Eds.; Springer: Boston, MA, USA, 2001; pp. 163–184.
14. Doroftei, M.; Anastasiu, P. Potential Impacts of Climate Change on Habitats and Their Effects on Invasive Plant Species in Danube Delta Biosphere Reserve, Romania. In *Managing Protected Areas in Central and Eastern Europe Under Climate Change: Advances in Global Change Research*; Rannow, S., Neubert, M., Eds.; Springer: Dordrecht, The Netherlands, 2014; pp. 267–278.
15. Hoveka, L.N.; Bezeng, B.S.; Yessoufou, K.; Boatwrigh, J.S.; Van der Bank, M. Effects of climate change on the future distributions of the top five freshwater invasive plants in South Africa. *S. Afr. J.* **2016**, *102*, 33–38. [CrossRef]
16. Hellmann, J.J.; Byers, J.E.; Bierwagen, B.G.; Dukes, J.S. Five potential consequences of climate change for invasive species. *Biol. Conserv.* **2008**, *22*, 534–543. [CrossRef] [PubMed]
17. Cronk, Q.C.B.; Fuller, J. Plant Invaders: The Threat to Natural Ecosystems. In *Plant Invaders*; Cronk, Q.C.B., Ed.; Earthscan Publications: London, UK, 1995.
18. Kolar, C.S.; Lodge, D.M. Progress in invasion biology: Predicting invaders. *Ecol. Evol.* **2001**, *16*, 199–204. [CrossRef]
19. Arim, M.; Abades, S.; Neill, P.; Lima, M.; Marquet, P. Spread dynamics of invasive species. *Proc. Natl. Acad. Sci. USA* **2006**, *103*, 374–378. [CrossRef] [PubMed]
20. Sebert-Cuvillier, E.; Paccaut, F.; Chabrerie, O.; Endels, P.; Goubet, O.; Decocq, G. Local population dynamics of an invasive tree species with a complex life-history cycle: A stochastic matrix model. *Ecol. Model.* **2007**, *201*, 127–143. [CrossRef]
21. Lind, E.M.; Parker, J.D. Novel Weapons Testing: Are Invasive Plants More Chemically Defended than Native Plants? *PLoS ONE* **2010**, *5*. [CrossRef]
22. Bennett, A.E.; Thomsen, M.; Strauss, S.Y. Multiple mechanisms enable invasive species to suppress native species. *Am. J. Bot.* **2011**, *98*, 1086–1094. [CrossRef] [PubMed]
23. Dormontt, E.E.; Lowe, A.J.; Prentis, P.J. Is rapid adaptive evolution important in successful invasions? In *Fifty Years of Invasion Ecology The Legacy of Charles Elton*; Richardson, D.M., Ed.; Wiley-Blackwell: Shichester, UK, 2011; pp. 175–193.
24. Sladonja, B.; Sušek, M.; Guillermic, J. Review on invasive tree of heaven (*Ailanthus altissima* (Mill.) Swingle) conflicting values: Assessment of its ecosystem services and potential biological threat. *J. Environ. Manag.* **2015**, *56*, 1009–1034. [CrossRef]

25. Millennium Ecosystem Assessment, Ecosystems and Human Well-being: Synthesis. Available online: https://www.millenniumassessment.org/documents/document.356.aspx.pdf (accessed on 13 October 2020).
26. Kavallieratos, N.G.; Stathas, G.J.; Athanassiou, C.G.; Tomanović, Ž. Aphid parasitoids (Hymenoptera: Braconidae: Aphidiinae) on citrus: Seasonal abundance, association with the species of host plant, and sampling indices. *Phytoparasitica* **2002**, *30*, 231. [CrossRef]
27. Cohen, Y.; Wang, W.; Ben-Daniel, B.H.; Ben-Daniel, Y. Extracts of *Inula viscosa* control downy mildew of grapes caused by *Plasmopara viticola*. *Phytopathology* **2006**, *96*, 417–424. [CrossRef]
28. Jaber, L.R.; Araj, S.E.; Qasem, J.R. Compatibility of endophytic fungal entomopathogens with plant extracts for the management of sweetpotato whitefly *Bemesia tabaci* Gennadius (Homoptera: Aleyrodidae). *Biol. Control.* **2018**, *117*, 164–171. [CrossRef]
29. Cobo, A.; González-Núñez, M.; Sánchez-Ramos, I.; Pascual, S. Selection of non-target tephritids for risk evaluation in classical biocontrol programmes against the olive fruit fly. *J. Appl. Entomol.* **2015**, *139*, 179–191. [CrossRef]
30. Kožuharova, E.; Lebanova, H.; Getov, I.; Benbassat, N.; Kochmarov, V. *Ailanthus altissima* (Mill.) Swingle-a terrible invasive pest in Bulgaria or potential useful medicinal plant? *Bothalia J.* **2014**, *44*, 213–229.
31. Macel, M.; de Vos, R.C.H.; Jansen, J.J.; van der Putten, W.H.; van Dam, N.M. Novel chemistry of invasive plants: Exotic species have more unique metabolomic profiles than native congeners. *Ecol. Evol.* **2014**, *4*, 2777–2786. [CrossRef] [PubMed]
32. Maoz, M.; Neeman, I. Effect of *Inula viscosa* extract on chitin synthesis in dermatophytes and *Candida albicans*. *J. Ethnopharmacol.* **2000**, *71*, 479–482. [CrossRef]
33. Bresch, C.; Mailleret, L.; Muller, M.M.; Poncet, C.; Parolin, P. Invasive plants in the Mediterranean basin: Which traits do they share? *J. Mediterr. Ecol.* **2013**, *12*, 13–19.
34. Al Hassan, M.; Chaura, J.; López-Gresa Pilar, M.; Borsai, O.; Daniso, E.; Donat, P.; Mayoral, O.; Vicente, O.; Boscaiu, M. Native-Invasive Plants vs. Halophytes in Mediterranean Salt Marshes: Stress Tolerance Mechanisms in Two Related Species. *Front. Plant Sci.* **2016**, *7*, 1438. [CrossRef] [PubMed]
35. Wacquant, P.J. Biogeographical and physiological aspects of the invasion by Dittrichia (ex-Inula) viscosa. In *A Ruderal Species in the Mediterranean Basin*; Greuter, W., Ed.; Kluwer: Dordrecht, The Netherlands, 1990; pp. 353–364.
36. Wacquant, J.P.; Bouab, N. Nutritional differentiation within the species *Dittrichia viscosa* W. Greuter, between a population from a calcareous habitat and another from an acidic habitat. In *Genetic Aspects of Plant Nutrition, Developments in Plant and Soil Sciences*; Sarić, M.R., Loughman, B.C., Eds.; Springer: Dordrecht, The Netherlands, 1983; pp. 285–291.
37. Parolin, P.; Scotta, I.M.; Bresch, C. Notes on the phenology of *Dittrichia viscosa*. *J. Mediterr. Ecol.* **2013**, *12*, 27–35.
38. Aronson, J.A. *HALOPH: A Data Base of Salt Tolerant Plants of the World*; Office of Arid Land Studies, University of Arizona: Tucson, AZ, USA, 1989.
39. Grašič, M.; Anžlovar, S.; Strgulc Krajšek, S. Germination rate of stinkwort (*Dittrichia graveolens*) and false yel-lowhead (*D. viscosa*) in relation to salinity. *Acta. Biol. Slov.* **2016**, *59*, 5–11.
40. De Laurentis, N.; Losacco, V.; Milillo, M.A.; Lai, O. Chemical investigations of volatile constituents of *Inula viscosa* (L.) Aiton (Asteraceae) from different areas of Apulia, Southern Italy. *Delpinoa* **2002**, *44*, 115–119.
41. Dor, E.; Hershenhorn, J. Allelopathic effects of *Inula viscosa* leaf extracts on weeds. *Allelopathy. J.* **2012**, *30*, 281–290.
42. Levizou, E.; Karageorgou, P.G.K.; Manetas, Y. Inhibitory effects of water soluble leaf leachates from *Dittrichia viscosa* on lettuce root growth, statocyte development and graviperception. *Flora-Morphol. Distrib. Funct. Ecol. Plants* **2002**, *197*, 152–157. [CrossRef]
43. Curadi, M.; Graifenberg, A.; Magnani, G.; Giustiniani, L. Growth and element allocation in tissues of *Inula viscosa* in sodic-saline conditions: A candidate for programs of desertification control. *Arid. Land Res. Manag.* **2005**, *19*, 257–265. [CrossRef]
44. Murciego, A.M.; Sanchez, A.G.; Gonzalez, M.R.; Gil, E.P.; Gordillo, C.T.; Fernandez, J.C.; Triguero, T.B. Antimony distribution and mobility in topsoils and plants (*Cytisus striatus, Cistus ladanifer* and *Dittrichia viscosa*) from polluted Sb-mining areas in Extremadura (Spain). *Environ. Pollut.* **2007**, *145*, 15–21. [CrossRef]
45. Perdikis, D.; Favas, C.; Lykouressis, D.; Fantinou, A. Ecological relationships between non-cultivated plants and insect predators in agroecosystems: The case of *Dittrichia viscosa* (Asteraceae) and *Macrolophus melanotoma* (Hemiptera: Miridae). *Acta Oecol.* **2007**, *31*, 299–306. [CrossRef]
46. Pest Rating Proposals and Final Rating for *Dittrichia viscosa*. Available online: https://blogs.cdfa.ca.gov/Section3162/?tag=dittrichia-viscosa (accessed on 28 May 2020).
47. Weeds Australia. Centre for Invasive Species Solutions, Canberra. Available online: https://profiles.ala.org.au/opus/weeds-australia/profile/Dittrichia%20viscosa (accessed on 18 June 2021).
48. Zurayk, R.; Khoury, N.; Talhouk, S.; Baalbaki, R. Salinity-Heavy Metal Interactions in Four Salt-tolerant Plant Species. *J. Plant Nutr.* **2001**, *24*, 1773–1786. [CrossRef]
49. Cafarchia, C.; De Laurentis, N.; Milillo, M.A.; Losacco, V.; Puccini, V. Fungistatic activity of a sesquiterpene lactone (tomentosin) isolated from fresh *Inula viscosa* (Asteraceae) flowers from the Puglia region. *Parasitologia* **2001**, *43*, 117–121.
50. Passalacqua, N.G.; Guarrera, P.M.; De Fine, G. Contribution to the knowledge of the folk plant medicine in Calabria region (Southern Italy). *Fitoterapia* **2007**, *78*, 52–68. [CrossRef]
51. Musthaba, S.M.; Athar, M.T.; Kamal, Y.T.; Baboota, S.; Javed, A.; Sayeed, A. Fast analysis and validation of rutin in anti-psoriatic ayurvedic formulation by HPLC. *J. Liq. Chromatogr. Relat. Technol.* **2011**, *34*, 446–455. [CrossRef]
52. Seca, A.; Grigore, A.; Pinto, D.; Silva, A. The genus Inula and their metabolites: From ethnopharmacological to medicinal uses. *J. Ethnopharmacol.* **2014**, *154*, 286–310. [CrossRef]

53. Merghoub, N.; El Btaouri, H.; Benbacer, L.; Gmouh, S.; Trentesaux, C.; Brassart, B.; Terryn, C.; Attaleb, M.; Madoulet, C.; Benjouad, A.; et al. *Inula Viscosa* Extracts Induces Telomere Shortening and Apoptosis in Cancer Cells and Overcome Drug Resistance. *Nutr. Cancer* **2016**, *68*, 131–143. [CrossRef] [PubMed]
54. Lounis, H.; Bergheim, I.; Bouhaimi, A.; Guigonis, J.M.; Belhamel, K. Anti-inflammatory and antioxidant activities of *Inula viscosa* and *Senecio anteuphorbium*. *Orient. Pharm. Exp. Med.* **2018**, *18*, 225–236. [CrossRef]
55. Askin, C.T.; Aslantürk, T.S. Evaluation of Cytotoxicity and Genotoxicity of *Inula viscosa* Leaf Extracts with Allium Test. *J. Biomed. Biotechnol.* **2010**, 8. [CrossRef]
56. Omezzine, F.; Rinez, A.; Ladhari, A.; Farooq, M.; Haouala, R. Allelopathic potential of *Inula viscosa* against crops and weeds. *Int. J. Agric. Biol.* **2011**, *13*, 841–849.
57. Oka, Y.; Ben-Daniel, B.H.; Cohen, Y. Nematicidal activity of powder and extracts of *Inula viscosa*. *J. Nematol.* **2001**, *3*, 735–742.
58. Mansour, F.; Azaizeh, H.; Saad, B.; Tadmor, Y.; Abo-Moch, F.; Said, O. The potential of middle eastern flora as a source of new safe bio-acaricides to control *Tetranychus cinnabarinus*, the carmine spider mite. *Phytoparasitica* **2004**, *32*, 66–72. [CrossRef]
59. Oka, Y.; Ben-Daniel, B.H.; Cohen, Y. Control of *Meloidogyne javanica* by formulations of *Inula viscosa* leaf extracts. *J. Nematol.* **2006**, *38*, 46. [CrossRef]
60. Alexenizer, M.; Dorn, A. Screening of medicinal and ornamental plants for insecticidal and growth regulating activity. *J. Pest Sci.* **2007**, *80*, 205–215. [CrossRef]
61. Sofou, K.; Isaakidis, D.; Spyros, A.; Büttner, A.; Giannis, A.; Katerinopoulos, H. Use of costic acid, a natural extract from *Dittrichia viscosa*, for the control of *Varroa destructor*, a parasite of the European honey bee. *Beilstein J. Org. Chem.* **2017**, *13*, 952–959. [CrossRef]
62. Cohen, Y.; Baider, A.; Ben-Daniel, B.; Ben-Daniel, Y. Fungicidal preparations from *Inula viscosa*. *Plant Prot. Sci.* **2002**, *38*, 629–630. [CrossRef]
63. Wang, W.; Ben-Daniel, B.; Cohen, Y. Control of Plant Diseases by Extracts of *Inula viscosa*. *Phytopathology* **2004**, *94*, 1042–1047. [CrossRef] [PubMed]
64. Talib, W.; Abu Zarga, M.; Mahasneh, A. Antiproliferative, Antimicrobial and Apoptosis Inducing Effects of Compounds Isolated from *Inula viscosa*. *Molecules* **2012**, *17*, 3291–3303. [CrossRef]
65. Al-Masri, M.I.; Sharawi, S.M.; Barakat, R.M. Effect of Clammy Inula (*Inula viscose*) Plant Extract in Combination with a Low Dose of the Fungicide Iprodione on *Botrytis cinerea* In Vitro and In Vivo. *Am. J. Plant Sci.* **2015**, *6*, 1519–1526. [CrossRef]
66. Dominguez, M.; Madejón, P.; Madejón, E.; Díaz Blanco, M. Novel energy crops for Mediterranean contaminated lands: Valorization of *Dittrichia viscosa* and *Silybum marianum* biomass by pyrolysis. *Chemosphere* **2017**, *186*, 976–986. [CrossRef] [PubMed]
67. Ater, M.; Lefèbvre, C.; Gruber, W.; Meerts, P. A phytogeochemical survey of the flora of ultramafic and adjacent normal soils in North Morocco. *Plant Soil* **2000**, *218*, 127–135. [CrossRef]
68. Swaileh, K.; Hussein, R.; Abu-Elhaj, S. Assessment of Heavy Metal Contamination in Roadside Surface Soil and Vegetation from the West Bank. *Arch. Environ. Contam. Toxicol.* **2004**, *23*, 22–30. [CrossRef]
69. Gisbert, C.; Almela, C.; Vélez, D.; López-Moya, J.R.; Serrano, R.; Montoro, R.; Navarro-Aviñó, J. Identification of As accumulation plant species growing on highly contaminated soils. *Int. J. Phytoremediation* **2008**, *10*, 183–184. [CrossRef]
70. Conesa, H.; María-Cervantes, A.; Alvarez-Rogel, J.; González, M. Influence of soil properties on trace element availability and plant accumulation in a Mediterranean salt marsh polluted by mining wastes: Implications for phytomanagement. *Sci. Total Environ.* **2011**, *409*, 4470–4479. [CrossRef]
71. Perez, C.; Martínez-Sánchez, M.; Martínez-López, S.; Bech, J.; Bolan, N. Distribution and bioaccumulation of arsenic and antimony in *Dittrichia viscosa* growing in mining-affected semiarid soils in southeast Spain. *J. Geochem. Explor.* **2012**, *123*, 128–135. [CrossRef]
72. Pistelli, L.; D'Angiolillo, F.; Morelli, E.; Basso, B.; Rosellini, I.; Posarelli, M.; Barbafieri, M. Response of spontaneous plants from an ex-mining site of Elba island (Tuscany, Italy) to metal (loid) contamination. *Environ. Sci. Pollut. Res. Int.* **2017**, *24*, 7809–7820. [CrossRef] [PubMed]
73. Barbafieri, M.; Dadea, C.; Tassi, E.; Bretzel, F.; Fanfani, L. Uptake of heavy metals by native species growing in a mining area in Sardenia, Italy: Discovering native flora for phytoremediation. *Int. J. Phytoremediation* **2011**, *13*, 985–997. [CrossRef]
74. Jimenez, M.N.; Bacchetta, G.; Casti, M.; Navarro, F.B.; Lallena, A.M.; Fernández-Ondoñoa, E. Potential use in phytoremediation of three plant species growing on contaminated mine-tailing soils in Sardinia. *Ecol. Eng.* **2011**, *37*, 392–398. [CrossRef]
75. Perez-Fernandez, M.A.; Calvo-Magro, E.; Ferrrer-Castan, D. Simulation of germination of pioneer species along an experimental drought gradient. *J. Environ. Biol.* **2006**, *27*, 679–685. [PubMed]
76. Gonzalez-Tejero, M.R.; Casares-Porcel, M.; Sánchez-Rojas, C.P.; Ramiro-Gutiérrez, J.M.; Mesa, J.; Pieroni, A.; Giusti, M.E.; Censorii, E.; de Pasquale, C.; Della, A.; et al. Medicinal plants in the Mediterranean area: Synthesis of the results of the project Rubia. *J. Ethnopharmacol.* **2008**, *116*, 341–357. [CrossRef] [PubMed]
77. Akkawi, M.; Abbasi, I.; Jaber, S.; Abouremeleh, Q.; Nasereddin, A.; Lutgen, P. Investigation of Traditional Palestinian Medicinal Plant *Inula viscosa* as Potential Anti-malarial Agent. *Br. J. Pharmacol.* **2014**, *5*, 156–162. [CrossRef]
78. Kawada, K.; Vovk, A.G.; Filatova, O.V.; Araki, M.; Nakamura, T.; Hayashi, I. Floristic composition and plant biomass production of steppe communities in the vicinity of Kharkiv, Ukraine. *Grassl. Sci.* **2005**, *51*, 205–213. [CrossRef]
79. Zhang, T.; Guo, R.; Gao, S.; Guo, J.; Sun, W. Responses of Plant Community Composition and Biomass Production to Warming and Nitrogen Deposition in a Temperate Meadow Ecosystem. *PLoS ONE* **2015**, *10*. [CrossRef] [PubMed]

80. Karageorgou, P.; Levizou, E.F.I.; Manetas, Y. The influence of drought, shade and availability of mineral nutrients on exudate phenolics of *Dittrichia viscosa*. *Flora* **2002**, *197*, 285–289. [CrossRef]
81. Terradas, J. Mediterranean woody plant growth-forms, biomass and production in the eastern part of the Iberian Peninsula. *Oecologia* **1990**, *10*, 337–349.
82. Brahmi-Chendouh, N.; Piccolella, S.; Crescente, G.; Pacifico, F.; Boulekbache-Makhlouf, L.; Hamri-Zeghichi, S.; Akkal, S.; Madani, K.; Pacifico, S. A nutraceutical extract from *Inula viscosa* leaves: UHPLC-HR-MS/MS based polyphenol profile, and antioxidant and cytotoxic activities. *J. Food Drug Anal.* **2019**, *27*, 692–702. [CrossRef] [PubMed]
83. Trimech, I.; Weiss, E.K.; Chedea, V.; Marin, D.; Detsi, A.; Ioannou, E.; Roussis, V.; Kefalas, P. Evaluation of Anti-oxidant and Acetylcholinesterase Activity and Identification of Polyphenolics of the Invasive Weed *Dittrichia viscosa*. *Phytochem. Anal.* **2014**, *25*, 421–428. [CrossRef] [PubMed]
84. Lattanzio, V.; Lattanzino, V.M.T.; Cardinali, A. Role of phenolics in the resistance mechanisms of plants against fungal pathogens and insects. In *Phytochemistry. Advances in Research*; Imperato, F., Ed.; Research Signpost: Trivandrum, India, 2006; pp. 23–67.
85. Mikulič-Petkovšek, M.; Štampar, F.; Veberič, R. Accumulation of phenolic compounds in apple in response to infection by the scab pathogen, *Venturia inaequalis*. *Physiol. Mol. Plant Pathol.* **2009**, *74*, 60–67. [CrossRef]
86. Orcaray, L.; Igal, M.; Zabalza, A.; Royuela, M. Role of Exogenously Supplied Ferulic and p-Coumaric Acids in Mimicking the Mode of Action of Acetolactate Synthase Inhibiting Herbicides. *J. Agric. Food Chem.* **2011**, *59*, 10162–10168. [CrossRef]
87. Dayan, F.E.; Duke, S.O. Natural compounds as next-generation herbicides. *J. Plant Physiol.* **2014**, *166*, 1090–1105. [CrossRef]
88. Pyšek, P.; Richardson, D.M. Invasive plants. In *Encyclopedia of Ecology*; Jørgensen, S.E., Ed.; Elsevier: Amsterdam, The Netherlands, 2008; pp. 2011–2020.
89. Pyšek, P.; Richardson, M.D.; Rejmanek, M.; Webster, G.L.; Williamson, M.; Kirschner, J. Alien plants in checklists and floras: Towards better communication between taxonomists and ecologists. *Taxon* **2004**, *53*, 131–143. [CrossRef]
90. James, L.; Evans, J.; Ralphs, M.; Child, R. *Noxious Range Weeds*; Westview Press: San Francisco, CA, USA, 1991; pp. 420–428.
91. Pyšek, P.; Richardson, D.M. Invasive Species, Environmental Change and Management, and Health. *Annu. Rev. Environ. Resour.* **2010**, *35*, 25–55. [CrossRef]
92. Niemelä, J. From systematics to conservation-carabidologists do it all. *Ann. Zool. Fenn.* **1996**, *33*, 1–4.
93. Jefferies, R. Allochthonous inputs: Integrating population changes and food-web dynamics. *Ecol. Evol.* **2000**, *15*, 19–22. [CrossRef]
94. Meiners, S.J. Native and exotic plant species exhibit similar population dynamics during succession. *Ecology* **2007**, *88*, 1098–1104. [CrossRef] [PubMed]
95. Burgiel, S.W.; Muir, A.A. *Invasive Species, Climate Change and Ecosystem-Based Adaptation: Addressing Multiple Drivers of Global Change*; Global Invasive Species Programme (GISP): Washington, DC, USA; Nairobi, Kenya, 2010.
96. Wijesundara, D. Can native plants become invasive? *Ceylon J. Sci.* **2017**, *46*, 1–2. [CrossRef]
97. Plieninger, T.; Hui, C.; Gaertner, M.; Huntsinger, L.L. The Impact of Land Abandonment on Species Richness and Abundance in the Mediterranean Basin: A Meta-Analysis. *PLoS ONE* **2014**, *9*. [CrossRef] [PubMed]
98. Mikulič, K.; Radović, A.; Kati, V.; Jelaska, S. Effects of land abandonment on bird communities of smallholder farming landscapes in post-war Croatia: Implications for conservation policies. *Community Ecol.* **2014**, *15*, 169–179. [CrossRef]
99. Benayas, J.M.R.; Artins, A.; Nicolau, J.M.; Schulz, J.J. Abandonment of agricultural land: An overview of drivers and consequences. *CAB. Rev. Perspect. Agric. Vet. Sci. Nutr. Nat. Resour.* **2007**, *2*, 14. [CrossRef]
100. Jackson, W.J.; Argent, R.M.; Bax, N.J.; Bui, E.; Clark, G.F.; Coleman, S.; Cresswell, I.D.; Emmerson, K.M.; Evans, K.; Hibberd, M.F.; et al. Overview: Invasive species are a potent, persistent and widespread threat to Australia's environment. In *Australia State of the Environment 2016*; Australian Government Department of the Environment and Energy: Canberra, Australia, 2016.
101. Yasumura, Y.; Crumpton-Taylor, M.; Fuentes, S.P.; Harberd, N. Step-by-Step Acquisition of the Gibberellin-DELLA Growth-Regulatory Mechanism during Land-Plant Evolution. *Curr. Biol.* **2007**, *17*, 1225–1230. [CrossRef] [PubMed]
102. Ratnayake, R.M.C.S. Why plant species become invasive? Characters Related to Successful Biological Invasion. In Proceedings of the Conference National Symposium on Invasive alien Species, Colombo, Sri Lanka, 11 November 2014.
103. Vilà, M.; Hulme, P. *Impact of Biological Invasions on Ecosystem Services*; Springer International Publishing: Cham, Switzerland, 2016.
104. Grice, A.C.; Friedel, M.H.; Marshall, N.A.; Van Klinken, R.D. Tackling Contentious Invasive Plant Species: A Case Study of Buffel Grass in Australia. *Environ. Manag.* **2012**, *49*, 285–294. [CrossRef]
105. Boari, A.; Vurro, M.; Calabrese, G.; Nesma, M.; Mahmoud, Z.; Cazzato, E.; Fracchiolla, M. Evaluation of *Dittrichia viscosa* (L.) Greuter Dried Biomass for Weed Management. *Plants* **2021**, *10*, 147. [CrossRef] [PubMed]

diversity

MDPI

Brief Report

Hitchhiking Exotic Clam: *Dreissena polymorpha* (Pallas, 1771) Transported via the Ornamental Plant Trade

Jiří Patoka [1,*] and Barbora Patoková [2]

1 Department of Zoology and Fisheries, Faculty of Agrobiology, Food and Natural Resources, Czech University of Life Sciences Prague, 165 00 Prague, Czech Republic
2 St. Ursula School, 110 00 Prague, Czech Republic; bsemrova@gmail.com
* Correspondence: patoka@af.czu.cz

Abstract: Ornamental aquaculture is one of the main sources of non-native species worldwide. Unintentionally transported "hitchhiking" organisms have been previously recorded; although most of these species are transported from tropical regions, here we report on the first accidental transport of the zebra mussel (*Dreissena polymorpha*) in a shipment of ornamental *Aegagropila linnaei* (Chlorophyta) from Russia to the Czech Republic. This invasive mussel is listed on the national blacklist of alien species in the Czech Republic and can be easily released in outdoor garden ponds together with *A. linnaei*. Since the Czech Republic is known to be a gateway for aquatic ornamental species from a European perspective, re-export to other European countries is also possible. Thus, the spread of *D. polymorpha* via this pathway cannot be excluded. This finding should be of importance to conservationists, traders, decision-makers and other stakeholders.

Keywords: zebra mussel; *Aegagropila linnaei*; Europe; aquarium; invasive species

Citation: Patoka, J.; Patoková, B. Hitchhiking Exotic Clam: *Dreissena polymorpha* (Pallas, 1771) Transported via the Ornamental Plant Trade *Diversity* **2021**, *13*, 410. https://doi.org/10.3390/d13090410

Academic Editor: José Luis García-Marín

Received: 20 July 2021
Accepted: 25 August 2021
Published: 27 August 2021

Publisher's Note: MDPI stays neutral with regard to jurisdictional claims in published maps and institutional affiliations.

Copyright: © 2021 by the authors. Licensee MDPI, Basel, Switzerland. This article is an open access article distributed under the terms and conditions of the Creative Commons Attribution (CC BY) license (https://creativecommons.org/licenses/by/4.0/).

In the era of globalization, biological invasions cause huge environmental and socio-economic losses worldwide, costing billions of dollars each year [1]. Ornamental aquaculture was identified as one of the main sources of non-native organisms [2–5] with thousands of individuals of thousands of animal and plant species transported per day both intra- and intercontinentally via the pet trade [6–10]. The occurrence of non-native biota is generally perceived as unwanted and harmful, even if there also exists certain examples of introduced endangered species such as *Arapaima gigas*, the fish of ornamental origin which is native and endangered in South America and recorded in the wild in Indonesia as a potential invader [11]. Therefore, decision-makers try to regulate this pathway via legislative restrictions, but the effectiveness is controversial and contentious in certain cases due to poor education of the general public in this regard [12].

When intentionally released or accidentally escaped, ornamental species can establish, multiply, and subsequently spread to the vicinity and behave as invasive, such as the marbled crayfish *Procambarus virginalis* in many European countries [13] and the apple snail *Pomacea* sp. in South Europe [14]. Certain ornamental species have been recorded in small quantities in the wild, such as the redclaw *Cherax quadricarinatus* [15] and the Mexican dwarf crayfish *Cambarellus patzcuarensis* [16] in Hungary, but their further spread cannot be excluded, at least in the case of the latter. Moreover, non-native pathogens, such as oomycete *Aphanomyces astaci* causing crayfish plague, are transmitted by their ornamental hosts [17,18] and can persist in the new environment by infecting native species even if their hosts do not survive.

Not only ornamental species are highlighted in this regard, but also associated biota have been recorded as being transported via this pathway [19,20]. These faunal assemblages are also known as "hitchhikers" and their ornamental hosts can be both animals and plants [21,22]. The predicted and discussed invasion potential of so-called hitchhikers is usually higher in comparison with ornamental species [23]. Some species are transported without close association with any of the ornamental species and can live without hosts freely when introduced to a new locality, and thus their invasion potential is high [22,24]. Improving knowledge about the spread of non-native species including hitchhikers via the international pet industry is crucial to establish effective management strategies to reduce introduction rates. The Czech Republic is known as one of the leading importers and exporters of aquatic ornamental species [2,8]. Most of these organisms are imported from tropical regions in South and South-east Asia, Africa and South America, while imports from other countries in temperate zones are mostly overlooked [2,7].

Therefore, we decided to inspect the shipment of dark green ornamental Marimo (also known as *Cladophora* balls or moss balls), which is a rare form of *Aegagropila linnaei* (Chlorophyta), to check on the possible presence of hitchhiking organisms. The shipment included 100 pcs of Marimo balls and was delivered by van from Moscow river in Russia, with a stop in Ukraine, to the Czech Republic in January 2021. Within the personal inspection we found one living and vital individual of freshwater zebra mussel *Dreissena polymorpha* [25]. This voucher individual had a shell length of 18 mm (Figure 1), which indicated it to be in the adult stage [26]. The voucher specimen was preserved in pure alcohol and deposited at the Department of Zoology and Fisheries, Czech University of Life Sciences Prague, the Patoka's Lab Collection, No. JP2021/01-001. The record is in line with anecdotal notes on *Dreissena polymorpha* associated with Marimo balls by hobbyists and aquarium owners [27]: https://eu.indystar.com/story/news/environment/2021/03/12/here-5-things-know-invasive-zebra-mussels-mossballs/4652002001/ (accessed on 19 July 2021).

Figure 1. The voucher specimen of zebra mussel *Dreissena polymorpha* transported via the ornamental plant trade from Russia to the Czech Republic; scale bar is equal to 10 mm.

Dreissena polymorpha (Bivalvia: Dreissenidae) is native to the Ponto-Caspian region [28], and thus Moscow river is out of its native range. *Dreissena polymorpha* appeared here from the middle of the 19th century [29,30]. The main method of dispersal of *D. polymorpha* is its 'natural' spread through channels and other artificial constructions, while drifting macrophytes and human transport can also be viewed as vectors of this mollusc dispersal [31]. For example, it was recently demonstrated that the overland transport of boats can facilitate dispersal of this species [32,33]. This clam is one of the most successful biofouling and prolific aquatic invasive species, jeopardizing native biota and entire ecosystems [34]. It is currently spread in 33 European countries (including the Czech Republic), 33 U.S. states and territories, and two provinces of Canada [35]. With the use of byssal threads, this mussel attaches to hard submerged surfaces and substrates, forming large colonies which may cause increased water turbidity, displacement of native mussel species, and alteration of nutrient cycling and trophic relationships [35–37]. Moreover, mutualistic interactions between *D. polymorpha* and certain invasive aquatic macrophytes from the genera *Elodea*, *Myriophyllum*, and *Potamogeton* have been recorded to enhance the biomass of the mentioned species [37]. Certain species from the first two mentioned genera are included on the list of invasive species of European Union concern (Regulation No. 1143/2014 on the prevention and management of the introduction and spread of invasive alien species).

Dreissena polymorpha was first recorded in the current territory of the Czech Republic in the early 1900s [38]. Subsequently, based on classification as invasive species with massive environmental impacts, it was added in the national blacklist of alien species [39]. Marimo balls are recommended as suitable species for ornamental outdoor pond planting/stocking (see: https://rybicky.net/atlasrostlin/cladophora_aegagropila [in Czech] (accessed on 19 July 2021)), a hobby which has increasing popularity in the Czech Republic [40]. Therefore, it is obvious that overlooked hitchhiking *D. polymorpha* can be released together with the Marimo balls and penetrate new waterbodies, for instance, when the ornamental pond is flooded.

Since the Czech Republic has been identified as a gateway for aquatic ornamental species from a European perspective, many imported animals and plants are subsequently re-exported abroad [2]. Thus, hitchhikers can be easily transported via this pathway to other European countries together with ornamental species.

From the aforementioned, it follows that *D. polymorpha* is a species of global concern and monitoring its introduction pathways is very important for improving the effectiveness of focused management and regulations. The first record of its introduction via international trade with ornamental aquatic plants should be of the attention of conservationists, traders, decision-makers and other stakeholders. Moreover, our finding highlighted the importance of monitoring the route of certain ornamental species transportation from non-tropical regions to the global market.

Author Contributions: Conceptualization, J.P.; methodology, J.P.; validation, J.P.; formal analysis, J.P.; investigation, J.P. and B.P.; resources, B.P.; data curation, J.P.; writing—original draft preparation, J.P.; writing—review and editing, B.P.; visualization, J.P.; supervision, J.P.; project administration, J.P.; funding acquisition, B.P. Both authors have read and agreed to the published version of the manuscript.

Funding: J.P. was funded by the Technology Agency of the Czech Republic within the project "DivLand", grant number SS02030018".

Institutional Review Board Statement: Not applicable.

Data Availability Statement: Not applicable.

Acknowledgments: We thank anonymous reviewers for their effort and time when evaluating our manuscript.

Conflicts of Interest: The authors declare no conflict of interest.

References

1. Vilà, M.; Hulme, P.E. *Impact of Biological Invasions on Ecosystem Services*; Springer International Publishing: Cham, Switzerland, 2017; 354p, ISBN 978-3-319-45121-3.
2. Kalous, L.; Patoka, J.; Kopecký, O. European hub for invaders: Risk assessment of freshwater aquarium fishes exported from the Czech Republic. *Acta Ichthyol. Piscat.* **2015**, *15*, 239–245. [CrossRef]
3. Nunes, A.L.; Tricarico, E.; Panov, V.E.; Cardoso, A.C.; Katsanevakis, S. Pathways and gateways of freshwater invasions in Europe. *Aquat. Invasions* **2015**, *10*, 359–370. [CrossRef]
4. Patoka, J.; Takdir, M.; Yonvitner; Aryadi, H.; Jerikho, R.; Nilawati, J.; Tantu, F.Y.; Bohatá, L.; Aulia, A.; Kamal, M.M.; et al. Two species of illegal South American sailfin catfish of the genus *Pterygoplichthys* well-established in Indonesia. *Knowl. Manag. Aquat. Ecosyst.* **2020**, *421*, 28. [CrossRef]
5. Haubrock, P.; Oficialdegui, F.J.; Yiwen, Z.; Patoka, J.; Yeo, D.C.J.; Kouba, A. The redclaw crayfish: A prominent aquaculture species with invasive potential in tropical and subtropical biodiversity hotspots. *Rev. Aquac.* **2021**, *13*, 1488–1530. [CrossRef]
6. Padilla, D.K.; Williams, S.L. Beyond ballast water: Aquarium and ornamental trades as sources of invasive species in aquatic ecosystems. *Front. Ecol. Environ.* **2004**, *2*, 131–138. [CrossRef]
7. Patoka, J.; Kalous, L.; Kopecký, O. Imports of ornamental crayfish: The first decade from the Czech Republic's perspective. *Knowl. Manag. Aquat. Ecosyst.* **2015**, *416*, 4. [CrossRef]
8. Evers, H.G.; Pinnegar, J.K.; Taylor, M.I. Where are they all from?—Sources and sustainability in the ornamental freshwater fish trade. *J. Fish Biol.* **2019**, *9*, 909–916. [CrossRef]

9. Akmal, S.G.; Zámečníková-Wanma, B.P.; Prabowo, R.E.; Khatami, A.M.; Novák, J.; Petrtýl, M.; Kalous, L.; Patoka, J. Marine ornamental trade in Indonesia. *Aquat. Living Resour.* **2020**, *33*, 25. [CrossRef]
10. Novák, J.; Kalous, L.; Patoka, J. Modern ornamental aquaculture in Europe: Early history of freshwater fish imports. *Rev. Aquac.* **2020**, *12*, 2042–2060. [CrossRef]
11. Marková, J.; Jerikho, R.; Wardiatno, Y.; Kamal, M.M.; Magalhães, A.L.B.; Bohatá, L.; Kalous, L.; Patoka, J. Conservation paradox of giant arapaima *Arapaima gigas* (Schinz, 1822) (Pisces: Arapaimidae): Endangered in its native range in Brazil and invasive in Indonesia. *Knowl. Manag. Aquat. Ecosyst.* **2020**, *421*, 47. [CrossRef]
12. Patoka, J.; Magalhães, A.L.B.; Kouba, A.; Faulkes, Z.; Jerikho, R.; Vitule, J.R.S. Invasive aquatic pets: Failed policies increase risks of harmful invasions. *Biodivers. Conserv.* **2018**, *27*, 3037–3046. [CrossRef]
13. Hossain, M.S.; Patoka, J.; Kouba, A.; Buřič, M. Clonal crayfish as biological model: A review on marbled crayfish. *Biologia* **2018**, *73*, 841–855. [CrossRef]
14. Lei, J.; Chen, L.; Li, H. Using ensemble forecasting to examine how climate change promotes worldwide invasion of the golden apple snail (*Pomacea canaliculata*). *Environ. Monit. Assess.* **2017**, *189*, 1–11. [CrossRef]
15. Weiperth, A.; Gál, B.; Kuříková, P.; Langrová, I.; Kouba, A. Risk assessment of pet-traded decapod crustaceans in Hungary with evidence of *Cherax quadricarinatus* (von Martens, 1868) in the wild. *North West J. Zool.* **2019**, *15*, 42–47.
16. Gál, B.; Kuříková, P.; Bláha, M.; Kouba, A.; Patoka, J. *Cambarellus patzcuarensis* in Hungary: The first dwarf crayfish established outside of North America. *Biologia* **2017**, *72*, 1529–1532.
17. Mrugała, A.; Kozubíková-Balcarová, E.; Chucholl, C.; Resino, S.C.; Viljamaa-Dirks, S.; Vukić, J.; Petrusek, A. Trade of ornamental crayfish in Europe as a possible introduction pathway for important crustacean diseases: Crayfish plague and white spot syndrome. *Biol. Invasions* **2015**, *17*, 1313–1326. [CrossRef]
18. Putra, M.D.; Bláha, M.; Wardiatno, Y.; Krisanti, M.; Jerikho, R.; Kamal, M.M.; Mojžišová, M.; Bystřický, P.K.; Kouba, A.; Kalous, L.; et al. *Procambarus clarkii* (Girard, 1852) and crayfish plague as new threats for biodiversity in Indonesia. *Aquat. Conserv.* **2018**, *28*, 1434–1440. [CrossRef]
19. Duggan, I.C.; Pullan, S.G. Do freshwater aquaculture facilities provide an invasion risk for zooplankton hitchhikers? *Biol. Invasions* **2017**, *19*, 307–314. [CrossRef]
20. Duggan, I.C.; Champion, P.D.; MacIsaac, H.J. Invertebrates associated with aquatic plants bought from aquarium stores in Canada and New Zealand. *Biol. Invasions* **2018**, *20*, 3167–3178. [CrossRef]
21. Patoka, J.; Bláha, M.; Devetter, M.; Rylková, K.; Čadková, Z.; Kalous, L. Aquarium hitchhikers: Attached commensals imported with freshwater shrimps via the pet trade. *Biol. Invasions* **2016**, *18*, 457–461. [CrossRef]
22. Patoka, J.; Bláha, M.; Kalous, L.; Vrabec, V.; Buřič, M.; Kouba, A. Potential pest transfer mediated by international ornamental plant trade. *Sci. Rep.* **2016**, *6*, 25896. [CrossRef]
23. Patoka, J.; Kopecký, O.; Vrabec, V.; Kalous, L. Aquarium molluscs as a case study in risk assessment of incidental freshwater fauna. *Biol. Invasions* **2017**, *19*, 2039–2046. [CrossRef]
24. Patoka, J.; Prabowo, R.E.; Petrtýl, M.; Reynolds, J.D.; Kuříková, P.; Zámečníková-Wanma, B.P.; Kalous, L. Marine hitchhikers: A preliminary study on invertebrates unintentionally transported via the international pet trade. *NeoBiota* **2020**, *61*, 33–36. [CrossRef]
25. Pallas, P.S. *Reise Durch Verschiedene Provinzen des Russischen Reichs. Theil 1. Physicalische Reise Durch Verschiedene Provinzen des Russischen Reichs im 1768- und 1769 sten Jahren*; Kaiserliche Akademie der Wissenschaften: St. Petersburg, Russia, 1771.
26. Mackie, G.L. Biology of the exotic zebra mussel, *Dreissena polymorpha*, in relation to native bivalves and its potential impact in Lake St. Clair. *Hydrobiologia* **1991**, *219*, 251–268. [CrossRef]
27. Bowman, S. Have an Aquarium? Here Are the 5 Things to Know about Invasive Zebra Mussels in Moss Balls. 2021. Available online: https://eu.indystar.com/story/news/environment/2021/03/12/here-5-things-know-invasive-zebra-mussels-moss-balls/4652002001/ (accessed on 19 July 2021).
28. Son, M.O. Native range of the zebra mussel and quagga mussel and new data on their invasions within the Ponto-Caspian Region. *Aquat. Invasions* **2007**, *2*, 174–184. [CrossRef]
29. Ludyanskiy, M.L.; McDonald, D.; MacNeill, D. Impact of the zebra mussel, a bivalve invader: *Dreissena polymorpha* is rapidly colonizing hard surfaces throughout waterways of the United States and Canada. *Bioscience* **1993**, *43*, 533–544. [CrossRef]
30. L'vova, A.A. On invasion of *Dreissena bugensis* (Bivalvia, Dreissenidae) in the Ucha Reservoir (Moscow oblast) and the Moscow River. *Zoologicheskiy Zhurnal* **2004**, *83*, 3–20.
31. Horvath, T.G.; Lamberti, G.A. Drifting macrophytes as a mechanism for zebra mussel (*Dreissena polymorpha*) invasion of lake-outlet streams. *Am. Midl. Nat.* **1997**, *138*, 29–36. [CrossRef]
32. Johnson, L.E.; Ricciardi, A.; Cartlon, J.T. Overland dispersal of aquatic invasive species: A risk assessment of transient recreational boating. *Ecol. Appl.* **2001**, *11*, 1789–1799. [CrossRef]
33. De Ventura, L.; Weissert, N.; Tobias, R.; Kopp, K.; Jokela, J. Overland transport of recreational boats as a spreading vector of zebra mussel *Dreissena polymorpha*. *Biol. Invasions* **2016**, *18*, 1451–1466. [CrossRef]
34. Ward, J.M.; Ricciardi, A. Impacts of *Dreissena* invasions on benthic macroinvertebrate communities: A meta-analysis. *Divers. Distrib.* **2007**, *13*, 155–165. [CrossRef]
35. Dölle, K.; Kurzmann, D.E. The freshwater mollusk *Dreissena polymorpha* (zebra mussel)—A review: Living, prospects and jeopardies. *Asian J. Environ. Ecol.* **2020**, *13*, 1–17. [CrossRef]

36. Ricciardi, A. Predicting the impacts of an introduced species from its invasion history: An empirical approach applied to zebra mussel invasions. *Freshw. Biol.* **2003**, *48*, 972–981. [CrossRef]
37. Crane, K.; Coughlan, N.E.; Cuthbert, R.N.; Dick, J.T.; Kregting, L.; Ricciardi, A.; MacIsaac, H.J.; Reid, N. Friends of mine: An invasive freshwater mussel facilitates growth of invasive macrophytes and mediates their competitive interactions. *Freshw. Biol.* **2020**, *65*, 1063–1072. [CrossRef]
38. Šefrová, H.; Laštůvka, Z. Catalogue of alien animal species in the Czech Republic. *Acta Univ. Agric. Silvic. Mendel. Brun.* **2014**, *53*, 151–170. [CrossRef]
39. Pergl, J.; Sádlo, J.; Petrusek, A.; Laštůvka, Z.; Musil, J.; Perglová, I.; Šanda, R.; Šefrová, H.; Šíma, J.; Vohralík, V.; et al. Black, Grey and Watch Lists of alien species in the Czech Republic based on environmental impacts and management strategy. *NeoBiota* **2016**, *28*, 1–37. [CrossRef]
40. Patoka, J.; Bláha, M.; Kalous, L.; Kouba, A. Irresponsible vendors: Non-native, invasive and threatened animals offered for garden pond stocking. *Aquat. Conserv.* **2017**, *27*, 692–697. [CrossRef]

diversity

MDPI

Article

Self-Compatibility and Reproductive Success of *Oenothera drummondii* subsp. *drummondii*: Is It Similar between Native and Non-Native Populations?

Juan B. Gallego-Fernández [1] and José G. García-Franco [1,2,*]

[1] Departamento de Biología Vegetal y Ecología, Universidad de Sevilla, 41012 Sevilla, Spain; galfer@us.es
[2] Red Ecología Funcional, Instituto de Ecología, Xalapa, Veracruz 91073, Mexico
* Correspondence: jose.garcia.franco@inecol.mx

Abstract: The mating system of plants widely distributed can change in native range but also in non-native habitats. *Oenothera drummondii,* native to the coastal dunes of the Gulf of Mexico, has been introduced to Europe, Africa, Asia and Oceania. Hand self- and cross-pollination were performed to determine compatibility and to compare fruit set, fruit weight, seed set and germination characteristics from natives and non-natives populations and a comprehensive integral reproductive success index (IRSI) was built. *Oenothera drummondii* exhibited high self-compatibility and mixed reproductive systems in all populations. Characteristics of fruits and seeds from self- and cross-pollination varied within and between native and non-native populations and some had a positive clinal variation in the native range. The IRSI was sensitive to changes of fruit set, seed set and final germination of both self- and cross-pollination, showing differences between native populations. Differences in characteristics of fruits and seeds in the native and non-native ranges suggest the occurrence of distinct selection factors. The mixed reproductive system of *O. drummondii* suggests it can take advantage of local visitors in the native range, but also can provide advantages for the establishment at non-native sites giving the opportunity to interact with local flower visitors.

Keywords: costal dunes; cross-pollination; reproductive success index; seed germination; incompatibility index; self-pollination

Citation: Gallego-Fernández, J.B.; García-Franco, J.G. Self-Compatibility and Reproductive Success of *Oenothera drummondii* subsp. *drummondii*: Is It Similar between Native and Non-Native Populations? *Diversity* **2021**, *13*, 431. https://doi.org/10.3390/d13090431

Academic Editor: Michael Wink

Received: 1 August 2021
Accepted: 3 September 2021
Published: 7 September 2021

Publisher's Note: MDPI stays neutral with regard to jurisdictional claims in published maps and institutional affiliations.

Copyright: © 2021 by the authors. Licensee MDPI, Basel, Switzerland. This article is an open access article distributed under the terms and conditions of the Creative Commons Attribution (CC BY) license (https://creativecommons.org/licenses/by/4.0/).

1. Introduction

Plant reproduction is central to the studies of population ecology, since it allows an understanding of the evolutionary processes of species throughout their distribution range [1,2]. Flowering plants possess exceptionally diverse reproductive strategies [3] and mating system type determines reproductive success [4]. Sexual reproduction in hermaphroditic plants generally includes self-pollination and cross-pollination, or a combination of both strategies [5,6]. Selfing resolves competition among individual plants for pollination services when there is a limitation of pollinators or mates [7], while outcrossing reduces inbreeding and increases genetic diversity. However, outcrossing is less efficient because it requires the presence of pollen vectors and neighbouring mates [4].

Studies in natural populations have indicated that mixed mating systems are frequent [5,8]. With this mating strategy, reproduction occurs by both self-fertilization (selfing) and mating with other individuals (outcrossing). One possible explanation is that autonomous pollination provides reproductive assurance for outcrossing species when pollinators are limited [5]. However, the rate of selfing can vary widely among populations throughout their natural range as a result of both biotic and environmental factors [2]. Some plant species show selfing ability, such as those with short lifetimes or that inhabit sites with frequent disturbances [9] and selfing species are more likely to have wider geographical ranges than those without this ability [10,11].

Human activities have not only caused major disturbances of natural habitats but have also led to species dispersal beyond their native ranges [12,13]. Several studies have shown that success in the establishment and spread of these non-native species depends on particular physiological, ecological and reproductive characteristics [14,15]. However, reproductive ability is important in the naturalization process because seed production and seed germination are necessaries to maintain populations [16]. Baker [17] stated that selfing plant species that arrive at new habitats after long-distance dispersal should have an adaptive advantage, since one or more individuals will have the opportunity to start a sexually reproductive population through the selfing reproductive mechanism. Known as Baker's law, this hypothesis has been widely discussed because of contradictory findings; however, recent studies indicate that self-fertilization is a common mating mechanism, mainly in species that colonize, both naturally following long-distance dispersal and with human assistance, as in the case of invasive species [11,18].

Phenotypic traits of native populations can evolve rapidly under the environmental conditions presented by novel habitats, leading to local adaptation of the individuals [19]. Differences between native and non-native populations can be observed by growing plants of populations of both origins in controlled uniform environments such as common gardens [15]. To date, few studies have compared the reproductive success of self- and cross-pollination from native and non-native populations [20–23] and none have included almost all of the parameters that define reproductive success (v.gr., fruit set, seed set, seed germination) and the range of the species distribution.

Oenothera drummondii subsp. *drummondii* Hook. naturally inhabits the back beaches and the first dune ridges of the coasts of the Gulf of Mexico and southeastern USA; although, during the last century it was introduced into parts of Europe, Asia, Africa, Australia and New Zealand [24–27]. This habitat is frequently disturbed by wind, burial by sand and impacted by waves during tropical storms. The species presents self-compatible herkogamous yellow flowers with anthesis at dusk, living for one night or for a few hours into the following morning in the native range [24,28,29]. The flowers are visited and pollinated by sphingid moths (e.g., *Hyles lineata*, *Manduca* sp.), although some hymenopterans (e.g., *Lasioglossum texaum*, *Apis mellifera*) also visit the flowers [24,29]. The seeds are very small with high germination capacity and some can germinate after floating in seawater for a period of time [30]. The genus *Oenothera* L. has been widely used for studies related to variation in pollination and breeding systems; some taxa are self-compatible and outcrossing, whereas others just self-compatible and autogamous [7,28,31]. Raven [32] considers that both cross- and self-pollination can occur in Onagraceae species that present positive herkogamy.

The current distribution of *O. drummondii*, with native and non-native populations distributed over a wide geographic range, with different ecological histories and environmental conditions, presents an opportunity to explore possible variation in mating systems and reproductive success as a result of both the evolutionary history in the native populations and the possible adaptation of non-native populations of *O. drummondii* to their novel environments. Faced with these scenarios, it is possible that native populations present a predominantly outcrossed mating system adapted to their native pollinators, while non-native populations present a predominantly self-mating system, since in their new habitats, flower visitors and pollinators can be not suitable for the flowers. On the other hand, since the native range of *O. drummondii* corresponds to a latitudinal gradient, we expected that the mating systems and fruit and seed characteristics will differ among these populations, especially among the most geographical distant populations, because change in pollinator presence (more abundant in the tropic climate than subtropical).

In this paper, we addressed the following specific questions: (1) What is the mating system (outcrossing, selfing or mixed) of *Oenothera drummondii*. (2) Does the mating system vary between native populations? (3) Do fruit and seed characteristics and seed germination vary between hand-pollination treatments in the native population range? (4) Does the success of self- or cross-pollination change according to the native or non-native status of populations? (5) Do the fruit and seed characteristics and seed germination of self- and cross-pollination vary among native and non-native populations? (6) Do the reproductive success and the self-compatibility index of non-native populations fall within the range of these parameters of native populations?

2. Materials and Methods

2.1. Taxon Studied

Oenothera drummondii subsp. *drummondii* is a short-lived shrub (up 4 years) that blooms from spring to autumn. In native populations, the flowers live very briefly (12–15 h) while in non-native populations, they can last for two or more days [29,33]. The flowers produce both nectar and pollen as floral rewards. Floral nectar is exuded by nectaries located at the base of tubular hypanthium. Pollen grains are connected to each other by viscin threads. The fruits are capsules with tetragonal dehiscence, which contain a large number of small seeds (ca. 200 seeds, each of 1.2 mm in length and 0.6 mm in width) that germinate readily under laboratory conditions [34].

2.2. Seed Collection and Plant Material

The study includes six native populations that together cover almost all of the geographic range of the natural distribution of the species in the Gulf of Mexico, four from Mexico and two from USA and four non-native populations corresponding to locations across its non-native distribution range (Spain, Israel, South Africa, Australia, Figure 1 and Supplementary Material Table S1 show the full list of the populations with their origin, countries, acronyms and geographic locations). Ripe fruits were collected during 2015 and 2016 from ca. 30 randomly selected mature-similar size plants in each population (1–5 fruits from each plant). Seeds of each population were obtained manually from fruits, pooled and stored in marked container. In December of 2016 and 2017, groups of seeds taken at random from each population were germinated in pots (15 × 25 cm) filled with a substrate composed of perlite (30%) and dune sand (70%) and regularly irrigated to field capacity. When the seedlings emerged, we removed all but the most vigorous individual in each pot, keeping a total of 18 plants from each population (180 plants in total). The plants were maintained under greenhouse conditions (temperature 25 °C, humidity 40% and natural daylight), with weekly irrigation of the substrate to field capacity, at the Centre of Technological Research and Innovation of the University of Seville (CITUS-US). In addition, the plants were fertilized every month with 200 mL of Hoagland solution (20%) throughout the study time in order not to limit their growth.

2.3. Pollination Treatments and Fruit and Seed Data

Flowering started nearly at the same time in all populations. Once the plants were flowering regularly, we performed the following hand-pollination treatments on the flowers of each population: (1) spontaneous selfing: the flower remained unmanipulated, (2) self-pollination: the flower was emasculated and pollinated with its own pollen. (3) cross-pollination: one flower was emasculated and pollinated using pollen from another emasculated flower of a different individual; we subsequently pollinated the latter flower with pollen from the former. All treatments were applied using fresh newly opened flowers with virgin styles that were pollen-saturated according to the pollination treatment. The plants produced different numbers of flowers, but enough to make all hand-pollination treatments on each one. Following application of the manipulation treatments, the flowers were carefully labelled and left unbagged, since the strict insect control practiced in the greenhouses ensured that no alien pollen could subsequently be deposited on the flowers.

The production and development of fruits were recorded regularly. The fruits were harvested when they were ripe, at ca. 6 weeks after pollination (fruit set) and pooled in paper bags per pollination treatment and population. With the number of fruits produced on each pollination treatment and population, we obtained the fruit set by means of the ratio of the number of flowers in the treatment and the number of fruits produced. Each fruit was weighed on an analytical balance (Ohaus Adventurer resolution 0.1 mg). The seeds of each fruit were then extracted manually and counted. With the number of seeds of each fruit developed, we calculate the seed set for each treatment and population, as the average number of seeds per fruit. Seeds of each pollination treatment and population were pooled and stored in label plastic containers in dry conditions until germination experiments.

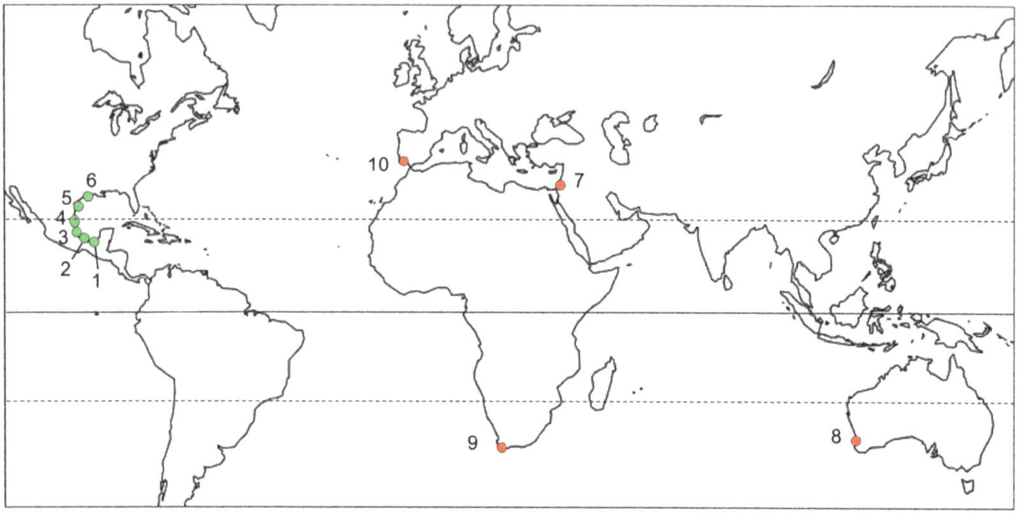

Figure 1. Geographical locations of the populations of *Oenothera drummondii* studied. Mexico: Native populations (green dots). Mexico: 1. Ojoshal (OJO); 2. Sontecomapan (SON); 3. La Mancha (MAN); 4. Tecolutla (TEC). US: 5. South Padre (SPA); 6. Bolivar (BOL). Non-native populations (red dots). Israel: 7. Rishon-Lezion (LEZ); Australia: 8. Mandurah (MAH). South Africa: 9. Muizenberg (MUI). Spain: 10. El Dique (DIQ). Line of Ecuador (solid black line) and intertropical region are shown (dashed black lines).

2.4. Germination Experiment

To evaluate the success of the hand-pollination methods, we conducted germination experiments with the seeds obtained. We only carry out germination experiments with the seeds produced by self-pollination and cross-pollination, because the seeds from the spontaneous pollination treatment were not well developed. Four sets of 100 randomly selected seeds were obtained from the seed mixture of each treatment and population (800 seeds from the two treatments per population; 8000 total experimental seeds). These seeds were placed in labelled Petri dishes with three layers of wet filter paper as a substrate and kept in germination chambers under controlled conditions (12 h light/darkness, temperature 25/20 °C). Previous studies indicate that *O. drummondii* seeds maintain high viability (≈90% after ca. 2 years) and do not require any pre-germinative treatment [29]. Daily, for a 90-day period, germinated seeds were quantified and removed and distilled water added as necessary to keep the papers moist. We considered a seed to have germinated when growth of the radicle was visible. For each treatment and population studied, we recorded the time to first germination in days (t_o), the percentage of total germination (final germination) and the germination rate. Germination rate was estimated using a modification of the Timson index [35] as follows,

$$\text{Germination rate} = \sum G/t, \tag{1}$$

where G is the percentage of germination accumulated at one-day intervals and t is the total number of days of the experiment. Once the germination period was completed, the crush test [36] was performed on the remaining seeds to determine if they were empty or if the embryo was still viable.

2.5. Data Analysis

Due to the low number of fruits and immature seeds in the spontaneous selfing treatment, all comparisons were made only between the manual self- and cross-pollination data; however, the data from spontaneous pollination are shown in Table A2. Prior to analysis, the normality of all data was tested (Shapiro-Wilk test) and germination data transformed (Arcsine). Since transformations did not produce normality of the data, non-parametric analyses were carried out. All analyses were performed using JMP (v9.0.1, SAS Institute Inc., Cary, NC, USA) and SPSS (IBM SPSS Statistics for Windows, version 25; North Castle, NY, USA).

The mating system of each population was evaluated with the self-compatibility index (SCI) [37], where:

$$\text{SCI} = \text{seed set from self-pollination}/\text{seed set from cross-pollination}, \quad (2)$$

in which values > 1 indicate full self-compatibility. Petanidou [22] note that the SCI allows comparison of data produced under different environmental growth conditions (i.e., different habitats, greenhouse). In our study, plants were obtained from seeds from populations of different geographic locations, but which were grown under the same greenhouse conditions.

We used a Wilcoxon t-test to determine whether the fruit sizes (weight, length and width), seed characteristics (seed set, seed set mass, individual seed weight), final germination and germination rate from each hand pollination treatment differ within each population (native and non-native). Subsequently, since the native populations studied are located along a latitudinal gradient, we explored whether fruit and seed characteristics obtained in each hand pollination treatment differed between the native populations by mean Wilcoxon/Kruskall—Wallis nonparametric analyses of variance and multiple Steel—Dwass comparisons. On another hand, linear regression analysis was performed to know any relation between the reproductive traits and the latitudinal distribution of the populations. Likewise, the relationships between latitude and final germination percentage and germination rate were compared by nonparametric ANOVA and explored through regression analysis. Since each native population represents a latitudinal position along the gradient of distribution of the study species (Figure 1), the charts presented in the results section include the population acronyms instead of latitudinal data. Population acronyms follow the order of latitudinal increase (see Table A1).

Latter we then explored whether reproductive success differed between the sets of native populations and non-native populations. Each data point pertaining to fruit and seed characteristics, as well as the final germination percentage and germination rate of all populations, was labelled and pooled according to origin (native and non-native) and pollination treatment (self- and cross-pollination) for inclusion in the analysis. The data were compared through Wilcoxon non-parametric analysis of variance and multiple Steel–Dwass comparisons, where the pollination treatment (self- and cross-) in each origin (native and non-native) was considered as a factor.

Plant reproductive success has been usually determined by the fruit set, the seed set, or the final germination, because each parameter shows the individual success of different reproductive phases in plants. As far as we know, however, there is no parameter that indicates the total reproductive success in plants. In order to represent the total reproductive success in a single value, we constructed the integral reproductive success index (IRSI), which was obtained as follows:

$$\text{IRSI} = \text{fruit set} \times \text{relative seed set} \times \text{final germination}, \quad (3)$$

where a value of 1 indicates very high success in all reproductive phases (e.g., 1.0 [fruit set] × 1.0 [relative seed set] × 1.0 [final germination] = 1); while values < 1 indicate that success could have been high in some of the reproductive phases, but low in others (e.g., 1.0 [fruit set] × 0.8 [relative seed set] × 0.5 [final germination] = 0.4). The relative seed set data for each hand pollination treatment (self- and cross-) of each population was obtained by dividing every seed set value by the highest seed set value recorded in all populations, regardless of the pollination treatment from which it was derived. We made the latter, based on the hypothesis that the highest seed set value represents that which could potentially achieved in any treatment and population. Finally, we performed a linear regression analysis, to explore whether the IRSI of each pollination treatment and the value of the SCI are related to the latitude at which each population is located.

3. Results

3.1. Compatibility, Fruit and Seed Characteristics among Native Populations

The fruit set of the two pollination treatments was high in all native populations (92–100%) and they therefore also presented a high level of compatibility (SCI 0.92–1.04) (Table A2). In general, fruit and seed characteristics did not differ between pollination treatments within the populations, but there was a tendency towards higher values in the cross-pollination treatment. The weight of fruits from cross-pollination were significantly greater in the MAN and BOL populations (Table A2).

The seeds of the two hand-pollination treatments presented a similar pattern to that of the fruits. Seed set did not differ significantly among the two hand-pollination treatment within almost all native populations; only in BOL seed set were significantly higher in the cross-pollination (Table A2). Time to first germination was similar between treatments in most native populations (4–5 d), in SON and TEC the time of first germination were greater in seeds from cross pollination (Table A3). Meanwhile, the germination rate and final germination was higher in seeds from cross pollination, although significant differing in SPA, MAN and TEC (Table A3).

On the other hand, the characteristics of fruits and seeds of each pollination treatment presented some differences between populations. The weight of the fruits derived from self-pollination did not differ significantly and showed less variability in the OJO and SON populations, but greater variability in SPA and BOL (Figure 2A), while the fruits of cross-pollination were significantly heavier in BOL and lighter in SON, but similar among the rest of the populations (Figure 2B).

The seeds of the two pollination treatments presented a similar pattern to that of the fruits. Seed set in the self-pollination treatment did not differ significantly among the populations (Figure 2C); however, seed set in the cross-pollination treatment did differ significantly, being higher in BOL and lower in SON and SPA (Figure 2D).

In general, seed germination varied among the hand-pollination treatments and populations. The populations BOL and SPA had the highest final germination percentages in both hand-pollination treatments, while OJO and SON had the lowest values in the two hand-pollination treatments (Figure 3), but only the final germination of seeds derived from the self-pollination treatment was positively related to latitude (Figure 3A,B). On the other hand, the germination rate (Timson Index) also differed among populations, showing the same pattern as that of final germination: low values in OJO and SON and higher values in SPA and BOL and again the germination rates obtained from seeds from the self-pollination treatment were significant positively related to latitude (Figure 3C).

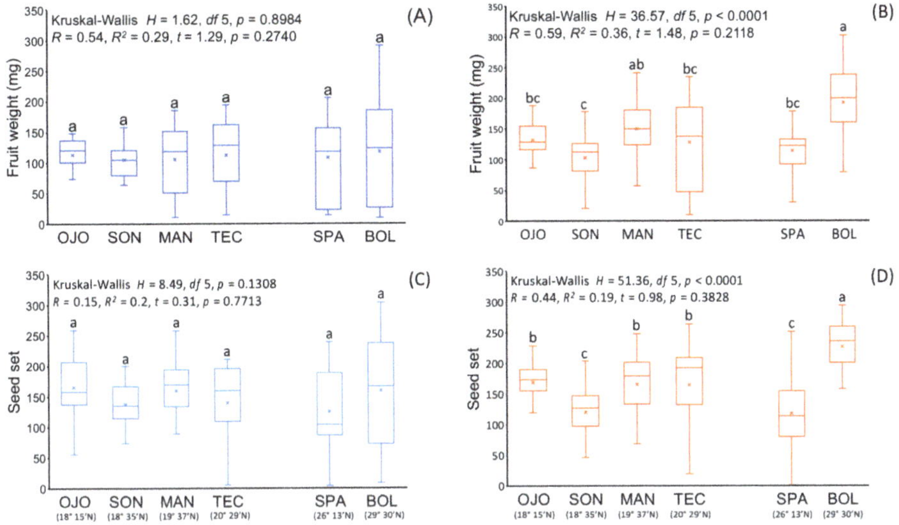

Figure 2. Fruit weight (**A**,**B**) and seed set (**C**,**D**) obtained by self- (blue boxes) and cross-pollination (orange boxes) in native populations of *Oenothera drummondii*. Comparison among populations (Kruskal–Wallis nonparametric ANOVA) and the relationship between the reproductive characteristic and latitude (linear regression analysis) are shown in each graph. Population acronyms are explained in Table A1. Latitudinal ubication is indicated below each acronymous. Different letters above the boxes indicate significant differences among populations (Steel–Dwass all pairs multiple comparisons $p < 0.001$). Boxes show the average (x), median (-) and quartiles.

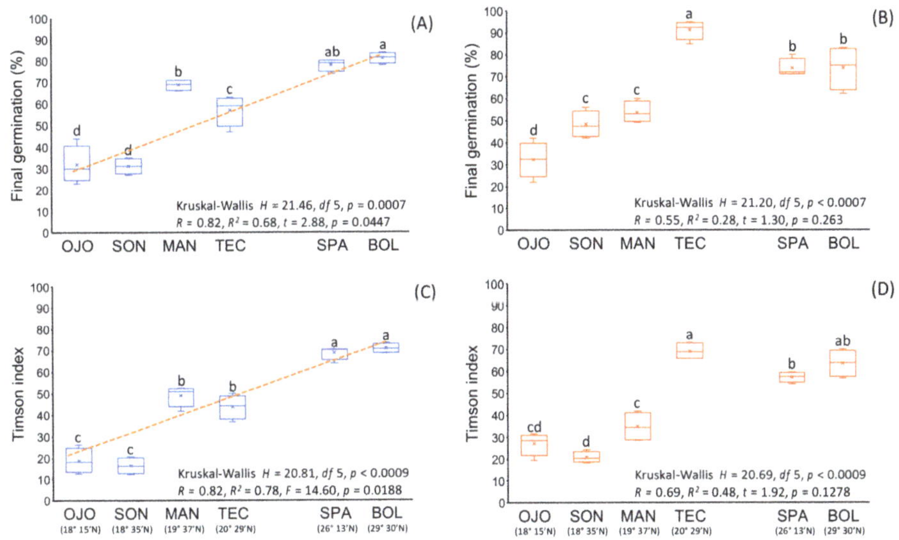

Figure 3. Final germination and Timson index of the seeds obtained by self- (**A**,**C**) and cross- (**B**,**D**) pollination in native populations of *Oenothera drummondii*. Comparison among populations (Kruskal–Wallis nonparametric ANOVA) and the relationship between the reproductive characteristic and latitude (linear regression analysis) are shown in each graph. Population acronyms are explained in Table A1. Latitudinal ubication is indicated below each acronymous. Dotted lines representing significant regression analyses but are not in scale with the latitudinal increase. Different letters above the boxes indicate significant differences among populations (Steel–Dwass all pairs multiple comparisons $p < 0.001$). Boxes show the average (x), median (-) and quartiles.

3.2. Fruit and Seed Characteristics of Non-Native Populations

As in the native populations, fruit set in the two pollination treatments of the non-native populations was high and SCI was therefore also high in all of these populations (Table A2). The fruit weight was not different between hand-pollination treatments within the non-native populations; (Table A2). Meanwhile, the seed set did not differ between hand-pollination in all non-native populations. On the other hand, time to first germination was very similar and did not differ between the treatments in each population, although the percentage of final germination was higher in the seeds derived from cross-pollination in the populations MAH and MUI (Table A3).

3.3. Comparison of Reproductive Characteristics between Native and Non-Native Populations

Fruit and seed characteristics differed significantly between the groups of native and non-native populations, but no clear general pattern was shown either by the origin of the populations (native and non-native) or by the pollination treatment (self- and cross-pollination) (Figure 4). The weight of the fruits from cross-pollination was significantly greater and was similar in the native and non-native populations, while fruit weight from selfing was lower but also similar among the two sets of populations (Figure 4A). However, seed set were both greater in the two pollination treatments of the non-native populations (Figure 4B), while the values of final germination and germination rate (Timson index) were higher in the cross-pollination treatment of the non-native populations (Figure 4C,D). Finally, the IRSI also were higher of both pollination treatments for the non-native group and the lowest was obtained for the self-pollination treatment of the native group (Figure 4E).

3.4. Integral Reproductive Success Index

The IRSI of the native populations show a positive tendency to latitudinal increase in both pollination treatments, although self-pollination was only marginally non-significant (Figure 5). This means that the populations located at a lower latitude (OJO and SON) presented the lowest reproductive success, while the population at the highest latitude (BOL) presented the highest reproductive success. Meanwhile, the compatibility index (SCI) presented a negative, but non-significant, relationship, being higher in the populations OJO and SON and decreasing to BOL (Figure 5). On another hand, the IRSI of both hand-pollination treatments and the SCI of all non-native populations were as high as of the native population located at the highest latitudinal-distribution (BOL).

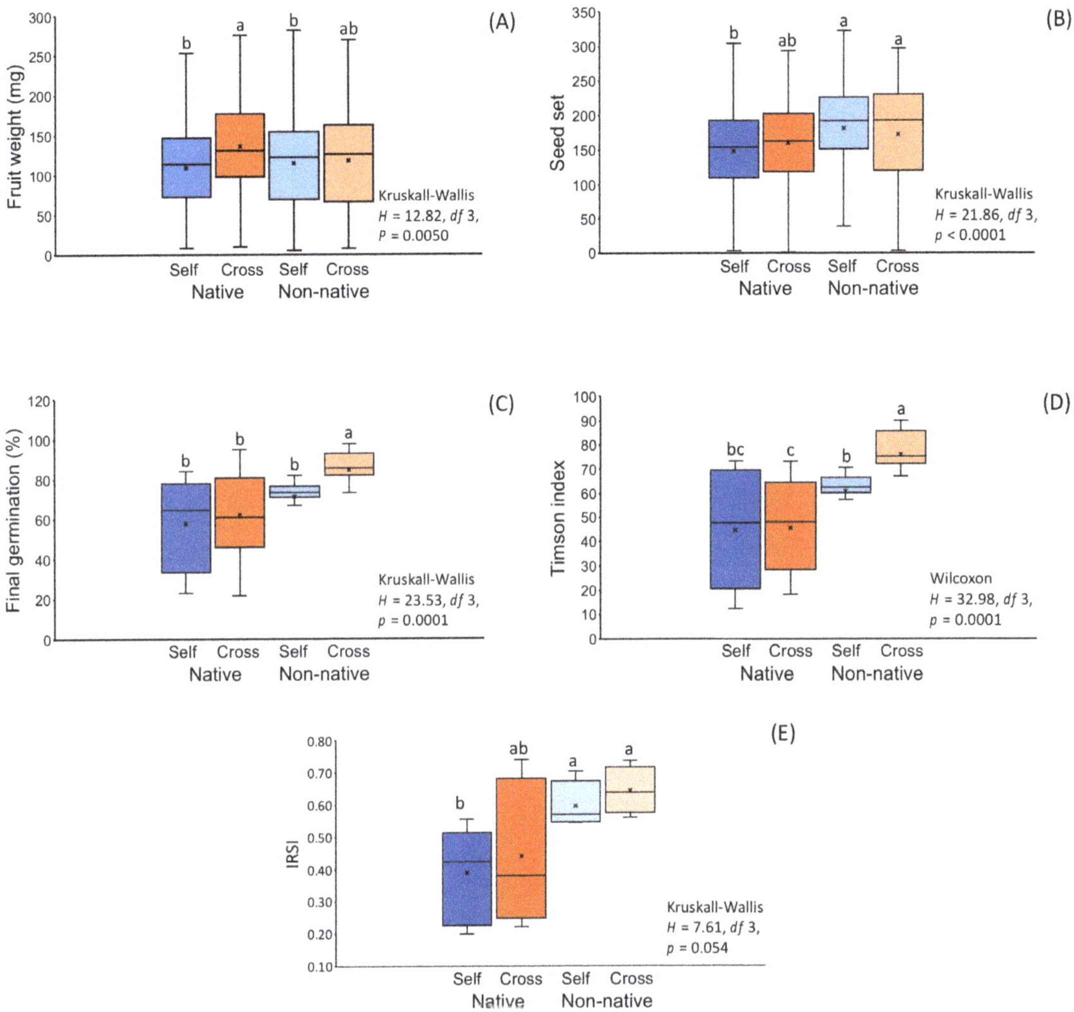

Figure 4. Comparison between groups of native and non-native populations of *Oenothera drummondii* of the characteristics of fruit weight (**A**), seed set (**B**), final and rate germination (**C**,**D**) and the integral reproductive success (IRSI) (**E**), obtained by self- (Self) and cross-pollination (Cross). The result of the Kruskal–Wallis nonparametric ANOVA is shown on each graph. Different letters above the boxes indicate significant differences (Steel–Dwass all pairs multiple comparisons, $p < 0.001$). Boxes show the average (x), median (-) and quartiles.

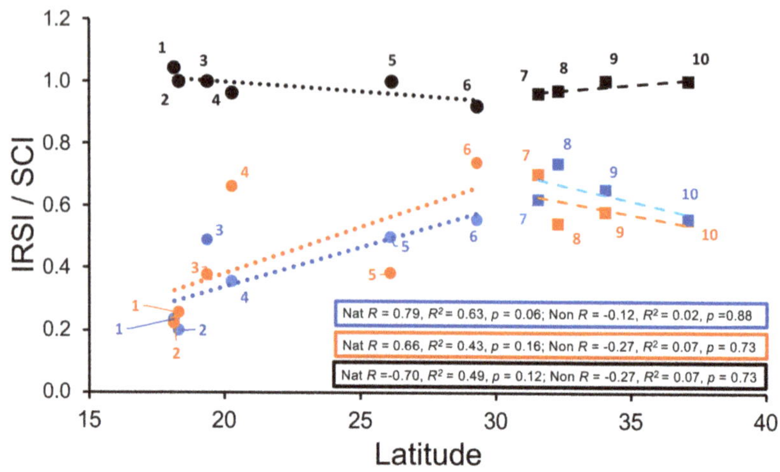

Figure 5. Relationship between self-compatibility index (SCI) (symbols and line black) and the integral reproductive success index (IRSI) of self- (symbols and line blue) and cross-pollination (symbols and line orange) with the latitudinal location of native and non-native populations of *Oenothera drummondii*. The boxes show the results of linear regression for each index for native (Nat) and non-native (Non) populations. Circles indicate native populations and squares indicate non-native populations. Numbers indicate acronyms of populations described in Figure 1.

4. Discussion

We have shown that *O. drummondii* presents a mixed system (self- and cross-pollination) with a high degree of self-compatibility, both in its native distribution range and in non-native populations. We also showed that the characteristics of the fruits and seeds produced by both self- and cross-pollination vary between native populations, with no clear pattern. This is also found when comparing these characteristics between the groups of native and non-native populations. However, in the native range, final accumulated germination of seeds and germination rate (Timson index) of self-compatibility treatment were positively related to latitudinal increase, whereas the self-compatibility index did so slightly negatively. In the non-native populations, these two parameters had values as high as those recorded in the native populations located at the highest latitudes. We also showed that the integral reproductive success index (IRSI) provides a better perspective of the true reproductive success of the *O. drummondii* populations, both native and non-native, since it incorporates the values of the most useful parameters of reproductive success (fruit set, seed set and germination).

4.1. Mating System and Fruit and Seed Characteristics in Native Populations

Studies in natural populations indicate wide variability in mating systems, but also show that mixed mating (self- and cross-pollination) are more common than expected [5,38,39]. The genus *Oenothera* is no exception, since wide variation is recognized in the species mating systems [7] and both cross- and self-pollination can occur in Onagraceae species that present, positive herkogamy [32]. Gallego-Fernández and García-Franco [34] noted positive herkogamy in *O. drummondii,* but it has not relationship with latitude increase. Our results show that *O. drummondii* is a species with a mixed mating system because of the high values of fruit set and seed set both by self- and cross-pollination.

Field observations in most of the studied native populations of *O. drummondii* indicate high fruit production [29]. The low fruit set recorded by spontaneous self-pollination in the greenhouse suggests low capacity of autogamy, so the high fruit set recorded in the field could be result of assisted pollination. The flowers of *O. drummondii* produce abundant nectar and pollen [24,29] and they are visited by sphingids and crepuscular bees, as occurs with other *Oenothera* species [28,40]. Observations made in one of the native populations studied (MAN) [29] show that bees (*Apis mellifera*) arrive to flowers shortly before sunset to collect pollen. The foraging activity is carried out in groups (2–5 bees) in one or two flowers of the same plant, limiting the pollination to selfing or geitonogamy. However, the sphingids (aff. *Manduca* sp.), which are nectar collectors with high energetic requirements, arrive shortly after dusk and fly around the population, visiting a large number of flowers of different individuals with no clear foraging sequence among plants, thus constituting potential outcrossing pollinators. This suggests that bees and sphingids both play important roles in the reproduction of *O. drummondii* in its native range of distribution. However, further observations and experiments are required in other native populations in the field in order to corroborate these findings.

Non-native populations also presented low fruit set and seed set by spontaneous self-pollination, but a high production of fruits and seeds in both hand-pollination treatments (self- and cross-pollination). Furthermore, opposed to the low number of fruits expected because of the alien environment in the non-native locations, it has been recorded a large production of fruits in one of our studied population (DIQ; [41]). On the other hand, in *O. drummondii* populations located in Europe and Israel (DIQ and LEZ), it has been recorded crepuscular anthesis and flower visits by local sphingids and bees [28,29,40]. This suggests that flowers of the non-native populations will have the same pattern of visits as those natives have. However, the flower lifespan in non-native populations is twice than in natives [29], which may be associated with limitation of pollinators [42]. Nevertheless, if pollinators are, in fact, limited, the constant coastal wind present probably shake violent the flowers and autogamy could thus be achieved.

The above indicates that *O. drummondii* presents those characteristics proposed by Baker [17] for colonizer species of new habitats. *Oenothera drummondii* has all of the benefits of being highly compatible and able to incorporate itself into the pollinator community, taking full advantage of its mixed mating system. Furthermore, there have been recorded low genetic variation in MAN (native) and DIQ (non-native) populations [43], which is evidence of colonization-extinction processes typical on coastal dunes [44,45].

4.2. *Fruit and Seed Characteristics of Native Populations*

The differences in most the characteristics of fruits and seeds among the native populations of *O. drummondii* showed a no relationship with latitudinal increase. However, there were a clear positive relationship in the fruit width and individual seed weight and a tendency in the seed-set weight to be smaller in lower latitudes and greater in those populations at higher latitudes. This pattern was also reflected in the final accumulated germination and in the germination rate (Timson index), since the populations located at low latitude had the lowest values of these parameters in the two pollination treatments studied.

Several studies have recorded floral trait variation patterns related to latitude increase, but rather than species following a general pattern, they respond differently to the conditions imposed by the latitudinal gradient [4]. In particular, the floral phenotypic features of *O. drummondii* differ among native and non-native populations and they do present a negative pattern with latitudinal increase [34]. The latitudinal positive relationship of individual seed weight in the two pollination treatments and of final germination and the germination rate in self-pollination suggest that the environmental conditions prevalent at higher latitudes can favour these reproductive characteristics of *O. drummondii*, since the wide contrasts found with the southern populations (OJO and SON). Reduction in seed set, seed size and germination are generally attributed to changes in local environmental conditions, due to the nutrient and water contents in the substrate, as well as pollen limitation (pollen load and quality) as a result of the absence or scarcity of pollinators and inbreeding [46,47]. Our study plants were maintained under controlled greenhouse conditions (in terms of nutrients, water, temperature and relative air humidity) throughout the experiment and all flowers were manually pollinated following a standardized method, so these factors did not influence the observed differences. The differences in seed size and germination recorded in the studied populations may therefore be the result of inbreeding. Such results have been recorded in self-compatible species and can be important in small and isolated populations [48,49]. Although we do not have information regarding the size of all studied populations in the field, we do know that those located at lower latitudes (OJO, SON, MAN and TEC) are small and relatively isolated; therefore, our results possibly reflect the effects of inbreeding taking place in the field.

4.3. Comparison between Native and Non-Native Populations

The characteristics of the fruits and seeds obtained by the hand-pollination treatments varied between the sets of native and non-native populations and, although in general they do not show a particular pattern, some tendencies towards higher values were observed in the germination of the non-native populations. This indicates that, even though non-native populations are isolated in geographically distant sites, they maintain a high level of self-compatibility and have not modified their potential for a mixed pollination system. This also suggests that these populations can take advantage of local pollinators [29,50], which could allow them to invade new sites [7]. Several *Oenothera* species have become successfully established in Europe and in other regions of the world, where some taxa are considered to be actively invasive species [7,44,51], similarities in environmental conditions between their native habitats and invaded sites must be important factors in the success of these species.

4.4. Integral Reproductive Success Index (IRSI)

Reproductive success in plants usually is measured by seed set (e.g., [52]), number and weight of fruits (e.g., [53]), fruit set, seed set, seed development [54,55]) or germination [56]. However, the most commonly used parameters are fruit set, seed set and germination, since these represent the final product of one reproductive stage process (pollination, fertilization, or seedling emergence). In the studied populations of *O. drummondii*, we tested all of the parameters that indicate reproductive success (Tables A2 and A3) but found that the individual responses were inconsistent. If we had considered the most common reproductive success parameters (fruit set, seed set and total germination) separately, interpretation of the reproductive success of *O. drummondii* can be biased in different ways. The fruit set was very high in the two pollination treatments of all populations, the seed set showed some differences among populations, while final germination in the native populations followed the latitudinal gradient and the non-native seeds showed similarity to the northernmost native populations in terms of final germination. When considering the three most common reproductive parameters, the proposed Integral Reproductive Success Index (IRSI) shows the full success of *O. drummondii* in each of the pollination treatments studied in the different populations, since the IRSI integrates the effects of each of the parameters in a final unique reproduction value for the species. The two populations located in the southernmost part of the native distribution had the lowest IRSI, but this positively increased in the higher-latitude populations. In addition, the values of the latter were similar to those of the non-native populations. This suggests that the southern native populations can be subject to a different selection of reproductive factors than those of the northern, while the great integral reproductive success of non-native populations of *O. drummondii* can be an important factor in their establishment success.

To our knowledge, there has been no reproductive index that integrates the different components of plant reproductive success. In our study, the IRSI and final accumulated germination values presented the same pattern. The latter had a strong effect on the value obtained by the former since neither fruit set nor seed set clearly differed between populations. However, under other situations, where each parameter can change, the IRSI can help to more clearly understand reproductive success.

5. Conclusions

Oenothera drummondii is a coastal species, which has ecological characteristics that allow it to live and colonize these ecosystems with stressful environmental conditions [30,41]. Our results showed that *O. drummondii* presents a mixed mating system. This can confer reproductive advantages both in native and non-native ranges, allowing the flowers to interact with the local fauna (bees and sphinxes). However, the environmental and biotic conditions present throughout the entire range of *O. drummondii* can impose contrasting selection pressures on its reproductive characteristics. Self-compatibility and germination responses (rate and final germination) are reduced in low latitude populations, while they are increased in higher latitude populations, including native and non-native. These reproductive traits may contribute to the successful establishment of *O. drummondii* outside the native range. However, field studies evaluating pollinators, mating and the same reproductive parameters studied here, including more both native and non-native populations, will allow us to understand the adaptation process of *Oenothera drummondii* in the new environments.

Supplementary Materials: The following are available online at https://www.mdpi.com/article/10.3390/d13090431/s1, Table S1: Characteristics of the fruits and seeds (mean ± SD) obtained by spontaneous, self- and cross-pollination (Self and Cross, respectively) in flowers of native and non-native populations of *Oenothera drummondii*.

Author Contributions: J.B.G.-F. and J.G.G.-F. planned, designed and performed the experiments and data analyses. In addition, both wrote and review the manuscript. Both authors have read and agreed to the published version of the manuscript.

Funding: This research was funded by the Ministerio de Economía y Competitividad (MINECO, Project CGL2015- 65058-R co-funded by FEDER.

Institutional Review Board Statement: Not applicable.

Data Availability Statement: The data presented in this study are openly available in the manuscript and Supplementary Material.

Acknowledgments: We thank Rusty Feagin, Mireia González, Pua Bar Kutiel and Ana Novoa for seed collection from some of native and non-native populations. We are grateful to the University of Seville Greenhouse General Services (CITIUS) for providing the facilities for growing and maintaining the plants. We also thank to Keith McMillan the review and edit of the English version. We appreciate the comments of two anonymous reviewers who improved the writing. The greenhouse work and the first draft of the manuscript was done during the sabbatical of JGGF at the University of Seville, supported by CONACYT, Mexico (CVU 7799).

Conflicts of Interest: The authors declare no conflict of interest. The funders had no role in the design of the study; in the collection, analyses, or interpretation of data; in the writing of the manuscript, or in the decision to publish the results.

Appendix A

Table A1. Geographic location of native and non-native populations of *Oenothera drummondii* studied. Numbers corresponding with geographic location in Figure 1.

	Origin	Country	Population	Acronym	Latitude	Longitude
1	Native	Mexico	Ojoshal	OJO	18°15′ N	93°59′ W
2	Native	Mexico	Sontecomapan	SON	18°33′ N	94°59′ W
3	Native	Mexico	La Mancha	MAN	19°37′ N	96°22′ W
4	Native	Mexico	Tecolutla	TEC	20°29′ N	97°01′ W
5	Native	USA	South Padre	SPA	26°13′ N	97°11′ W
6	Native	USA	Bolivar	BOL	29°30′ N	94°30′ W
7	Non-native	Israel	Rishon-Lezion	LEZ	31°59′ N	34°43′ W
8	Non-native	Australia	Mandurah	MAH	32°32′ S	115°41′ E
9	Non-native	South Africa	Muizenberg	MUI	34°06′ S	18°28′ E
10	Non-native	Spain	Dique	DIQ	37°09′ N	06°54′ W

Table A2. Characteristics of the fruits and seeds (mean ± SD) obtained by spontaneous, self- and cross-pollination (Self and Cross, respectively) in flowers of native and non-native populations of *Oenothera drummondii*. The last row of each trait of fruits and seeds show the comparison (Wilcoxon test) between self- and cross-pollination data of each population (ns = not significant; * $p = 0.05$; ** $p = 0.01$; *** $p = <0.001$). Population acronyms provided in Figure 1and Table A1.

Characteristics of Fruits and Seeds	Pollination Treatment	Native Populations						Non-Native Populations			
		OJO	SON	MAN	TEC	SPA	BOL	LEZ	MAH	MUI	DIQ
Fruit set (%)	Spontaneous	0	12	20	15	2	19	5	4	0	7
	Self	96	100	96	96	100	92	96	97	100	100
	Cross	92	100	96	100	100	100	100	100	100	100
Self-Compatibility Index		1.04	1.00	1.00	0.96	1.00	0.92	0.96	0.97	1.00	1.00
Fruit weight (mg)	Spontaneous	11.4 ± 2.1	18.1 ± 14.6	34.2 ± 29.4	24.3 ± 24.1	22.9 ± 17.2	35.8 ± 48.6	18.0 ± 12.2	22.9 ± 5.3	11.9 ± 12.9	17.8 ± 24.9
	Self	112.7 ± 28.5	104.6 ± 37.6	105.6 ± 56.8	112.4 ± 56.9	107.8 ± 68.5	117.5 ± 85.0	112.6 ± 64.0	139.0 ± 81.7	97.7 ± 35.5	109.0 ± 48.7
	Cross	131.3 + 51.4	102.8 + 41.5	149.2 + 50.7	127.9 + 70.9	113.8 + 44.0	192.8 ± 66.5	96.3 ± 61.6	176.1 ± 67.5	116.6 ± 37.3	84.4 ± 56.6
		ns	ns	***	ns	ns	***	ns	ns	*	ns
Seed set	Spontaneous	0.0 ± 0.0	46.7 ± 38.8	42.0 ± 36.6	54.8 ± 83.2	62.0 ± 35.4	89.8 ± 109.9	58.0 ± 0.0	10.0 ± 0.0	0.0 ± 0.0	72 ± 100.4
	Self	165.5 ± 47.1	137.7 ± 34.7	159.6 ± 58.7	139.7 ± 61.8	137.8 ± 66.1	159.8 ± 95.2	215.7 ± 59.0	165.6 ± 96.8	161.5 ± 35.8	190.5 ± 60.3
	Cross	168.4 ± 67.5	120.2 ± 44.7	164.9 ± 65.4	164.6 ± 75.4	117.9 ± 62.8	227.1 ± 45.3	176.8 ± 94.8	175.2 ± 87.8	173.2 ± 49.7	161.7 ± 83.3
		ns	ns	ns	ns	ns	**	ns	ns	ns	ns

Table A3. Seed germination characteristics and the integral reproductive success index (IRSI) of self- and cross-pollination treatments in flowers of native and non-native populations of *Oenothera drummondii*. The last row of each germination parameter shows the comparison (Wilcoxon test) between self- and cross-pollination data of each population (ns = not significant; * $p = 0.05$). Population acronyms provided in Figure 1and Table A1.

	Pollination Treatment	Native Populations						Non-Native Populations			
		OJO	SON	MAN	TEC	SPA	BOL	LEZ	MAH	MUI	DIQ
First, germination (d)	Self	4.0 ± 0.0	4.5 ± 1.0	5.0 ± 0.8	5.5 ± 1.0	5.0 ± 1.2	4.0 ± 0.0	4.5 ± 1.0	4.0 ± 0.0	4.0 ± 0.0	4.0 ± 0.0
	Cross	4.0 ± 0.0	7.3 ± 1.5	5.0 ± 0.8	4.0 ± 0.0	5.75 ± 0.50	4.0 ± 0.0	4.0 ± 0.0	5.0 ± 1.2	4.0 ± 0.0	5.3 ± 1.5
		ns	*	ns	*	ns	ns	ns	ns	ns	ns
Timson Index	Self	18.8 ± 5.9	16.3 ± 3.9	49.0 ± 4.9	43.7 ± 5.6	68.8 ± 3.2	70.9 ± 2.2	61.6 ± 3.6	65.2 ± 1.8	62.9 ± 3.7	54.5 ± 15.4
	Cross	27.1 ± 5.3	20.9 ± 2.5	43.8 ± 6.7	69.2 ± 3.8	57.2 ± 2.3	63.6 ± 6.2	72.0 ± 3.8	87.6 ± 1.6	76.7 + 4.5	67.6 ± 12.9
		ns	ns	*	*	*	ns	*	*	*	ns
Final germination (%)	Self	31.8 ± 8.9	31.0 ± 3.7	68.8 ± 2.6	57.0 ± 7.1	78.0 ± 2.8	81.3 ± 2.5	73.0 ± 4.3	72.8 ± 2.1	77.3 ± 4.4	62.8 ± 15.5
	Cross	32.3 ± 8.2	48.3 ± 6.1	53.8 ± 4.9	91.25 ± 4.4	73.8 ± 9.9	73.8 ± 9.9	79.8 ± 4.9	95.3 ± 1.9	85.3 ± 3.9	78.3 ± 15.8
		ns	*	*	*	ns	ns	ns	*	*	ns
Integral Reproductive Success	Self	0.23	0.20	0.49	0.35	0.50	0.50	0.70	0.54	0.58	0.55
Index (IRSI)	Cross	0.22	0.26	0.38	0.66	0.38	0.74	0.62	0.74	0.65	0.56

References

1. Pierre-Olivier, C. The Evolution of Plant Mating System: Is It Time for a Synthesis? In *Studies in Population Genetics*; Fusté, M.C., Ed.; IntechOpen: Rijeka, Croatia, 2012; pp. 17–38.
2. Whitehead, M.R.; Lanfear, R.; Mitchell, R.J.; Karron, J.D. Plant mating systems often vary widely among populations. *Front. Ecol. Evol.* **2018**, *6*, 38. [CrossRef]
3. Barrett, S.C.H. Mating strategies in flowering plants: The outcrossing–selfing paradigm and beyond. *Philos. Transect. R. Soc. B* **2003**, *358*, 991–1004. [CrossRef]
4. Barrett, S.C.H.; Harder, L.D. The ecology of mating and its evolutionary consequences in seed plants. *Annu. Rev. Ecol. Evol. Syst.* **2017**, *48*, 135–157. [CrossRef]
5. Goodwillie, C.; Kalisz, S.; Eckert, C.G. The evolutionary enigma of mixed mating systems in plants: Occurrence, theoretical explanations, and empirical evidence. *Ann. Rev. Eco. Evol. Syst.* **2005**, *36*, 47–79. [CrossRef]
6. Gamble, D.E.; Bontrager, M.; Angert, A.L. Floral trait variation and links to climate in the mixed-mating annual *Clarkia pulchella*. *Botany* **2018**, *96*, 425–435. [CrossRef]
7. Antoń, S.; Denisow, B. Pollination biology and breeding system in five nocturnal species of *Oenothera* (Onagraceae): Reproductive assurance and opportunities for outcrossing. *Plant Syst. Evol.* **2018**, *304*, 1231–1243. [CrossRef]
8. Randle, A.M.; Slyder, J.B.; Kalisz, S. Can differences in autonomous selfing ability explain differences in range size among sister-taxa pairs of *Collinsia* (Plantaginaceae)? An extension of Baker's Law. *New Phytol.* **2009**, *183*, 618–629. [CrossRef]
9. Munoz, F.; Violle, C.; Cheptou, P.-O. CSR ecological strategies and plant mating systems: Outcrossing increases with competitiveness but stress-tolerance is related to mixed mating. *Oikos* **2016**, *125*, 1296–1303. [CrossRef]
10. Grossenbacher, D.L.; Runquist, R.; Goldberg, E.; Brandvain, Y. Geographic range size is predicted by plant mating system. *Ecol. Lett.* **2015**, *18*, 706–713. [CrossRef]
11. Razanajatovo, M.; Maurel, N.; Dawson, W.; Essl, F.; Kreft, H.; Pergl, J.; Pyšek, P.; Weigelt, P.; Winter, M.; van Kleunen, M. Plants capable of selfing are more likely to become naturalized. *Nat. Commun.* **2016**, *7*, 13313. [CrossRef]
12. Eckert, C.; Kalisz, S.; Geber, M.A.; Sargent, R.; Elle, E.; Cheptou, P.O.; Goodwillie, C.; Johnston, M.O.; Kelly, J.K.; Moeller, D.A.; et al. Plant mating systems in a changing world. *Trends Ecol. Evol.* **2009**, *25*, 35–43. [CrossRef]
13. Hargreaves, A.L.; Eckert, C.G. Evolution of dispersal and mating systems along geographic gradients: Implications for shifting ranges. *Funct. Ecol.* **2014**, *28*, 5–21. [CrossRef]
14. Pyšek, P.; Richardson, D.M. Traits associated with invasiveness in alien plants: Where do we stand? In *Biological Invasions, Section II*; Nentwig, W., Ed.; Springer: Berlin, Germany, 2007; pp. 97–125.
15. Van Kleunen, M.; Bossdorf, O.; Dawson, W. The ecology and evolution of alien plants. *Ann. Rev. Ecol. Evol. Syst.* **2018**, *49*, 25–47. [CrossRef]
16. Richardson, D.M.; Pišek, P.; Rejmánek, M.; Barbour, M.G.; Panetta, F.D.; West, C.J. Naturalization and invasion of alien plants: Concepts and definitions. *Divers. Distrib.* **2000**, *6*, 93–107. [CrossRef]
17. Baker, H.G. Self-compatibility and establishment after 'long-distance' dispersal. *Evolution* **1955**, *9*, 347–348. [CrossRef]
18. Cheptou, P.O. Does the evolution of self-fertilization rescue populations or increase the risk of extinction? *Ann. Bot.* **2019**, *123*, 337–345. [CrossRef] [PubMed]
19. Van Boheemen, L.A.; Atwater, D.Z.; Hodgins, K.A. Rapid and repeated local adaptation to climate in an invasive plant. *New Phytol.* **2019**, *222*, 614–627. [CrossRef] [PubMed]
20. Stout, J.C. Pollination of invasive *Rhododendron ponticum* (Ericaceae) in Ireland. *Apidologie* **2007**, *38*, 198–206. [CrossRef]
21. Ollerton, J.; Watts, S.; Connerty, S.; Lock, J.; Parker, L.; Wilson, I.; Schueller, S.; Nattero, J.; Cocucci, A.A.; Izhaki, I.; et al. Pollination ecology of the invasive tree tobacco *Nicotiana glauca*: Comparisons across native and non-native ranges. *J. Pollinat. Ecol.* **2012**, *9*, 85–95. [CrossRef]
22. Petanidou, T.; Godfree, R.C.; Song, D.S.; Kantsa, A.; Dupont, Y.L.; Waser, N. Self-compatibility and plant invasiveness: Comparing species in native and invasive ranges. *Perspect. Plant Ecol. Evol. Syst.* **2012**, *14*, 3–12. [CrossRef]
23. Issaly, E.A.; Sérsic, A.N.; Pauw, A.; Cocucci, A.A.; Travest, A.; Benítez-Vieyra, S.M.; Paiaro, V. Reproductive ecology of the bird-pollinated *Nicotiana glauca* across native and introduced ranges with contrasting pollination environments. *Biol. Invasions* **2020**, *22*, 485–498. [CrossRef]
24. Dietrich, W.; Wagner, W.L. Systematics of *Oenothera* Section *Oenothera* Subsection *Raimannia* and Subsection *Nutantigemma* (Onagraceae). *Syst. Bot. Monogr.* **1988**, *24*, 1–91. [CrossRef]
25. Frean, M.; Balkwill, K.; Gold, C.; Burt, S. The expanding distributions and invasiveness of *Oenothera* in southern Africa. *S. Afr. J. Bot.* **1997**, *63*, 449–458. [CrossRef]
26. Heenan, P.B.; de Lange, P.J.; Cameron, E.K.; Champion, P.D. Checklist of dicotyledons, gymnosperms, and pteridophytes naturalised or casual in New Zealand: Additional records 1999–2000. *N. Z. J. Bot.* **2002**, *40*, 155–174. [CrossRef]
27. Heyligers, P.C. Flora of the Stockton and Port Hunter sandy foreshores with comments on fifteen notable introduced species. *Cunninghamia* **2008**, *10*, 493–511.
28. Gregory, D.P. Hawkmoth pollination in the Genus *Oenothera*. *Aliso* **1964**, *5*, 385–419. [CrossRef]

29. Gallego-Fernández, J.B.; Morales-Sánchez, J.A.; Martínez, M.L.; García-Franco, J.G.; Zunzunegui, M. Recovery of beach-foredune vegetation after disturbance by storms. *J. Coast. Res.* **2020**, *95* (Suppl. S1), 34–38. [CrossRef]
30. Gallego-Fernández, J.B.; (University of Sevilla, Sevilla, Andalucia, Spain); García-Franco, J.G.; Instituto de Ecología, A.C., Xalapa, Veracruz, Mexico). Personal communication, 2019.
31. Theiss, K.E.; Holsinger, K.E.; Evans, M.E.K. Breeding system variation in 10 evening primroses (*Oenothera* sections *Anogra* and *Kleinia*; Onagraceae). *Am. J. Bot.* **2010**, *97*, 1031–1039. [CrossRef] [PubMed]
32. Raven, P.H. A survey of reproductive biology in Onagraceae. *N. Z. J. Bot.* **1979**, *17*, 575–593. [CrossRef]
33. Dietrich, W.; Wagner, W.L.; Raven, P.H. Systematics of *Oenothera* Section *Oenothera* Subsection *Oenothera* (Onagraceae). *Syst. Bot. Monogr.* **1997**, *50*, 1–234. [CrossRef]
34. Gallego-Fernández, J.B.; García-Franco, J.G. Floral trait variation in *Oenothera drummondii* subsp. drummondii across a wide latitudinal range of native and non-native populations. *Flora* **2021**, *280*, 151851. [CrossRef]
35. Khan, M.A.; Ungar, I.A. The effect of salinity and temperature on the germination of polymorphic seeds and growth of *Atriplex triangularis* Willd. *Am. J. Bot.* **1984**, *71*, 481–489. [CrossRef]
36. Sawma, J.T.; Mohler, C.L. Evaluating seed viability by an unimbued seed crush test in comparison with the tetrazolium test. *Weed Technol.* **2002**, *16*, 781–786. [CrossRef]
37. Lloyd, D.G.; Schoen, D.J. Self-fertilization and cross-fertilization in plants. 1. Functional dimensions. *Int. J. Plant Sci.* **1992**, *153*, 358–369. [CrossRef]
38. Vogler, D.W.; Kalisz, S. Sex among the flowers: The distribution of plant mating systems. *Evolution* **2001**, *55*, 202–204. [CrossRef] [PubMed]
39. Fornoni, J.; Ordano, M.; Pérez-Ishiwara, R.; Boege, K.; Domínguez, C.A. A comparison of floral integration between selfing and outcrossing species: A meta-analysis. *Ann. Bot.* **2016**, *117*, 299–306. [CrossRef]
40. Raguso, R.A.; Kelber, A.; Pfaff, M.; Levin, R.A.; McDade, L.A. Floral biology of North American *Oenothera* sect. Lavauxia (Onagraceae): Advertisements, rewards, and extreme variation in floral depth. *Ann. Mo. Bot. Gard.* **2007**, *94*, 236–257. [CrossRef]
41. Gallego-Fernández, J.B.; Martínez, M.L.; García-Franco, J.G.; Zunzunegui, M. Multiple seed dispersal modes of an invasive plant species on coastal dunes. *Biol. Invasions* **2021**, *23*, 111–127. [CrossRef]
42. Ashman, T.L. Flower Longevity. In *Plant Cell Death Processes*; Nooden, L.D., Ed.; Elsevier Aademic Press: San Diego, CA, USA, 2004; pp. 349–362.
43. Hernández-Espinosa, R.; González-Astorga, J.; Espinosa de los Monteros, A.; Cabrera-Toledo, D.; Gallego-Fernández, J.B. Transferability of microsatellite markers developed in *Oenothera* spp. to the invasive ipecies *Oenothera drummondii* Hook. (Onagraceae). *Diversity* **2020**, *12*, 387. [CrossRef]
44. Gallego-Fernández, J.B.; Martínez, M.L.; García-Franco, J.G.; Zunzunegui, M. The impact on plant communities of an invasive alien herb, *Oenothera drummondii*, varies along the beach-coastal dunes gradient. *Flora* **2019**, *260*, 151466. [CrossRef]
45. Obeso, J.R.; Aedo, C. Plant-species richness and extinction on isolated dunes along the rocky coast of Northwestern Spain. *J. Veg. Sci.* **1992**, *3*, 129–132. [CrossRef]
46. Hirayama, K.; Ishida, K.; Tomaru, N. Effects of pollen shortage and self-pollination on seed production of an endangered tree, *Magnolia stellata*. *Ann. Bot.* **2005**, *95*, 1009–1015. [CrossRef]
47. Huang, Q.; Burd, M. The effect of pollen limitation on the evolution of mating system and seed size in hermaphroditic plants. *Am. Nat.* **2019**, *193*, 447–457. [CrossRef]
48. Weller, S.G. The relationship of rarity to plant reproductive biology. In *Restoration of Endangered Species*; Bowles, M.L., Whelan, C.J., Eds.; Cambridge University Press: Cambridge, UK, 1994; pp. 90–117.
49. Weekley, C.W.; Race, T. The breeding system of *Ziziphus celata* Judd and D. W. Hall (Rhamnaceae), a rare endemic plant of the Lake Wales Ridge, Florida, USA: Implications for recovery. *Biol. Conserv.* **2001**, *100*, 207–213. [CrossRef]
50. Eisikowitch, D.; Lazar, Z. Flower change in *Oenothera drummondii* Jooker as a response to pollinators' visits. *Bot. J. Linn. Soc.* **1987**, *95*, 101–111. [CrossRef]
51. Mihulka, A.; PysÏek, P. Invasion history of *Oenothera* congeners in Europe: A comparative study of spreading rates in the last 200 years. *J. Biogeogr.* **2001**, *28*, 597–609. [CrossRef]
52. Ye, Z.-M.; Jin, X.-F.; Wang, Q.-F.; Yang, C.-F.; Inouye, D.W. Pollinators shift to nectar robbers when florivory occurs, with effects on reproductive success in *Iris bulleyana* (Iridaceae). *Plant Biol.* **2017**, *19*, 760–766. [CrossRef] [PubMed]
53. Herbertsson, L.; Jönsson, A.M.; Anderson, G.K.S.; Seibel, K.; Rubdiöf, M.; Ekroos, J.; Stjernman, M.; Olsson, O.; Smith, H.G. The impact of sown flower strips on plant reproductive success in Southern Sweden varies with landscape context. *Agric. Ecosyst. Environ.* **2018**, *259*, 127–134. [CrossRef]
54. Sáyago, R.; Quesada, M.; Aguilar, R.; Ashworth, L.; Lopezaraiza-Mikel, M.; Martén-Rodríguez, S. Consequences of habitat fragmentation on the reproductive success of two *Tillandsia* species with contrasting life history strategies. *AoB Plants* **2018**, *10*, ply038. [CrossRef]
55. Delnevo, N.; van Etten, E.J.; Byrne, M.; Stock, W.D. Floral display and habitat fragmentation: Effects on the reproductive success of the threatened mass-flowering *Conospermum undulatum* (Proteaceae). *Ecol. Evol.* **2019**, *9*, 11494–11503. [CrossRef] [PubMed]
56. Wickert, K.L.; O'Neal, E.S.; Davis, D.D.; Kasson, M.T. Seed production, viability, and reproductive limits of the invasive *Ailanthus altissima* (Tree-of-Heaven) within invaded environments. *Forests* **2017**, *8*, 226. [CrossRef]

diversity

Article

Systematics of the Arboreal Neotropical '*thorellii*' Clade of *Centruroides* Bark Scorpions (Buthidae) and the Efficacy of Mini-Barcodes for Museum Specimens

Aaron M. Goodman [1,2,*], Lorenzo Prendini [2] and Lauren A. Esposito [1]

1 Institute for Biodiversity Science and Sustainability, California Academy of Sciences, 55 Music Concourse Drive, San Francisco, CA 94118, USA; lesposito@calacademy.org
2 Division of Invertebrate Zoology, American Museum of Natural History, Central Park West at 79th St., New York, NY 10024, USA; lorenzo@amnh.org
* Correspondence: agoodman@amnh.org; Tel.: +1-415-672-0493

Abstract: Fragmented and degraded DNA is pervasive among museum specimens, hindering molecular phylogenetics and species identification. Mini-barcodes, 200–300-base-pair (bp) fragments of barcoding genes, have proven effective for species-level identification of specimens from which complete barcodes cannot be obtained in many groups, but have yet to be tested in arachnids. The present study investigated the efficacy of mini-barcodes combined with longer sequences of the Cytochrome *c* Oxidase Subunit I (COI) gene in the systematics of the arboreal Neotropical '*thorellii*' clade of *Centruroides* Marx, 1890 bark scorpions (Buthidae, C.L. Koch 1837), the species of which have proven to be difficult to identify and delimit due to their similar morphology. The phylogeny of 53 terminals, representing all nine species of the clade and representative species belonging to related clades of *Centruroides*, rooted on *Heteroctenus junceus* (Herbst, 1800) and based on up to 1078 base pairs of COI and 112 morphological characters, is presented to test the monophyly of the clade and the limits of its component species. The results support the recognition of nine species of the '*thorellii*' clade, in accordance with a recent taxonomic revision, and highlight the efficacy of mini-barcodes for identifying morphologically similar cryptic species using specimens of variable age and preservation.

Keywords: scorpions; taxonomy; phylogenetics; species delimitation; COI; genetic degradation; morphology

Citation: Goodman, A.M.; Prendini, L.; Esposito, L.A. Systematics of the Arboreal Neotropical '*thorellii*' Clade of *Centruroides* Bark Scorpions (Buthidae) and the Efficacy of Mini-Barcodes for Museum Specimens *Diversity* **2021**, *13*, 441. https://doi.org/10.3390/d13090441

Academic Editor: Luc Legal

Received: 8 July 2021
Accepted: 10 August 2021
Published: 16 September 2021

Publisher's Note: MDPI stays neutral with regard to jurisdictional claims in published maps and institutional affiliations.

Copyright: © 2021 by the authors. Licensee MDPI, Basel, Switzerland. This article is an open access article distributed under the terms and conditions of the Creative Commons Attribution (CC BY) license (https://creativecommons.org/licenses/by/4.0/).

1. Introduction

Although natural history collections constitute one of the largest repositories of genetic information and are paramount for biological research, successful amplification of DNA from museum specimens often proves difficult due to degradation and fragmentation [1,2]. Many museum specimens were not collected with the intention (or knowledge) of future genomic studies, resulting in varying methods of preservation both among and within taxa and varying rates of subsequent molecular degradation among specimens of similar age [2]. Several factors, including time, dehydration, environmental exposure, and the presence of bacterial or fungal contamination, markedly affect the quality of DNA in collections of vertebrate and invertebrate taxa [2–4]. Although DNA barcoding is cost-effective for species identification, the utility of this method for museum material is severely limited by the failure to amplify complete Cytochrome *c* Oxidase Subunit I (COI) sequences and the cost in finances, time, and effort to repeat amplifications for adequate coverage. Overcoming the obstacle of sequencing museum specimens is critical, as natural history collections provide the basis for studies of historical populations, the discovery of new (and often cryptic) species, and the verification of barcoded specimens to types [5–8].

Research suggests that barcodes as short as 200 nucleotide base pairs (bp), the length at which DNA from degraded museum material stabilizes [2,8], are sufficient to acquire 95% confidence in species-level identification, thus promoting the use of 'mini-barcodes', primer sets designed to amplify 200–300-base-pair (bp) fragments of barcoding genes [8]. Mini-barcodes have proven effective for species-level identification from degraded museum specimens in a variety of systematics studies concerning fish, reptiles, mammals, insects, and plants, as well as for the identification of processed or traditional food products [8–15].

The Neotropical '*thorellii*' clade of *Centruroides* Marx, 1890 bark scorpions (Buthidae C.L. Koch 1837)—mostly inhabiting the forest canopy and characterized by their relatively small size, mottled coloration, and elongated metasoma and telson, more pronounced in the adult males—represents an ideal system for testing the efficacy of mini-barcodes in arachnids (Figure 1). Historically, *Centruroides hoffmanni* Armas, 1996, *Centruroides rileyi* Sissom, 1995, and *Centruroides schmidti* Sissom, 1995 were assigned together with *Centruroides thorellii* Kraeplin 1891 to the informal '*thorellii*' species group based on their similar morphology, habitat, and distribution in the forests of southern Mexico and northern Central America [16–20]. However, it was recently demonstrated that the '*thorellii*' species group is a paraphyletic assemblage, with *C. thorellii* being more closely related to a clade of dark, large-bodied species of *Centruroides*, suggesting that the apparent morphological similarity to other species of the '*thorellii*' group is convergent [21,22].

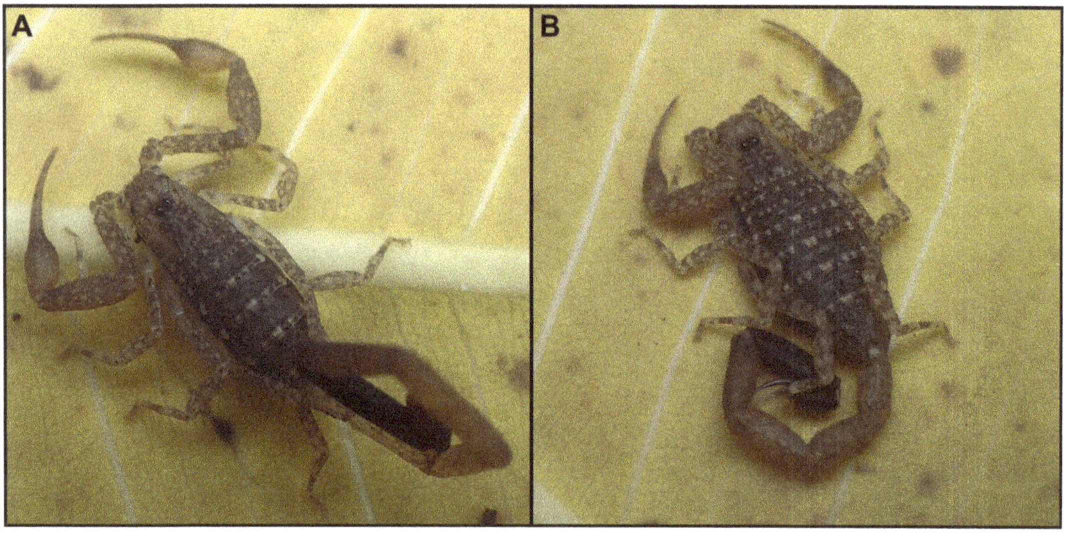

Figure 1. *Centruroides berstoni* (Goodman et al., 2021), a representative species of the arboreal Neotropical '*thorellii*' clade of *Centruroides* Marx, 1890 bark scorpions (Buthidae C.L. Koch, 1837), habitus in life: ♂ (**A**) and ♀ (**B**), Biotopo Chocón Machacas, Município Livingston, Guatemala.

The difficulty of collecting these canopy-dwelling scorpions appears to have resulted in their diversity and distribution being severely undersampled. Many collection records are represented by singletons. The comparative rarity of these scorpions in collections resulted in a tripling of the known diversity to a total of nine species in a recent revision [23], wherein six new species, previously identified as operational taxonomic units (OTUs) on the basis of DNA sequence data [22], were described. However, insufficient material prevented a decision as to whether the OTUs identified were new species or variable populations of more widespread species [22]. A larger sample of specimens from across the distribution permitted a more thorough assessment of species limits within the clade, resulting in the descriptions of *Centruroides berstoni* Goodman et al., 2021, *Centruroides catemacoensis* Goodman et al., 2021, *Centruroides chanae* Goodman et al., 2021,

Centruroides cuauhmapan Goodman et al., 2021, *Centruroides hamadryas* Goodman et al., 2021, and *Centruroides yucatanensis* Goodman et al., 2021.

The present study aimed to test the species limits and phylogenetic relationships among the species of the '*thorellii*' clade recognized by Goodman et al. (2021) using COI mini-barcodes of individuals from disparate localities combined with longer COI sequences and morphological characters, and in so doing demonstrating the efficacy of mini-barcodes for species-level identification of old and/or poorly preserved arachnid specimens from natural history collections [23].

2. Materials and Methods

Taxon Sampling

The ingroup comprised nine terminals, representing all nine species of the '*thorellii*' clade [23]. Each species was represented by three to seven individuals from one to five localities, including the type localities of all species except *C. hoffmanni*, *C. rileyi*, and *C. schmidti*, the identifications of which were verified based on morphology [16,19]. Exemplar species representing the North American, Central American, and Caribbean clades of *Centruroides* [21] were included as outgroups. Three species per clade were included to represent the genetic variability within each. Specimens of *C. thorellii* were also included to test its phylogenetic position. Trees were rooted on *Heteroctenus junceus* (Herbst, 1800), representing *Heteroctenus* Herbst, 1800, the sister taxon of *Centruroides* [21]. The complete taxon sample comprised nine ingroup species and ten outgroup species, considered satisfactory for testing the monophyly of the '*thorellii*' clade.

Material Examined

Specimens and tissue samples used for molecular and morphological analyses were obtained from the American Museum of Natural History (AMNH), New York; the California Academy of Sciences (CAS), San Francisco; the Colección Nacional de Arácnidos, Instituto de Biología (CNAN), Universidad Nacional Autónoma de México, Mexico City; and the Oxford University Museum of Natural History (OUMNH), UK.

Field-collected specimens from localities in Veracruz, Mexico, and southern Guatemala were detected using ultraviolet light, euthanized by submersion in 95% ethanol, and subsequently injected with ethanol to improve internal preservation. A total of 53 tissue samples, stored at −20 °C at the CAS and the Ambrose Monell Cryo Collection at the AMNH, were used for DNA extraction (Appendix A).

DNA Sequencing

Genomic DNA was extracted from muscle tissue from the fourth leg of well-preserved specimens using the spin column extraction protocol of the Qiagen DNeasy Blood & Tissue Kit (Valencia, CA, USA). DNA extracts were stored in 50–200 μL of elution (AE) buffer depending on the DNA concentration assessed using a Nanodrop 2000C Spectrophotometer (ThermoFisher Scientific, Waltham, MA, USA).

PCR amplification of complete mitochondrial and nuclear genes proved impossible for many older samples due to degradation and fragmentation of the DNA. Therefore, a 125-bp hypervariable region of the COI gene was partially amplified using the mini-barcode forward primer Uni-MinibarF1, 5′-TCCACTAATCACAARGATATTGGTAC-3′, and the reverse primer Uni-MinibarR1, 5′-GAAAATCATAATGAAGGCATGAGC-3′ [8]. Partial COI sequences, ca. 650 bp in length, were successfully amplified for field-collected specimens using the primers LCO1490 (5′-GGTCAACAAATCATAAAGATATTGG-3′) and HCO2198 (5′-TAAACTTCAGGGAGACCAAAAAATCA-3′) [24]. Complete COI sequences (1078 bp) were incorporated into the dataset from Esposito (2011) using the primers HCOEXTERNA (5′-GAAGTTTATATTTTAATTTTACCTGG-3′) and HCOEXTERNB (5′-CCTATTGAWARAACATARTGAAAATG-3′) [25].

Each PCR reaction was conducted in a 25-microliter volume, comprising 2.5 μL of 10× PCR buffer, 0.5 μL of dNTPs (10 mM stock), 1.0 μL of $MgCl_2$ (50 mM stock), 1.0 μL of each primer (25 mM stock), 1.0 μL of bovine serum albumin, 0.25 μL of Invitrogen

(Waltham, MA, USA) Taq polymerase enzyme (5 units/μL stock), 7.75 μL of Millipore (Burlington, MA, USA) deionized H_2O, and 10 μL of DNA template. Reactions with the LCO1490/HCO2198 primers used 2 μL of DNA template with 15.75 μL of Millipore deionized H_2O. PCR amplifications, performed on a BioRad MyCycler thermocycler (Bio-Rad Laboratories, Hercules, CA, USA) at the CAS Center for Comparative Genomics (CCG) and an Epicenter thermocycler (Eppendorf, Hamburg, Germany) at the AMNH Sackler Institute for Comparative Genomics (SICG), included an initial denaturing step at 95 °C for 2 min; 40 cycles of denaturing at 95 °C for 30 s, annealing at 47 °C for 30 s, and extension at 72 °C for 60 s; and a final extension at 72 °C for 10 min. PCR products were checked using gel electrophoresis on 1% agarose gel with ethidium bromide, and 1–2 μL of ExoSAP-IT (USB Scientific, Cleveland, OH, USA) was used to clean PCR products.

Sanger dideoxy sequencing was conducted using fluorescent-labeled BigDye Terminator 3.1 (Applied Biosystems, Waltham, MA, USA). Each 10-microliter reaction included 5.45 μL of Millipore deionized H_2O, 1.5 μL of 5× BigDye buffer, 0.3 μL of primer (10 mM stock), 0.75 μL of BigDye, and 2 μL of cleaned PCR product. The STeP50 Program (Platt et al., 2007) was used for cycle sequencing on a BioRad MyCycler thermocycler at the CAS CCG and an Epicenter thermocycler (Eppendorf, Hamburg, Germany) at the AMNH SICG. The labeled single-stranded DNA was precipitated by adding 2.5 μL of 125 mM di-NaEDTA to each sample, followed immediately by washing and centrifuging in 100% and 70% ethanol. Samples were then dried in a 65 °C incubator for 8 min, and 10 μL of HiDi formamide (Applied BioSystems) was added to each pellet, denatured at 94 °C for 2 min, and cooled on ice for 5 min. The denatured and fluorescent-labeled DNA pellets were sequenced on an ABI 3130x Genetic Analyzer (Applied BioSystems) at the CAS CCG and a Prism™ 3730x (Applied BioSystems) at the AMNH SICG.

Newly generated sequences were edited, forward and reverse primers were removed, and complementary strands were assembled into consensus sequences using Geneious v. 11.0.4 [26], by reference to a 1078-bp fragment of COI for *C. rileyi* from GenBank [21]. The dataset comprised 53 sequences, 14 mini-barcodes (127–134 bp), 12 partial (430–659 bp) sequences, 14 previously unpublished complete (1078 bp) sequences [21], and 13 complete sequences from Esposito and Prendini (2019) (Appendices 2 and 3) [22]. Sequences less than 150 bp in length were uploaded to the Dryad digital repository (doi:10.5061/dryad.fttdz08t2), whereas sequences greater than 150 bp in length were submitted to GenBank (Appendix B).

Morphological Character Matrix

Adult specimens from which DNA sequences were generated were scored using relevant characters from Esposito (2011), Esposito et al. (2017), and other sources (Table 1) [22,27]. The data matrix comprised 112 characters, including 43 (35%) from the prosoma, 38 (33%) from the mesosoma, and 30 (29%) from the metasoma, 41% derived from carination and surface macrosculpture, 25% from shape and morphometrics, 24% from coloration, and 7% from other character systems, e.g., macrosetae, trichobothria, and internal anatomy (Appendix C). Morphological examination of specimens was conducted using a Leica M125 stereomicroscope with an ocular micrometer calibrated at 10× magnification, with an LED-6WD UV spotlight to enhance the visualization of granulation and carination of the exoskeleton under ultraviolet light. Character states were scored from 0 to 5, as unknown (?) for missing data, or as polymorphic if two or more states were exhibited. The morphological terminology follows Hjelle (1990) and Sissom (1990), except for carapace and metasomal carination, which follows Vachon (1952); trichobothria, which follows Vachon (1974); tergite and pedipalp carination, which follows Prendini (2000); book lung anatomy, which follows Kamenz and Prendini (2008); and ovariuterine anatomy, which follows Volschenk et al. (2008) [28–34]. The matrix is deposited in Morphobank (http://morphobank.org/permalink/?P4047; accessed on 28 June 2021).

Phylogenetic Analysis

Sequences were aligned using the ClustalW method [35,36] in Mesquite v. 3.51 [37] and checked by eye. The average nucleotide composition of the aligned COI sequences was 17.7% A, 13% C, 25.2% G, and 44% T, and 725 sites were identical (67.4%), with 40.7% missing data as sequences varied greatly in length (125–1078 bp). Gaps in the dataset were treated as missing data and codon positions were optimized to reduce stop codons.

Phylogenetic relationships were inferred with a combined analysis of the molecular and morphological datasets using Bayesian inference (BI) in MrBayes v. 4.3.6 [38] and only the molecular dataset using maximum likelihood (ML) in RAxML v. 8.0.0 [39] on the CIPRES supercomputing cluster [40]. PartitionFinder2 [41] was used to determine the evolutionary models for BI and the partition definitions for ML. PartitionFinder2 obtained the highest AIC and BIC scores for the HKY+G model (for codons 1, 2, and 3); hence, that model was used for all subsequent analyses. Morphological characters were incorporated into the ML analysis and optimized using the gamma model of rate of heterogeneity and corrected for ascertainment bias using a Lewis correction type [42].

Table 1. Distribution of 112 morphological characters scored for phylogenetic analysis of the arboreal Neotropical '*thorellii*' clade of *Centruroides* Marx, 1890 bark scorpions (Buthidae C.L. Koch 1837) and outgroup species of *Centruroides* and *Heteroctenus junceus* (Herbst, 1800). Character states are scored from 0 to 5, as unknown (?), or as polymorphic [].

Outgroup
Heteroctenus junceus
2101131122 1120211321 0100011011 1212101013 0001112120 2000121000
0111020011 01120021?? 0101021011 1011001001 1111211110 11
Centruroides arctimanus
1010010011 0120221221 0000010?1? 111111210? 00111???20 2121000010
0110102?10 0111000??? ???1100000 1210000100 0110011110 00
Centruroides bani
0110021222 1120011121 0001010001 1011112000 0110011222 2112110000
0110101010 01110031?? ???1101001 1[01]10000 0011001111 11201
Centruroides exilicauda
0100030202 1120200010 0000010111 0000200000 0010113110 2110220011
0000201010 11110130?? 0111021111 1011101010 1101211110 11
Centruroides gracilis
1000011112 1120201321 0100111111 1212202000 0110003120 2120220011
0110201020 1102100100 0111001111 101?001011 1110011012 11
Centruroides hentzi
2010030122 0120200020 0000014110 1010102000 0101101112 2111010010
0111202020 00110000?? ???1101011 1210000000 1101211000 01
Centruroides infamatus
1000031212 1120200011 0000010111 1010102000 0010111111 2110121010
0110100010 01010020?? ???1001111 1210101011 1101011?12 11
Centruroides ochraceus
0000031202 1100211111 0100010?1 0000102000 00102???22 2110221011
0110101?21 01010100?? ???1101111 1010011011 1101211110 11
Centruroides thorellii
0000021101 1120200000 0100010110 0101102000 0010112301 2100120011
0110201020 01011001?? ???1100011 1301100010 0110000002 01
Centruroides tuxtla
0100131222 1120100021 0000010111 1011102005 0110202322 2111120011
0101102010 1112010??? ???1000011 1210100110 0010000012 01

Table 1. *Cont.*

Ingroup
Centruroides berstoni
1110010111 1000000010 00010101?0 1010012115 0110002111 2112220011
0?00202?10 01110100?? ???1100000 0110000000 01110101?0 01
Centruroides catemacoensis
1110010222 1021200010 0001010110 1010010115 0010202111 2110020010
0110202120 00100100?? ???1100000 0110000000 0111210000 01
Centruroides chanae
1110010111 1021200000 0001314110 1111010005 00112???12 2102220011
0?10202120 0111010??? ???1110000 0010000000 0111211010 01
Centruroides cuauhmapan
1110010110 1120200010 0000314110 1010012010 0111202121 2103120011
0?10202010 01110100?? ???1100000 0010000000 0111010000 01
Centruroides hamadryas
1110010111 1000000010 00010101?0 1010012115 0110002111 2112220011
0?00202?10 01110100?? ???1100000 0110000000 01110101?0 01
Centruroides hoffmanni
0100010111 1121000010 0000010110 1111012005 1110112110 2103220011
0?00202020 01110100?? ???1100000 0210000010 0110211010 01
Centruroides rileyi
1110010110 1120200010 0000314110 1010012010 0111202121 2103120011
0?10202010 01110100?? ???1100000 0010000000 0111010000 01
Centruroides schmidti
1110010111 0120200010 0101010110 1010012010 0100012122 2102220011
0000202120 11110101?? 0111100000 0010000000 0101211010 00
Centruroides yucatanensis
1110010111 1120000010 0001014110 1010012105 0010012021 2103020011
0?10201120 0111010??? ???1100000 0010000000 0110010112 01

A rapid bootstrap analysis was run with 1000 iterations. MrBayes was run twice on four threads for 50,000,000 generations, with Markov chains sampled every 1000 generations and the standard 25% burn-in calculated. Branches with posterior probability values $\geq$0.95 were considered strongly supported [43]. A 50% majority-rule consensus tree was created, and nonparametric bootstraps were estimated for the optimal nucleotide substitution model. Nodes with bootstrap values $\geq$70 were considered strongly supported [43].

Species Delimitation

The limits of putative species were evaluated on the majority-rule consensus tree from the Bayesian 95% posterior probability of the concatenated dataset of morphological characters and aligned DNA sequences using the species delimitation plugin in Geneious [26,44]. Clusters of sequences were evaluated based on their association with individuals collected at the type locality. Specimens which were monophyletic with those collected at the type locality or within the vicinity thereof with greater than 0.95 posterior probability were assigned to the species in question. Several metrics were investigated for assessing species delimitation in Geneious, including P (AB) for reciprocal monophyly [45]; P (RD), which measures the probability of an observed clade's degree of distinctiveness [46] and values for the probability of population identification of a hypothetical sample based on the groups being tested; P ID (Strict), and P ID (Liberal).

3. Results

Phylogenetic analyses with BI and ML produced almost identical tree topologies with well-supported terminal nodes but weakly supported internal nodes (Figures 2 and 3). Nine well-supported reciprocally monophyletic species-level clades were recovered in both analyses, but relationships within each clade received low support (Figures 2 and 3). *Centruroides hoffmanni*, *C. rileyi*, and *C. schmidti* received high posterior probabilities and likelihood values for monophyly (0.99/100, 0.95/100, and 0.99/100, respectively). *Centruroides thorellii* was placed outside the '*thorellii*' clade (Figures 2 and 3). All recently described species were well-supported: *C. berstoni* (0.99/99); *C. catemacoensis* (1.00/100); *C. chanae* (1.00/100); *C. cuauhmapan* (0.99/100); *C. hamadryas* (0.98/82); *C. yucatanensis* (1.00/100) (Figures 2 and 3). The species delimitation analysis identified nine distinct species, all of which received values for Rosenberg's P (AB) lower than 0.01, supporting the results of the BI and ML analyses (Table 2).

4. Discussion

Systematics

The present study is the first analysis of scorpions in which COI mini-barcodes from museum specimens were used for species identification. Three nominal species and six OTUs, identified by Esposito (2011) and described as new species of the '*thorellii*' clade by Goodman et al. (2021), were confirmed by the analyses presented here, highlighting the efficacy of mini-barcodes for identifying morphologically similar cryptic species using specimens of variable age and preservation [22,23].

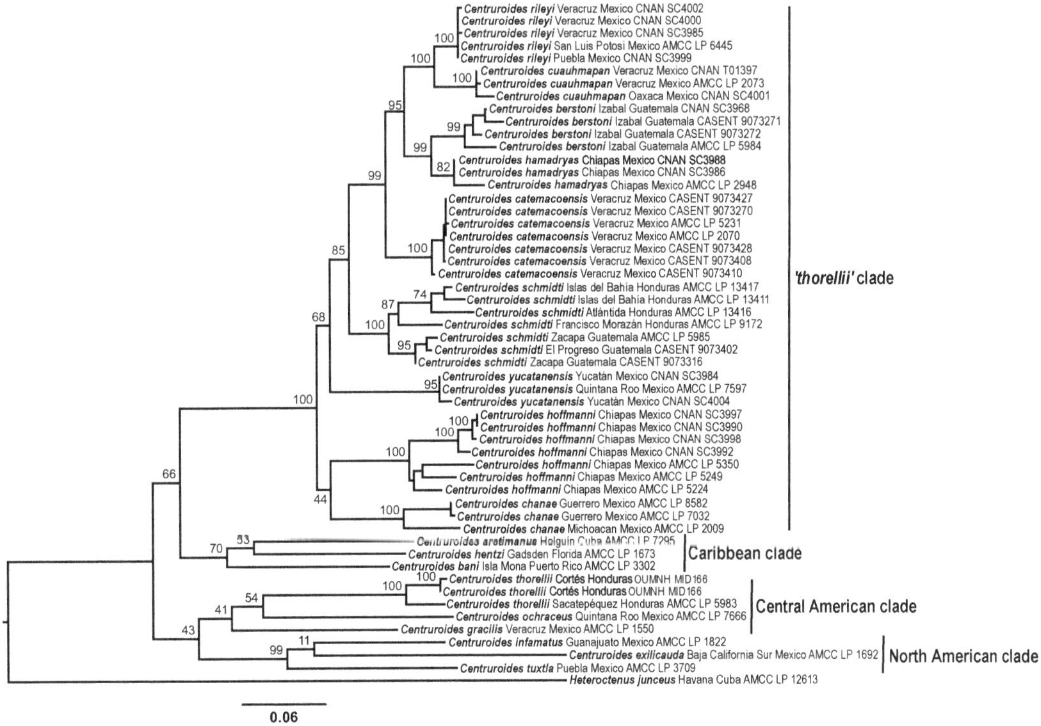

Figure 2. Phylogeny of the arboreal Neotropical '*thorellii*' clade of *Centruroides* Marx, 1890 bark scorpions (Buthidae C.L. Koch 1837) based on maximum likelihood analysis of 112 morphological characters and 1078 base pairs of DNA sequence from the mitochondrial Cytochrome *c* Oxidase Subunit I gene for 53 specimens of the ingroup with outgroup species of *Centruroides* and *Heteroctenus junceus* (Herbst, 1800). Material was deposited in the following collections: Ambrose Monell Cryo Collection (AMCC) at the American Museum of Natural History, New York; Colección Nacional de Arácnidos, Instituto de Biología (CNAN), Universidad Nacional Autónoma de Mexico, Mexico City; California Academy of Sciences (CASENT), San Francisco; Oxford University Museum of Natural History (OUMNH), UK.

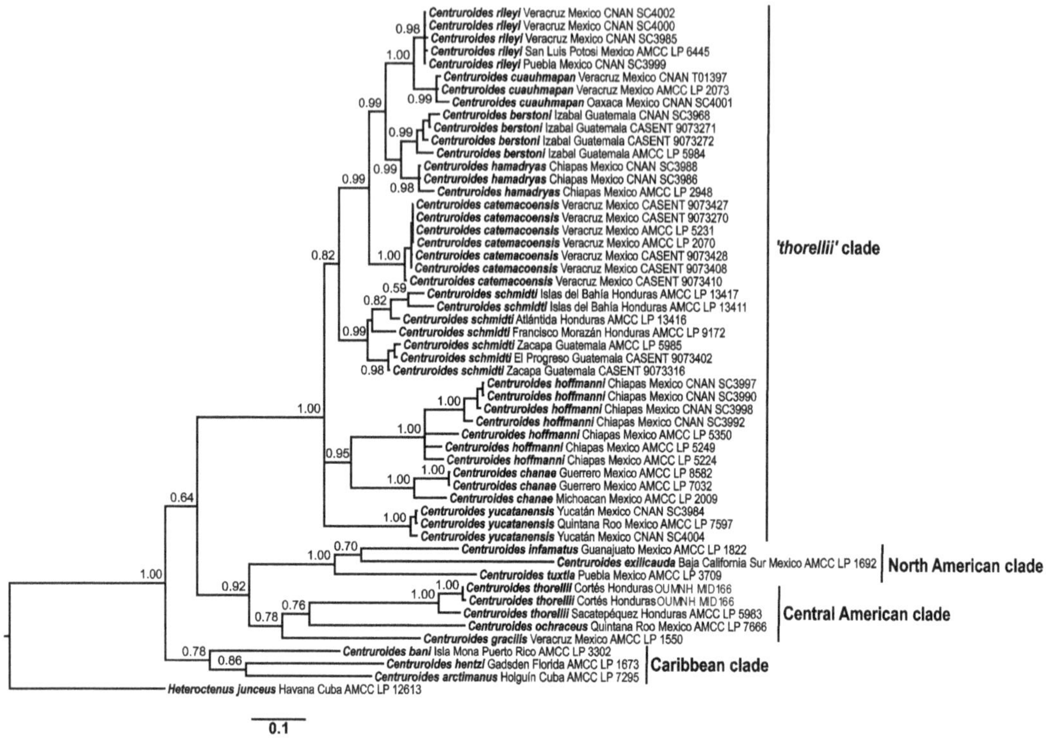

Figure 3. Phylogeny of the arboreal Neotropical '*thorellii*' clade of *Centruroides* Marx, 1890 bark scorpions (Buthidae C.L. Koch 1837) based on Bayesian analysis of 112 morphological characters and 1078 base pairs of DNA sequence from the mitochondrial Cytochrome *c* Oxidase Subunit I gene for 53 specimens of the ingroup with outgroup species of *Centruroides* and *Heteroctenus junceus* (Herbst, 1800). Material was deposited in the following collections: Ambrose Monell Cryo Collection (AMCC) at the American Museum of Natural History, New York; Colección Nacional de Arácnidos, Instituto de Biología (CNAN), Universidad Nacional Autónoma de Mexico, Mexico City; California Academy of Sciences (CASENT), San Francisco; Oxford University Museum of Natural History (OUMNH), UK.

Although this study demonstrates the effectiveness of mini-barcodes in sequencing specimens preserved in suboptimal conditions, several caveats must be taken into consideration.

Mitochondrial genes such as COI are regarded as the gold standard for DNA barcoding as their high mutation rates allow the sorting of specimens to species level with relative ease, low cost, and efficiency [46,47]. However, recent studies demonstrated that short mini-barcodes (<200 bp) perform more poorly than barcodes of medium length (>200 bp), necessitating other lines of evidence to differentiate species [45]. The addition of slow-mutating nuclear genes is necessary to resolve deep basal evolutionary relationships [21,22,27]. The incorporation of both mitochondrial and nuclear genes with differing mutation rates is desirable as it improves the support and stability of nodes.

The ML and BI trees presented here received low support for the internal nodes, most likely due to the absence of a nuclear gene. Successful amplification of nuclear genes proved fruitless, even when attempting to amplify shorter fragments such as the Internal Transcribed Spacer (ITS2, 350 bp) and Histone 3 (H3, 300 bp) [48,49]. The shorter, 125-bp mini-barcode primers would have provided insufficient support for species delimitation in the absence of other lines of evidence. The inclusion and analysis of a morphological character matrix with the COI data allowed an integrative approach to species delimitation in the '*thorellii*' clade.

Table 2. Genetic distances and species delimitation metrics among and within species of the arboreal Neotropical '*thorellii*' clade of *Centruroides* Marx, 1890 bark scorpions (Buthidae C.L. Koch 1837). Definitions modified from Masters et al. (2011): Intra Dist: average pairwise tree distance among species; Inter Dist: average pairwise tree distance within species; Intra/Inter ratio of Intra Dist to Inter Dist; P ID (Strict): mean probability, with 95% confidence interval (CI), of correct identification of unknown specimen using placement on tree; P ID (Liberal): mean probability, with 95% confidence interval (CI), of correct identification of unknown specimen using BLAST (best sequence alignment), DNA Barcoding (closest genetic distance), or placement on tree; Av (MRCA-tips): mean distance between most recent common ancestor of species and its conspecific members; P (RD): probability that clades have observed degree of distinctiveness; P (AB): probability that clades are reciprocally monophyletic.

Putative Species	Specimens	Closest Species	Intra Dist	Inter Dist	Intra/Inter	P ID (Strict)	P ID (Liberal)	Av (MRCA-Tips)	P (RD)	P (AB)
1: *C. rileyi*	5	**2:** *C. cuauhmapan*	0.007	0.055	0.12	0.85 (0.73, 0.98)	0.97 (0.87, 1.0)	0.003	0.05	0.01
2: *C. cuauhmapan*	3	**1:** *C. rileyi*	0.002	0.055	0.36	0.55 (0.37, 0.73)	0.80 (0.66, 0.95)	0.010	0.19	0.01
3: *C. berstoni*	4	**4:** *C. hamadryas*	0.035	0.075	0.47	0.55 (0.41, 0.70)	0.83 (0.72, 0.94)	0.022	0.93	0.01
4: *C. hamadryas*	3	**3:** *C. berstoni*	0.016	0.075	0.22	0.65 (0.47, 0.82)	0.88 (0.74, 1.0)	0.081	0.13	0.01
5: *C. catemacoensis*	7	**1:** *C. rileyi*	0.007	0.110	0.06	0.92 (0.81, 1.0)	0.98 (0.92, 1.0)	0.010	0.05	5.50×10^{-7}
6: *C. schmidti*	7	**5:** *C. catemacoensis*	0.073	0.161	0.46	0.71 (0.61, 0.82)	0.90 (0.84, 0.96)	0.045	0.59	4.50×10^{-8}
7: *C. hoffmanni*	7	**5:** *C. chanae*	0.069	0.251	0.28	0.81 (0.71, 0.92)	0.93 (0.87, 0.99)	0.05	1.0	1.85×10^{-3}
8: *C. chanae*	3	**9:** *C. schmidti*	0.058	0.251	0.23	0.64 (0.46, 0.81)	0.88 (0.73, 1.0)	0.042	1.0	1.85×10^{-3}
9: *C. yucatanensis*	3	**8:** *C. schmidti*	0.014	0.197	0.07	0.74 (0.57, 0.92)	0.97 (0.82, 1.0)	0.008	0.05	4.20×10^{-6}

Importantly, in most cases, multiple specimens of each species were represented by both partial COI and mini-barcodes, allowing more accurate alignments. Future analyses using mini-barcodes should include additional datasets for species identification if possible [46,50].

Diversification

The complex topography of the Mexican and Central American landscapes together with the ecological specialization of these scorpions are hypothesized to have driven diversification in the '*thorellii*' clade, which appears to display high levels of endemism. The arboreal habits of the clade confer a unique niche safe from predation by larger ground-dwelling scorpion species as well as other ground-dwelling predators [51]. Species such as *C. berstoni*, *C. catemacoensis*, and *C. hamadryas* exhibit very specific temperature, humidity, and substrate preferences [52]. Evidence of population structure among habitats but genetic admixture within habitats suggests these species are not constrained by dispersal within a given habitat but have narrow ecological requirements preventing them from dispersing between habitats.

The analyses identified geographically structured intraspecific genetic variation, despite morphological conservatism. For example, *C. hoffmanni* displayed a genetic structure among specimens collected less than 50 km apart. This species occurs predominantly within the Chiapas Depression, which acts as a corridor for some species and a barrier for others [53]. The results of the species delimitation analyses suggested all individuals of *C. hoffmanni* are conspecific, as genetic differentiation is minimal. The specialized arboreal microhabitat of *C. hoffmanni* suggests dispersal is limited across its broad geographical range, resulting in genetic divergence among populations occupying heterogeneous habitats, from evergreen and broadleaf forests to savannahs and scrub, across the Chiapas Depression.

A similar pattern was apparent with *C. schmidti*, in which divergence was evident among specimens from Guatemala and Honduras. Localities within the neighboring El Progresso and Zacapa departments of Guatemala share contiguous pine savannah and thorn forests. Localities within Honduras originated from Atlántida and Islas del Bahía, tropical coastlines with offshore islands characterized by broadleaf forests which receive high rainfall.

A third example is provided by *C. chanae*, in which divergence was evident between specimens separated by the Balsas Depression, an arid basin covering part of southwestern Guerrero and most of northwestern Michoacán in southern Mexico, which represents a barrier for other arachnid species [54–57]. Future studies should sample the entire ranges of *C. chanae*, *C. hoffmanni*, and *C. schmidti* to better understand the population structure within these species and investigate the possibility of additional cryptic species.

In contrast to *C. chanae*, *C. hoffmanni*, and *C. schmidti*, little genetic divergence was evident among specimens of *C. rileyi* distributed over a large area of northeastern Mexico, in the states of San Luis Potosí and Tamaulipas, suggesting unrestricted gene flow among populations. This may be due to the absence of barriers to dispersal across the homogenous Tamaulipan mezquital, a subhumid xeric shrubland biome which encompasses most of the lowlands of both states [58].

5. Conclusions

Systematics has reached a critical point in the 21st century where the rate of species description must outpace the decline of biodiversity [46]. Specimens must be rapidly sorted to the species level to infer large-scale biodiversity patterns. Such baseline work is essential to determine changes in species richness caused by climate change and habitat destruction. Luckily, vast numbers of specimens, including many undescribed species, are already housed within natural history collections; the challenge comes in developing faster, cheaper, and more efficient molecular sorting methods to harness data for their delimitation and identification, regardless of the age or preservation of the specimen [47].

Mini-barcodes have been lauded for species-level identification in degraded museum material due to their low cost on short-read sequencing platforms (e.g., Illumina) and their similar performance to complete barcodes [47]. This study presents the first use of mini-barcodes to delimit arachnid species from museum material and highlights the potential of mini-barcodes for species delimitation in arthropods more broadly. With roughly 28 million species remaining to be described [59], mini-barcodes provide another tool in the taxonomist's toolkit to assist with documenting the world's biodiversity while the opportunity remains.

Author Contributions: Conceptualization, L.A.E. and A.M.G.; methodology, L.A.E., L.P. and A.M.G.; software, L.A.E.; validation, L.A.E. and L.P.; formal analysis, A.M.G.; investigation, A.M.G. and L.A.E.; resources, L.A.E. and L.P.; data curation, A.M.G. and L.P.; writing—original draft preparation, A.M.G.; writing—review and editing, A.M.G., L.A.E. and L.P.; visualization, A.M.G., L.A.E. and L.P.; supervision, L.A.E. and L.P.; project administration, L.A.E.; funding acquisition, A.M.G., L.A.E. and L.P. All authors have read and agreed to the published version of the manuscript.

Funding: The research presented herein comprised part of the MSc thesis of A.M. at San Francisco State University and the CAS, partially supported by a grant from the Vincent Roth Fund for Systematic Research of the American Arachnological Society and a Collections Study Grant from the AMNH. Some data presented herein emanated from the PhD dissertation of L.A.E. at the City University of New York (CUNY) and the AMNH, supported by a U.S. National Science Foundation (NSF) GK-12 Fellowship, a CUNY/NSF AGEP Grant, a CUNY Presidential Fellowship, a CUNY College Now Fellowship, and an NSF Postdoctoral Fellowship (1003087). Additional funding for this research was provided by a grant from the Theodore Roosevelt Memorial Fund of the AMNH to L.A.E.; an Ernst Mayr Award from the Museum of Comparative Zoology, Harvard University, to L.A.E.; an NSF Doctoral Dissertation Improvement Grant (DEB 0910147) to L.P. and L.A.E.; an NSF grant (DEB 0413453) to L.P., and a grant from the Richard Lounsbery Foundation to L.P.

Institutional Review Board Statement: Ethical review and approval were waived for this study, due to experimentation on nonliving invertebrate museum specimens.

Informed Consent Statement: Not applicable.

Data Availability Statement: The data presented in this study are openly available in the Dryad digital repository (datadryad.org) at (doi:10.5061/dryad.fttdz08t2; accessed on 18 June 2021).

Acknowledgments: We thank the following for assisting with fieldwork or donating material used in the study: L. Allen, A.J. Ballesteros, D. Barrales-Alcalá, M.A. Barrios-Izás, P. Berea, G. Bonilla, J.L. Castelo, D. Chibras, M. Cordóva, M. Escalante, T. Gearheart, E. González-Santillán, J. Gorneau, J.H. Huff, A. Jaimes, M.K. Lippey, S. Longhorn, R. Monjaraz-Ruedas, H. Montaño, G. Montiel-Parra, D. Ortíz, R. Paredes, J. Ponce-Saavedra, M. Ramírez, M.E. Soleglad, A. Tietz, A. Valdez, M. van Dam, G. Villegas, C. Víquez, and H. Yamaguti; R. Coates Lutes for use of the Los Tuxtlas Field Station; E.S. Volschenk for providing unpublished morphological characters; D. Casellato, P. Rubi, and T. Sharma for generating some of the DNA sequence data at the AMNH; A. Lam, M. van Dam, L. Bonomo, and S.F. Loria for assistance with DNA sequencing and phylogenetic analysis at the CAS; P. Colmenares for logistical support with collections at the AMNH, and four reviewers for constructive comments on a previous draft of the manuscript.

Conflicts of Interest: The authors declare no conflict of interest.

Appendix A

Specimens and tissue samples from which DNA was extracted and morphology was examined for phylogenetic analysis of the arboreal Neotropical '*thorellii*' clade of *Centruroides* Marx, 1890 bark scorpions (Buthidae C.L. Koch 1837) and outgroup species of *Centruroides* and *Heteroctenus junceus* (Herbst, 1800). Material was deposited in the following collections: Ambrose Monell Cryo Collection (AMCC) at the American Museum of Natural History, New York; Colección Nacional de Arácnidos, Instituto de Biología (CNAN), Universidad Nacional Autónoma de México, Mexico City; California Academy of Sciences (CASENT), San Francisco; Oxford University Museum of Natural History (OUMNH), UK.

Outgroup

Centruroides arctimanus **Armas, 1976: CUBA:** *Holguín*: Município Holguín: 20°53′32.8″ N 76°17′04.7″ W, 3.x.2007, A. Tietz, 2 juv. (AMCC [LP 7295]).
Centruroides bani **Armas and Marcano Fondeur, 1987: U.S.A.:** *Puerto Rico:* Isla Mona, road between Sardiniera and airport: 18°03′48.4″ N 67°53′14.3″ W, 18.x.2009, L.A. Esposito and H. Yamaguti, 1 ♂ (AMCC [LP 3302]).
Centruroides exilicauda **(Wood, 1863): MEXICO:** *Baja California Sur:* Município Los Cabos: Cabo San Lucas, 15 mi. E, 12°53′23″ N 109°54′56″ W, 1.vi.1999, M.E. Soleglad, 1 ♂ (AMCC [LP 1692]).
Centruroides gracilis **(Latreille, 1804): MEXICO:** *Veracruz*: Município Actopan: Puente Nacional, Los Idolos, 19°25′44.9″ N 96°32′12.4″ W, 5.v.2006, O.F. Francke, P. Berea, and A. Ballesteros, 1 ex. (AMCC [LP 1550]).
Centruroides hentzi **(Banks, 1900): U.S.A.:** *Florida*: Gadsden County, 16.v.2000, T. Gearheart, 1 ♂ (AMCC [LP 1673]).
Centruroides infamatus **(C.L. Koch, 1844): MEXICO:** *Guanajuato*: Município Acámbaro: 2 km N Pueblo San Luis de los Agustinos, 20°12′27.9″ N 100°41′22.2″ W, 25.III.2006, O.F. Francke, A. Valdez, G. Villegas, H. Montaño, and C. Santibañez, 2 ♂ (AMCC [LP 1822]).
Centruroides ochraceus **(Pocock, 1898): MEXICO:** *Quintana Roo*: Município Benito Juárez: Puerto Morelos, Botanical Gardens, 20°50′42.1″ N 86°54′12.9″ W, 4.VIII.2007, G. Montiel and R. Paredes, 1 subad. ♂ (AMCC [LP 7666]).
Centruroides thorellii **(Kraepelin, 1891): GUATEMALA:** *Sacatepéquez*: Município Antigua: Parque Senderos de Alux, 15–20 km W of Guatemala City, 14°36′38.9″ N 90°38′20.8″ W, 30.VII.2008, J. Huff, C. Víquez, E. Agreda, and D. Ortíz, 1 ♀, 4 ex. (AMCC [LP 5983]). **HONDURAS:** *Cortés*: Município San Pedro Sula: Parque Nacional Cusuco, Cantilles Site, CA2 (Transect 2, subsite 3), 15°30′48″ N 88°14′23″ W, M. D'Sousa, K. Sagastume, and S. Longhorn, 2 ♀ (OUMNH [MID166]).
Centruroides tuxtla **Armas, 1999: MEXICO:** *Chiapas*: Município Suchiapa: La Vuelta del Alacran, 4.8 km from km 27 on road from Ocozocuautla to Villa Flores, 16°32′35.2″ N 93°12′09.8″ W, x.2004, R. Paredes, O.F. Francke, and G. Villegas, 1 ♂, 1 ♀ (AMCC [LP 3709]).
Heteroctenus junceus **(Herbst, 1800): CUBA:** *Guantánamo*: Município Baracoa: Alejandro Humboldt National Park, near El Yunque de Baracoa, 20°20′42.1″ N 74°33′59.1″ W, 370 m, 5.iv.2012, CarBio team, 1 ♂ (AMCC [LP 12613]).

Ingroup

Centruroides berstoni **Goodman et al., 2021: GUATEMALA:** *Departamento Izabal*: Município Livingston: Biotopo Chocón Machacas, 15°44′05.3″ N 88°54′57.2″ W, 15 m, 25.ix.2019, A.M. Goodman, 8 ♂, 1 ♀, 1 juv. ♂ (CASENT 9073271); Río Dulce, Hotel Tijax, 15°40′12.2″ N 89°00′27″ W, 49 m, 8.vii.2006, J. Huff, C. Víquez, and D. Ortíz, collected along trails through old secondary growth tropical forest using UV at night, 1 ♂ (AMCC [LP 5984]); 15°39′51.2″ N 89°00′14.6″ W, 17 m, 24.ix.2019, A.M. Goodman, collected along gravel road of Hacienda Tijax Parking Lot, flanked by bamboo groves and live fencing, 1 ♂ (CASENT 9073272); 15°44′05.3″ N 88°54′57.2″ W, 15 m, 25.ix.2019, A.M. Goodman, 8 ♂, 1 ♀, 1 juv. ♂ (CASENT 9073271). Município Morales: Morales, Finca Fiymeza, Sendero Anfibio, 15°24′24.1″ N 88°41′46.8″ W, 595 m, 17.viii.2017, D. Barrales and R. Monjaraz, 1 juv. ♀ (CNAN SC3968).
Centruroides catemacoensis **Goodman et al., 2021: MEXICO:** *Veracruz*: Município Catemaco: Estacion Biología Los Tuxtlas, UNAM, 18°35′05.6″ N 95°04′29.9″ W, 134 m, 26.viii.2005, O.F. Francke, M. Córdova, A. Jaimes, A. Valdez, and H. Montaño, 1 ♂ (AMCC [LP 5231]); 18°34′54″ N 95°04′54.6″ W, 134 m, 19.vii.2002, J. Ponce and O.F. Francke, 1 ♀ (AMCC [LP 2070]), 74–416 m, 17.vii–25.vii.2018, A.M. Goodman, J. Gorneau, and M.K. Lippey, 2 ♂ (CASENT 9073270, 9073427), 4 juv. ♀ (CASENT 9073408, 9073410, 9073428).
Centruroides chanae **Goodman et al., 2021: MEXICO:** *Guerrero*: Município Copala: Microondas Fogos, 16°33′59.5″ N 98°53′18.1″ W, 103 m, 22.vi.2007, O.F. Francke, M. Es-

calante, H. Montaño, and A. Ballesteros, 1 ♂ (AMCC [LP 7032]), 1 ♀ (AMCC [LP 8582]). *Michoacán*: Município Aquila: Faro de Bucerias, 18°21′08.3″ N 103°30′20.9″ W, 13 m, 10.iii.2002, J. Ponce, low deciduous forest, 3 ♂ (CNAN SC4005); 18°35′50.5″ N 103°30′04.3″ W, 221 m, 14.iv.2002, J. Ponce and E. González, low deciduous forest, 1 ♂ (AMCC [LP 2009]).

***Centruroides cuauhmapan* Goodman et al., 2021: MEXICO:** *Oaxaca*: Municípo San Juan Bautista Tuxtepec: 17 km from San Juan Bautista Tuxtepec, Cerro del Oro Dam, 17°59′55″ N 96°15′47.2″ W, 74 m, 23.v.1990, E. Barrera and A. Cadena, 1 ♂ (CNAN SC4001). *Veracruz*: Municípo Amatlán de los Reyes: Cañada Blanca, 18°55′43.5″ N 96°51′26″ W, 555 m, 18.vii.2002, E. González, found in coffee plantation in lowland rainforest, collected at night with UV light, 1 ♂ (AMCC [LP 2073]). Municípo Actopan: Los Idolos, 19°25′44.9″ N 96°32′12.4″ W, 112 m, 5.v.2006, O.F. Francke, P. Berea, and A.J. Ballesteros, collected with UV detection, 2 ♂paratypes (CNAN T01397).

***Centruroides hamadryas* Goodman et al., 2021: MEXICO:** *Chiapas*: Municipío Ocosingo: La Galleta, 2 km SE of Frontera Corozal, 16°48′12.7″ N 90°52′11.1″ W, 132–150 m, 28.iv.2004, R. Paredes and J.L. Castelo, collected with UV light detection, 2 ♂ (AMCC [LP 2948]); 16°49′55″ N 90°56′08″ W, 146 m, 7.iv.2005, A. Valdez, O.F. Francke, and A. Ballesteros, collected at night with UV lamp, 1 juv. ♂ (CNAN SC3986); 16°48′18.5″ N 90°54′25″ W, 114 m, 28.iv.2005, A. Valdez, O.F. Francke, and A. Ballesteros, urban area towards blue water bridge, collected with UV light detection, 2 ♂, 1 ♀, 1 juv. ♂, 1 juv. ♀ (CNAN SC3988).

***Centruroides hoffmanni* Armas, 1996: MEXICO:** *Chiapas*: Municípo Ángel Albino Corzo: 8 km from Siltepec, 18°48′33″ N 92°40′30.6″ W, 663 m, 17.viii.2007, C. Mayorga, G. Ortega, and L. Cervantes, 1 juv. ♀ (CNAN SC3990). Municípo Comitan: Parque Nacional Lagunas de Montebelo, 16°17′17″ N 91°56′16″ W, 1473 m, 3.ix.2005, O.F. Francke, M. Córdova, A. Jaimes, A. Valdez, and H. Montaño, 1 juv. ♂ (CNAN SC3992). Municípo La Concordia: Villa Corzo La Tigrilla, San Julián, Revolución Mexicana, 16°00′00″ N 92°50′47″ W, 544 m, 17.iii.2007, C. Mayorga, G. Ortega, and L. Cervantes; 1 ♂, 1 ♀ (CNAN SC3998). Municípo Tuxtla Gutiérrez: Las Delicias, 16°45′31.1″ N 93°06′26.4″ W, 529 m, 2.iii.2005, O.F. Francke, M. Córdova, A. Jaimes, A. Valdez, and H. Montaño, 1 ♀ (AMCC [LP 5224]); Gutiérrez, San Julián, Revolución Mexicana, 16°11′41″ N 93°01′16″ W, 544 m, 16.iii.2007, G. Ortega and A. Cervantes, 2 ♂, 3 ♀ (CNAN SC3997). Municípo Tzimol: Carretera [Hwy] Comitán–Tzimol Santa Rosa, 16°11′03.4″ N 92°16′59.3″ W, 632–730 m, 2.ix.2005, O.F. Francke, M. Córdova, A. Jaimes, A. Valdez, and H. Montaño, 1 ♂, 1 ♀ (AMCC [LP 5249]). Municípo Villaflores: Reserva de La Biosfera La Sepultura, 1 km SE of Ejido California, 16°15′14.2″ N 93°35′46.4″ W, 1009–1132 m, 30.viii.2005, O.F. Francke, M. Córdova, A. Jaimes, A. Valdez, and H. Montaño, 1 ♂ (AMCC [LP 5350]).

***Centruroides rileyi* Sissom, 1995: MEXICO:** *Puebla*: Municípo Cuetzalan el Progreso: Cuetzalan, Santiago Yancuitlalpan, 18°54′42.2″ N 98°35′15.3″ W, 2554 m, 19.v.1995, G. Oclogaig Barrzia, juvs (CNAN SC3999). *San Luis Potosí*: Municípo Axtlan de Terrazas: Axtlan de Terrazas, 21°25′34.9″ N 98°52′42″ W, 100 m, 28.iv.2006, O.F. Francke, A. Valdez, G. Villegas, and R. Paredes, 1 ♂ (AMCC [LP 6445]). *Veracruz*: Municípo Paplanta: Papantla 20°27′24.1″ N 97°18′56.1″ W, 2197 m, iii.2000, J.L. Castelo, 1 ♂ (CNAN SC4000). Municípo Tamiahua: Moralillo, Cerro Azul, 21°11′03.8″ N 97°44′49.6″ W, 153 m, 27.ii.2007, E. Barrera and L. Cervantes, 2 ♀ (CNAN SC3985), 1 juv. ♀ (CNAN SC4002).

***Centruroides schmidti* Sissom, 1995: GUATEMALA:** *Departamento El Progreso*: Municípo Rio Hondo: San Francisco Zapotitlan: Finca El Olvido, Las Minas, 15°02′04.8″ N 89°52′26.7″ W, 1214 m, 18.ix.2019, A.M. Goodman, L.A. Esposito, and L. Allen, 5 ♀, 1 juv. ♂, 1 juv. ♀ (CASENT 9073402). *Departamento Zacapa*: Municípo Rio Hondo: Bosque Pino, Guadalupe, Manta de Golpeo, 14°58′04.7″ N 89°24′47″ W, 751 m, 20.ix.2019, A.M. Goodman, M. Barrios, and M. van Dam, UV hand collection, found on oak and pine trees, 2–3 m high, 7 ♂, 1 ♀, 1 juv. ♂, 1 juv. ♀ (CASENT

9073278); Aldea Casas de Pinto, near turnoff for Zacapa at Rio Hondo, 15°01′38.2″ N 89°36′57.2″ W, 77 m, 13.vii.2006, J.H. Huff, semi-arid region with scrub forest and cacti, collected under rocks in shaded areas and at night using UV, 1 ad. (AMCC [LP 5985]). **HONDURAS:** *Departamento Atlántida*: Município La Ceiba: Parque Nacional Pico Bonito, Pico Bonito, trails from Visitor Centre and park entrance, 14°43′30.6″ N 86°44′11.5″ W, 184 m, 30.viii.2013, S. Longhorn, dense wet lowland tropical forest near large river, sweeping and beating, may have been on or under wood, day search, 1 juv. ♂ (AMCC [LP 13416]). *Departmento Francisco Morazán*: Município San Antonio de Oriente: E.A.P. Zamorano, Monte Redondo, Acuacultura, 13°39′59.6″ N 86°59′21″ W, 773 m, 23.ix.2008, C. Víquez, UV at night, 1 ♀ (AMCC [LP 9172]). *Departamento Islas de la Bahía*: Município Roatán: Cayos Cochinos, Cayos Menor, forest trails, 15°57′26.9″ N 86°30′03.3″ W, 101 m, 2.viii.2012, S. Longhorn, scrub oak forest, 1 ♀ (AMCC [LP 13411]); Isla Utila, 16°06′22.1″ N 86°54′08.1″ W, 12 m, 21.vii.2012, S. Longhorn, scrub forest/wet savannah, 1 ♀ (AMCC LP [13417]).

Centruroides yucatanensis **Goodman et al., 2021: MEXICO:** *Quintana Roo*: Município Benito Juarez: Puerto Morelos, Jardin Botanico Alfredo Barrera, 20°50′42.1″ N 86°54′12.9″ W, 38 m, 4.vii.2007, G. Montiel, R. Paredes, M. Ramírez, D. Chibras, and G. Bonilla, 1 ♂ (AMCC [LP 7597]), 2 juv. ♂ (CNAN SC3984). *Yucatán*: Município Felipe Carrillo Puerto: Cenote Chac-ha, 3.5 km N and 3 km E of Kalacmul, 20°04′40.3″ N 88°08′27.9″ W, 23 m, 9.vii.2007, R. Paredes, D. Chibras, and G. Montiel, 1 ♀ (CNAN SC4004).

Appendix B

Tissue samples, base pair lengths, localities, and GenBank accession codes of DNA sequences from the mitochondrial Cytochrome *c* Oxidase Subunit I gene used for phylogenetic analysis of the arboreal Neotropical '*thorellii*' clade of *Centruroides* Marx, 1890 bark scorpions (Buthidae C.L. Koch 1837) and outgroup species of *Centruroides* and *Heteroctenus junceus* (Herbst, 1800). Material was deposited in the following collections: Ambrose Monell Cryo Collection (AMCC) at the American Museum of Natural History, New York; Colección Nacional de Arácnidos, Instituto de Biología (CNAN), Universidad Nacional Autónoma de México, Mexico City; California Academy of Sciences (CASENT), San Francisco; Oxford University Museum of Natural History (OUMNH), UK. Sequences less than 150 bp in length were deposited in the Dryad digital repository (doi:10.5061/dryad.fttdz08t2). Sequences of 127–134 bp are mini-barcodes, 430–659 bp are partial COI, and sequences of 1078 bp are complete COI.

Species	Coll. Number	Locality	Length	GenBank Code
Outgroup				
H. junceus	AMCC [LP 12613]	Cuba: Guantánamo: Humboldt N. P.	1078	KY982192.1
C. bani	AMCC [LP 3302]	Puerto Rico: Isla Mona	1078	MK479164.1
C. exilicauda	AMCC [LP 1692]	Mexico: Baja California Sur: Cabo San Lucas	1078	KY982179.1
C. gracilis	AMCC [LP 1550]	Mexico: Veracruz: Los Idolos	1078	MK479175.1
C. hentzi	AMCC [LP 1673]	U.S.: Florida: Gadsden Co.	1078	MK479177.1
C. infamatus	AMCC [LP 1822]	Mexico: Guanajuato: Acámbaro	1078	KY982181.1
C. ochraceus	AMCC [LP 7666]	Mexico: Morelos: Puerto Morelos Bot. Gard.	1078	MK479194.1
C. thorellii	AMCC [LP 5983]	Guatemala: Sacatepéquez: Parque Alux	1078	MK479208.1
	OUMNH [MID166]	Honduras: Cortés: San Pedro Sula	642	MZ366335
			658	MZ366336
C. tuxtla	AMCC [LP 3709]	Mexico: Chiapas: La Vuelta de Alacran	1078	MK479209.1
Ingroup				
C. berstoni	CASENT 9073271	Guatemala: Izabal: Biotopo Chocón Machacas	430	MZ366346
	CASENT 9073272	Guatemala: Izabal: Hotel Tijax	621	MZ366345
	CNAN SC3968	Guatemala: Izabal: Morelos	658	MZ366344

Appendix B
Cont.

Species	Coll. Number	Locality	Length	GenBank Code
Ingroup				
C. catemacoensis	AMCC [LP 2070]	Mexico: Veracruz: Los Tuxtlas	1078	MZ429054
	AMCC [LP 5231]		1078	MZ429055
	CASENT 9073270		659	MZ366343
	CASENT 9073408		648	MZ366342
	CASENT 9073410		658	MZ366341
	CASENT 9073427		659	MZ366340
	CASENT 9073428		648	MZ366339
C. chanae	AMCC [LP 2009]	Mexico: Michoacán: Faro de Bucerias	1078	MZ429056
	AMCC [LP 7032]		1078	MZ429057
	AMCC [LP 8582]	Mexico: Guerrero: Microondas Fogos	1078	MZ429058
C. cuauhmapan	AMCC [LP 2073]	Mexico: Veracruz: Cañada Blanca	1078	MZ429059
	CNAN SC4001	Mexico: Oaxaca: Cerro del Oro	40	-
	CNAN T01397	Mexico: Veracruz: Los Idolos	127	-
C. hamadryas	AMCC [LP 2948]	Mexico: Chiapas: La Galleta	1078	MZ429060
	CNAN SC3986		127	-
	CNAN SC3988		127	-
C. hoffmanni	AMCC [LP 5224]	Mexico: Chiapas: Las Delicias	1078	MZ429061
	AMCC [LP 5249]	Mexico: Chiapas: Santa Rosa	1078	MZ429062
	AMCC [LP 5350]	Mexico: Chiapas: Res. Biosfera Sepultura	1078	MK479178.1
	CNAN SC3990	Mexico: Chiapas: Siltepec	134	-
	CNAN SC3992	Mexico: Chiapas: P. N. Lagunas de Montebelo	137	-
	CNAN SC3997	Mexico: Chiapas: Gutierrez	134	-
	CNAN SC3998	Mexico: Chiapas: Villa Corzo	138	-
C. rileyi	AMCC [LP 6445]	Mexico: San Luis Potosí: Axtlan de Terrazas	1078	KY982183.1
	CNAN SC3985	Mexico: Veracruz: Cerro Azul	127	-
	CNAN SC3999	Mexico: Puebla: Cuetzalan	127	-
	CNAN SC4000	Mexico: Veracruz: Papantla	127	-
	CNAN SC4002	Mexico: Veracruz: Cerro Azul	127	-
C. schmidti	AMCC [LP 13416]	Honduras: Atlántida: Pico Bonito	127	-
	AMCC [LP 13411]	Honduras: Isla del Bahía: Cayos Menor	1078	MZ429064
	AMCC [LP 13417]		1078	MZ429065
	AMCC [LP 5985]	Guatemala: Zacapa: Aldea casas de Pinto	1078	MZ429063
	AMCC [LP 9172]	Honduras: Francisco Morazán: E.A.P. Zamorano	1078	KY982184.1
	CASENT 9073316	Guatemala: Zacapa: Guadalupe	606	MZ366338
	CASENT 9073402	Guatemala: Zacapa: Las Minas	620	MZ366337
C. yucatanensis	AMCC [LP 7597]	Mexico: Quintana Roo: Puerto Morelos	1078	MK479201.1
	CNAN SC3984		127	-
	CNAN SC4004	Mexico: Yucatán: Cenote Chac-ha	127	-

Appendix C

Morphological characters and character states used in phylogenetic analysis of the arboreal Neotropical '*thorellii*' clade of *Centruroides* Marx, 1890 bark scorpions (Buthidae C.L. Koch 1837) and outgroup species of *Centruroides* and *Heteroctenus junceus* (Herbst, 1800). Morphological terminology follows Hjelle (1990) and Sissom (1990), except for carapace and metasomal carination, which follows Vachon (1952); trichobothria, which follows Vachon (1974); tergite and pedipalp carination, which follows Prendini (2000); book lung anatomy, which follows Kamenz and Prendini (2008); and ovariuterine anatomy, which follows Volschenk et al. (2008).

Carapace

1. Lateral ocular carinae: 0, present, distinct; 1, reduced to several granules; 2, absent (E.S. Volschenk and L. Prendini, unpublished data; Esposito et al. 2017, 2018).
2. Centrolateral carinae: 0, present; 1, absent (E.S. Volschenk and L. Prendini, unpublished data; Esposito et al., 2017, 2018).
3. Anterior centrosubmedian carinae: 0, present; 1, absent (E.S. Volschenk and L. Prendini, unpublished data; Esposito et al., 2017, 2018).
4. Posterior centrosubmedian carinae: 0, present; 1, absent (E.S. Volschenk and L. Prendini, unpublished data; Esposito et al., 2017, 2018).

5. Anteromedian notch: 0, present; 1, absent (Esposito, 2011).
6. Surface granulation density: 0, smooth; 1, sparsely granular; 2, densely granular medially; 3, densely granular throughout (Esposito, 2011).
7. Surface granulation texture: 0, weakly granular, shagreened; 1, large, rounded granules; 2, large, conical granules (Esposito, 2011).
8. Anterior median ocular sulcus: 0, absent; 1, wide; 2, narrow, deep (Esposito, 2011).
9. Median ocular sulcus: 0, absent; 1, wide; 2, narrow, deep (Esposito, 2011).
10. Posteromedian sulcus: 0, absent; 1, wide; 2, narrow, deep (Esposito, 2011).
11. Posteromarginal sulci: 0, absent; 1, present (Esposito, 2011).
12. Posterolateral sulci: 0, absent; 1, present (Esposito, 2011).
13. Anterior margin, carina: 0, absent; 1, smooth; 2, granular (Esposito, 2011).
14. Lateral margins, carina: 0, absent; 1, smooth; 2, granular (Esposito, 2011).
15. Posterior margin, carina between posterior centrosubmedian carinae: 0, absent; 1, smooth; 2, granular (Esposito, 2011).

Pedipalps

16. Chela prodorsal carina: 0, granular; 1, smooth; 2, absent (E.S. Volschenk and L. Prendini, unpublished data; Esposito, 2011; Esposito et al., 2017, 2018).
17. Chela retrodorsal carina: 0, granular; 1, smooth; 2, absent (E.S. Volschenk and L. Prendini, unpublished data; Esposito, 2011).
18. Chela retromedian carinae: 0, granular; 1, smooth; 2, absent; 3, vestigial, reduced to several sparse granules; ?, unknown (E.S. Volschenk and L. Prendini, unpublished data; Esposito, 2011; Esposito et al., 2017, 2018).
19. Chela retroventral accessory carina: 0, complete; 1, reduced; 2, absent (E.S. Volschenk and L. Prendini, unpublished data; Esposito, 2011).
20. Chela retroventral carinae: 0, granular; 1, smooth; ?, unknown (Esposito, 2011; Esposito et al., 2017, 2018).
21. Chela retrodorsal accessory carinae: 0, present; 1, absent (Esposito, 2011; Esposito et al., 2017, 2018).
22. Chela prodorsal accessory carinae: 0, present; 1, absent (Esposito, 2011; Esposito et al., 2017, 2018).
23. Patella dorsomedian carina: 0, present; 1, absent (Esposito, 2011).
24. Femur retromedian carinae: 0, small granules; 1, large, conical granules (Esposito, 2011).
25. Fixed finger, number of median denticle subrows: 0, eight; 1, nine; 2, ten or more; 3, seven plus fused proximal subrow; 4, six plus fused proximal subrow (Esposito, 2011; Esposito et al., 2017, 2018).
26. Fixed and movable fingers, supernumary granules; 0, absent; 1, present (Soleglad and Fet, 2003; Esposito, 2011; Esposito et al., 2017, 2018).
27. Movable finger, number of median denticle subrows: 0, eight; 1, nine; 2, eleven; 3, thirteen or more; 4, seven plus fused proximal subrow (Soleglad and Fet, 2003; Prendini, 2004; Esposito, 2011; Esposito et al., 2017, 2018).
28. Chela shape (♂): 0, incrassate (bulbous or swollen); 1, slender (Prendini, 2001, 2004; Esposito, 2011; Esposito et al., 2017, 2018).
29. Chela shape (♀): 0, incrassate (bulbous or swollen); 1, slender (Prendini, 2001; Esposito, 2011; Esposito et al., 2017, 2018).
30. Chela movable finger, proximal lobe (♂): 0, present; 1, absent (Prendini, 2001; Esposito, 2011; Esposito et al., 2017, 2018).
31. Patella prolateral surface, setation: 0, long, dense setae; 1, short, sparse setae (E.S. Volschenk and L. Prendini, unpublished data; Esposito, 2011).

Legs

32. Leg I, tarsal setation: 0, short, dense setae; 1, long, dense setae; 2, sparse setae (E.S. Volschenk and L. Prendini, unpublished data; Esposito, 2011; Esposito et al., 2017, 2018).
33. Leg IV, tarsal setation: 0, short, dense setae; 1, long, dense setae; 2, sparse setae (Esposito, 2011; Esposito et al., 2017, 2018).
34. Legs I–IV, trochanter lateral margin carina: 0, absent; 1, smooth; 2, granular (Esposito, 2011).
35. Telotarsal ungues: 0, long, slightly curved; 1, hooked (Esposito, 2011).

Pectines

36. Pectinal tooth shape: 0, elongate; 1, rounded, spade-like (Esposito, 2011; Esposito et al., 2017, 2018).
37. Proximal teeth, nodules on dorsal surface: 0, one; 1, multiple; 2, absent (Esposito, 2011; Esposito et al., 2017, 2018).
38. Dorsal fulcra: 0, present; 1, reduced (Esposito, 2011).
39. Tympanum-like expansion between lamellae and teeth: 0, absent; 1, present (Esposito, 2011).
40. Proximal fulcra, setal count: 0, one; 1, two; 2, three; 3, four; 4, six or more; 5, none; ?, unknown. (Esposito, 2011; Esposito et al., 2017, 2018).
41. Pectinal plate anterior margin, sulcus (♂): 0, present; 1, absent (Esposito, 2011; Esposito et al., 2017, 2018).
42. Pectinal plate, posterior margin (♂): 0, straight; 1, convex; 2, concave (Esposito, 2011; Esposito et al., 2017, 2018).
43. Median pectinal plate depression (♂): 0, present; 1, absent (Esposito, 2011; Esposito et al., 2017, 2018).
44. Lateral pectinal plate depression (♂): 0, present; 1, absent (Esposito, 2011).
45. Pectinal plate shape (♂): 0, square; 1, rectangular; 2, trapezoidal (Esposito, 2011).
46. Pectinal plate anterior margin, sulcus (♀): 0, present; 1, absent; ?, unknown (Esposito, 2011).
47. Pectinal plate posterior margin (♀): 0, straight; 1, slightly convex; 2, prominently rounded; 3, concave; ?, unknown (Esposito, 2011).
48. Pectinal plate depressions (♀): 0, absent; 1, single wide median depression; 2, paired lateral depressions; 3, single small, deep median depression (pinhole); ?, unknown (Esposito, 2011).

Sternites

49. Sternite VI, ventromedian carina: 0, absent; 1, granular; 2, smooth (E.S. Volschenk and L. Prendini, unpublished data; Esposito, 2011; Esposito et al., 2017, 2018).
50. Sternite V, setation (♂): 0, absent; 1, present, setal base not situated in pits (surface smooth); 2, present, setal base situated in pits (surface punctate) (Esposito, 2011).
51. Sternite VI, ventrolateral carinae: 0, absent; 1, reduced to single granule; 2, present, more than one granule (E.S. Volschenk and L. Prendini, unpublished data; Esposito, 2011; Esposito et al., 2017, 2018).

Tergites

52. Tergites III–VI, dorsolateral carinae: 0, present; 1, absent (E.S. Volschenk and L. Prendini, unpublished data; Esposito, 2011; Esposito et al., 2017, 2018).
53. Tergites III–VI, dorsosubmedian carinae: 0, absent; 1, vestigial; 2, distinct (Prendini, 2004; Esposito, 2011; Esposito et al., 2017, 2018).
54. Tergite VII, median carina: 0, narrow, granular carina; 1, broad, granular carina; 2, broad, smooth carina; 3, vestigial (E.S. Volschenk and L. Prendini, unpublished data; Esposito, 2011; Esposito et al., 2017, 2018).

Metasoma

55. Segment II, median lateral carinae: 0, complete; 1, posteriorly restricted; 2, absent (Esposito, 2011; Esposito et al., 2017, 2018).
56. Segment III, median lateral carinae: 0, complete; 1, posteriorly restricted; 2, absent (Esposito, 2011; Esposito et al., 2017, 2018).
57. Segment III, dorsolateral carinae, posterior granules: 0, similar to preceding granules; 1, larger than preceding granules, acuminate (E.S. Volschenk and L. Prendini, unpublished data; Esposito, 2011; Esposito et al., 2017, 2018).
58. Segment IV, median lateral carinae: 0, absent or obsolete; 1, present (Esposito, 2011; Esposito et al., 2017, 2018).
59. Segment V, posterior margin (anal rim) granulation: 0, present; 1, absent (Esposito, 2011; Esposito et al., 2017, 2018).
60. Segment V, dorsolateral carinae: 0, present; 1, absent (Esposito, 2011; Esposito et al., 2017, 2018).
61. Segment V, median lateral carinae: 0, present; 1, absent (Esposito 2011; Esposito et al., 2017, 2018).
62. Segment V, ventrolateral carinae: 0, present; 1, absent; ?, unknown (Esposito, 2011).
63. Segment V, ventromedian carinae: 0, absent or obsolete; 1, present (Esposito, 2011; Esposito et al., 2017, 2018)
64. Segment V, ventrosubmedian carinae: 0, absent or obsolete; 1, present (Esposito, 2011; Esposito et al., 2017, 2018).
65. Segment V, ratio of segment length to width (♂): 0, slightly elongated, length less than 2x width; 1, moderately elongated, length 2.5–3x width; 2, markedly elongated, length more than 3x width (Esposito, 2011; Esposito et al., 2017, 2018).
66. Segments I–IV, width: 0, narrowing posteriorly, segment I slightly wider than IV; 1, slightly widening posteriorly, segment I slightly narrower than IV; 2, markedly widening posteriorly, segment I much narrower than IV (Esposito, 2011; Esposito et al., 2017, 2018).
67. Metasoma length relative to length of prosoma and mesosoma (♂): 0, similar or slightly greater; 1, 1.5–2x; 2, more than 2x (Esposito, 2011; Esposito et al., 2017, 2018).
68. Metasoma length relative to length of prosoma and mesosoma (♀): 0, similar or slightly greater; 1, 1.5–2x; 2, more than 2x; ?, unknown (Lamoral, 1978; Prendini 2001, 2003; Esposito, 2011; Esposito et al., 2017, 2018).

Telson

69. Telson shape and length (♂): 0, spherical, length similar to width; 1, slightly ovate, length approximately 1.5x width; 2, ovate, length more than 2x width (Esposito, 2011; Esposito et al., 2017, 2018).
70. Telson vesicle, width in relation to width of metasomal segment V (♂): 0, similar; 1, narrower; 2, much narrower, less than half (Lamoral, 1978; Prendini 2001, 2003; Esposito, 2011; Esposito et al., 2017, 2018).
71. Telson vesicle, ventromedian carinae: 0, present; 1, absent (E.S. Volschenk and L. Prendini, unpublished data; Esposito, 2011; Esposito et al., 2017, 2018).
72. Telson vesicle, ventrolateral carinae: 0, present; 1, absent (E.S. Volschenk and L. Prendini, unpublished data; Esposito, 2011; Esposito et al., 2017, 2018).
73. Telson aculeus angle: 0, angled approximately 90° to vesicle; 1, angled less than 90° to vesicle (Esposito, 2011).
74. Telson vesicle, proximal margin, notch: 0, absent; 1, present, unmodified; 2, present, projecting vertically (Esposito, 2011).
75. Telson vesicle, lateral lobes: 0, absent; 1, present (Esposito, 2011).
76. Telson ventral surface: 0, granular; 1, smooth (Esposito, 2011).
77. Telson subaculear tubercle: 0, distinct, singular; 1, distinct, bifurcate; 2, obsolete, slight protuberance; 3, absent (Lamoral, 1980; Stockwell, 1989; Prendini, 2001, 2003; Esposito, 2011; Esposito et al., 2017, 2018).

Total length

78. Total length, sexual dimorphism: 0, male shorter than or similar to female; 1, male much longer than female; ?, unknown (Esposito, 2011; Esposito et al., 2017, 2018).

Ovariuterus

79. Number of loops ('cells'): 0, eight; 1, nine; ?, unknown (Volschenk et al., 2008; Esposito, 2011; Esposito et al., 2017, 2018).
80. Loop shape: 0, simple; 1, complex bridged; ?, unknown (Volschenk et al., 2008; Esposito, 2011; Esposito et al., 2017, 2018).

Book lungs

81. Lamellar surface: 0, slender venation; 1, ribbed venation; ?, unknown (Kamenz and Prendini, 2008; Esposito, 2011; Esposito et al., 2017, 2018).
82. Lamellar edge: 0, thorns; 1, smooth or wrinkled; ?, unknown (Kamenz and Prendini, 2008; Esposito, 2011; Esposito et al., 2017, 2018).
83. Spiracle, posterior margin: 0, hilocks; 1, subconical; ?, unknown (Kamenz and Prendini, 2008; Esposito, 2011; Esposito et al., 2017, 2018).

Color and infuscation

84. Cheliceral manus, reticulate infuscation: 0, present; 1, laterally restricted; 2, absent (Esposito, 2011).
85. Pedipalp segments, color pattern: 0, chela manus and patella similar to femur; 1, chela manus darker than femur; 2, chela manus and patella darker than femur; 3, chela manus paler than femur (Esposito, 2011).
86. Pedipalp femur, mottled infuscation: 0, present; 1, absent (Esposito, 2011).
87. Pedipalp patella, mottled infuscation: 0, present; 1, absent (Esposito, 2011).
88. Pedipalp chela manus, mottled infuscation: 0, present; 1, absent (Esposito, 2011).
89. Pedipalp chela fingers, mottled infuscation: 0, present; 1, absent (Esposito, 2011).
90. Tergites I–VI, mottled infuscation: 0, present; 1, absent (Esposito, 2011).
91. Tergites I–VI, lateral band of infuscation, shape: 0, absent; 1, narrow; 2, distinct, rectangular; 3, wide, almost touching lateral margins (Esposito, 2011).
92. Tergites I–VI, lateral band of infuscation, intensity: 0, absent; 1, faint; 2, mottled; 3, distinct (Esposito, 2011).
93. Tergites I–VI, infuscation: 0, eye-shaped pattern; 1, absent (Esposito, 2011).
94. Tergites I–VI, lateral margins infuscation: 0, distinct black line; 1, absent; ?, unknown (Esposito, 2011).
95. Tergites I–VI, median stripe of infuscation: 0, absent; 1, present (Esposito, 2011).
96. Tergite VII, color pattern: 0, similar to other tergites; 1, paler than other tergites (Esposito, 2011).
97. Carapace, mottled infuscation: 0, present; 1, absent (Esposito, 2011).
98. Carapace, bands of infuscation: 0, absent; 1, two broad bands; 2, four narrow lines (Esposito, 2011).
99. Carapace, interocular triangle infuscation: 0, present; 1, absent (Esposito, 2011).
100. Metasomal segments I–V, ventral surfaces, mottled infuscation: 0, present; 1, absent (Esposito, 2011).
101. Metasomal segments I–V, lateral surfaces, mottled infuscation: 0, present; 1, absent (Esposito, 2011).
102. Metasomal segments I–V, ventromedian stripe of infuscation: 0, present; 1, absent (Esposito, 2011).
103. Metasomal segment V, color pattern: 0, similar to preceding segments; 1, darker than preceding segments (Esposito, 2011; Esposito et al., 2017, 2018).
104. Telson, lateral bands of infuscation: 0, present; 1, absent (Esposito, 2011).
105. Telson, median stripe of infuscation: 0, complete; 1, posteriorly confined; 2, absent (Esposito, 2011).
106. Telson, mottled infuscation: 0, present; 1, absent (Esposito, 2011).
107. Sternites III–VI, mottled infuscation: 0, present; 1, absent (Esposito, 2011).

108. Sternite V, pale surface (♂): 0, present; 1, absent; ?, unknown (Esposito, 2011).
109. Sternite V, pale surface (♀): 0, present; 1, absent; ?, unknown (Esposito, 2011).
110. Sternite VII, color pattern: 0, similar to other sternites; 0, paler than preceding sternites; 2, darker than preceding sternites (Esposito, 2011).
111. Legs I–IV, dorsal surfaces, mottled infuscation: 0, present; 1, absent (Esposito, 2011).
112. Legs I–IV, ventral surfaces, mottled infuscation: 0, present; 1, absent (Esposito, 2011).

References

1. Wandeler, P.; Hoeck, P.E.; Keller, L.F. Back to the future: Museum specimens in population genetics. *Trends Ecol. Evol.* **2007**, *22*, 634–642. [CrossRef] [PubMed]
2. Zimmermann, J.; Hajibabaei, M.; Blackburn, D.C.; Hanken, J.; Cantin, E.; Posfai, J.; Evans, T.C. DNA damage in preserved specimens and tissue samples: A molecular assessment. *Front. Zool.* **2008**, *5*, 18. [CrossRef]
3. Dean, M.D.; Ballard, J.W. Factors affecting mitochondrial DNA quality from museum preserved *Drosophila simulans*. *Entomol. Exp. Appl.* **2001**, *98*, 279–283. [CrossRef]
4. Evans, T. DNA damage. *NEB Expr.* **2007**, *2*, 1–3.
5. Cárdenas, P.; Moore, J.A. First records of *Geodia* demosponges from the New England seamounts, an opportunity to test the use of DNA mini-barcodes on museum specimens. *Mar. Biodivers.* **2019**, *49*, 163–174. [CrossRef]
6. Francis, C.M.; Borisenko, A.V.; Ivanova, N.V.; Eger, J.L.; Lim, B.K.; Guillén-Servent, A.; Kruskop, S.V.; Mackie, I.; Hebert, P.D. The role of DNA barcodes in understanding and conservation of mammal diversity in Southeast Asia. *PLoS ONE* **2010**, *5*, e12575. [CrossRef] [PubMed]
7. Hebert, P.D.; Penton, E.H.; Burns, J.M.; Janzen, D.H.; Hallwachs, W. Ten species in one: DNA barcoding reveals cryptic species in the neotropical skipper butterfly *Astraptes fulgerator*. *Proc. Natl. Acad. Sci. USA* **2004**, *101*, 14812–14817. [CrossRef] [PubMed]
8. Meusnier, I.; Singer, G.A.; Landry, J.F.; Hickey, D.A.; Hebert, P.D.; Hajibabaei, M. A universal DNA mini-barcode for biodiversity analysis. *BMC Genom.* **2008**, *9*, 214. [CrossRef]
9. Alberdi, A.; Garin, I.; Aizpurua, O.; Aihartza, J. The foraging ecology of the mountain long-eared bat *Plecotus macrobullaris* revealed with DNA mini-barcodes. *PLoS ONE* **2012**, *7*, e35692. [CrossRef]
10. Bagley, J.C.; De Aquino, P.D.; Breitman, M.F.; Langeani, F.; Colli, G.R. DNA barcode and minibarcode identification of freshwater fishes from Cerrado headwater streams in central Brazil. *J. Fish Biol.* **2019**, *95*, 1046–1060. [CrossRef]
11. Dong, W.; Liu, H.; Xu, C.; Zuo, Y.; Chen, Z.; Zhou, S. A chloroplast genomic strategy for designing taxon specific DNA mini-barcodes: A case study on ginsengs. *BMC Genet.* **2014**, *15*, 138. [CrossRef]
12. Dubey, B.; Meganathan, P.R.; Haque, I. DNA mini-barcoding: An approach for forensic identification of some endangered Indian snake species. *Forensic Sci. Int. Genet.* **2011**, *5*, 181–184. [CrossRef] [PubMed]
13. Sultana, S.; Ali, M.E.; Hossain, M.M.; Naquiah, N.; Zaidul, I.S. Universal mini COI barcode for the identification of fish species in processed products. *Food Res. Int.* **2018**, *105*, 19–28. [CrossRef]
14. Wannell, G.J.; Griffiths, A.M.; Spinou, A.; Batista, R.; Mendonça, M.B.; Wosiacki, W.B.; Fraser, B.; Wintner, S.; Papadopoulos, A.I.; Krey, G.; et al. A new minibarcode assay to facilitate species identification from processed, degraded or historic ray (Batoidea) samples. *Conserv. Genet. Resour.* **2020**, *18*, 659–668. [CrossRef]
15. Yang, F.; Ding, F.; Chen, H.; He, M.; Zhu, S.; Ma, X.; Jiang, L.; Li, H. DNA barcoding for the identification and authentication of animal species in traditional medicine. *Evid.-Based Complement. Altern. Med.* **2018**, *2018*, 5160254. [CrossRef] [PubMed]
16. de Armas, L.F. Presencia de *Centruroides schmidti* Sissom en el sureste de Mexico y descripcion de dos especies nuevas (Scorpiones: Buthidae). *Rev. Nicar. Entomol.* **1996**, *36*, 21–33.
17. Hoffmann, C.C. Monografias para la entomología médica de México. Monografia No. 2, Los escorpiones de México. Segunda parte: Buthidae. *An. Inst. Biol. Univ. Nac. Autón. México* **1932**, 3, 243–361.
18. Martín-Frías, E.; de Armas, L.F.; Paniagua-Solís, J.F. Redescription of the Mexican scorpion *Centruroides hoffmanni* Armas, 1996 (Scorpiones: Buthidae). *Euscorpius* **2005**, *2005*, 1–7. [CrossRef]
19. Sissom, W.D. Redescription of the scorpion *Centruroides thorelli* Kraepelin (Buthidae) and description of two new species. *J. Arachnol.* **1995**, *23*, 91–99.
20. Ponce-Saavedra, J.; Moreno-Barajas, R.J. El género *Centruroides* Marx 1890 (Scorpiones: Buthidae) en México. *Biológicas* **2005**, *7*, 42–51.
21. Esposito, L.A.; Prendini, L. Island ancestors and New World biogeography: A case study from the scorpions (Buthidae: Centruroidinae). *Sci. Rep.* **2019**, *9*, 3500. [CrossRef]
22. Esposito, L. Systematics and Biogeography of the New World Scorpion Genus *Centruroides* Marx, 1890 (Scorpiones: Buthidae). Ph.D. Dissertation, City University of New York, New York, NY, USA, 2011.
23. Goodman, A.; Prendini, L.; Francke, O.F.; Esposito, L.A. Systematic revision of the arboreal Neotropical "*thorellii*" clade of *Centruroides* Marx, 1890, bark scorpions (Buthidae C.L. Koch, 1837) with descriptions of six new species. *Bull. Am. Mus. Nat. Hist.* **2021**, *452*, 1–92.

24. Folmer, R.H.; Nilges, M.; Folkers, P.J.; Konings, R.N.; Hilbers, C.W. A model of the complex between single-stranded DNA and the single-stranded DNA binding protein encoded by gene V of filamentous bacteriophage M13. *J. Mol. Biol.* **1994**, *240*, 341–357. [CrossRef] [PubMed]
25. Arango, C.P.; Wheeler, W.C. Phylogeny of the sea spiders (Arthropoda, Pycnogonida) based on direct optimization of six loci and morphology. *Cladistics* **2007**, *23*, 255–293. [CrossRef]
26. Kearse, M.; Moir, R.; Wilson, A.; Stones-Havas, S.; Cheung, M.; Sturrock, S.; Thierer, T. Geneious Basic: An integrated and extendable desktop software platform for the organization and analysis of sequence data. *J. Bioinform.* **2012**, *28*, 1647–1649. [CrossRef]
27. Esposito, L.A.; Yamaguti, H.Y.; Pinto-da-Rocha, R.; Prendini, L. Plucking with the plectrum: Phylogeny of the New World buthid scorpion subfamily Centruroidinae Kraus, 1955 (Scorpiones: Buthidae) reveals evolution of three pecten-sternite stridulation organs. *Arthropod Syst. Phylogeny* **2018**, *76*, 87–122.
28. Hjelle, J.T. Anatomy and morphology. In *The Biology of Scorpions*; Polis, G.A., Ed.; Stanford University Press: Stanford, CA, USA, 1990; pp. 9–63.
29. Sissom, W.D. Systematics, biogeography and paleontology. In *The Biology of Scorpions*; Polis, G.A., Ed.; Stanford University Press: Stanford, CA, USA, 1990; pp. 64–160.
30. Vachon, M. *Études sur les Scorpions*; Institut Pasteur d'Algérie: Algiers, Algeria, 1952; 482p.
31. Vachon, M. Étude des caractères utilisés pour classer les familles et les genres de scorpions (Arachnides). 1. La trichobothriotaxie en arachnologie. Sigles trichobothriaux et types de trichobothriotaxie chez les scorpions. *Bull. Mus. Hist. Nat. Paris* **1974**, *140*, 857–958.
32. Prendini, L. Phylogeny and classification of the superfamily Scorpionoidea Latreille 1802 (Chelicerata, Scorpiones): An exemplar approach. *Cladistics* **2000**, *16*, 1–78. [CrossRef]
33. Kamenz, C.; Prendini, L. An atlas of book lung fine structure in the order Scorpiones (Arachnida). *Bull. Am. Mus. Nat. Hist.* **2008**, *316*, 1–359. [CrossRef]
34. Volschenk, E.S.; Mattoni, C.I.; Prendini, L. Comparative anatomy of the mesosomal organs of scorpions (Chelicerata, Scorpiones), with implications for the phylogeny of the order. *Zool. J. Linn. Soc.* **2008**, *154*, 651–675. [CrossRef]
35. Larkin, M.A.; Blackshields, G.; Brown, N.P.; Chenna, R.; McGettigan, P.A.; McWilliam, H.; Thompson, J.D. Clustal W and Clustal X version 2.0. *J. Bioinform.* **2007**, *23*, 2947–2948. [CrossRef]
36. Thompson, J.D.; Higgins, D.G.; Gibson, T.J. CLUSTAL W: Improving the sensitivity of progressive multiple sequence alignment through sequence weighting, position-specific gap penalties and weight matrix choice. *Nucleic Acids. Res.* **1994**, *22*, 4673–4680. [CrossRef] [PubMed]
37. Maddison, W.P.; Maddison, D.D. Mesquite: A Modular System for Evolutionary Analysis. Version 3.4. Available online: https://www.mesquiteproject.org (accessed on 20 June 2020).
38. Ronquist, F.; Huelsenbeck, J.P. MrBayes, 3, Bayesian phylogenetic inference under mixed models. *J. Bioinform.* **2003**, *19*, 1572–1574. [CrossRef]
39. Stamatakis, A. RAxML version, 8, A tool for phylogenetic analysis and post-analysis of large phylogenies. *J. Bioinform.* **2014**, *30*, 1312–1313. [CrossRef]
40. Miller, M.A.; Pfeiffer, W.; Schwartz, T. Creating the CIPRES Science Gateway for inference of large phylogenetic trees. In Proceedings of the Gateway Computing Environments Workshop (GCE), New Orleans, LA, USA, 14 November 2010; pp. 1–8.
41. Lanfear, R.; Frandsen, P.B.; Wright, A.M.; Senfeld, T.; Calcott, B. PartitionFinder, 2: New methods for selecting partitioned models of evolution for molecular and morphological phylogenetic analyses. *Mol. Biol. Evol.* **2017**, *34*, 772–773. [CrossRef] [PubMed]
42. Lewis, P.O. A likelihood approach to estimating phylogeny from discrete morphological character data. *Syst. Biol.* **2001**, *50*, 913–925. [CrossRef] [PubMed]
43. Alfaro, M.E.; Zoller, S.; Lutzoni, F. Bayes or bootstrap? A simulation study comparing the performance of Bayesian Markov Chain Monte Carlo sampling and bootstrapping in assessing phylogenetic confidence. *Mol. Biol. Evol.* **2003**, *20*, 255–266. [CrossRef]
44. Masters, B.C.; Fan, V.; Ross, H.A. Species delimitation—A Geneious plugin for the exploration of species boundaries. *Mol. Ecol. Resour.* **2011**, *11*, 154–157. [CrossRef]
45. Rosenberg, N.A. Statistical tests for taxonomic distinctiveness from observations of monophyly. *J. Evol.* **2007**, *61*, 317–323. [CrossRef]
46. Moritz, C.; Cicero, C.; Godfray, C. DNA barcoding: Promise and pit-falls. *PLoS Biol.* **2004**, *2*, e354. [CrossRef]
47. Yeo, D.; Srivathsan, A.; Meier, R. Longer is not always better: Optimizing barcode length for large-scale species discovery and identification. *Syst Biol.* **2020**, *69*, 999–1015. [CrossRef] [PubMed]
48. Larena, I.; Salazar, O.; Gonzalez, V.; Julian, M.C.; Rubio, V. Design of a primer for ribosomal DNA internal transcribed spacer with enhanced specificity for ascomycetes. *J. Biotechnol.* **1999**, *75*, 187–194. [CrossRef]
49. Colgan, D.J.; McLauchlan, A.; Wilson, G.D.F.; Livingston, S.P.; Edgecombe, G.D.; Macaranas, J.; Cassis, G.; Gray, M.R. Histone H3 and U2 snRNA DNA sequences and arthropod molecular evolution. *Aust. J. Zool.* **1998**, *46*, 419–437. [CrossRef]
50. Johnson, N.K.; Cicero, C. New mitochondrial DNA data affirm the importance of Pleistocene speciation in North American birds. *Evolution* **2004**, *58*, 1122–1130. [CrossRef]
51. Rodrigo, A.; Bertels, F.; Heled, J.; Noder, R.; Shearman, H.; Tsai, P. The perils of plenty: What are we going to do with all these genes? *Philos. Trans. R. Soc. B* **2008**, *363*, 3893–3902. [CrossRef]

52. Goodman, A.; Esposito, L. Niche partitioning in congeneric scorpions. *Invertebr. Biol.* **2020**, *139*, e12280. [CrossRef]
53. Johnson, J.D.; Mata-Silva, V.; García-Padilla, E.; Wilson, L.D. The herpetofauna of Chiapas, Mexico: Composition, physiographic distribution, and conservation status. *Mesoam. Herpetol.* **2015**, *2*, 272–329.
54. Towler, W.I. Phylogenetic Structure of Two Central Mexican *Centruroides* Species Complexes. Master's Thesis, Marshall University, Huntington, WV, USA, 2002.
55. Ponce Saavedra, J.; Francke, O.F. Una nueva especie de alacrán del género *Centruroides* Marx 1890 (Scorpiones, Buthidae) de la depresión del Balsas, México. *Acta Zool. Mex.* **2004**, *20*, 221–232.
56. Saavedra, J.P.; Sissom, W.D. A new species of the genus *Vaejovis* (Scorpiones, Vaejovidae) endemic to the Balsas basin of Michoacán, Mexico. *J. Arachnol.* **2004**, *32*, 539–544. [CrossRef]
57. Mendoza, J.; Francke, O.F. Systematic revision of *Brachypelma* red-kneed tarantulas (Araneae: Theraphosidae), and the use of DNA barcodes to assist in the identification and conservation of CITES-listed species. *Invert. Syst.* **2017**, *31*, 157–179. [CrossRef]
58. Meyer, S.E.; García-Moya, E.; Lagunes-Espinoza, L.D.C. Topographic and soil surface effects on gypsophile plant community patterns in central Mexico. *J. Veg. Sci.* **1992**, *3*, 429–438. [CrossRef]
59. Ødegaard, F. How many species of arthropods? Erwin's estimate revised. *Biol. J. Linn. Soc.* **2000**, *71*, 583–597. [CrossRef]

diversity

MDPI

Article

Characterization and Gene Expression of Vitellogenesis-Related Transcripts in the Hepatopancreas and Ovary of the Red Swamp Crayfish, *Procambarus clarkii* (Girard, 1852), during Reproductive Cycle

Chiara Manfrin [1,*,†], Moshe Tom [2,†], Massimo Avian [1], Silvia Battistella [1], Alberto Pallavicini [1] and Piero Giulio Giulianini [1]

1 Department of Life Sciences, University of Trieste, Via Licio Giorgieri 5, 34127 Trieste, Italy; avian@units.it (M.A.); battiste@units.it (S.B.); pallavic@units.it (A.P.); giulianini@units.it (P.G.G.)
2 Israel Oceanographic and Limnological Research, P.O.B 9753, Haifa 3109701, Israel; tom@ocean.org.il
* Correspondence: cmanfrin@units.it
† These authors contributed equally to this work.

Citation: Manfrin, C.; Tom, M.; Avian, M.; Battistella, S.; Pallavicini, A.; Giulianini, P.G. Characterization and Gene Expression of Vitellogenesis-Related Transcripts in the Hepatopancreas and Ovary of the Red Swamp Crayfish, *Procambarus clarkii* (Girard, 1852), during Reproductive Cycle *Diversity* **2021**, *13*, 445. https://doi.org/10.3390/d13090445

Academic Editors: Michael Wink and Gary H. Dickinson

Received: 9 August 2021
Accepted: 10 September 2021
Published: 16 September 2021

Publisher's Note: MDPI stays neutral with regard to jurisdictional claims in published maps and institutional affiliations.

Copyright: © 2021 by the authors. Licensee MDPI, Basel, Switzerland. This article is an open access article distributed under the terms and conditions of the Creative Commons Attribution (CC BY) license (https://creativecommons.org/licenses/by/4.0/).

Abstract: The major component of the animal egg yolk is the lipoglycoprotein vitellin, derived from its precursor vitellogenin (VTG), which is produced species-specifically in decapod crustaceans in the hepatopancreas and/or in the ovary of reproductive females. Previous studies on *Procambarus clarkii* vitellogenesis report the existence of two single VTGs. Here, from a multiple tissue transcriptome including ovaries and hepatopancreas of *P. clarkii*, we characterized four different VTG and two VTG-like transcriptomes encoding for the discoidal lipoprotein-high density lipoprotein/β-glucan binding protein (*dLp/HDL-BGBP*). The relative expression of the various genes was evaluated by quantitative Real-Time PCR in both the ovary and hepatopancreas of females at different reproductive stages (from immature until fully mature oocytes). These studies revealed tissue-specificity and a reproductive stage related expression for the VTGs and a constitutive expression in the hepatopancreas of *dLp/HDL-BGBP* independent from the reproductive stage. This study may lead to more detailed study of the vitellogenins, their transcription regulation, and to the determination of broader patterns of expression present in the female hepatopancreas and ovary during the vitellogenesis. These findings provide a starting point useful for two different practical aims. The first is related to studies on *P. clarkii* reproduction, since this species is highly appreciated on the market worldwide. The second is related to the study of new potential interference in *P. clarkii* reproduction to delay or inhibit the worldwide spread of this aggressively invasive species.

Keywords: *Procambarus clarkii*; Crustacea; Decapoda; vitellogenesis; digital gene expression analysis; discoidal lipoprotein; β-glucan binding protein

1. Introduction

Ovarian development is characterized by the accumulation of a major egg-yolk protein precursor, vitellogenin (VTG), that serves as a food reserve for the embryo. Since the 1970s, the formation of a lipoprotein complex or vitellogenin has been known to be a prerequisite to the constitution of the protein yolk in myriapods, crustaceans, insects, amphibians, and birds. This vitellogenin is associated with other prosthetic groups and synthetized outside the ovary, transported by the hemolymph and sequestered by the oocytes [1]. In crustaceans the development of the oocyte comprises two distinguished stages, named "primary vitellogenesis" characterized by glycoprotein granule accumulation and "secondary vitellogenesis" or vitellogenesis strictly speaking, that occurs solely during reproduction [2]. This latest phase, in crustaceans, is mainly heterosynthetic since the vitellogenin is carried in the hemolymph and sequestered by the oocytes [3].

The source of VTG in crustaceans has been controversial for many years. Several authors in the late 1960s suggested hemocytes as the site of VTG synthesis [4], but the two main tissues of VTG production are now globally accepted to be the ovaries as shown in *Callinectes sapidus* [5] and *Penaeus semisulcatus* [6], or the hepatopancreas, as found in *Charybdis feriata* [7], *Macrobrachium rosenbergii* [8], *Oziothelphusa senex* [9], and *Pandalus hypsinotus* [10]. In some other crustacean species, the VTG synthesis sites are both the hepatopancreas and ovary, as shown in *Carcinus maenas* [11], *Eriocheir sinensis* [12], *Macrobrachium nipponense* [13], *Marsupenaeus japonicus* [14], *Penaeus monodon* [15], and *Scylla paramamosain* [16], although the contribution of the ovary is relatively less than that of the hepatopancreas.

Another controversial question concerns the number of VTGs discovered in a single crustacean species, or rather whether one or more homologous VTGs participate together in crustacean reproduction. In the Arthropoda, there are three homologous genes that arose from ancient insect vitellogenin duplications and are known as *VTG-like-A*, *-B*, and *-C* in *Apis mellifera* [17] and four copies in *Solenopsis invicta* [18]. The function of these homologous VTGs is unclear [17]. During evolution, it seemed that crustaceans reduced these genes up to two VTG copies, which have been described in a variety of species [19–29].

In the present study, we used *Procambarus clarkii* (Girard, 1852) as a model organism to identify and investigate the expression of VTGs and VTG-like transcripts. The red swamp crayfish, native to Mexico and the United States, is considered a ubiquitous invasive species worldwide, while representing a valuable aquaculture resource on international markets [30–32]. It is an r-selected species with a ductile life cycle that breeds year-round, with variable recruitment peaks in summer, early winter, and spring [33,34]. A constant 100% of mature males observed in the North Italy population from July to September indicates possible successful mating during this season which may extend to the end of October at these latitudes [35].

Many studies focusing on VTG and *P. clarkii* reproduction refer to one single VTG in this decapod species [36–39]. To investigate this aspect, a comprehensive *P. clarkii* transcriptome assemblage consisting of 12 different tissues was previously constructed by one of us (Manfrin communication) and used here to identify the available *P. clarkii* VTG transcripts and those of the high-density lipoprotein/β glucan binding protein (*dLp/HDL-BGBP*, hereinafter referred to as *BGBP*). The later gene is responsible for transport of lipids and is also fundamental for the innate immune response of crustaceans [40,41]. The transcript characterization was followed by examination of their transcription patterns in the female ovary and hepatopancreas during the ovarian development period, observing their transcription level during the ovarian developmental period.

2. Materials and Methods

2.1. Ethical Note

The following experimental procedures are in accordance with current Italian law. No special permits were required for this study, as no endangered or protected species were involved. Individuals were kept under appropriate laboratory conditions to ensure their welfare and responsiveness. After completion of the experiments, the crayfish were euthanized by hypothermia.

2.2. Transcripts Identification

The identification of VTGs and general lipid carriers which share domains with VTG followed a similarity-based process based on the use of conserved domains, both in public repositories (i.e., NCBI) and using a comprehensive transcriptome assembly (hereafter defined as ATLAS). The ATLAS consisted of 12 different *P. clarkii* tissues, namely brain, heart, ventral ganglia, eyestalk, green gland, ovary, testis, hepatopancreas, muscle, Y-organ, gill, and hemocyte, Illumina sequenced (depth 2 × 100 bp) (unpublished results). CLC Genomics Workbench v.12 (Qiagen, Hilden, Germany) was used to map the reads from each tissue to the assembly for initial identification of tissues of interest, using the RNA-seq analysis tool with the following parameter settings: mismatch cost 2, gap insertion and deletion cost 3, end gaps 0, length fraction 0.95, similarity fraction 0.98. The protein molecular weight was estimated by Expasy-Compute pI/Mw and the presence of cleavage site was evaluated with SignalP 5.0.

2.3. Comparative Studies

The phylogeny of VTGs was inferred using the online tool NGPhylogeny.fr [42]. Briefly, the VTGs corresponding to all available complete proteins from the group "Crustacean" at GenBank, were aligned through MAFFT v.7 [43] and the alignment was curated by BMGE [44]. PhyML was used to infer the tree [45].

The ovaries of females were removed to photographically record the external morphology at various stages of development; photos were taken with an Olympus BX50.2.4.

Total RNA was extracted from tissues (ovary and hepatopancreas) frozen in liquid nitrogen using the TriReagent RNA isolation kit (Sigma-Aldrich, Saint Louis, MO, USA) according to the manufacturer's instructions. The resulting RNAs were further purified using the E.Z.N.A.™ MicroElute RNA Clean-up Kit (Omega Bio-Tek, Norcross, GA, USA). The amount of RNA was quantified spectrophotometrically using Nanodrop 2000 (Thermo Fisher Scientific, Waltham, MA, USA) and its quality was analyzed by capillary electrophoresis BioAnalyzer 2100 (Agilent, Santa Clara, CA, USA).

2.4. Transcript Expression by qRT-PCR

The relative gene expression of selected *VTGs* and *BGBPs* in *P. clarkii* was examined by quantitative Real-Time PCR (qRT-PCR), in both ovary and hepatopancreas of twenty females containing oocytes from the 2nd to 6th developmental stages [46]. Preparation of cDNAs was performed using M-MLV reverse transcriptase (Promega) and oligoT(18) primer starting from 1 μg of total RNA and following the manufacturer's instructions. Specific primers were designed using Primer3Web version 4.0.0 [47] and Oligo Calculator version 3.26 [48] to predict possible secondary structure and hairpin formation, as shown in Table 1. A dilution of 1/5 of the initial cDNA was used for qRT-PCR, which was performed in triplicate on the CFX96 Real-Time PCR detection system (Bio Rad, Hercules, CA, USA), using the following thermal profile: denaturation at 95 °C for 1′, 40 cycles at 95 °C for 15″ and 60 °C for 20″ and a final melting curve analysis from 65 °C to 95 °C with an increment of 0.5 °C every 5″. The 15 μL reaction mix contained the SsoAdvanced universal probes supermix (BioRad, Hercules, CA, USA), and the reverse and forward primers were used at the final concentration of 0.5 μM each.

Table 1. Primer sequences used for VTGs and *BGBPs* qRT-PCR experiments. PcVTG1-4 amplify VTGs, PcBGBP1 and 2 amplify discoidal lipoprotein-high density-β-glucan binding transcripts. PcEF1α: Elongation factor 1α. Pcβ-Actin: beta-Actin. PcGAPDH: Glyceraldehyde 3-phosphate dehydrogenase. The two letters Pc stands for *Procambarus clarkii*.

Primer ID	Forward 5′-3′	Reverse 5′-3′
PcVTG1	TCACCAGTCAACAGAGCAGC	TTCTCAGCACACCGAACTGC
PcVTG2	GAGGGTGGAAAGTCAGCTCC	ACAGTTCATCGCTCCTTCGG
PcVTG3	GTCGGACTGCAGATGAAGGG	AACAAAGCCTTCGGTTTGCG
PcVTG4	TCTGTTGAGAAAGCCGAGCC	TCTAGGCGTACTAGACCCAGC
PcBGBP1	CACACAAGACGAAGTGCTGC	TAAACGGTGCTAAGGGCTGG
PcBGBP2	CCCCTAGCATTAGCAACCCC	ACAACTCGGCGTCTTTCTCG
PcEF1α	AGATCTGAAACGTGGTTTTGTT	TCAATCTTTTCCAGAAGTTCGT
Pcβ-Actin	AGGGCGTGATGGTTGGTAT	CCGTGCTCAATGGGATATTT
PcGAPDH	CTCCATCTTTGACGCTAAGGC	GCACTATCCACCTTCTGCATG

Primer efficiencies were calculated using LinReg v.12.18 [49] for all primer sets used. As putative housekeeping genes, Elongation factor 1 alpha (*EF1α*), β-Actin and Glyceraldehyde 3-phosphate dehydrogenase (*GAPDH*) were tested by qRT-PCR following the same thermal cycle profile as the target transcripts. Their expression in all the experimental samples was evaluated using the BestKeeper [50], NormFinder [51] and geNorm [52] software to select at least the two most suitable and stable reference genes. Expression values of all examined transcripts for each experimental group and gene expression analysis were performed using Bio Rad CFX Manager v. 3.1 software.

3. Results

3.1. Characterization of VTG-like Transcripts

The full length VTG of *Cherax quadricarinatus* (GenBank identifier AAG17936; [19]) was used as initial reference to characterize the identified VTGs from the *P. clarkii* ATLAS transcriptome library. This reference sequence is 7944 nucleotides long and encodes for a protein of 2584 aa. Four domains were found by CDD [53] at NCBI: a vitellogenin_N domain (pfam 01347) in the interval 42–585 aa, a domain of unknown function (DUF1943, pfam09172) in the range 617–918 aa, a von Willebrand factor type D domain (2345–2491 aa) and a C8 domain (pfam08742) in the range 2530–2576 aa.

The four identified VTGs from *P. clarkii* (Figure S1) had an open reading frame length between 6024 bp (*VTG3*) and 8361 bp (*VTG1*) encoding for proteins ranging from 2007 to 2777 aa (Figure 1A), while the two identified BGBPs ranged for proteins of 4234 aa and 4806 aa (Figure 1B).

All the transcripts contained both a Vitellogenin_N super family and DUF1943 domains and various domains such as von Willebrand factor, type D domain shared among VTG from *C. quadricarinatus*, VTG1, VTG4 and BGBP2 from *P. clarkii* or C8 domain present in VTG from *C. quadricarinatus* and BGBP2.

The *VTG1* (GenBank ID OK142726) in *P. clarkii* consisted of an open reading frame of 8361 bp and a deduced protein of 2777 amino acids with a predicted molecular weight (MW) of 315 kDa. It has a putative cleavage site between position 19 and 20: VRA-AP (probability: 0.93) indicating the secretory nature of the molecule. The domains alongside their positions on the transcripts are indicated in Figure 1.

The *VTG2* (GenBank ID OK142727) consisted of an open reading frame of 6075 bp encoding for a protein of 2024 aa with a predicted MW of 270 KDa. A cleavage site was identified between position 18 and 19: ARA-AP (probability: 0.90).

The *VTG3* (GenBank ID OK142728) consisted of an open reading frame of 6024 bp encoding for a protein of 2007 aa with an estimated MW of 223 KDa. The cleavage site was identified between position 18 and 19: ARA-AP (probability: 0.90).

The *VTG4* (GenBank ID OK142729) consisted of an open reading frame of 7746 bp encoding for a peptide of 2571 aa (MW 289 KDa). A cleavage site was present between aa in position 18 and 19: ARA-AP (probability: 0.90).

The *VTG1* resulted the longest transcript followed by *VTG4*, whereas *VTG2*, and 3 had roughly the same length (Figure S2).

The high-density lipoprotein/β glucan binding protein (*BGBP*) is a pattern recognition protein responsible for the transport of lipids which is also fundamental for the innate immune response of crustaceans [40,41].

The *BGBP1* (GenBank ID OK142730) and the *BGBP2* (GenBank ID OK142731) were found to be longer than *VTGs* with 14,661 bp (encoding for a protein of 4886 aa) and 12,705 bp (protein of 4234 aa), respectively. DUF 1943 domain is the only domain shared with the VTGs and a VWF domain, type D is shared between VTG1 and BGBP2.

Phylogenetic inference (Figure 2) suggests that all vitellogenins descend from two lineages that hold Cladocera, Isopoda and Copepoda together (cluster in light grey in Figure 2) and Decapoda apart (cluster in dark grey in Figure 2). As expected, all VTGs found in *P. clarkii* belonged to the Decapoda lineage and formed a cluster with the VTG of *C. quadricarinatus* and *H. americanus*.

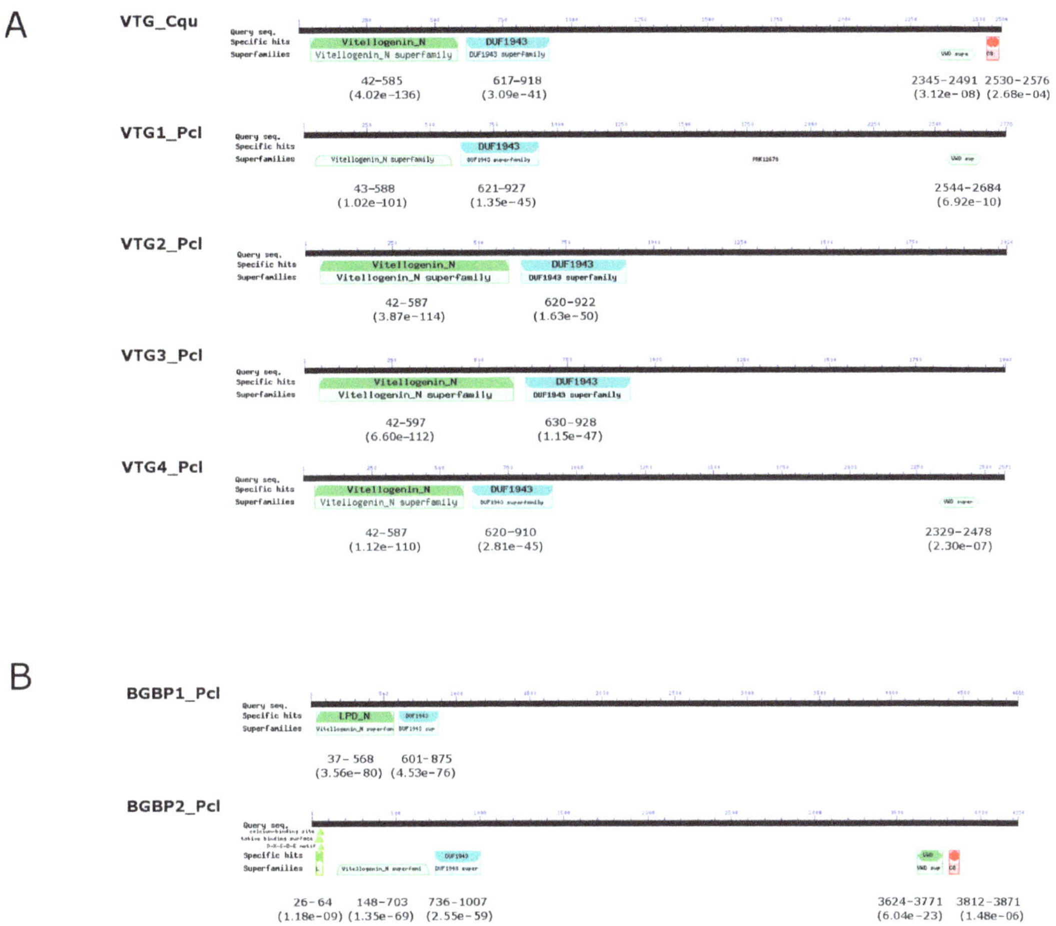

Figure 1. (**A**) Protein structure of VTGs and BGBPs from *P. clarkii* (Pcl) along with the VTG used as reference from *C. quadricarinatus* (Cqu). Domains are shown alongside their position in each protein and between brackets is reported the E-value obtained in CDD (NCBI). (**B**) Protein structure of BGBPs from *P. clarkii*.

3.2. Expression of VTG-Like Transcripts during Ovarian Development

The expression pattern of the 4 VTGs and the 2 *BGBPs* in the ovary and in the hepatopancreas of twenty *P. clarkii* females expressing all stages of oocyte development from stage 2 to 6 (Figure 3) was investigated. The oocyte development staging followed Alcorlo et al. [46] criteria and the expression level was evaluated by qRT-PCR.

Figure 4 shows the different *VTG*s and *BGBPs* expression profiles at the diverse ovarian stages. Three different gene expression patterns have been observed. The first one is characteristic of the *VTG1* with a gradual increase of its expression from ovarian development stage 4 exclusively in the ovary. The second one is characteristic of *VTG2-4*, with a marked expression in the hepatopancreas only at ovarian development stage 6. The third one concerns *BGBP1-2*, whose expression appears to be homogeneous and stable among all ovarian stages, but only in the hepatopancreas tissue.

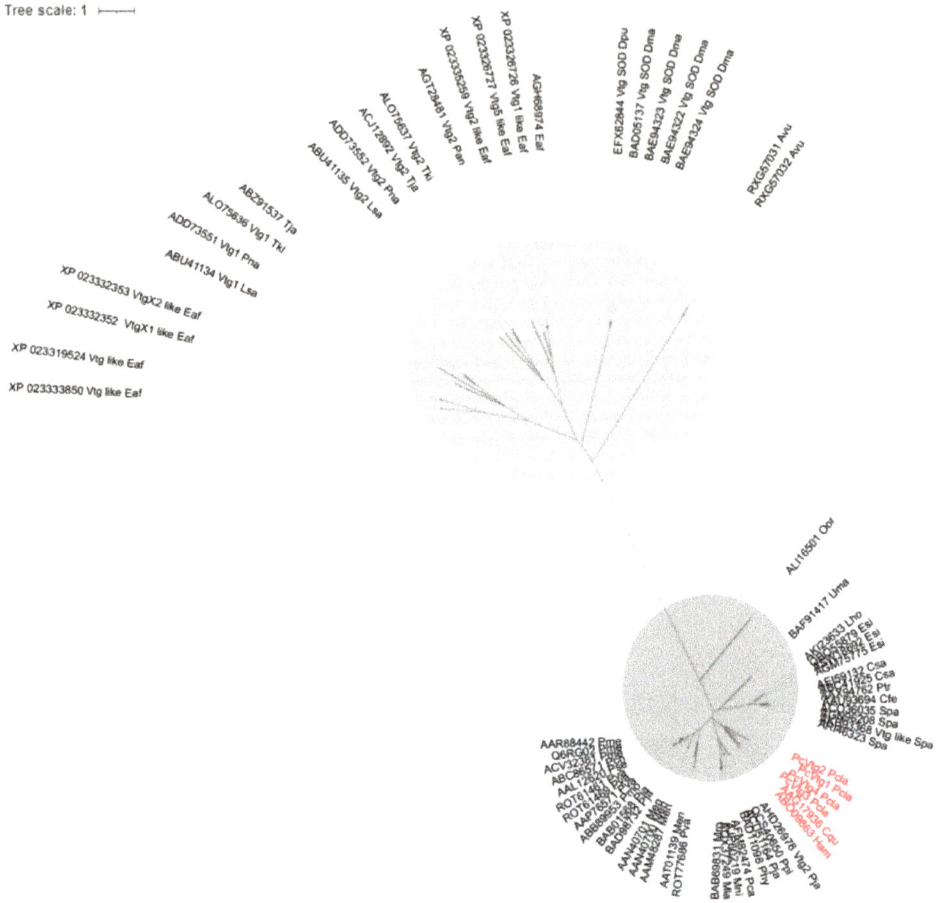

Figure 2. PhylML tree of complete VTGs stored in GenBank (until 4 July 2021). Each sequence is labelled with the corresponding GenBank ID followed by a three-character code: the first letter represents the genus name and the other two the species. *P. clarkii* VTGs characterized in the present study are indicated in red. In light grey is selected the cluster of Cladocera, Isopoda and Copepoda, while in dark grey the cluster of Decapoda.

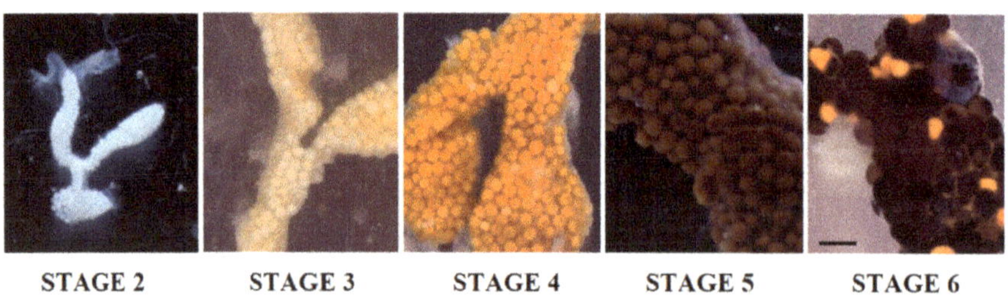

Figure 3. Appearance and coloration of the ovaries of *P. clarkii* in relation to their stage of maturity. The ovaries at different stages are shown at the same magnification. Stage 2 represents immature oocytes until stage 6, when the ovary is fully mature and active. Scale bar 2.5 mm.

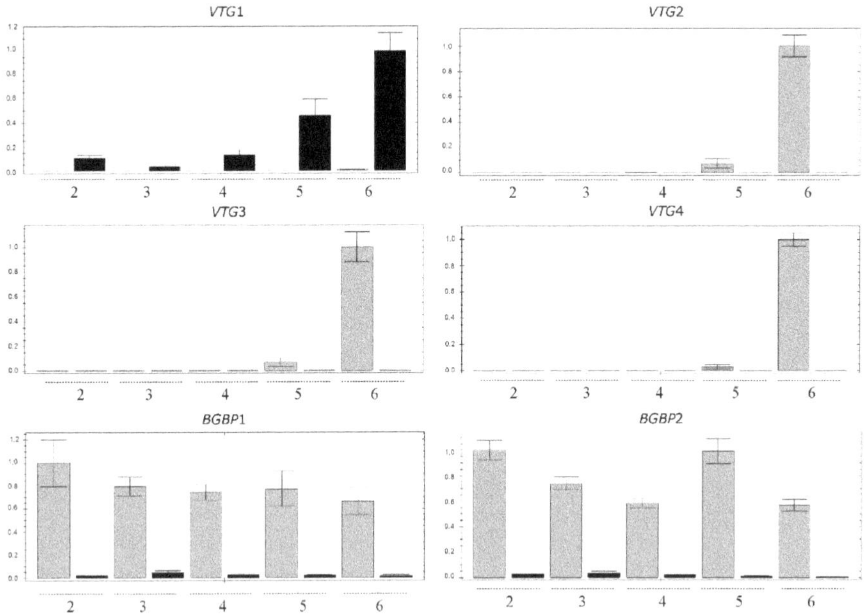

Figure 4. Relative VTGs and *BGBPs* expressions in *P. clarkii* during ovarian development assessed through qRT-PCR, using *GAPDH* and *EF1α* as reference genes. Results are mean ± SD of three technical replicates. The *y* axis of each graph is scaled based on the highest level of expression and indicates the relative expression of each target transcript. Dark bars represent the ovary and grey bars the hepatopancreas tissues; 2, 3, 4, 5, and 6 are the ovary developmental stages.

4. Discussion

The study investigated the number of VTGs orthologs present in the red swamp crayfish *P. clarkii*, since up to two VTGs were found in previous studies [19–29]. From the ATLAS library derived from the sequencing of 12 different tissues, four VTGs and two BGBP (VTG-like) transcripts were assembled and mapped (Figure 1), and this novel phenomenon was probably the result of multiplicity of VTG genes and/or alternative spliced forms.

The full length of the single VTG cDNA in crustaceans is about 8 kilobases (kb) in size and encodes 2500–2600 amino acid residues. The VTG sizes identified in *P. clarkii* are in line with the work by Avarre and colleagues [54].

To validate the bioinformatic results, phylogenetic analyses on all the complete VTGs available at present from public repositories were analyzed, identifying a separation between Decapoda and a group clustering Copepoda, Isopoda and Cladocera (Figure 2). The presence of an N-terminal lipid binding domain and a DUF 1943 domain suggests the relationship with the large lipid transfer proteins [41]. Two putative dibasic furin cleavage sites (with the motifs RAKR and RARR, respectively) were identified bordering the sequence of the BGBP. A similar protein with identical domain architecture was found in the prawn *Macrobrachium rosenbergii* suggesting a conserved structure among crustacean species [41].

The present study was aimed at two additional aspects of the vitellogenesis, the differential contribution of the hepatopancreas and the ovary in the process and the relationships of the *dLp/HDL-BGBP* protein to the vitellogenic process in view of its mutual domains with the VTG and its role as lipid carrier. The *dLp/HDL-BGBP* was found to not to be affected by ovarian development regulation pathways, hence not specifically contributing to ovarian development. Its expression in both organs is constitutive, higher in the hepatopancreas in comparison to the ovary. This high hepatopancreatic expression may be generally required to carry lipid products from this metabolic organ to other tissues.

The major novel finding of this work is the multiplicity of VTG transcripts in *P. clarkii* above the previously recognized two genes, and their organ-specific transcription in the ovary or the hepatopancreas with different expression scheduling. It starts in the ovary at stage 4/5 in which the ovary quickly enlarges and accumulates reserve materials. The hepatopancreas contributes to VTG production only at stage 6 probably to reinforce VTG production and boost maturation (Figure 4). These results support the idea that multiple VTGs are involved in ovarian maturation and that the contribution comes from both the hepatopancreas and the ovary, especially at stage 6 (complete ovarian maturation) in red swamp crayfish. Monitoring the four VTGs identified here could expand studies of reproduction in this species and shed light on potential methods that interfere ovarian maturation to develop new methods to contrast the spread of *P. clarkii*. Conversely, an in-depth analysis of the expression of VTGs could also be useful for aquaculture. It may be suggested that the ovary is the primary site of VTG production, required by the oocytes located in the ovary. However, the need for fast development of bigger ova led to the evolutionary need for an external reinforcement of the production by the major crustacean metabolic organ, the hepatopancreas. Only one VTG is expressed in the ovary, VTG1, while the others (VTG2-4) are expressed only in the hepatopancreas (Figure 4).

Cambarid crayfish complete their larval development within the ova and consequently have relatively large ova containing sufficient amounts of reserve materials to complete the development without external feeding. Future studies may focus on the full sequencing of all involved genes, attempting at the elucidation of mutual or different regulatory upstream sequences which may explain the different transcription patterns.

Supplementary Materials: The following figures are available online at https://www.mdpi.com/article/10.3390/d13090445/s1, Figure S1: Tracks of each reconstructed VTGs and lipid carriers, Figure S2: Schematic alignment of the 4 VTGs retrieved from the ATLAS of *P. clarkii*.

Author Contributions: Conceptualization, C.M., M.T., M.A., S.B., A.P. and P.G.G.; methodology, C.M. and M.T.; software, C.M. and M.T.; validation, C.M. and M.T.; formal analysis, C.M. and M.T.; resources, A.P. and P.G.G.; data curation, C.M., A.P. and P.G.G.; writing—original draft preparation, C.M. and M.T.; writing—review and editing, C.M., M.T., M.A., S.B., A.P. and P.G.G.; visualization, C.M., and P.G.G.; supervision, P.G.G.; All authors have read and agreed to the published version of the manuscript.

Funding: This research received no external funding.

Institutional Review Board Statement: Not applicable.

Data Availability Statement: The data presented in this study, in particular the sequences of the identified 4 VTGs and 2 BGPBs are openly available in GenBank.

Acknowledgments: We thank Luca Peruzza and Federica Piazza for their contribution in collecting the crayfish specimens and Marco Gerdol and Samuele Greco for bioinformatics support.

Conflicts of Interest: The authors declare no conflict of interest. The funders had no role in the design of the study; in the collection, analyses, or interpretation of data; in the writing of the manuscript, or in the decision to publish the results.

References

1. Busson-Mabillot, S. Données récentes sur la vitellogenèse. *Ann. Biol.* **1969**, *8*, 199–228.
2. Dhainaut, A.; De Leersnyder, M. Etude cytochimique et ultrastructurale de l'évolution ovocytaire du crabe *Eriocheir sinensis*. I. Ovogenèse naturelle. *Arch. Biol.* **1976**, *87*, 261–282.
3. Meusy, J.J. Vitellogenin, the extraovarian precursor of the protein yolk in Crustacea: A review. *Reprod. Nutr. Dévelop.* **1980**, *20*, 1–21. [CrossRef] [PubMed]
4. Kerr, M.S. The hemolymph proteins of the blue crab, *Callinectes sapidus*. II. A lipoprotein serologically identical to oocyte lipovitellin. *Develop. Biol.* **1969**, *20*, 1–17. [CrossRef]

5. Lee, C.Y.; Watson, D. In Vitro study of vitellogenesis in the blue crab (*Callinectes sapidus*): Site and control of vitellin synthesis. *J. Exp. Zool.* **1995**, *271*, 364–372. [CrossRef]
6. Browdy, C.L.; Fainzilber, M.; Tom, M.; Loya, Y.; Lubzens, E. Vitellin synthesis in relation to oogenesis in vitro incubated ovaries of *Penaeus semisulcatus*. *J. Exp. Zool.* **1990**, *255*, 205–215. [CrossRef]
7. Mak, A.S.; Choi, C.L.; Tiu, S.H.; Hui, J.H.; He, J.G.; Tobe, S.S.; Chan, S.M. Vitellogenesis in the red crab *Charybdis feriatus*: Hepatopancreas-specific expression and farnesoic acid stimulation of vitellogenin gene expression. *Mol. Reprod. Dev.* **2005**, *70*, 288–300. [CrossRef]
8. Yang, W.J.; Ohira, T.; Tsutsui, N.; Subramoniam, T.; Huong, D.T.T.; Aida, K.; Wilder, M.N. Determination of amino acid sequence and site of mRNA expression of four vitellins in the giant freshwater prawn, *Macrobrachium rosenbergii*. *J. Exp. Zool.* **2000**, *287*, 413–422. [CrossRef]
9. Girish, B.P.; Swetha, C.; Reddy, P.S. Hepatopancreas but not ovary is the site of vitellogenin synthesis in female fresh water crab, *Oziothelphusa senex*. *Biochem. Biophy. Res. Commun.* **2014**, *447*, 323–327. [CrossRef]
10. Tsutsui, N.; Saido-Sakanaka, H.; Yang, W.J.; Jayasankar, V.; Jasmani, S.; Okuno, A.; Ohira, T.; Okumura, T.; Aida, K.; Wilder, M.N. Molecular characterization of a cDNA encoding vitellogenin in the coonstriped shrimp, *Pandalus hypsinotus* and site of vitellogenin mRNA expression. *J. Exp. Zool.* **2004**, *301*, 802–814. [CrossRef]
11. Ding, X.; Nagaraju, G.P.C.; Novotney, D.; Lovett, D.L.; Borst, D.W. Yolk protein expression in the green crab, *Carcinus maenas*. *Aquaculture* **2010**, *298*, 325–331. [CrossRef]
12. Li, K.; Chen, L.; Zhou, Z.; Li, E.; Zhao, X.; Guo, H. The site of vitellogenin synthesis in Chinese mitten-handed crab *Eriocheir sinensis*. *Comp. Biochem. Physiol. B* **2006**, *143*, 453–458. [CrossRef]
13. Bai, H.; Qiao, H.; Li, F.; Fu, H.; Sun, S.; Zhang, W.; Jin, S.; Gong, Y.; Jiang, S.; Xiong, Y. Molecular characterization and developmental expression of vitellogenin in the oriental river prawn *Macrobrachium nipponense* and the effects of RNA interference and eyestalk ablation on ovarian maturation. *Gene* **2015**, *562*, 22–31. [CrossRef]
14. Okumura, T.; Yamano, K.; Sakiyama, K. Vitellogenin gene expression and hemolymph vitellogenin during vitellogenesis, final maturation, and oviposition in female kuruma prawn, *Marsupenaeus japonicus*. *Comp. Biochem. Physiol. A* **2007**, *147*, 1028–1037. [CrossRef]
15. Hiransuchalert, R.; Thamniemdee, N.; Khamnamtong, B.; Yamano, K.; Klinbunga, S. Expression profiles and localization of vitellogenin mRNA and protein during ovarian development of the giant tiger shrimp *Penaeus monodon*. *Aquaculture* **2013**, *412–413*, 193–201. [CrossRef]
16. Jia, X.; Chen, Y.; Zou, Z.; Lin, P.; Wang, Y.; Zhang, Z. Characterization and expression profile of Vitellogenin gene from *Scylla paramamosain*. *Gene* **2013**, *520*, 119–130. [CrossRef]
17. Salmela, H.; Stark, T.; Stucki, D.; Fuchs, S.; Freitak, D.; Dey, A.; Kent, C.F.; Zayed, A.; Dhaygude, K.; Hokkanen, H.; et al. Ancient Duplications Have Led to Functional Divergence of Vitellogenin-Like Genes Potentially Involved in Inflammation and Oxidative Stress in Honey Bees. *Genome Biol. Evol.* **2016**, *8*, 495–506. [CrossRef]
18. Wurm, Y.; Wang, J.; Riba-Grognuz, O.; Corona, M.; Nygaard, S.; Hunt, B.G.; Ingram, K.K.; Falquet, L.; Nipitwattanaphon, M.; Gotzek, D.; et al. The genome of the fire ant *Solenopsis invicta*. *Proc. Natl. Acad. Sci. USA* **2011**, *108*, 5679–5684. [CrossRef]
19. Abdu, U.; Davis, C.; Khalaila, I.; Sagi, A. The vitellogenin cDNA of *Cherax quadricarinatus* encodes a lipoprotein with calcium binding ability, and its expression is induced following the removal of the androgenic gland in a sexually plastic system. *Gen. Comp. Endocrinol.* **2002**, *127*, 263–272. [CrossRef]
20. Avarre, J.-C.; Michelis, R.; Tietz, A.; Lubzens, E. Relationship between vitellogenin and vitellin in a marine shrimp (*Penaeus semisulcatus*) and molecular characterization of vitellogenin complementary DNAs. *Biol. Reprod.* **2003**, *69*, 355–364. [CrossRef]
21. Okuno, A.; Yang, W.J.; Jayasankar, V.; Saido-Sakanaka, H.; Huong, D.T.T.; Jasmani, S.; Atmomarsono, M.; Subramoniam, T.; Tsutsui, N.; Ohira, T.; et al. Deduced Primary Structure of Vitellogenin in the Giant Freshwater Prawn, *Macrobrachium rosenbergii*, and Yolk Processing During Ovarian Maturation. *J. Exp. Zool.* **2002**, *292*, 417–429. [CrossRef]
22. Jeon, J.M.; Lee, S.O.; Kim, K.S.; Baek, H.J.; Kim, S.; Kim, I.K.; Mykles, D.L.; Kim, H.W. Characterization of two vitellogenin cDNAs from a Pandalus shrimp (*Pandalopsis japonica*): Expression in hepatopancreas is down-regulated by endosulfan exposure. *Comp. Biochem. Physiol. B* **2010**, *157*, 102–112. [CrossRef]
23. Phiriyangkul, P.; Utarabhand, P. Molecular Characterization of a cDNA Encoding vitellogenin in the banana shrimp, *Penaeus (Litopenaeus) merguiensis* and sites of vitellogenin mRNA Expression. *Molec. Reprod. Dev.* **2006**, *73*, 410–423. [CrossRef]
24. Raviv, S.; Parnes, S.; Segall, C.; Davis, C.; Sagi, A. Complete sequence of *Litopenaeus vannamei* (Crustacea: Decapoda) vitellogenin cDNA and its expression in endocrinologically induced sub-adult females. *Gen. Comp. Endocrinol.* **2006**, *145*, 39–50. [CrossRef] [PubMed]
25. Tiu, S.H.K.; Hui, J.H.L.; Mak, A.S.C.; He, J.G.; Chan, S.M. Equal contribution of hepatopancreas and ovary to the production of vitellogenin (PmVg1) transcripts in the tiger shrimp, *Penaeus monodon*. *Aquaculture* **2006**, *254*, 666–674. [CrossRef]
26. Tsang, W.S.; Quackenbush, S.L.; Chow, B.K.C.; Tiu, S.H.K.; He, J.G.; Chan, S.M. Organization of the shrimp vitellogenin gene: Evidence of multiple genes and tissue specific expression by the ovary and hepatopancreas. *Gene* **2003**, *303*, 99–109. [CrossRef]
27. Tsutsui, N.; Kawazoe, I.; Ohira, T.; Jasmani, S.; Yang, W.J.; Wilder, M.N.; Aida, K. Molecular characterization of a cDNA encoding vitellogenin and its expression in the hepatopancreas and ovary during vitellogenesis in the kuruma prawn, *Penaeus japonicus*. *Zool. Sci.* **2000**, *17*, 651–660. [CrossRef] [PubMed]

28. Xie, S.; Sun, L.; Liu, F.; Dong, B. Molecular characterization and mRNA transcript profile of vitellogenin in Chinese shrimp, *Fenneropenaeus chinensis*. *Molec. Biol. Rep.* **2009**, *36*, 389–397. [CrossRef]
29. Yang, F.; Xu, H.T.; Dai, Z.M.; Yang, W.J. Molecular characterization and expression analysis of vitellogenin in the marine crab *Portunus trituberculatus*. *Comp. Biochem. Physiol. B* **2005**, *142*, 456–464. [CrossRef] [PubMed]
30. Wang, Q.; Cheng, L.; Liu, J.; Li, Z.; Xie, S.; Silva, S.S.D. Freshwater aquaculture in PR China: Trends and prospects. *Rev. Aquac.* **2015**, *7*, 283–302. [CrossRef]
31. Putra, M.D.; Bláha, M.; Wardiatno, Y.; Krisanti, M.; Yonvitner; Jerikho, R.; Kamal, M.M.; Mojžišová, M.; Bystřický, P.K.; Kouba, A.; et al. Procambarus clarkii (Girard, 1852) and crayfish plague as new threats for biodiversity in Indonesia. *Aquat. Conserv.* **2018**, *28*, 1434–1440. [CrossRef]
32. Haubrock, P.J.; Oficialdegui, F.J.; Zeng, Y.; Patoka, J.; Yeo, D.C.J.; Kouba, A. The red claw crayfish: A prominent aquaculture species with invasive potential in tropical and subtropical biodiversity hotspots. *Rev. Aquac.* **2021**, *13*, 1488–1530. [CrossRef]
33. Gherardi, F.; Barbaresi, S.; Salvi, G. Spatial and temporal patterns in the movement of *Procambarus clarkii*, an invasive crayfish. *Aquat. Sci.* **2000**, *62*, 179–193. [CrossRef]
34. Scalici, M.; Gherardi, F. Structure and dynamics of an invasive population of the red swamp crayfish (*Procambarus clarkii*) in a Mediterranean wetland. *Hydrobiologia* **2007**, *583*, 309–319. [CrossRef]
35. Peruzza, L.; Piazza, F.; Manfrin, C.; Bonzi, L.C.; Battistella, S.; Giulianini, P.G. Reproductive plasticity of a *Procambarus clarkii* population living 10 °C below its thermal optimum. *Aquat. Inv.* **2015**, *10*, 199–208. [CrossRef]
36. Silveyra, G.R.; Silveyra, P.; Vatnick, I.; Medesani, D.A.; Rodríguez, E.M. Effects of atrazine on vitellogenesis, steroid levels and lipid peroxidation, in female red swamp crayfish *Procambarus clarkii*. *Aquat. Toxicol.* **2018**, *197*, 136–142. [CrossRef]
37. Guan, Z.B.; Yin, J.; Chen, K.; Shui, Y.; Cai, Y.J.; Liao, X.R. The hepatopancreas and ovary are the sites of vitellogenin synthesis in female red swamp crayfish *Procambarus clarkii* (Girard, 1852) (Decapoda: Astacoidea: Cambaridae). *J. Crust. Biol.* **2016**, *36*, 637–641. [CrossRef]
38. Kang, P.F.; Mao, B.; Fan, C.; Wang, Y.F. Transcriptomic information from the ovaries of red swamp crayfish (*Procambarus clarkii*) provides new insights into development of ovaries and embryos. *Aquaculture* **2019**, *505*, 333–343. [CrossRef]
39. Zhong, Y.; Zhao, W.; Tang, Z.; Huang, L.; Zhu, X.; Liang, X.; Yan, A.; Lu, Z.; Yu, Y.; Tang, D.; et al. Comparative transcriptomic analysis of the different developmental stages of ovary in red swamp crayfish *Procambarus clarkii*. *BMC Genom.* **2021**, *22*, 199. [CrossRef]
40. Duvic, B.; Söderhäll, K. Purification and partial characterization of a beta-1,3-glucanbinding-protein membrane receptor from blood cells of the crayfish *Pacifastacus leniusculus*. *Eur. J. Biochem.* **1992**, *207*, 223–228. [CrossRef]
41. Stieb, S.; Roth, Z.; Dal Magro, C.; Fischer, S.; Butz, E.; Sagi, A.; Khalaila, I.; Lieb, B.; Schenk, S.; Hoeger, U. One precursor, three apolipoproteins: The relationship between two crustacean lipoproteins, the large discoidal lipoprotein and the high density lipoprotein/β-glucan binding protein. *Biochim. Biophys. Acta* **2014**, *1841*, 1700–1708. [CrossRef]
42. Lemoine, F.; Correia, D.; Lefort, V.; Doppelt-Azeroual, O.; Mareuil, F.; Cohen-Boulakia, S.; Gascuel, O. NGPhylogeny.fr: New generation phylogenetic services for non-specialists. *Nucleic Acids Res.* **2019**, *47*, 260–265. [CrossRef]
43. Katoh, K.; Standley, D.M. MAFFT Multiple Sequence Alignment Software Version 7: Improvements in Performance and Usability. *Mol. Biol. Evol.* **2013**, *30*, 772–780. [CrossRef]
44. Criscuolo, A.; Gribaldo, S. BMGE (Block Mapping and Gathering with Entropy): A new software for selection of phylogenetic informative regions from multiple sequence alignments. *BMC Evol. Biol.* **2010**, *10*, 210. [CrossRef] [PubMed]
45. Guindon, S.; Dufayard, J.F.; Lefort, V.; Anisimova, M.; Hordijk, W.; Gascuel, O. New Algorithms and Methods to Estimate Maximum-Likelihood Phylogenies: Assessing the Performance of PhyML 3.0. *Syst. Biol.* **2010**, *59*, 307–321. [CrossRef] [PubMed]
46. Alcorlo, P.; Geiger, W.; Otero, M. Reproductive biology and life cycle of the invasive crayfish *Procambarus clarkii* (Crustacea: Decapoda) in diverse aquatic habitats of South-Western Spain: Implications for population. *Fundam. Appl. Limnol.* **2008**, *173*, 197–212. [CrossRef]
47. Untergasser, A.; Cutcutache, I.; Koressaar, T.; Ye, J.; Faircloth, B.C.; Remm, M.; Rozen, S.G. Primer3-new capabilities and interfaces. *Nucleic Acids Res.* **2012**, *40*, e115. [CrossRef]
48. Ruijter, J.M.; Ramakers, C.; Hoogaars, W.M.; Karlen, Y.; Bakker, O.; van den Hoff, M.J.; Moorman, A.F. Amplification efficiency: Linking baseline and bias in the analysis of quantitative PCR data. *Nucleic Acids Res.* **2009**, *37*, e45. [CrossRef]
49. Kibbe, W.A. OligoCalc: An online oligonucleotide properties calculator. *Nucleic Acids Res.* **2007**, *35*, 43–46. [CrossRef]
50. Pfaffl, M.W.; Tichopad, A.; Prgomet, C.; Neuvians, T.P. Determination of stable housekeeping genes, differentially regulated target genes and sample integrity: BestKeeper—Excel-based tool using pairwise correlations. *Biotech. Lett.* **2004**, *26*, 509–515. [CrossRef]
51. Andersen, C.L.; Jensen, J.L.; Ørntoft, T.F. Normalization of real-time quantitative reverse transcription-PCR data: A model-based variance estimation approach to identify genes suited for normalization, applied to bladder and colon cancer data set. *Cancer Res.* **2004**, *64*, 5245–5250. [CrossRef]
52. Vandesompele, J.; De Preter, K.; Pattyn, F.; Poppe, B.; Van Roy, N.; De Paepe, A.; Speleman, F. Accurate normalization of real-time quantitative RT-PCR data by geometric averaging of multiple internal control genes. *Genome Biol.* **2002**, *3*, research0034.1. [CrossRef] [PubMed]

53. Marchler-Bauer, A.; Bryant, S.H. CD-Search: Protein domain annotations on the fly. *Nucl. Acids Res.* **2004**, *32*, 327–331. [CrossRef] [PubMed]
54. Avarre, J.C.; Lubzens, E.; Babin, P.J. Apolipocrustacein, formerly vitellogenin, is the major egg yolk precursor protein in decapod crustaceans and is homologous to insect apolipophorin II/I and vertebrate apolipoprotein B. *BMC Evol. Biol.* **2007**, *7*, 3. [CrossRef] [PubMed]

diversity

MDPI

Review

Shattercane (*Sorghum bicolor* (L.) Moench Subsp. *Drummondii*) and Weedy Sunflower (*Helianthus annuus* L.)—Crop Wild Relatives (CWRs) as Weeds in Agriculture

Panagiotis Kanatas [1,*], Ioannis Gazoulis [2], Stavros Zannopoulos [3], Alexandros Tataridas [2], Anastasia Tsekoura [2], Nikolaos Antonopoulos [2] and Ilias Travlos [2]

[1] Department of Crop Science, University of Patras, P.D. 407/80, 30200 Mesolonghi, Greece

[2] Laboratory of Agronomy, Agricultural University of Athens, 75, Iera Odos str., 11855 Athens, Greece; giangazoulis@gmail.com (I.G.); a.tataridas@gmail.com (A.T.); natasa_tse@hotmail.com (A.T.); nikolasantwno@gmail.com (N.A.); travlos@aua.gr (I.T.)

[3] Ministry of Rural Development and Food, Hellenic Food Authority, 124, Kifisias str., 11526 Athens, Greece; sgeoponos@gmail.com

* Correspondence: pakanatas@gmail.com; Tel.: +30-697-7568-815

Abstract: Shattercane (*Sorghum bicolor* (L.) Moench subsp. *drummondii*) and weedy sunflower (*Helianthus annuus* L.) are two examples of crop wild relatives (CWRs) that have become troublesome weeds in agriculture. Shattercane is a race belonging to a different subspecies than domesticated sorghum (*Sorghum bicolor* (L.) Moench subsp. *bicolor*). Weedy sunflower populations are natural hybrids between wild and domesticated sunflower (*Helianthus annuus* L.). Both species have key weedy characteristics, such as early seed shattering and seed dormancy, which play an important role in their success as agricultural weeds. They are widely reported as important agricultural weeds in the United States and have invaded various agricultural areas in Europe. Shattercane is very competitive to sorghum, maize (*Zea mays* L.), and soybean (*Glycine max* (L.) Merr.). Weedy sunflower causes severe yield losses in sunflower, maize, soybean, pulse crops, and industrial crops. Herbicide resistance was confirmed in populations of both species. The simultaneous presence of crops and their wild relatives in the field leads to crop–wild gene flow. Hybrids are fertile and competitive. Hybridization between herbicide-tolerant crops and wild populations creates herbicide-resistant hybrid populations. Crop rotation, false seedbed, cover crops, and competitive crop genotypes can suppress shattercane and weedy sunflower. Preventative measures are essential to avoid their spread on new agricultural lands. The development of effective weed management strategies is also essential to prevent hybridization between sorghum, sunflower, and their wild relatives and to mitigate its consequences.

Keywords: seed shattering; yield loss; herbicide resistance; hybrid fitness; weed management; preventative measures; cultural practices

Citation: Kanatas, P.; Gazoulis, I.; Zannopoulos, S.; Tataridas, A.; Tsekoura, A.; Antonopoulos, N.; Travlos, I. Shattercane (*Sorghum bicolor* (L.) Moench Subsp. *Drummondii*) and Weedy Sunflower (*Helianthus annuus* L.)—Crop Wild Relatives (CWRs) as Weeds in Agriculture *Diversity* **2021**, *13*, 463. https://doi.org/10.3390/d13100463

Academic Editor: Michael Wink

Received: 6 September 2021
Accepted: 23 September 2021
Published: 25 September 2021

Publisher's Note: MDPI stays neutral with regard to jurisdictional claims in published maps and institutional affiliations.

Copyright: © 2021 by the authors. Licensee MDPI, Basel, Switzerland. This article is an open access article distributed under the terms and conditions of the Creative Commons Attribution (CC BY) license (https://creativecommons.org/licenses/by/4.0/).

1. Introduction

Crop wild relatives (CWRs) are wild plant species closely related to domesticated crops. According to Maxted et al. [1], the genetic relationships between crops and CWRs are described by the following taxa groups: TG1a—crop taxon; TG1b—the same species of crop; TG2—the same series or section of crop; TG3—the same subgenus of crop; TG4—the same genus of crop; and TG5—the same tribe, but different genus of crop. The species in taxa groups TG1a, TG1b, TG2, and TG3 are of unique interest from both plant breeding and weed science perspectives because they belong in the primary gene pool of a genus (GP–1) and can successfully interbreed [2–5].

These wild taxa are valuable genetic resources that should be explored for use in plant breeding programs. They can increase genetic diversity in cultivated species through hybridization and they can transfer beneficial traits such as resistance to biotic and abiotic stress factors. CWRs were successfully used to confer resistance traits to soil salinity, drought, and bacterial leaf blight in durum wheat (*Triticum durum* Desf.), barley (*Hordeum vulgare* L.), and rice (*Oryza sativa* L.), respectively [6–8]. Adaptation of crops to stress conditions through the use of CWRs leads to improved crop yields, yield stability over time, and the improved quality of agricultural products. For example, in soybean (*Glycine max* (L.) Merr.), a wild relative of the crop was reported to have candidate genes that improve 1000-seed weight and thus soybean seed yield [9]. In processing tomato (*Solanum lycopersicum* L.), CWRs belonging to the same genus contain genes that can improve fruit quality traits, such as total soluble solids content, sugar content, and the fruit dry weight to fruit fresh weight ratio [10]. Similar results were reported for cereals and legumes [11]. The use of CWRs is also recommended for crop improvement in dominant multipurpose crops such as sorghum (*Sorghum bicolor* (L.) Moench subsp. *bicolor*) and sunflower (*Helianthus annuus* L.) [2,12].

However, wild species that are closely related to domesticated crops can occur as weeds on agricultural lands. A special group of agricultural weeds includes weedy relatives of some crops that belong to the same species as domesticated plants. In particular, weedy rice (*Oryza sativa* L.), shattercane (*Sorghum bicolor* (L.) Moench subsp. *drummondii*), and weedy sunflower (*Helianthus annuus* L.) are prominent examples of weedy relatives of rice, sorghum, and sunflower, respectively [13–15]. All these species are competitive and have undesirable agronomic characteristics, such as early seed shattering and seed dormancy, which play an important role in their success as weeds in agriculture [15–17]. Although the seeds can be assumed to be edible, they cannot be harvested because seed shattering occurs before crop maturity [14,18,19]. In addition, seed dormancy allows these wild plants to form large seed banks in the soil and become persistent in agricultural areas [15,20,21].

Shattercane and weedy sunflower are among the most competitive weeds against their closely related domesticated crop species, namely sorghum and sunflower, respectively [22,23]. Furthermore, both are troublesome species in a variety of crops. In particular, shattercane was reported to cause significant yield losses in maize (*Zea mays* L.) and soybean (*Glycine max* (L.) Merr.); weedy sunflower competition was reported to limit the productivity of maize, soybean, pulse crops, and industrial crops [24–28]. On the contrary, weedy rice exclusively infests rice fields and causes severe yield losses everywhere in the world where direct-seeded rice is cultivated [15,19]. Apart from their competitive ability, populations of these weedy crop relatives have developed resistance to common herbicides used to control weeds in summer field crops [29]. In addition, strong genetic and botanical ties favor hybridization between weedy and cultivated plants, leading to more complex weed problems. It should be noted that wild plants contribute to crop improvement when gene flow occurs from CWRs to domesticated plants, under controlled conditions. On the other hand, if there is natural gene flow from the crop to its wild relatives in the field, this leads to the development of fertile hybrid populations that have comparable fitness to their wild and domesticated parents [3,4]. In cases where the domesticated parent is a herbicide-tolerant crop, the crop–weed gene flow can lead to the development of herbicide-resistant hybrid populations [30,31].

Gene flow from crops to their wild relatives occurs by cross-pollination [32]. The pollen parents of the first outcrossing may be domesticated plants in cultivation, crop volunteers that arose after the harvest of the previous crop, or feral populations that escaped cultivation. The terms 'crop volunteers' and 'feral populations' are explained in later sections, as is the crucial role of such plant populations as genetic bridges for the success of the crop–weed gene flow under certain circumstances [3,33,34]. In outcrosses of sorghum × shattercane and rice × weedy rice, pollen is transferred by wind from the crop plants to their wild relatives [35–37]. In contrast, cross-pollination between domesticated and wild forms of *H. annuus* occurs by insects, especially honeybees (*Apis mellifera* L.), since *H. annuus* is an insect-pollinated species [38,39]. The hybridization process begins when desiccated pollen grains (from the flowers of the domesticated plants) land on the stigma (in the flowers of the crop's wild relatives). The male gametophytes (e.g., pollen grains) rapidly rehydrate and begin to germinate [40]. Subsequently, a pollen tube grows through the pistil tissues of the stigma and style, across the surface of the placenta, and then through the micropyle of the ovule to reach the female gametophyte in the embryo sac [40,41]. The growth of the pollen tube stops, and two gametes are released [40]. It should be noted that pollen tube growth is both polar and directional. Cytosolic Ca^{2+} ions are thought to play an important role in pollen tube formation, growth, and polarity as secondary messengers [40,42,43]. In any case, the physical distance between crop plants and their wild relatives and the synchrony of their flowering times are crucial factors affecting hybridization rates [18,37].

The current study summarizes information on weedy relatives of crops that are problematic weeds in agriculture because they compete with a wide range of crops, have high invasive potential, and can also successfully interbreed with their closely related domesticated crops and generate complex weed problems [32]. Regarding species selection, it should be noted that shattercane and weedy sunflower are problematic species in a wider range of crops compared to weedy rice. Moreover, the interactions between weedy rice and rice and the appropriate strategies to control weedy rice in direct-seeded rice fields were recently studied [15,19]. Therefore, the present study focused on the weedy relatives of sorghum and sunflower, i.e., shattercane and weedy sunflower, respectively. Although the selected species are known to be important weeds mainly in the United States, there is much evidence that they have also invaded various agricultural areas in Europe [44,45].

First, we present information on the origin of these species and also on their important morphological and ecological traits. Then, we summarize information on their competitive ability against their closely related domesticated species and also against other important summer field crops. In addition, we include information about their occurrence in Europe where they have the potential to become serious invaders in the future. Cases are presented where herbicide resistance was confirmed. Evidence of successful hybridization between crops and their wild relatives is included along with information on the fitness of the hybrids produced; we also include cases where crop–wild gene flow led to the development of herbicide-resistant hybrids. Weed management strategies that can be effective in controlling these species are discussed. Emphasis is also placed on the role of weed management in preventing gene flow from crops to their wild relatives.

2. Shattercane [*Sorghum bicolor* (L.) Moench Subsp. *drummondii*]

2.1. Origin

The genus *Sorghum* is divided in five subgenera, namely *Eu–sorghum*, *Chaetosorghum*, *Heterosorghum*, *Parasorghum* and *Stiposorghum*. The *Eu–sorghum* subgenus includes the following species: *Sorghum bicolor* (L.) Moench, *Sorghum propinquum* (Kunth) Hitchc., *Sorghum halepense* (L.) Pers., and *Sorghum almum* Parodi [46].

Sorghum bicolor (L.) Moench is divided into three subspecies whose members are all diploids (2n = 20): (1) *Sorghum bicolor* (L.) Moench subsp. *bicolor* which contains all cultivated sorghum lines classified by Harlan and De Wet [47] into five races (*bicolor*, *guinea*, *caudatum*, *kafir*, and *durra*), (2) *Sorghum bicolor* (L.) Moench subsp. *verticilliflorum* (Steud.) de Wet ex Wiersema & J. Dahlb. which contains wild progenitors of cultivated sorghums classified into four races, namely *aethiopicum*, *arundinaceum*, *verticilliflorum*, and *virgatum* [48] and (3) *S. bicolor* ssp. *drummondii* (Nees ex Steud.) De Wet ex Davidse which includes two races called sudangrass and shattercane [48]. This subspecies is the product of natural hybridization between *S. bicolor* subsp. *bicolor* × *S. bicolor* subsp. *verticilliflorum* [46]. While sudangrass can be grown as a forage crop, shattercane is a weedy relative of sorghum that is a considered a serious weed whose agronomic importance has increased over the years [14,26,27,31,46,49–51].

As for the other species of the *Eu–sorghum* subgenera, *S. propinquum* is a diploid (2n = 20), rhizomatous, biennial to perennial, wild species [52]. *S. halepense* is another rhizomatous perennial wild species, which is tetraploid (2n = 40), also known as johnsongrass [52]. This species is thought to have arisen either by natural hybridization between *S. bicolor* × *S. propinquum* or by chromosome duplication in *S. propinquum* [14,53]. Regarding *S. almum*, it is a tetraploid (2n = 40), rhizomatous, perennial species and is a natural hybrid between *S. bicolor* × *S. halepense* [54]. Of the perennial species presented, *S. propinquum* and *S. almum* are not reported as troublesome weeds [16]. In contrast, johnsongrass is one of the most common and noxious weeds in agriculture, which can also reduce biodiversity due to its high invasive potential [55].

This article could focus on both shattercane and johnsongrass, as both species are troublesome weeds that can hybridize with the crop [2,3]. However, the present study focuses on species that belong to the primary gene pool (GP–1) of a genus and have very strong genetic links to the crop. Species in this primary gene pool readily interbreed and produce fertile hybrids [16]. In contrast, species belonging to the secondary gene pool (GP–2) of a genus can also interbreed with the crop, but successful gene transfer between these two gene pools can be difficult in some situations. Since shattercane belongs to the primary gene pool (GP–1) of sorghum while johnsongrass belongs to the secondary gene pool (GP–2) of the genus [2], further information is provided only on shattercane. In addition, the main aspects of johnsongrass biology and ecology, as well as its negative impacts on agriculture and biodiversity, were already summarized in a previous study [5].

2.2. Morphological and Ecological Traits

Shattercane is a warm-season annual grass that originated in Africa [46]. The plants have erect, unbranched stems and can grow 1–4 m tall. This weedy race has some key characteristics that explain its evolution into a troublesome weed. First of all, plant height cannot be regulated in shattercane because it lacks a dwarfing trait that is controlled in cultivated sorghum by four recessive dwarfing genes [56]. Therefore, the increased canopy height results in lower harvest index values and makes mechanical harvesting an impossible task [21]. It should also be noted that the great height of shattercane improves its ability to compete with tall cereals such as maize (*Zea mays* L.) and increases its ability to disperse seeds over long distances [57].

As for seed dispersal, it is an ecological trait playing a central role in the success of this species as an agricultural weed. Seed dispersal is rapid, and the explanation lies in the abscission layer that forms at the base of the spikelet at the stage of physiological seed maturity. This abscission layer allows the seed to detach from the panicle and immediately fall to the soil surface. It is worth mentioning that only a light breeze (e.g., a wind moving at a very low speed of 7–12 km h^{-1}) is adequate to cause seed shattering before the cultivated crop can be harvested [14]. In addition, the shattered seeds can stay dormant for a long time in the soil and remain viable. Burnside et al. [58] reported a seed survival period of up to 13 years in the United States while Fellows and Roeth [59] found that the dormancy period can be further extended if the seeds are tightly enclosed in the glumes. As for the reproductive ability of shattercane, plants typically produce 1–6 panicles with each panicle producing 500–1500 seeds [57]. An interesting fact is that shattercane has an extended emergence window since the seeds can germinate late in the growing season. These later-emerging weeds may exhibit aggressive growth rates, reach maturity, and produce seeds that enrich the species' seedbank dynamics in the soil [14].

2.3. Competitive Ability and Distribution

Shattercane populations can establish on agricultural land, field margins, and marginal areas in various regions across the world. Its presence as a weed was reported in North America, in Africa where it is believed to have originated, in Asia, and also in Europe [16,27,44,46,49,60]. Shattercane infestations result in significant yield loss in important summer field crops including grain sorghum, maize, and soybean. All reports of yield loss due to shattercane competition are from field trials conducted in the United States, with the exception of the case study by Raey et al. [27], which was conducted in Iran, Asia (Table 1).

Table 1. Yield losses of summer field crops due to shattercane (*Sorghum bicolor* (L.) Moench subsp. *drummondii*) interference. Results presented are from field trials repeated in time or space.

Crop	Shattercane Density	Yield Loss	Reference
Grain Sorghum	5.6 Plants m^{-2}	73–82%	[61]
Maize	13–20 Plants m^{-1} of Row	22%	[49]
Maize	20 Plants m^{-2}	43–85%	[51]
Maize	6.6 Plants m^{-2}	19%	[26]
Maize	40 Plants m^{-2}	34%	[62]
Soybean	3.3 Plants m^{-1} of Row	60%	[63]
Soybean	12 Plants m^{-2}	57%	[27]

Especially in the United States, shattercane is one of the most common and problematic weeds in grain sorghum [50]. There are also case studies from this continent showing the competitive ability of shattercane against domesticated sorghum and other important summer field crops. In sorghum, early studies revealed that 5.6 shattercane plants m^{-2}, spaced 45 cm apart, caused a 73–82% yield loss in grain sorghum [61]. The competitive advantage of shattercane compared to grain sorghum growth was also recently highlighted in greenhouse studies [21,31,36,64,65]. In most of the case studies mentioned, shattercane exhibited a more aggressive growth compared to grain sorghum and the weeds were significantly taller than the domesticated plants. Shattercane is also reported as a strong competitor to maize and soybean. In particular, Beckett and Stoller [49] found that 13 to 20 shattercane plants m^{-1} of row resulted in a 22% grain yield loss in maize. Season-long shattercane interference (from 20 plants m^{-2}) reduced grain yield by 43–85% in the study by Hans and Johnson [51]. The same authors also observed significant yield reductions when shattercane was left uncontrolled until it was 31 cm tall. At a density of 6.6 plants m^{-2}, Deines et al. [26] predicted a grain yield loss of 19%. King and Hagood [64] found that shattercane competition (at a density of 40 plants m^{-2}) resulted in up to 34% grain yield loss. In soybean (*Glycine max* (L.) Merr.), seed yield decreased by more than 60% due to full-season competition by 3.3 shattercane plants m^{-1} of the row [65]. There is also evidence from Asia showing that 50 soybean plants m^{-2} were outcompeted by 12 shattercane plants m^{-2} and suffered a 57% loss in seed yield [27].

There are not many official reports on the presence of shattercane in Europe. However, it should be noted that in the context of climate change, the resilient and versatile sorghum has gained importance as a multipurpose crop in Europe [66]. Sorghum acreage has increased in all European sorghum producing countries, namely France, Italy, Hungary, Romania, Bulgaria, Austria and Greece [67]. Defelice [14] pointed out that shattercane can spread anywhere in the world where domesticated sorghum is grown. Therefore, it is possible that populations of shattercane have developed in the European countries mentioned above, although this is not officially reported. The U.S. Department of Agriculture (USDA) has conducted a weed risk assessment for this weed species and concluded that the presence of shattercane in sorghum producing countries is underreported because it is difficult to distinguish shattercane from sorghum [68]. Berenji and Dahlberg [44], Dahlberg et al. [69] and Schwartz–Lazaro and Gage [70] mention that there are at least two distinct areas where shattercane populations were reported, namely southeastern Hungary and northeastern Serbia. Dahlberg et al. [69] also included a photograph of a shattercane population growing in a broom corn field in their study. Broomcorn is a cultivated race of sorghum whose panicles are used as raw material for making natural corn brooms [71]. The morphology of shattercane is very similar to broom corn. Since Europe, especially Hungary, Romania and Serbia, are the main producers of broom and broom corn in the world [44], it is logical to assume that populations of broom corn may have developed in these areas but are not yet reported due to the morphological similarities between broom corn and shattercane. In view of this situation, a research goal of weed scientists in Europe should be to carefully survey sorghum fields to detect populations of shattercane and take action to control this weed before it becomes established in Europe.

2.4. Herbicide Resistance

Research has shown that consecutive applications of ALS (acetolactate synthase)-inhibiting herbicides in a particular field inevitably result in the selection of ALS-resistant shattercane populations (Table 2).

Table 2. Cases of herbicide resistance in shattercane (*Sorghum bicolor* (L.) Moench subsp. *drummondii*) populations. Results presented are from temporally and spatially replicated dose-response experiments.

Crop	Herbicide	Mode of Action	Chemical Family	Reference
Maize	Primisulfuron–Methyl	ALS Inhibitor	Sulfonylurea	[72]
Maize	Primisulfuron–Methyl Nicosulfuron Imazethapyr	ALS Inhibitor ALS Inhibitor ALS Inhibitor	Sulfonylurea Sulfonylurea Imidazolinone	[73]
Maize–Soybean Rotation	Primisulfuron–Methyl Nicosulfuron Imazethapyr	ALS Inhibitor ALS Inhibitor ALS Inhibitor	Sulfonylurea Sulfonylurea Imidazolinone	[74]
Maize	Imazethapyr	ALS Inhibitor	Imidazolinone	[75]
Maize	Nicosulfuron Imazethapyr Imazapyr	ALS Inhibitor ALS Inhibitor ALS Inhibitor	Sulfonylurea Imidazolinone Imidazolinone	[64]
Maize–Soybean Rotation	Nicosulfuron Imazethapyr	ALS Inhibitor ALS Inhibitor	Sulfonylurea Imidazolinone	[76]

Anderson et al. [72] reported shattercane resistance to primisulfuron–methyl in a biotype collected from a maize field treated with primisulfuron–methyl and nicosulfuron for three consecutive growing seasons. In the study by Lee et al. [73], shattercane populations from 12 fields were resistant to primisulfuron–methyl and nicosulfuron. In the same study, another population was susceptible to primisulfuron–methyl and nicosulfuron but resistant to imazethapyr. The presence of a biotype with noticeable levels of resistance to primisulfuron and cross-resistance to nicosulfuron and imazethapyr was also confirmed [74]. Resistance evolved after 10 years of use of ALS–inhibiting herbicides in a field where maize was rotated with soybean. Zelaya and Owen [75] observed that one population was 29 times more resistant to imazethapyr compared to a sensitive population. These authors noted that resistance occurred in an environment where the use of ALS-inhibiting herbicides was an important component of the selection pressure. In another study, the continuous use of nicosulfuron for weed control in silage maize resulted in the selection of a shattercane population that was resistant to nicosulfuron and exhibited cross-resistance to imazethapyr and imazapyr [64]. Werle et al. [76] screened 190 shattercane populations and observed five and four populations that were resistant to imazethapyr and nicosulfuron, respectively, and two populations that were cross-resistant to nicosulfuron and imazethapyr. All of these cases of herbicide resistance in shattercane were reported in the United States. However, if shattercane becomes a serious invader in European fields, crop rotation and herbicide rotation practices should be used to prevent the development of herbicide-resistant populations.

2.5. Hybridization with Domesticated Sorghum

Both shattercane and sorghum belong to the primary gene pool of the genus, they are sexually compatible, and can be wind pollinated. Therefore, these sympatric species can successfully outcross under favorable field conditions and produce fertile hybrids [21,36,60]. Schmidt et al. [37] highlighted flowering duration of sorghum and flowering overlap between the two species as important factors determining hybridization rates in the field. The same authors also emphasized the crucial role of wind speed and direction in the outcome of the hybridization process [37]. Moreover, hybridization rates tend to increase when the distance between interacting populations becomes smaller [3,37]. Another noteworthy point is that gene transfer from the crop to its wild relatives is more frequent than gene transfer in the opposite direction [77]. One possible explanation is that populations of domesticated plants in agricultural fields are usually much larger, and the domesticated plants, therefore, produce larger amounts of pollen compared to their wild relatives [77].

In any case, hybrids between sorghum and shattercane can be competitive, as shown by case studies where successful hybridization was reported. Sahoo et al. [36] found that grain sorghum × shattercane hybrids produced 31% more biomass and were 56–61% taller compared to grain sorghum. They also found that the hybrids produced 40–63% and 42–61% more spikelets per panicle and seeds per plant, respectively, compared to their domesticated parents. In this study, hybrid relative fitness was similar to shattercane as also observed in the study by Schmidt et al. [37]. In the study by Magomere et al. [78], F_1 hybrids produced 1509 more seeds than their parent plants, while the mean seed weight of the hybrids was 41% higher than that of grain sorghum. Similar observations were made for aboveground biomass production and tillering capacity, indicating a competitive advantage of the F_1 hybrids over their domesticated parents [78]. Schmidt et al. [21] revealed also that F_2 hybrids are characterized by lower vegetative growth and fecundity than shattercane but their relative fitness can be comparable to that of grain sorghum. In particular, these authors reported no significant differences between grain sorghum and grain sorghum × shattercane F_2 hybrids in the number of panicles per plant, aboveground biomass production, and seed production [21]. In the pot experiments by Werle et al. [31], F_1 hybrids outcompeted an ALS–resistant grain sorghum inbred line and caused a biomass yield loss of 75–95%. Aside from their competitive ability, seed dormancy is another characteristic of these hybrids that might enable them to be highly persistent on agricultural lands. Indeed, there is evidence that seed dormancy is similar to shattercane and seeds can survive in the soil for many years [3,21,36].

Another consequence of hybridization between domesticated sorghum and its wild relative, shattercane, is the emergence of herbicide-resistant hybrids under certain circumstances. First, it should be noted that in the past, germplasm from shattercane populations with resistance to ALS-inhibiting herbicides was used to develop the 'Inzen' technology, i.e., to develop ALS-tolerant grain sorghum populations [79]. Werle et al. [63] revealed that most of herbicide-resistant shattercane populations have evolved independently and resistance is not the result of pollen-mediated gene flow between ALS-tolerant grain sorghum and shattercane. However, there is evidence that possible outcrossing between the crop and its wild relative may indeed result in the creation of ALS-resistant grain sorghum × shattercane hybrids. For instance, Werle et al. [23] found that shattercane × ALS–tolerant grain sorghum hybrids were tolerant to ALS-inhibiting herbicides and herbicide application did not reduce hybrid growth. Adugna and Bekele [60] also reported that such hybrids can be tolerant to herbicides and at the same time competitive against grain sorghum and exhibit similar fitness to shattercane. In another study conducted under greenhouse and real field conditions, the creation of ALS-tolerant hybrids was confirmed as the hybrids were not affected by the application of a nicosulfuron plus rimsulfuron mixture and maintained their competitive advantage over their ALS-tolerant grain sorghum parents [31].

3. Weedy Sunflower (*Helianthus annuus* L.)

3.1. Origin

The genus *Helianthus* is native to the temperate zones of North America and includes 52 species and 19 subspecies with 14 annuals and 39 perennials. The basal chromosome number is n = 17. All 14 of the annual species are diploid (2n = 34), while in the group of perennial species there are 26 diploid, 3 tetraploid (2n = 68), 7 hexaploid (2n = 102) and 3 mixaploid species [80,81]. Taxonomically, there are four distinct sections in the genus, namely the annual polyphyletic section *Helianthus*, the annual monophyletic section

Agrestis, the perennial polyphyletic section *Ciliares* with two races, and the perennial polyphyletic section *Divaricati* with four races [82]. The species *Helianthus annuus* L. of the section *Helianthus* includes the domesticated sunflower (*Helianthus annuus* L. var. *macrocarpus*) cultivated for its oil seeds and also its weedy or wild forms [83].

In an early study by Heiser [84], it was suggested that there are three subspecies of *H. annuus*, namely *H. annuus* subsp. *lenticularis*, *H. annuus* subsp. *texanus*, and *H. annuus* subsp. *annuus*, the last subspecies being emphasized as the weedy sunflower. However, there are still no official names for the subspecies. In another study, Heiser [83] emphasized that *H. annuus* exhibits high morphological variability, so that its wild and weedy relatives cannot be adequately classified into separate subspecies. This is in contrast to the genus *S. bicolor*, where shattercane is a race belonging to a different subspecies than the domesticated sorghum. Indeed, there is strong evidence that the weedy forms of *H. annuus* are not represented by a specific subspecies but are the natural result of hybridization with domesticated sunflower. There is evidence of crop introgression in weedy sunflowers since they combine wild and domesticated traits in proportions that vary between wild and domesticated plants [4,85–88]. In some recent studies, the various forms of *H. annuus* are divided into the domesticated sunflower, the weedy sunflowers, which include the "agrestal" biotypes, and the wild sunflowers, which include the "ruderal" biotypes. The term "agrestal" is used to describe plants evolved under selection pressure on agricultural land while the term "ruderal" refers to plants inhabiting naturally disturbed sites [89].

For the species *H. annuus*, the weedy "agrestal" biotypes are considered natural crop–wild hybrids [4,13,20,22]. The initial invasions of such biotypes on agricultural land might be attributed to importations of contaminated sunflower seed lots. The importations of contaminated seed from the United States were the dominant hypothesis for the spread of weedy sunflowers in European fields [38,86,87]. As for the wild "ruderal" biotypes, their spread into non–crop areas such as roadsides, water channels, firebreaks, etc., is thought to be promoted by anthropogenic activities [39,85,90,91]. The ruderal biotypes can hybridize recurrently with the domesticated plants leading to the spread of highly competitive hybrids in the field [4,39,92]. Although seed transport by humans is considered to explain the invasion of weedy sunflowers in South America, the role of ruderal biotypes in the spread of weedy forms in these regions and also in North America is highlighted [88,92–94]. For instance, Kane and Rieseberg [94] attributed the development of multiple weedy sunflower populations in the United States to the presence of ruderal populations near cultivated sunflower fields. Several factors favor the hybridization process including the overlapping flowering periods of domesticated and wild sunflower, the self-incompatibility trait of wild sunflower, and the presence of shared pollinators under real field conditions [34]. In addition, pollen transfer from the crop to wild plants can occur even from 1 km away [18].

3.2. *Morphological and Ecological Traits*

Various forms of *H. annuus* occur as domesticated sunflowers, as weeds in agriculture and as wild plants on naturally disturbed, uncultivated sites. Domesticated sunflowers have unbranched stems of 1.2–2.0 m tall, topped by a single, large-diameter yellow-colored head. In addition, anthocyanins are not present in the plant tissues [83,84]. However, weedy sunflowers have taller stems characterized by apical or full branching. Unlike domesticated plants, weedy sunflowers form several heads per plant, usually between 17 and 34. Head diameter, seeds per head, 1000 seed weight, and seed oil content are significantly lower compared to cultivated sunflowers. Head color can be red or yellow. Anthocyanins are found in the stem, petioles, and stigma. Research has shown that the morphology of weedy sunflowers is intermediate between wild biotypes and domesticated sunflowers [4,13,39,86,88]. The wild trait of self-incompatibility and the domesticated trait of male-sterility can be also observed [34,86].

Seed dormancy is an important ecological trait of weedy sunflowers that enables seed bank formation on agricultural lands. In greenhouse tests conducted by Presotto et al. [17]

with five weedy sunflower populations, seed dormancy reached 77% when no stratification treatments were applied. In another recent study, weedy sunflower seeds remained viable and dormant in the soil for 42 months, suggesting that such biotypes form persistent seedbanks and even establish on agricultural lands outside their native range [20]. Seed dormancy and seedbank formation are traits that originated in wild populations and were transferred to weedy sunflowers through crop–wild hybridization [17,20,86,95,96]. On top of seed dormancy, the seed shattering ability of weedy sunflowers contributes to their success as agricultural weeds. The seeds are easily detached from the heads due to the anatomy of the disks, which are characterized by a lower depth–width ratio compared to domesticated sunflowers, replenishing the seed bank of weedy sunflower in the field [4,92]. As for seed production, it can range between 2200 and 6460 seeds per plant [22,97,98]. Presotto et al. [30] found that the fitness and seed production of weedy sunflowers can be significantly reduced compared to their domesticated and wild parents. However, the same authors found that relative fitness and fertility of plants tended to increase when weedy sunflowers were backcrossed with cultivated or wild sunflower populations.

3.3. Competitive Ability and Distribution

Weedy sunflowers were reported as agricultural weeds in their native range, i.e., in North America, South America and particularly Argentina, and also in several countries in Europe [22,25,86,98,99]. Their competitive ability is attributed to their early-season vigor, rooting, and vegetative growth, plant height, and allelopathic potential [13,100,101]. There are several reports highlighting the detrimental effect of weedy sunflower interference on the yield performance of summer field crops (Table 3).

In the three-year field trials conducted by Casquero et al. [13] in Argentina, sunflower seed yield loss surpassed 50% due to weedy sunflower interference at the density of 4 plants m^{-2}. At higher density, i.e., 10.7 plants m^{-2}, weedy sunflower reduced sunflower seed numbers per plant, 1000 seed weight and seed yield per plant by 66, 41, and 80%, respectively [22]. As for the presence of weedy sunflowers as agricultural weeds in the United States, Deines et al. [26] found that weedy sunflower was 11 times more competitive than shattercane and predicted a yield loss of 46% for maize due to competition from 4 weedy sunflower plants m^{-2}. In the study by Falkenberg et al. [99], competition 20–25 plants m^{-2} reduced maize net return by 66–68% compared to the case where weedy sunflower was controlled by herbicide application. In soybean, the presence of 3 plants m^{-2} reduced seed yield by 47–72% compared to weed-free conditions [24]. Geier et al. [100] noticed that weedy sunflower interference at a density of 4.6 plants m^{-2} resulted in almost complete seed yield loss. In another study, weedy sunflower caused a 94% reduction in seed yield under real–field conditions and reduced soybean height and biomass under greenhouse conditions [101]. In pulse crops, Mesbah et al. [102] observed that 1.5 weedy sunflower plants per m of row reduced the dry bean (*Phaseolus vulgaris* L.) seed yield by 27–34% and also that weedy sunflower was far more competitive than green foxtail (*Setaria viridis* (L.) Beauv.). Moreover, cowpea [*Vigna unguiculata* (L.) Walp.] biomass was reported to decrease by 77–82% in the presence of 6 weeds m^{-2} [28]. In cotton (*Gossypium hirsutum* L.), season-long interference resulted in complete yield loss at densities of 5, 10, 20, and 50 weedy sunflower plants m^{-2} [25]. As for another industrial crop, competition from 6, 12, 18, and 24 plants per 30 m of row was reported to reduce the root yield of sugar beet (*Beta vulgaris* L.) by 40, 52, 67, and 73%, respectively [103]. These authors also found that weedy sunflower was more competitive than velvetleaf (*Abutilon theophrasti* Medic.). In northeastern Mexico, Rosales–Robles et al. [104] recorded a grain yield loss of 27, 49, 60, 71 and 75% for spring wheat (*Triticum aestivum* L.), in the presence of 2, 4, 8, 16, and 32 weedy sunflower plants m^{-2}.

Table 3. Yield losses of summer field crops due to weedy sunflower (*Helianthus annuus* L.) interference. Results presented are from field trials repeated in time or space.

Crop	Weedy Sunflower Density	Yield Loss	Reference
Sunflower	4 Plants m^{-2}	50%	[13]
Sunflower	10.7 Plants m^{-2}	80%	[22]
Sunflower	12–15 Plants m^{-2}	35–60%	[86]
Maize	4 Plants m^{-2}	34%	[26]
Spring Wheat	2–32 Plants m^{-2}	27–75%	[104]
Soybean	3 Plants m^{-2}	47–72%	[24]
Soybean	4.6 Plants m^{-2}	97%	[100]
Soybean	220 Heads m^{-2}	94%	[101]
Dry Bean	1.5 Plants m^{-1} of Row	27–34%	[102]
Cowpea	6 Plants m^{-2}	77–82%	[28]
Cotton	5–50 Plants m^{-2}	100%	[25]
Sugar Beet	6–30 Plants m^{-1} of Row	40–73%	[103]

There is also evidence that weedy sunflower has invaded European fields in recent years. Infestations were observed mainly in the Mediterranean and Balkan Peninsula countries. In France, Muller et al. [86] recorded significant losses in seed yield (35–60%) of sunflower when grown in competition with 12–15 weedy sunflower plants m^{-2}. The same authors found 12 weedy sunflower populations in a total of 300 sunflower fields studied in Andalusia, Spain. In the same prefecture, Poverene and Cantamutto [105] detected weedy sunflower infestations at a density of 5–7 plants per 100 m^2 in a sunflower field and also detected weedy sunflower patches in uncultivated areas near sunflower fields. In Central Italy, weedy sunflower plants were found in sunflower, maize, sugar beet, processing tomato, alfalfa (*Medicago sativa* L.), and tobacco (*Nicotiana tabacum* L.) fields. The most severe infestations were observed on the moist margins of arable fields where tillage and herbicide treatments were limited or absent [106]. There are no official reports of weedy sunflower in Greece. However, farmers have recently complained about the presence of weedy sunflower plants in sunflower fields in the sunflower growing area of Domokos in Central Greece. According to these unofficial descriptions, the weedy plants are present at densities of 3–6 plants m^{-2} and exhibit typical weedy characteristics, such as branching and the formation of multiple heads with smaller diameters compared to cultivated sunflower hybrids (personal communication; unpublished data). Field surveys will be conducted at these sites to further investigate the development of weedy sunflower populations and also to quantify the effects of competition from weedy sunflowers on sunflower productivity under Greek soil and climatic conditions.

Elsewhere in the Balkan Peninsula, Saulic et al. [107] observed three weedy sunflower populations in northern Serbia, and the different populations showed variability in several morphological characteristics. Bozic et al. [108] conducted field experiments at two sites in Central Serbia where weedy sunflower populations occurred. These authors found that crop-to-weed gene flow was possible and depended on flowering time overlap, wind speed and direction, and also on the distance between the domesticated and wild plants. Stojićević et al. [45] demonstrated that weedy sunflower is a highly invasive species in Serbia, occurring at almost 200 sites with sunflower, maize and spring wheat. These authors found heavy infestations at some sites (20–30 plants m^{-2}) and reported that weedy sunflower can produce about 50–100 small-sized heads per plant (10,000–20,000 seeds per plant). Vrbnicanin et al. [98] studied three populations collected from Central Serbia and found that two populations were potentially resistant to nicosulfuron. According to Bozic et al. [108] and Vrbnicanin et al. [98], this species is also considered invasive in Croatia, Romania and Hungary. As for its occurrence on Central Europe, this weed was detected in sunflower fields and adjacent uncultivated areas on Czech Republic [109].

3.4. Herbicide Resistance

In addition to their competitive ability, weedy sunflower populations have developed resistance to several herbicides (Table 4).

Table 4. Cases of herbicide resistance in weedy sunflower [*Helianthus annuus* L.] populations. Results presented are from temporally and spatially replicated dose-response experiments.

Crop	Herbicide	Mode of Action	Chemical Family	Reference
Soybean	Imazethapyr	ALS Inhibitor	Imidazolinone	[110]
Soybean	Imazethapyr Imazamox Thifensulfuron–Methyl Chlorimuron–Ethyl	ALS Inhibitor ALS Inhibitor ALS Inhibitor ALS Inhibitor	Imidazolinone Imidazolinone Sulfonylurea Sulfonylurea	[111]
Soybean	Imazethapyr Imazaquin Imazamox Chlorimuron–Ethyl Cloransulam–Methyl Flumetsulam	ALS Inhibitor ALS Inhibitor ALS Inhibitor ALS Inhibitor ALS Inhibitor ALS Inhibitor	Imidazolinone Imidazolinone Imidazolinone Sulfonylurea Triazolopyrimidine Triazolopyrimidine	[112]
Soybean	Imazethapyr Chlorimuron–Ethyl	ALS Inhibitor ALS Inhibitor	Imidazolinone Sulfonylurea	[113]
Soybean	Imazethapyr Chlorimuron–Ethyl	ALS Inhibitor ALS Inhibitor	Imidazolinone Sulfonylurea	[75]
Maize	Glyphosate	EPSPS Inhibitor	Glycine	[114]
Sunflower	Imazamox	ALS Inhibitor	Imidazolinone	[30]
Sunflower	Imazapyr	ALS Inhibitor	Imidazolinone	[115]

The herbicide–resistant populations may be naturally selected following consecutive applications of herbicides with the same mode of action in a particular field. Resistance may also occur as a result of gene flow between herbicide–tolerant domesticated sunflower and its wild relatives.

3.4.1. Natural Selection of Herbicide–Resistant Weedy Sunflower Populations

In the USA, resistance to imazethapyr was confirmed in a population found in a soybean field treated with this herbicide for seven consecutive years [110]. Baumgartnen et al. [111] reported that these biotypes exhibited cross-resistance to imazamox, thifensulfuron–methyl, and chlorimuron–ethyl. Allen et al. [112] observed reduced sensitivity to imazethapyr, imazaquin, imazamox, chlorimuron–ethyl, cloransulam–methyl, and flumetsulam. These populations were collected from a soybean field where chlorimuron–ethyl was consecutively applied to control weedy sunflower in the past. White et al. [113] found a population that was 9 and 39 times more resistant to chlorimuron–ethyl and imazethapyr, respectively, compared to a sensitive population. This population was collected from a field where these herbicides were applied for eight years in rotation for the control of weedy sunflowers in soybean. Zelaya and Owen [75] noticed that a population was 36 and 43 times more resistant to imazethapyr and chlorimuron–ethyl, respectively, compared to a sensitive population. In addition, seven weedy sunflower populations were recently reported to have evolved resistance to glyphosate in fields where glyphosate–resistant maize and cotton were planted for several growing seasons [114].

It should be noted that the cases of herbicide resistance mentioned above were reported from the United States. As for Europe, Vrbnicanin et al. [98] collected two sunflower populations in Serbia from fields treated with nicosulfuron in consecutive years. These authors found that the application of nicosulfuron at the recommended field dose had no effect on the relative fitness and fecundity of the two potentially resistant populations. Although this is not an official case where herbicide resistance was confirmed in dose-response experiments, these results suggest that herbicide-resistant weedy sunflower may be evolving in Europe.

3.4.2. Herbicide Resistance as a Gene–Flow Consequence in *H. annuus*

Following the introduction of "Clearfield" technology, there is increasing consideration of the spread of imidazolinone-resistant weedy sunflowers in the USA and Europe. This technology was developed in 2003 to create sunflower hybrids with resistance to imidazolinone herbicides and to allow farmers to selectively control broadleaf weeds in the crop; imazamox is the only active ingredient registered for this purpose in the USA, while imazamox and imazapyr are approved in Europe [115]. However, there is evidence that these herbicide–resistant sunflower genotypes can successfully interbreed with wild populations that are present near a cultivated field, leading to the creation of imidazolinone-resistant weedy sunflower populations. Resistance to imazamox, for example, was reported by Massinga et al. [116] in the United States, while Presotto et al. [30] confirmed resistance to imazapyr in Argentina. In such populations, seed dormancy is not affected by hybridization. Seed production, although low in some cases, can increase rapidly when weedy sunflowers backcross with domesticated and wild sunflowers [30,98]. Another consequence of backcrossing is the successful transfer of herbicide resistance traits from weedy sunflowers to wild populations. These herbicide-resistant wild populations can encroach on new cultivated sunflower fields, hybridize with the crop, and generate new populations of herbicide-resistant weedy sunflowers [116].

4. Management of Shattercane [*Sorghum bicolor* (L.) Moench subsp. *drummondii*] and Weedy Sunflower (*Helianthus annuus* L.)

4.1. Proactive Strategies

Weed management should initially rely on the introduction of proactive strategies that prevent the spread of weeds to new agricultural lands [117,118]. Although the spread and establishment of these species is primarily facilitated by early seed shattering, late–emerging individuals may reach maturity at crop harvest [4,14]. Given the morphological and phenological overlap between these crops and their weedy relatives, weed seeds may be harvested when crops are harvested, resulting in seed lot contamination. As a result, shattercane and weedy sunflower can enter new sorghum and sunflower fields, respectively, as seed lot contaminants [39,63]. The machines used for seedbed preparation, sowing and harvesting, and threshing of grains and seeds should be carefully cleaned before moving them from one field to another [13,19]. In addition, systematic scouting of sorghum and sunflower fields for early detection of shattercane and weedy sunflower is crucial when weed density is low. When weedy populations are well established, their control is almost impossible [63,86]. Such proactive strategies prevent the spread of both species, their hybridization with domesticated sorghum and sunflower, and mitigate the consequences of gene flow between crops and their weedy relatives [76,119].

Another important measure to prevent gene flow is the management of crop volunteers and feral populations along field margins and in non–crop areas. To define the two terms: volunteers are crop plants derived from the unintentional loss of seeds during harvest [34]. The germination of these seeds creates populations of crop volunteers that can either grow in subsequent crops in the same field or migrate into field margins and adjacent non–crop areas. In the latter case, populations of a domesticated crop that escape from the field, survive, and successfully reproduce in unmanaged ecosystems are referred to as feral populations [33]. Feral sorghum and sunflower populations can successfully interbreed with shattercane and weedy sunflower, respectively, if they are located at the edge of an infested field [3,34]. The gene flow that occurs from feral to weedy individuals can be very problematic. In sorghum and sunflower fields infested with shattercane and weedy sunflower, respectively, gene flow may be reduced or not occur at all if there is no overlap in flowering time between crops and their weedy relatives. In such cases, flowering overlap may occur between feral populations in field margins and weedy populations growing in the agricultural field. Consequently, gene flow continues to occur. In other words, feral populations derived from volunteer crop plants can potentially serve as genetic bridges for gene transfer between crop plants and their weedy relatives [34].

Herbicide application is the most effective practice to control shattercane and weedy sunflower in field margins and non–crop areas. Glyphosate may be the most effective active ingredient enabling broad spectrum weed control in marginal areas [5]. However, overreliance should be avoided to prevent the development of glyphosate resistant weeds as recently observed in weedy sunflower populations [114]. To maintain its efficacy over time, alternative weed control options in non–crop areas should gain interest. For instance, recent research has shown that natural, environmentally friendly, non–selective herbicides can be effective on annual weeds if applied repeatedly in early weed growth stages [120].

4.2. *Reactive Strategies*

Once shattercane and weedy sunflower infestations are observed on agricultural land, reactive strategies for their management include the use of cultural practices, herbicides, and mechanical methods. Effective weed management is essential to avoid yield loss in a variety of summer field crops (including sorghum and sunflower) and also to prevent crop–weed gene flow in sorghum and sunflower fields.

4.2.1. Cultural Practices

Crop rotation is a cultural practice that increases crop diversity in an agricultural area since a series of crops are sequentially grown over time on the same land. In crop rotation systems, crop mimics such as shattercane are subjected to diverse agronomic practices and are affected by alterations in fundamental crop management practices, i.e., tillage, fertilization, irrigation regimes becoming less adaptable and competitive [121]. In addition, crop rotation is accompanied by the rotation of herbicides with different modes of action delaying the selection of herbicide-resistant populations [122]. The importance of crop rotation for the management of shattercane and weedy sunflower was highlighted in the case studies by Werle et al. [63] and Presotto et al. [30], respectively. Diversifying the corn-soybean rotation with cool-season crops such as winter wheat (*Triticum aestivum* L.) and canola (*Brassica napus* L.) resulted in significantly lower weedy sunflower infestation in the study by Anderson [123] especially under no-till conditions.

Growing a cover crop before the establishment of the main cash crop is another cultural practice that can be used for shattercane and weedy sunflower suppression. In the study by Whalen et al. [124] where shattercane was one of the dominant weeds in a soybean field, a cover crop mixture of cereal rye (*Secale cereale* L.) and hairy vetch (*Vicia villosa* Roth) resulted in 83% lower weed biomass; weed suppression increased when the use of cover crops was combined with the application of pre–emergence herbicides with soil residual activity. Sunn hemp (*Crotolaria juncea* L.) is a cover crop with aggressive growth recently reported to have suppressed weedy sunflower emergence and growth in the subsequent cash crop [125]. Intercropping, narrow row spacing, increased seeding rates, fertilization, and irrigation management should also be investigated for the suppression of shattercane and weedy sunflower. There is evidence that such practices contribute to weed management in summer field crops where shattercane and weedy sunflower are troublesome weeds [126–130]. The selection of competitive hybrids and cultivars was also reported to suppress shattercane in maize and weedy sunflowers in summer legumes such as cowpea [28,64]. In addition, the biological cycle of a particular crop genotype may result in no flowering overlap between the crop and its weedy relatives. Therefore, hybrid and cultivar selection may be an option to prevent crop–weed gene flow in sorghum and sunflower. The same is noted for manipulations in crop sowing dates [3,119].

The preparation of a firm seedbed, the use of germinable crop seed, sowing date and sowing depth selection are also cultural practices ensuring optimal crop growth and can lead to the suppression of noxious weeds such as shattercane and weedy sunflower [117]. False seedbed is another cultural, non–chemical, practice recommended for the control of shattercane and weedy sunflower in a great variety of summer crops including sorghum and sunflower. To apply this practice, the conventional tillage practices used for seedbed preparation are not followed by crop establishment. On the contrary, weeds are left to emerge. At this time, irrigations are encouraged because they stimulate greater weed emergence. After approximately 2 weeks, when the main flush of emergence has passed, weeds are controlled by shallow tillage. Weed control is followed by crop sowing [131]. If shattercane and weedy sunflower populations continue to occur, they can be controlled by subsequent cultivations between crop rows [132].

4.2.2. Herbicides and Mechanical Methods

The strong botanical ties between crops and their weedy relatives precludes, in most cases, selective herbicide use to control shattercane in sorghum fields and weedy sunflower in sunflower fields. The selective control of shattercane and weedy sunflower is possible only when "Inzen" sorghum and "Clearfield" sunflower are treated with ALS-inhibiting herbicides. However, crop–wild gene flow is very likely to result in the spread of herbicide-resistant hybrids in the field [30,63]. In any case, herbicide application is more preferable to be carried out before crop sowing under the concept of stale seedbed. Stale seedbed includes the same actions as false seedbed apart from the weed control method. In stale seedbeds, weed control is carried out by the application of a non–selective herbicide [131]. Both glyphosate and pelargonic acid, a natural contact type non–selective herbicide, have been recently reported to provide sufficient control of annual weeds in summer crops [133].

There are more selective herbicide options in crops which are not genetically related to shattercane and weedy sunflower. However, herbicides with different modes of action should be rotated or applied in mixtures to avoid the development of herbicide-resistant populations [118]. As for mechanical methods, cultivation between crop rows can effectively control both species [134,135]. Mechanical operations may need to be repeated; a general recommendation is to increase the number of interrow cultivations to increase the efficacy of mechanical weed control [126]. There is also evidence that multiple mowing operations between crop rows can also provide solutions in shattercane control [136]. The same author denoted that mowing can be effectively combined with herbicide application. Such practices should also be tested against weedy sunflower.

5. Conclusions

Shattercane and weedy sunflower are two examples of CWRs that have become troublesome weeds in agriculture. Key weedy characteristics such as early seed shattering and seed dormancy play an important role in their success as agricultural weeds. Both species are very competitive to their closely related domesticated crops. Moreover, they can cause severe yield losses in a wide variety of summer field crops. Both species are widely reported as important agricultural weeds in the United States and have invaded various agricultural areas in Europe. Resistance to herbicides was confirmed in both shattercane and weedy sunflower populations. Crop rotation, false seedbed, cover crops, and competitive crop genotypes are valuable cultural practices for suppressing both species. In addition, preventative measures should be also adopted to avoid their spread to new agricultural land. The development of effective weed management strategies is also essential to prevent hybridization between sorghum, sunflower and their wild relatives and mitigate its consequences.

Author Contributions: Conceptualization, P.K., I.G., and I.T.; methodology, P.K.; software, P.K. and I.G.; validation, P.K., I.G., A.T. (Alexandros Tataridas), A.T. (Anastasia Tsekoura), and N.A.; writing—original draft preparation, P.K., I.G., A.T. (Alexandros Tataridas), N.A., and I.T.; writing—review and editing, P.K., I.G., and S.Z.; visualization, P.K. and S.Z.; supervision, P.K.; project administration, P.K. and I.T. All authors have read and agreed to the published version of the manuscript.

Funding: This research received no external funding.

Institutional Review Board Statement: Not applicable.

Informed Consent Statement: Not applicable.

Data Availability Statement: Not applicable.

Conflicts of Interest: The authors declare no conflict of interest.

References

1. Maxted, N.; Ford-Lloyd, B.V.; Jury, S.; Kell, S.; Scholten, M. Towards a definition of a crop wild relative. *Biodivers. Conserv.* **2006**, *15*, 2673–2685. [CrossRef]
2. Ananda, G.K.S.; Myrans, H.; Norton, S.L.; Gleadow, R.; Furtado, A.; Henry, R.J. Wild sorghum as a promising resource for crop improvement. *Front. Plant Sci.* **2020**, *11*, 1108. [CrossRef] [PubMed]
3. Ohadi, S.; Hodnett, G.; Rooney, W.; Bagavathiannan, M. Gene flow and its consequences in *Sorghum* spp. *Crit. Rev. Plant Sci.* **2017**, *36*, 367–385. [CrossRef]
4. Presotto, A.; Fernández-Moroni, I.; Poverene, M.; Cantamutto, M. Sunflower crop-wild hybrids: Identification and risks. *Crop Prot.* **2011**, *30*, 611–616. [CrossRef]
5. Travlos, I.S.; Montull, J.M.; Kukorelli, G.; Malidza, G.; Dogan, M.N.; Cheimona, N.; Antonopoulos, N.; Kanatas, P.J.; Zannopoulos, S.; Peteinatos, G. Key aspects on the biology, ecology and impacts of johnsongrass [*Sorghum halepense* (L.) Pers] and the role of glyphosate and non-chemical alternative practices for the management of this weed in Europe. *Agronomy* **2019**, *9*, 717. [CrossRef]
6. Honsdorf, N.; March, T.J.; Berger, B.; Tester, M.; Pillen, K. High-throughput phenotyping to detect drought tolerance QTL in wild barley introgression lines. *PLoS ONE* **2014**, *9*, e97047.
7. Munns, R.; James, R.A.; Xu, B.; Athman, A.; Conn, S.J.; Jordans, C.; Byrt, C.S.; Haere, R.A.; Tyerman, S.D.; Tester, M.; et al. Wheat grain yield on saline soils is improved by an ancestral Na^+ transporter gene. *Nat. Biotechnol.* **2012**, *30*, 360–364. [CrossRef]
8. Zhou, Y.L.; Uzokwe, V.N.E.; Zhang, C.H.; Cheng, L.R.; Wang, L.; Chen, K.; Gao, X.Q.; Sun, Y.; Chen, J.J.; Zhu, L.H.; et al. Improvement of bacterial blight resistance of hybrid rice in China using the *Xa23* gene derived from wild rice (*Oryza rufipogon*). *Crop Prot.* **2011**, *30*, 637–644. [CrossRef]
9. Hu, D.; Zhang, H.; Du, Q.; Hu, Z.; Yang, Z.; Li, X.; Wang, J.; Huang, F.; Yu, D.; Wang, H.; et al. Genetic dissection of yield-related traits via genome-wide association analysis across multiple environments in wild soybean (*Glycine soja* Sieb. and Zucc.). *Planta* **2020**, *251*, 1–17. [CrossRef]
10. Capel, C.; Del Carmen, A.F.; Alba, J.M.; Lima-Silva, V.; Hernández-Gras, F.; Salinas, M.; Boronat, A.; Angosto, T.; Botella, M.A.; Fernández-Muñoz, R.; et al. Wide-genome QTL mapping of fruit quality traits in a tomato RIL population derived from the wild-relative species *Solanum pimpinellifolium* L. *Theor. Appl. Genet.* **2015**, *128*, 2019–2035. [CrossRef]
11. Swamy, B.M.; Sarla, N. Yield-enhancing quantitative trait loci (QTLs) from wild species. *Biotechnol. Adv.* **2008**, *26*, 106–120. [CrossRef] [PubMed]
12. Kaya, Y.; Jocic, S.; Miladinovic, D. Sunflower. In *Technological Innovations in Major World Oil Crops*; Gupta, S.K., Ed.; Springer: New York, NY, USA, 2012; Volume 1, pp. 85–129.
13. Casquero, M.; Presotto, A.; Cantamutto, M. Exoferality in sunflower (*Helianthus annuus* L.): A case study of intraspecific/interbiotype interference promoted by human activity. *Field Crops Res.* **2013**, *142*, 95–101. [CrossRef]
14. Defelice, M.S. Shattercane, *Sorghum bicolor* (L.) Moench ssp. *drummondii* (Nees ex Steud.) De Wet ex Davidse—Black sheep of the family. *Weed Technol.* **2006**, *20*, 1076–1083.
15. Nadir, S.; Xiong, H.-B.; Zhu, Q.; Zhang, X.-L.; Xu, H.-Y.; Li, J.; Dongchen, W.; Henry, D.; Guo, X.-Q.; Khan, S. Weedy rice in sustainable rice production. A review. *Agron. Sustain. Dev.* **2017**, *37*, 46. [CrossRef]
16. Ejeta, G.; Grenier, C. Sorghum and its weedy hybrids. In *Crop Ferality and Volunteerism*; Gressel, J., Ed.; CRC Press: Boca Raton, FL, USA, 2005; pp. 123–135.
17. Presotto, A.; Poverene, M.; Cantamutto, M. Seed dormancy and hybridization effect of the invasive species, *Helianthus annuus*. *Ann. Appl. Biol.* **2014**, *164*, 373–383. [CrossRef]
18. Arias, D.M.; Rieseberg, L.H. Gene flow between cultivated and wild sunflowers. *Theor. Appl. Genet.* **1994**, *89*, 655–660. [CrossRef] [PubMed]
19. Chauhan, B.S. Strategies to manage weedy rice in Asia. *Crop Prot.* **2013**, *48*, 51–56. [CrossRef]
20. Presotto, A.; Hernández, F.; Casquero, M.; Vercellino, R.; Pandolfo, C.; Poverene, M.; Cantamutto, M. Seed bank dynamics of an invasive alien species, *Helianthus annuus* L. *J. Plant Ecol.* **2020**, *13*, 313–322. [CrossRef]
21. Schmidt, J.J.; Yerka, M.K.; Pedersen, J.F.; Lindquist, J.L. Growth, fitness, and overwinter survival of a shattercane (*Sorghum bicolor* ssp. *drummondii*) × grain sorghum (*Sorghum bicolor* ssp. *bicolor*) F_2 population. *Weed Sci.* **2018**, *66*, 634–641.
22. Casquero, M.; Cantamutto, M. Interference of the agrestal *Helianthus annuus* biotype with sunflower growth. *Weed Res.* **2016**, *56*, 229–236. [CrossRef]
23. Werle, R.; Bernards, M.L.; Sattler, S.E.; Lindquist, J.L. Susceptibility of shattercane × ALS-resistant sorghum hybrids and their parents to rimsulfuron and nicosulfuron. In Proceedings of the 53rd Annual Meeting of Weed Science Society of America, Baltimore, MD, USA, 4–7 February 2013; Weed Science Society of America: Champaign, IL, USA, 2013. Abstract 331.
24. Allen, J.R.; Johnson, W.G.; Smeda, R.J.; Kremer, R.J. ALS resistant *Helianthus annuus* interference in *Glycine max*. *Weed Sci.* **2000**, *48*, 461–466. [CrossRef]

25. Charles, G.W.; Sindel, B.M.; Cowie, A.L.; Knox, O.G. Determining the critical period for weed control in high-yielding cotton using common sunflower as a mimic weed. *Weed Technol.* **2019**, *33*, 800–807. [CrossRef]
26. Deines, S.R.; Dille, J.A.; Blinka, E.L.; Regehr, D.L.; Staggenborg, S.A. Common sunflower (*Helianthus annuus*) and shattercane (*Sorghum bicolor*) interference in corn. *Weed Sci.* **2004**, *52*, 976–983. [CrossRef]
27. Raey, Y.; Ghassemi-Golezani, K.; Javanshir, A.; Alyari, H.; Mohammadi, S.A. Interference between shattercane (*Sorghum bicolor*) and soybean (*Glycine max*). *N. Z. J. Crop Hortic. Sci.* **2005**, *33*, 53–58. [CrossRef]
28. Wang, G.; McGiffen, M.E.; Ehlers, J.D.; Marchi, E.C. Competitive ability of cowpea genotypes with different growth habit. *Weed Sci.* **2006**, *54*, 775–782. [CrossRef]
29. Heap, I. The International Herbicide-Resistant Weed Database. Available online: www.weedscience.org (accessed on 2 September 2021).
30. Presotto, A.; Ureta, M.S.; Cantamutto, M.; Poverene, M. Effects of gene flow from IMI resistant sunflower crop to wild *Helianthus annuus* populations. *Agric. Ecosyst. Environ.* **2012**, *146*, 153–161. [CrossRef]
31. Werle, R.; Schmidt, J.J.; Laborde, J.; Tran, A.; Creech, C.F.; Lindquist, J.L. Shattercane × ALS-tolerant sorghum F_1 hybrid and shattercane interference in ALS-tolerant sorghum. *J. Agric. Sci.* **2014**, *6*, 159–165. [CrossRef]
32. Ellstrand, N.C.; Prentice, H.C.; Hancock, J.F. Gene flow and introgression from domesticated plants into their wild relatives. *Annu. Rev. Ecol. Evol. Syst.* **1999**, *30*, 539–563. [CrossRef]
33. Gressel, J. Introduction—The challenges of ferality. In *Crop Ferality and Volunteerism*; Gressel, J., Ed.; CRC Press: Boca Raton, FL, USA, 2005; pp. 1–7.
34. Reagon, M.; Snow, A.A. Cultivated *Helianthus annuus* (Asteraceae) volunteers as a genetic "bridge" to weedy sunflower populations in North America. *Am. J. Bot.* **2006**, *93*, 127–133. [CrossRef]
35. Engku, A.K.; Norida, M.; Juraimi, A.S.; Rafii, M.Y.; Abdullah, S.N.A.; Alam, M.A. Gene flow from Clearfield® rice to weedy rice under field conditions. *Plant Soil Environ.* **2016**, *62*, 16–22.
36. Sahoo, L.; Schmidt, J.J.; Pedersen, J.F.; Lee, D.J.; Lindquist, J.L. Growth and fitness components of wild × cultivated *Sorghum bicolor* (Poaceae) hybrids in Nebraska. *Am. J. Bot.* **2010**, *97*, 1610–1617. [CrossRef] [PubMed]
37. Schmidt, J.J.; Pedersen, J.F.; Bernards, M.L.; Lindquist, J.L. Rate of shattercane × sorghum hybridization in situ. *Crop Sci.* **2013**, *53*, 1677–1685. [CrossRef]
38. Faure, N.; Serieys, H.; Bervillé, A. Potential gene flow from cultivated sunflower to volunteer, wild *Helianthus* species in Europe. *Agric. Ecosyst. Environ.* **2002**, *89*, 183–190. [CrossRef]
39. Ureta, M.S.; Carrera, A.D.; Cantamutto, M.A.; Poverene, M.M. Gene flow among wild and cultivated sunflower, *Helianthus annuus* in Argentina. *Agric. Ecosyst. Environ.* **2008**, *123*, 343–349. [CrossRef]
40. Taiz, L.; Zeiger, E.; Moller, I.M.; Murphy, A. *Plant Physiology and Development*, 6th ed.; Sinauer Associates: Sunderland, CT, USA, 2015.
41. Taylor, L.P.; Hepler, P.K. Pollen germination and tube growth. *Annu. Rev. Plant Biol.* **1997**, *48*, 461–491. [CrossRef] [PubMed]
42. Steinhorst, L.; Kudla, J. Calcium—A central regulator of pollen germination and tube growth. *Biochim. Biophys. Acta* **2013**, *1833*, 1573–1581. [CrossRef] [PubMed]
43. Zheng, R.H.; Su, S.D.; Xiao, H.; Tian, H.Q. Calcium: A Critical factor in pollen germination and tube elongation. *Int. J. Mol. Sci.* **2019**, *20*, 420. [CrossRef]
44. Berenji, J.; Dahlberg, J. Perspectives of sorghum in Europe. *J. Agron. Crop Sci.* **2004**, *190*, 332–338. [CrossRef]
45. Stojićević, D.; Ilić, A.; Sekulić, T.; Stupar, V.; Božić, D.; Vrbničanin, S. Distribution of weedy sunflower on territory of Republic of Serbia and potential risks for agriculture. *J. Hortic. For. Biotechnol.* **2017**, *21*, 132–137.
46. De Wet, J.M.J. Systematics and evolution of *Sorghum* Sect. *Sorghum* (Gramineae). *Am. J. Bot.* **1978**, *65*, 477–484. [CrossRef]
47. Harlan, J.R.; de Wet, J.M.J. A simplified classification of cultivated sorghum. *Crop Sci.* **1972**, *12*, 172–176. [CrossRef]
48. Wiersema, J.H.; Dahlberg, J. The nomenclature of *Sorghum bicolor* (L.) Moench (Gramineae). *Taxon* **2007**, *56*, 941–946. [CrossRef]
49. Beckett, T.H.; Stoller, E.W.; Wax, L.M. Interference of four annual weeds in corn (*Zea mays*). *Weed Sci.* **1988**, *36*, 764–769. [CrossRef]
50. Dille, J.A.; Stahlman, P.W.; Thompson, C.R.; Bean, B.W.; Soltani, N.; Sikkema, P.H. Potential yield loss in grain sorghum (*Sorghum bicolor*) with weed interference in the United States. *Weed Technol.* **2020**, *34*, 624–629. [CrossRef]
51. Hans, S.R.; Johnson, W.G. Influence of shattercane [*Sorghum bicolor* (L.) Moench.] interference on corn (*Zea mays* L.) yield and nitrogen accumulation. *Weed Technol.* **2002**, *16*, 787–791. [CrossRef]
52. Celarier, R.P. Cytotaxonomic notes on the subsection *Halepensia* of the genus *Sorghum*. *Bull. Torrey Bot. Club* **1958**, *85*, 49–62. [CrossRef]
53. Paterson, A.H.; Schertz, K.F.; Lin, Y.R.; Liu, S.C.; Chang, Y.L. The weediness of wild plants: Molecular analysis of genes influencing dispersal and persistence of johnsongrass, *Sorghum halepense* (L.) Pers. *Proc. Natl. Acad. Sci. USA* **1995**, *92*, 6127–6131. [CrossRef] [PubMed]
54. Dillon, S.L.; Lawrence, P.K.; Henry, R.J.; Price, H.J. *Sorghum* resolved as a distinct genus based on combined ITS1, ndh F and Adh 1 analyses. *Plant Syst. Evol.* **2007**, *268*, 29–43. [CrossRef]
55. Klein, P.; Smith, C.M. Invasive johnsongrass, a threat to native grasslands and agriculture. *Biologia* **2021**, *76*, 413–420. [CrossRef]
56. Quinby, J.R.; Karper, R.E. Inheritance of height in sorghum. *Agron. J.* **1954**, *46*, 211–216. [CrossRef]
57. Horak, M.J.; Moshier, L.J. Shattercane (*Sorghum bicolor*) biology and management. *Rev. Weed Sci.* **1994**, *6*, 133–149.

58. Burnside, O.C.; Wicks, G.A.; Fenster, C.R. Longevity of shattercane seed in soil across Nebraska. *Weed Res.* **1977**, *17*, 139–143. [CrossRef]
59. Fellows, G.M.; Roeth, F.W. Factors influencing shattercane (*Sorghum bicolor*) seed survival. *Weed Sci.* **1992**, *40*, 434–440. [CrossRef]
60. Adugna, A.; Bekele, E. Morphology and fitness components of wild × crop F1 hybrids of *Sorghum bicolor* (L.) in Ethiopia: Implications for survival and introgression of crop genes in the wild pool. *Plant Genet. Resour.* **2013**, *11*, 196–205. [CrossRef]
61. Vesecky, J.F.; Feltner, K.C.; Vanderlip, R.L. Wild cane and forage sorghum competition in grain sorghum. *Weed Sci.* **1973**, *21*, 28–32. [CrossRef]
62. King, S.R.; Hagood, E.S. Herbicide programs for the control of ALS-resistant shattercane (*Sorghum bicolor*) in corn (*Zea mays*). *Weed Technol.* **2006**, *20*, 416–421. [CrossRef]
63. Fellows, G.M.; Roeth, F.W. Shattercane (*Sorghum bicolor*) interference in soybean (*Glycine max*). *Weed Sci.* **1992**, *40*, 68–73. [CrossRef]
64. Hoffman, M.L.; Buhler, D.D. Utilizing sorghum as a functional model of crop-weed competition. I. Establishing a competitive hierarchy. *Weed Sci.* **2002**, *50*, 466–472. [CrossRef]
65. Werle, R.; Tenhumberg, B.; Lindquist, J.L. Modeling shattercane dynamics in herbicide-tolerant grain sorghum cropping systems. *Ecol. Modell.* **2017**, *343*, 131–141. [CrossRef]
66. Schaffasz, A.; Windpassinger, S.; Friedt, W.; Snowdon, R.; Wittkop, B. Sorghum as a novel crop for Central Europe: Using a broad diversity set to dissect temperate-adaptation. *Agronomy* **2019**, *9*, 535. [CrossRef]
67. Popescu, A. Sorghum production in the EU–28 in the period 2008–2019 and its forecast for 2020–2014 horizon. *Sci. Pap. Ser. Manag. Econ. Eng. Agric. Rural. Dev.* **2020**, *20*, 479–488.
68. United States Department of Agriculture (USDA). Weed Risk Assessment for *Sorghum bicolor* (L.) Moench Nothosubsp. *drummondii* (Steud.) de Wet ex Davidse (Poaceae). Available online: https://www.aphis.usda.gov/plant_health/plant_pest_info/weeds/downloads/wra/sorghum-bicolor-drummondii.pdf (accessed on 18 September 2021).
69. Dahlberg, J.; Berenji, J.; Sikora, V.; Latković, D. Assessing sorghum [*Sorghum bicolor* (L.) Moench] germplasm for new traits: Food, fuels & unique uses. *Maydica* **2012**, *56*, 85–92.
70. Schwartz-Lazaro, L.M.; Gage, K.L. Sustainable weed control in grain sorghum. In *Weed Control: Sustainability, Hazards, and Risks in Cropping Systems Worldwide*; Korres, N.E., Burgos, N.R., Duke, S.O., Eds.; CRC Press: Boca Raton, FL, USA, 2018; pp. 262–275.
71. Berenji, J.; Dahlberg, J.; Sikora, V.; Latkovi, D. Origin, history, morphology, production, improvement, and utilization of broomcorn [*Sorghum bicolor* (L.) Moench] in Serbia. *Econ. Bot.* **2011**, *65*, 190–208. [CrossRef]
72. Anderson, D.D.; Roeth, F.W.; Martin, A.R. Discovery of a primisulfuron-resistant shattercane (*Sorghum bicolor*) biotype. *Weed Technol.* **1998**, *12*, 74–77. [CrossRef]
73. Lee, C.D.; Martin, A.R.; Roeth, F.W.; Johnson, B.E.; Lee, D.J. Comparison of ALS inhibitor resistance and allelic interactions in shattercane accessions. *Weed Sci.* **1999**, *47*, 275–281. [CrossRef]
74. Brenly-Bultemeier, T.L.; Stachler, J.; Harrison, S.K. Confirmation of shattercane (*Sorghum bicolor*) resistance to ALS-inhibiting herbicides in Ohio. *Plant Health Prog.* **2002**, *3*, 1. [CrossRef]
75. Zelaya, I.A.; Owen, M.D.K. Evolved resistance to acetolactate synthase-inhibiting herbicides in common sunflower (*Helianthus annuus*), giant ragweed (*Ambrosia trifida*), and shattercane (*Sorghum bicolor*) in Iowa. *Weed Sci.* **2004**, *52*, 538–548. [CrossRef]
76. Werle, R.; Jhala, A.J.; Yerka, M.K.; Dille, J.A.; Lindquist, J.L. Distribution of herbicide-resistant shattercane and johnsongrass populations in sorghum production areas of Nebraska and Northern Kansas. *Agron. J.* **2016**, *108*, 321–328. [CrossRef]
77. Mutegi, E.; Sagnard, F.; Labuschagne, M.; Herselman, L.; Semagn, K.; Deu, M.; de Villiers, S.; Kanyenji, B.M.; Mwongera, N.; Traore, P.C.S.; et al. Local scale patterns of gene flow and genetic diversity in a crop-wild-weedy complex of sorghum (*Sorghum bicolor* (L.) Moench) under traditional agricultural field conditions in Kenya. *Conserv. Genet.* **2012**, *13*, 1059–1071. [CrossRef]
78. Magomere, T.O.; Obukosia, S.D.; Shibairo, S.I.; Ngugi, E.K.; Mutitu, E. Evaluation of relative competitive ability and fitness of *Sorghum bicolor* × *Sorghum halepense* and *Sorghum bicolor* × *Sorghum sudanense* F_1 hybrids. *J. Biol. Sci.* **2015**, *15*, 1–15. [CrossRef]
79. Tuinstra, M.R.; Al-Khatib, K. Acetolactate Synthase Herbicide Resistant Sorghum. U.S. Patent Application 20080216187, 4 September 2008.
80. Schilling, E.E. Helianthus. In *Flora of North America*; Flora of North America Editorial Committee, Ed.; Oxford University Press: New York, NY, USA; Oxford, UK, 2006; Volume 21, pp. 141–169.
81. Stebbins, J.C.; Winchell, C.J.; Constable, J.V.H. *Helianthus winteri* (Asteraceae), a new perennial species from the southern Sierra Nevada foothills, California. *Aliso* **2013**, *31*, 19–24. [CrossRef]
82. Timme, R.E.; Simpson, B.B.; Linder, C.R. High-resolution phylogeny for *Helianthus* (Asteraceae) using the *18S–26S* ribosomal DNA external transcribed spacer. *Am. J. Bot.* **2007**, *94*, 1837–1852. [CrossRef] [PubMed]
83. Heiser, C.B., Jr. Taxonomy of *Helianthus* and origin of domesticated sunflower. In *Sunflower Science and Technology*; Carter, J.F., Ed.; American Society of Agronomy, Crop Science Society of America, Soil Science Society of America: Madison, WI, USA, 1978; Agronomy Monographs; Volume 19, pp. 31–53.
84. Heiser, C.B., Jr. Variation and subspeciation in the common sunflower, *Helianthus annuus*. *Am. Midl. Nat.* **1954**, *51*, 287–305. [CrossRef]
85. Cantamutto, M.; Poverene, M. The transgenic sunflower. In *Genetics, Genomics and Breeding of Sunflower*; Hu, J.G., Seiler, G., Kole, C., Eds.; Science Publishers: Enfield, CT, USA, 2010; pp. 279–312.

86. Muller, M.H.; Delieux, F.; Fernandez-Martinez, J.M.; Garric, B.; Lecomte, V.; Anglade, G.; Leflon, M.; Motard, C.; Segura, R. Occurrence, distribution and distinctive morphological traits of weedy *Helianthus annuus* L. populations in Spain and France. *Genet. Resour. Crop Evol.* **2009**, *56*, 869–877. [CrossRef]
87. Muller, M.H.; Latreille, M.; Tollon, C. The origin and evolution of a recent agricultural weed: Population genetic diversity of weedy populations of sunflower (*Helianthus annuus* L.) in Spain and France. *Evol. Appl.* **2011**, *4*, 499–514. [CrossRef]
88. Presotto, A.; Hernández, F.; Díaz, M.; Fernández-Moroni, I.; Pandolfo, C.; Basualdo, J.; Cuppari, S.; Cantamutto, M.; Poverene, M. Crop-wild sunflower hybridization can mediate weediness throughout growth-stress tolerance trade-offs. *Agric. Ecosyst. Environ.* **2017**, *249*, 12–21. [CrossRef]
89. Lincoln, R.J.; Boxshall, G.A.; Clarck, P.F. *A Dictionary of Ecology, Evolution and Systematics*, 2nd ed.; Cambridge University Press: Cambridge, UK, 1998.
90. Cantamutto, M.; Torres, L.; Presotto, A.; Gutierrez, A.; Ureta, S.; Poverene, M. Migration pattern suggested by terrestrial proximity as possible origin of wild annual *Helianthus* populations in central Argentina. *Biol. Invasions* **2010**, *12*, 541–551. [CrossRef]
91. Poverene, M.; Cantamutto, M.; Seiler, G.J. Ecological characterization of wild *Helianthus annuus* and *Helianthus petiolaris* germplasm in Argentina. *Plant Genet. Resour.* **2009**, *7*, 42–49. [CrossRef]
92. Burke, J.M.; Gardner, K.A.; Rieseberg, L.H. The potential for gene flow between cultivated and wild sunflower (*Helianthus annuus*) in the United States. *Am. J. Bot.* **2002**, *89*, 1550–1552. [CrossRef] [PubMed]
93. Cantamutto, M.; Presotto, A.; Moroni, I.F.; Alvarez, D.; Poverene, M.; Seiler, G. High infraspecific diversity of wild sunflowers (*Helianthus annuus* L.) naturally developed in central Argentina. *Flora* **2010**, *205*, 306–312. [CrossRef]
94. Kane, N.C.; Rieseberg, L.H. Genetics and evolution of weedy *Helianthus annuus* populations: Adaptation of an agricultural weed. *Mol. Ecol.* **2008**, *17*, 384–394. [CrossRef]
95. Alexander, H.M.; Schrag, A.M. Role of soil seed banks and newly dispersed seeds in population dynamics of the annual sunflower, *Helianthus annuus. J. Ecol.* **2003**, *91*, 987–998. [CrossRef]
96. Moody-Weis, J.; Alexander, H.M. The mechanisms and consequences of seed bank formation in wild sunflowers (*Helianthus annuus*). *J. Ecol.* **2007**, *95*, 851–864. [CrossRef]
97. Mercer, K.L.; Wyse, D.L.; Shaw, R.G. Effects of competition on the fitness of wild and crop-wild hybrid sunflower from a diversity of wild populations and crop lines. *Evolution* **2006**, *60*, 2044–2055. [CrossRef]
98. Vrbnicanin, S.P.; Bozic, D.M.; Pavlovic, D.M.; Saric-Krsmanovic, M.M.; Stojicevic, D.; Uludag, A. Fitness studies on invasive weedy sunflower populations from Serbia. *Rom. Biotechnol. Lett.* **2017**, *22*, 12464–12472.
99. Falkenberg, N.R.; Cogdill, T.J.; Rister, M.E.; Chandler, J.M. Economic evaluation of common sunflower (*Helianthus annuus*) competition in field corn. *Weed Technol.* **2012**, *26*, 137–144. [CrossRef]
100. Geier, P.W.; Maddux, L.D.; Moshier, L.J.; Stalman, P.W. Common sunflower (*Helianthus annuus*) interference in soybean (*Glycine max*). *Weed Technol.* **1996**, *16*, 787–791. [CrossRef]
101. Irons, S.M.; Burnside, O.C. Competitive and allelopathic effects of sunflower (*Helianthus annuus*). *Weed Sci.* **1982**, *30*, 372–377. [CrossRef]
102. Mesbah, A.O.; Miller, S.D.; Koetz, P.J. Common sunflower (*Helianthus annuus*) and green foxtail (*Setaria viridis*) interference in dry bean. *Weed Technol.* **2004**, *18*, 902–907. [CrossRef]
103. Schweizer, E.E.; Bridge, L.D. Sunflower (*Helianthus annuus*) and velvetleaf (*Abutilon theophrasti*) interference in sugarbeets (*Beta vulgaris*). *Weed Sci.* **1982**, *30*, 514–519. [CrossRef]
104. Rosales-Robles, E.; Salinas-García, J.R.; Sánchez-de-la-Cruz, R.; Rodríguez-del-Bosque, L.A.; Esqueda-Esquivel, V. Interference and control of wild sunflower (*Helianthus annuus* L.) in spring wheat (*Triticum aestivum* L.) in northeastern México. *Cereal Res. Commun.* **2002**, *30*, 439–446. [CrossRef]
105. Poverene, M.; Cantamutto, M. A comparative study of invasive *Helianthus annuus* populations in their natural habitats of Argentina and Spain. *Helia* **2010**, 33, 63–74. [CrossRef]
106. Vischi, M.; Cagiotti, M.E.; Cenci, C.A.; Seiler, G.J.; Olivieri, A.M. Dispersal of wild sunflower by seed and persistant basal stalks in some areas of Central Italy. *Helia* **2006**, *29*, 89–94. [CrossRef]
107. Saulic, M.; Stojicevic, D.; Matkovic, A.; Bozic, D.; Vrbnicanin, S. Population variability of weedy sunflower as invasive species. In Proceedings of the 4th ESENIAS Workshop: International Workshop on IAS in Agricultural and Non-Agricultural Areas in ESENIAS Region, Çanakkale, Turkey, 16–17 December 2013; Uludağ, A., Trichkova, T., Tomov, R., Eds.; Çanakkale Onsekiz Mart University, Turkey—East and South European Network for Invasive Alien Species (ESENIAS)—Institute of Biodiversity and Ecosystem Research, BAS, Bulgaria: Çanakkale, Turkey, 2013; pp. 79–85.
108. Bozic, D.; Pavlovic, D.; Bregola, V.; Di Loreto, A.; Bosi, S.; Vrbnicanin, S. Gene flow from herbicide-resistant sunflower hybrids to weedy sunflower. *J. Plant Dis. Prot.* **2015**, *122*, 183–188. [CrossRef]
109. Holec, J.; Soukup, J.; Cerovska, M.; Novakova, K. Common sunflower (*Helianthus annuus* var. *annuus*)—Potential threat to coexistence of sunflower crops in Central Europe. In Proceedings of the 2nd European Conference on Co-Existence between GM and Non-GM Based Agricultural Supply Chain, Montpellier, France, 14–15 November 2005; Messean, S., Ed.; Agropolis Productions: Montpellier, France, 2005; pp. 271–272.
110. Al-Khatib, K.; Baumgartner, J.R.; Peterson, D.E.; Currie, R.S. Imazethapyr resistance in common sunflower (*Helianthus annuus*). *Weed Sci.* **1998**, *46*, 403–407. [CrossRef]

111. Baumgartner, J.R.; Al-Khatib, K.; Currie, R.S. Cross-resistance of imazethapyr-resistant common sunflower (*Helianthus annuus*) to selected imidazolinone, sulfonylurea, and triazolopyrimidine herbicides. *Weed Technol.* **1999**, *13*, 489–493. [CrossRef]
112. Allen, J.R.; Johnson, W.G.; Smeda, R.J.; Wiebold, W.J.; Massey, R.E. Management of acetolactate synthase (ALS)–resistant common sunflower (*Helianthus annuus* L.) in soybean (*Glycine max*). *Weed Technol.* **2001**, *15*, 571–575. [CrossRef]
113. White, A.D.; Owen, M.D.; Hartzler, R.G.; Cardina, J. Common sunflower resistance to acetolactate synthase-inhibiting herbicides. *Weed Sci.* **2002**, *50*, 432–437. [CrossRef]
114. Singh, V.; Etheredge, L.; McGinty, J.; Morgan, G.; Bagavathiannan, M. First case of glyphosate resistance in weedy sunflower (*Helianthus annuus*). *Pest Manag. Sci.* **2020**, *76*, 3685–3692. [CrossRef]
115. Tan, S.; Evans, R.R.; Dahmer, M.L.; Singh, B.K.; Shaner, D.L. Imidazolinone-tolerant crops: History, current status and future. *Pest Manag. Sci.* **2005**, *61*, 246–257. [CrossRef]
116. Massinga, R.A.; Al-Khatib, K.; Amand, P.S.; Miller, J.F. Gene flow from imidazolinone-resistant domesticated sunflower to wild relatives. *Weed Sci.* **2003**, *51*, 854–862. [CrossRef]
117. Gazoulis, I.; Kanatas, P.; Papastylianou, P.; Tataridas, A.; Alexopoulou, E.; Travlos, I. Weed management practices to improve establishment of selected lignocellulosic crops. *Energies* **2021**, *14*, 2478. [CrossRef]
118. Norsworthy, J.K.; Ward, S.M.; Shaw, D.R.; Llewellyn, R.S.; Nichols, R.L.; Webster, T.; Bradley, K.; Frisvold, G.; Powles, S.; Burgos, N.R.; et al. Reducing the risks of herbicide resistance: Best management practices and recommendations. *Weed Sci.* **2012**, *60*, 31–62. [CrossRef]
119. Roumet, M.; Noilhan, C.; Latreille, M.; David, J.; Muller, M.-H. How to escape from crop-to-weed gene flow: Phonological variation and isolation-by-time within weedy sunflower population. *New Phytol.* **2013**, *197*, 642–654. [CrossRef] [PubMed]
120. Kanatas, P.; Antonopoulos, N.; Gazoulis, I.; Travlos, I.S. Screening glyphosate-alternative weed control options in important perennial crops. *Weed Sci.* **2021**, 1–15. [CrossRef]
121. Reeves, D.W. Cover crops and rotations. In *Crops Residue Management (Advances in Soil Science)*, 1st ed.; Hatfield, J.L., Stewart, B.A., Eds.; Lewis Publishers: Boca Raton, FL, USA, 1994; pp. 125–172.
122. Kanatas, P. Mini-Review: The role of crop rotation, intercropping, sowing dates and increased crop density towards a sustainable crop and weed management in arable crops. *Agraarteadus* **2020**, *31*, 22–27.
123. Anderson, R.L. Crop sequence and no-till reduce seedling emergence of common sunflower (*Helianthus annuus*) in following years. *Weed Technol.* **2007**, *21*, 355–358. [CrossRef]
124. Whalen, D.M.; Shergill, L.S.; Kinne, L.P.; Bish, M.D.; Bradley, K.W. Integration of residual herbicides with cover crop termination in soybean. *Weed Technol.* **2020**, *34*, 11–18. [CrossRef]
125. Soti, P.; Racelis, A. Cover crops for weed suppression in organic vegetable systems in semiarid subtropical Texas. *Org. Agric.* **2020**, *10*, 429–436. [CrossRef]
126. Kanatas, P.J.; Gazoulis, I. The integration of increased seeding rates, mechanical weed control and herbicide application for weed management in chickpea (*Cicer arietinum* L.). *Phytoparasitica* **2021**, 1–13. [CrossRef]
127. Kanatas, P.; Gazoulis, I.; Travlos, I. Irrigation timing as a practice of effective weed management in established alfalfa (*Medicago sativa* L.) crop. *Agronomy* **2021**, *11*, 550. [CrossRef]
128. Kanatas, P.; Travlos, I.; Kakabouki, I.; Papastylianou, P.; Gazoulis, I. Yield of organically grown maize hybrids as affected by two green manure crops in Greece. *Chil. J. Agric. Res.* **2020**, *80*, 334–341. [CrossRef]
129. Iqbal, N.; Manalil, S.; Chauhan, B.S.; Adkins, S.W. Effect of narrow row-spacing and weed crop competition duration on cotton productivity. *Arch. Agron. Soil Sci.* **2020**, 1–13. [CrossRef]
130. Rad, S.V.; Valadabadi, S.A.R.; Pouryousef, M.; Saifzadeh, S.; Zakrin, H.R.; Mastinu, A. Quantitative and qualitative evaluation of *Sorghum bicolor* L. under intercropping with legumes and different weed control methods. *Horticulturae* **2020**, *6*, 78. [CrossRef]
131. Travlos, I.; Gazoulis, I.; Kanatas, P.; Tsekoura, A.; Zannopoulos, S.; Papastylianou, P. Key factors affecting weed seeds' germination, weed emergence, and their possible role for the efficacy of false seedbed technique as weed management practice. *Front. Agron.* **2020**, *2*, 1. [CrossRef]
132. Del Pino, A.M.; Pannacci, E.; Di Michele, A.; Bravi, E.; Marconi, O.; Tei, F.; Palmerini, C.A. Selective inhibition of wild sunflower reproduction with mugwort aqueous extract, tested on cytosolic Ca^{2+} and germination of the pollen grains. *Plants* **2021**, *10*, 1364. [CrossRef] [PubMed]
133. Kanatas, P.; Travlos, I.; Papastylianou, P.; Gazoulis, I.; Kakabouki, I.; Tsekoura, A. Yield, quality and weed control in soybean crop as affected by several cultural and weed management practices. *Not. Bot. Horti Agrobot. Cluj-Napoca* **2020**, *48*, 329–341. [CrossRef]
134. Burnside, O.C. Shattercane control in narrow-row soybeans. *Agron. J.* **1980**, *72*, 753–757. [CrossRef]
135. Muller, M.H.; Lecomte, V.; Garric, B.; Jouffret, P.; Leflon, M.; Pourageaux, F.; Ségura, R. Weedy sunflowers in France: Prevalence and first inferences on their origin. In Proceedings of the 17th International Sunflower Conference, Cordoba, Spain, 8–12 June 2008; pp. 685–690.
136. Donald, W.W. Control of both winter annual and summer annual weeds in no-till corn with between-row mowing systems. *Weed Technol.* **2007**, *21*, 591–601. [CrossRef]

diversity

MDPI

Article

FloCan—A Revised Checklist for the Flora of the Canary Islands

Carl Beierkuhnlein [1,2,3,*], Anna Walentowitz [1] and Walter Welss [4]

[1] Department of Biogeography, University of Bayreuth, Universitätsstr. 30, 95440 Bayreuth, Germany; anna.walentowitz@uni-bayreuth.de
[2] Geographical Institute Bayreuth (GIB), 95440 Bayreuth, Germany
[3] Bayreuth Center of Ecology and Environmental Science (BayCEER), 95440 Bayreuth, Germany
[4] Botanical Garden, Friedrich-Alexander-University Erlangen-Nürnberg, Loschgestraße 1, 91054 Erlangen, Germany; walter.welss@fau.de
* Correspondence: carl.beierkuhnlein@uni-bayreuth.de; Tel.: +49-921-552270

Abstract: The flora of the Canary Islands has been subject to botanical studies for more than 200 years. Several biodiversity databases are available for the archipelago. However, there are various drivers of change in real biodiversity and the knowledge about it constantly needs to be kept track of. Island floras are both: exposed to species loss and to species introductions, either through natural processes or by anthropogenic drivers. Additionally, the evolution of endemic plant species plays a substantial role. Endemic species are sensitive to population decline due to small population sizes and possible low competitiveness against incoming species. Additionally, there is continuous progress in systematics and taxonomy. Species names or their taxonomic attribution can be modified. Here, we check published plant lists for the Canary Islands and literature, and compile currently accepted taxa into an updated checklist. For this FloCan checklist, several sources were compiled, checked for completeness and quality, and their taxonomy was updated. We illustrate how far plant names are considered in regional or global databases. This work represents the current state of knowledge on Canary Island plant diversity, including introduced and recently described taxa. We provide a comprehensive and updated basis for biogeographical and macroecological studies. Particularly, the number of non-native species is being extended substantially. The adaptation to standard international nomenclature supports integration into large-scale studies.

Keywords: alien species; archipelago; biodiversity; databases; endemism; evolutionary arena; GBIF; invasive species; island biogeography; island biota; Macaronesia; macroecology; non-native alien species; plants; TRY; EU Biodiversity Strategy

Citation: Beierkuhnlein, C.; Walentowitz, A.; Welss, W. FloCan—A Revised Checklist for the Flora of the Canary Islands *Diversity* **2021**, *13*, 480. https://doi.org/10.3390/d13100480

Academic Editors: Tibor Magura and Michael Wink

Received: 25 August 2021
Accepted: 23 September 2021
Published: 29 September 2021

Publisher's Note: MDPI stays neutral with regard to jurisdictional claims in published maps and institutional affiliations.

Copyright: © 2021 by the authors. Licensee MDPI, Basel, Switzerland. This article is an open access article distributed under the terms and conditions of the Creative Commons Attribution (CC BY) license (https://creativecommons.org/licenses/by/4.0/).

1. Introduction

The Flora of the Canary Islands archipelago has attracted botanists for centuries. Alexander von Humboldt, for example, spent a week on Tenerife in the year 1799. During this short stay, he described the elevational distribution of plant species at the slope of Mount Teide. The iconic figure displaying the altitudinal distribution key species was published after his return to Europe in 1826 [1]. In the same year as Humboldt's ascent to Mont Teide, in 1799, the first natural history book on the Canary Islands comprising species lists, and written by Viera y Clavijo, was released [2].

Knowing about Humboldt's experiences, Charles Darwin was keen to see the vegetation of Tenerife in January 1832 when the Beagle arrived offshore. However, nobody was allowed to leave the Beagle at the harbour of Santa Cruz de Tenerife due to quarantine restrictions because of the Cholera epidemic in London at that time. Darwin depicts the situation in his report published in 1839 [3] "*Oh misery, misery—we were just preparing to drop our anchor within 1/2 a mile of Santa Cruz when a boat came alongside bringing with it our death-warrant. The consul declared we must perform a rigorous quarantine of twelve days. Those who have never experienced it can scarcely conceive what a gloom it cast on every one: Matters were soon decided by the Captain ordering all sail to be set & make a course for the Cape Verd Islands. We have left perhaps one of the most interesting places in the world, just at the moment when we were near enough for every object to create, without satisfying, our utmost curiosity"*. Captain Fitzroy, manoeuvring the Beagle, wrote about this situation "*this was a great disappointment to Mr Darwin, who had cherished a hope of visiting the peak. To see it, to anchor and be on the point of landing, yet be obliged to turn away without the slightest prospect of beholding Tenerife again, was indeed to him a real calamity"*. Having a strong background in botany, Darwin might have had an eye-opening experience if this incidence had not prevented him from exploring the flora of the Canary Islands.

The middle of the 19th century saw a strong stimulation of botanical assessments with the outstanding "Histoire naturelle des Iles Canaries" written by Philip Barker Webb and Sabin Berthelot (and Afred Moquin-Tandon) [4]. This book became the most important landmark for a complete flora of the archipelago. A few decades later, Hermann Christ [5,6] published another list of plant species in the archipelago. During the 20th century, an extensive series of floristic studies and species lists were edited with the pioneering work of Charles-Josef Pitard [7]. Increasingly new contributions by botanists such as Bornmüller, Buch, Sventenius, Broussonet, Masferrer, Burchard, Hansen, Kunkel, Sunding, Bramwell, Santos Guerra, and Schönfelder, to name but a few, increased the knowledge about the Canarian flora [8–13]. Additionally, nowadays the archipelago continues to attract the attention of international naturalists and scientists. This long legacy of botanical research evokes the impression that the plant species of the islands are well known, and it may be one of the best investigated regions of the planet. However, even though the Canary Islands were colonised and settled by Europeans centuries ago and have become an attractive destination for tourists, no complete survey of the entire islands could ever be conducted. The steep terrain of remote mountain slopes, inaccessible gorges (barrancos), and rugged cliffs at their coastline are restricting human investigations and hamper accessibility [14,15]. Substantial parts of the archipelago cannot be reached and are not even accessible to climbers due to the loose parent material of young volcanic rock.

In recent years, several new plant species have been described in the archipelago [16–21]. However, the publication of a new species does not translate directly into being incorporated and accepted in standard international databases. This process takes time. Additionally, expert knowledge exists about species that exist in nature but has not yet been addressed in official and accepted scientific publications. Several species are known and mentioned by experts but not officially described yet. This even applies to woody species of considerable size and with clear morphological distinction (Figure 1).

Furthermore, plant species recorded in the past have been reported to have disappeared (e.g., *Hypecoum procumbens* L. or *Grammitis quaerenda* Bolle), to have extinguished populations, or to have become extinct [22]. Some endemic plant species are currently at the brink of extinction (e.g., *Lotus eremiticus* A. Santos). This is a common syndrome on islands, where species populations can be small and viable population sizes can easily be undershot. Comparable examples exist in the Galapagos islands for previously important key species from the endemic genus *Scalesia* [23] or on Mauritius, exhibiting the prominent example of *Sideroxylon grandiflorum* [24]. Others are likely to have disappeared in the past and might only be reflected as genera in rare studies of pollen records [25]. Many islands have an extensive legacy of biodiversity loss. However, there is little quantitative evidence for this due to the limited preservation of plant remains.

Because of the continuous progress in plant systematics and taxonomy, there is a need to update and unify nomenclature, particularly for those genera or families under debate and for those experiencing considerable upheaval. Understandably, it is mostly those groups that have undergone rapid radiation and diversification in the archipelago, respectively, where the identification of species and their relatedness is work in progress [26–28].

(a) (b)

Figure 1. On the island of La Palma (Canary Islands) just one species of the genus *Carlina* L. with narrow leaves is listed in most floras. This species (*C. falcata* Svent.) (**a**) is relatively abundant and widespread on the island. However, in remote parts of northern cliffs a subspecies of *Carlina canariensis* (L.f.) Cav. was recorded recently (**b**), which has not been scientifically described up to now. A regular publication of this taxon following the rules of ICN (International Code of Nomenclature for algae, fungi, and plants) is in preparation. This illustrates that even for well-known genera of woody species, new species' descriptions are still to be expected.

During recent decades, several floras and species lists of the Canary Islands have been published and updated as online databases reflecting the state of knowledge and its rapid development. Here, we can just mention prominent examples of this vast literature [22,29–35]. It is difficult to decide on one work as a standard. Some are mainly rooted in regional and local knowledge; others are better related to international taxonomic standards; some are more recently published; others follow a more rigorous understanding of systematics; some are continuously updated online databases; others are printed books. When looking closer at some taxa, there is disagreement in many details, including the acceptance of species and differences in the perceptions of their systematic and taxonomic status. However, there is a general need to unify and update taxonomic and spatial information on species [36] to reflect real biodiversity at its best and to enable, e.g., inter-island comparison.

Regional flora and checklists of the Canary Islands are increasingly used in macroecological and biogeographical studies [37–39], which, however, may reflect just a subsection of the real species diversity of the archipelago [40,41]. The resulting findings may be questionable if a substantial part of the existing species is being ignored [42]. Furthermore, outdated taxonomy might inhibit or weaken studies at larger scales.

Openly available public webpages on endemic plants of the Canary Islands, for instance in Wikipedia, differ strongly in content between languages. The Spanish site lists 122 endemic plant species [43]; the English version linked to the same page informs about only 68 endemic plant species [44]! This illustrates that there is not a clear common ground on this topic. A reason for these differences might be the definition of endemism. It is by its nature scale-dependent [45,46] and may be subjectively defined if the area it is related to is not clearly limited [47]. Additionally, there is human bias to be considered meaning that people might be used to certain species names or taxa that have a high value in nature conservation, and that persist even if there is scientific evidence that such names can no longer be accepted and must be updated.

The advancement of knowledge and confusion due to new findings for the Canary Island flora is understandable at best with the example of the dragon tree (*Dracaena draco* (L.) L.). The dragon tree is maybe the most iconic plant species in the archipelago. Alexander von Humboldt was already fascinated by the impressive life form of several specimen in the valley of Orotava, Tenerife, back in his days. Then, 200 years later in 1997, a small population of a subspecies of *Dracaena draco* was found in the Moroccan High Atlas Mountains (subsp. *ajgal* Benabid & Cuzin). In consequence, *Dracaena draco* was no longer a Macaronesian endemic species (to the Canary Islands and Cabo Verde) but an endemic subspecies (*D. draco* subsp. *draco*) for these islands. One year later, in 1998, another *Dracaena* species *Dracaena tamaranae* A. Marrero, R.S. Almeida & M. Gonzalez-Martin was found on the island of Gran Canaria, again adding an endemic *Dracaena* species to the archipelago [48]. Such surprises and discoveries would rather be expected for less famous plants or for cases where the deterministic traits are not so obvious but are astonishing for the most well-known plant taxa of the islands.

Another prominent example is the Canary Island laurel tree, a key species of the laurel forest, which is best preserved in this archipelago. *Laurus novocanariensis* Rivas Mart., Lousã, Fern. Prieto, E. Días, J.C. Costa & C. Aguiar was named *Laurus azorica* (Seub.) Franco before and even further back in time was named *Laurus canariensis* Webb & Berthel. non Willd. To add to the confusion, *Laurus canariensis* Willd. is a published but invalid synonym for another Lauracean species of the island: *Apollonias barbujana* (Cav.) Bornm. In summary, *Laurus novocanariensis*, as it is currently named, was an archipelago endemic species first, became a Macaronesian endemic, and back again an archipelago endemic taxon. The name has changed due to an increasing understanding of phylogenetic relatedness. However, it may happen that in the future this taxon could be understood as a synonym to *Laurus nobilis* L. from the Mediterranean which would mean that one more classified endemic species would be lost from the list of plant species just in the human understanding of biological taxonomy—without any consequences for nature [49].

Changes in the number of species that are described for islands are mostly due to newly introduced and even invasive species. These species are being introduced by past and present trade and travelling activities that connect the islands to other parts of the world [50]. This applies strongly to the Canary Islands that have been used as a testing ground for the acclimatisation of exotic species from the New World that were intended to be introduced to the Mediterranean as early as in the 16th century [51]. The subtropical and oceanic climate supports the establishment of many species across tropical and subtropical biogeographic realms. The differentiation of climatic conditions within the archipelago and even within topographically diverse islands with pronounced elevational zones and differences in precipitation and moisture regimes between leeward and windward sides adds to the spectrum of available habitats for the establishment of non-native biota [52]. Many plants have been introduced for ornamental purposes. Those species may remain confined to gardens or parks, but they may also start reproducing and spreading after a certain time lag and establish a legacy of offspring. However, the same mechanism applies for newly introduced species as for unknown endemics; they must first be detected before they can be registered in any database or list. Too often knowledge about cultivated ornamental plants that were established outside their natural range has been ignored until such species turned "wild" and created problems and damage.

Another process that is modifying the diversity of islands is extinction. Many of these extinction events may have occurred unnoticed since invasive species, unknown herbivores such as rabbits or goats, or other predators and pathogens have been introduced. However, the pressure of introduced herbivores on island species that have not evolved defence mechanisms is still pertinent [53]. Many endemic plant species have dramatically declined in distribution and abundance with the consequence of becoming hyper-endemic [46], surviving as remnant populations at the brink of extinction. In the case of *Lotus eremiticus* A. Santos only one "population" of few specimens remains in nature, most probably built up by a single clone [54]. One single (local) disturbance event could erase such a species globally. Considering the vulnerability of such hyper-endemics, it becomes evident that the human impact has clearly reduced intraspecific diversity. Generally, it is not a given fact that island taxa exhibit low genetic diversity as concluded from their small populations [55,56].

The genus *Aeonium* is iconic to the archipelago and is also considered as an example of adaptive radiation. Of the 42 species of the genus, 36 are endemic to the Canary Islands. Surprisingly, the phylogenetically oldest species do not originate from the continent

(Morocco, Eastern Africa, Yemen), but the continental species instead represent the young branches of the phylogenetic tree [57]. The earth history of "Paleo-Macaronesia" is one explanation for this pattern [58].

Many more volcanic islands more existed in this tectonically complex area at the edge of the oceanic and the continental crust during the last 80 Mio. years. These former islands have disappeared from the ocean surface due to erosion. However, they can be detected with the help of bathymetry as guyots at the sea level of the Last Glacial Maximum 200 (LGM), which was their basis for erosion [59]". Furthermore, the spatial fluctuation of island area during the Pleistocene had an influence on plant species richness [60].

Increasingly, modern molecular methods allow for new insights into the distinction of species and to their attribution to higher phylogenetic units, influencing nomenclature, systematics, and biogeography [61–64]. Realistically, this process is a continuous one that will not end soon because of the immense diversity of plant populations and the historical focus on mere morphological traits for classification. Hidden relations between taxa need to be uncovered, and cryptic species that resemble other published species need to be identified. Consequently, data sources for regional and trans-regional assessments need to be updated continuously and adapted to international standards in nomenclature to avoid artefacts just through deviating terminologies.

The Canary Islands are an outstanding example of an essential field for ecological research, the preservation of endemic taxa, and for biogeographical studies. Here, we offer an overview of the current knowledge and perception of the flora of the Canary Islands, knowing that near future developments in phylogeny and discoveries of new species might modify this picture. Nevertheless, we identify that the customary reference to one specific list of plant species has an influence on the scientific statements made. Based on the comparison of established approaches, databases, and lists of the Canary Islands, we identified common general taxonomic agreement but also inconsistencies. Based on this overview, we present an updated and revised flora for the archipelago that is open for additions and corrections.

2. Materials and Methods

We first reviewed and compiled published species lists for the Canary Islands [8–13,22,29–35]. The resulting plant species list was then complemented with individual publications on specific taxonomic groups for specific regions or islands. In addition, documented cultivated ornamental plants and crops were included. This study is focused on the 7 major islands of the Canary Islands archipelago: El Hierro, La Palma, La Gomera, Tenerife, Gran Canaria, Fuerteventura, and Lanzarote. With very few exceptions, all taxa are attributed to their occurrence on these islands.

Small islets in the close vicinity of Lanzarote and Fuerteventura were excluded as they have been covered in a recent study that provided a checklist for these islets [65]. However, there were no additional species listed in comparison with our checklist.

Criteria for accepting a taxon in this new FloCan plant checklist for the Canary Islands were the reliability of records and whether the respective species or subspecies name is accepted or considered a synonym following international standards of nomenclature and systematics. In the case of questionable records, additional proof was explicitly searched, and if a record for a given species on an individual island or for the entire archipelago was not found, this species was deleted from the list. The same applied if a plant population was considered an independent subspecies (or species), but this taxonomic categorization did not align with international standards. However, such synonyms or erroneously reported species are also maintained in an appendix to allow checking for the existence of these taxa in the future. For highly debated taxa, we also consider current scientific literature on plant phylogenetics (e.g., [66] for the genus *Micromeria*).

The resulting list was then compared to international standard taxonomic checklists prioritising those species that are listed in Plants of the World online (POWO) [67]. As a result, published species names in one list or dataset can translate into subspecies and vice versa.

Additionally, we consulted World Flora Online (WFO) [68] for all taxa, which is the replacement of The Plant List (TPL) [69] that has not been updated since 2013. Therefore, TPL was not considered explicitly in this checklist to avoid redundance and outdated viewpoints. We further compared all taxa with Catalogue of Life [70], which generally accepts more taxa as valid names than World Flora Online. We do not suggest that one of these databases is superior to another one. However, for standardisation we decided to follow the suggested plant names of one checklist, generally Plants of the World Online, indicating nevertheless whether other databases share the same name for a respective plant species, or suggest a synonym as the accepted name.

Additionally, we checked names in GBIF [71], which is not a taxonomic database, but is frequently used and to be considered a standard to obtain species' distribution data. Furthermore, we screened the TRY [72,73] database for plant functional traits. These international databases differ in terms of scientific scrutiny and their general philosophy. For example, TRY does not provide author citation, which is a substantial part of botanical names.

In this study, databases are not evaluated. Rather it shall be illustrated which taxa are considered under the name given in this checklist (or under a synonym) in these pre-valent databases. We follow, with only very few exceptions, in the naming for plant species, genus, family, and subspecies the standards of POWO. Other infraspecific units (e.g., varieties) are only considered in a few cases. There is no clear agreement across checklists to which degree infraspecific taxa should be considered below the level of subspecies.

Deviating nomenclature or missing representation in other plant lists is protocolled. This option in our checklist illustrates which taxa are generally accepted and where no consistent opinion across published plant species lists can be seen. There are cases where the same species' name was published by several authors. Confusion could result from missing author citations if one of these apparently equitable names is defined as a synonym of another accepted species.

For list comparison we used a semi-automatised approach and compared our species list with global databases via the "taxize" package [74] applied in R [75]. Every species for which several or no results were given were rechecked manually. However, this filter was complemented by individual reviews and revisions for all taxa, to overcome, for example, deviating spelling of names. Botanical publications that are based on taxa should follow the International Code of Nomenclature for algae, fungi, and plants (ICN) as a global compilation of published plant names in scholarly publications serves the International Plant Names Index (IPNI) [76]. However, IPNI is not aiming to provide the latest state of knowledge for the progress in taxonomy and its reflection in botanical nomenclature.

Island biota and endemic species, but also recently published data on taxa with revised nomenclature, are likely to be underrepresented in global databases. This needs to be considered to avoid data bias in trans-regional biodiversity studies. In our checklist, the current state of representation of Canary Island plants in these databases is protocolled, even though such data repositories will further develop towards higher completeness and representation. This comparison serves as orientation on the reliability of research that uses such open data sources without scrutinising and comparing every single species.

In addition to published floras and plant lists, we reviewed the current botanical literature, focusing on studies about recently documented established non-native species and taxonomic studies for selected species groups [77–108]. These studies are sources for additional species records and changes in plant names, which are not yet included in global or regional databases.

To indicate taxa that are highly debated compared to others, we checked the number of published synonyms in POWO. Infraspecific taxa such as subspecies and varieties are also listed if they are officially published. We separate our analyses for different taxonomic levels. In some genera (e.g., *Aeonium*, *Micromeria*) a series of hybrids are described. Hybrids are listed in a separate table in order not to ignore this aspect of biodiversity, but also to show that such organisms, that might appear with a certain regularity, have not evolved (yet) to accepted species.

Finally, we suggest an updated plant list including information on the spatial distribution of taxa across the Canary Islands and their status. We classify native species as probably native and surely native. Non-native species were categorised as probably non-native in cases where this is not certain, surely non-native and invasive non-native (i.e., intruding into and substantially modifying natural ecosystems). The term "non-native" is being used as a synonym of the term "alien". Our study aims to improve the picture of the currently existing taxa on the island, but also to acknowledge that numerous taxa are unanimous or under debate.

3. Results

This new FloCan checklist informs about the current state of knowledge of the flora of the Canary Islands considering regional floras and international taxonomic databases as well as specific publications (Supplement Materials Part S1–S6). It aims at providing a transparent overview of the acceptance of species and infraspecific names, suggesting a revised checklist. The total number of accepted taxa in this checklist sums up to 2812 (1781 native, 1031 non-native), comprising 2416 species (1452 native, 964 non-native) and 396 infraspecific taxa (329 native, 67 non-native) (Supplement Materials Part S1, Main Table for species and Infraspecific taxa)".

There are substantial differences between islands in the number and proportion of native and non-native taxa, species, and infraspecific taxa (A species list can be found in Supplement Materials Part S2). The proportion of native taxa is high for the arid islands with less topographic diversity in the eastern part of the archipelago (Fuerteventura and Lanzarote) (Figure 2). Islands with a large human population such as Tenerife or Gran Canaria posess a large number of non-native species. Generally, there are not many non-native infraspecific taxa, which reflects that such biota are an indicator of ongoing speciation.

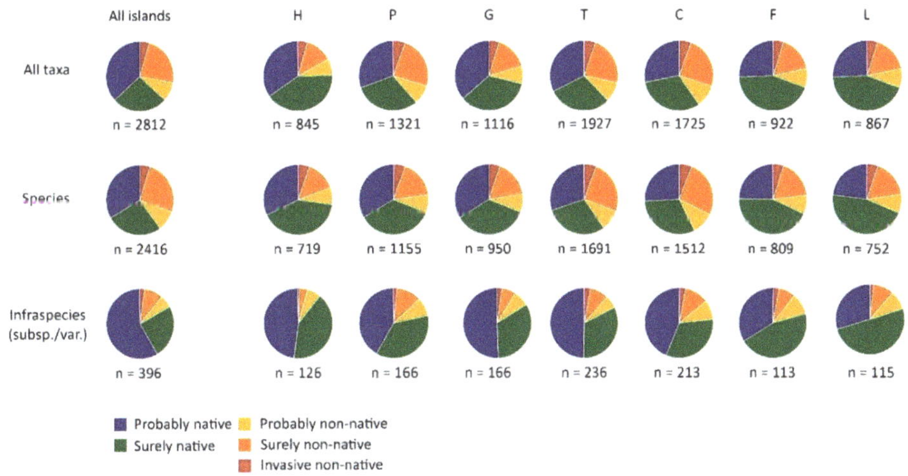

Figure 2. Categories of native and non-native taxa in the Canary Islands and for the individual islands El Hierro (H), La Palma (P), La Gomera (G), Tenerife (T), Gran Canaria (C), Fuerteventura (F) and Lanzarote (L) based on the here presented checklist. Proportions are given at the level of all taxa, and separately for species and infra specific taxa (sub-species and varieties). Absolute numbers of taxa are given below the pie charts. Generally, the proportion of accepted native infraspecific units (subspecies, varieties) is higher compared with the accepted species. Highest numbers of taxa are listed for the large islands with pronounced topography and diverse climatic conditions.

Within the native taxa (Figure 3), we find a consistent proportion of endemic taxa between approx. 20 and 40 per cent. The proportion of single-island endemic species (SIE) is only higher in comparison with multi-island (archipelago) endemics (MIE) for the sum of all taxa and species across all islands. On individual islands, the proportion of MIE is always larger than that of SIE. The arid islands with less pronounced relief again show comparable patterns, dominated by native species that are non-endemic. Surprisingly, the proportion (and absolute number) of SIE is very small also for El Hierro, whereas the other mountainous islands exhibit a consistently high proportion of SIE. Infraspecific taxa exhibit higher proportions of SIE and MIE compared to species or taxa level consistently for all individual islands but show the opposite pattern for the entire archipelago.

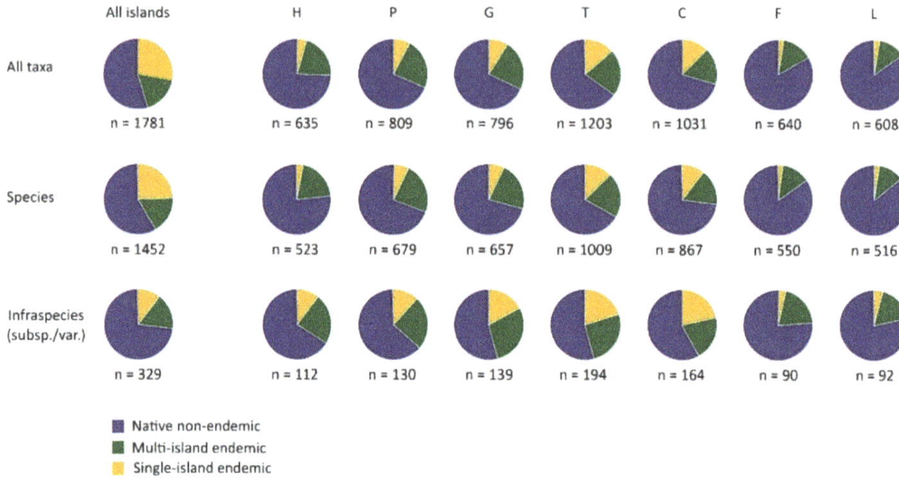

Figure 3. Categories for native taxa subdivided into native non-endemic species (blue), multi-island endemics (green) and single island endemics (yellow). Proportions are given at the level of all taxa, and separately for species and infraspecific taxa (sub-species and varieties). Absolute numbers of taxa are given below the pie charts (n). The relations are illustrated for the entire Canary Islands archipelago (All islands) and for the individual islands El Hierro (H), La Palma (P), La Gomera (G), Tenerife (T), Gran Canaria (C), Fuerteventura (F) and Lanzarote (L). Surprisingly, approximately one quarter of all species in the entire archipelago are SIE. The highest proportions of endemic taxa relate to the islands with pronounced topography and diverse climatic conditions (H, P, G, T, C).

As this FloCan checklist provides an updated list of plant taxa, including their representation in other existing floras and databases, we want to illustrate to what degree other lists are deviating from the here suggested taxonomy (Figure 4). Uncertainty is generally high for infraspecific taxa. The reflection of accepted taxa is good in GBIF. However, many accepted taxa, mainly endemic and infraspecific taxa are not well represented in the TRY database, which can cause bias in ecological studies.

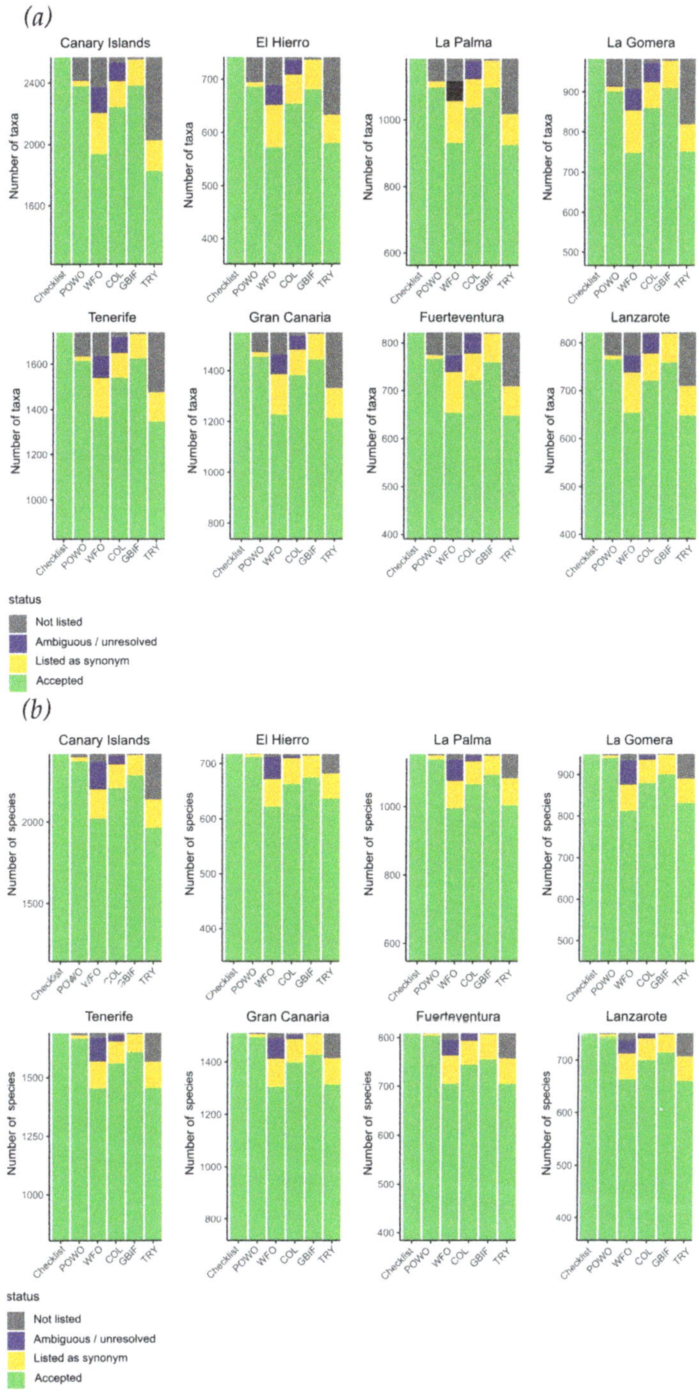

Figure 4. *Cont.*

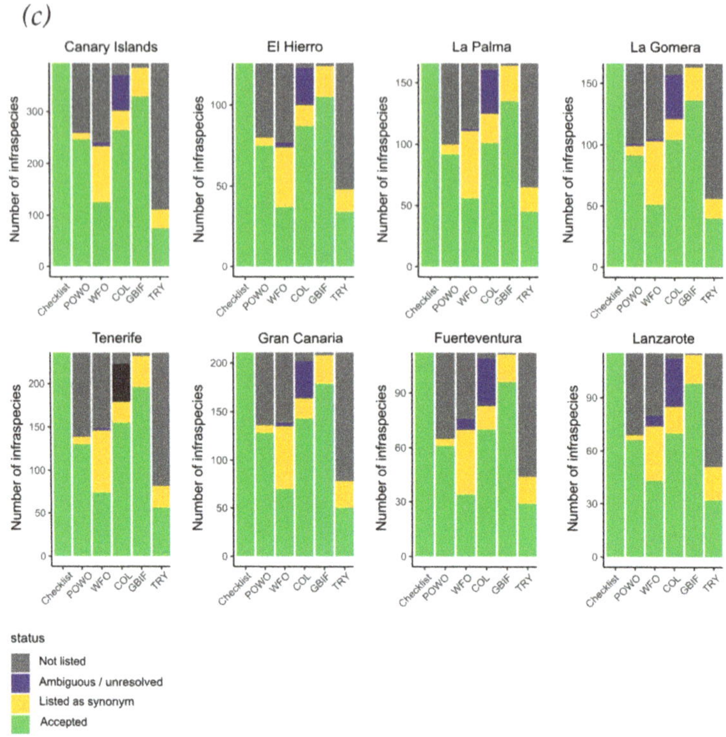

Figure 4. Representation of the accepted taxa suggested in this checklist for the Canary Islands in global databases. (**a**) for all taxa, (**b**) for species, and (**c**) for infraspecific taxa (subspecies and varieties). Note that scales for (**a**,**b**) do not begin with 0 taxa/species but start at 50% of each individual y-axis to work out differences. Axis (**c**) for infraspecific taxa start in 0. Numbers of accepted taxa diverge between POWO and WFO, reflecting that POWO was taken as a reference. COL numbers are close to those of POWO, which supports the decision to select this database for reference. The patterns for individual islands resemble the pattern of the entire archipelago.

Generally, the compiled list of taxa is well reflected in two currently applied reference lists for the flora of the Canary Islands [22,34] (Figure 5). However, a substantial number of taxa appear in these lists as not-accepted synonyms, and also in international reference databases (Figure 4).

In addition to the main checklist, we add a list of currently not considered taxa that were published for the Canary Islands before, illustrating the reason for exclusion such as being a synonym of another accepted taxon or an obviously erroneous record or misunderstanding (Supplement Materials Part S5). As it does not make sense to list all synonyms ever published, which can sum up to more than 100 for a single species, we report those synonyms in detail in Supplement Materials Part S6 that are used in the above-mentioned datasets and floras referred to in this checklist to guide readers towards accepted names in this checklist.

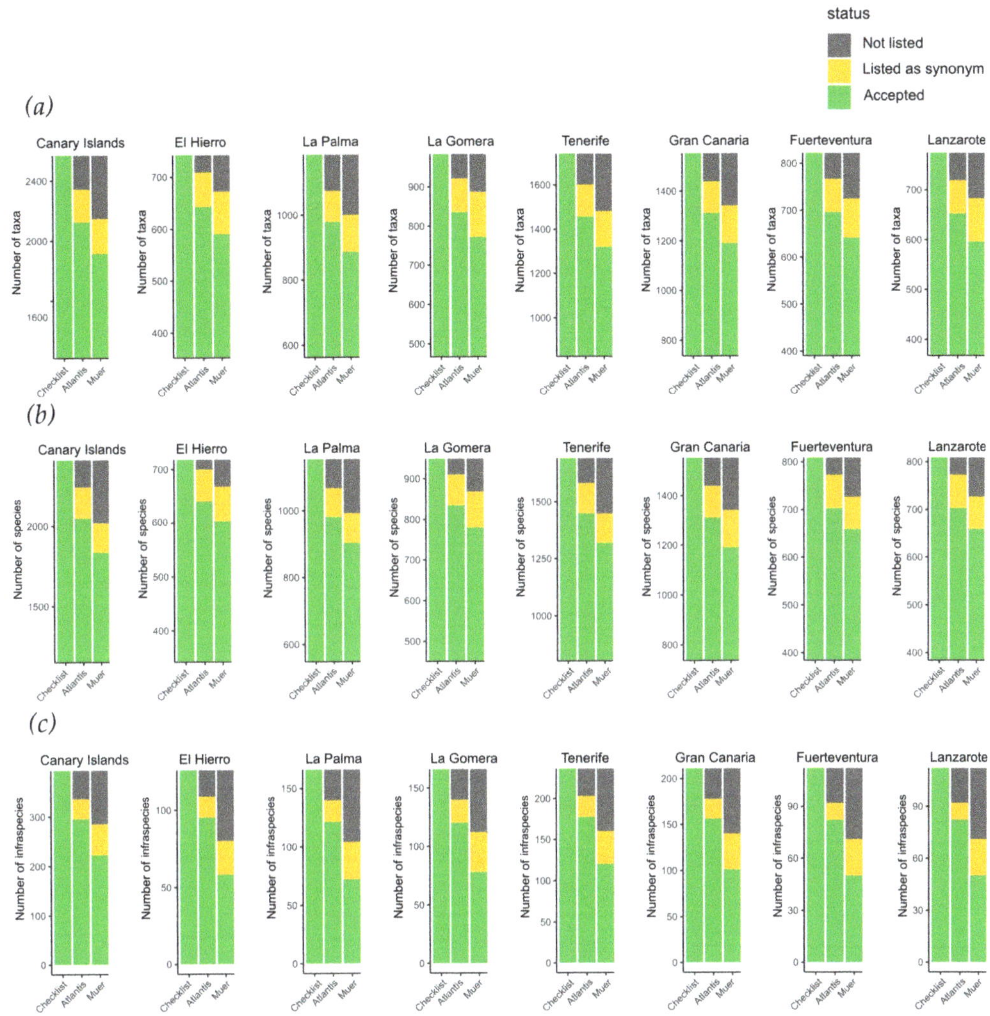

Figure 5. Comparing numbers of the here presented FloCan Checklist (without hybrid taxa, synonyms, and non-naturalized ornamental garden and park plants) with the Atlantis database [22] and with the taxa of Muer et al. [34] separately for (**a**) taxa, (**b**) species and (**c**) infraspecies (subspecies, varieties). Categories given are "not listed" in the respective database, "listed as synonym" and "accepted". This comparison illustrates how many of the accepted taxa in FloCan are also reflected in other established references.

Published hybrids are listed in addition, in a separate list (Supplement Materials Part S3), as this field is highly likely to be incomplete, and less clearly regulated. Particularly in specific genera (e.g., *Micromeria*) several hybrids are described. As hybrid speciation can be an important process, hybrids contribute to the biodiversity of the Canary Islands.

Furthermore, we add a list of not (yet) naturalized plants planted in gardens and parks (Supplement Materials Part S4), focusing on recorded perennial, long-lived plants because those species might naturalise and establish in the future. This list contains non-native exotic species that have not been found and reported yet to produce natural offspring. The species of this list are not included in the here presented analyses as we could not compile an exhaustive list of these species given the information available.

In Table 1, we provide a comparison for the taxa accepted in the FloCan Checklist (Supplement Materials Part S1, Main Table) with other modern plant lists such as Atlantis Biota or Muert et al. Major differences relate to the consideration of recently recorded non-native species, but also to taxonomic revisions.

Table 1. Numbers of accepted plant taxa (families, genera, species, subspecies, varieties) in the FloCan checklist for the Canary Island archipelago in comparison with other current plant lists (Atlantis, Muer) and with reference to the accepted taxa in international taxonomic data bases Plants of the World Online (POWO), Catalogue of Life (COL), World Flora Online (WFO), GBIF and TRY.

Taxonomic Level	FloCan	Atlantis Biota	Muer et al.	POWO	COL	WFO	GBIF	TRY
Year	2021	2021	2016	2021	2021	2021	2021	2021
Families	171	162	155	171	166	164	170	166
Genera	863	747	699	854	815	775	832	788
Species	2416	2050	1834	2370	2207	2022	2284	1965
Infraspecifics	396	295	222	247	264	125	329	74

Differences in the number of families are related to deviating reflection of changes in plant taxonomy and systematics. The splitting up of the family Scrophulariaceae is one of several examples of fundamental changes in plant families. Additionally, the increasing consideration and recording of exotic species contributes to an increase in plant families, respectively.

Progress in the exploration of species and in systematics reveals that a flora (plant list) is a moving target due to processes such as the increasing identification of endemism (Table 2) through records of new species or molecular identification of their taxonomic separation. Another process contributing to the ongoing changes in regional biodiversity is the establishment of non-native species (Table 3). Both processes are considered in the updated FloCan checklist.

Table 2. Numbers of endemic species and endemic infraspecific taxa (subspecies, varieties) in current plant lists of the Canary Island archipelago.

Taxonomic Level	FloCan	Atlantis Biota	Muer et al.
Year	2021	2021	2016
Species	608	541	499
Infraspecifics	197	147	152

Table 3. Numbers of non-native species and non-native infraspecific taxa (subspecies, varieties) in current plant lists of the Canary Island archipelago.

Taxonomic Level	FloCan	Atlantis Biota	Muer et al.
Year	2021	2021	2016
Species	964	781	662
Infraspecifics	67	44	16

4. Discussion

This new checklist reflects the current state of knowledge on the Flora of the Canary Islands, one of the most important evolutionary arenas in the holarctic realm [109]. This knowledge is under constant change and development. Therefore, it seems impossible to provide one final product, even in the case of the Canary Islands archipelago, which has been subject to botanical studies since the beginning of the 19th century. New species are still being found that were not known to science before, while small endemic populations of species are at the brink of extinction. Additionally, new non-native species are introduced as crops, ornamental plants or accidentally, and then may become established in the natural or semi-natural environment close to settlements or along roadsides. Some of these species become invasive, intruding into natural communities, and modifying ecosystems and their functioning, partly resulting in a deterioration of ecosystem services. Such ongoing and even accelerated changes resulting from the mostly undersaturated floras of islands combined with increasing connectivity due to transport and traffic make island biota a moving target.

For only very few taxa, no occurrence records could be attributed to specific islands, although these taxa occur or have occurred in the archipelago. Such species are listed in some sources as occurring in the archipelago, but without a precise location on specific islands or without confirmation during the last decades. One endemic plant species (*Solanum nava* Webb. & Berthel.) is likely completely extinct now, although this was already thought to be the case in the 1970s. One species has most likely become regionally extinct (*Grammitis quaerenda* Bolle), and three others have not been recorded for years (*Glinus lotoides* L., *Hypecoum procumbens* L., *Picris hieracioides* L.). Nevertheless, such taxa were not excluded as there is a chance of rediscovery. Other species with no clear local records in this list are part of complex groups that require specialist knowledge (e.g., *Taraxacum campylodes* G.E. Haglund). Currently missing clear local records also apply to some ornamental plants such as *Acanthus spinosus* L., *Amphilophium crucigerum* (L.) L.G. Lohmann, or *Syagrus weddelliana* (H.Wendl.) Becc. and to some tree species that have been planted in forestry (e.g., *Pinus sylvestris* L.).

In Supplement Materials Part S4 we provide an additional list of plant species recorded in gardens and parks. This list aims to create awareness about possible additional invasion processes even if the specimens do not yet show natural regeneration and dispersal. We encourage, however, the monitoring of these species, as some of these have been recorded to become invasive in other places of the world (e.g., *Artemisia absinthium* L.). Some herbaceous species on this list, such as *Sanguisorba minor* L. could naturalise rapidly without necessarily becoming invasive. As several of these garden and park species are trees and shrubs, they might invade and modify natural ecosystems. Nevertheless, such a list can only be incomplete as there is a constant import of ornamental plants. For the reasons of non-proven natural regeneration and incompleteness, this list was separated from the general checklist. Consequently, the main FloCan checklist does not comprise these ornamental garden plants. Here, we exclude them from the analyses, even if they can become non-native members of the natural vegetation quite rapidly.

Progress in taxonomy and systematics results in modified attribution of organisms to species, subspecies and even genera and families. This may be uncomfortable for practitioners that are used to specific terms and may even affect legal regulations with fixed terminology and nomenclature, but it is an intrinsic and essential condition of botanical sciences. Because of this continuously ongoing struggle to improve the understanding of nature, there is no complete agreement on all facets of taxonomy within the scientific community. Even if there are clear rules and regulations for accepting a scientific name, it may take time until such insights are generally accepted and translated into floristic inventories. However, deviating viewpoints will always exist.

Global databases are "work in progress" constantly being subject to change. Moreover, they can be incomplete, particularly for island biota. This applies mainly to the TRY database [72], which includes only very few infraspecific taxa, does not inform about species names' authors (which can result in errors due to synonyms), and misses many endemic species. However, the frequent use of such sources in large-scale studies implies the necessity to clarify to which degree island biota are covered and which restrictions need to be considered.

A global invader can serve as an example for a possible confusion of plant names. *Pennisetum setaceum* (Forssk.) Chiov. appears in GBIF [71] as *Pennisetum setaceum* (Forssk.) Chiov. as well as *Cenchrus setaceus* (Forssk.) Morrone, with deviating occurrence records for the Canary Islands. A total of 8970 records are provided for *C. setaceus* for all Canary Islands but only 4473 records are given for *P. setaceum*, including no reference for the island La Gomera (status 28 June 2021). This illustrates that a critical screening of available data is recommended because this is just one species, and a fully automated data mining can hardly recognise such errors that can even occur for widespread species when the nomenclature has changed or synonyms have been published. This species is listed in our checklist as *Cenchrus setaceus* (Forssk.) Morrone. This is also the name of the species in the Atlantis Biota data base of the Cabildo Insular of the Canary Islands [22]. However, it appears under a synonym in Muer et al. [34]. Our reference database for international standards, Plants of the World Online [67], also accepts this name, as well as Catalogue of Life [70], whereas it is seen as ambiguous in World Flora Online [68]. Comparable thorough screening across databases was done for all taxa in our checklist.

If endemic plants on oceanic islands are seen as a reflection of ongoing evolutionary processes, infraspecific variation cannot be ignored. Subspecies or varieties may not be very precisely defined and can be seen as beyond the biological species concept. However, speciation has many facets, including hybrid speciation and apomixis. Hence, biogeographical assessments which are aiming to characterise the moving target of evolution cannot ignore such infraspecific units. Again, it was Charles Darwin who was very aware of this fact. In August 1857 he wrote to J.D. Hooker: "*I am got extremely interested in tabulating according to mere size of genera, the species having any varieties marked by greek letters or otherwise: the result as far as I have yet gone seems to me one of the most important arguments I have yet met with, that varieties are only small species—or species only strongly marked varieties. The subject is in many ways so very important for me; I wish much you would think of any well-worked Floras with from 1000–2000 species, with the varieties marked. It is good to have hair-splitters & lumpers*" [110]. Hopefully, Charles Darwin would have been happy with the plant list provided herewith.

Today, big data algorithms are applied in biogeographic research relying on the correctness of species names, records and occurrence. (e.g., [38–42,60,111,112]). Such analyses are dependent on the expert knowledge fed into databases and their maintenance and quality control. However, there are many sources of error in databases because real-time control cannot be implemented and autotomized, or AI (artificial intelligence) algorithms are not (yet) able to replace expert knowledge, including specific challenges for certain taxa that exhibit, for instance, apomictic processes for reproduction. Additionally, the progress in phylogenetic research is not equal across taxonomic groups, with some being more thoroughly scrutinised because there is a larger scientific community working with these. In contrast, others are more neglected with the consequence of a possibly outdated nomenclature.

The Canary Islands are of outstanding importance to biodiversity covered by the European Union and should be considered strongly in the implementation of the EU Biodiversity Strategy for 2030 [113]. Ongoing land use changes, pressures related to tourism, climate change and additionally, the negative impact of introduced alien herbivores [114] are calling for a reinforced commitment in nature conservation [115]. The designation of protected areas is a common tool in conservation, requiring a sound knowledge of the uniqueness of biota, including infraspecific taxa. Databases and checklists are an important basis for such strategies [116]. The Canary Island protected area network is a work in progress. Besides the management of protected areas, the entire archipelago should be seen as a cradle of nature [109]. To achieve the objective to preserve its biodiversity of the future, laws and regulations need to be efficiently implemented, but in addition, enhanced support in human resources and financial endowment is required.

This FloCan checklist is the result of an in-depth survey on regional literature and web-based platforms, including own experience, data recording in the field and screening of international standard databases. It reflects the current understanding of taxa and the recent information about species records. Each flora, however, is a work in progress or can even be seen as a "moving target". Additional non-native species can be expected to establish, and even endemic species not yet known to science can be discovered in remote places. Additionally, cryptic taxa can be hidden under the disguise of morphological similarity and will be detectable with molecular methods. Here, we update the state of knowledge in the present moment. Nevertheless, our transparent approach to illustrate converging, but also deviating perceptions and points of view in common plant lists enables more critical and realistic biogeographical assessments.

5. Conclusions

The Canary Island archipelago is a hotspot of plant endemism and a safe site for remnant populations of plant taxa that have become extinct on neighbouring continents during Pleistocene and even Holocene climatic fluctuations. The oceanic climate, combined with pronounced topography, offers suitable habitats for a wide range of species and plant functional groups. However, the total species richness is, like on all islands, relatively low due to dispersal filters. Non-native species have become abundant since humans contributed to their transport, establishment, and provision of disturbed and anthropogenic habitats. Processes that are contributing to phytodiversity, such as invasion, extinction, or evolution, are progressing with varying momentum and different speeds. Furthermore, they do not proceed equally on all islands or in all ecosystems. The global importance of the Canary Islands requires a continuous survey and monitoring of biodiversity. The FloCan checklist aims to reflect the state of knowledge in July 2021 and is very likely to be amended and adapted in the future. Progress in phylogenetics may modify the status of well-known taxa. Still, new species are being discovered, and more and more non-native species are likely to become establish and detected. Additionally, ornamental plants may start regenerating after a lag period or develop possible invasive population dynamics under climate change. Therefore, this study explicitly includes many non-native plants that were missing in previous lists. Being adapted to the current international taxonomy standards, this list can be used for trans-regional or even global biogeographical studies.

Supplementary Materials: The following material is available online at https://www.mdpi.com/article/10.3390/d13100480/s1, Part S1: FloCan Main List (Accepted plant species and infraspecific taxa); Part S2: FloCan Plant Species List; Part S3: Published hybrids (not included in biogeographical analyses); Part S4: Recorded non-naturalized garden plants and plants in public places such as parks (not included in biogeographical analyses); Part S5: List of previously reported plants that were excluded and explanation for exclusion; Part S6: Synonyms used in currently published plant lists (but not accepted in FloCan).

Author Contributions: Conceptualization, C.B.; methodology, C.B. and A.W.; validation, C.B. and W.W.; formal analysis, A.W.; investigation, C.B. and A.W.; resources, C.B. and W.W.; data curation, C.B., A.W. and W.W.; writing—original draft preparation, C.B.; writing—review and editing, A.W. and W.W.; visualization, A.W.; supervision, C.B.; project administration, C.B.; All authors have read and agreed to the published version of the manuscript.

Funding: This research received no external funding.

Institutional Review Board Statement: Not applicable.

Informed Consent Statement: Not applicable.

Data Availability Statement: The data provided by this study are documented in the Supplementary Materials.

Acknowledgments: This work was greatly supported by Kira Meyer in data compilation, and by Reinhold Stahlmann for all technical aspects of data handling. We are grateful to Felix Medina, from the Cabildo Insular on the Island of La Palma, Spain, for his support in field research. Thanks to Alessandro Chiarucci, Bologna, Italy, Ole Vetaas, Bergen, Norway, Severin Irl, Frankfurt, Germany, Carsten Hobohm, Flensburg, Germany, and Richard Field, Nottingham, UK, for their company in the field and stimulating discussions that generated the need for improved botanical data availability.

Conflicts of Interest: The authors declare no conflict of interest.

References

1. Humboldt, A.V.; Bonpland, A. *Viage á Las Regiones Equinocciales del Nuevo Continente*; Casa de Rosa: Paris, France, 1826.
2. Viera y Clavijo, D.J. *Diccionario de Historia Natural de las Islas Canarias, o Índice Alfabético Descriptivo de sus Tres Reinos, Animal, Vegetal y Mineral*; Excma. Mancomunidad de Cabildos de Las Palmas, Plan Cultural: Las Palmas de Gran Canaria, Spain, 1799.
3. Darwin, C. Voyage of the Beagle. In *The Narrative of the Voyages of H.M. Ships Adventure and Beagle*; Henry Colburn: London, UK, 1839; Volume 3.
4. Webb, P.B.; Berthelot, S. Histoire naturelle des Iles Canaries. In *Phytogeographia Canariensis*; Béthune: Paris, France, 1844; 1430p.
5. Christ, H. Vegetation und Flora der Canarischen Inseln. *Engler's Bot. Jahrbücher* **1885**, *6*, 458–526.
6. Christ, H. Specilegium canariense. *Engler's Bot. Jahrbücher* **1887**, *9*, 86–172.
7. Pitard, J.; Proust, L. Les Iles Canaries. In *Flore de l'Archipel*; Béthune: Paris, France, 1908; 503p.
8. Eriksson, O.; Hansen, A.; Sunding, P. *Flora of Macaronesia: Checklist of Vascular Plants*; Sommerfeltia: Umea, Sweden, 1974.
9. Kunkel, G. *Flora y Vegetación del Archipiélago Canario. Tratado Florístico. 2ª Parte. Dicotiledóneas*; Ed. Edirca: Las Palmas de Gran Canaria, Spain, 1991; 312p.
10. Kunkel, G. *Die Kanarischen Inseln und ihre Pflanzenwelt*; Gustav Fischer Verlag: Stuttgart, Jena, Germany; New York, NY, USA, 1993.
11. Bramwell, D. *Flora de las Islas Canarias*; Rueda: Madid, Spain, 1997.
12. Schönfelder, P.; Schönfelder, I. *Die Kosmos-Kanarenflora*; Kosmos: Stuttgart, Germany, 2012; 319p.
13. Santos Guerra, A. *Vegetacion y Flora de La Palma*; Editorial Interinsular Canaria: Santa Cruz de Tenerife, Spain, 1983; 348p.
14. Fernandez-Palacios, J.M.; Martin Esquivel, J.L. *Naturaleza de las Islas Canarias*; Editorial Interinsular Canaria: Santa Cruz de Tenerife, Spain, 2002.
15. Pott, R.; Hüppe, J.; Wildpret de la Torre, W. *Die Kanarischen Inseln—Natur- und Kulturlandschaften*; Ulmer: Stuttgart, Germany, 2003.
16. Montelongo, V.; Bramwell, D.; Fernández-Palacios, O. *Parolinia glabriuscula* (Brassicaceae), una nueva especie para Gran Canaria (Islas Canarias, España). *Bot. Macaronésica* **2003**, *24*, 67–72.
17. Gil Gonzalez, J.; Gil Gonzales, M.L.; Morales Mateos, J.B.; Mesa Coello, R. *Vicia vulcanorum* (Fabaceae) a new species from the island of Lanzarote (Canary Islands). *Collect. Bot.* **2012**, *31*, 19–27. [CrossRef]
18. Bañares Baudet, Á.; Acevedo Rodríguez, A.; Rebolé Beaumont, Á. *Monanthes subrosulata*, a new species of *M.* sect. Sedoidea (Crassulaceae) from La Palma, Canary Islands, Spain. *Willdenowia* **2013**, *43*, 25–31. [CrossRef]
19. Crawford, D.J.; Mort, M.E.; Archibald, J.K. *Tolpis santosii* (Asteraceae, Chichorieae), a new species from La Palma, Canary Islands. *Vieraea* **2013**, *41*, 169–175.
20. Gil Gonzalez, J.; Morales Mateos, J.B.; Gil Gonzales, M.L.; Mesa Coello, R. *Vicia voggenreiteriana* (Fabaceae) a new species from the island of La Gomera (Canary Islands). *Viaeraea* **2013**, *41*, 189–201.
21. Friesen, N.; Herden, T.; Schönfelder, P. *Allium canariense* (Amaryllidaceae), a species endemic of the Canary Islands. *Phytotaxa* **2015**, *221*, 1–20. [CrossRef]
22. Atlantis Biota Database. Available online: https://www.biodiversidadcanarias.es/biota/ (accessed on 5 July 2021).
23. Mauchamp, A.; Atkinson, R. *Rapid, Recent and Irreversible Habitat Loss: Scalesia Forest in the Galapagos Islands*; Galapagos Report 2009–2010; GNPD; GCREG; CDF; GC: Galápagos, Ecuador, 2010; pp. 108–112.
24. Wittmer, M.C. The dodo and the tambalacoque tree: An obligate mutualism reconsidered. *Oikos* **1991**, *61*, 133–137. [CrossRef]
25. de Nascimento, L.; Willis, K.J.; Fernandez-Palacios, J.M.; Criado, C.; Whittaker, R.J. The long-term ecology of the lost forests of La Laguna, Tenerife (Canary Islands). *J. Biogeogr.* **2009**, *36*, 499–514. [CrossRef]
26. Bañares Baudet, Á.; Marrero Gómez, M.V.; Scholz, S. Taxonomic and nomenclatural notes on Crassulaceae of the Canary Islands, Spain. *Willdenowia* **2008**, *38*, 475–489. [CrossRef]
27. Peroni, A.; Peroni, G. Considerazioni tassonomiche sul genere *Cystopteris* Bernh. (Athyriaceae: Pteridophyta) di El Hierro (Isole Canarie). *Ann. Mus. Cov. Rovereto Sez. Arch. St. Sc. Nat.* **2006**, *21*, 207–210.
28. Vitales, D.; Garnatje, T.; Pellicer, J.; Vallès, J.; Santos-Guerra, A.; Sanmartin, I. The explosive radiation of *Cheirolophus* (Asteraceae, Cardueae) in Macaronesia. *BMC Evol. Biol.* **2014**, *14*, 118. [CrossRef]
29. Hansen, A.; Sunding, P. *Flora of Macaronesia: Checklist of Vascular Plants*, 4th ed.; Sommerfeltia: Umea, Sweden, 1993; Volume 17, pp. 1–295.
30. Hohenester, A.; Welss, W. *Exkursionsflora für die Kanarischen Inseln. Mit Ausblicken auf ganz Makaronesien*; Eugen Ulmer: Stuttgart, Germany, 1993.
31. Bramwell, D.; Bramwell, Z. *Flores Silvestres de las Islas Canarias*, 4th ed.; Rueda Ed.: Madrid, Spain, 2001; 437p.
32. Izquierdo, I.; Martín, J.L.; Zurita, N.; Arechavaleta, M. (Eds.) Lista de especies silvestres de Canarias (hongos, plantas y animales terrestres). In *Consejería de Medio Ambiente y Ordenación Territorial*; Gobierno de Canarias: Las Palmas de Gran Canaria, Spain, 2004; 500p.
33. Arechavaleta, M.; Rodriguez, S.; Zurita, N.; Garcia, A. *Lista de Especies Silvestres de Canarias. Hongos, Plantas y Animales Terrestres*; Gobierno de Canarias: Las Palmas de Gran Canaria, Spain, 2010; 579p.
34. Muer, T.; Sauerbier, H.; Cabrera Calixto, F. *Die Farn- und Blütenpflanzen der Kanarischen Inseln*; Markgraf Publishers: Weikersheim, Germany, 2016.
35. Flora de Canarias (Gil). Available online: http://www.floradecanarias.com/ (accessed on 5 July 2021).

36. Freiberg, M.; Winter, M.; Gentile, A.; Zizka, A.; Muellner-Riehl, A.N.; Weigelt, A.; Wirth, C. LCVP, The Leipzig catalogue of vascular plants, a new taxonomic reference list for all known vascular plants. *Sci. Data* **2020**, *7*, 416. [CrossRef]
37. Kier, G.; Kreft, H.; Lee, T.G.; Jetz, W.; Ibisch, P.L.; Nowicki, C.; Mutke, J.; Barthlott, W. A global assessment of endemism and species richness across island and mainland regions. *Proc. Natl. Acad. Sci. USA* **2009**, *106*, 9322–9327. [CrossRef]
38. Steinbauer, M.; Beierkuhnlein, C. Characteristic pattern of species diversity on the Canary Islands. *Erdkunde* **2010**, *64*, 57–71. [CrossRef]
39. Lenzner, B.; Weigelt, P.; Kreft, H.; Beierkuhnlein, C.; Steinbauer, M. The general dynamic model of island biogeography revisited at the level of major flowering plant families. *J. Biogeogr.* **2017**, *44*, 1029–1040. [CrossRef]
40. Fernández-Palacios, J.M.; Otto, R.; Borregard, M.K.; Kreft, H.; Price, J.P.; Steinbauer, M.J.; Weigelt, P.; Whittaker, R.J. Evolutionary winners are ecological losers among oceanic island plants. *J. Biogeogr.* **2021**, *48*, 2186–2198. [CrossRef]
41. Cutts, V.; Hanz, D.M.; Barjas-Barbosa, M.P.; Algar, A.C.; Steinbauer, M.J.; Irl, S.D.H.; Kreft, H.; Weigelt, P.; Fernandez-Palacios, J.M.; Field, R. Scientific floras can be reliable sources for some trait data in a system with poor coverage in global trait databases. *J. Veg. Sci.* **2021**, *32*, e12996. [CrossRef]
42. Hortal, J.; Lobo, J.M.; Jiménez-Valverde, A. Limitations of Biodiversity Databases: Case Study onSeed-Plant Diversity in Tenerife, Canary Islands. *Conserv. Biol.* **2007**, *21*, 853–863. [CrossRef] [PubMed]
43. Wikipedia Spanish on Categoría: Flora endémica de Canarias. Available online: https://es.wikipedia.org/wiki/Categor%C3%ADa:Flora_end%C3%A9mica_de_Canarias (accessed on 28 June 2021).
44. Wikipedia English on Endemic Plants of the Canary Islands. Available online: https://en.wikipedia.org/wiki/Category:Endemic_flora_of_the_Canary_Islands (accessed on 28 June 2021).
45. Hobohm, C.; Janišová, M.; Steinbauer, M.; Landi, S.; Field, R.; Vanderplank, S.; Beierkuhnlein, C.; Grytnes, J.A.; Vetaas, O.R.; Fidelis, A.; et al. Global endemics-area relationships of vascular plants. *Perspect. Ecol. Conserv.* **2019**, *17*, 41–49. [CrossRef]
46. Hohbohm, C.; Moro-Richter, M.; Beierkuhnlein, C. Distribution and Habitat Affinity of Endemic and Threatened Species—Global Assessment. In *Perspectives for Biodiversity and Ecosystems*; Hobohm, C., Ed.; Springer: Berlin/Heidelberg, Germany, 2021.
47. Thompson, J.D. *Plant Evolution in the Mediterranean: Insights for Conservation*; Oxford University Press: Oxford, UK, 2020.
48. Marreiro, A.; Almeida, R.S.; Gonzalez-Martin, M. A new species of wild dragon tree, *Dracaena* (Dracaenaceae) from Gran Canaria and its taxonomic and biogeographic implications. *Bot. J. Linn. Soc.* **1998**, *128*, 291–314.
49. Rodríguez-Sánchez, F.; Guzmán, B.; Valido, A.; Vargas, P.; Arroyo, J. Late Neogene history of the laurel tree (*Laurus*, L.; Lauraceae) based on phylogeographical analyses of Mediterranean and Macaronesian populations. *J. Biogeogr.* **2009**, *36*, 1270–1281. [CrossRef]
50. Pyšek, P.; Jarosik, V.; Hulme, P.E.; Hejda, M.; Schaffner, U.; Vila, M.A. Global assessment of invasive plant impacts on resident species, communities and ecosystems: The interaction of impact measures, invading species traits and environment. *Glob. Chang. Biol.* **2011**, *18*, 1725–1737. [CrossRef]
51. Ríos-Mesa, D.; Pereira-Lorenzo, S.; González-Díaz, A.; Hernádez-González, J.; González-Diaz, E.; Saúco, V.G. The status of Chestnut cultivation and utilization in the Canary Islands. *Adv. Hortic. Sci.* **2011**, *25*, 90–98.
52. Irl, S.; Schweiger, A.; Steinbauer, M.; Ah-Peng, C.; Arévalo, J.R.; Beierkuhnlein, C.; Chiarucci, A.; Daehler, C.C.; Fernandez-Palacios, J.M.; Flores, O.; et al. Human impact, climate and dispersal strategies determine plant invasion on islands. *J. Biogeogr.* **2021**. early view. [CrossRef]
53. Irl, S.; Steinbauer, M.; Messinger, J.; Blume-Werry, G.; Palomares-Martínez, A.; Beierkuhnlein, C.; Jentsch, A. Burned and devoured—Introduced herbivores, fire and the endemic flora of the high elevation ecosystem on La Palma, Canary Islands. *Arct. Antarct. Alp. Res.* **2014**, *46*, 859–869. Available online: http://www.gobiernodecanarias.org/boc/2006/237/011.html (accessed on 28 October 2020). [CrossRef]
54. Frankham, R. Do island populations have less genetic variation than mainland populations? *Heredity* **1997**, *78*, 311–327. [CrossRef] [PubMed]
55. Fernández-Mazuecos, M.; Vargas, P. Genetically depauperate in the continent but rich in oceanic islands: *Cistus monspeliensis* (Cistaceae) in the Canary Islands. *PLoS ONE* **2011**, *6*, e17172. [CrossRef] [PubMed]
56. García-Verdugo, C.; Calleja, J.A.; Vargas, P.; Silva, L.; Moreira, O.; Pulido, F. Polyploidy and microsatellite variation in the relict tree Prunus lusitanica L.: How effective are refugia in preserving genotypic diversity of clonal taxa? *Mol. Ecol.* **2013**, *22*, 1546–1557. [CrossRef] [PubMed]
57. Mort, M.E.; Soltis, D.E.; Soltis, P.S.; Francisco-Ortega, J.; Santos-Guerra, A. Phylogenetics and evolution of the Macaronesian clade of Crassulaceae inferred from nuclear and chloroplast sequence data. *Syst. Bot.* **2002**, *27*, 271–288.
58. Fernández-Palacios, J.M.; de Nascimento, L.; Otto, R.; Delgado, J.D.; García-del-Rey, E.; Arévalo, J.R.; Whittaker, R.J. A reconstruction of Palaeo-Macaronesia, with particular reference to the long-term biogeography of the Atlantic island laurel forests. *J. Biogeogr.* **2011**, *38*, 226–246. [CrossRef]
59. Van den Boogard, P. The origin of the Canary Island Seamount Province—New ages of old seamounts. *Sci. Rep.* **2017**, *3*, 2107. [CrossRef] [PubMed]
60. Weigelt, P.; Steinbauer, M.; Cabral, J.S.; Kreft, H. Late Quaternary climate change shapes island biodiversity. *Nature* **2016**, *532*, 99–102. [CrossRef] [PubMed]
61. Aparicio, A.; Martín-Hernanz, S.; Parejo-Farnés, C.; Arroyo, J.; Lavergne, S.; Yeşilyurt, E.B.; Yeşilyurt, M.-L.; Rubio, E.; Albaladejo, R.G. Phylogenetic reconstruction of the genus *Helianthemum* (Cistaceae) using plastid and nuclear DNA-sequences: Systematic and evolutionary inferences. *Taxon* **2017**, *66*, 868–885. [CrossRef]

62. Sánchez Meseguer, A.; Aldasoro, J.J.; Sanmartín, I. Bayesian inference of phylogeny, morphology and range evolution reveals a complex evolutionary history in St. John's wort (*Hypericum*). *Mol. Phylogenetics Evol.* **2013**, *67*, 379–403. [CrossRef] [PubMed]
63. Graham, R.E.; Reyes-Betancourt, J.A.; Chapman, M.A. Inter-island differentiation and contrasting patterns of diversity in the iconic Canary Island sub-alpine endemic *Echium wildpretii* (Boraginaceae). *Syst. Biodivers.* **2021**, *19*, 507–521. [CrossRef]
64. Schmitt, T. *Molekulare Biogeographie*; Haupt Verlag: Bern, Germany, 2020; 504p.
65. Fernández-Palacios, J.M.; Negrín, Z.; Fernández Lugo, S.; Arévalo, J.R.; de Nascimento, L. Terrestrial Biota Checklist of the Chinijo Archipelago and Lobos (Canary Islands). *Rev. Sci. Insul.* **2018**, *1*, 51–86. [CrossRef]
66. Puppo, P.; Curto, M.; Gusmão-Guedes, J.; Cochofel, J.; Pérez de Paz, P.L.; Bräuchler, C.; Meimberg, H. Molecular phylogenetics of *Micromeria* (Lamiaceae) in the Canary Islands, diversification and inter-island colonization patterns inferred from nuclear genes. *Mol. Phylogenetics Evol.* **2015**, *89*, 160–170. [CrossRef] [PubMed]
67. Plants of the World Online (Kew). Available online: http://www.plantsoftheworldonline.org/ (accessed on 5 June 2021).
68. World Flora Online. Available online: http://www.worldfloraonline.org/ (accessed on 5 June 2021).
69. The Plant List. Available online: http://www.theplantlist.org/ (accessed on 5 June 2021).
70. Roskov, Y.; Abucay, L.; Orrell, T.; Nicolson, D.; Bailly, N.; Kirk, P.M.; Bourgoin, T.; DeWalt, R.E.; Decock, W.; De Wever, A.; et al. (Eds.) *Species 2000 & ITIS Catalogue of Life, 2018 Annual Checklist. Species 2000*; Naturalis: Leiden, The Netherlands, 2018; Available online: www.catalogueoflife.org/annual-checklist/2018 (accessed on 5 June 2021).
71. GBIF. Available online: https://www.gbif.org/ (accessed on 5 June 2021).
72. TRY. Available online: https://www.try-db.org/TryWeb/Home.php (accessed on 5 June 2021).
73. Kattge, J.; Bönisch, G.; Díaz, S.; Lavorel, S.; Prentice, I.C.; Leadley, P.; Tautenhahn, S.; Werner, G.D.A.; Aakala, T.; Abedi, M.; et al. TRY plant trait database—Enhanced coverage and open access. *Glob. Chang. Biol.* **2020**, *26*, 119–188. [CrossRef]
74. Chamberlain, S.; Szocs, E. Taxize—Taxonomic search and retrieval in R. *F1000Research* **2013**, *2*, 191. [CrossRef]
75. R Core Team. *R: A Language and Environment for Statistical Computing*; R Foundation for Statistical Computing: Vienna, Austria, 2020.
76. International Plant Names Index IPNI. Available online: https://www.ipni.org/ (accessed on 5 June 2021).
77. Nezadal, W.; Lindacher, R.; Welss, W. Lokalendemiten und Phytodiversität der westkanadischen Inseln La Palma und La Gomera. *Feddes Repert.* **1999**, *110*, 19–30. [CrossRef]
78. Barone, R. Notas corológicas sobre dos especies de *Atriplex*, L. (Chenopodiaceae) en Tenerife, Islas Canarias. *Botánica Macaronésica* **2003**, *24*, 161–164.
79. Leibbach, R.; Peter, R. Bemerkungen zu den Orchideen von La Palma (Kanarische Inseln). *J. Eur. Orch.* **2007**, *39*, 55–70.
80. Otto, R.; Scholz, H.; Scholz, S. Supplements to the flora of the Canary Islands, Spain: Poaceae. *Willdenowia* **2008**, *38*, 491–496. [CrossRef]
81. Prina, A.O.; Martinez-Laborde, J.B. A taxonomic revision of *Crambe* section *Dendrocrambe* (Brassicaceae). *Bot. J. Linn. Soc.* **2008**, *156*, 291–304. [CrossRef]
82. Verloove, F.; Reyes-Betancort, J. Additions to the flora of Tenerife (Canary Islands, Spain). *Collect. Bot.* **2011**, *30*, 63–78. [CrossRef]
83. Sánchez de Lorenzo Cáceres, J. Sobre la presencia de *Balanites aegyptiaca* (L.) Del. (Zygophyllaceae) en Tenerife, Islas Canarias. *Bouteloua* **2012**, *10*, 82–84.
84. Reisigl, H. Gefährdete endemische Blütenpflanzen der Trockeninsel Fuerteventura: Herkunft, Ökologie, Gesellschaft. *Bauhinia* **2013**, *24*, 39–52.
85. Verloove, F. New xenophytes from Gran Canaria (Canary Islands, Spain), with emphasis on naturalized and (potentially) invasive species. *Collect. Bot.* **2013**, *32*, 59–82. [CrossRef]
86. Santos Guerra, A.; Reyes Betancort, J.A.; Padrón Mederos, M.A.; Mesa Coello, R. Plantas poco o nada conocidas de la flora vascular silvestre de las Islas Canarias. *Bot. Complut.* **2013**, *37*, 99–108. [CrossRef]
87. Verloove, F.; Guiggi, A. Some new xenophytes from Fuerteventura (Canary Islands, Spain). *Bouteloua* **2013**, *13*, 13,–42.
88. Santos Guerra, A.; Reyes-Betancort, J. Nuevas adiciones y citas de interés para la flora autóctona de las islas Canarias. *Vieraea* **2014**, *42*, 249–258.
89. Otto, R.; Verloove, F. New xenophytes from La Palma (Canary Islands, Spain), with emphasis on naturalized and (potentially) invasive species. *Collect. Bot.* **2016**, *35*, e001.
90. Marrero Rodríguez, Á. Eucaliptos en Gran Canaria, identificación y corología. Hacia una reseña histórica. *Botánica Macaronésica* **2016**, *29*, 91–137.
91. Salas-Pascual, M.; Quintana Vega, G. *Salvinia molesta* D. S. Mitch. (Salviniaceae), nueva cita para Canarias y España. *Botánica Macaronésica* **2016**, *29*, 73–81.
92. Verloove, F.; Alves, P. New vascular plant records for the western part of the Iberian Peninsula (Portugal and Spain). *Folia Bot. Extrem.* **2016**, *10*, 5–23.
93. Galan de Mera, A.; Linares-Perea, E.; Vicente-Orellana, J.A. *Taraxacum* (Asteraceae) in the Azores, Madeira and the Canary Island. *Ann. Bot. Fenn.* **2017**, *54*, 273–285. [CrossRef]
94. Verloove, F. New xenophytes from the Canary Islands (Gran Canaria and Tenerife; Spain). *Acta Bot. Croat.* **2017**, *76*, 120–131. [CrossRef]

95. Verloove, F.; Ojeda-Land, E.; Smith, G.F.; Guiggi, A.; Reyes-Betancort, J.A.; Samarín, C.; González Hernández, A.; Barone, R. New records of naturalised and invasive cacti (Cactaceae) from Gran Canaria and Tenerife, Canary Islands, Spain. *Bradleya* **2017**, *35*, 58–79. [CrossRef]
96. Otto, R.; Verloove, F. New xenophytes from La Palma (Canary Islands, Spain), with emphasis on naturalized and (potentially) invasive species—Part 2. *Collect. Bot.* **2018**, *37*, e005.
97. Expósito, A.B.; Siverio, A.; Bermejo, L.A.; Sobrino-Vesperinas, E. Checklist of alien plant species in a natural protected area: Anaga Rural Park (Tenerife, Canary Islands) effect of human infrastructures on their abundance. *Plant Ecol. Evol.* **2018**, *151*, 142–152. [CrossRef]
98. Verloove, F.; Salas-Pascual, M.; Marrero Rodríguez, Á. New records of alien plants for the flora of Gran Canaria (Canary Islands, Spain). *Flora Mediterr.* **2018**, *28*, 119–135.
99. Portero Alvarez, A.M.; Martin-Carbajal Gonzalez, J.; Reyes-Betancort, J.; Mesa Coello, R.A. *Lotus gomerythus* (Fabaceae-Loteae) Spec. *Nova. Botanica Macaronesica* **2019**, *30*, 89–98.
100. Marrero, A. Vicia tenoi (Fabaceae) una nueca especie del macizo de Teno, Tenerife (Islas Canarias). *Bot. Macaronesica* **2019**, *30*, 153–156.
101. Arango Toro, O. *Aeoniun lioi* (Crassulaceae): Una nueva especie de Tenerife, Islas Canarias. *Bot. Macaronesica* **2019**, *30*, 7–22.
102. Verloove, F.; Thiede, J.; Marrero Rodríguez, Á.; Salas-Pascual, M.; Reyes-Betancort, J.; Ojeda-Land, E.; Smith, G. A synopsis of feral Agave and Furcraea (Agavaceae, Asparagaceae s. lat.) in the Canary Islands (Spain). *Plant Ecol. Evol.* **2019**, *152*, 470–498. [CrossRef]
103. Marrero Rodríguez, Á. Adiciones corológicas a la flora vascular de Gran Canaria, especies xenófitas, ocasionales o potenciales invasoras. *Botánica Macaronésica* **2019**, *30*, 121–142.
104. Otto, R.; Verloove, F. New xenophytes from La Palma (Canary Islands, Spain), with emphasis on naturalized and (potentially) invasive species—Part 3. *Collect. Bot.* **2020**, *39*, e001.
105. Verloove, F.; Déniz Suárez, E.; Salas Pascual, M. New records of non-native vascular plants in Gran Canaria (Spain, Canary Islands). *Flora Mediterr.* **2020**, *30*, 121–136.
106. White, O.W.; Reyes-Betancort, J.A.; Chapman, M.A.; Carine, M.A. Geographical isolation, habitat shifts and hybridisation in the diversification of the Macaronesian endemic genus *Argyranthemum* (Asteraceae). *New Phytol.* **2020**, *228*, 1953–1971. [CrossRef] [PubMed]
107. Verloove, F. New records in vascular plants alien to Tenerife (Spain, Canary Islands). *Biodivers. Data J.* **2021**, *9*, e62878. [CrossRef] [PubMed]
108. White, O.W.; Reyes-Betancort, J.A.; Chapman, M.A.; Carine, M.A. Recircumscription of the Canary Island endemics *Argyranthemum broussonetii* and *A. callichrysum* (Asteraceae: Anthemideae) based on evolutionary relationships and morphology. *Willdenowia* **2021**, *51*, 129–139. [CrossRef]
109. Nürk, N.M.; Linder, H.P.; Onstein, R.E.; Larcombe, M.J.; Hughes, C.E.; Fernández, L.P.; Schlüter, P.M.; Valente, L.; Beierkuhnlein, C.; Cutts, V.; et al. Diversification in evolutionary arenas—Assessment and synthesis. *Ecol. Evol.* **2020**, *10*, 6163–6182. [CrossRef] [PubMed]
110. Darwin, C. Letter no. 2130. To, J.D. Hooker, 1 August 1857. Darwin Correspondence Project. Available online: https://www.darwinproject.ac.uk/ (accessed on 5 July 2021).
111. Irl, S.; Schweiger, A.; Medina, F.M.; Fernández-Palacios, J.M.; Harter, D.; Jentsch, A.; Provenzale, A.; Steinbauer, M.J.; Beierkuhnlein, C. An island view of endemic rarity—Environmental drivers and consequences for nature conservation. *Divers. Distrib.* **2017**, *23*, 1132–1142. [CrossRef]
112. García-Verdugo, C.; Baldwin, B.G.; Fay, M.F.; Caujapé-Castells, J. Life history traits and patterns of diversification in oceanic archipelagos: A meta-analysis. *Bot. J. Linn. Soc.* **2014**, *174*, 334–348. [CrossRef]
113. EU Biodiversity Strategy for 2030. Available online: https://www.eea.europa.eu/policy-documents/eu-biodiversity-strategy-for-2030-1 (accessed on 17 July 2021).
114. Gangoso, L.; Donázar, J.A.; Scholz, S.; Palacios, C.J.; Hiraldo, F. Contradiction in conservation of island ecosystems: Plants, introduced herbivores and avian scavengers in the Canary Islands. *Biodivers. Conserv.* **2006**, *15*, 2231–2248. [CrossRef]
115. Fernández-Palacios, J.M.; Martin Esquivel, J.L. *Naturaleza de las Islas Canarias—Ecologia y Conservation*; Publicaciones Turquesa: Santa Cruz de Tenerife, Spain, 2001; 474p.
116. Reyes-Betancort, J.A.; Santos-Guerra, A.; Guma, I.R.; Humphries, C.J.; Carine, M.A. Diversity, rarity and the evolution and conservation of the Canary Islands endemic flora. *An. Jardín Botánico Madr.* **2008**, *65*, 25–45. [CrossRef]

diversity

MDPI

Article

Overlooked Species Diversity and Distribution in the Antarctic Mite Genus *Stereotydeus*

Claudia Brunetti [1,*], Henk Siepel [2], Peter Convey [3,4], Pietro Paolo Fanciulli [1], Francesco Nardi [1] and Antonio Carapelli [1,*]

[1] Department of Life Sciences, University of Siena, Via A. Moro 2, 53100 Siena, Italy; paolo.fanciulli@unisi.it (P.P.F.); francesco.nardi@unisi.it (F.N.)

[2] Radboud Institute for Biological and Environmental Sciences (RIBES), Animal Ecology and Physiology, Radboud University, P.O. Box 9100, 6500 GL Nijmegen, The Netherlands; henk.siepel@ru.nl

[3] British Antarctic Survey, NERC, High Cross, Madingley Road, Cambridge CB3 0ET, UK; pcon@bas.ac.uk

[4] Department of Zoology, University of Johannesburg, P.O. Box 524, Auckland Park 2006, South Africa

* Correspondence: clabrunetti12@gmail.com (C.B.); antonio.carapelli@unisi.it (A.C.); Tel.: +39-057-723-4410 (A.C.)

Abstract: In the harsh Antarctic terrestrial ecosystems, invertebrates are currently confined to sparse and restricted ice free areas, where they have survived on multi-million-year timescales in refugia. The limited dispersal abilities of these invertebrate species, their specific habitat requirements, and the presence of geographical barriers can drastically reduce gene flow between populations, resulting in high genetic differentiation. On continental Antarctica, mites are one of the most diverse invertebrate groups. Recently, two new species of the free living prostigmatid mite genus *Stereotydeus* Berlese, 1901 were discovered, bringing the number of Antarctic and sub-Antarctic species of this genus up to 15, of which 7 occur along the coast of Victoria Land and in the Transantarctic Mountains. To examine the biodiversity of *Stereotydeus* spp., the present study combines phylogenetic, morphological and population genetic data of specimens collected from nine localities in Victoria Land. Genetically distinct intraspecific groups are spatially isolated in northern Victoria Land, while, for other species, the genetic haplogroups more often occur sympatrically in southern Victoria Land. We provide a new distribution map for the *Stereotydeus* species of Victoria Land, which will assist future decisions in matters of the protection and conservation of the unique Antarctic terrestrial fauna.

Keywords: Victoria Land; molecular phylogeny; *cox1*; *28S*; biogeography; terrestrial invertebrates; acari; *Stereotydeus* spp.

Citation: Brunetti, C.; Siepel, H.; Convey, P.; Fanciulli, P.P.; Nardi, F.; Carapelli, A. Overlooked Species Diversity and Distribution in the Antarctic Mite Genus *Stereotydeus* *Diversity* **2021**, *13*, 506. https://doi.org/10.3390/d13100506

Academic Editor: Michael Wink

Received: 24 August 2021
Accepted: 13 October 2021
Published: 19 October 2021

Publisher's Note: MDPI stays neutral with regard to jurisdictional claims in published maps and institutional affiliations.

Copyright: © 2021 by the authors. Licensee MDPI, Basel, Switzerland. This article is an open access article distributed under the terms and conditions of the Creative Commons Attribution (CC BY) license (https://creativecommons.org/licenses/by/4.0/).

1. Introduction

Due to Antarctica's isolation and extreme environmental conditions, the continent's terrestrial biota has limited species level diversity and many higher taxonomic groups are completely missing or very poorly represented [1,2]. As a result of the climatic factors and the typically low availability of organic nutrients in soils, lichens and mosses are the only macroscopic flora present on the continent [1,3–6]. Similarly, the Antarctic terrestrial fauna consists of a small number of microarthropod species (mites and springtails) as well as other microscopic invertebrates (nematodes, tardigrades and rotifers), making the continental region amongst the simplest ecosystems on Earth [2,7].

The challenging environmental conditions, isolation and the patchy distribution of ice free areas have been recognized as the main factors affecting and defining populations of the Antarctic terrestrial invertebrate fauna, both physiologically and genetically [8]. As a consequence, in order to survive the harsh Antarctic conditions, these terrestrial animals have evolved impressive biochemical and physiological adaptations, to tolerate prolonged periods of freezing and dry conditions, amongst other severe stresses [9–12]. Behavioral strategies also play a role. For instance, continental Antarctic springtails (Collembola) and mites (Acari) are often found concentrated under rocks, where the environment tends to be moister, rich in organic carbon and with low salinity [13], and where microbial diversity is also present, stabilizing mineral soils and allowing colonization by both micro-invertebrates and flora [2]. Although temperature plays an important role in regulating microarthropod life cycles, the major factor regulating their survival and growth remains the availability of liquid water [5,14]. An additional challenge for microarthropod survival derives from the bottleneck caused by their dispersal abilities, especially over longer distances.

Studies have suggested that rafting on the surface of melt water streams is a possible route for dispersal [15–18], as is the use of animal vectors (zoochory; e.g., on bird plumage or in nesting materials) [19–23] and, also, human mediated transport [22]. A further mechanism is dispersal by wind (anemochory). However, although the latter is known to be an effective dispersal strategy in, for instance, some oribatid mites [24], it may not be effective for Antarctic microarthropods, at least over longer distances/timescales, due to the risk of desiccation and the lack of an anhydrobiotic dispersal stage [7,25,26]. In order to understand the dispersal, over short and long distances, of microarthropods in Victoria Land, molecular studies have been conducted on different springtail species [22,27–31]. These have identified that the presence of glacial barriers strongly influences species distributions, and that these have likely limited gene flow between restricted and isolated refugia during various glacial maxima [22,28,32]. Analogous biogeographical patterns have been reported for the prostigmatid mite *Stereotydeus mollis* by Womersley and Strandtmann, 1963, in Victoria Land [33–36], although with higher genetic divergence, possibly due to higher activity levels and shorter generation time [33,37] and/or to a longer evolutionary history than for the springtails. As the evolution of these microarthropods in Antarctica has taken place over many millions of years, they represent suitable subjects to test speciation hypotheses and identify evolutionary trends and patterns of Antarctic fauna [33,38,39].

Free living mites are one of the most abundant and widespread microarthropod groups in Antarctica [40] and, among these, the best represented groups are the suborders Prostigmata and Oribatida and the order Mesostigmata. Within the Prostigmata, one of the most diverse families is the Penthalodidae, which includes the cosmopolitan genus *Stereotydeus* Berlese, 1860 [7]. However, while many studies have been conducted on the morphological and, more recently, genetic characteristics of springtails [27–29,31,35,41,42] present in Victoria Land, very few particularly genetic studies have investigated the biodiversity of Antarctic mites generally, and specifically *Stereotydeus*. Indeed, after early morphological studies in the 1960s [43–45], few studies on the physiology and ecology of the genus have been conducted [11–13,40,46,47], these are particularly focusing on *S. mollis*. Very recently, two new *Stereotydeus* species (*S. ineffabilis* and *S. nunatakis*) have been described from an area of Victoria Land [48], bringing the number of known Antarctic and sub-Antarctic members of the genus to 15 [48]. Focusing on Victoria Land, five species (*S. delicatus* Strandtmann, 1967, *S. punctatus* Strandtmann, 1967, *S. belli* Trouessart, 1902, *S. ineffabilis* Brunetti and Siepel, 2021 and *S. nunatakis* Brunetti, 2021) are currently known from North Victoria Land and two (*S. mollis* and *S. shoupi* Strandtmann, 1967) from South Victoria Land and the central Transantarctic Mountains [36]. Given the harsh field conditions and the small size and cryptic characters of members of this genus, the precise taxonomic determination of specimens in situ is challenging. In the laboratory, the combination of genetic and morphological approaches provides a powerful tool for detecting different levels of diversity. During the last two decades, the development of barcoding techniques using the mitochondrial cytochrome *c* oxidase subunit I (*cox1*) gene in combination with different nuclear markers has helped to discriminate cryptic species and determine the origin of morphological variation in multiple taxa [31,49,50]. However, over the period since this technology has become available, only three genetic studies have been conducted on Antarctic representatives of the genus, focusing exclusively on *S. mollis* in Southern Victoria Land [33,34,36] and giving a tantalizing hint of the high level of diversity hidden within and between different populations of this single species.

At the same time, given the recent discovery of the two new *Stereotydeus* species in Northern Victoria Land in a study that also reviewed the morphological characters relevant to the identification of Antarctic *Stereotydeus* species [48], the question of a possible overlap between these new taxa with the species already known from the area (*S. belli*, *S. punctatus* and *S. delicatus*) and with *S. mollis* from Southern Victoria Land has to be addressed. In addition to that, the current lack of genetic knowledge of a species morphologically described more than fifty years ago needs addressing, not only for the systematic understanding of the genus, but also to contribute to the future development and implementation of sustainable conservation planning in Antarctica. Although Antarctica is often assumed to be a pristine continent, it is increasingly clear that Antarctic ecosystems and biodiversity are facing the same threats as in the rest of the world, particularly from climate change, pollution, biological invasions and an increase in direct human impacts and activities [51–53]. In this context, the poor existing knowledge of species diversity and their dispersal ability are considered limiting factors to their effective management and conservation [31,54,55].

In the current study, we investigated, using a combined taxonomic approach, the distribution, phylogenetic relationships and the population genetics of the genus representatives of the *Stereotydeus* present in Victoria Land, with the support of morphological characteristics fundamental for species identification. In the Antarctic Conservation Biogeographic Regions system (ACBRs [52,56,57]), Victoria Land is divided into Northern and Southern Victoria Land. Nevertheless, the area between Mount Melbourne and the Drygalski Ice Tongue has been singled out for its unusual biogeographic connections and possible role in the promotion of the genetic differentiation of terrestrial taxa in numerous studies targeting Collembola [27,58,59]. As such, this region, named "Central" for convenience, has been separated from the northern ACBR in our analyses. Furthermore, we provide more than 150 new sequences for the mitochondrial barcode region *cox1*, and the nuclear *28S*, of five different *Stereotydeus* species from Victoria Land.

2. Materials and Methods

2.1. Sample Collection

Stereotydeus specimens were collected from nine different localities in Victoria Land (Figure 1; Table 1) during the 2017–2018 and 2018–2019 austral summer expeditions of the Italian National Antarctic Research Program (PNRA: PNRA16_00234), and were immediately preserved in >99.5% ethanol and stored at −80 °C. A total of 159 individuals were used for the molecular analyses. Of these, the whole body of 137 specimens was used for the genetic analyses (Table 1; see Section 2.2). The remaining 22 individuals (see Section 2.3) were used in the morphological investigation, with only 2–4 legs used for the DNA extraction.

Table 1. Coordinates and altitudes of sampling localities and ID codes for the different populations sampled; the numbers of individuals (*n.*) extracted and used for the molecular analyses and the species found at each locality, are given.

ID	Locality	Victoria Land	Lat (S)	Long (E)	Altitude	*n.*	Species
CHA	*Cape Hallett (Adelie Cove)*	North	72°26′25″	169°56′32″	140 m	10	*S. belli*
CCI	*Crater Cirque*	North	72°37′52″	169°22′22″	200 m	14	*S. belli; S. punctatus*
CJO	*Cape Jones*	North	73°16′38″	169°12′54″	310 m	17	*S. belli*
KAY	*Kay Island*	North	74°04′14″	165°18′60″	140 m	10	*S. belli*
CIC	*Campo Icaro*	Central *	74°42′45″	164°06′21″	70 m	35	*S. ineffabilis; S. delicatus*
VEG	*Vegetation Island*	Central *	74°47′00″	163°37′00″	120 m	10	*S. delicatus*
INE	*Inexpressible Island*	Central *	74°53′39″	163°43′44″	30 m	10	*S. ineffabilis; S. delicatus*
PRI	*Prior Island*	South	75°41′31″	162°52′34″	130 m	17	*S. ineffabilis; S. nunatakis*
SNU	*Starr Nunatak*	South	75°53′57″	162°35′08″	60 m	14	*S. ineffabilis; S. nunatakis*

* CIC, VEG and INE have been considered as "Central" to facilitate the division of the sampling area based on geography, although they all formally lie within the defined ACBR North Victoria Land [52,57].

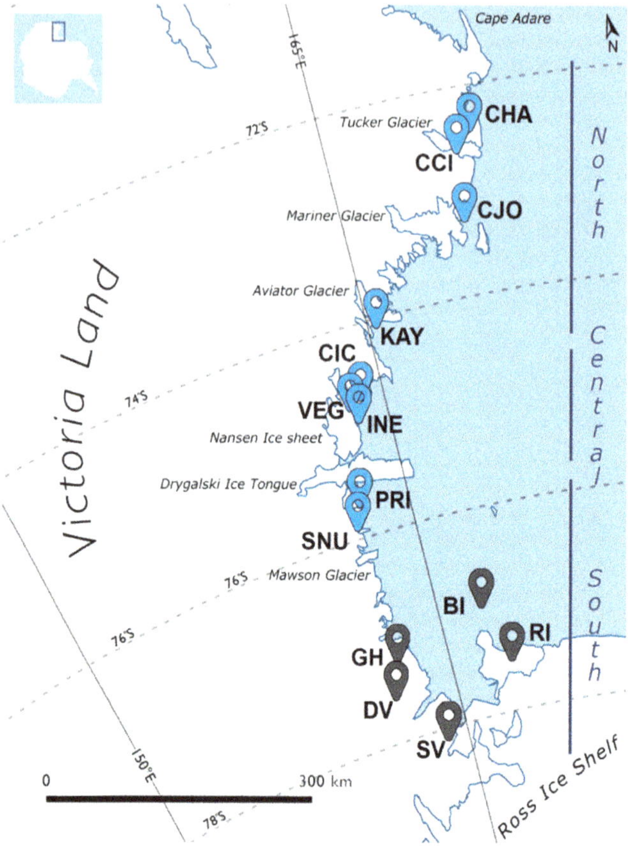

Figure 1. Map of sampling localities for the *Stereotydeus* spp. samples analyzed in this study (blue; see Table 1 for locality abbreviations) and in published studies of *S. mollis* [33,34,36] (dark grey): DV = McMurdo Dry Valleys (Taylor, Wright and Victoria Valleys and vicinity), SV = southern Dry Valleys (Garwood, Marshall and Miers Valleys and vicinity), BI = Beaufort Island; RI = Ross Island and GH = Granite Harbour (coastlines from ADD Simple Basemap, NPI/Quantarctica 3 [60]).

2.2. Molecular Dataset

Total genomic DNA was extracted from 137 whole individuals from the nine collection sites (Table 1) and the outgroup specimen, the winter grain mite *Penthaleus major* (Acari: Penthaleidae; Accession number *cox1*: MZ350753; Accession number *28S*: MZ442288; Table S1 in the Supplementary Materials) using the Wizard® SV genomic DNA Purification System (Promega, Madison, WI, USA) and eluting in 50 μL ddH_2O.

Region II of mtDNA cytochrome *c* oxidase subunit I (*cox1*) was amplified using the mite specific primers COI-2F (5′-TTYGAYCCIDYIGGRGGAGGAGATCC-3′) and COI-2R (5′-GGRTARTCWGARTAWCGNCGWGGTAT-3′) [61]. A preliminary amplification of the 28S gene was performed on a restricted pool of five *Stereotydeus* individuals from each of six localities (CHA, CCI, CJO, CIC, INE and SNU) and including all the species, with the primer pair D1a (5′-CCCSCGTAAYTTAAGCATAT-3′) and D5b1 (5′-ACACACTCCTTAGCGGA-3′) [62]. A new specific primer pair (Ste-28S-F (5′-GGACGTGAAACCGCTTGTA-3′) and Ste-28S-R (5′-TCTGACGATCGATTTGCAC-3′)) was designed in conserved regions (750 bp) and used to amplify all the remaining *Stereotydeus* specimens and the outgroup. PCRs were performed in 25 μL reaction volume containing: 2.5 μL of genomic DNA, 0.5 mM of each primer, 0.2 mM of each deoxynucleotide triphosphates (dNTPs), 2.5 mM of $MgCl_2$, 5 μL of

Green GoTaq Flexi buffer and 0.625 U of GoTaq Flexi DNA Polymerase (Promega, Madison, WI, USA). The amplifications were performed in a GeneAmp® PCR System 2700 (Applied Biosystems, Foster City, CA, USA) thermal cycler. The initial denaturation step was set at 95 °C for 5 min, followed by 35 cycles at 95 °C for 1 min, 45 °C (for *cox1*) or 50 °C (for the *28S*) for 1 min, and 72 °C for 90 s, and a final extension step at 72 °C for 7 min. PCR products were then purified using the kit Wizard® SV Gel and PCR Clean-up (Promega, Madison, WI, USA) and sequenced on both strands (with the same primers used for PCRs) with a DNA Analyzer ABI 3730, at the core facility of the Bio-Fab Research Lab (Rome, Italy). The sequences were assembled and manually corrected using the MacVector™ software (MacVector, Inc., USA; version 16.0.8-[63]).

In addition to the new samples extracted for this study, all 56 publicly available *cox1* sequences for the genus *Stereotydeus* were downloaded from GenBank (Table S2 in Supplementary Materials) and included in the analyses. These included 50 of *S. mollis*, 2 of *S. shoupi*, 1 of *S. belli*, 1 of *S. villosus* and 2 of *Stereotydeus* sp. together with a second outgroup, another eupodid mite *Eriorhynchus* sp. (Acari: Eriorhynchidae; see Table S2).

The two haplotypes, DQ305366 (S2-[34]) and DQ305388 (B-[33]), were excluded from this analysis because they are homonyms of DQ305362 and DQ305389, respectively. An error in naming them may have occurred when deposited in GenBank, therefore, following the analyses of Demetras et al. [36], the latter two were used in our analyses. Although we included all the remaining deposited haplotypes, some incongruences are noted in three other sequences: (i) for DQ305362 (S2-[34]), coordinates are missing because the precise sampling site in Wright Valley is not clear (W3 and/or W5); (ii) for DQ305382 (S20–V11 from Victoria Valley [34]), coordinates were not included in the original article [34]; (iii) DQ305367 (S6-[34]) was used in Demetras et al. [36] but is missing in the original article of McGaughran et al. [34], therefore, the coordinates are not shown (see Table S2). For the specimens from Demetras et al. [36], only the generic location of southern Dry Valleys (i.e., Garwood, Marshall and Miers Valleys, Shangri La and vicinity, according to Collins et al. [64]) was given, but not the exact coordinates, so they are not shown in this study.

2.3. Combined Morphological Analysis

In parallel to this study, morphological analyses have been performed on numerous specimens (between 20–50 for each sampled species, data not published). The morphological comparisons clearly defined the boundaries between all the *Stereotydeus* species occurring on Victoria Land, as recently published in Brunetti et al. [48], where not only the new species of *S. ineffabilis* and *S. nunatakis* are described, but also all the characters so far used to describe and distinguish the Antarctic *Stereotydeus* species are reviewed (see [48] Tables A1–A7), and the keys to identification are provided. Unfortunately, the lack of specimens of *S. mollis, S. shoupi* and *S. villosus* from accessible localities prevented us from improving the original descriptions with the new characters studied in these species and, therefore, were not available for combined morphological analyses.

In addition, after a quick molecular screening, we decided to deeply investigate the morphological aspects of *S. delicatus* and *S. ineffabilis* in relation to their genetic differentiation. We focused our attention on Campo Icaro, Inexpressible Island, Prior Island and Starr Nunatak, due to the presence at those localities of the new species described (*S. ineffabilis* and *S. nunatakis*). We also questioned the exact correspondence of previously published sequences to specific *Stereotydeus* taxa. In this respect, the combination of morphological and molecular analyses performed on the same specimens, collected in the central and southern sites of our sampling area, and the recent taxonomic description of new species of the genus (i.e., *S. ineffabilis*), challenged the attribution of some haplotypes to *S. mollis*.

Due to the small size of the specimens and, consequently, of the characteristics useful for an accurate taxonomic determination, 22 adult individuals (13 *S. ineffabilis* from four localities and 9 *S. delicatus* from Campo Icaro; Table 2) were selected for the joint morphological/molecular investigation and also used in all the molecular analyses. Only adult specimens were considered in the morphological comparison because, at the nymphal stages, most of the characteristics useful for the positive identification of *Stereotydeus* species are not yet developed (e.g., small size, sex structures not developed, division of the femora absent or incomplete, reduced number of aggenital and genital setae, and reduced number of rhagidial organs; see [48]).

From each specimen, 2–4 legs were removed (to perform the genetic analyses) while the remainder of the body was incubated on a slide with few drops of lactic acid (20%) at 37–45 °C for 30 min to clear the samples, which were then observed under a Leica DM RBE microscope for morphological analysis. The morphological characters considered for identification of *S. delicatus* and *S. ineffabilis* were: (a) the length (μm); (b) the division of the femora (presence/absence); (c) the position of the anal pore; (d) the number of aggenital and (e) the number of genital setae; (f) the length of the 4th segment of the pedipalp compared to the 3rd segment; (g) the shape of the epirostrum; and (h) the disposition of the rhagidial organs on tarsi I and II.

Table 2. New specimens extracted for the haplotypic and morphological analyses. Sampling localities with their ID codes, date of collection and the slide codes, and the sex and species of the new *Stereotydeus* individuals are given.

Locality	ID	Date	Slide	Sex	Species
Campo Icaro	CIC	28 January 2019	CI1	M	*S. delicatus*
			CI3	F	*S. ineffabilis*
			CI5	F	*S. delicatus*
			CI7	M	*S. delicatus*
		24 December 2017	CI9	M	*S. delicatus*
			CI10	M	
			CI11	F	
			CI12	F	
			CI13	F	
			CI14	M	
Inexpressible Island	INE	21 January 2019	I1	F	*S. ineffabilis*
			I2	M	
			I3	F	
			I4	M	
			I5	F	
Prior Island	PRI	11 January 2019	P1	M	*S. ineffabilis*
			P2	M	
			P3	F	
			P5	M	
Starr Nunatak	SNU	11 January 2018	S1	M	*S. ineffabilis*
			S2	M	
			S5	F	

2.4. Phylogenetic Analyses

For both mitochondrial and nuclear markers, 159 sequences were obtained and the datasets were separately aligned using the online tool Clustal Omega [65]; and manually corrected and trimmed (147 bp and 54 bp were trimmed for the *cox1* and *28S* respectively) using the MacVector™ software (MacVector, Inc., Cary, NC, USA; version 16.0.8-[63]). The resulting *cox1* dataset was then aligned with the two outgroups, while the *28S* dataset was aligned only with the *P. major* outgroup, due to the lack of the ribosomal DNA sequence in Genbank for *Eriorhynchus* sp. The outgroups were selected from mite families related to ingroups in order to reduce the phylogenetic distance with the Antarctic *Stereotydeus* spp. In detail, the species *P. major* (from a closely related family to that of ingroups) was selected as outgroup both for combined and single locus analyses. In addition, the *cox1* sequence of *Eryorinchus* sp. was also included as outgroup because it has been widely used in previous studies on Antarctic *Stereotydeus* spp.

The *cox1* dataset was concatenated to the *28S* alignment to generate a multilocus dataset through FaBox [66], with the online tool Fasta alignment joiner (available at https://users-birc.au.dk/palle/php/fabox/alignment_joiner.php; accessed on 18 September 2020).

The multilocus alignment was then run on the Gblocks server 0.91b ([67]; available at http://molevol.cmima.csic.es/castresana/Gblocks_server.html; accessed on 18 September 2020) under strict settings and the hypervariable regions of the *28S* alignment were discarded. After the run, 1034 positions, out of the 1171 of the initial dataset (88%), were kept. Ultimately, the four single- and the multilocus alignments used for the phylogenetic and population genetics analyses were: (i) *cox1* with outgroup; (ii) *cox1* all haplotypes; (iii) combined *cox1-28S*; and (iv) combined *cox1-28S* with associated morphological information (Table 3).

Table 3. List of the datasets (single and multilocus), number of new sequences obtained and used in each dataset (*n.*), markers, reference sequences (Ref.) and outgroups used for the analyses and models of nucleotide evolution that best fitted, divided according to the partition applied and to the respective tree search optimization criteria.

		n.	Single/Multi Locus	*cox1*	*28S*	Ref.	Outgroups	Best Model 1st	2nd	3rd	Non-Cod
i	*cox1 with outgroups*	159	single	x	-	*S. shoupi* (2) *S. villosus* *Stereotydeus* sp. (2) *S. belli*	*Eriorhynchus* sp. *P. major*	K81UF+I+G	GTR+I	F81+I	-
ii	*cox1 all haplotypes*	159	single	x	-	*S. shoupi* (2) *S. villosus* *Stereotydeus* sp. (2) *S. belli* *S. mollis* (50)	*Eriorhynchus* sp. *P. major*	K81UF+G	GTR+I+G	F81+I	-
iii	*combined cox1-28S*	159	multi	x	x	-	*P. major*	K81UF+I+G	TRN+I	F81+I	GTR+G
iv	*combined cox1-28S with morphologica information*	99	multi	x	x	-	*P. major*	HKY+I+G	TIM+G	F81+G	TVM+G

To identify the haplotypes and their frequencies within populations, all the alignments were run with the online software DNA-Collapser [66]. The sequences of the resulting haplotypes were used to calculate the genetic distances between the haplotypes using the software R 3.6.1 [68] with the "ape 5.3" package [69]. The best evolutionary models were selected before the tree search (Table 3), partitioning the datasets with the software PartitionFinder 2.1.1 [70] based on Akaike's information criterion (AIC) and a greedy strategy: 1st, 2nd and 3rd codon positions for the *cox1* protein-encoding gene and one single partition were considered for the *28S* (Table 3). Bayesian analysis was performed with MrBayes 3.2.7 software [71], applying four chains (three hot and one cold) for 10^6 generations, with a sampling frequency of one tree every 1000 iterations and with 25% of the tree topologies discarded (burn in step) from the final result. For better visualization, the resulting phylogenetic trees were then zoomed and expanded and the node labels (posterior probabilities) were added with FigTree 1.4.4 software [72]. The new *Stereotydeus* mitochondrial and nuclear sequences were deposited in GenBank (*cox1* Accession numbers: MZ350724-MZ350752; *28S* Accession numbers: MZ442270-MZ442287; Table S1 in Supplementary Materials).

2.5. Population Structure Analyses

The population genetics study was performed using the *cox1* dataset without the outgroups applied for the phylogenetic analysis. *S. mollis* sequences were not included in the analysis. This was due to: (i) the incongruences found in the Genbank sequences (see Section 2.2. and Table S2 in Supplementary Materials), (ii) the fact that no morphological investigations were performed on these individuals, and (iii) because new *S. mollis* specimens were not available for a morphological analysis during this study. Haplotype frequencies were obtained using the online tool DNA collapser [66]. The network clade analysis was performed on TCS 1.21 [73] using a connection limit of 98% and visualized with the online tool tcsBU ([74]; available at https://cibio.up.pt/software/tcsBU/; accessed on 28 November 2020) to estimate the haplotype networks for each species. To investigate the genetic characteristics of populations and to test for the presence of population structure, Arlequin version 3.11 [75] was used for each species separately. The haplotype (h) and nucleotide (π) diversity indices [76], as well as the mean number of pairwise differences (θ) and segregating sites (θS), were computed at the population level. Analysis of molecular variance (AMOVA; [77]) was used to measure the extent to which genetic variance could be assigned to the hierarchical structure of population organization (testing them with the structure according to the populations: "Cape Hallett", "Crater Cirque", "Cape Jones" and "Kay Island" for *S. belli*; "Campo Icaro", "Vegetation Island" and "Inexpressible Island" for *S. delicatus*; "Campo Icaro", "Inexpressible Island", "Prior Island" and "Starr Nunatak" for *S. ineffabilis* and "Prior Island" and "Starr Nunatak" for *S. nunatakis*), with the statistical significance of variance components tested with 16,000 permutations. Pairwise differences between haplotypes (Φ_{ST} values) were calculated using simple distances and these were used to look for significant relationships between population genetic distance (Φ_{ST}).

3. Results

Using the *cox1* haplotypes of the 50 *S. mollis* specimens already available on GenBank as templates, 495 bp of a uniform and unambiguous alignment from 159 sequenced individuals were used for all genetic analyses. For *28S*, 1034 positions of the 159 sequenced individuals, together with the outgroup *P. major*, were used for phylogenetic analyses.

For each *Stereotydeus* species, between 2–14 *cox1* and 1–9 *28S* haplotypes were found (Table 4) while, for each locality, between 1–11 *cox1* and 1–4 *28S* haplotypes were found. Most *28S* haplotypes were unique at the species level, with the only exception being RX1 from CIC, shared by both *S. delicatus* and *S. ineffabilis*. In addition, for the combined set of *cox1* and *28S*, from 3–16 and from 2–9 haplotypes were found for the *Stereotydeus* species and the localities, respectively. The number of *Stereotydeus* species identified per site ranged from 1–2 (Table 5).

Thirty-six unique haplotypes for *cox1*, ranging in divergence from 0.2 to 2.5% and 18 unique haplotypes for *28S* (from 0.2 to 9.0%), were identified. The compiled matrix of percentage genetic distances (Table 6) showed a gradient of arbitrarily estimated comparisons corresponding to intraspecific distances (0% to 8.48%), intermediate values between intra- and interspecific distances (8.49% to 10.7%), and interspecific distances (10.8% to 16.8%).

Table 4. Number of specimens analyzed per species and number of haplotypes detected within the species for the mitochondrial and nuclear markers and the combined set of the *cox1* and *28S* (combined).

Species	Specimens	Haplotypes		
		cox1	*28S*	Combined
S. belli	39	10	9	14
S. punctatus	12	4	1	4
S. ineffabilis	59	14	3	16
S. delicatus	39	6	2	10
S. nunatakis	10	2	2	3

Table 5. Sampling locality codes (ID), number of sequenced individuals per area (*n.*), number of species per area (N.) and their names, and list of all haplotypes for each species. Haplotype code: the first letter indicates the marker (M: mitochondrial; R: nuclear ribosomal DNA) and the genus (S: *Stereotydeus*) in the combined haplotypes; the second letter is the initial of the species name (B/b = *belli*; P/p = *punctatus*; D/d = *delicatus*; I/i = *ineffabilis*; N/n = *nunatakis*; RX identifies the haplotype only present in the Campo Icaro (CIC) area and found in both *S. delicatus* and *S. ineffabilis*) followed by the progressive number of the haplotype.

ID	*n.*	N.	Species	Haplotypes		
				cox1	*28S*	Combined
CHA	10	1	*S. belli*	MB1(10)	RB1(9), RB2(1)	Sb1(9), Sb2(1)
CCI	14	2	*S. belli*	MB10(2)	RB8(1), RB9(1)	Sb12(1), Sb13(1)
			S. punctatus	MP1(3), MP2(7), MP3(1), MP4(1)	RP1(12)	Sp1(7), Sp2(3), Sp3(1), Sp4(1)
CJO	17	1	*S. belli*	MB2(2), MB3(3), MB4(3), MB5(8), MB6(1)	RB3(16), RB4(1)	Sb3(8), Sb4(2), Sb5(3), Sb6(3), Sb7(1)
KAY	10	1	*S. belli*	MB7(1), MB8(3), MB9(6)	RB5(7), RB6(1), RB7(1), RB8(1)	Sb8(1), Sb9(2), Sb10(5), Sb11(1), Sb14(1)
CIC	45	2	*S. delicatus*	MD1(18), MD2(1), MD3(2), MD4(1), MD5(6)	RD1(12), RD2(2), RX1(14)	Sd1(1), Sd2(1), Sd3(9), Sd4(2), Sd5(8), Sd6(1), Sd9(1), Sd10(4), Sd11(1)
			S. ineffabilis	MI1(6), MI2(1), MI3(1), MI5(4), MI12(4), MI13(1)	RX1(17)	Si1(1), Si2(4), Si5(4), Si11(1), Si12(6), Si13(1)
VEG	10	1	*S. delicatus*	MD5(9), MD6(1)	RD1(10)	Sd8(1), Sd10(9)
INE	15	2	*S. delicatus*	MD6(1)	RD2(1)	Sd7(1)
			S. ineffabilis	MI6(3), MI7(3), MI8(6), MI12(1), MI14(1)	RI1(14)	Si3(1), Si6(3), Si7(3), Si8(6), Si14(1)
PRI	21	2	*S. ineffabilis*	MI4(15), MI9(1), MI10(1), MI11(2)	RI3(19)	Si4(2), Si9(1), Si10(1), Si16(15)
			S. nunatakis	MN1(2)	RN1(2)	Sn1(2)
SNU	17	2	*S. ineffabilis*	MI4(5), MI9(3), MI11(1)	RI2(1), RI3(8)	Si4(1), Si9(3), Si15(1), Si16(4)
			S. nunatakis	MN1(7), MN2(1)	RN1(7), RN2(1)	Sn1(6), Sn2(1), Sn3(1)

Table 6. Matrix of percentage genetic distances between *Stereotydeus cox1* sequences. Cells were colored in a gradient from green (0%-low divergence) to red (16.8%-high divergence) to facilitate interpretation (see Table 5 for the haplotype labels).

	MI4	MI1	MI2	MI3	MI14	MI5	MI9	MI10	MI6	MI8	MI7	MD6	MD1	MD4	MD3	MD2	MD5	MP3	MP4	MP2	MP1	MB10	MB9	MB8	MB7	MB1, *S. belli*	MB6	MB5	MB3	MB4	MB2	MN1	MN2	MI12	MI13	MI11	*Stereotydeus* sp.a	*Stereotydeus* sp.b	*S. villosus*	*S. shoupi*	*S. shoupi* 1
MI4	0																																								
MI1	0.4	0																																							
MI2	0.61	0.2	0																																						
MI3	0.61	0.2	0.4	0																																					
MI14	0.61	0.2	0.4	0.4	0																																				
MI5	2.42	2.42	2.63	2.63	2.22	0																																			
MI9	2.42	2.83	3.03	3.03	2.63	1.41	0																																		
MI10	2.42	2.83	3.03	3.03	2.63	1.82	0.4	0																																	
MI6	2.63	2.63	2.83	2.83	2.42	0.2	1.62	2.02	0																																
MI8	2.63	2.63	2.83	2.83	2.42	1.01	1.62	2.02	0.81	0																															
MI7	2.83	2.83	3.03	3.03	2.63	1.21	1.82	2.22	1.01	0.2	0																														
MD6	1.82	1.82	2.02	2.02	1.62	1.41	2.42	2.83	1.62	1.62	1.82	0																													
MD1	2.22	2.22	2.42	2.02	2.02	1.62	2.42	2.83	1.82	1.82	2.02	1.41	0																												
MD4	2.42	2.42	2.63	2.22	2.22	1.82	2.63	3.03	2.02	2.02	2.22	1.62	0.2	0																											
MD3	2.42	2.42	2.63	2.22	2.22	1.82	2.63	3.03	1.62	1.62	1.82	1.62	0.2	0.4	0																										
MD2	2.63	2.63	2.83	2.42	2.42	2.02	2.83	3.23	2.02	2.02	2.22	1.82	0.4	0.61	0.4	0																									
MD5	8.28	8.69	8.89	8.89	8.89	7.88	8.48	8.69	7.88	8.28	8.08	8.48	8.48	8.69	8.48	8.69	0																								
MP3	6.46	6.87	6.87	7.07	7.07	8.28	7.88	8.28	8.48	8.28	8.08	7.68	7.68	7.88	7.88	8.08	5.86	0																							
MP4	6.67	7.07	7.07	7.27	7.27	8.08	7.68	8.08	7.88	7.68	7.47	7.47	7.88	8.08	7.68	8.08	5.66	1.21	0																						
MP2	6.67	7.07	7.07	7.27	7.27	8.08	7.68	8.08	8.28	8.08	7.88	7.47	7.88	8.08	8.08	8.28	5.66	0.2	1.01	0																					
MP1	6.87	7.27	7.27	7.47	7.47	8.28	7.88	8.28	8.48	8.28	8.08	7.68	8.08	8.28	8.28	8.48	5.45	0.4	1.21	0.2	0																				
MB10	9.49	9.9	9.9	10.1	10.1	9.9	9.7	9.9	9.9	9.7	9.9	9.7	9.9	10.1	9.9	10.1	8.89	6.46	6.26	6.26	6.46	0																			
MB9	9.29	9.7	9.7	9.9	9.9	9.7	9.49	9.7	9.7	9.49	9.7	9.49	9.7	9.9	9.7	9.9	9.09	6.67	6.46	6.46	6.67	0.2	0																		
MB8	9.49	9.9	9.9	10.1	10.1	9.9	9.7	9.9	9.9	9.7	9.9	9.7	9.9	10.1	9.9	10.1	8.48	6.06	5.86	5.86	6.06	0.4	0.61	0																	
MB7	10.7	11.1	11.1	11.3	11.3	11.1	10.9	11.1	11.1	10.9	11.1	10.9	11.1	11.3	11.1	11.1		7.07	6.87	6.87	6.87	4.04	4.24	3.64	0																
MB1, *S. belli*	10.5	10.9	10.9	11.1	11.1	10.7	11.3	11.5	10.5	10.7	10.5	10.7	11.1	11.3	10.9	11.3	9.09	8.08	8.08	7.88	8.08	5.05	5.25	5.05	5.45	0															
MB6	10.3	10.7	10.7	10.9	10.9	10.7	10.5	10.7	10.5	10.3	10.5	10.5	10.7	10.9	10.5	10.9	10.3	8.28	7.68	8.08	8.28	3.43	3.64	3.43	4.24	5.45	0														
MB5	10.7	11.1	11.1	10.9	11.3	11.1	10.9	11.1	10.9	10.7	10.9	10.9	10.7	10.9	10.5	10.5	10.5	8.48	7.68	8.28	8.48	3.84	4.04	3.84	4.65	5.45	1.01	0													
MB3	10.7	11.1	11.1	10.9	11.3	11.1	10.9	11.1	10.9	10.7	10.9	10.9	10.7	10.9	10.5	10.9	10.1	8.08	7.27	7.88	8.08	3.64	3.84	3.23	4.04	5.66	0.81	0.61	0												
MB4	10.7	11.1	11.1	10.9	11.3	11.1	10.9	11.1	10.9	10.7	10.9	10.9	10.7	10.9	10.5	10.9	10.1	8.08	7.27	7.88	8.08	3.84	4.04	3.43	4.24	5.45	1.01	0.4	0.2	0											
MB2	11.1	11.5	11.5	11.3	11.7	11.1	11.3	11.5	10.9	11.1	11.3	11.3	11.1	11.3	10.9	11.3	10.1	8.48	7.68	8.28	8.48	3.84	4.04	3.43	4.24	5.45	1.41	1.21	0.61	0.81	0										
MN1	11.3	11.5	11.5	11.7	11.3	12.7	12.3	11.9	12.9	12.9	13.1	11.7	12.3	12.5	12.5	12.7	13.7	9.49	9.9	9.29	9.49	11.92	11.9	11.9	12.1	11.72	12.3	12.5	12.7	12.5	12.9	0									
MN2	11.5	11.7	11.7	11.9	11.5	12.9	12.5	12.1	13.1	13.1	13.3	11.9	12.5	12.7	12.7	12.9	13.5	9.7	10.1	9.49	9.29	12.12	12.1	12.1	12.1	11.92	12.5	12.7	12.9	12.7	13.1	0.2	0								
MI12	13.3	13.7	13.7	13.5	13.9	14.1	13.7	14.1	14.1	14.1	14.3	13.7	13.3	13.1	13.3	13.5	13.5	12.5	12.3	12.5	12.7	12.93	12.7	12.7	13.3	13.94	14.6	14.1	13.9	13.7	14.1	14.6	14.8	0							
MI13	13.9	14.1	14.1	14.3	14.3	14.8	13.9	14.1	14.6	14.1	13.9	14.3	14.3	14.1	14.1	14.6	11.9	11.7	11.3	11.9	12.1	12.12	12.3	11.9	11.7	13.13	12.3	12.7	12.3	12.5	12.9	16.6	16.8	13.1	0						
MI11	14.3	14.8	15	14.6	15	14.8	15	15.2	14.8	15	15.2	14.8	14.8	14.6	14.8	14.6	12.9	12.1	11.5	11.9	11.9	12.12	12.3	12.5	13.1	13.94	13.5	12.9	13.5	13.3	13.5	12.9	12.9	15.8	14.1	0					
Stereotydeus sp.a	12.3	12.7	12.7	12.9	12.9	13.7	13.3	13.3	13.7	13.5	13.3	13.9	13.1	13.3	13.1	13.1	12.1	9.49	9.7	9.7	9.7	11.11	11.1	10.7	10.9	12.32	11.9	11.7	11.7	11.5	11.9	13.7	13.7	13.3	15.4	14.1	0				
Stereotydeus sp.b	12.3	12.7	12.7	12.9	12.9	13.7	13.3	13.3	13.7	13.5	13.3	13.9	13.1	13.3	13.1	13.1	12.5	9.49	9.7	9.7	9.7	11.52	11.5	11.1	10.9	12.73	11.9	11.7	11.7	11.5	11.9	13.7	13.7	13.9	15.6	13.9	1.21	0			
S. villosus	12.9	12.9	12.9	13.1	13.1	13.1	12.7	13.1	12.9	12.7	12.9	13.1	12.9	13.1	12.7	13.1	13.9	10.5	10.1	10.5	10.7	11.31	11.3	10.9	11.3	12.12	11.1	11.3	11.1	10.9	11.5	14.3	14.6	15.6	15.8	16.4	11.11	10.71	0		
S. shoupi	14.3	14.8	14.8	14.6	15	14.6	14.3	14.6	14.3	14.3	14.1	15	14.6	14.8	14.3	14.8	14.6	12.5	12.1	12.3	12.5	12.93	13.1	12.5	12.1	13.13	13.3	12.9	12.7	12.5	12.5	15.2	15.4	16	15	15.8	12.53	11.92	13.54	0	
S. shoupi 1	14.6	15	15	14.8	15.2	14.8	14.6	14.8	14.6	14.6	14.3	15.2	14.8	15	14.6	15	14.8	12.7	12.3	12.5	12.7	12.73	12.9	12.7	12.3	13.33	13.1	12.7	12.9	12.7	12.7	15	15.2	16.6	15	15.2	12.73	12.12	13.74	0.61	0

Green frame—*S. ineffabilis*; orange frame—*S. delicatus*; magenta frame—*S. punctatus*; blue frame—*S. belli*; light blue frame—*S. nunatakis*.

3.1. Haplotype Network Analyses

The total number of nucleotide substitutions (absolute changes) ranged from 1 (*S. nunatakis* in SNU) to 117(*S. ineffabilis* in CIC) within all the populations of the five different taxa. Four subnetworks were found for *S. belli*, with two single haplotypes not connected with any other haplotype: MB1 and MB7, from CHA and KAY, respectively. Within the species and within the clusters, the number of nucleotide substitutions ranged from a minimum of nine, recorded in CJO, to a maximum of 21, in KAY (mean 7.50 ± 25.48) (Figure 2). For *S. punctatus*, one single network was observed where all haplotypes were connected with each other within an upper range of seven nucleotide changes (Figure 2). Three clusters were found for *S. delicatus*, with two single haplotypes not connected with any other haplotype: MD5 (VEG and CIC) and MD6 (VEG and INE). The number of nucleotide substitutions ranged from 42, in VEG, to 46, in CIC, in this species and within the populations (mean 29.33 ± 9.95) (Figure 2). Six networks were found for *S. ineffabilis*, with three single haplotypes not connected with any other haplotype: MI11 (SNU and PRI), MI12 (CIC and INE) and MI13 (CIC). These haplotypes are also placed together in a different position in the phylogenetic trees, with the respect to the other conspecific haplotypes (see Section 3.2). The differences within both species and populations ranged from 79, in SNU, to 117, in CIC (mean 89 ± 18.67) (Figure 2). For *S. nunatakis*, only two haplotypes were observed, differing by a single substitution (mean 0.50 ± 0.71) (Figure 2).

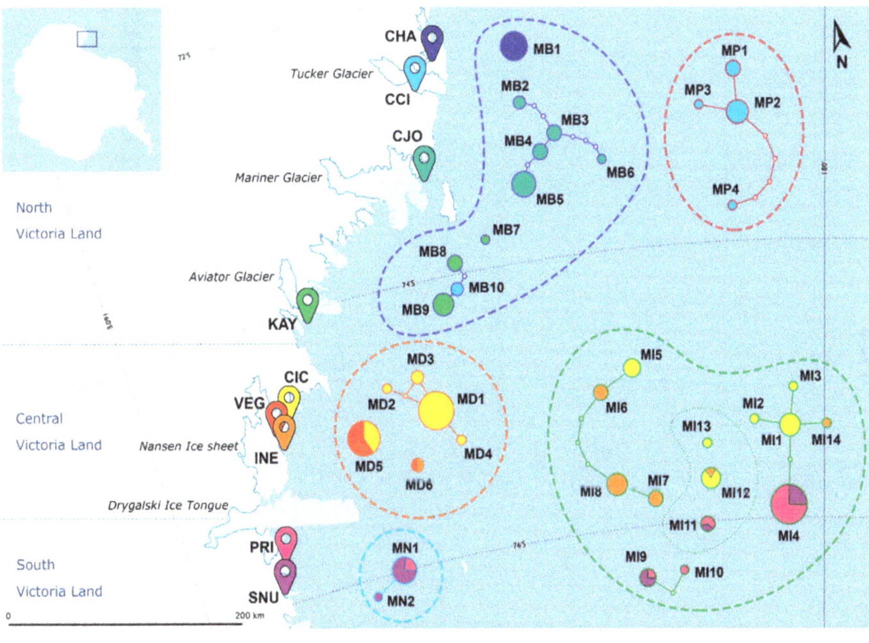

Figure 2. Haplotype networks of *cox1* for the five *Stereotydeus* species in Victoria Land (from 72 °S to 76 °S). Collection sites are indicated by the pie chart colors; the species are identified by the outlines of the networks together with the haplotype ID and the dashed lines around the clusters (coastlines from ADD Simple Basemap, NPI/Quantarctica 3 [60]).

3.2. Phylogenetic Analyses

- *cox1* with outgroups

For this single locus analysis, a total of 165 *Stereotydeus* sequences and two outgroups (*Eriorhynchus* sp. and *P. major*) were used. Before the addition of the outgroups, two unrooted analyses were also performed (Table S3 and Figure S1 in Supplementary Materials). One monophyletic group was formed by the haplotypes of *S. belli* (MB1-10) and includes 29 specimens from Northern Victoria Land (CHA, CJO and two from CCI), all those of the

KAY population (Central Victoria Land) and also the single sequence of *S. belli* (specimen from Cape Hallett) (Figures 2 and 3). Another monophyletic group included all 12 *S. punctatus* sequences (MP1-4) belonging to the CCI population. One paraphyletic group included the *S. delicatus* specimens (MD1-6) and the S. *ineffabilis* specimens (MI1-10, 14), with individuals from Southern Victoria Land (CIC, VEG, INE for *S. delicatus* and CIC, INE, PRI and SNU for *S. ineffabilis*) (Figures 2 and 3). Three haplotypes of *S. ineffabilis* were not included in the latter group, but they were clustered together, although with low statistical support (Figures 2 and 3). These three sequences, together with the branch that carries the two *S. nunatakis* haplotypes, did not cluster with the remaining ingroup, due to the insertion of three sequences of other species, although with medium statistical support (pp = 0.74 and 0.87) (Figure 3).

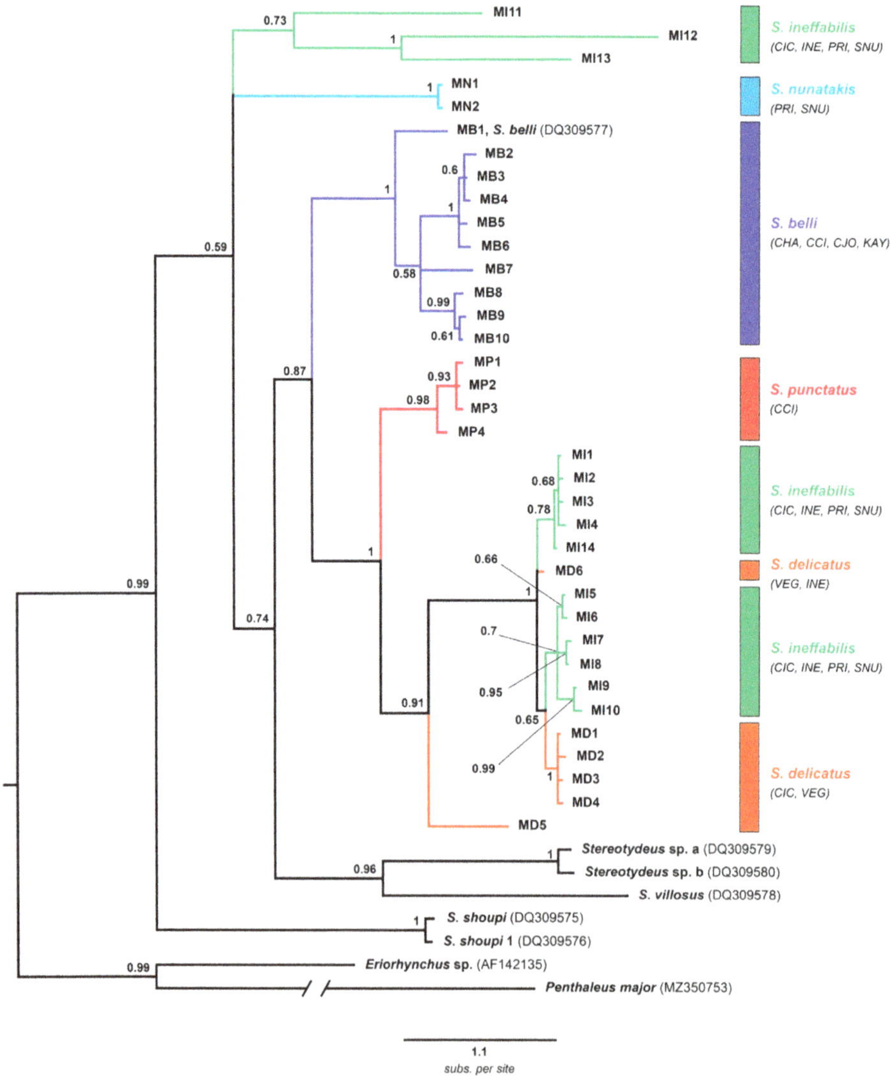

Figure 3. The phylogenetic tree of the *cox1* dataset of *Stereotydeus* specimens from Victoria Land (locality IDs are shown in brackets). Posterior probabilities are shown at the nodes. Labels indicate the ID of the haplotypes (detailed description of haplotypes in Table 5). Accession numbers for reference sequences and *Erioryнchus* sp. [33,34,36] and *P. major* outgroups are also shown.

- *cox1* all haplotypes

Fifty previously published *S. mollis* reference sequences were included for this analysis (Table S2). Despite the *S. ineffabilis* haplotypes being spread throughout the entire phylogenetic tree and not all nodes being statistically well supported, the two monophyletic groups of *S. belli* and *S. punctatus* were still distinct from the remaining species with good support at nodes (pp = 0.95 and 1, respectively). *S. delicatus* was, once again, recovered as a paraphyletic group: one cluster of four haplotypes (MD1-4) and two separated branches (MD5 and MD6), although with low support at nodes. The cluster of two haplotypes for *S. nunatakis*, together with MI11 and two *S. mollis* haplotypes (Sm49 and Sm50), was collapsed with the other three sequences at the base of the main cluster. Six (MI1, 4, 6, 9, 11, 12) out of the fourteen *S. ineffabilis* haplotypes were identical to previously published sequences (L, K, J, Sm44, R, O, respectively) originally assigned to *S. mollis* before the identification and description of *S. ineffabilis* as a new species [48] (Table 7). After the morphological identification of the specimens related to these haplotypes (Table 7), these sequences are now considered as *S. ineffabilis*. In addition, when sequences initially assigned to *S. mollis* clustered together with the *S. ineffabilis* haplotypes and were statistically well supported (pp > 0.85), we tentatively considered them as belonging to *S. ineffabilis* (e.g., Sm50, P; Figure 4).

Table 7. *S. ineffabilis* specimens used for the haplotypic and morphological analyses (Slide) with *cox1* haplotypes (*cox1*) identical to previously published sequences of *S. mollis* (haplo.). Sampling localities with their ID codes where the specimens were found and accession numbers (Acc. num.) of the *S. mollis* haplotypes are also provided.

	Slide	**ID**	***cox1***	**Haplo.**	**Acc. Num.**
S. ineffabilis	CI3	CIC	M1	L	DQ305390
	P1, 2, 5; S5	PRI, SNU	MI4	K	DQ305385
	I2, 4	INE	MI6	J	DQ305397
	P3; S1	PRI, SNU	MI9	Sm44	HM537086
	S2	SNU	MI11	R	DQ309574
	I3	INE	MI12	O	DQ309572

- Combined *cox1-28S*

Following the phylogenetic analyses of the combined dataset of *cox1* and *28S* sequences (1034 bp), four phylogroups were detected: three monophyletic groups (*S. belli*, *S. punctatus* and *S. nunatakis*, although with variable support, pp = 0.55–1) and one paraphyletic clade (statistically low support, pp = 0.66, including *S. ineffabilis* and *S. delicatus* as mutually para/polyphyletic groups). The combination of the two datasets generated 14 unique haplotypes for *S. belli* from northern Victoria Land (CHA, CJO and CCI) and central Victoria Land (KAY), 4 unique haplotypes for *S. punctatus* from northern Victoria Land (CCI), 3 unique haplotypes for *S. nunatakis* from southern Victoria Land (PRI and SNU), 11 unique haplotypes for *S. delicatus* from southern Victoria Land (CIC, VEG and INE) and 16 unique haplotypes for *S. ineffabilis* from southern Victoria Land (CIC, INE, PRI and SNU) (Figure 5).

- Combined *cox1-28S* with morphology

In order to further clarify the paraphyletic relationships, a table of some morphological characteristics was linked to the combined *cox1-28S* tree, restricted to *S. ineffabilis* and *S. delicatus* sequences. All the nodes clustering the deepest branches together were statistically well supported, with the exception of that separating the Si 4 haplotype from the main cluster (pp = 0.64) (Figure 6).

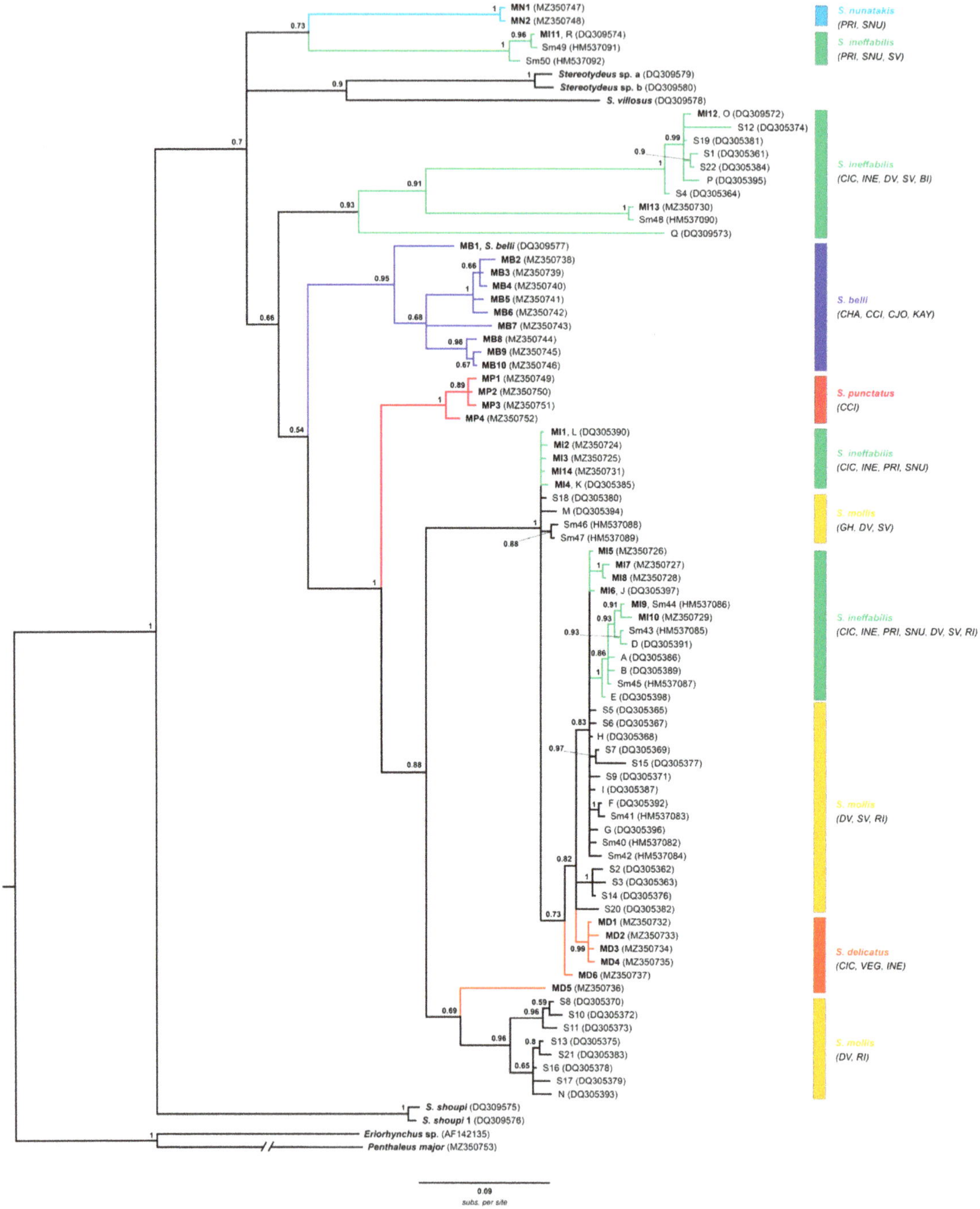

Figure 4. Phylogenetic tree of *cox1* haplotypes of *Stereotydeus* specimens from Victoria Land and the McMurdo and southern Dry Valleys [33,34,36]. Posterior probabilities are shown at the nodes. New haplotypes (in bold): labels indicate the ID of the haplotype; accession numbers are shown in brackets.

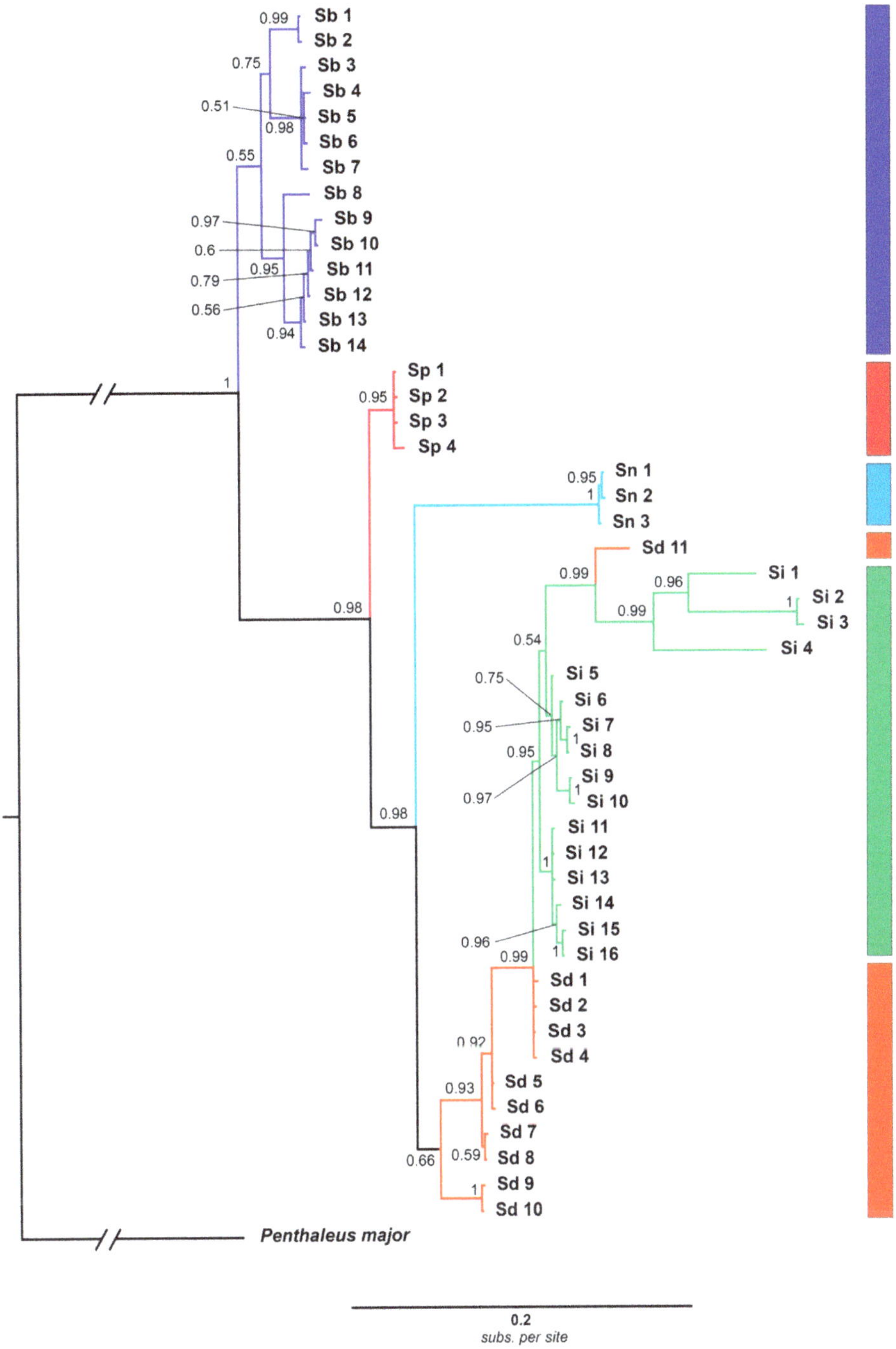

Figure 5. Phylogenetic tree of the combined mitochondrial and nuclear sequences of *Stereotydeus* specimens from Victoria Land. Posterior probabilities are shown at the nodes; labels indicate the ID of the haplotypes. For detailed description of haplotypes, see Table 5.

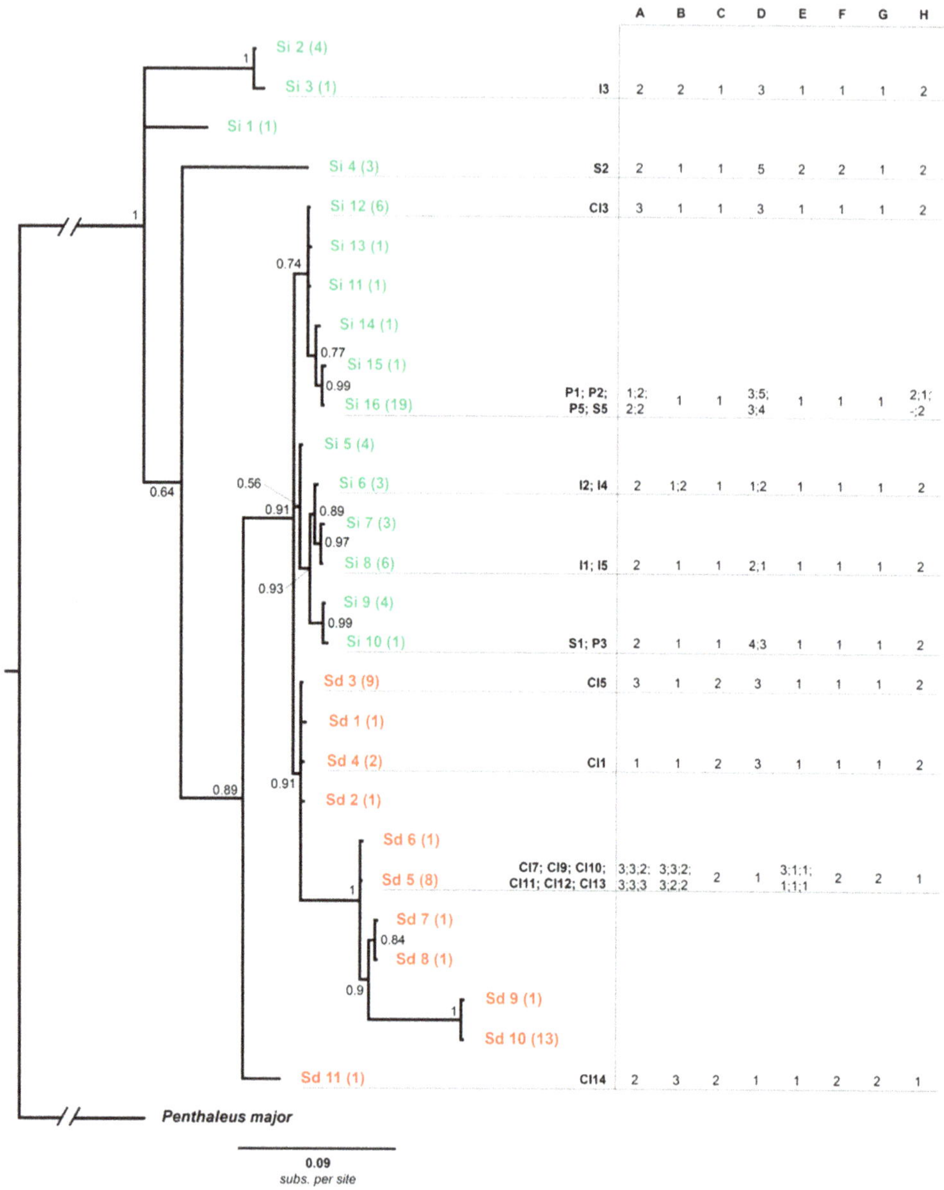

Figure 6. Phylogenetic tree of the combined mitochondrial and nuclear sequences of *S. delicatus* (orange labels) and *S. ineffabilis* (green labels) and table with codes of the morphological characters (see Table 8). Posterior probabilities are shown at the nodes of the phylogenetic tree. For the specimen ID (black, bold), see Table 2.

The *S. delicatus* specimens had a mean length of 451.83 µm (± 27.39 µm), ranging from CI1 (389.99 µm) to CI7 (481.55 µm). The femora were divided in CI7, CI9, CI11 (Sd 5) and CI14 (Sd 11), undivided in CI1 (Sd 4) and CI5 (Sd 3) and with partial division in CI10, CI12 and CI13 (Sd 5). The position of the anal pore was always apical. In all the specimens observed with haplotype Sd 5 and CI14 (Sd 11), there were four pairs of aggenital setae, while CI1 and CI5 had five pairs. Six pairs of genital setae were present in all the specimens, with the exception of CI7, which had seven pairs. The length of the 4th segment of the pedipalp was longer than the 3rd in all the specimens with haplotypes Sd 5 and CI14 (Sd 11), while in CI1 and CI5 the two segments were comparable in length. The trilobe shaped epirostrum was weakly developed in CI1 and CI5, while, in the remaining specimens, it was evident and strongly developed. The three rhagidial organs on tarsi I and II showed an axis of symmetry in all specimens, with the exception, again, of CI1 and CI5.

Table 8. Morphological characters considered for the identification of *S. delicatus* and *S. ineffabilis*. Every row of a character is represented by a number (1–5) used to link them to the *combined morphology* phylogenetic tree (Figure 6). A. Length (μm); B. Femora; C. Position of the anal pore; D. Aggenital setae; E. Genital setae; F. Length of the 4th segment of pedipalp compared to the 3rd; G. Epirostrum; H. Disposition of the rhagidial organs on tarsi I and II.

Code	A	B	C	D	E	F	G	H
1	<400	undivided	ventral	4/4	6/6	IV = III	weak	symmetry
2	401–450	barely divided	apical	4/5	6/7	IV > III	evident	no symmetry
3	451–489	divided		5/5	7/7			
4	>490			5/6				
5				6/6				

The *S. ineffabilis* specimens had a mean length of 427.62 μm (± 18.61 μm), ranging from P1 (386.62 μm) to CI3 (460.44 μm). The femora were undivided except for individuals I3 (Si 3) and I4 (Si 6), where the division was only partial. The anal pore was always ventral in the terminal portion (see [48] Figures 1b and 5b). The number of aggenital setae was variable: two specimens (I2, I5) had four pairs, five (I3, CI3, P1, P3 and P5) had five pairs, two (S2, P2) had six pairs, while four had an intermediate number (I1 and I4 had 9 setae; S1 and S5 had 11 setae). Six pairs of genital setae were present in all specimens with the exception of S2 (Si 4), which presented an asymmetry with 13 setae. The length of the two terminal segments of the pedipalps was comparable in all the specimens examined except in S2 (Si 4), where the 4th segment was longer than the 3rd. The trilobed shape of the epirostrum was weakly developed in all specimens. The three rhagidial organs on tarsi I and II showed an axis of symmetry only in P2 (Si 16). P5 legs I and II were missing, so it was not possible to determine the positions of the rhagidial organs (for the morphological features see [48], Figures 1–5).

Although character C seems the only listed character that sharply sorts out the two species, when few exceptions of specimens are not considered, the list of characters increases (see [48] for the keys and the synoptic Tables A1–7 of the Antarctic *Stereotydeus* species).

3.3. Population Structure Analyses

Haplotype diversity (*h*) for *cox1* in *S. belli* ranged from 0 to 0.743 (mean 0.336). Within populations, CJO had the highest haplotype diversity and CIIA and CCI the lowest. Nucleotide diversity (π) was low for all four populations, with the highest value being in the KAY population (0.010) (Table 9). The values of mean nucleotide pairwise differences $\theta(\pi)$ and mean number of segregating sites $\theta(S)$ ranged from 0 to 5.200 (mean 1.194 ± 2.486) and from 0 to 7.423 (mean 2.521 ± 3.501), respectively. The KAY population had the highest values of both $\theta(\pi)$ and $\theta(S)$, while CHA and CCI had the lowest. For *S. delicatus*, *h* ranged from 0 to 0.553 (mean 0.384), with the highest values in CIC (0.553) and the lowest in INE. Measures of π showed a similar pattern to haplotype diversity, with the highest values found in CIC (0.030). The highest values of $\theta(\pi)$ and $\theta(S)$ were recorded in CIC (14.966) and in VEG, respectively, while the INE population had the lowest values for both parameters. In *S. ineffabilis* populations, *h* ranged from 0.380 to 0.801 (mean 0.647). Within the populations, CIC, again, had the highest haplotype diversity, while PRI had the lowest. Reflecting the *h* measures, π had the highest value in CIC (0.071), with the lowest recorded in INE (0.026). The values of $\theta(\pi)$ and $\theta(S)$ ranged from 13.121 to 35.375 (mean 21.460 ± 9.829) and from 22.317 to 29.579 (mean 26.202 ± 3.497), respectively. The CIC population had the highest values of both $\theta(\pi)$ and $\theta(S)$, while INE and PRI had the lowest. These parameters were also calculated for the two *S. nunatakis* populations. However, because, in PRI, only one haplotype was detected, all parameters for this population were 0, while in SNU the values were 0.250 for h and $\theta(\pi)$ and 0.001 and 0.386 for π and $\theta(S)$, respectively.

Table 9. Population genetic parameters for *cox1* in *S. belli*, *S. delicatus*, *S. ineffabilis* and *S. nunatakis* sampled across Victoria Land (Area). n, number of individuals; N_H, number of haplotypes within the populations and their frequencies; h, haplotype diversity; π, nucleotide diversity; $\theta(\pi)$, mean number of pairwise differences; θ(S), mean number of segregating sites; haplotypes shared between populations are indicated in italics (see Table 5 for details).

			Stereotydeus belli			
Area	**n.**	**N_H**	**h ± σ**	**π ± σ**	**θ(π) ± σ**	**θ(S) ± σ**
CHA	10	MB1(10)	0.000 ± 0.000	0.000 ± 0.000	0.000 ± 0.000	0.000 ± 0.000
CCI	2	MB10(2)	0.000 ± 0.000	0.000 ± 0.000	0.000 ± 0.000	0.000 ± 0.000
CJO	17	MB2(2), MB3(3), MB4(3), MB5(8), MB6(1)	0.743 ± 0.086	0.005 ± 0.003	2.559 ± 1.616	2.662 ± 1.247
KAY	10	MB7(1), MB8(3), MB9(6)	0.600 ± 0.130	0.010 ± 0.006	5.200 ± 3.108	7.423 ± 3.330
			Stereotydeus delicatus			
Area	**n**	**N_H**	**h ± σ**	**π ± σ**	**θ(π) ± σ**	**θ(S) ± σ**
CIC	28	MD1(18), MD2(1), MD3(2), MD4(1), *MD5(6)*	0.553 ± 0.093	0.030 ± 0.015	14.966 ± 7.682	11.307 ± 3.860
VEG	10	*MD5(9), MD6(1)*	0.200 ± 0.154	0.017 ± 0.010	8.400 ± 4.807	14.846 ± 6.322
INE	1	*MD6(1)*	0.000 ± 0.000	0.000 ± 0.000	0.000 ± 0.000	0.000 ± 0.000
			Stereotydeus ineffabilis			
Area	**n**	**N_H**	**h ± σ**	**π ± σ**	**θ(π) ± σ**	**θ(S) ± σ**
CIC	17	MI1(6), MI2(1), MI3(1), MI5(4), *MI12(4)*, MI13(1)	0.801 ± 0.060	0.071 ± 0.037	35.375 ± 18.143	29.579 ± 10.687
INE	14	MI6(3), MI7(3), MI8(6), *MI12(1)*, MI14(1)	0.769 ± 0.083	0.026 ± 0.014	13.121 ± 7.058	24.213 ± 9.242
PRI	19	*MI4(15), MI9(1)*, MI10(1), *MI11(2)*	0.380 ± 0.134	0.033 ± 0.017	16.316 ± 8.502	22.317 ± 7.940
SNU	9	*MI4(5), MI9(3), MI11(1)*	0.639 ± 0.126	0.042 ± 0.023	21.028 ± 11.648	28.699 ± 12.242
			Stereotydeus nunatakis			
Area	**n**	**N_H**	**h ± σ**	**π ± σ**	**θ(π) ± σ**	**θ(S) ± σ**
PRI	2	*MN1(2)*	0.000 ± 0.000	0.000 ± 0.000	0.000 ± 0.000	0.000 ± 0.000
SNU	8	*MN1(7)*, MN2(1)	0.250 ± 0.180	0.001 ± 0.001	0.250 ± 0.355	0.386 ± 0.386

As in [27,35,54], to establish the best combination for the population structure, AMOVA screenings were run for three species testing different combinations of population clusters: 10 runs were performed for *S. belli* (four populations: CHA, CCI, CJO and KAY), 3 for *S. delicatus* (three populations: CIC, VEG and INE) and 9 for *S. ineffabilis* (four populations: CIC, INE, SNU and PRI). As *S. nunatakis* was found only in two populations (PRI, SNU), the AMOVA was not calculated. For *S. belli*, the best resulting asset was (CHA vs. CCI+KAY vs. CJO), for *S. delicatus* (VEG vs. CIC+INE) and for *S. ineffabilis* (CIC vs. INE vs. SNU+PRI).

When group structure was assigned to populations for each species, the AMOVA analysis revealed more variation among groups and within populations (for *S. ineffabilis*) than among populations within groups (Table 10). In particular, for *S. belli* and *S. delicatus*, the Φ_{CT} values were similar (10.48068 and 9.51162, respectively). while for *S. ineffabilis* the value was only 2.94891. In contrast, Φ_{ST} values were higher in *S. ineffabilis* (10.89525) than in *S. belli* and *S. delicatus* (1.25345 and 6.66210, respectively).

Table 10. Percentage of variation (%) of molecular variance (AMOVA) of different levels of hierarchical population structure for *Stereotydeus* spp. for the mtDNA cytochrome *c* oxidase subunit *I* (*cox1*). The test was carried out with structure enforced according to geographical regions (see Section 3.3. for details).

Species		**Among Groups** Φ_{CT}	**Among Populations within Groups** Φ_{SC}	**Within Populations** Φ_{ST}
S. belli	Variance component	10.48068	0.05397	1.25345
	p	(0.16735 ± 0.00273)	(0.45057 ± 0.00422)	(0.00000 ± 0.00000)
	%	88.91	0.46	10.63
S. delicatus	Variance component	9.51162	0.28149	6.66210
	p	(0.33383 ± 0.00347)	(0.24403 ± 0.00340)	(0.0006 ± 0.00006)
	%	57.80	1.71	40.49
S. ineffabilis	Variance component	2.94891	−0.55777	10.89525
	p	(0.16135 ± 0.00259)	(0.62355 ± 0.00382)	(0.00056 ± 0.00018)
	%	22.19	−4.20	82.00

4. Discussion

This study provides over 150 new sequences representing all species of the mite genus *Stereotydeus* from Victoria Land. Combined with the morphological assessments that we provided, this information sheds light on an understudied taxon and provides a good starting point for further taxonomic studies of the species of the genus (Figure 7).

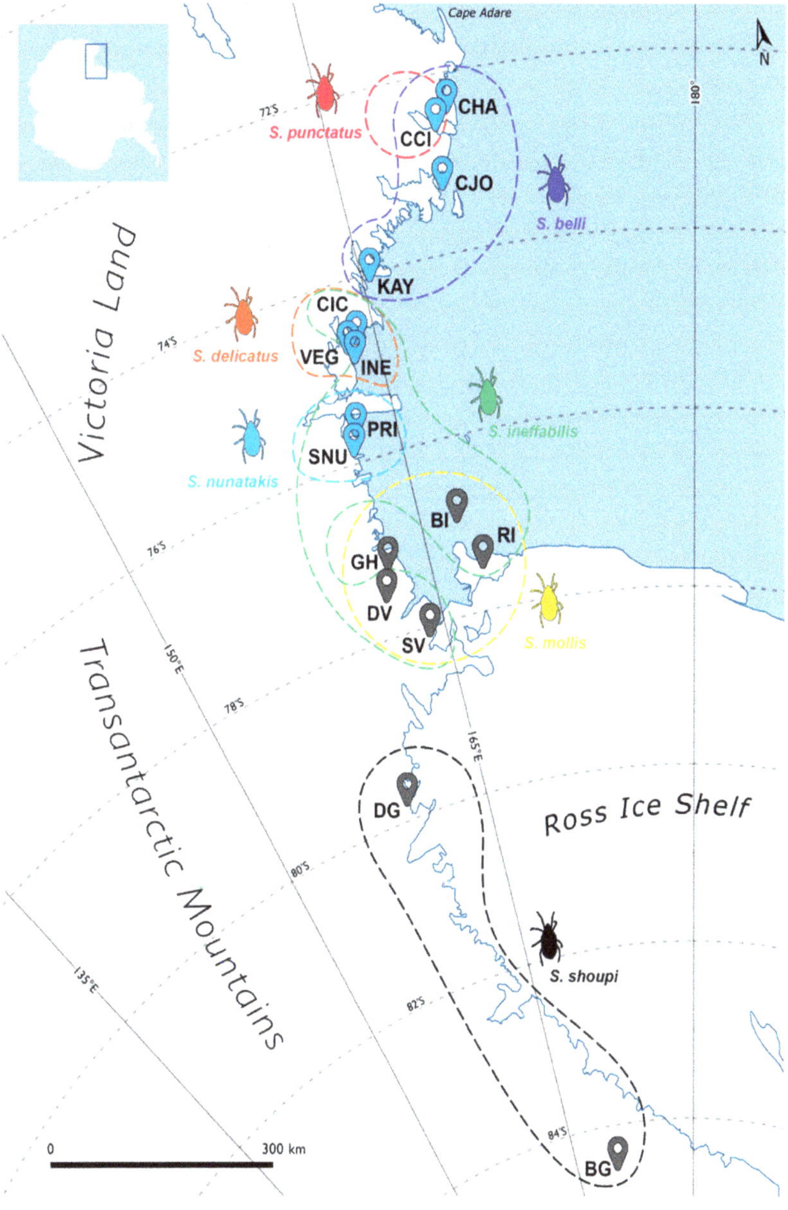

Figure 7. Updated map of the distribution of the *Stereotydeus* spp. of Victoria Land, based on the combination of new morphological and molecular data obtained this study and [48] (blue) and previous molecular data from [33,34,36] (grey). (coastlines from ADD Simple Basemap, NPI/Quantarctica 3 [60]).

4.1. North Victoria Land Taxa

Based on the analyses performed in this study, we found a latitudinal pattern in the distribution of *Stereotydeus* species in the Victoria Land coastal region. The presence of *S. belli* characterizes all populations from Cape Hallett (CHA) to Kay Island (KAY), while *S. punctatus* has, so far, been detected only in Crater Cirque (CCI). This is the first genetic study to be conducted on the latter species and, although only comprising a limited number of samples (12 individuals analyzed resulting in 4 *cox1* and 1 *28S* unique haplotypes), the presence of genetic variability is already evident. In addition, this is an easy species to identify morphologically due to the peculiar dorsal position of the anal opening that does not occur in any other *Stereotydeus* species. Early records of this taxon were reported by Strandtmann [44] and Gressitt and Shoup [40], also from Cape Adare and Cape Hallett.

For *S. belli* it is possible that historical events, such as habitat fragmentation due to glacial events, divergence in isolation and subsequent range expansion [32–34,78–80], are responsible for the patchy distribution of these populations and their genetic isolation. All the *S. belli* populations were clearly distinct, with KAY and CCI clustered together and separated from both CHA and CJO. This fragmented and apparently disjointed distribution is consistent with reports from other terrestrial invertebrate species in north Victoria Land (e.g., [31,64]). The same studies have reported that invertebrate populations in the region of the Tucker Glacier are genetically more closely related to populations in central–south Victoria Land, compared to others in relatively closer geographical proximity. Recent studies of springtail species endemic to Victoria Land, including *Cryptopygus cisantarcticus* [64] and *Friesea gretae* and *F. propria* [31] (*F. grisea* in Collins et al. [64]), highlighted the important role of the Tucker Glacier as an insurmountable barrier leading to high levels of genetic divergence between populations from either side of the glacier, plausibly representing distinct species. Combining the inferences made in the current study with previous springtail studies specific to northern Victoria Land [31,64], a comparably important role may be played by Crater Cirque, where *S. belli* and *S. punctatus* occur in sympatry.

4.2. Central-South Victoria Land Taxa

This study presents the first record in the central part of Victoria Land of *S. delicatus,* which was originally discovered and described by Strandtmann [44] (although with only one individual from each location) from Cape Adare and Edisto Inlet; thus, our new data considerably expand the known distribution of this species southwards. Our sampling area is located in a part of Terra Nova Bay that is affected by powerful winds, while the Hells Gate moraine creates an abrupt interruption between Inexpressible Island (INE) and the peninsula of the Northern Foothills where Campo Icaro (CIC) is located. The distribution of the haplotypes found in this area suggests a possible role played by Vegetation Island (VEG) in acting as a bridge to connect CIC and INE. It is plausible that gene connectivity bypassed the inhospitable Hells Gate channel by using VEG as a midpoint between CIC and INE, although further intermediate steps may have been available at different points in the past.

Considering the phylogenetic relationship of *S. delicatus* with the other *Stereotydeus* taxa reviewed in this study, the link with the newly described species *S. ineffabilis* is immediately evident. Even though the species are morphologically distinct (Figure 6, Table 8; see also [44,48] for species descriptions), individuals of the two species from the CIC locality share the same unique haplotype (RX1) for the nuclear marker *28S*, although the combined analyses of the latter with the mitochondrial marker *cox1* and the morphological characteristics provided a good resolution of the boundary between the two species. A possible explanation for these results is that these taxa have "recently" undergone a speciation process and, because of the different resolutions of the two markers, it is possible that the large ribosomal subunit may not yet have accumulated sufficient mutations to enable distinguishing between the two sympatric species. A "slow" nucleotide substitution rate in *28S* is not unusual and has recently also been recorded in *Friesea* lineages from Victoria

Land [31]. Specifically in relation to understanding the geographic distribution and genetic diversity of *S. delicatus*, it is now crucial to expand the sampling and study effort to include the north Victoria Land locations of Cape Adare and Edisto Inlet, where the species was first recorded and described by Strandtmann [44].

While *S. delicatus* shows a well defined pattern of distribution, that of *S. ineffabilis* appears to be more complex. As for *S. delicatus*, the presence of the Hells Gate moraine isolates the populations north of the Drygalski Ice Tongue but, observing the haplotype networks, it is possible that, in the past, the two areas were linked, with the populations starting to differentiate only when the connection was broken. It is notable that the two *S. ineffabilis* populations south of the Drygalski Ice Tongue show a genetic connection to the population north of the glacier, although also showing some differentiation. As the Drygalski Ice Tongue is considered the geographical barrier that sharply delimits the faunas of north and south Victoria Land, our data provide a first indication of geneflow between north and south Victoria Land, and the first record of a terrestrial microarthropod species shared between the two regions.

In comparing the genetic diversity present in north and south Victoria Land, this study included also *Stereotydeus* spp. *cox1* haplotypes reported in previous studies [33,34,36] in the phylogenetic analyses performed. A striking outcome of these analyses is the strong link that emerged between *S. mollis* and *S. ineffabilis* sequences. The great genetic variability of the *cox1* marker alone proved ineffective in drawing a clear distinction between the two taxa. In order to stabilize the phylogenetic signal of the mitochondrial marker, it will be crucial to include one or more nuclear markers in future studies, as well as combining genetic and morphological approaches. In the absence of nuclear DNA sequence data from the *Stereotydeus* specimens, several morphological characteristics (e.g., the smaller size of the adults, the asymmetry in the tarsal rhagidial organs, the position of the solenidia on the tibiae and the genua, the number of the aggenital setae; see [48] for more details) were useful in identifying boundaries between *S. mollis* and *S. ineffabilis*. A high level of genetic diversity of recent origin (see branching pattern on Figures 4 and 5) is generally interpreted as an indication of recent demographic expansion. However, the present distributions of the *S. ineffabilis*, *S. delicatus* and *S. mollis* phylogroups may best be interpreted as being the result of alternative and temporally disjunct colonization events and speciation processes that occurred several times and started from different glacial refugia over a time interval of more than 10 Myr.

Together with *S. ineffabilis*, *S. nunatakis* was also present in the Prior Island (PRI) and Starr Nunatak (SNU) sampling locations [48]. Although the number of samples for genetic and morphological analyses was low, some variability and divergence was apparent. Based on the combined mitochondrial and nuclear phylogenetic analysis and the computation of genetic distances, *S. nunatakis* appears to be more closely related to *S. punctatus*, from north Victoria Land, than to the other species from south Victoria Land, *S. ineffabilis*, *S. mollis* and *S. shoupi*.

4.3. Speciation in Action

The patterns of diversity observed today in many Antarctic species can be traced back to historic events, such as habitat fragmentation, divergence in isolation and subsequent range expansion, that influenced the distribution of species particularly at local scales [32–34,81]. The resulting patterns of genetic variation can be used to infer ecological factors (e.g., effective population size, dispersal capacity), as well as those affecting speciation processes. Allopatric speciation in populations that are geographically separated appears to be characteristic for populations of many terrestrial invertebrate species native to Victoria Land, and is considered the result of the different fragmentation and isolation events of ancestral and widespread lineages [19,20,27]. As for these other invertebrates, we suggest that, due to their limited dispersal abilities and the presence of physiological barriers such as low tolerance to desiccation and abiotic barriers, our resulting populations also started to differentiate independently. However, especially for the southern Victoria Land species of *Stereotydeus*, the scenario appears to be more complex, due to the presence of four geographically and genetically closely related species. In recent decades, the suggestion that speciation might also occur in populations that are not geographically isolated (i.e., sympatric speciation [82–84]) has become increasingly accepted. It is possible, for example, that, when limited resources are available to members of sympatric populations, interactions through both direct (i.e., interference) and indirect (i.e., exploitation) competition could lead groups of individuals, especially those belonging to populations of large size, to adopt different behaviors, select different habitats, establish temporal shifts of activity patterns or avoid mating or generating hybrids with low fitness. Thus, ecologically based barriers to gene flow evolve between populations resulting in an "ecological selection" [83,85,86]. This selection can occur under different geographic conditions [83], so it cannot be excluded that this process may also have contributed to the current patterns of variability and distribution of *Stereotydeus* species in Antarctica.

Although the biogeographical patterns of springtails and mites in coastal Victoria Land share some similarities [28,33,35], their intra- and interspecific genetic distances are not entirely comparable. Interspecific genetic distances calculated between species of Acari are generally greater than those observed in comparisons between Collembola (e.g., [33,34]), and it is not possible to exclude this being influenced by the different survival strategies and/or life histories of free living mites [12]. It is possible that all aspects of the life history strategy of Antarctic terrestrial invertebrates (e.g., including generation time, life cycles, physiology and metabolism), in combination with environmental conditions, could be major factors influencing evolutionary rates (nucleotide substitutions). However, it is also not clear, in general, how rates of evolution differ across species or, if they do, what factors drive these differences. The factors responsible for the high levels of divergence shown by mites have previously been suggested to include the smaller size of the animals, their shorter generation time and higher activity levels [37] and their greater recolonization/dispersal abilities [33] in comparison with springtail taxa. However, these hypotheses have not been explicitly tested. Prostigmatid mites lack an impermeable cuticle, and behavioral strategies, such as microhabitat selection, along with physiological acclimatization [12] are likely to play a fundamental role in the isolation of populations and their survival. As suggested by Demetras et al. [36], some behavioral differences may have a role in increasing genetic divergence, as has also been noted for some Antarctic springtail species [87]. Thus, through combining morphological, genetic and ecological studies of terrestrial fauna, we can better understand the evolutionary origins, dispersal history and current distribution of Antarctic invertebrates.

Due to the close phylogenetic relationships between the central and southern species (*S. ineffabilis*, *S. delicatus* and *S. mollis*), in the future it will be fundamental to carry out and implement new combined taxonomical studies and enlarge the number of specimens in the analyses. The inclusion of a more recent revision of the original materials used for the first description of *S. delicatus* and *S. mollis* will help to identify additional morphological characters, if any, necessary to distinguish these species with respect to *S. ineffabilis*. In fact, when the amount of divergence at inter- and intraspecific level is overlapping, morphology is important to identify species boundaries. In addition, the genetic differentiation of species of "recent" origin may be less variable with respect to more ancient ones. Thus, the combination of new morphological analyses and a deeper genetic screening through the incorporation of more nuclear markers and/or genome comparisons will be the starting point to better define some of the phylogenetic relationships of all the Victoria Land *Stereotydeus* species.

In summary, the contemporary distributions of species of *Stereotydeus* occurring in Victoria Land follow defined latitudinal patterns, including two major features. These are characterized by, first, a more genetically defined cluster in the north Victoria Land populations of *S. belli* and *S. punctatus* and, second, a more complex, in terms of species composition, cluster including populations in south Victoria Land.

Supplementary Materials: The following are available online at https://www.mdpi.com/article/10.3390/d13100506/s1, Figure S1: Unrooted phylogenetic trees of *Stereotydeus* specimens with posterior probabilities shown at nodes, Table S1: Accession numbers of the *cox1* and *28S* sequences of *Stereotydeus* species and *Penthaleus major* deposited on GenBank and included in the analyses, Table S2: Accession numbers of the *cox1* sequences of *Stereotydeus* species and one Eriorhynchidae mite downloaded from GenBank and included in the analyses, Table S3: List of the datasets, number of new sequences obtained and used in each dataset, markers, reference sequences and outgroups used for the analyses and models of nucleotide evolution that best fitted, divided according to the partition applied and to the respective tree search optimization criteria.

Author Contributions: Conceptualization, C.B. and A.C.; methodology, C.B., A.C. and F.N.; validation, A.C., F.N., H.S. and P.P.F.; formal analysis, C.B.; investigation, C.B.; resources, A.C., P.P.F., F.N. and H.S.; data curation, A.C., F.N. and C.B.; writing—original draft preparation, C.B.; writing—review and editing, C.B., A.C., F.N., P.C., H.S. and P.P.F.; visualization, C.B.; supervision, A.C., H.S., P.C. and F.N.; project administration, C.B. and A.C.; funding acquisition, A.C., F.N. and P.P.F. All authors have read and agreed to the published version of the manuscript.

Funding: This research was funded by the Italian Program of Research in Antarctica (PNRA16_00234) to A.C. Partial support was also provided by the University of Siena. P.C. is supported by NERC core funding to the British Antarctic Surveys "Biodiversity, Ecosystems and Adaptation" Team. The paper also contributes to the SCAR "State of the Antarctic Ecosystem" (AntEco) international program.

Institutional Review Board Statement: Not applicable.

Data Availability Statement: Data are contained within the article. The sequences are available on NCBI under the Accession Numbers MZ350724-MZ350753 (*cox1*) and MZ442270-MZ442288 (*28S*) (see also Table S1 in Supplementary Materials).

Acknowledgments: We would like to thank Hatef Servatibeiragh, Bernardo Stockl Junior, Veronica Cannucci, Chiara Leo and Claudio Cucini for their help with the lab work. We wish to thank also the four anonymous reviewers for their helpful comments.

Conflicts of Interest: The authors declare no conflict of interest. The funders had no role in the design of the study; in the collection, analyses, or interpretation of data; in the writing of the manuscript; or in the decision to publish the results.

References

1. Block, W. Terrestrial Microbiology, Invertebrates and Ecosystems. In *Antarctic Ecology*; R. M. Laws: London, UK, 1984.
2. Convey, P. Antarctic Ecosystems. In *Encyclopedia of Biodiversity*; Levin, S., Ed.; Academic Press: San Diego, CA, USA, 2017; Volume 1, pp. 179–187.
3. Spain, A.V. Some Aspects of Soil Conditions and Arthropod Distribution in Antarctica. *Pac. Insects Monogr.* **1971**, *25*, 21–26.
4. Schwarz, A.-M.J.; Green, J.D.; Green, T.G.A.; Seppelt, R.D. Invertebrates Associated with Moss Communities at Canada Glacier, Southern Victoria Land, Antarctica. *Polar Biol.* **1993**, *13*, 157–162. [CrossRef]
5. Kennedy, A.D. Water as a Limiting Factor in the Antarctic Terrestrial Environment: A Biogeographical Synthesis. *Arct. Alp. Res.* **1993**, *25*, 308–315. [CrossRef]
6. Skotnicki, M.L.; Selkirk, P.M. Plant Biodiversity in an Extreme Environment. In *Trends in Antarctic Terrestrial and Limnetic Ecosystems: Antarctica as a Global Indicator*; Bergstrom, D.M., Convey, P., Huiskes, A.H.L., Eds.; Springer: Dordrecht, The Netherlands, 2006; pp. 161–175. ISBN 978-1-4020-5277-4.
7. Marshall, D.J.; Pugh, P.J.A. Origin of the Inland Acari of Continental Antarctica, with Particular Reference to Dronning Maud. *Zool. J. Linn.* **1996**, *118*, 101–118. [CrossRef]
8. Convey, P.; Peck, L.S. Antarctic Environmental Change and Biological Responses. *Sci. Adv.* **2019**, *5*, eaaz0888. [CrossRef]
9. Cannon, R.J.C.; Block, W. Cold Tolerance of Microarthropods. *Biol. Rev.* **1988**, *63*, 23–77. [CrossRef]
10. Block, W.; Baust, J.G.; Franks, F.; Johnston, I.A.; Bale, J. Cold Tolerance of Insects and Other Arthropods. *Philos. Trans. R. Soc. Lond. B Biol. Sci.* **1990**, *326*, 613–633.
11. Sinclair, B.J.; Sjursen, H. Terrestrial Invertebrate Abundance across a Habitat Transect in Keble Valley, Ross Island, Antarctica. *Pedobiologia* **2001**, *45*, 134–145. [CrossRef]
12. Sjursen, H.; Sinclair, B.J. On the Cold Hardiness of *Stereotydeus mollis* (Acari: Prostigmata) from Ross Island, Antarctica. *Pedobiologia* **2002**, *46*, 188–195. [CrossRef]
13. Caruso, T.; Bargagli, R. Assessing Abundance and Diversity Patterns of Soil Microarthropod Assemblages in Northern Victoria Land (Antarctica). *Polar Biol.* **2007**, *30*, 895–902. [CrossRef]
14. Janetschek, H. Arthropod Ecology of South Victoria Land. *Antarct. Res. Ser.* **1967**, *10*, 205–293.
15. Gressitt, J.L. Introduction: Dispersal. In *Entomology of Antarctica*; Gressitt, J.L., Ed.; Antarctic Research Series; American Geophysical Union: Washington, DC, USA, 1967; Volume 10, pp. 25–27.

16. Coulson, S.J.; Hodkinson, I.D.; Webb, N.R.; Harrison, J.A. Survival of Terrestrial Soil-Dwelling Arthropods on and in Seawater: Implications for Trans-Oceanic Dispersal. *Funct. Ecol.* **2002**, *16*, 353–356. [CrossRef]
17. Hawes, T.C.; Worland, M.R.; Bale, J.S.; Convey, P. Rafting in Antarctic Collembola. *J. Zool.* **2008**, *274*, 44–50. [CrossRef]
18. Hawes, T.C. Rafting in the Antarctic Springtail, *Gomphiocephalus hodgsoni*. *Antarct. Sci.* **2011**, *23*, 456–460. [CrossRef]
19. Pugh, P.J.A. Acarine Colonisation of Antarctica and the Islands of the Southern Ocean: The Role of Zoohoria. *Polar Rec.* **1997**, *33*, 113–122. [CrossRef]
20. Strong, J. Ecology of Terrestrial Arthropods at Palmer Station, Antarctic Peninsula. In *Entomology of Antarctica*; Gressitt, J.L., Ed.; Antarctic Research Series; American Geophysical Union: Washington, DC, USA, 1967; Volume 10, pp. 357–371. ISBN 978-1-118-66869-6.
21. Tilbrook, P.J. Arthropod Ecology in the Maritime Antarctic. In *Entomology of Antarctica*; Gressitt, J.L., Ed.; Antarctic Research Series; American Geophysical Union: Washington, DC, USA, 1967; Volume 10, pp. 331–356. ISBN 978-1-118-66869-6.
22. Stevens, M.I.; Hogg, I.D. Long-Term Isolation and Recent Range Expansion from Glacial Refugia Revealed for the Endemic Springtail *Gomphiocephalus hodgsoni* from Victoria Land, Antarctica. *Mol. Ecol.* **2003**, *12*, 2357–2369. [CrossRef]
23. Wallwork, J.A. Distribution Patterns of Cryptostigmatid Mites (Arachnida: Acari) in South Georgia. *Pac. Insects* **1972**, *14*, 615–625.
24. Lehmitz, R.; Russell, D.; Hohberg, K.; Christian, A.; Xylander, W.E.R. Wind Dispersal of Oribatid Mites as a Mode of Migration. *Pedobiologia* **2011**, *54*, 201–207. [CrossRef]
25. Pugh, P.J.A. Have Mites (Acarina: Arachnida) Colonised Antarctica and the Islands of the Southern Ocean via Air Currents? *Polar Rec.* **2003**, *39*, 239–244. [CrossRef]
26. Hawes, T.C.; Worland, M.R.; Convey, P.; Bale, J.S. Aerial Dispersal of Springtails on the Antarctic Peninsula: Implications for Local Distribution and Demography. *Antarct. Sci.* **2007**, *19*, 3–10. [CrossRef]
27. Fanciulli, P.P.; Summa, D.; Dallai, R.; Frati, F. High Levels of Genetic Variability and Population Differentiation in *Gressittacantha terranova* (Collembola, Hexapoda) from Victoria Land, Antarctica. *Antarct. Sci.* **2001**, *13*, 246–254. [CrossRef]
28. Frati, F.; Spinsanti, G.; Dallai, R. Genetic Variation of MtCOII Gene Sequences in the Collembolan *Isotoma klovstadi* from Victoria Land, Antarctica: Evidence for Population Differentiation. *Polar Biol.* **2001**, *24*, 934–940. [CrossRef]
29. Torricelli, G.; Carapelli, A.; Convey, P.; Nardi, F.; Boore, J.L.; Frati, F. High Divergence across the Whole Mitochondrial Genome in the "Pan-Antarctic" Springtail *Friesea grisea*: Evidence for Cryptic Species? *Gene* **2010**, *449*, 30–40. [CrossRef]
30. Torricelli, G.; Frati, F.; Convey, P.; Telford, M.; Carapelli, A. Population Structure of *Friesea grisea* (Collembola, Neanuridae) in the Antarctic Peninsula and Victoria Land: Evidence for Local Genetic Differentiation of Pre-Pleistocene Origin. *Antarct. Sci.* **2010**, *22*, 757–765. [CrossRef]
31. Carapelli, A.; Greenslade, P.; Nardi, F.; Leo, C.; Convey, P.; Frati, F.; Fanciulli, P.P. Evidence for Cryptic Diversity in the "Pan-Antarctic" Springtail *Friesea antarctica* and the Description of Two New Species. *Insects* **2020**, *11*, 141. [CrossRef] [PubMed]
32. Stevens, M.I.; Greenslade, P.; Hogg, I.D.; Sunnucks, P. Southern Hemisphere Springtails: Could Any Have Survived Glaciation of Antarctica? *Mol. Biol. Evol.* **2006**, *23*, 874–882. [CrossRef] [PubMed]
33. Stevens, M.I.; Hogg, I.D. Contrasting Levels of Mitochondrial DNA Variability between Mites (Penthalodidae) and Springtails (Hypogastruridae) from the Trans-Antarctic Mountains Suggest Long-Term Effects of Glaciation and Life History on Substitution Rates, and Speciation Processes. *Soil Biol. Biochem.* **2006**, *38*, 3171–3180. [CrossRef]
34. McGaughran, A.; Hogg, I.D.; Stevens, M.I. Patterns of Population Genetic Structure for Springtails and Mites in Southern Victoria Land, Antarctica. *Mol. Phylogenet. Evol.* **2008**, *46*, 606–618. [CrossRef]
35. McGaughran, A.; Torricelli, G.; Carapelli, A.; Frati, F.; Stevens, M.I.; Convey, P.; Hogg, I.D. Contrasting Phylogeographical Patterns for Springtails Reflect Different Evolutionary Histories between the Antarctic Peninsula and Continental Antarctica. *J. Biogeogr.* **2010**, *37*, 103–119. [CrossRef]
36. Demetras, N.J.; Hogg, I.D.; Banks, J.C.; Adams, B.J. Latitudinal Distribution and Mitochondrial DNA (COI) Variability of *Stereotydeus* spp. (Acari: Prostigmata) in Victoria Land and the Central Transantarctic Mountains. *Antarct. Sci.* **2010**, *22*, 749–756. [CrossRef]
37. Martin, A.P.; Palumbi, S.R. Body Size, Metabolic Rate, Generation Time, and the Molecular Clock. *Proc. Natl. Acad. Sci. USA* **1993**, *90*, 4087–4091. [CrossRef]
38. Sinclair, B.J.; Stevens, M.I. Terrestrial Microarthropods of Victoria Land and Queen Maud Mountains, Antarctica: Implications of Climate Change. *Soil Biol. Biochem.* **2006**, *38*, 3158–3170. [CrossRef]
39. Convey, P.; Biersma, E.M.; Casanova-Katny, A.; Maturana, C.S. Chapter 10—Refuges of Antarctic diversity. In *Past Antarctica*; Oliva, M., Ruiz-Fernández, J., Eds.; Academic Press: London, UK, 2020; pp. 181–200. ISBN 978-0-12-817925-3.
40. Gressitt, J.L.; Shoup, J. Ecological Notes on Free-Living Mites in North Victoria Land. In *Entomology of Antarctica*; Gressitt, J.L., Ed.; Antarctic Research Series; American Geophysical Union: Washington, DC, USA, 1967; Volume 10, pp. 307–320.
41. Greenslade, P. An Antarctic Biogeographical Anomaly Resolved: The True Identity of a Widespread Species of Collembola. *Polar Biol.* **2018**, *41*, 969–981. [CrossRef]
42. Greenslade, P. A New Species of *Friesea* (Collembola: Neanuridae) from the Antarctic Continent. *J. Nat. Hist.* **2018**, *52*, 2197–2207. [CrossRef]
43. Womersley, H.; Strandtmann, R.W. On Some Free Living Prostigmatic Mites of Antarctica. *Pac. Insects* **1963**, *5*, 451–472.
44. Strandtmann, R.W. Terrestrial Prostigmata (trombidiform mites). In *Antarctic Research Series*; Gressitt, J.L., Ed.; American Geophysical Union: Washington, DC, USA, 1967; pp. 51–80. ISBN 978-1-118-66869-6.

45. Pittard, D.A. A Comparative Study of the Life Stages of the Mite, *Stereotydeus mollis* W. & S. (Acarina). *Pac. Insects Monogr.* **1971**, *25*, 1–14.
46. Fitzsimons, J.M. Temperature and Three Species of Antarctic Arthropods. *Pac. Insects Monogr.* **1971**, *25*, 127–135.
47. Block, W. Ecological and Physiological Studies of Terrestrial Arthropods in the Ross Dependency 1984-85. *Brit. Antarct. Surv. Bull.* **1985**, *68*, 115–122.
48. Brunetti, C.; Siepel, H.; Fanciulli, P.P.; Nardi, F.; Convey, P.; Carapelli, A. Two New Species of the Mite Genus *Stereotydeus* Berlese, 1901 (Prostigmata: Penthalodidae) from Victoria Land, and a Key for Identification of Antarctic and Sub-Antarctic Species. *Taxonomy* **2021**, *1*, 116–141. [CrossRef]
49. Xiao, J.-H.; Wang, N.-X.; Li, Y.-W.; Murphy, R.W.; Wan, D.-G.; Niu, L.-M.; Hu, H.-Y.; Fu, Y.-G.; Huang, D.-W. Molecular Approaches to Identify Cryptic Species and Polymorphic Species within a Complex Community of Fig Wasps. *PLoS ONE* **2010**, *5*, e15067. [CrossRef]
50. Jörger, K.M.; Schrödl, M. How to Describe a Cryptic Species? Practical Challenges of Molecular Taxonomy. *Front. Zool.* **2013**, *10*, 59. [CrossRef] [PubMed]
51. Chown, S.L.; Lee, J.E.; Hughes, K.A.; Barnes, J.; Barrett, P.J.; Bergstrom, D.M.; Convey, P.; Cowan, D.A.; Crosbie, K.; Dyer, G.; et al. Challenges to the Future Conservation of the Antarctic. *Science* **2012**, *337*, 158–159. [CrossRef] [PubMed]
52. Lee, J.R.; Raymond, B.; Bracegirdle, T.J.; Chadès, I.; Fuller, R.A.; Shaw, J.D.; Terauds, A. Climate Change Drives Expansion of Antarctic Ice-Free Habitat. *Nature* **2017**, *547*, 49–54. [CrossRef] [PubMed]
53. Bergami, E.; Rota, E.; Caruso, T.; Birarda, G.; Vaccari, L.; Corsi, I. Plastics Everywhere: First Evidence of Polystyrene Fragments inside the Common Antarctic Collembolan *Cryptopygus antarcticus*. *Biol. Lett.* **2020**, *16*, 20200093. [CrossRef] [PubMed]
54. Carapelli, A.; Convey, P.; Frati, F.; Spinsanti, G.; Fanciulli, P.P. Population Genetics of Three Sympatric Springtail Species (Hexapoda: Collembola) from the South Shetland Islands: Evidence for a Common Biogeographic Pattern. *Biol. J. Linn. Soc.* **2017**, *120*, 788–803. [CrossRef]
55. Wauchope, H.S.; Shaw, J.D.; Terauds, A. A Snapshot of Biodiversity Protection in Antarctica. *Nat. Commun.* **2019**, *10*, 946. [CrossRef]
56. Terauds, A.; Chown, S.L.; Morgan, F.; Peat, H.J.; Watts, D.J.; Keys, H.; Convey, P.; Bergstrom, D.M. Conservation Biogeography of the Antarctic. *Divers. Distrib.* **2012**, *18*, 726–741. [CrossRef]
57. Terauds, A.; Lee, J.R. Antarctic Biogeography Revisited: Updating the Antarctic Conservation Biogeographic Regions. *Divers. Distrib.* **2016**, *22*, 836–840. [CrossRef]
58. Hawes, T.C.; Torricelli, G.; Stevens, M.I. Haplotype Diversity in the Antarctic Springtail *Gressittacantha terranova* at Fine Spatial Scales - A Holocene Twist to a Pliocene Tale. *Antarct. Sci.* **2010**, *22*, 766–773. [CrossRef]
59. Carapelli, A.; Leo, C.; Frati, F. High Levels of Genetic Structuring in the Antarctic Springtail *Cryptopygus terranovus*. *Antarct. Sci.* **2017**, *29*, 311–323. [CrossRef]
60. Matsuoka, K.; Skoglund, A.; Roth, G. Quantarctica [Dataset]. *Nor. Polar Inst.* **2018**, *10*. Available online: https://www.npolar.no/quantarctica/ (accessed on 25 June 2021).
61. Otto, J.C.; Wilson, K.J. Assessment of the Usefulness of Ribosomal 18S and Mitochondrial COI Sequences in Prostigmata Phylogeny. In *Acarology: Proceedings of the 10th International Congress*; CSIRO Publishing: Melbourne, Australia, 2001; Volume 100.
62. Friedrich, M.; Tautz, D. An Episodic Change of RDNA Nucleotide Substitution Rate Has Occurred during the Emergence of the Insect Order Diptera. *Mol. Biol. Evol.* **1997**, *14*, 644–653. [CrossRef] [PubMed]
63. MacVector, Inc. *MacVector*; MacVector, Inc.: Apex, NC, USA, 2018.
64. Collins, G.E.; Hogg, I.D.; Convey, P.; Barnes, A.D.; McDonald, I.R. Spatial and Temporal Scales Matter When Assessing the Species and Genetic Diversity of Springtails (Collembola) in Antarctica. *Front. Ecol. Evol.* **2019**, *7*, 76. [CrossRef]
65. Sievers, F.; Wilm, A.; Dineen, D.; Gibson, T.J.; Karplus, K.; Li, W.; Lopez, R.; McWilliam, H.; Remmert, M.; Söding, J.; et al. Fast, Scalable Generation of High-quality Protein Multiple Sequence Alignments Using Clustal Omega. *Mol. Syst. Biol.* **2011**, *7*, 539. [CrossRef] [PubMed]
66. Villesen, P. FaBox: An Online Toolbox for Fasta Sequences. *Mol. Ecol. Notes* **2007**, *7*, 965–968. [CrossRef]
67. Castresana, J. Selection of Conserved Blocks from Multiple Alignments for Their Use in Phylogenetic Analysis. *Mol. Biol. Evol.* **2000**, *17*, 540–552. [CrossRef]
68. R Core Team R: The R Project for Statistical Computing. 2016. Available online: https://oasishub.co/dataset/the-r-project-for-statistical-computing (accessed on 16 March 2021).
69. Paradis, E.; Schliep, K. Ape 5.0: An Environment for Modern Phylogenetics and Evolutionary Analyses in R. *Bioinformatics* **2019**, *35*, 526–528. [CrossRef] [PubMed]
70. Lanfear, R.; Frandsen, P.B.; Wright, A.M.; Senfeld, T.; Calcott, B. PartitionFinder 2: New Methods for Selecting Partitioned Models of Evolution for Molecular and Morphological Phylogenetic Analyses. *Mol. Biol. Evol.* **2016**, *34*, 772–773. [CrossRef]
71. Ronquist, F.; Teslenko, M.; van der Mark, P.; Ayres, D.L.; Darling, A.; Höhna, S.; Larget, B.; Liu, L.; Suchard, M.A.; Huelsenbeck, J.P. MrBayes 3.2: Efficient Bayesian Phylogenetic Inference and Model Choice Across a Large Model Space. *Syst. Biol.* **2012**, *61*, 539–542. [CrossRef]
72. Rambaut, A. FigTree v 1.4. 2012. Available online: http://tree.bio.ed.ac.uk/software/figtree/ (accessed on 21 April 2021).
73. Clement, M.; Posada, D.; Crandall, K. TCS: A Computer Program to Estimate Gene Genealogies. *Mol. Ecol.* **2000**, *9*, 1657–1659. [CrossRef]

74. Múrias dos Santos, A.; Cabezas, M.P.; Tavares, A.I.; Xavier, R.; Branco, M. TcsBU: A Tool to Extend TCS Network Layout and Visualization. *Bioinformatics* **2016**, *32*, 627–628. [CrossRef] [PubMed]
75. Excoffier, L.; Laval, G.; Schneider, S. Arlequin (Version 3.0): An Integrated Software Package for Population Genetics Data Analysis. *Evol. Bioinform.* **2005**, *1*, 47–50. [CrossRef]
76. Nei, M. *Molecular Evolutionary Genetics*; Columbia University Press: New York, NY, USA, 1987.
77. Excoffier, L.; Smouse, P.E.; Quattro, J.M. Analysis of Molecular Variance Inferred from Metric Distances among DNA Haplotypes: Application to Human Mitochondrial DNA Restriction Data. *Genetics* **1992**, *131*, 479–491. [CrossRef] [PubMed]
78. Wise, K.A.J. Collembola (springtails). In *Entomology of Antarctica*; Gressitt, J.L., Ed.; Antarctic Research Series; American Geophysical Union: Washington, DC, USA, 1967; pp. 123–148.
79. Knowles, L.L. Did the Pleistocene Glaciations Promote Divergence? Tests of Explicit Refugial Models in Montane Grasshoppers. *Mol. Ecol.* **2001**, *10*, 691–701. [CrossRef] [PubMed]
80. Rowe, K.C.; Heske, E.J.; Brown, P.W.; Paige, K.N. Surviving the Ice: Northern Refugia and Postglacial Colonization. *Proc. Natl. Acad. Sci. USA* **2004**, *101*, 10355–10359. [CrossRef] [PubMed]
81. Huiskes, A.H.L.; Convey, P.; Bergstrom, D.M. Trends in Antarctic Terrestrial and Limnetic Ecosystems: Antarctica as a Global Indicator. In *Trends in Antarctic Terrestrial and Limnetic Ecosystems: Antarctica as a Global Indicator*; Bergstrom, D.M., Convey, P., Huiskes, A.H.L., Eds.; Springer: Dordrecht, The Netherlands, 2006; pp. 1–13. ISBN 978-1-4020-5277-4.
82. Via, S. Sympatric Speciation in Animals: The Ugly Duckling Grows Up. *Trends Ecol. Evol.* **2001**, *16*, 381–390. [CrossRef]
83. Nosil, P. *Ecological Speciation*; Oxford University Press: Oxford, UK, 2012.
84. Foote, A.D. Sympatric Speciation in the Genomic Era. *Trends Ecol. Evol.* **2018**, *33*, 85–95. [CrossRef]
85. Rundle, H.D.; Nosil, P. Ecological Speciation. *Ecol. Lett.* **2005**, *8*, 336–352. [CrossRef]
86. Schluter, D. Evidence for Ecological Speciation and Its Alternative. *Science* **2009**, *323*, 737–741. [CrossRef]
87. McGaughran, A.; Redding, G.P.; Stevens, M.I.; Convey, P. Temporal Metabolic Rate Variation in a Continental Antarctic Springtail. *J. Insect Physiol.* **2009**, *55*, 130–135. [CrossRef]

diversity

MDPI

Article

A New Shrimp Genus (Crustacea: Decapoda) from the Deep Atlantic and an Unusual Cleaning Mechanism of Pelagic Decapods

Alexander Vereshchaka *, Dmitry Kulagin and Anastasiia Lunina

Shirshov Institute of Oceanology, Russian Academy of Sciences, Nakhimovski Prospekt 36, 117997 Moscow, Russia; kulagin.dima@gmail.com (D.K.); lunina@ocean.ru (A.L.)
* Correspondence: alv@ocean.ru

Abstract: The deep sea is the largest biome on Earth and hosts the majority of as yet undescribed species; description of these may trigger a new mindset about evolution and function of characters. We describe and diagnose a new genus and species *Sclerodora crosnieri* sp. nov. belonging to the superfamily Oplophoroidea. We examined and coded 81 characters for morphological analyses and used four gene markers for molecular analyses involving the new taxon and representatives of all other genera of Oplophoroidea. Retrieved morphological and molecular trees were similar and suggested that the new genus is a sister group to *Hymenodora* and both form a clade sister to the rest of Acanthephyridae. We provide an amended key to all genera of Oplophoroidea. We found an unusual chelate structure on the dactyl of the fifth pereopod, tested and confirmed a hypothesis that this structure is common for the whole family Acanthephyridae. We suggest that this derived structure is linked to an active cleaning of branchia—a function associated with chelipeds in some other carid shrimps. Convergent chelate structures are likely efficient for cleaning branchia, whichever appendage is adapted for these functions. In Oplophoridae (sister to Acanthephyridae), cleaning function is carried out by well-developed epipods.

Keywords: Caridea; phylogeny; morphology; new taxon; Oplophoroidea

Citation: Vereshchaka, A.; Kulagin, D.; Lunina, A. A New Shrimp Genus (Crustacea: Decapoda) from the Deep Atlantic and an Unusual Cleaning Mechanism of Pelagic Decapods *Diversity* **2021**, *13*, 536. https://doi.org/10.3390/d13110536

Academic Editor: Michael Wink

Received: 7 October 2021
Accepted: 22 October 2021
Published: 26 October 2021

Publisher's Note: MDPI stays neutral with regard to jurisdictional claims in published maps and institutional affiliations.

Copyright: © 2021 by the authors. Licensee MDPI, Basel, Switzerland. This article is an open access article distributed under the terms and conditions of the Creative Commons Attribution (CC BY) license (https://creativecommons.org/licenses/by/4.0/).

1. Introduction

The deep sea (i.e., below 200 m in depth) is the largest biome on Earth; the deep-pelagic domain accounts for nearly 94% of the habitable volume of the World Ocean [1], whereas only 16% of all named species on Earth are marine [2]. The deep-sea is suggested to host the majority of as yet undescribed species, which results in continuous discovery of new taxa from this environment. This process, which is usually a routine in zoology, occasionally yields taxa triggering a new mindset about evolution of characters and their functions.

In fact, while examining the deep pelagic fauna of the Central Atlantic, we found an unusual shrimp of the superfamily Oplophoroidea [3], which could not be attributed to any of the oplophoroid genera. Further sequencing of gene markers confirmed results of morphological analyses and the generic status of the new taxon. Here we examine and code 81 characters for morphological analyses and use four gene markers for molecular analyses to map the taxon on the phylogenetic tree.

Morphological examinations of the new genus resulted in a reanalysis of cleaning mechanism of the whole superfamily Oplophoroidea. Cleaning and grooming of branchia is an important function and a significant challenge for decapods and involves various mechanisms [4]. In Caridea, one of mechanism (a passive one) is linked to setobranchs and a hooked epipod unique to this group. The epipod hook of one appendage fits around the bases of the setobranch setae on the appendage posterior to it. During limb movements, when the coxae of these two limbs move apart, the setobranch setae are drawn down over the gill lamellae. When the coxae move toward each other, the setobranch setae are guided back to the gills through the epipod hook. When the epipod hook is displaced from the setobranch, the setae of the latter lose their location with respect to the gills [4]. An alternative mechanism (an active one) is linked to grooming chelipeds: one pair is generally used in body grooming and cleaning the gills when epipod-setobranch complexes have been lost [5,6]. Generally, each of these mechanisms is conservative at the genus and family level in the Caridea and the active and passive cleaning do not occur together [4].

In our specimen, neither of the described mechanisms was possible: epipods on the last two pairs of the pereopods (fourth and fifth) were absent and no gill-cleaning structures on the chelipeds were observed. Instead, the specimen has a very specialized dactyl of the fifth pereopods: short and forming a very characteristic chelate structure. We hypothesized that this character may mirror an alternative active cleaning mechanism involving the fifth pereopod, not the chelipeds as in other carids. In order to test this hypothesis, we checked structure of epipods and fifth pereopods in all other species of the superfamily Oplophoroidea, ran phylogenetic analyses, and mapped these characters on the resulting trees.

Oplophoroidea hitherto included 70 valid species within the two families, Oplophoroidea and Acanthephyridae; Oplophoridae encompass three genera (*Janicella* Chace, 1986, *Oplophorus* H. Milne Edwards, 1837, and *Systellaspis* Spence Bate, 1888) and are considered as a sister clade to Acanthephyridae [7], which includes *Acanthephyra* A. Milne-Edwards, 1881, *Ephyrina* Smith, 1885, *Heterogenys* Chace, 1986, *Hymenodora* G.O. Sars, 1877, *Kemphyra* Chace, 1986, *Meningodora* Smith, 1882, and *Notostomus* A. Milne-Edwards, 1881. Most genera are widely distributed and have been explored in numerous publications of the 19th and 20th centuries (e.g., [8]). Oplophoroidea was recently revised on the basis of both morphological and molecular analyses ([7,9–11]) and the finding of an undescribed genus and species belonging to this superfamily is surprising.

2. Methods

2.1. Morphological Analysis

We chose outgroups from Pasiphaeoidea and Bresilioidea, both representing the sister clade to Oplophoroidea ([12], Figure 1). In analysis 1 we used *Pasiphaea sivado* (Risso, 1816), the type species of *Pasiphaea*, as the outgroup. In Analysis 2, we used *Alvinocaris longirostris* Kikuchi and Ohta, 1995 as the outgroup. In addition to a new species, we included as the ingroups representatives of all valid species of *Hymenodora* (four species), and representatives of all other genera of Oplophoroidea: three genera of Oplophoridae and seven genera of Acanthephyridae (Table 1).

Table 1. Individuals used in morphological analyses. MNHN - National Museum of Natural History (Paris, France); ZMUK—National History Museum, Copenhagen, Denmark; IO RAN—Institute of Oceanology, Russian Academy of Scienses, USNM -National Museum of Natural History, Smithsonian Institution.

Species	Coordinates	Other Information	Museum, Number
Acanthephyra quadrispinosa	29°39′ S, 44°16′ E	Expedition ATIMO VATAE. SUD MADAGASCAR, Sud Pointe Barrow. Chaultier "Nosy Be 11", Stn. CP 3596, 986–911 m. 12.05.2010.	MNHN-IU-2010-4285
Acanthephyra acutifrons	14°43′ N, 45° 02′ W	"Professor Logatchev" 39 cruise St 215 RT, RTAK	IO RAN 39L 215 RT №1
Ephyrina ombango	10°23,17′ N, 46°45,34′ W	DEMERABY, CP07, chalutage 4850 m. 20.09.80	MNHN-IU-2018-1579
Ephyrina ombango	9°18′ S, 11°10′ E	"Ombango", C14, St.325,midwater traul, 0–725 m, 02.03.1961, 23h00–23h15	MNHN-IU-2014-11098

Table 1. *Cont.*

Species	Coordinates	Other Information	Museum, Number
Heterogenis microphtalma	No data	Collection de S.A.S.le Prince de Monaco, Station 7/3. №12h, 16–19.8.96. Chal 4360 m	MNHN-IU-2018-1578
Hymenodora acanthitelsonis	45°18′ N, 125°43′ W–45°17′ N, 125°49′ W	Pacific Ocean, Unated States, Oregon, W of Pacific City. Yaqina BMT.189, 18.03.1970.	USNM 137500
Hymenodora glacialis	02°03′ S, 118°45′ E	Indonesie, CORINDON -Makassar. St CH286, 1710–1730 m	Na 10655
Hymenodora glacialis	73°28′ N, 10°07′ W	Mer de Norvege, Campagne NORBI, N.O. "Jean Charcot", Stn CP16, 2937 m, 07.08.1975	MNHN-IU-2008-16833
Hymenodora gracilis	37°39′ S, 77°26′ E	Ile Amsterdam, Campagne Jasus (MD 50), N.O. "Marion Dufresne", Stn CP193, 2800–3075 m. 27.06.1986	MNHN-IU-2008-16839
Hymenodora frontalis	15° N, 45° W	ROV "Vityaz", 59 th cruise, st. 7497, № 271. 18.06.1976, 1500–2500 m	
Janicella spinicauda	1°28′ S, 48°06′ E	ROV "Vityaz", 17th cruise, St. 2604,13.11.88, 670–690 m	ZMUK
Janicella spinicauda	8°44′ S, 43°54′ E	Dana Expedition, St. 3939-1, 23.12.1929, 500 meter wire	ZMUK
Kemphyra corallina	37°54′ S, 77°22′ E	Iles St Paul et Amsterdam, "Marion Dufresne" Cne MD Jasus Stn CP 56. 2280–2310 m. 14.07.1986. 20h02–22h31	MNHN-IU-2018-1581
Kemphyra corallina	33°59′ S, 43°55′ E	Indian Ocean: Walters shoal, Plaine Sud. N.O. "Marion Dufresne", Campagne MD208(Walters Shoal). Stn CP49156 1865–2058 m, 12.05.2017	MNHN-IU-2016-9402
Meningodora longiscula	9°55′ N, 142°00′ E	Nouvelle-Caledonie, Campagne Caride V. Stn 15, 1000 m. 12.09.1969	MNHN-IU-2011-5635
Notostomus elegans		37 cruise RV Logatchev, St 156 TS	IO RAN
Oplophorus gracilirostris	25°11′ N, 122°35′ E	Dana Expedition, St. 3722-3, 300 m wire	ZMUK
Oplophorus gracilirostris	20°08′ N, 82°59′ W	Dana Expedition, St. 1218, 800 m wire	ZMUK
Oplophorus gracilirostris	12°30′ S, 48°16′ E	ROV "Vityaz", 17th cruise, St. 2597, 12.11.88, 360–555 m wire.	ZMUK
Oplophorus gracilirostris	22°06′ N, 84°58′ W	Dana Expedition, St. 1223, 500 m wire	ZMUK
Pasiphaea sivado	35°47′ N, 05°17′ W	Detroit de Gibraltar, N.O. "Cryos", BALGIM, St. CP150, 280–300 m, 18.06.1984	MNHN-IU-2018-1611
Sclerodora crosnieri **sp.nov.**	16° N, 46° W	39th Cruise of R/V "Professor Logatchev", March 2018	ZMUK
Systellaspis pellucida	12°30′ S, 48°16′ E	Indian Ocean. North end of Madagascar. ROV "Vityaz", 17th cruise, St. 2597, 360–555 m	IO RAN
Systellaspis pellucida	25°11′ N, 122°35′ E	North Western Pacific Ocean. S.E. and E. of Formosa. Dana Expedition 3722(2) 29.05.1929, 600 mw	ZMUK
Systellaspis pellucida	25°11′ N, 122°35′ E	North Western Pacific Ocean. S.E. and E. of Formosa. Dana Expedition 3722(1) 29.05.1929, 1000 mw	ZMUK

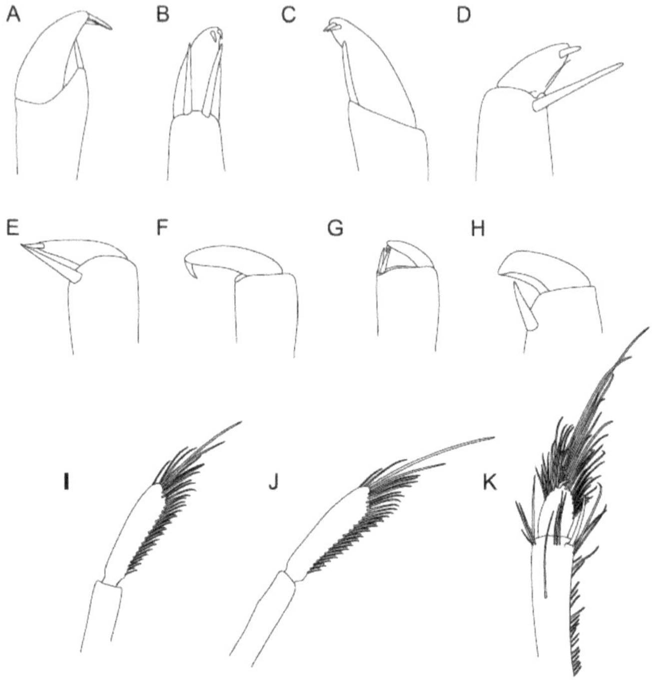

Figure 1. Terminal part of the fifth pereopod in Oplophoroidea (schematic, most non-robust setae removed): (**A**)—*Sclerodora crosnieri*, sp. nov., (**B**)—*Hymenodora frontalis*, side view, (**C**)—*Hymenodora frontalis*, inner view, (**D**)—*Acanthephyra acutifrons*, (**E**)—*Notostomus elegans*, (**F**)—*Kemphyra corallina*, (**G**)—*Meningodora longisulca*, (**H**)—*Heterogenys microphthalma*, (**I**)—*Systellaspis debilis*, (**J**)—*Oplophorus gracilirostris*, (**K**)—*Janicella spinicauda*.

For each included taxon we identified and encoded 81 morphological characters (not weighted, Supporting Information, File S1. The dataset (File S2) was handled and analyzed using a combination of programs using maximum parsimony settings: WIN-CLADA/NONA and TNT [13,14]. Trees were generated in TNT with 30,000 trees in memory, under the 'implicit enumeration' algorithm. Relative stability of clades was assessed by standard bootstrapping (sample with replacement) with 10,000 pseudoreplicates and by Bremer support (algorithm TBR, saving up to 10,000 trees up to 8 steps longer). In all analyses, clades were considered robust if they had synchronously Bremer support $\geq$3 and bootstrap support $\geq$70.

2.2. Molecular Analyses

In order to resolve the phylogenetic position of the new species within the superfamily Oplophoroidea, we selected two mitochondrial (COI, 16S) and two nuclear genes (18S, H3) owing to their phylogenetic utility and different inheritance patterns. Outgroups and ingroups were the same as in the morphological analysis. NCBI GenBank accession numbers of sequences taken for phylogenetic analysis are listed in Table 2.

Table 2. Details of the analyzed species and sequences used in the study. Newly retrieved sequences are highlighted in bold; 'N'—missing data.

Taxon	GenBank Accession Numbers				Source
	COI	16S	18S	H3	
Outgroup taxa					
Pasiphaeoidea Dana, 1852					
Pasiphaeidae Dana, 1852					
Pasiphaea sivado (Risso, 1816)	KP759487	KP725629	KP725826	MF279416	Aznar-Cormano et al., 2015; Liao et al., 2017
Bresilioidea Calman, 1896					
Alvinocarididae Christoffersen, 1986					
Alvinocaris longirostris Kikuchi & Ohta, 1995	KP215329	KP215285	KP215300	KP215342	Aznar-Cormano et al., 2015
Ingroup taxa					
Oplophoroidea Dana, 1852					
Oplophoridae Dana, 1852					
Janicella spinicauda (A. Milne-Edwards, 1883)	MH572546	KP075932	MH100869	MH107256	Wilkins and Bracken-Grissom, 2020 (GenBank); Wong et al., 2015; Lunina et al., 2019
Oplophorus gracilirostris A. Milne-Edwards, 1881	KP076150	KP075920	KP075847	KP076072	Wong et al., 2015
Systellaspis pellucida (Filhol, 1884)	JQ306184	KP075925	JF346250	KP076077	Matzen da Silva et al., 2011; Wong et al., 2015; Li et al., 2011
Acanthephyridae Spence Bate, 1888					
Acanthephyra quadrispinosa Kemp, 1939	KP759363	KP725479	KP725677	KP726051	Aznar-Cormano et al., 2015
Ephyrina ombango Crosnier & Forest, 1973	MW043004	MW043448	MW043463	MW052289	Lunina et al., 2020
Heterogenys microphthalma (Smith, 1885)	KP076183	KP075898	KP075787	KP076124	Wong et al., 2015
Kemphyra corallina (A.Milne-Edwards, 1883)	MW043006	MW043450	MW043465	MW052291	Lunina et al., 2020
Meningodora longisulca Kikuchi, 1985	MW043007	MW043451	MW043466	MW052292	Lunina et al., 2020
Notostomus elegans A. Milne-Edwards, 1881	MW043011	MW043455	MW043470	MW052296	Lunina et al., 2020
Hymenodora frontalis Rathbun, 1902	DQ882080	N	N	N	Costa et al., 2007
Hymenodora glacialis (Buchholz, 1874)	FJ602519	GQ131896	GQ131915	N	Bucklin et al., 2010; Chan et al., 2010
Hymenodora gracilis Smith, 1886	MH572613	MH542891	KP075827	KP076134	Wilkins and Bracken-Grissom, 2020 (GenBank); Wong et al., 2015
Sclerodora crosnieri gen. nov., sp. nov.	**OK382996**	**OK382953**	**OK382952**	**OK424597**	This study

Total genomic DNA was extracted from the fifth pleopod using the Qiagen DNeasy® Blood and Tissue Kit in accordance with the manufacturer's protocol. Polymerase chain reaction (PCR) amplification of the COI gene was performed with the primers COI-acant-for2a (5′-GGDGTNGGNACDGGNTGRAC-3′) and/COH6 [15]. We have designed a new internal primer within the barcoding region, since all attempts at amplification with LCO 1490 [16] or COL6 [17] primers were unsuccessful. The length of the resulting fragment was 397 bp. The mitochondrial large subunit 16S rRNA was amplified by 16L2/16H3 primers (~550 bps, [18,19]), the nuclear small subunit 18S rRNA was amplified by A/L, C/Y, O/B primers (~1800 bps, [20]), and H3 gene fragment was amplified by H3A/H3B primers (~330 bps, [21]). A pre-made PCR mix (ScreenMix-HS) from Evrogene™ (1 × ScreenMix-HS, 0.4 μM of each primer, 1–1.5 μL of DNA template, and completed with milliQ H_2O to make up a total volume of 20 μL) was used for the amplification. The thermal profile used an initial denaturation for 3 min at 95 °C followed by 35–40 cycles of 20 s at 94 °C, 30 s at 47–56 °C depending on primer pair, 1 min at 72 °C and a final extension of 7 min at 72 °C. PCR products were purified by ethanol precipitation and sequenced in both directions using BigDye Terminator v3.1 (Applied Biosystems, Foster City, CA, USA). Each sequencing reaction mixture, including 0.5 μL of BigDye Terminator v3.1, 0.8 μL of 1 μM primer, and 1–2 μL of purified PCR template, was run for 30 cycles of 96 °C (10 s), 50 °C (5 s), and 60 °C (4 min). Sequences were purified by ethanol precipitation to remove unincorporated primers and dyes. Products were re-suspended in 14 μL formamide and electrophoresed in ABI Prism-3500 sequencer (Applied Biosystems) at the joint usage center 'Methods of molecular diagnostics' of the IEE RAS. The nucleotide sequences were cleaned and assembled using CodonCode Aligner version 7.1.1. Protein-coding sequences (COI, H3) were checked for indels and stop codons to prevent the inclusion of pseudogenes. All sequences were then compared to genes reported in GenBank using BLAST (National Center for Biotechnology Information, NCBI) to check for potential contamination.

For each gene-fragment, the sequences were aligned using MUSCLE [22] implemented in MEGA version X [23], and the alignment accuracy was adjusted by eye. Missing data were designated with a "?" for any incomplete sequences. All obtained sequences were submitted to the NCBI GenBank database (Table 2).

In order to assess phylogenetic relationships between species, we run Bayesian Inference (BI) and Maximum likelihood (ML) analyses. We ran ML analysis in RAxML GUI v2.0 [24,25] applying the GTR + G model. Bootstrap resampling with 1000 replicates was made using the thorough bootstrap procedure to assign support to branches in the ML tree. Trees were generated for each individual gene dataset and examined for conflicting topologies. Final ML tree was generated using the partitioned by gene and codone dataset of all concatenated genes.

The BI analysis was conducted in MrBayes v3.2.6 [26] for the concatenated dataset of all genes. The combined dataset was partitioned and analyzed using models selected by PartitionFinder2 [27]. AICc metric implemented in PartitionFinder2 was used to obtain the optimal partitioning scheme. Two independent runs, each consisting of four chains, were executed for this analysis. A total of 10,000,000 generations were performed for the combined dataset, with sampling every 1000 generations, and the first 25% trees (i.e., 2500 trees for combined dataset) were discarded as "burn-in". A 1% average standard deviation of split frequencies was reached after about 0.75 million generations.

We considered the clades statistically supported if they had a synchronous support of posterior probabilities ≥0.9 on the BI tree and bootstrap value ≥70% on the ML tree.

To quantify COI genetic distances between species/genera of Oplophoroidea, we used the Kimura 2-parameter model [28] implemented in MEGA X.

3. Results

3.1. Morphological Analyses and Supporting Synapomorphies

Examination of epipods and fifth pereopods in all species of Oplophoroidea revealed a great conformity between both characters:

Family Acanthephyridae, all species: epipods on the fourth pereopods absent; dactyli of the fifth pereopods short, greatly modified in a chelate structure (Figure 1A–H).

Family Oplophoridae, all species: epipods of the fourth pereopod well-developed with a prominent hook serving for cleaning of posterior branchia; dactyli of the fifth pereopods long and not greatly specialized (Figure 1I–K).

Phylogenetic Analysis 1 with *Pasiphaea sivado* as outgroup retrieved a single most parsimonious (MP) tree (Figure 2A, Files S3 and S4) with a score of 101 (Ci = 80, Ri = 84). The tree showed three major clades: Oplophoridae, *Hymenodora* + *Sclerodora*, and the rest of Acanthephyridae. Oplophoridae was sister group to Acanthephyridae, and *Hymenodora* + *Sclerodora* was sister group to the rest of Acanthephyridae. *Hymenodora* was sister group to *Sclerodora*.

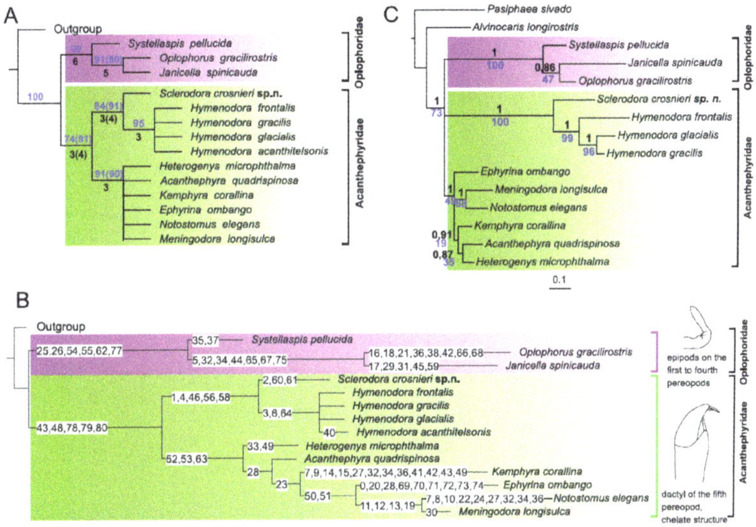

Figure 2. Phylogenetic trees with *Pasiphaea sivado* and *Alvinocaris longirostris* as outgroups. (**A**)—morphological MP tree; only clades supported by both Bremer values (black, below branches) and bootstrap values (blue, above branches) are shown; if support values in analyses differed, values retrieved in Analysis 2 are given in parentheses. (**B**)—synapomorphies, above branches, see coding in File S1. (**C**)—molecular BI and ML tree, only supported clades are shown. The horizontal scale bar marks the number of expected substitutions per site. Statistical support indicated as Bayesian posterior probabilities (black, above branches) and ML bootstrap with 1000 replicates (blue, below branches).

Analysis 2 with *Alvinocaris longirostris* as outgroup also retrieved a single MP tree (Figure 2A, File S3) with a score of 100 (Ci = 81, Ri = 85). Tree topology was the same as in Analysis 1.

The clade *Hymenodora* + *Sclerodora* was supported by five synapomorphies (Figure 2B, File S5): the presence of dorsal subuliform teeth (1) and the loss of subtriangular teeth (4) on the rostrum; a left mandible with the molar process compressed and sub-bilinear (46), a second maxilla with the proximal endite elongate, without submarginal papilla and lamina (56), and a first maxilliped with the endopod two-segmented, greatly overreaching endites (58). Within this clade, *Sclerodora* was supported by the presence of dorsal subuliform teeth both on the rostrum and carapace extending from the dorsal ridge (2), a second maxilliped with the terminal segment subtriangular and attached transversely (60) and bearing robust terminal setae (61). *Hymenodora* was supported by the presence of dorsal subuliform teeth only on the rostrum (3), a reticulum of carinae on membranous carapace (6), and a second maxilliped with the subovoid terminal segment attached diagonally (64).

3.2. *Molecular Analyses*

A total of 15 species representing all genera of the superfamily Oplophoroidea and two outgroup species were put in the data matrix. In addition, all species of *Hymenodora* deposited in GenBank (three out of four) were also added to the data matrix. Prior to the analyses, all sequences from GenBank were checked for contamination or possible misidentification using BLAST search and preliminary phylogenetic reconstruction with each gene separately. ML trees generated for each individual gene dataset revealed no conflicting topologies between genes, at least in branching with bootstrap values $\geq$60% (File S6). The concatenated four-marker dataset comprised 3321 bp. Results from PartitionFinder2 recommended a 7-partition scheme by gene and codon (H3, COI), which was used in the final analyses (File S7).

Molecular analyses (Figure 2C, File S8) showed that the new species was a sister group to *Hymenodora,* and both formed a common robust clade. The rest of Acanthephyridae and Oplophoridae also formed robust clades; deeper nodes within Oplophoroidea remained unresolved.

Genetic K2P distances between the new species and three *Hymenodora* species ranged from 31.9% to 32.9% in COI gene (File S9). These values significantly exceeded K2P distances between all *Hymenodora* species (9.4–27.0%) as well as K2P distances between representatives of six genera of Acanthephyridae (17.9–28.4%) and three genera of Oplophoridae (23.3–28.9%).

4. Discussion

4.1. *Taxonomic Implication*

Results of morphological and molecular analyses were very similar and suggested the same position of *Sclerodora* on the phylogenetic tree. This taxon was sister to *Hymenodora,* and, along with *Hymenodora,* formed a robust clade sister to the rest of the Acanthephyridae. Calculations of genetic K2P distances suggested a generic status of the new taxon: *Sclerodora* was more distant from the sister *Hymenodora* than any pair of genera within Acanthephyridae or Oplophoridae from each other. In addition to a significant genetic distance, *Sclerodora* was supported by remarkable synapomorphies linked to the carapace (the presence of dorsal subuliform teeth extending from the dorsal ridge) and mouthparts (shape and articulation of the terminal segment of the second maxilliped, unique in the superfamily Oplophoroidea). Both molecular and morphological evidences suggest the generic status of *Sclerodora* and its position within the clade *Hymenodora* + *Sclerodora* and within the major clade Acanthephyridae.

In order to encapsulate results of morphological and molecular analyses in the phylogenetic classification, we here erect and diagnose the new genus and provide an amended key to all genera of Oplophoroidea.

4.1.1. *Sclerodora* gen. nov.

Emended diagnosis: Integument robust; rostrum overreaching eye cornea, armed with subuliform dorsal teeth, no ventral teeth; carapace with dorsal ridge armed with subuliform teeth in anterior part; antennal angle rounded, branchiostegal spine rudimentary, no hepatic spine, no uninterrupted lateral carina extending from near orbit to near posterior margin, hepatic and branchiostegal carinae weak; abdomen with all somites dorsally rounded and lacking teeth; 6th somite longer than 5th. Eyes with cornea narrower than eyestalk; antennal scale without lateral teeth; mandibles dissimilar, molar process with transverse distal surface triangular on right member of pair and compressed and sub-bilinear on left member, incisor process toothed along entire opposable margin; 1st maxilla with endopod bearing distal prominence with a single robust seta; 2nd maxilla with proximal endite elongate, lacking papilla and submarginal lamina; 1st maxilliped with two-segmented endopod greatly overreaching endites; 2nd maxilliped with distal segment subtriangular, attached transversely to preceding segment and bearing terminal robust setae; 3rd maxilliped and

1st pereopod with exopods not unusually broad or rigid; pereopods with neither ischium nor merus broadly compressed, fourth pair without epipod.

Species included: *Sclerodora crosnieri* sp. nov.

Type species: *Sclerodora crosnieri* sp. nov. (type by monotypy).

Etymology: From Greek 'σκληρόσ', firm, hard, and 'δορα', integument; a reference to the integument of the new species, which is firmer than that in the sister genus *Hymenodora*.

Remarks: *Sclerodora* is similar to *Hymenodora* and both differ from other Oplophoroidea in the replacement of usual subtriangular teeth on the rostrum with subuliform structures spaced from each other; in having an unusual molar process (compressed and sub-bilinear) on the left mandible; in a unique elongate proximal endite of the second maxilla lacking submarginal papilla and lamina; in a two-segmented endopod greatly overreaching endites of the first maxilliped. At the same time, *Sclerodora* differs from *Hymenodora* in a presence of the dorsal subuliform teeth extending from the common carina both on the rostrum and the carapace. Such a character of *Sclerodora* as a subtriangular terminal segment of the second maxilliped, attached transversely and bearing robust terminal setae, is unique and not found in other Oplophoroidea.

4.1.2. *Sclerodora crosnieri* sp. nov.

Material: *Holotype*, female, 26 mm carapace length, 80 mm total length (telson broken); 39th Cruise of R/V "Professor Logachev"; 2018, March 2; 15° N, 45° W; Isaacs-Kidd midwater trawl, oblique tow 0–2500 m; kept in the Natural History Museum, Copenhagen University, Denmark.

Description: Carapace smooth, 1.73 times as long as high, suprabranchial and hepatic ridges prominent (Figure 3A); dorsal carina with eight small irregular teeth in 1/4 anterior part; rostrum with four dorsal teeth (Figure 3B). Abdomen with sixth somite twice as long as fifth; telson (broken) with dorsolateral spines (Figure 3C).

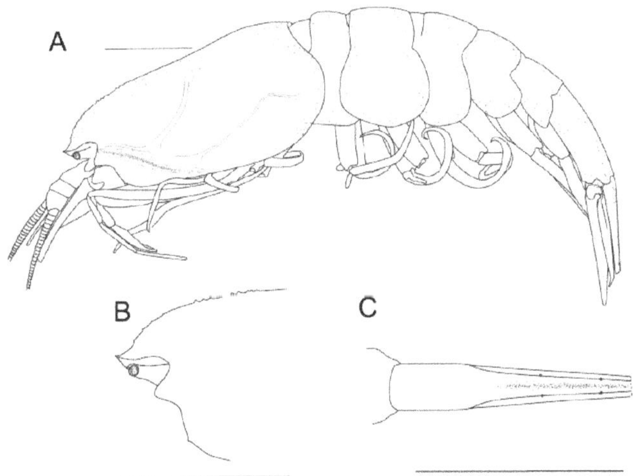

Figure 3. Body of *Sclerodora crosnieri* sp. nov., holotype: (**A**)—general view. (**B**)—anterior part of carapace, lateral view. (**C**)—telson, dorsal view. All scales: 10 mm.

Mandible (Figure 4A) with 2-segmented palp; first maxilla with distal endite bearing two rows of robust setae (Figure 4B); second maxilla with two distal endites subequal (Figure 4C); first maxilliped with distal segment of endopod nearly twice as long as basal segment (Figure 4D); second maxilliped with distal segment bearing five terminal stout setae (Figure 4E); third maxilliped with well-developed hook-bearing epipod, distal segment densely covered with setae over entire margin (Figure 4F). First pereopod with carpus bearing distal tooth, propodus densely covered with setae over flexor margin, bearing large terminal and tiny subterminal spines, inner margins of chela rifled (Figure 5A–C); second pereopod with propodus bearing large terminal and tiny subterminal spines, inner margins of chela rifled (Figure 5D–F); third pereopod with ischium armed with three spines and dactyl bearing seven robust setae on flexor margin (Figure 5G); fourth pereopod with ischium armed with a single spine and dactyl bearing seven robust setae on flexor margin (Figure 5H); fifth pereopod with propodus covered with rifled setae over flexor margin and a single terminal robust seta in the chelate structure, dactyl curved and bearing two terminal robust setae (Figure 5I–J).

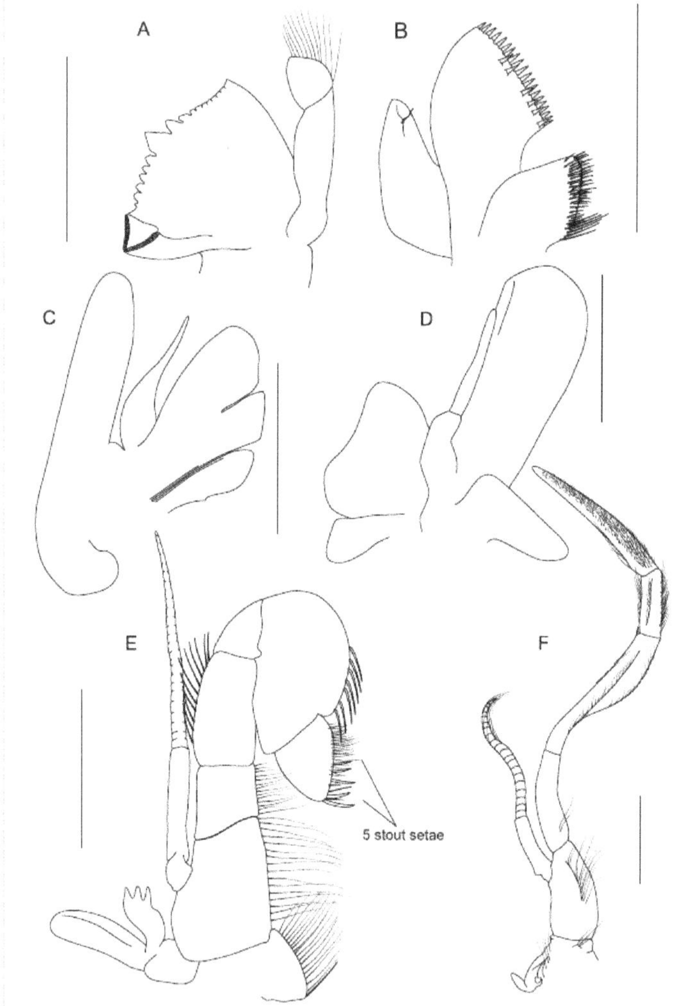

Figure 4. Mouthparts of *Sclerodora crosnieri* sp. nov., holotype, right parts: (**A**)—mandible. (**B**)—first maxilla, inner view. (**C**)—second maxilla, setae removed. (**D**)—first maxilliped, setae removed. (**E**)—second maxilliped. (**F**)—third maxilliped. All scales: 3 mm.

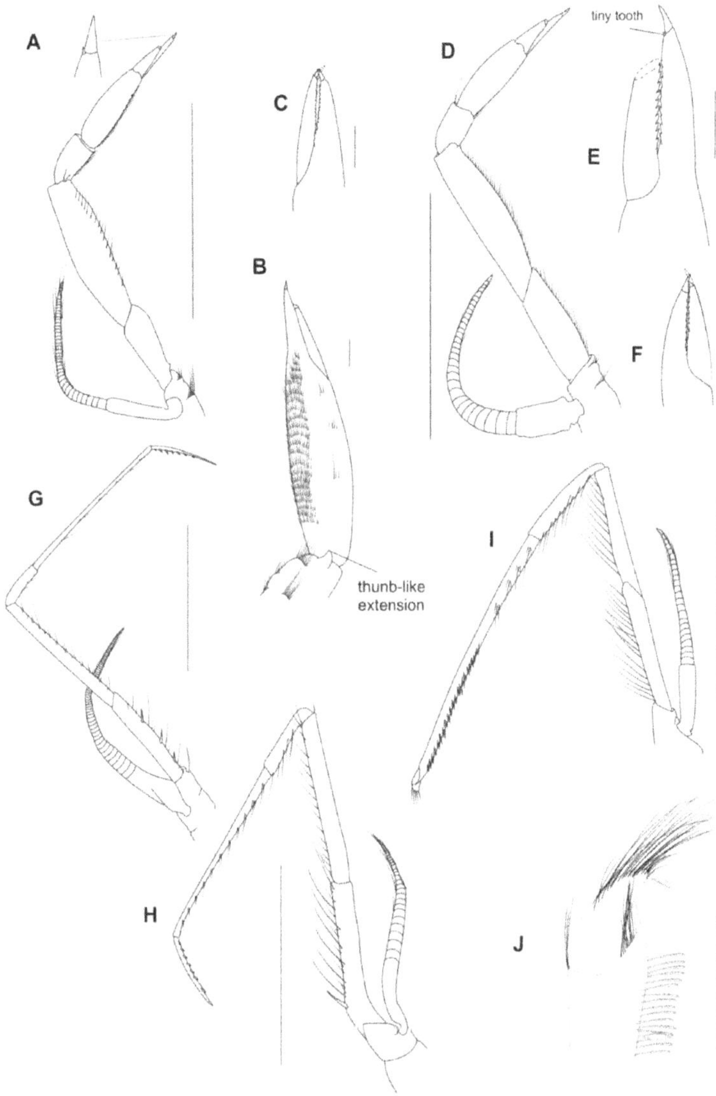

Figure 5. Pereopods of *Sclerodora crosnieri* **sp. nov**., holotype: (**A**)—right first pereopod. (**B**)—right first chela, inner view. (**C**)—left first chela. (**D**)—right second pereopod. (**E**)—right second chela. (**F**—left second chela. (**G**)—right third pereopod. (**H**)—left fourth pereopod. (**I**)—right fifth pereopod. (**J**)—terminal part of right fifth pereopod. All scales for entire pereopods: 10 mm, all scales for their tips: 1 mm.

Etymology: named after the late Alain Crosnier, prominent carcinologist greatly contributed to taxonomy of decapods and, in particular, oplophoroid shrimps.

4.1.3. Key to Genera of Oplophoroidea

1. Sixth abdominal somite with distinct dorsal carina 2
- Sixth abdominal somite dorsally smooth 6
2. Hepatic spine present .... *Kemphyra* Chace, 1986
- Hepatic spine absent 3
3. Third abdominal somite with long dorsal tooth overreaching fourth somite .... *Heterogenys* Chace, 1986
- Tooth on third abdominal somite, if present, not overreaching fourth somite 4
4. Carapace dorsally denticulate over nearly entire length; first abdominal somite dorsally carinate .... *Notostomus* A. Milne-Edwards, 1881
- Carapace dorsally not denticulate on posterior half; first abdominal somite smooth 5
5. A single continuous lateral carina on carapace (extending from near orbit to near posterior margin on carapace .... *Meningodora* Smith, 1882
- None or two continuous lateral carinae on carapace (one extending from near orbit, another extending from near branchiostegal spine) .... *Acanthephyra* A. Milne-Edwards, 1881
6. Rostrum unarmed. Meri and ischia of pereopods greatly wide and compressed .... *Ephyrina* Smith, 1885
- Rostrum denticulate. Meri and ischia of pereopods not greatly wide and compressed 7
7. Rostral teeth subuliform, spaced from each other. Cornea subequal or narrower than eyestalk 8
- Rostral teeth subtriangular, extending from a common crest. Cornea wider than eyestalk 9
8. Dorsal subuliform teeth only on rostrum .... *Hymenodora* G.O. Sars, 1877
- Dorsal subuliform teeth both on rostrum and anterior part of carapace .... *Sclerodora* gen.nov.
9. Carapace strongly chitinized, subtriangular in cross-section. Abdomen with sixth somite not longer than fifth, third to fourth somites with strong dorsomedial spines (at least $\frac{1}{2}$ of segment length) 10
- Carapace moderately chitinized, suboval in cross-section. Abdomen with sixth somite nearly twice as long as fifth, third to fourth somites without strong dorsomedial spines .... *Systellaspis* Spence Bate, 1888
10. Second abdominal somite with strong dorsomedial spine. .... *Janicella* Chace, 1986
- Second abdominal somite without strong dorsomedial spine .... *Oplophorus* H. Milne Edwards

4.2. A New Suggested Cleaning and Grooming Mechanism

Examination of epipods and fifth pereopods in all species of Oplophoroidea reveals a remarkable co-evolution between both characters. In this superfamily, reduction of the fourth epipod is associated with development of a chelate structure on the fifth pereopod. This structure is morphologically similar in all genera (Figure 1) and remarkably resembles grooming chelae in other carids as illustrated in [4]. When we map these synapomorphies on a phylogenetic tree (Figure 2B), we can see that a chelate structure on the fifth pereopod (linked to a lost epipod on the fourth pereopod and indicating active grooming) is a derived structure, whereas a long dactyl not forming chelate structure (linked to a full set of the epipods and passive grooming) occurs basally on the morphological tree and likely plesiomorphic as suggested by Bauer [4].

We suggest that Acanthephyridae evolved an active cleaning mechanism, which is a derived one and alternative to that described by Bauer [4,5]: posterior branchiae are groomed and cleaned by the fifth pereopods instead of the chelipeds. Convergent chelate structures suggest that the chela is especially efficient for cleaning and grooming branchiae, whichever appendage is adapted for these functions. In Oplophoridae, which are basal on the phylogenetic morphological tree, the cleaning function is carried out passively by well-developed epipods.

Our results confirm Bauer's [4,5] statement that the major type of gill-cleaning method is generally a characteristic at the family level and that the active cleaning is more derived than the passive one. Interestingly, in Oplophoridae the last three pereopods likely take another function and act as a holding structure during mating [10], which may favor copulation in the turbulent water column [9].

Supplementary Materials: The following are available online at https://www.mdpi.com/article/10.3390/d13110536/s1, File S1: Character list, File S2: Character state, File S3: Retrieved morphological trees. Bremer support, File S4: Retrieved morphological trees. Bootstrap support, File S5: Synapomorphies, File S6: Maximum likelihood (RAxML) phylograms for each individual gene dataset, File S7: Partitioning scheme and best models selected by PartitionFinder2, File S8: Molecular BI and ML trees, File S9: Estimates of evolutionary divergence between species.

Author Contributions: A.L. analyzed the specimens morphologically; D.K. ran the genetic analyses; A.V., A.L. and D.K. wrote the paper and participated in the revisions of it. All authors have read and agreed to the published version of the manuscript.

Funding: This research was supported by RSF Project No. 18-14-00231.

Institutional Review Board Statement: Not applicable.

Data Availability Statement: Data is contained within the article and supplementary material.

Acknowledgments: We thank S.G. Kobylyansky for material from the 39th Cruise of the R/V "Professor Logatchev"; H. Braken-Grissom, Laura Corbari, J. Olesen and T. Sutton for possibility to examine specimens of Oplophoroidea.

Conflicts of Interest: None of the authors have any financial competing interests.

References

1. Dawson, M.N. Species richness, habitable volume, and species densities in freshwater, the sea, and on land. *Front. Biogeogr.* **2012**, *4*. [CrossRef]
2. Costello, M.J.; Chaudhary, C. Marine biodiversity, biogeography, deep-sea gradients, and conservation. *Curr. Biol.* **2017**, *27*, 511–527. [CrossRef] [PubMed]
3. WoRMS Editorial Board. World Register of Marine Species. 2021. Available online: http://www.marinespecies.org (accessed on 4 October 2021). [CrossRef]
4. Bauer, R.T. Decapod crustacean grooming: Functional morphology, adaptive value, and phylogenetic significance. In *Functional Morphology of Feeding and Grooming in Crustacea*; CRC Press: Boca Raton, FL, USA, 2020.
5. Bauer, R.T. Antifouling adaptations of caridcan shrimp (Decapoda: Caridea): Gill cleaning mechanisms and grooming of brooded embryos. *Zool. J. Linn. Soc.* **1979**, *65*, 281–303. [CrossRef]
6. Bauer, R.T. Grooming behavior and morphology in the decapod Crustacea. *J. Crust. Biol.* **1981**, *1*, 153–173. [CrossRef]
7. Wong, J.M.; Pérez-Moreno, J.L.; Chan, T.Y.; Frank, T.M.; Bracken-Grissom, H.D. Phylogenetic and transcriptomic analyses reveal the evolution of bioluminescence and light detection in marine deep-sea shrimps of the family Oplophoridae (Crustacea: Decapoda). *Mol. Phylogenet. Evol.* **2015**, *83*, 278–292. [CrossRef]
8. Chace, F.A. *The Caridean Shrimps (Crustacea: Decapoda) of the Albatross Philippine Expedition, 1907–1910, Part 4: Families Oplophoridae and Nematocarcinidae*; Smithsonian Institution Press: Washington, DC, USA, 1986; p. 81.
9. Lunina, A.; Vereshchaka, A. The role of the male copulatory organs in the colonization of the pelagic by shrimp-like eucarids. *Deep Sea Res. Part II Top. Stud. Oceanogr.* **2017**, *137*, 327–334. [CrossRef]
10. Lunina, A.A.; Kulagin, D.N.; Vereshchaka, A.L. Oplophoridae (Decapoda: Crustacea): Phylogeny, taxonomy and evolution studied by a combination of morphological and molecular methods. *Zool. J. Linn. Soc.* **2019**, *186*, 213–232. [CrossRef]
11. Lunina, A.A.; Kulagin, D.N.; Vereshchaka, A.L. Phylogenetic revision of the shrimp genera Ephyrina, Meningodora and Notostomus (Acanthephyridae: Caridea). *Zool. J. Linn. Soc.* **2020**, zlaa161. [CrossRef]
12. Li, C.P.; De Grave, S.; Chan, T.Y.; Lei, H.C.; Chu, K.H. Molecular systematics of caridean shrimps based on five nuclear genes: Implications for superfamily classification. *Zool. Anz.* **2011**, *250*, 270–279. [CrossRef]
13. Nixon, K. The parsimony ratchet, a new method for rapid parsimony analysis. *Cladistics* **1999**, *15*, 407–414. [CrossRef]
14. Goloboff, P.; Farris, S.; Nixon, K. TNT: Tree Analysis Using New Technology. 2000. Available online: http://www.lillo.org.ar/phylogeny/tnt (accessed on 24 October 2021).
15. Schubart, C.D.; Huber, M.G.J. Genetic comparisons of German populations of the stone crayfish, Austropotamobius torrentium (Crustacea: Astacidae). *Bull. Français Pêche Piscic.* **2006**, *380–381*, 1019–1028. [CrossRef]
16. Folmer, O.; Black, M.; Hoeh, W.; Lutz, R.; Vrijenhoek, R. DNA primers for amplification of mitochondrial cytochrome c oxidase subunit I from diverse metazoan invertebrates. *Mol. Mar. Biol. Biotechnol.* **1994**, *3*, 294–299. [PubMed]
17. Schubart, C.D. Mitochondrial DNA and decapod phylogenies: The importance of pseudogenes and primer optimization. *Crustacean* **2009**, *18*, 47–65.

18. Schubart, C.D.; Cuesta, J.A.; Felder, D.L. Glyptograpsidae, a new brachyuran family from Central America: Larval and adult morphology, and a molecular phylogeny of the Grapsoidea. *J. Crust. Biol.* **2002**, *22*, 28–44. [CrossRef]
19. Reuschel, S.; Schubart, C.D. Phylogeny and geographic differentiation of Atlanto–Mediterranean species of the genus Xantho (Crustacea: Brachyura: Xanthidae) based on genetic and morphometric analyses. *Mar. Biol.* **2006**, *148*, 853–866. [CrossRef]
20. Apakupakul, K.; Siddall, M.E.; Burreson, E.M. Higher level relationships of leeches (Annelida: Clitellata: Euhirudinea) based on morphology and gene sequences. *Mol. Phylogenetics Evol.* **1999**, *12*, 350–359. [CrossRef]
21. Colgan, D.J.; McLauchlan, A.; Wilson, G.D.F.; Livingston, S.P.; Edgecombe, G.D.; Macaranas, J.; Gray, M.R. Histone H3 and U2 snRNA DNA sequences and arthropod molecular evolution. *Aust. J. Zool.* **1998**, *46*, 419–437. [CrossRef]
22. Edgar, R.C. MUSCLE: Multiple sequence alignment with high accuracy and high throughput. *Nucleic Acids Res.* **2004**, *32*, 1792–1797. [CrossRef] [PubMed]
23. Kumar, S.; Stecher, G.; Li, M.; Knyaz, C.; Tamura, K. MEGA X: Molecular Evolutionary Genetics Analysis across computing platforms. *Mol. Biol. Evol.* **2018**, *35*, 1547–1549. [CrossRef] [PubMed]
24. Stamatakis, A. RAxML version 8: A tool for phylogenetic analysis and post-analysis of large phylogenies. *Bioinformatics* **2014**, *30*, 1312–1313. [CrossRef]
25. Edler, D.; Klein, J.; Antonelli, A.; Silvestro, D. raxmlGUI 2.0 beta: A graphical interface and toolkit for phylogenetic analyses using RAxML. *Methods Ecol. Evol.* **2021**, *12*, 373–377. [CrossRef]
26. Ronquist, F.; Teslenko, M.; Van Der Mark, P.; Ayres, D.L.; Darling, A.; Höhna, S.; Huelsenbeck, J.P. MrBayes 3.2: Efficient Bayesian phylogenetic inference and model choice across a large model space. *Syst. Biol.* **2012**, *61*, 539–542. [CrossRef] [PubMed]
27. Lanfear, R.; Frandsen, P.B.; Wright, A.M.; Senfeld, T.; Calcott, B. PartitionFinder 2: New methods for selecting partitioned models of evolution for molecular and morphological phylogenetic analyses. *Mol. Biol. Evol.* **2016**, *34*, 772–773. [CrossRef] [PubMed]
28. Kimura, M. A simple method for estimating evolutionary rate of base substitutions through comparative studies of nucleotide sequences. *J. Mol. Evol.* **1980**, *16*, 111–120. [CrossRef] [PubMed]

diversity

MDPI

Article

Establishment of a New Filamentous Cyanobacterial Genus, *Microcoleusiopsis* gen. nov. (Microcoleaceae, Cyanobacteria), from Benthic Mats in Open Channel, Jiangxi Province, China

Ruozhen Geng [1,2], Wenke Li [2,3], Aimin Chao [4], Xiaoyu Guo [2,3], Hua Li [2], Gongliang Yu [2,*] and Renhui Li [1,*]

1 College of Life and Environmental Sciences, Wenzhou University, Wenzhou 325039, China; 20210438@wzu.edu.cn
2 Key Laboratory of Algal Biology, Institute of Hydrobiology, Chinese Academy of Sciences, Wuhan 430072, China; liwenke@ihb.ac.cn (W.L.); guoxiaoyu@ihb.ac.cn (X.G.); lih@ihb.ac.cn (H.L.)
3 University of Chinese Academy of Sciences, Beijing 100039, China
4 Zhejiang Key Laboratory of Ecological and Environmental Monitoring, Forewarning and Quality Control, Zhejiang Environmental Monitoring Center, Hangzhou 310012, China; chaoaimin@126.com
* Correspondence: yugl@ihb.ac.cn (G.Y.); renhui.li@wzu.edu.cn (R.L.)

Abstract: Cyanobacterial taxonomic studies performed by using the modern approaches always lead to creation of many new genera and species. During the field survey for cyanobacterial resources in China, a filamentous cyanobacterial strain was successfully isolated from a microbial mat attached to rock surfaces of the Ganfu Channel, Jiangxi Province, China. This strain was morphologically similar to the cyanobacterial taxa belonging to the genera *Microcoleus* and *Phormidium*. The phylogenetic analyses based on 16S rRNA gene sequences showed that this strain formed a well-supported clade, close to the filamentous genera *Microcoleus*, *Tychonema*, and *Kamptonema*. The maximum similarity of 16S rRNA gene sequence of this strain with the related genera was 95.04%, less than the threshold for distinguishing bacterial genus. The ITS secondary structures also distinguish this strain from the related cyanobacterial genera. Therefore, combined with morphology, 16S rRNA gene sequence, and ITS secondary structures, a novel cyanobacterial genus here as *Microcoleusiopsis* was established, with the species type as *Microcoleusiopsis ganfuensis*.

Keywords: filamentous cyanobacteria; *Microcoleusiopsis ganfuensis*; polyphasic; taxonomy

Citation: Geng, R.; Li, W.; Chao, A.; Guo, X.; Li, H.; Yu, G.; Li, R. Establishment of a New Filamentous Cyanobacterial Genus, *Microcoleusiopsis* gen. nov. (Microcoleaceae, Cyanobacteria), from Benthic Mats in Open Channel, Jiangxi Province, China *Diversity* **2021**, *13*, 578. https://doi.org/10.3390/d13110548

Academic Editor: Michael Wink

Received: 29 September 2021
Accepted: 26 October 2021
Published: 29 October 2021

Publisher's Note: MDPI stays neutral with regard to jurisdictional claims in published maps and institutional affiliations.

Copyright: © 2021 by the authors. Licensee MDPI, Basel, Switzerland. This article is an open access article distributed under the terms and conditions of the Creative Commons Attribution (CC BY) license (https://creativecommons.org/licenses/by/4.0/).

1. Introduction

In the past decade, the molecular biological methods have shown a powerful solution to taxonomic problems in many cyanobacterial categories [1]. The polyphasic approach—based on the combination of morphological, cytomorphological, ecological, and molecular characteristics—has been widely used in characterization and integrated to solve the taxonomic problems of cyanobacteria have been accepted by more and more cyanobacterial researchers, leading to much progress in studies on cyanobacterial diversity [2,3]. The classification criteria based on only morphological observation gradually lost their original utility, and the morphological boundaries among many related genera became even more blurred. The problem that morphological characteristics could not be well integrated with phylogeny was so evident that it became urgent to revise the existing classification system of cyanobacteria from a more phylogenetic perspective. Thus, based on the polyphasic method, Komárek et al. proposed the eight-order system, later the ten-order system, resolving some phylogenetic issues [2,4–8].

The genus *Microcoleus* Desmazières ex Gomont was first described in 1892 [9], and this genus contains a group of filamentous cyanobacteria widely existing in various ecological niches, and was considered as one of the largest genera in the family Microcoleaceae. The type species *Microcoleus vaginatus* (Vaucher) Gomont ex Gomont, was characterized with many bright blue-green trichomes per colorless and unlamellated sheath, with specific ecology (soil biotope) [10,11]. As currently defined, there are 112 species of *Microcoleus* including aquatic and terrestrial species in all database, only 55 species have been accepted taxonomically based on the Algaebase Database up to 2017 (www.algaebase.org, accessed on 13 May 2021). Most species of this genus have typical characteristics of usually simple filaments, densely packed trichomes, isodiametric vegetative cells, strongly constricted cross walls, no calyptra, end cells typically longer than wide, sheaths open at the apex, and crosswise cell division [12].

For a long time, the phylogenetic evidence has indicated the genus *Microcoleus* to be polyphyletic. Its taxonomic revisions were continuously performed, mainly by separating several species in the genus away from the type species *M. vaginatus*. Boyer et al. (2002) summarized the 31 strains of *Microcoleus* as two morphological species (*M. vaginatus* and *M. steenstrupii*) falling into two distinct clades which were regarded as two genera [13]. Similarly, Siegesmund et al. (2008) proposed another important species within *Microcoleus*, *M. chthonoplastes*, in the new genus/species *Coleofasciculus chthonoplastes* based on its genetic distance to the type species [14]. Strunecky et al. (2013) targeted the morphological and molecular criteria for the revision of the genus *Microcoleus* through extensive examination of 92 strains of *M. vaginatus* and *Phormidium autumnale* from a wide range of regions and biotopes, and they further established the new family of Microcoleaceae and more than 10 new combination species of *Microcoleus* by transferring from species formerly placed in the genus *Phormidium* and *Oscillatoria* [15]. Niiyama and Tuji (2019) also described a new species, *Microcoleus pseudautumnalis*, producing both 2-methylisoborneol (2-MIB) and geosmin based on the polyphasic approach [16]. Similarly, Kimberly et al. (2020) also proposed a novel anatoxin-a and dihydroanatoxin-a producing species, *M. anatoxicus*, and these two recent studies even provided some new clues revealing the new species of *Microcoleus* related to some environmental issues [17]. However, the further revisions for the taxonomy of the genus *Microcoleus* are required, which will lead to more new genera and species during the revisionary course.

In recent years, the construction of water diversion projects has become an important measure in China to solve the problems for the increasing demand of water resources in water shortage areas, leading to a large number of new artificial channel with flowing water biotopes. Filamentous cyanobacteria accounting for a large proportion of microbial mats growing on both sides of the channels are mainly composed of Oscillatorean cyanobacteria such as *Microcoleus*, *Oscillatoria*, *Phormidium*, *Lyngbya*, and *Tychonema*. In this study, one filamentous cyanobacterial strain with *Microcoleus*-like morphology was isolated from the Ganfu Channel in Jiangxi Province, China. The polyphasic method based on morphological and molecular and phylogenetic analyses was used to characterize this new isolated cyanobacterium, and results revealed that it represents a novel genus of the family Microcoleaceae. Thus, the new genus as *Microcoleusiopsis* gen. nov and type species as *Microcoleusiopsis ganfuensis* sp. nov. were described.

2. Materials and Methods

2.1. Sampling and Cultivation

Benthic mat samples were separated in August 2019 from Ganfu channel, Jiangxi Province, China (28°33′7.48″ N, 115°56′44.62″ E). For strain isolation, mats were scraped off using a circular knife and live material was washed thoroughly in sterile liquid CT medium [18]. Sub-samples were coated onto the surface of sterile solid CT plate and the Pasteur pipette washing method was used to obtain unialgal filaments or single cells under 40× microscope (Olympus CKX31, Tokyo, Japan), kept at 25 °C under cool white fluorescence light on a 12:12 h L:D photoperiod with a photon flux density of 40 μmol m^{-2} s^{-1}. Finally, a filamentous strain (named as CHAB 4138) was isolated and transferred into several 25 mL flasks containing 15 mL of CT medium. These strains were stored in the culture collection of Harmful Algae Biology laboratory (CHAB) in the Institute of Hydrobiology, Wuhan, China.

2.2. Morphological and Ultrastructural Characterization

Cell morphological observation was investigated with a Nikon Eclipse 80i microscope (Nikon, Japan). Filaments and vegetative cells were measured more than 100 individuals with a DS-Ri1 digital camera (Nikon, Japan). Microphotographs taken at 400 times were analyzed by using Nikon software NIS-Elements D 3.2. For ultrastructure examination, fresh samples were fixed using 2.5% glutaraldehyde in 0.1 M phosphate buffer at a pH 7.2 and 4 °C for three days. Then, these samples were washed using 0.1 M phosphate buffer after which they were post-fixed using 1% osmium tetroxide for 2 h, and washed again using 0.1 M phosphate buffer to remove osmium tetroxide after which they were dehydrated using a sequential ethanol gradient (30, 50, 70, 90, and 100%) and embedded in Spurr's resin [19]. Uranyl acetate (2%) and lead citrate were used to stain the sections. Finally, the specimens were examined with an HT7700 (Japan) transmission electron microscope under 80 kV on Hitachi TEM system control (Hitachi, Tokyo, Japan).

2.3. DNA Extraction and PCR Amplification

To avoid extra bacteria contamination, fresh material of strain CHAB 4138 was collected by filtering onto Millipore filter (3.0 μm aperture, Merck Millipore, Darmstadt, Germany) and was further cleaned with sterile CT medium for two to three times, collected in clean EP tubes. Total genomic DNA from this strain was extracted using the modified cetyltrimethylammonium bromide (CTAB) method [20]. DNA was quantified using a NanoDrop™ 1000 Spectrophotometer (Thermo Scientific, Waltham, MA, USA).

The primers PA [21] and B23S [22] were used to amplify segments including the 16S rRNA gene and the 16S–23S internal transcribed spacer (ITS). Each PCR amplification was performed using a BIO-RAD Thermal Cycler (Bio-Rad, Hercules, CA, USA) with total PCR reaction volume of 20 μL consisted of 1 μL of genomic DNA (100 ng μL^{-1}), 0.5 μL of each primer (10 μmol L^{-1}), 8 μL of sterile water and 10 μL of 2× PCR mix with Taq polymerase (Beijing Tsingke Biotech Co., Ltd., Beijing, China). The program for 16S rRNA gene ran for one cycle of 3 min at 94 °C; 34 cycles of 30 s at 94 °C, 30 s at 58 °C (30 s at 55 °C for ITS), and 1 min at 72 °C (30 s for ITS) and then a final elongation step at 72 °C for 5 min. The PCR products were purified by the Qiaquik PCR purification columns (Qiagen, Germany) using TSINGKE DNA Gel Extraction Kit (Beijing Tsingke Biotech Co., Ltd., Beijing, China), cloned to the pMDTM18-T vector (TaKaRa, TaKaRa BioInc., Otsu, Japan) and inserted into *Escherichia coli* trans5α cells. Finally, the positive clones including target fragment were sequenced bidirectionally using an ABI 3730 Automated Sequencer (PerkinElmer, Waltham, MA, USA). At least three clones were sequenced for each target fragment.

2.4. Detection for Cyanotoxin Synthesis Genes

Genomic DNA from strain CHAB 4138 was detected for the cyanotoxin synthesis genes such as microcystins, paralytic shellfish toxins, cylindrospermopsin, and anatoxin-a. The corresponding primers and PCR procedures refer to the methods of previous studies by Jungblut and Neilan [23], Al-Tebrineh et al. [24], McGregor and Sendall [25], and Rantala-Ylinen et al. [26], respectively.

2.5. Phylogenetic Analyses

The 16S rRNA gene sequences obtained from a single clone of strain CHAB 4138 were initially screened at the NCBI Website (BLAST), and higher similar reference sequences were downloaded from GenBank database to construct the molecular phylogeny of these two strains. Using MAFFT v7.312 software we obtained a matrix of 162 sequences with 1237 nucleotide sites [27] after multiple sequence alignment. The standard selection nucleic acid substitution model (GTR+I+G) based on the Akaike information criterion (AIC) for Bayesian analysis (BI) and maximum likelihood analysis (ML) were selected to analyzed the alignments, and then particular parameters were individually estimated by MrBayes v3.2.6 [28] and PhyML 3.0 [29]. The Kimura–2 model was selected with 1000 bootstrap replicates to perform neighbor joining (NJ) analysis using MEGA software v7.0 [30]. Both

ML and Bayesian phylogenetic trees were viewed and edited in FigTree v1.4.3 (http://tree.bio.ed.ac.uk/software/figtree/), and all obtained phylogenetic trees were edited by Tree View 1.6.6 software [31]. Similarity matrix of the 16S rRNA was established via MEGA software v7.0 to calculate p-distance with pairwise deletion of gaps.

2.6. Construction of Secondary Structure of 16S–23S Internal Transcribed Spacer (ITS)

The 16S–23S rRNA ITS secondary structures of D1–D1′, Box–B and V3 helices of this strain and other closed species were determined using RNA structure, version 5.6 [32]. The 16S–23S rRNA gene nucleotide sequences obtained in this study have been deposited in the GenBank database, and the accession numbers are OK422506 and OK422507.

3. Results

3.1. Morphological Description

Microcoleusiopsis R. Geng et G. Yu gen. nov.

Diagnosis: This genus appears morphologically similar to the genera of *Microcoleus* and *Phormidium*. The phylogenetic relationship was close to members of the family Microcoleaceae.

Description: In nature, colonies macroscopic, usually forming algal mats attached to the rock surface on freshwater rivers. Filaments long, straight, or slightly curved, blue-green to yellow-brown, surrounded by hyaline, colorless envelopes. Trichomes cylindrical, isopolar, not attenuated toward ends. Vegetative cells discoid, isopolar, always broader than long. Reproduction by motile hormogonia formed by necridia. Thylakoids radially arranged.

Type species: *Microcoleusiopsis ganfuensis* R. Geng et G. Yu sp. nov.

Etymology: The name of new genus "*Microcoleusiopsis*" was chosen because it was closely related to genus *Microcoleus*.

Microcoleusiopsis ganfuensis R. Geng et G. Yu sp. nov. (Figure 1).

Diagnosis: This species appears morphologically similar to the genera of *Microcoleus*-like. Filaments are long, not attenuated towards ends, and not or slightly constricted at the cross-walls. Apical cell rounded, without calyptra or thickened outer cell wall. Phylogenetic analysis suggested that this species formed a separated clade which was close to members of the families Microcoleaceae, such as *Microcoleus*, *Tychonema*, and *Kamptonema*.

Description: In nature, colonies usually form cyanobacterial mats attached to the surface of wet rocks on freshwater rivers and channels. Filaments long, unbranched, straight or slightly curved, blue-green, green when young, and yellow-brown when old, surrounded by hyaline, colorless sheaths. Trichomes isopolar, cylindrical, not attenuated towards ends, not or slightly constricted at the cross-walls. Vegetative cells usually discoid, isopolar, 2.28–(3.09)–4.27 μm long, 4.52–(5.69)–6.18 μm broad, width: length ratio 1.8, with granular content, not aerotopes. Apical cell rounded, without calyptra or thickened outer cell wall. Sheath finer, colorless, hyaline, not diffluent, and always open at the apex. Reproduction by motile hormogonia formed by necridia. Heterocytes and akinetes were not observed. Thylakoids radially arranged (Figure 2).

Reference strain: CHAB 4138.

Type locality: In Ganfu open channel, Jiangxi Province, China. (August 2019, 28°33′7.48″ N, 115°56′44.62″ E).

Holotype here designated: Dry material of this strain CHAB 4138 with no. JXGF201902, stored at Freshwater Algae Biology Herbarium (HBI), Institute of Hydrobiology, Chinese Academy of Science, Wuhan, Hubei Province, China.

Etymology: The name of species "*ganfuensis*" was chosen because this strain was separated from the Ganfu open channel.

Habitat: Attached on wet rock surfaces.

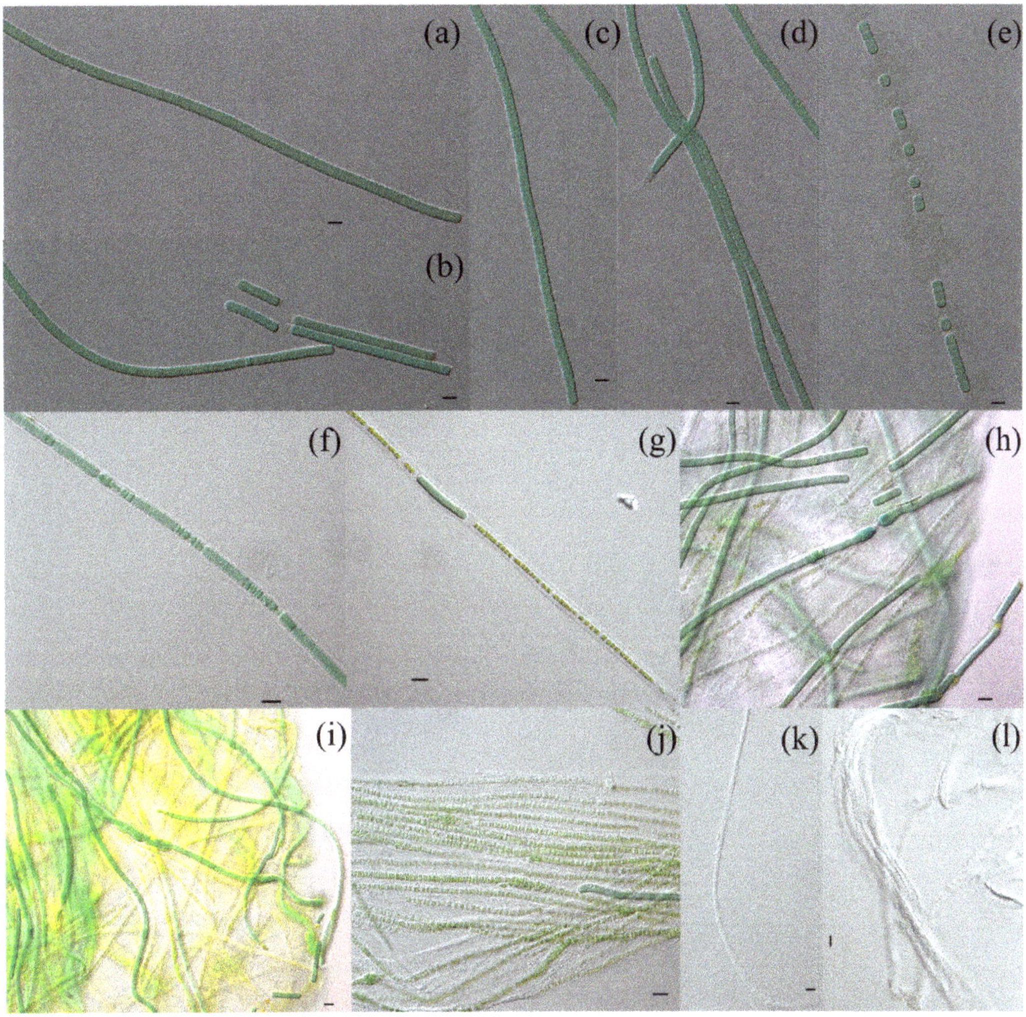

Figure 1. Light microscopy of *Microcoleusiopsis ganfuensis* strains. (**a–e**) Immature filaments without sheaths. (**f**) Immature filaments with colorless sheaths. (**g,h**) Trichome fragmentation and formation of necridia. (**i,j**) Old filaments of 3–month–old. (**k,l**) Decline filaments with lamellated sheaths. Scale bars: 10 μm.

3.2. Molecular and Phylogeny Analyses

Through single sequencing, we obtained two 16S rRNA gene clones (1494bp) of strain CHAB 4138 which shared 99.91% similarities with each other. The 16S rRNA gene phylogenetic trees based on NJ, ML, and Bayesian methods with 162 sequences of family Microcoleaceae and Oscillatoriaceae strains downloaded from the NCBI database (Figure 3) indicated that the two clones of CHAB 4138 clustered a well-supported independent cluster (cluster A), supported by NJ/ML/BI approaches with high bootstrap values of 99%/100%/1.00. This unique clade was close to those formed by the filamentous genera *Microcoleus* (cluster B), *Tychonema* (cluster C), *Kamptonema* (cluster D), and *Heteroleibleinia* (cluster E), with a maximum similarity as 95.04%, probably representing a novel genus of filamentous cyanobacteria (sharing similarities to *Microcoleus*, *Kamptonema*, *Heteroleibleinia*, *Tychonema*, *NeoLyngbya*, *Lyngbya*, *Okeania*, *Hydrocoleum*, *Dapis*, *Moorea*, *Symploca*, *Caldora*, *Wilmottia*, *Laspinema*, *Trichodesmium*, *Coleofasciculus*, *Oscillatoria*, *Aerosakkonema*, and

Phormidium were 94.09–95.04%, 94.43–94.52%, 94.43–94.52%, 94.26–94.35%, 94.09–94.17%, 93.83–93.91%, 93.13–93.22%, 92.96–93.04%, 92.78–92.87%, 92.70–92.78%, 92.61–92.70%, 92.52–92.61%, 92.43–92.52%, 92.43–92.52%, 92.09–92.17%, 92.09–92.17%, 92.00–92.09%, 92.00–92.09%, and 91.74–91.83%, respectively) (Table 1).

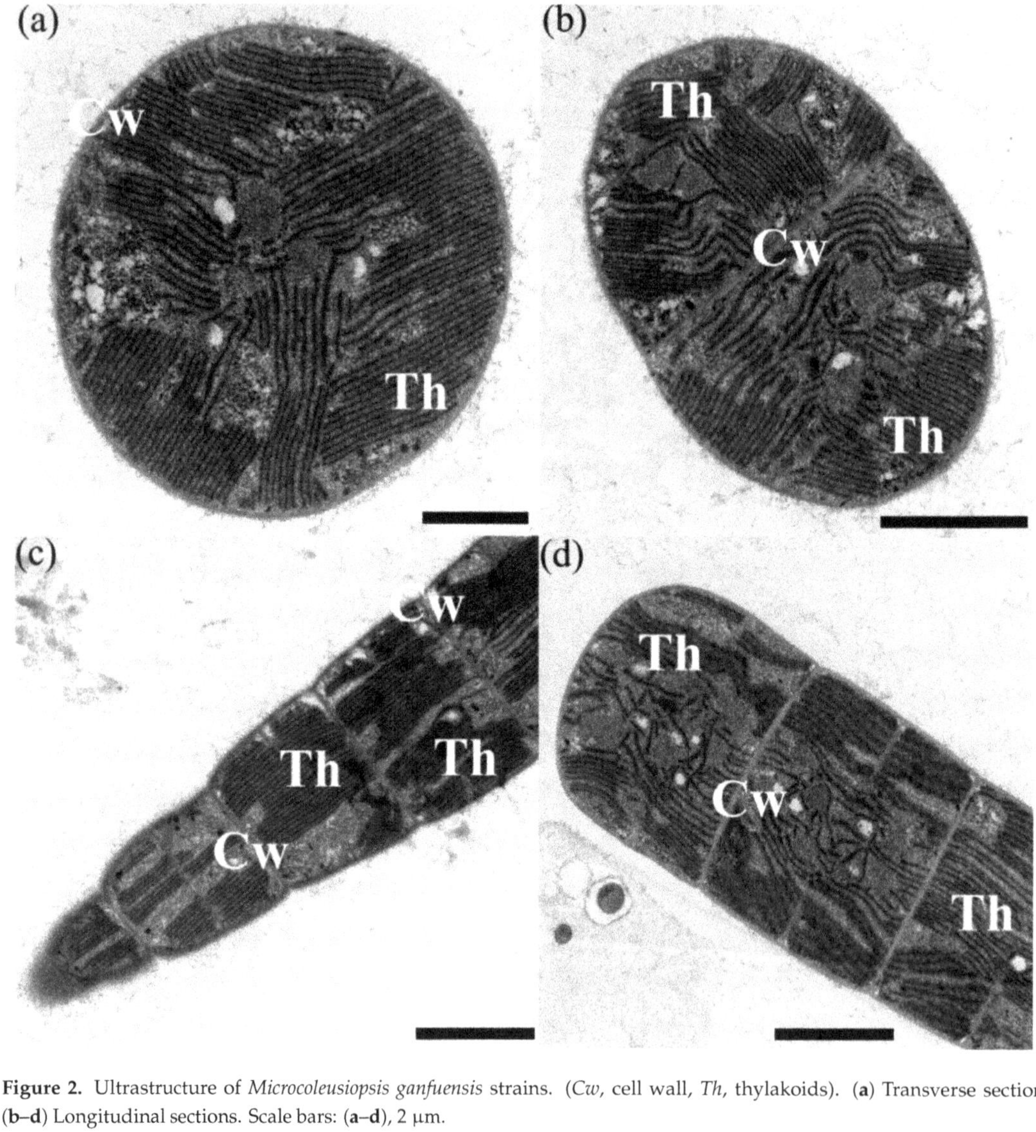

Figure 2. Ultrastructure of *Microcoleusiopsis ganfuensis* strains. (*Cw*, cell wall, *Th*, thylakoids). (**a**) Transverse section. (**b–d**) Longitudinal sections. Scale bars: (**a–d**), 2 μm.

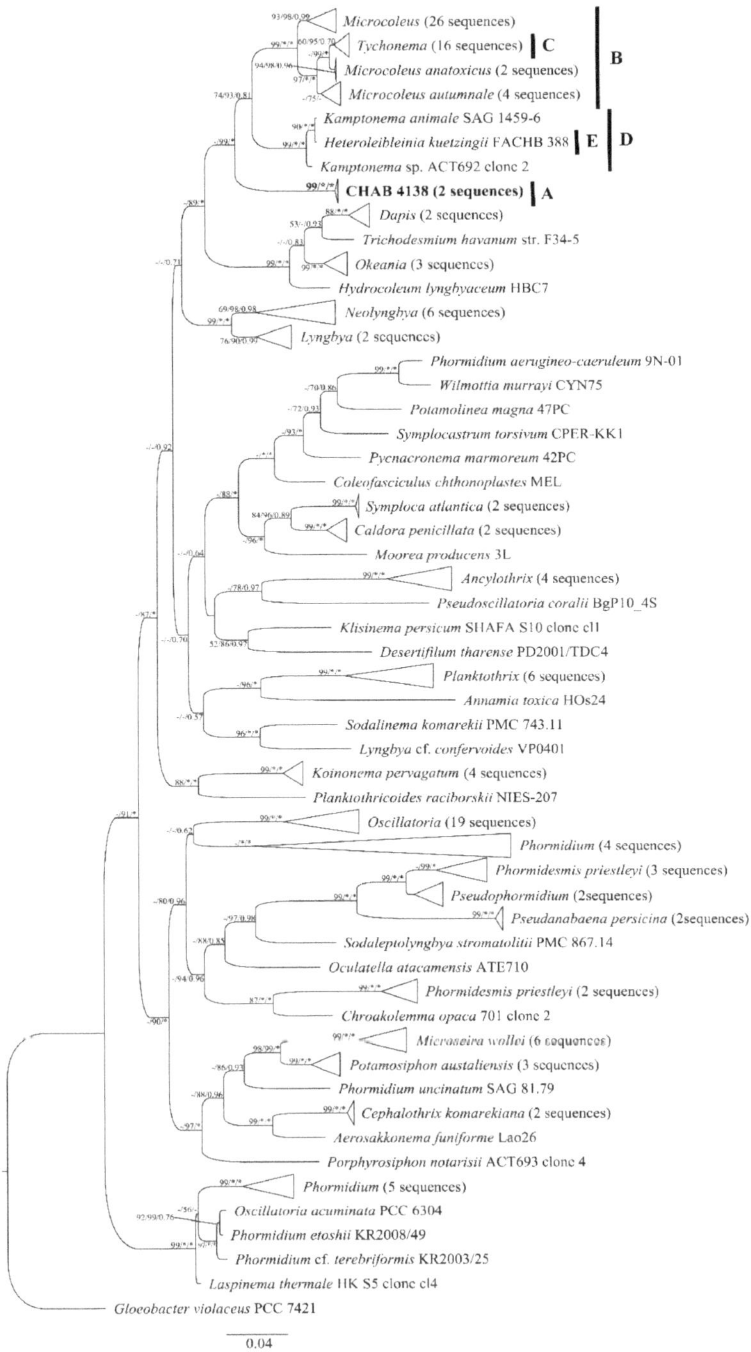

Figure 3. Bayesian inference (BI) phylogenetic tree of *Microcoleusiopsis ganfuensis* CHAB 4138 based on 16S rRNA gene sequences. Bootstrap values greater than 50% are showed on the BI tree for NJ/ML methods and Bayesian posterior probabilities. **A–E** represent *Microcoleusiopsis ganfuensis* strains, *Microcoleus* strains, *Tychonema* strains, *Kamptonema* strains and *Heteroleibleinia* strain, respectively. "*" indicates bootstrap values of 100 in ML and NJ and BI posterior probabilities of 1.00. The novel filamentous strains of this study indicate in bold. Bar, 0.04.

Table 1. Sequence similarity comparison of the 16S rRNA gene between *Microcoleusiopsis ganfuensis* strains and its closed species and genera. Similarity = [1 − (p-distance)] * 100.

Strains	1	2	3	4	5	6	7	8	9	10	11	12	13	14	15	16	17	18	19	20	21	22	23	24
1. *Microcoleusiopsis ganfuensis* CHAB 4138 clone 1																								
2. *Microcoleusiopsis ganfuensis* CHAB 4138 clone 2	99.91																							
3. *Microcoleus pseudautumnalis* Ak1609	94.70	94.78																						
4. *Microcoleus vaginatus* CSU-U-KK1	94.96	95.04	99.39																					
5. *Microcoleus anatoxicus* PTRS-2	94.26	94.35	98.70	98.35																				
6. *Microcoleus autumnale* SAG 78.79	94.09	94.17	98.52	98.00	99.13																			
7. *Oscillatoria princeps* CCALA 1115 clone F3	92.00	92.09	93.04	92.61	92.78	92.52																		
8. *Phormidium etoshii* KR2008/49	91.74	91.83	92.70	92.78	92.70	92.78	92.61																	
9. *Tychonema bourrellyi* FEM GT529	94.26	94.35	98.61	98.09	99.74	99.22	92.78	92.52																
10. *Tychonema bornetii* NIVA-CYA 60	94.26	94.35	98.61	98.09	99.74	99.22	92.78	92.52	100.00															
11. *Kamptonema animale* SAG 1459-6	94.43	94.52	96.17	95.74	95.91	95.57	93.39	91.57	95.74	95.74														
12. *Heteroleibleinia kutzingii* FACHB 388	94.43	94.52	96.17	95.74	95.91	95.57	93.39	91.57	95.74	95.74	100.00													
13. *Lyngbya hieronymusii* CN4-3	93.83	93.91	93.39	93.65	93.30	93.65	91.91	93.30	93.30	93.30	93.30	93.30												
14. *Neolyngbya granulosa* ALCB 114393	94.09	94.17	93.13	93.39	92.35	92.52	91.22	92.78	92.17	92.17	92.61	92.61	96.26											
15. *Coleofasciculus chthonoplastes* MEL	92.09	92.17	92.52	92.78	92.26	92.26	92.26	91.48	92.09	92.09	93.83	93.83	93.39	93.39										
16. *Hydrocoleum lyngbyaceum* HBC7	92.96	93.04	93.74	93.48	92.96	93.13	91.57	91.74	92.96	92.96	93.57	93.57	93.39	93.30	92.52									
17. *Okeania plumata* NAC8-45	93.13	93.22	94.09	93.65	93.83	94.00	90.87	91.74	93.83	93.83	94.35	94.35	93.39	92.96	92.43	97.30								
18. *Symploca atlantica* PCC 8002	92.61	92.70	93.30	92.96	93.13	92.61	92.61	91.39	92.87	92.87	93.48	93.48	92.87	92.52	94.00	92.09	92.70							
19. *Wilmottia murrayi* CYN75	92.43	92.52	93.04	93.22	93.22	92.87	91.91	90.78	92.96	92.96	93.22	93.22	92.52	92.26	94.09	91.65	92.70	93.74						
20. *Aerosakkonema funiforme* Lao26	92.00	92.09	93.04	93.30	92.09	92.26	92.70	90.78	92.00	92.00	92.78	92.78	92.09	92.09	92.87	92.35	91.65	92.43	92.09					
21. *Dapis pnigousa* BCBC12-12	92.78	92.87	93.91	93.48	93.65	93.83	91.22	91.13	93.65	93.65	93.91	93.91	93.13	92.17	92.70	97.04	97.04	92.09	92.09	91.39				
22. *Moorea producens* 3L	92.70	92.78	91.48	91.04	91.30	91.30	91.65	90.87	91.30	91.30	93.04	93.04	92.17	92.00	93.57	91.74	92.17	94.26	92.17	90.78	91.57			
23. *Caldora penicillata* HMC13-9	92.52	92.61	92.78	92.35	92.26	92.26	92.87	92.09	92.17	92.17	93.30	93.30	92.52	92.78	94.87	92.09	92.26	96.09	93.65	92.17	92.43	94.43		
24. *Laspinema thermale* HK S5 clone cl4	92.43	92.52	93.30	93.57	93.04	93.13	93.22	98.96	92.87	92.87	92.26	92.26	93.65	93.48	91.57	92.52	92.09	91.91	91.13	91.39	91.39	91.57	92.52	
25. *Trichodesmium havanum* str. F34-5	92.09	92.17	93.65	93.22	93.04	93.39	90.70	91.57	93.04	93.04	93.13	93.13	92.35	92.35	91.57	96.52	97.22	91.57	91.57	90.87	97.22	90.78	91.83	91.83

Besides, four type of cyanotoxin genes were not detected in *Microcoleusiopsis ganfuensis* CHAB 4138 and we did not obtain any PCR products by using the primers responsible for the synthesis genes for these toxins (microcystins, cylindrospermopsin, paralytic shellfish toxins and anatoxins).

3.3. Analyses of ITS between 16S and 23S rRNA Gene and Secondary Structures

The partial 16S–23S ITS sequences of *Microcoleusiopsis ganfuensis* CHAB 4138 were obtained with a total length of 761 bp in this study (Table 2), and they were used, together with seven species clones from three genera including *Microcoleus*, *Oscillatoria* and *Coleofasciculus* downloaded from NCBI, to construct the ITS secondary structures. In general, all sequences contained both tRNA ^{Ile} and tRNA ^{Ala} (Table 2). As the most conserved structure, the D1–D1′ helix (Figure 4) of strain CHAB 4138 was similar to those of several species of close genera like *Microcoleus vaginatus* CSU-U-KK1, *Microcoleus vaginatus* PTRS-2, *Microcoleus autumnale* SAG 78.79, and *Oscillatoria princeps* CCALA 1115 in basal and apical stem–loop, but significantly different from those of *Microcoleus pseudautumnalis* Ak1609 and *Coleofasciculu chthonoplastes* MEL. In the basal stem of strain CHAB 4138 and other six species mentioned above, there was a 4-bp helix (a 6-bp helix in *C. chthonoplastes* MEL), followed by a small unidirectional bulge, and the apical structures contained a 4-bp helix (5-bp in *M. pseudautumnalis* Ak1609, *M. vaginatus* PTRS-2 and *M. autumnale* SAG 78.79; 3-bp in *C. chthonoplastes* MEL) with a 15-bp loop (5-bp in *M. pseudautumnalis* Ak1609 and *C. chthonoplastes* MEL; 14-bp in *M. vaginatus* CSU-U-KK1 and *M. autumnale* SAG 78.79; 16-bp and 17-bp in *M. vaginatus* PTRS-2 and *O. princeps* CCALA 1115, respectively).

The Box–B (Figure 5) and V3 (Figure 6) helices of CHAB 4138 were conspicuously different from those of other related genera in sequence length and stem–loop structures (Table 2). CHAB 4138 had its own unique Box–B helix, consisting of one 4-bp helix, two 3-bp helices, two 6-bp helices, two small unidirectional bulges, one 1:1 bp base bilateral bulge, one 2:4 bp base bilateral bulge, and one 4-bp apical loop. Whereas the other six related species had five Box–B helices types, especially the genus *Microcoleus* could be divided into three types, represented by *M. pseudautumnalis* Ak1609, *M. vaginatus* CSU-U-KK1, and *M. vaginatus* PTRS-2 with *M. autumnale* SAG 78.79, respectively. No regular patterns were found for V3 helices between CHAB 4138 and other seven filamentous species. The studied strain CHAB 4138 only had a 5-bp helix followed by a 6-bp apical loop, which significantly differed from other species.

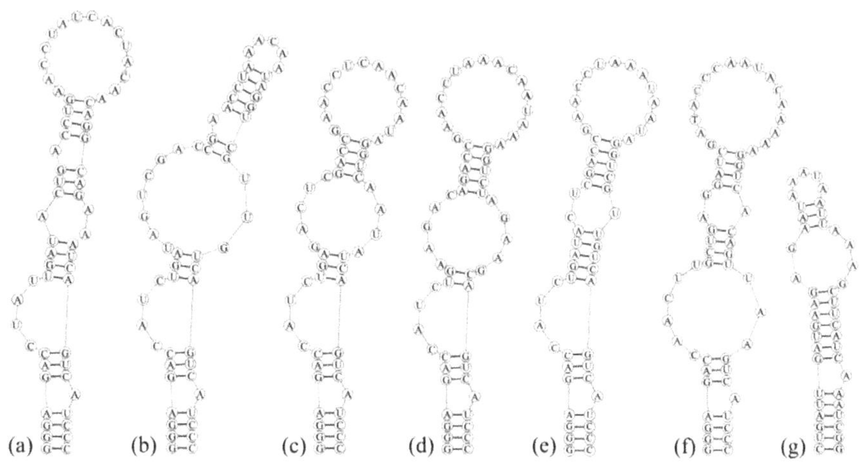

Figure 4. D1–D1′ helix in *Microcoleusiopsis ganfuensis* and other six related species. (**a**). *Microcoleusiopsis ganfuensis* CHAB 4138. (**b**). *Microcoleus pseudautumnalis* AK1609. (**c**). *Microcoleus vaginatus* CSU-U-KK1. (**d**). *Microcoleus vaginatus* PTRS-2. (**e**). *Microcoleus autumnale* SAG 78.79. (**f**) *Oscillatoria princeps* CCALA 1115. (**g**). *Coleofasciculu chthonoplastes* MEL.

Table 2. Analyses on ITS of 16S–23S region for *Microcoleusiopsis ganfuensis* strains.

Organisms	GenBank	ITS Total Length (nt)	D1–D1′ Helix Length (nt)	D2 Region	$tRNA^{Ile}$	$tRNA^{Ala}$	Box B Helix Length (nt)	Box A Spacer	V3 Helix Length (nt)
Microcoleusiopsis ganfuensis **CHAB 4138**	OK422506	761	60	CTTTCAAACTAG	+	+	58	GAACCTTGAAAA	16
Microcoleus pseudautumnalis Ak1609	LC486302	545	58	CTTTCAAACTAT	+	+	38	GAACCTTGAAAA	39
Microcoleus vaginatus CSU-U-KK1	EF667962	586	60	CTTTCAAACTAT	+	+	40	GAACCTTGAAAA	40
Microcoleus anatoxicus PTRS-2	MT013208	548	63	CTTTCAAACTAT	+	+	37	GAACCTTGAAAA	33
Oscillatoria princeps CCALA 1115 clone F3	MG255277	746	60	CTTTCAAACTAA	+	+	37	GAACCTTGAAAA	62
Microcoleus autumnale SAG 78.79	AM778717	573	58	CTTTCAAACTAT	+	+	53	GAACCTTGAAAA	31
Coleofasciculus chthonoplastes MEL	EF654038	526	44	CTTTCAAACTGG	+	+	27	GAACCTTGAAAA	37

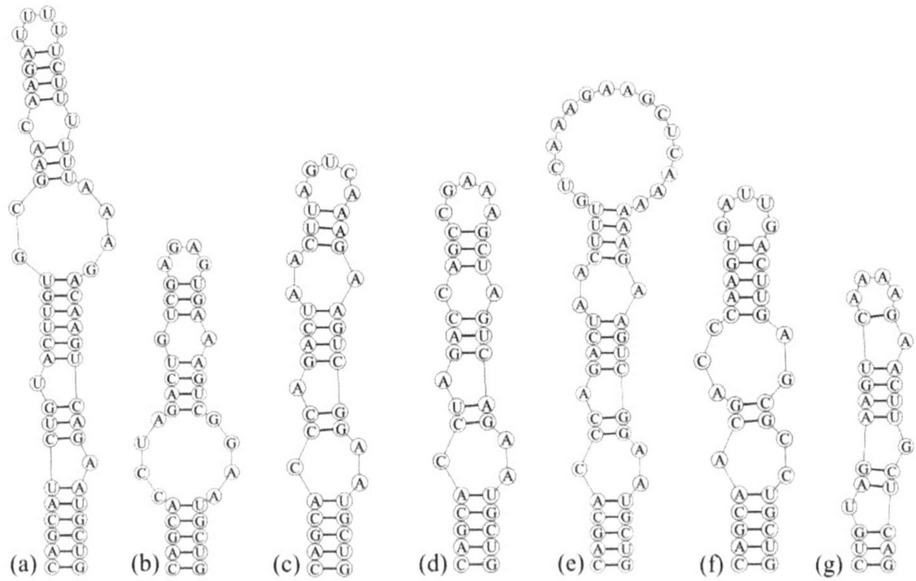

Figure 5. Box–B helix in *Microcoleusiopsis ganfuensis* and other six related species. (**a**). *Microcoleusiopsis ganfuensis* CHAB 4138. (**b**). *Microcoleus pseudautumnalis* AK1609. (**c**). *Microcoleus vaginatus* CSU-U-KK1. (**d**). *Microcoleus vaginatus* PTRS-2. (**e**). *Microcoleus autumnale* SAG 78.79. (**f**) *Oscillatoria princeps* CCALA 1115. (**g**). *Coleofasciculu chthonoplastes* MEL.

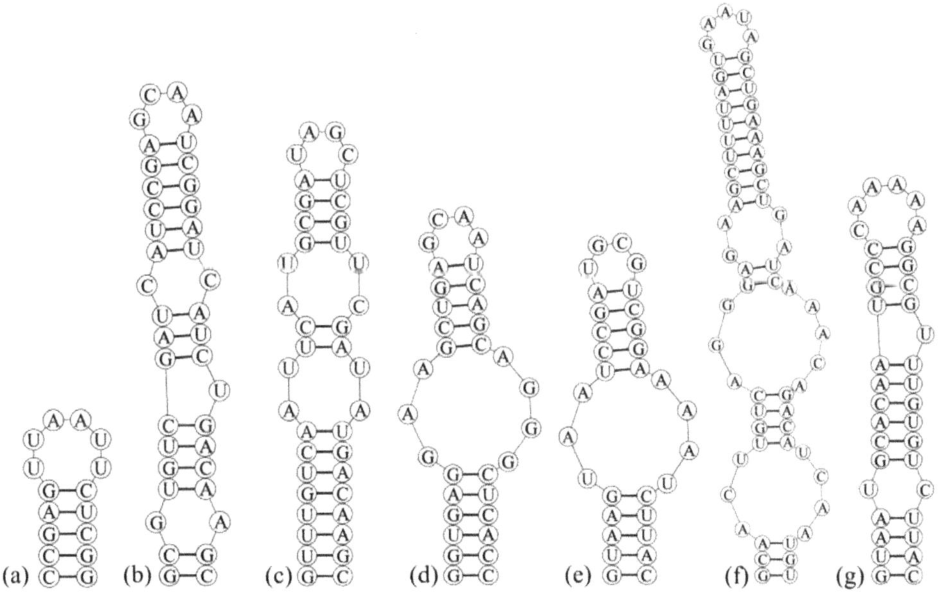

Figure 6. V3 helix in *Microcoleusiopsis ganfuensis* and other six related species. (**a**). *Microcoleusiopsis ganfuensis* CHAB 4138. (**b**). *Microcoleus pseudautumnalis* AK1609. (**c**). *Microcoleus vaginatus* CSU-U-KK1. (**d**). *Microcoleus vaginatus* PTRS-2. (**e**). *Microcoleus autumnale* SAG 78.79. (**f**) *Oscillatoria princeps* CCALA 1115. (**g**). *Coleofasciculu chthonoplastes* MEL.

4. Discussion

Benthic cyanobacteria can grow in patches on the attached substrates to form algal mats, and they are important primary producers in river and lake communities. Those mat-forming taxa mainly include *Chroococcus*-like cyanobacteria containing single cell and colonies with mucilage, a considerable number of filamentous *Oscillatoria*-like cyanobacteria without cell differentiation, *Nostoc*-like cyanobacteria with cell differentiation, *Stigonema*-like cyanobacteria with true branches, and *Chamaesiphon*-like cyanobacteria forming endospores [33,34]. During the field investigation, Oscillatorean cyanobacteria were found to be the main dominant species in the algal mats, and their characterization and correct identification based on the modern taxonomic system should be emphasized. It is expected that the ideal cyanobacteria genera and species in the current cyanobacterial taxonomy should be monophyletic, which means the need to make constant revisions to have this goal achieved [8,14,35–38].

Previous taxonomy of cyanobacteria was morphology based system, especially at a high rank, by using morphological characteristics such as the size of filaments and cells, polarity and branch types of filaments [34]. However, with the introduction of molecular biology methods, typical characteristics were proven to appear and lose many times during evolutionary process of cyanobacteria, making the distinction between some species of critical morphological characteristics increasingly blurred [34,39]. In this study, the benthic filamentous cyanobacterium isolated from the algal mats of the Ganfu Channel was difficult to be accurately classified based on morphological characteristics only such as the shapes of cells and filaments, types of end cells, and facultative presence of sheaths. Therefore, the polyphasic approach exhibited its power to determine the correct taxonomic attribution and phylogenetic relationship of this novel filamentous cyanobacterium.

The phylogenetic tree based on the 16S rRNA gene sequences indicated that the position of *Microcoleusiopsis ganfuensis* CHAB 4138 was close to the filamentous genera *Microcoleus*, *Tychonema*, *Kamptonema*, and *Heteroleibleinia*; however, the strains within the genus *Microcoleus* formed three small clades, and further clustered into a large clade with the strains of *Tychonema* (Figure 3 clade B). *Kamptonema*, originally described as "*Oscillatoria animalis*" (Figure 3 clade D), is a newly established filamentous cyanobacterial genus of family Microcoleaceae, by separating from genus *Phormidium* in recent years [40]. In addition, a geosmin producer [41]—*Heteroleibleinia kuetzingii* FACHB 388 (one filamentous strain originally identified as *Lyngbya kuetzingii* at the FACHB Culture Collection)—was shown to be clustered with *Kamptonema* strains in family Microcoleaceae (Figure 3 clade E), supported by NJ/ML/BI approaches as 99%/100%/1.00, and such a result implied that this strain may need to be re-identified as belonging to the genus *Kamptonema*. Comparison of 16S rRNA sequences showed that the two clones of CHAB 4138 clustered a well-supported independent cluster (cluster A), with a maximum similarity of 16S rRNA sequences as 95.04% to the existing cyanobacterial taxa, below the threshold of bacterial genus; therefore, this strain probably represents a new cyanobacterial taxon [42–44].

As one of the effective tools to distinguish cyanobacterial species, the secondary structures of ITS including D1–D1′, Box–B, and V3 helices can also distinguish *Microcoleusiopsis ganfuensis* from other filamentous cyanobacteria [45–49]. The D1–D1′ (Figure 4), Box–B (Figure 5), and V3 (Figure 6) helices of *M. ganfuensis* were significantly different from other related genera (*Microcoleus*, *Oscillatoria*, and *Coleofasciculus*) in stem–loop structures. It is worth mentioning that there were three configurations of the stem–loop structure of Box–B helix in multiple strains of the genus *Microcoleus* in this study, one as *M. pseudautumnalis* Ak1609, one as *M. vaginatus* CSU-U-KK1, and the third as *M. vaginatus* PTRS-2 and *M. autumnale* SAG 78.79—implying some relationship between ITS secondary structures and the ability of secondary metabolites.

Nowadays, the biological proliferation dominated by benthic filamentous cyanobacteria in rivers, lakes, and channels worldwide is frequently increasing, and the harmful effects caused by benthic cyanobacteria has gradually become a problem which cannot be ignored [50–52]. *Microcoleus* and *Tychonema* species were widely reported as toxigenic cyanobacteria since they were found to produce neurotoxic anatoxin-a/homoanatoxin-a in USA [17,53], Italy [54,55], and Germany [56]. Species *Kamptonema formosum*, a member of the newly established genus, was even found to form microcystins, anatoxin-a/homoanatoxin-a, and other anatoxin congeners in a recent published paper [57]. However, in this study, *Microcoleusiopsis ganfuensis* was shown to lack the synthesis genes of four type of cyanotoxins, indicating that it may not be a potential producer of cyanotoxins. Furthermore, the morphological observation based on both field sample and the cultured strain showed no bundle formation of trichomes covered by a sheath, confirming the distinction of *M. ganfuensis* from the type species of *Microcoleus*. Thus, the establishment of the new genus/species *Microcoleusiopsis ganfuensis* was well supported by the combination of morphology, 16S rRNA gene sequence, and 16S–23S ITS secondary structures.

Author Contributions: Conceptualization, R.G. and W.L.; methodology, R.G.; software, R.G.; formal analysis, W.L. and X.G.; investigation, R.G.; resources, A.C.; data curation, R.G.; writing—original draft preparation, R.G.; writing—review and editing, R.L.; visualization, H.L.; funding acquisition, G.Y. And R.L. All authors have read and agreed to the published version of the manuscript.

Funding: This study is funded by the NSFC projects (NSFC no. 51779247 and NSFC no. 31970219).

Institutional Review Board Statement: Not applicable.

Data Availability Statement: Not applicable.

Acknowledgments: We would like to thank the calculating resources supported by the Wuhan Branch, Supercomputing Center, Chinese Academy of Sciences, China for the computing resources. The first author thanks Zhenfei Xing and Yuan Xiao for technical assistance with TEM images.

Conflicts of Interest: The authors declare no conflict of interest.

References

1. Garcia–Pichel, F.; Lopez–Cortes, A.; Nubel, U. Phylogenetic and morphological diversity of cyanobacteria in soil desert crusts from the Colorado plateau. *Appl. Environ. Microbiol.* **2001**, *67*, 1902–1910. [CrossRef] [PubMed]
2. Komárek, J. Several problems of the polyphasic approach in the modern cyanobacterial system. *Hydrobiologia* **2018**, *811*, 7–17. [CrossRef]
3. Mareš, J.; Strunecký, O.; Bučinská, L.; Wiedermannová, J. Evolutionary patterns of thylakoid architecture in cyanobacteria. *Front. Microbiol.* **2019**, *10*, 277. [CrossRef]
4. Comte, K.; Holland, D.P.; Walsby, A.E. Changes in cell turgor pressure related to uptake of solutes by *Microcystis* sp. strain 8401. *FEMS Microbiol. Ecol.* **2007**, *61*, 399–405. [CrossRef] [PubMed]
5. Sciuto, K.; Rascio, N.; Andreoli, C.; Moro, I. Polyphasic characterization of ITD–01, a cyanobacterium isolated from the Ischia Thermal District (Naples, Italy). *Fottea.* **2011**, *11*, 31–39. [CrossRef]
6. Komárek, J.; Kaštovský, J.; Mareš, J.; Johansen, J.R. Taxonomic classification of cyanoprokaryotes (*Cyanobacterial genera*) 2014, using a polyphasic approach. *Preslia* **2014**, *86*, 295–335.
7. Dvořák, P.; Poulíčková, A.; Hašler, P.; Belli, M.; Casamatta, D.A.; Papini, A. Species concepts and speciation factors in cyanobacteria, with connection to the problems of diversity and classification. *Biodivers. Conserv.* **2015**, *24*, 739–757. [CrossRef]
8. Komárek, J.; Johansen, J.R.; Šmarda, J.; Strunecký, O. Phylogeny and taxonomy of *Synechococcus*-like cyanobacteria. *Fottea* **2020**, *20*, 171–191. [CrossRef]
9. Gomont, M. Monographie des Oscillariées (*Nostocacées homocystées*). *Annal. Sci. Nat. Bot.* **1892**, *16*, 91–264.
10. Geitler, L. Schizophyta (Klasse Schizophyceae). In *Natürliche Pflanzenfamilien*; Engler, A., Prantl, K., Eds.; Duncker & Humblot: Berlin, Germany, 1942; pp. 1–232.
11. Drouet, F. Revision of the classifification of the Oscillatoriaceae. *Monogr. Acad. Nat. Sci. Phil.* **1968**, *15*, 1–370.
12. Anagnostidis, K.; Komárek, J. Modern approach to the classification system of cyanophytes 3-Oscillatoriales. *Algol. Stud.* **1988**, *50–53*, 327–478.
13. Boyer, S.L.; Johansen, J.R.; Flechtner, V.R. Phylogeny and Genetic Variance in Terrestrial *Microcoleus* (Cyanophyceae) Species Based on Sequence Analysis of the 16S rRNA Gene and Associated 16S–23S ITS Region. *J. Phycol.* **2002**, *38*, 1222–1235. [CrossRef]
14. Siegesmund, M.A.; Johansen, J.R.; Karsten, U.; Friedl, T. *Coleofasciculus* gen. nov. (Cyanobacteria): Morphological and molecular criteria for revision of the genus *Microcoleus* Gomont. *J. Phycol.* **2008**, *44*, 1572–1585. [CrossRef] [PubMed]
15. Strunecký, O.; Komárekjeffrey, J.; Johansenjeffrey, R. Molecular and morphological criteria for revision of the genus *Microcoleus* (Oscillatoriales, Cyanobacteria). *J. Phycol.* **2013**, *49*, 1167–1180. [CrossRef] [PubMed]
16. Niiyama, Y.; Tuji, A. *Microcoleus pseudautumnalis* sp. nov. (Cyanobacteria, Oscillatoriales) producing 2–methylisoborneol. *Bull. Natl. Mus. Nat. Sci. Ser. B* **2019**, *45*, 93–101.

17. Conklin, K.Y.; Stancheva, R.; Otten, T.G. Molecular and morphological characterization of a novel dihydroanatoxin-a producing *Microcoleus* species (cyanobacteria) from the Russian River, California, USA. *Harmful Algae* **2020**, *93*, 101767. [CrossRef]
18. Ichimura, T. Isolation and culture methods of algae. *Methods Phycol. Stud.* **1979**, *2*, 294–305.
19. Spurr, A.R. A low–viscosity epoxy resin embedding medium for electron microscopy. *J. Ultrastruct. Res.* **1969**, *26*, 31–43. [CrossRef]
20. Neilan, B.A.; Jacobs, D.; Goodman, A.E. Genetic diversity and phylogeny of toxic cyanobacteria determined by DNA polymorphisms within the phycocyanin locus. *Appl. Environ. Microbiol.* **1995**, *61*, 3875–3883. [CrossRef]
21. Edwards, U.; Rogall, T.; Blöcker, H.; Emde, M.; Böttger, E.C. Isolation and direct complete nucleotide determination of entire genes. Characterization of a gene coding for 16S ribosomal RNA. *Nucleic Acids Res.* **1989**, *17*, 7843–7853. [CrossRef]
22. Gkelis, S.; Rajaniemi, P.; Vardaka, E.; Moustaka–gouni, M.; Lanaras, T.; Sivonen, K. *Limnothrix redekei* (Van Goor) Meffert (Cyanobacteria) strains from Lake Kastoria, Greece form a separate phylogenetic group. *Microb. Ecol.* **2005**, *49*, 176–182. [CrossRef] [PubMed]
23. Jungblut, A.D.; Neilan, B.A. Molecular identification and evolution of the cyclic peptide hepatotoxins, microcystin and nodularin, synthetase genes in three orders of cyanobacteria. *Arch. Microbiol.* **2006**, *185*, 107–114. [CrossRef]
24. Al-Tebrineh, J.; Pearson, L.A.; Yasar, S.A.; Neilan, B.A. A multiplex qPCR targeting hepato- and neurotoxigenic cyanobacteria of global significance. *Harmful Algae* **2012**, *15*, 19–25. [CrossRef]
25. Mcgregor, G.B.; Sendall, B.C. Phylogeny and toxicology of *Lyngbya wollei* (Cyanobacteria, Oscillatoriales) from north-eastern Australia, with a description of *Microseira* gen. nov. *J. Phycol.* **2015**, *51*, 109–119. [CrossRef]
26. Rantala-Ylinen, A.; Känä, S.; Wang, H. Anatoxin-a synthetase gene cluster of the cyanobacterium *Anabaena* sp. strain 37 and molecular methods to detect potential producers. *Appl. Environ. Microbiol.* **2011**, *77*, 7271–7278. [CrossRef]
27. Katoh, K.; Standley, D.M. MAFFT multiple sequence alignment software version 7: Improvements in performance and usability. *Mol. Biol. Evol.* **2013**, *30*, 772–780. [CrossRef] [PubMed]
28. Ronquist, F.; Teslenko, M.; Van Der Mark, P.; Ayres, D.L.; Darling, A.; Höhna, S.; Huelsenbeck, J.P. MrBayes 3.2: Efficient Bayesian phylogenetic inference and model choice across a large model space. *Syst. Biol.* **2012**, *61*, 539–542. [CrossRef]
29. Guindon, S.; Dufayard, J.F.; Lefort, V.; Anisimova, M.; Hordijk, W.; Gascuel, O. New Algorithms and Methods to Estimate Maximum–Likelihood Phylogenies: Assessing the Performance of PhyML 3.0. *Syst. Biol.* **2010**, *59*, 307–321. [CrossRef]
30. Kumar, S.; Stecher, G.; Tamura, K. MEGA7: Molecular Evolutionary Genetics Analysis version 7.0 for bigger datasets. *Mol. Biol. Evol.* **2016**, *33*, 1870–1874. [CrossRef]
31. Page, R.D.M. TreeView: An application to display phylogenetic trees on personal computers. *Comput. Appl. Biosci.* **1996**, *12*, 357–358. [PubMed]
32. Mathews Lab. RNAstructure Version 5.6. Available online: http://rna.urmc.rochester.edu/RNAstructure.html (accessed on 13 May 2021).
33. Rippka, R.; Deruelles, J.; Waterbury, J.B.; Herdman, M.; Stanier, R.Y. Generic assignments, strain histories and properties of pure cultures of cyanobacteria. *J. Gen. Microbiol.* **1979**, *111*, 1–61. [CrossRef]
34. Komárek, J. Cyanoprokaryota—3.Teil/ 3rd Part: Heterocytous Genera. In *Süsswasserflora von Mitteleuropa Freshwater Flora of Central Europe 19/2*; Büdel, B., Krienitz, L., Gärtner, G., Schagerl, M., Eds.; Springer/Spektrum: Heidelberg, Germany, 2013.
35. Anagnostidis, K.; Komárek, J. Modern approach to the classification system of cyanophytes. 1-Introduction. *Algol. Stud. Arch. Für Hydrobiol.* **1985**, *38–39*, 291–302.
36. Johansen, J.R.; Casamatta, D.A. Recognizing cyanobacterial diversity through adoption of a new species paradigm. *Algol. Studies* **2005**, *117*, 71–93. [CrossRef]
37. Casamatta, D.A.; Johansen, J.R.; Vis, M.L. Molecular and morphological characterization of ten polar and near–polar strains with the Oscillatoriales (Cyanobacteria). *J. Phycol.* **2010**, *41*, 421–438. [CrossRef]
38. Iii, R.; Johansen, J.R.; Kovácik, L. A unique Pseudanabaenalean (cyanobacteria) genus *Nodosilinea* gen. nov. based on morphological and molecular data1. *J. Phycol.* **2011**, *47*, 1397–1412.
39. Gugger, M.; Molica, R.; Le, B.B. Genetic diversity of *Cylindrospermopsis* strains (Cyanobacteria) isolated from four continents. *Appl. Environ. Microbiol.* **2005**, *71*, 1097–1100. [CrossRef] [PubMed]
40. Strunecký, O.; Komárek, J.; Šmarda, J. *Kamptonema* (Microcoleaceae, Cyanobacteria), a new genus derived from the polyphyletic *Phormidium* on the basis of combined molecular and cytomorphological markers. *Preslia* **2014**, *86*, 193–207.
41. Wang, Z.J.; Song, G.F.; Li, Y.G. The diversity, origin, and evolutionary analysis of geosmin synthase gene in cyanobacteria. *Sci. Total. Environ.* **2019**, *689*, 789–796. [CrossRef] [PubMed]
42. Wayne, L.G.; Brenner, D.J.; Colwell, R.R.; Grimont, P.A.D.; Kandler, O.; Krichevsky, M.I.; Starr, M.P. Report of the ad hoc committee on reconciliation of approaches to bacterial systematics. *Int. J. Syst. Evol. Microbiol.* **1987**, *37*, 463–464. [CrossRef]
43. Stackebrandt, E.; Goebel, B.M. Taxonomic note: A placefor DNA–DNA reassociation and 16S rRNA sequence analysisin the present species definition in bacteriology. *Int. J. Syst. Bacteriol.* **1994**, *44*, 846–849. [CrossRef]
44. Stackebrandt, E.; Ebers, J. Taxonomic parameters revisited: Tarnished gold standards. *Microbiology* **2006**, *33*, 152–155.
45. Iteman, I.; Rippka, R.; Marsac, N.T.D.; Herdman, M. Comparison of conserved structural and regulatory domains within divergent 16s rRNA–23s rRNA spacer sequences of cyanobacteria. *Microbiology* **2000**, *146*, 1275–1286. [CrossRef] [PubMed]
46. Vaccarino, M.A.; Johansen, J.R. *Brasilonema angustatum* sp. nov. (Nostocales), a new filamentous cyanobacterial species from the Hawaiian Islands. *J. Phycol.* **2012**, *48*, 1178–1186. [CrossRef] [PubMed]

47. Osorio–Santos, K.; Pietrasiak, N.; Bohunická, M.; Miscoe, L.H.; Kováčik, L.; Martin, M.P.; Johansen, J.R. Seven new species of *Oculatella* (Pseudanabaenales, Cyanobacteria): Taxonomically recognizing cryptic diversification. *Eur. J. Phycol.* **2014**, *49*, 450–470. [CrossRef]
48. Pietrasiak, N.; Mühlsteinová, R.; Siegesmund, M.A.; Johansen, J.R. Phylogenetic placement of *Symplocastrum* (Phormidiaceae, Cyanophyceae) with a new combination *S. californicum* and two new species: *S. flechtnerae* and *S. torsivum*. *Phycologia* **2014**, *53*, 529–541. [CrossRef]
49. Mareš, J. Multilocus and SSU rRNA gene phylogenetic analyses of available cyanobacterial genomes, and their relation to the current taxonomic system. *Hydrobiologia* **2018**, *811*, 19–34. [CrossRef]
50. Quiblier, C.; Wood, S.; Echenique-Subiabre, I.; Heath, M.; Villeneuve, A.; Humbert, J.-F. A review of current knowledge on toxic benthic freshwater cyanobacteria—Ecology, toxin production and risk management. *Water Res.* **2013**, *47*, 5464–5479.
51. Fetscher, A.E.; Howard, M.D.; Stancheva, R.; Kudela, R.M.; Stein, E.D.; Sutula, M.A.; Busse, L.B.; Sheath, R.G. Wadeable streams as widespread sources of benthic cyanotoxins in California, USA. *Harmful Algae* **2015**, *49*, 105–116. [CrossRef]
52. Echenique-Subiabre, I.; Dalle, C.; Duval, C.; Heath, M.W.; Coute, A.; Wood, S.A.; Humbert, J.-F.; Quiblier, C. Application of a spectrofluorimetric tool (bbe BenthoTorch) for monitoring potentially toxic benthic cyanobacteria in rivers. *Water Res.* **2016**, *101*, 341–350. [CrossRef]
53. Bouma-Gregson, K.; Olm, M.R.; Probst, A.J.; Anantharaman, K.; Power, M.E.; Banfield, J.F. Impacts of microbial assemblage and environmental conditions on the distribution of anatoxin-a producing cyanobacteria within a river network. *ISME J.* **2019**, *13*, 1618–1634. [CrossRef] [PubMed]
54. Shams, S.; Capelli, C.; Cerasino, L.; Ballot, A.; Dietrich, D.R.; Sivonen, K.; Salmaso, N. Anatoxin-a producing *Tychonema* (cyanobacteria) in European waterbodies. *Water Res.* **2015**, *69*, 68–79. [CrossRef] [PubMed]
55. Nico, S.; Leonardo, C.; Adriano, B.; Camilla, C. Planktic *Tychonema* (cyanobacteria) in the large lakes south of the alps: Phylogenetic assessment and toxigenic potential. *Fems Microbiol. Ecol.* **2016**, *92*, fiw155.
56. Jutta, F.; Camilla, B.; Britta, G.; Anja, H.; Roswitha, K.; Kinga, T. Fatal neurotoxicosis in dogs associated with tychoplanktic, anatoxin-a producing *Tychonema* sp. in mesotrophic lake tegel, Berlin. *Toxins* **2018**, *10*, 60.
57. Blahova, L.; Sehnal, L.; Lepsova-Skacelova, O.; Szmucova, V.; Babica, P.; Hilscherova, K.; Teikari, J.; Sivonen, K.; Blaha, L. Occurrence of cylindrospermopsin, anatoxin-a and their homologs in the southern Czech Republic–Taxonomical, analytical, and molecular approaches. *Harmful Algae* **2021**, *108*, 102101. [CrossRef] [PubMed]

diversity

MDPI

Article

Issi saaneq gen. et sp. nov.—A New Sauropodomorph Dinosaur from the Late Triassic (Norian) of Jameson Land, Central East Greenland †

Victor Beccari [1,2,3,*,‡], Octávio Mateus [1,2,‡], Oliver Wings [4,‡], Jesper Milàn [5,‡] and Lars B. Clemmensen [6,‡]

1 GeoBioTec, Department of Earth Sciences, Faculdade de Ciência e Tecnologia, Universidade Nova de Lisboa, 2829-516 Lisbon, Portugal; omateus@fct.unl.pt
2 Museu da Lourinhã, 2530-158 Lourinhã, Portugal
3 SNSB—Bayerische Staatssammlung für Paläontologie und Geologie, 80333 Munich, Germany
4 Natural Sciences Collections (ZNS), Martin Luther University Halle-Wittenberg, 06108 Halle, Germany; oliver.wings@zns.uni-halle.de
5 Geomuseum Faxe, Østsjællands Museum, 4640 Faxe, Denmark; jesperm@oesm.dk
6 Department of Geosciences and Natural Resource Management, University of Copenhagen, 1165 Copenhagen, Denmark; larsc@ign.ku.dk
* Correspondence: victor.beccari@gmail.com
† urn:lsid:zoobank.org:act:76F3ECA8-8B98-4735-8354-398B53A455EA; urn:lsid:zoobank.org:act:8DF8A871-C8D7-4944-A8A9-338BF82AB5CA; urn:lsid:zoobank.org:pub:8AB4D333-BD8E-40B3-B978-37AD481C20E3.
‡ urn:lsid:zoobank.org:author:CF510830-CFDF-4955-8874-2C09C24A123C (V.B.); urn:lsid:zoobank.org:author:9310955D-ACA9-4090-8699-3BB795F7B067 (O.M.); urn:lsid:zoobank.org:author:00B8FDE4-3E97-47B6-9D60-9BBD260615A6 (O.W.); urn:lsid:zoobank.org:author:76740A35-5E64-413B-A561-EAEEB024F79E (J.M.); urn:lsid:zoobank.org:author:04906BD1-B206-448B-94F1-ABBC6388EB10 (L.B.C.).

Citation: Beccari, V.; Mateus, O.; Wings, O.; Milàn, J.; Clemmensen, L.B. *Issi saaneq* gen. et sp. nov.—A New Sauropodomorph Dinosaur from the Late Triassic (Norian) of Jameson Land, Central East Greenland *Diversity* **2021**, *13*, 561. https://doi.org/10.3390/d13110561

Academic Editor: Eric Buffetaut

Received: 29 September 2021
Accepted: 29 October 2021
Published: 3 November 2021

Publisher's Note: MDPI stays neutral with regard to jurisdictional claims in published maps and institutional affiliations.

Copyright: © 2021 by the authors. Licensee MDPI, Basel, Switzerland. This article is an open access article distributed under the terms and conditions of the Creative Commons Attribution (CC BY) license (https://creativecommons.org/licenses/by/4.0/).

Abstract: The Late Triassic (Norian) outcrops of the Malmros Klint Formation, Jameson Land (Greenland) have yielded numerous specimens of non-sauropod sauropodomorphs. Relevant fossils were briefly reported in 1994 and were assigned to *Plateosaurus trossingensis*. However, continuous new findings of early non-sauropod sauropodomorphs around the globe facilitate comparisons and allow us to now revise this material. Here, the non-sauropod sauropodomorph *Issi saaneq* gen. et sp. nov. is described based on two almost complete and articulated skulls. The two skulls represent a middle-stage juvenile and a late-stage juvenile or subadult. *Issi saaneq* differs from all other sauropodomorphs by several unique traits: (1) a small foramen at the medial surface of the premaxilla; (2) an anteroposteriorly elongated dorsoposterior process of the squamosal; (3) a relatively high quadrate relative to rostrum height; (4) a well-developed posterodorsal process of the articular. These features cannot be explained by taphonomy, ontogeny, or intraspecific variation. *Issi saaneq* shows affinities to Brazilian plateosaurids and the European *Plateosaurus*, being recovered as the sister clade of the latter in our phylogenetic analysis. It is the northernmost record of a Late Triassic sauropodomorph, and a new dinosaur species erected for Greenland. *Issi saaneq* broadens our knowledge about the evolution of plateosaurid sauropodomorphs.

Keywords: sauropodomorph; Triassic; plateosaurid; dinosaur; Greenland

1. Introduction

The non-sauropod sauropodomorph *Plateosaurus engelhardti* [1], was the first dinosaur to be named outside the UK [2]. Numerous specimens were since assigned to the genus *Plateosaurus*, and new species proposed. Such is the case of *Plateosaurus ingens* [3], formerly *Gresslyosaurus ingens* [4], *Plateosaurus erlenbergiensis* [5], *Plateosaurus gracilis* [2], formerly *Sellosaurus gracilis* [6], and *Plateosaurus trossingensis* [7]. The validity of these specimens has been debated over the last few decades, with one main alternative of *Plateosaurus* classification being the most accepted. Most authors recognize three valid species, *Pl. gracilis* as a sister taxon of *Pl. engelhardti* and *Pl. ingens* [8–11]. However, the material assigned to *Pl. ingens* is under preparation for a redescription, as it possibly represents a new genus [12]. The main issue with *Plateosaurus* taxonomy arises due to the fragmentary nature of the holotype. Therefore, Galton, [13] proposed that the specimen SMNS 13200, a complete skeleton including cranial and post-cranial material, assigned to *Pl. trossingensis*, should be the neotype of *Plateosaurus*, which was accepted by the decision of the International Commission on Zoological Nomenclature [14].

The interrelationships of early sauropodomorphs and the phylogenetic relationship of *Plateosaurus* are receiving increasing attention with novel taxa described for the Late Triassic of South America, Africa and Europe in the past few decades but it is still plagued by issues that are not yet resolved [12,15–19]. McPhee et al. [18] listed the major issues with early sauropodomorph taxonomy, such as disagreeing and poorly understood character conceptions; fragmentary material and missing data for key specimens; lack of complete descriptions and restricted access to several Chinese taxa; and the inclusion of chimeric specimens as operational taxonomic units (OTUs). Among the first attempts to assess *Plateosaurus* interrelationships was the phylogenetic analysis done by Yates [15], where *Plateosaurus* was recovered at the base of Plateosauria, defined as the least inclusive clade containing *Plateosaurus* and Sauropoda (*sensu* [15]). The clade that includes species closer to *Plateosaurus trossingensis* than to Sauropoda was referred to as Plateosauridae in the same work [15] and is recovered in most cladistic analyses with *Unaysaurus tolentinoi* [20] as the sister taxon to *Plateosaurus* [11,12,15,18,21]. However, this comes as no surprise as the dataset used to assess non-sauropod sauropodomorphs usually derives from the work of Yates [15].

McPhee et al. [18] tested the position of *U. tolentinoi* as the sister taxon to *Plateosaurus* and, although relatively constant in his analysis, the clade Plateosauridae is only supported by two unambiguous synapomorphies of the humerus: a non-convex humeral head (character 220, state 1 in the dataset of [18]) and the length of humerus being three times that of the transverse width of its distal condyles (226 (0) in the dataset of [18]). The latter, however, is shared with several other non-sauropod taxa. The introduction of *Macrocollum itaquii* [22] in recent phylogenetic analyses placed *U. tolentinoi* as a sister taxon to *Mac. itaquii*, both forming with the Indian sauropodomorph *Jaklapallisaurus asymmetrica* [23] the clade Unaysauridae [19,22]. Unaysauridae was first recovered as a sister clade to Plateosauridae and then shifted to a more derived position inside of Massopoda (*sensu* [15]) [19,22]. Although close relationship to the contemporaneous sauropodomorphs from Brazil is expected, the validity of Unaysauridae suffers from the missing data and fragmentary nature of specimens. For example, one synapomorphy of unaysaurids is related to the astragalus medial end length ratio to the anteroposterior length of the lateral end, a which has not been confirmed in *U. tolentinoi*, whereas a second synapomorphy, the presence of a promaxillary fenestra, cannot be observed in *J. asymmetrica* which lacks the required cranial remains [23].

Jenkins et al. [24] reported for the first time the presence of an early sauropodomorph in Late Triassic strata (Norian) of the Fleming Fjord Group (formerly Fleming Fjord Formation, see [25]), of Jameson Land in central East Greenland. Jenkins and colleagues [24] briefly described a skull (NHMD 164741, formerly MCZ Field no. 61/91G) from the Malmros Klint Formation (formerly Malmros Klint Member, see [25]). This specimen was collected in Summer of 1991 and assigned to *Plateosaurus engelhardti* (=*Pl. trossingensis*) due to its dental structure and number of teeth, a single dorsal process of the premaxilla, and a Y-shaped quadratojugal [24]. Unfortunately, the lack of a thorough description rendered the allocation of this specimen to *Pl. trossingensis* as tentative, at most, and recent studies advise that this classification should be taken with caution [26]. Furthermore, other Norian sauropodomorphs described in the last decades added new relevant information about the taxonomy of the Greenland specimens.

More new specimens were collected from the same locality: at least three additional specimens were recovered and are yet to be formally described. Among them is an almost complete skull of a juvenile sauropodomorph (NHMD 164758, formerly 1/G95 or 1/95/G, collected in 1995) and two unpublished individuals with cranial and postcranial material (NHMD 164734, formerly 4.88.G and GM.V 2013-683, collected in 2012; and NHMD 164775, a yet unprepared specimen collected in 2012).

Here we focus on the reassessment and thorough description of the skull NHMD 164741, regarded as *Pl. engelhardti* (=*Pl. trossingensis*) by Jenkins et al. [24], and the first description of specimen NHMD 164758, another skull from the same locality, using μCT-scan for the evaluation of hard-to-access features. New features shared by both specimens also allow us to provide an updated phylogeny of early Sauropodomorpha.

2. Geological Setting

Late Triassic sediments are well exposed in the Jameson Land Basin, located in central East Greenland at about 71° N at the present-day land areas of Jameson Land and Scoresby Land (Figure 1). The Jameson Land Basin is bounded by the N-S stretching Liverpool Land to the east and the Stauning Alper to the west. To the north and south, the basin was demarcated by a fracture zone in the Kong Oscars Fjord and the Scoresby Sund respectively [25,27]. During the Late Triassic, the basin was located at the northern rim of the Pangaea supercontinent at approximately 43° N [28]. This position placed the basin in a transition zone between the relatively dry interior of Pangaea and the more humid peripheral part of this continent [29,30], or well inside the humid temperate belt [28].

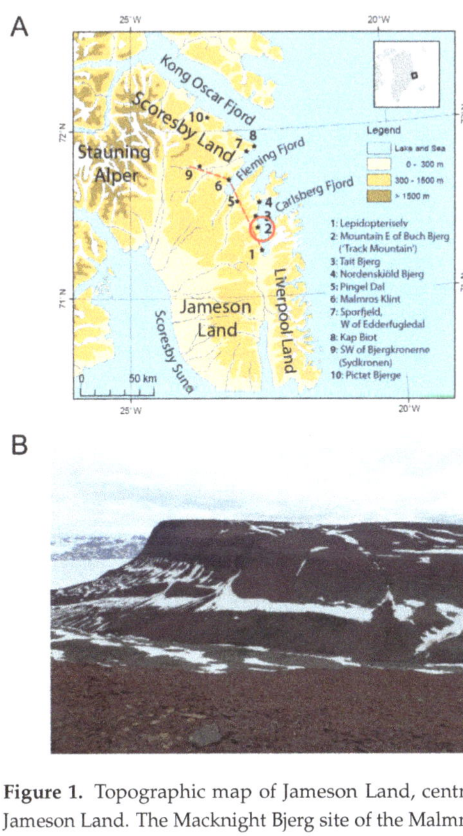

Figure 1. Topographic map of Jameson Land, central East Greenland. (**A**) Topographic map of Jameson Land. The Macknight Bjerg site of the Malmros Klint Formation is marked by the red circle and number 2. Modified from [25]. (**B**) Photograph of the outcrops in Buch Bjerg "Track-site".

The Fleming Group consists of three formations, a lowermost Edderfugledal Formation, a middle Malmros Klint Formation, and an uppermost Ørsted Dal Formations [28]. The Edderfugledal and Malmros Klint formations formed in shallow lacustrine/playa lake environments, whereas a large part of the Ørsted Dal Formation records lake and mudflat deposition [25]. The Malmros Klint Formation is exposed in impressive cliff exposures at Carlsberg Fjord near the eastern margin of the basin, where it has a typical thickness of about 125 m (Figure 2).

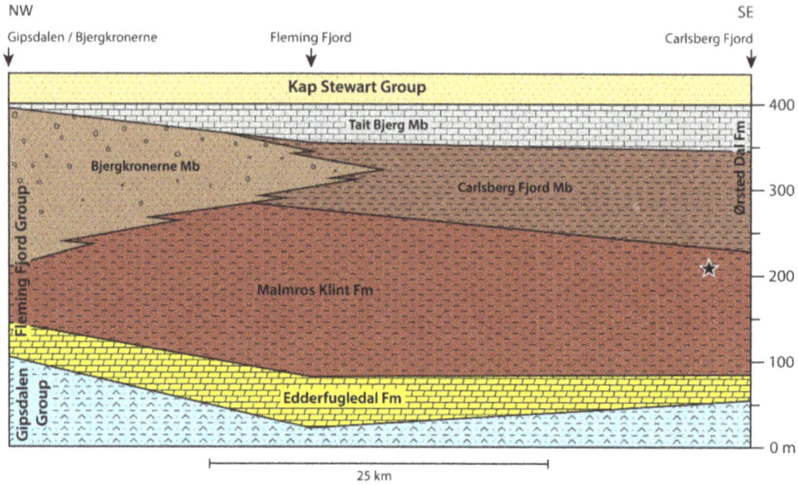

Figure 2. Cross-section of the lithostratigraphic units in the Fleming Fjord Group. Schematic cross-section of the lithostratigraphic units in the Fleming Fjord Group, Jameson Land. The star marks the uppermost Macknight Bjerg site where the sauropodomorph specimens were collected. Modified from [25].

The Malmros Klint and Ørsted Dal Formations comprise a diverse faunal assemblage, containing chondrichthyans and actinopterygian fishes, dipnoans such as *Ceratodus tenuensis* [31], theropod and sauropodomorph dinosaurs, temnospondyls, turtles, aetosaurs, phytosaurs and pterosaurs [24,26,31–36]. Furthermore, the fauna includes teeth and bony elements of early mammaliaforms [24,37,38].

New palaeomagnetic work [28] indicates that the Fleming Fjord Group was deposited between 220 and 209 Ma. Specimens NHMD 164741 and NHMD 164758 were recovered from the same locality, the "Iron Cake" site, in the north side of Macknight Bjerg. This material in the "Iron Cake" site comes from a 1 m thick lake-shore sandstone interval in the uppermost part of the Malmros Klint Formation, 25 m below its boundary to the Ørsted Dal Formation (Figure 2). According to the age model of Kent and Clemmensen [28], this site has an age of 214 Ma (mid Norian), whereas the site in uppermost Carlsberg Fjord Member with other skeletal remains of a sauropodomorph has an age of 211 Ma.

3. Materials and Methods

3.1. Specimens in This Study

NHMD 164741 (Figure 3A) represents a nearly complete, partially articulated, laterally compressed skull with mandible and teeth. NHMD 164758 (Figure 3B) consists of a nearly complete, articulated and laterally compressed skull with mandible and teeth. Postcranial elements of possibly the same individual (NHMD 164758) were collected in 2012 but are yet unprepared and not described in the current work. Both skulls are permanently housed at the Natural History Museum of Denmark (NHMD). Specimen NHMD 164758 was broken into at least two pieces, separating the anterior and posterior regions of the skull, and glued back together using Paraloid® B-72 (Kremer Pigmente, München, Bavaria, Germany) dissolved in acetone (equal parts/50%). The anatomy of both specimens was compared to other basal sauropodomorph dinosaurs described in the literature (Table 1).

Figure 3. Photograph of the skulls NHMD 164741 and NHMD 164758. (**A**) NHMD 164741 in left lateral view. (**B**) NHMD 164758 in left lateral view.

Table 1. List of specimens used in this study for purposes of comparative anatomy.

Taxon	Specimen(s)	Source(s)
Bagualosaurus agudoensis	UFRGS-PV-1099-T	[39]
Buriolestes schultzi	CAPPA/UFSM 0035; ULBRA-PVT280	[40–42]
Coloradisaurus brevis	PVL 3967	[43,44]

Table 1. *Cont.*

Taxon	Specimen(s)	Source(s)
Efraasia minor	SMNS 12667	[2,45,46]
Leyesaurus marayensis	PVSJ 706	[11]
Lufengosaurus huenei	IVPP V15	[47]
Macrocollum itaquii	CAPPA/UFSM 0001a; CAPPA/UFSM 0001b	[19,22]
Massospondylus carinatus	BP/1/5241; BP/1/4934	[16]
Massospondylus kaalae	SAM-PK-K1325	[48]
Ngwevu intlokoi	BP/1/4779	[17]
Pampadromaeus berberenai	ULBRA-PVT016	[49]
Panphagia protos	PVSJ 874	[50]
Plateosaurus trossingensis	AMNH FARB 6810; GPIT-PV-30704; MB.R.1937; MSF 07.M; MSF 08.M; MSF 08.H; MSF 09.2; MSF 11.4; MSF 12.3; MSF 15.4; MSF 15.8; MSF 16.1; MSF 17.4; MSF 23; MSF 33; NAAG_00011238; NAAG_ 00011239; SMA 09.1; SMNS 12949; SMNS 12950; SMNS 13200; SMNS 52968	[2,3,51–53]
Plateosaurus gracilis	GPIT 18318a	[2]
Saturnalia tupiniquim	MCP 3845-PV	[54]

3.2. *Digitization and Image Treatment*

The specimens were digitized using both photogrammetry and μCT-scanning, followed by the creation of a texturized 3D model and observations of internal structures. The photogrammetric method used was based on the Walk-Around Method [55] using a D3500 camera (Nikon, Tokyo, Japan). The pictures were taken from the full 360° of each specimen with a 10° interval between each picture. The 3D photogrammetry models were created with the commercial software Metashape v.1.71 (Agisoft, St. Petersburg, Russia). The first alignment was set to High accuracy, then the point cloud was cleaned manually and using gradual selection, then a dense point cloud was generated set to High quality. The meshes were generated using the High Face Count setting and textures were created in 7680/1 Texture Size/Count setting.

Both specimens were CT-scanned at CENIEH (Burgos, Spain), using MicroCT V | Tome | X s 240 by GE Sensing & Inspection Technologies Phoenix X-Ray (Hürth, North Rhine-Westphalia, Germany). The resulting high-resolution stack contains 2848 .tiff images of 0.08999975 mm voxel size and 1922 × 562 × 2636 resolution (NHMD 164741) and a stack of 2821 .tiff images of 0.0679998 mm voxel size and1810 × 756 × 2821 resolution (NHMD 164758). The raw files are available to download at MorphoSource (ark:/87602/m4/393344, accessed on 29 September 2021). Image Segmentation was carried out with the commercial software Avizo v9.1 (Thermo Fisher, Waltham, MA, USA). Due to the low contrast between bones, the process of segmentation was done mainly using the brush selection tool, slice by slice, and applying interpolation when needed. The "Remove Island" feature of Avizo was applied for islands smaller than 30 pixels to remove the excessive noise of the meshes. The image segmentation process resulted in a total of 65 meshes for NHMD 164741 and 73 meshes for NHMD 164758. The meshes were generated using no smooth operator and exported as wavefront (.obj) files.

All meshes were treated and rendered with the free open-source software Blender v2.92. The segmented meshes were smoothed using the Smooth Laplacian modifier with Lambda factor = 1 and 5 repeats and then decimated to 20% of the original face count for rendering purposes. All the meshes (in .obj file format) are available at MorphoSource (ark:/87602/m4/393381, accessed on 29 September 2021). The measurements were taken both on the physical specimens and digitally in Blender. All renders were done using Cycles as a render engine and with an accurate scale bar in the software. All pictures and renders of the specimens were handled in the commercial imaging softwares Adobe Lightroom 2021 and Adobe Photoshop CC 2021 (Adobe, Mountain View, CA, USA).

3.3. Skull Reconstruction

The digital reconstruction of the specimens was done by retrodeforming, moving, and mirroring the best-preserved elements in the software Blender v2.92. This allowed for tentative volume visualization and natural placement of bones of the laterally crushed specimens. The measurements taken in these skeletal reconstructions are approximated and the digital reconstruction was not used in the description and comparative anatomy section of this manuscript.

3.4. Phylogenetic Analysis

The specimens NHMD 164741 and NHMD 164758 were scored separately in the data matrix of Rauhut et al. [12], containing 67 OTUs and 382 characters (120 cranial and 262 post-cranial) (Supplementary Materials, Data S1). *Mac. itaquii* was scored in this dataset, following the descriptive work of Müller et al. [22] and Müller [19] to assess the validity of the clade Unaysauridae and its relationship to Plateosauridae. The phylogenetic analysis was conducted using the free software TNT v1.5 [56]. The trees were recovered using Traditional Search, with 1000 Wagner trees replicates, holding 20 trees per replicate, with TBR algorithm and 1 random seed, collapsing the trees after the search. Consistency and retention indexes and Bremer Support were obtained using a premade script. Bootstrap was calculated using absolute frequencies and 100 replicates. All characters were treated with the same weight, and characters 8, 13, 19, 23, 40, 57, 62, 69, 92, 102, 117, 121, 122, 129, 132, 148, 150, 151, 158, 168, 170, 171, 178, 210, 211, 213, 232, 237, 254, 263, 268, 282, 295, 316, 322, 330, 352, 365, 368, 370, 375, and 380 were treated as ordered.

3.5. Institutional Abbreviations

AMNH FARB, American Museum of Natural History, Fossil Amphibian, Reptile and Bird Collection, New York, NY, USA; BP, Bernard Price Institute for Palaeontological Research, University of the Witwatersrand, Johannesburg, South Africa; BRSMG, Bristol City Museum and Art Galleries, Bristol, UK; CAPPA/UFSM, Centro de Apoio à Pesquisa Paleontológica da Quarta Colônia da Universidade Federal de Santa Maria, Rio Grande do Sul, Brazil; GPIT, Institut und Museum für Geologie und Paläontologie der Universität Tübingen, Germany; IVPP, Institute of Vertebrate Paleontology and Paleoanthropology, Beijing, People's Republic of China; MB.R., Museum für Naturkunde, collection of fossil Reptilia, Berlin, Germany; MCP, Museu de Ciências e Tecnologia Pontifícia Universidade Católica do Rio Grande do Sul, Porto Alegre, Brazil; MSF, Sauriermuseum Frick, Frick, Switzerland; NHMD, GeoCenter Møns Klint, Møns Klint, Denmark, PVL, Paleontolgía de Vertebrados, Instituto Muíguel Lillo, Tucumán, Argentina; SAM-PK, Iziko South African Museum, Cape Town, South Africa; SMA, Sauriermuseum Aathal, Aathal-Seegräben, Switzerland; SMNS, Staatliches Museum für Naturkunde, Stuttgart, Germany; UFRGS-PV, Paleovertebrate Collection of the Laboratório de Paleovertebrados da Universidade Federal do Rio Grande do Sul, Porto Alegre, Brazil; UFSM, Universidade Federal de Santa Maria, Brazil; ULBRA-PVT, Universidade Luterana do Brasil, Coleção de Paleovertebrados, Canoas, Brazil.

3.6. Nomenclature Acts

The electronic edition of this article conforms to the requirements of the amended International Code of Zoological Nomenclature (ICZN), and hence the new names contained herein are available under that Code from the electronic edition of this article. This published work and the nomenclatural acts it contains have been registered in ZooBank, the online registration system for the ICZN. The ZooBank LSIDs (Life Science Identifiers) can be resolved, and the associated information viewed through any standard web browser by appending the LSID to the prefix http://zoobank.org/. The LSID for this publication is: (urn:lsid:zoobank.org:pub:8AB4D333-BD8E-40B3-B978-37AD481C20E3). The electronic edition of this work was published in a journal with an ISSN and has been archived and is available from the following digital repositories: PubMed Central, LOCKSS.

4. Results

4.1. Systematic Palaeontology

Dinosauria [57]
Saurischia [58]
Sauropodomorpha [59]
Plateosauridae [60]
Issi saaneq gen. et sp. nov.

4.2. Etymology

From Kalaallisut, "issi" meaning cold and "saaneq" meaning bone. Pronounced 'is-y sa-ah-neq'. In reference to the conditions in which the fossils were recovered. We have selected a name in Inuit language to honor the local culture.

4.3. Holotype

NHMD 164741, (Figures 3A, 4 and 5) a nearly complete and partially articulated skull of a late-stage juvenile to sub-adult specimen, missing the anteriormost region of the premaxillae and dentaries and missing most of the right elements. The skull preserves the left premaxilla, both maxillae, both nasals, the left lacrimal, incomplete jugals, incomplete prefrontals, incomplete left postorbital, left squamosal, left quadratojugal, left quadrate, both frontals, the distal part of the left parietal, parts of the braincase (i.e., fragments of the basisphenoid, a fragment of the left laterosphenoid and the left paroccipital process), both pterygoids, the left ectopterygoid, fragments of the palatines, fragments of the left vomer, both dentaries, the left coronoid process, left splenial, left angular, left surangular, left prearticular, left articular and teeth.

4.4. Paratype

NHMD 164758, (Figures 3B, 6 and 7) a nearly complete and articulated skull of a medium-stage juvenile specimen, with lateral deformation, missing most of the posterodorsal skull elements (i.e., squamosals, most of the parietals, supraoccipital and prootic). The skull preserves both premaxillae, both maxillae, both nasals, both lacrimals, both prefrontals, the left postorbital, both jugals, the condylar area of both quadrates, both frontals, the anterior part of the left parietal, parts of the braincase (i.e., the left orbitosphenoid, the left laterosphenoid, right basipterygoid process and the parasphenoid process of the basisphenoid), the complete palatal region and the complete mandibles.

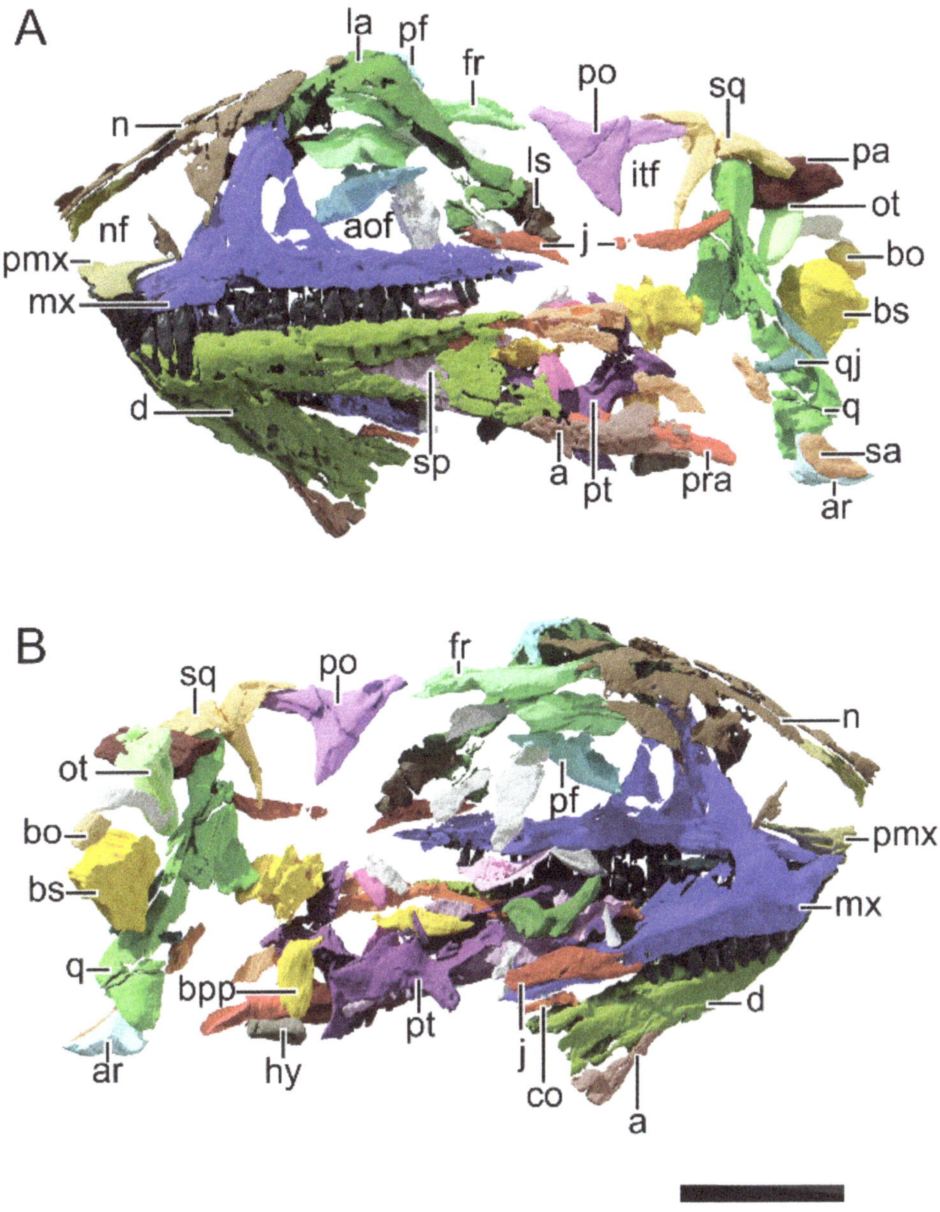

Figure 4. Digital reconstruction of the skull of NHMD 164741. (**A**) Left lateral view. (**B**) Right lateral view. Abbreviations: a, angular; aof, antorbital fenestra; ar, articular; bo, basioccipital; bpp, basipterygoid process; bs, basisphenoid; co, coronoid; d, dentary; fr, frontal; hy, hyoid; itf, infratemporal fenestra; j, jugal; la, lacrimal; ls, laterosphenoid; mx, maxilla; n, nasal; nf, narial fenestra; ot, otoccipital; pa, parietal; pf, prefrontal; pmx, premaxilla; po, postorbital; pra, prearticular; pt, pterygoid; q, quadrate; qj, quadratojugal; sa, surangular; sp, splenial; sq, squamosal. Scale bar = 50 mm.

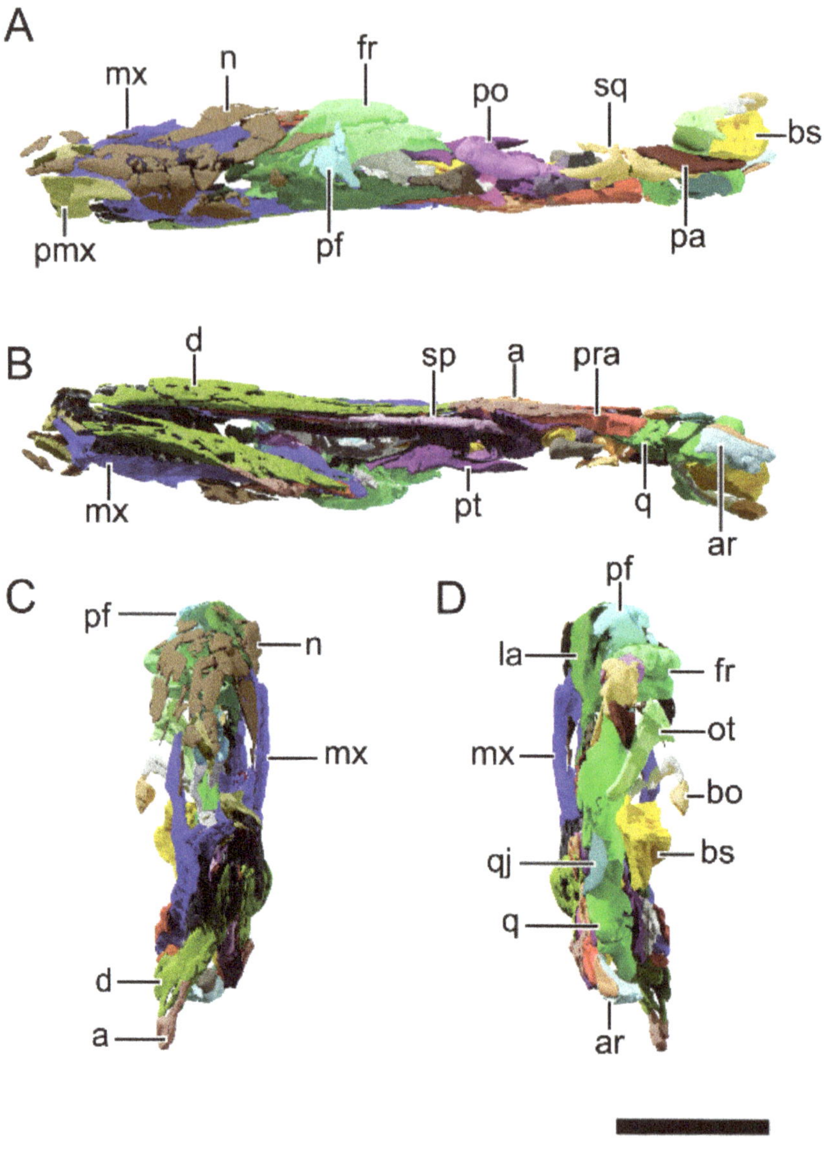

Figure 5. Digital reconstruction of the skull of NHMD 164741. (**A**) Dorsal view. (**B**) Ventral view. (**C**) Anterior view. (**D**) Posterior view. Abbreviations: a, angular; ar, articular; bo, basioccipital; bs, basisphenoid; d, dentary; fr, frontal; la, lacrimal; ls, laterosphenoid; mx, maxilla; n, nasal; ot, otoccipital; pa, parietal; pf, prefrontal; pmx, premaxilla; po, postorbital; pra, prearticular; pt, pterygoid; q, quadrate; qj, quadratojugal; sa, surangular; sp, splenial; sq, squamosal. Scale bar = 50 mm.

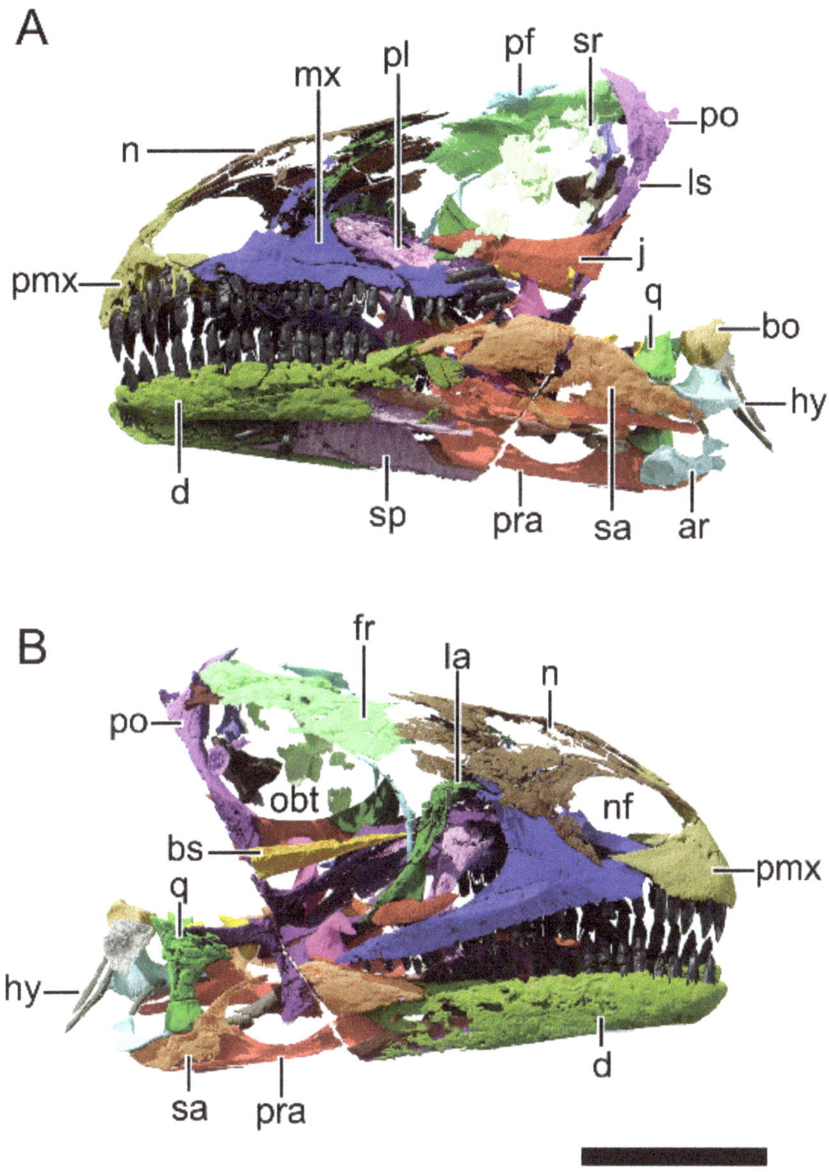

Figure 6. Digital reconstruction of the skull of NHMD 164758. (**A**) Left lateral view. (**B**) Right lateral view. Abbreviations: ar, articular; bo, basioccipital; bs, basisphenoid; d, dentary; fr, frontal; hy, hyoid; j, jugal; la, lacrimal; ls, laterosphenoid; mx, maxilla; n, nasal; nf, narial fenestra; obt, orbit; pf, prefrontal; pl, palatine; pmx, premaxilla; po, postorbital; pra, prearticular; pt, pterygoid; q, quadrate; sa, surangular; sp, splenial; sr, sclerotic ring. Scale bar = 50 mm.

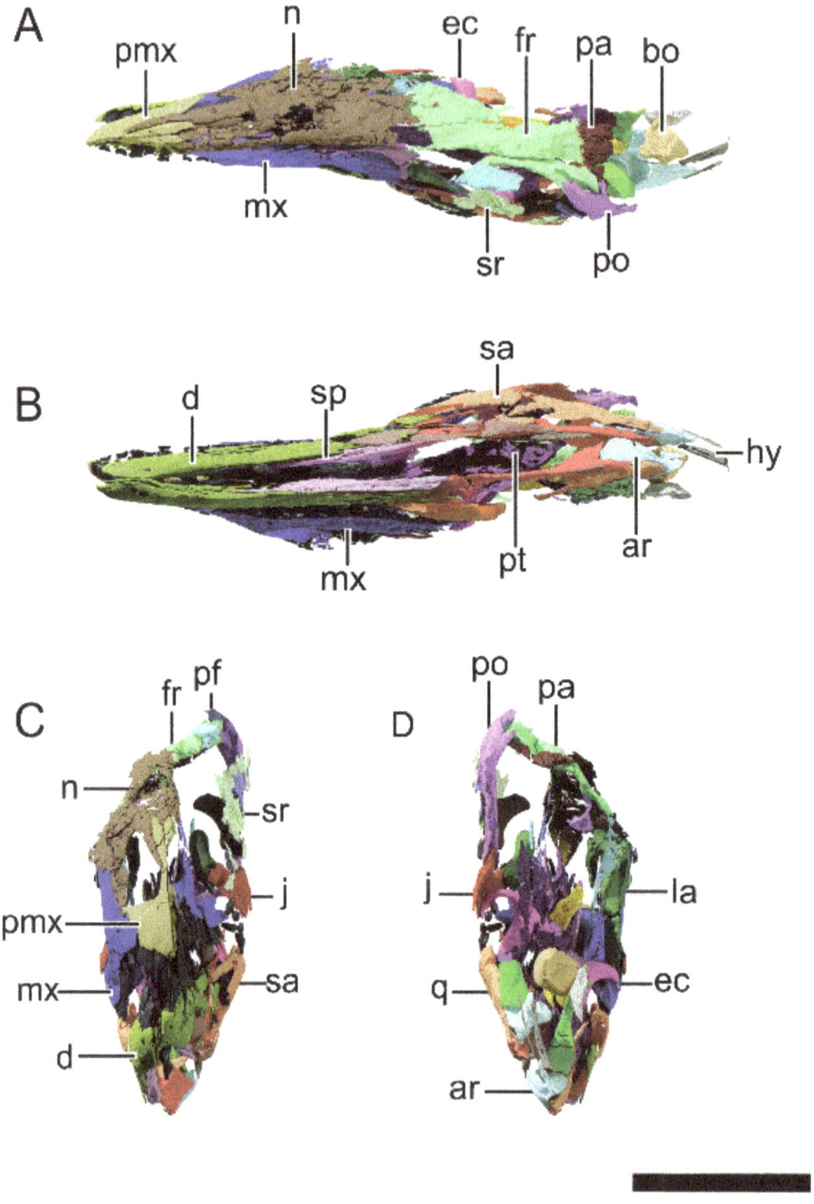

Figure 7. Digital reconstruction of the skull of NHMD 164758. (**A**) Dorsal view. (**B**) Ventral view. (**C**) Anterior view. (**D**) Posterior view. Abbreviations: ar, articular; bo, basioccipital; d, dentary; ec, ectopterygoid; fr, frontal; hy, hyoid; j, jugal; la, lacrimal; mx, maxilla; n, nasal; pa, parietal; pf, prefrontal; pmx, premaxilla; po, postorbital; pra, prearticular; pt, pterygoid; q, quadrate; sa, surangular; sp, splenial; sr, sclerotic ring. Scale bar = 50 mm.

4.5. Locality

NHMD 164741 and NHMD 164758 were collected at Macknight Bjerg site, Jameson Land, central East Greenland, of coordinates 71°23.010′ N, 22°34.114′ W and 71°22.993′ N, 22°33.972′ W, respectively (Figure 1).

4.6. Horizon and Age

NHMD 164741 and NHMD 164758 were collected at Malmros Klint Formation, Fleming Fjord Group, of mid-Norian stage of the Late Triassic [28,29,61].

4.7. Diagnosis

Issi saaneq can be distinguished from other basal sauropodomorphs on the basis of a unique trait combination comprising six phylogenetic synapomorphies (i) and four cranial autapomorphies (ii). (i) weakly developed narial fossa (character 10, state 0); small subnarial foramen (12, 1); anterior margin of the external naris anterior to the mid-length of the premaxilla (17, 0); anteroposterior length of the antorbital fossa less than that of the orbit (28, 1); antorbital fossa ending before the ventral process of the lacrimal (41, 1); strongly curved jugal process of the ectopterygoid (86, 1). (ii) the presence of a small foramen at the medial surface of the premaxilla at the base of the lateral process of the premaxilla; an anteroposteriorly elongated dorsoposterior process of the squamosal; a quadrate relatively tall in comparison to the rostrum height; a well-developed posterodorsal process of the articular, square-shaped in lateral view. *Issi saaneq* possesses features thought to be autapomorphic to other plateosaurids, i.e., five teeth in the premaxilla (as in *Plateosaurus*), a promaxillary fenestra (as in the Brazilian plateosaurids *Mac. itaquii* and *U. tolentinoi*), a lateral sheet of bone in the lacrimal covering the posterodorsal corner of the antorbital fenestra (as in *Pl. trossingensis*), and a secondary fossa ventral to the Meckelian groove (as in *U. tolentinoi*).

4.8. Description and Comparative Anatomy

4.8.1. Generalities

NHMD 164741 consists of a partially complete and almost fully articulated skull. The specimen is almost twice as long as tall (length is measured from the anteriormost preserved region to the end of the squamosal and height is measured from the ventral surface of the left dentary to the apex of the left lacrimal), and due to a lateral crushing, its overall width is around a fifth of its height (see Table 2 for the skull general measurements). The anteriormost region of the snout (anterior part of left premaxilla and whole right premaxilla) and anterior region of both dentaries are missing. Thus, the total length of the skull cannot be precisely measured, although it would not be much longer than preserved. Most elements in the skull of NHMD 164741 are preserved in three dimensions, with little individual deformation, even though the skull is crushed laterally. This lateral compression displaced most bones from the right side of the skull and disarticulated some elements (i.e., left and right frontals, left postorbital, left quadratojugal). The left elements of the skull are mostly present, showing some fractures, but preserving the overall shape of the skull. The orbital region, however, is poorly preserved due to compaction, with the overall orbital shape only tentatively recovered as semi-circular, according to a slight anterodorsal expansion over the caudal margin of the lacrimal. The right jaw joint region of the skull is missing, whereas the left elements of this region are slightly ventrally displaced, but mostly still in association (i.e., quadratojugal, quadrate, articular and the posterior part of the surangular). The braincase and occipital region are mostly disarticulated, its bones missing or too fragmented for precise identification.

Table 2. General skull measurements for the Greenland specimens NHMD 164741 and NHMD 164758. Asterisk (*) indicates maximum preserved measurement.

Measurements (in mm)	NHMD 164741	NHMD 164758
Skull anteroposterior length (from the anterior tip of the premaxilla to posterior margin of occipital condyle)	243.7 *	167.9
Skull maximum dorsoventral height at orbit including the mandible	112.7	93.6
Skull dorsoventral height at orbit excluding mandible	75.2	58.3
Rostrum dorsoventral height (measured at the posterior margin of the external naris level)	57.9	40.9
External naris maximum anteroposterior length	35.1	25.1
Orbit length	52.5 *	49.9
Orbit height	46.5 *	41.0
Premaxilla maximum anteroposterior length	-	36.9
Premaxilla alveolar anteroposterior length	-	24.3
Premaxilla maximum dorsoventral height	40.3	37.8
Narial fossa anteroposterior length	28.7	24.9
Maxilla anteroposterior maximum length	129.9	95.5
Maxilla anteroposterior alveolar length	118.8	85.7
Maxilla dorsoventral height (from the dorsal tip of the dorsal process to ventral margin of maxilla)	57.0	42.0
Antorbital fossa maximum anteroposterior length	46.5	34.8
Nasal anteroposterior length	98.7	80.0
Lacrimal dorsoventral height	64.5	41.9
Lacrimal maximum length of dorsal region	43.1	37.5
Prefrontal dorsoventral maximum height	-	32.3
Frontal maximum anteroposterior length	51.6	42.3
Frontal maximum mediolateral width	49.1	18.7
Postorbital maximum length	48.6	25.5
Postorbital maximum height	40.8	41.6
Squamosal maximum length	49.5	-
Squamosal maximum height	38.9	-
Jugal maximum anteroposterior length	57.8	48.9
Jugal height under the orbit	11.3	10.5
Quadrate dorsoventral height	91.8	-
Quadrate mediolateral width at condylar region	13.9	10.8
Quadrate pterygoid flange anteroposterior length	24.3	17.8
Pterygoid maximum anteroposterior length	92.4	83.8
Pterygoid maximum dorsoventral height	52.4	46.6
Ectopterygoid maximum length of medial flange	29.8	20.5
Ectopterygoid maximum mediolateral width	22.4	21.6
Palatine maximum anteroposterior length	31.3	39.8
Palatine maximum mediolateral width	9.1	13.2
Vomer maximum anteroposterior length	-	42.6
Mandible maximum length	210.1 *	167.1

Table 2. *Cont.*

Measurements (in mm)	NHMD 164741	NHMD 164758
Mandible maximum height	45.6	33.8
Dentary maximum length	118.0 *	100.1
Dentary maximum height	34.4	22.1
Surangular maximum length	118.1	79.8
Surangular maximum height	44.8	25.5

The NHMD 164758 skull is relatively smaller (0.69 in length) than that of NHMD 164741, but its elements show less lateral compaction than NHMD 164741, resulting in a better-preserved palatal region. Similar to NHMD 164741, the skull is relatively dorsoventrally taller than in other plateosaurids (see Table 2). The right elements of the snout are better preserved, as the lateral surface of the left snout elements was eroded. The orbit in NHMD 164758 is subcircular and preserves the left sclerotic ring.

4.8.2. Premaxilla

NHMD 164741 preserves the posterior half of the left premaxilla (Figure 8). NHMD 164758 bears the only complete premaxilla of both specimens (right element, Figure 9). The premaxilla is triangular in lateral profile, encompassing most of the narial fenestra. The main body of the premaxilla is slightly anteroposteriorly longer than dorsoventrally tall and contains five alveoli, the first of which is adjacent to the rostral tip of the bone. A 5-tooth premaxilla is also seen in *Plateosaurus* (except for specimens with 6 premaxillary teeth, i.e., HMN XXIV, SMNS 12949 and SMNS 13200), and the position of the first premaxillary tooth is close to the anterior margin of the premaxilla. An anteriorly located first premaxillary tooth is similar to that of the juvenile *Pl. trossingensis* (MSF 15.8B) and is thought to be related to ontogeny [53], as in mature specimens of *Pl. trossingensis* (i.e., MSF 11.4, MSF 15.8, and 16.1) there is usually a gap between the anterior tip of the premaxilla and the first premaxillary tooth.

The main body is laterally perforated by a small foramen at the base of the dorsal process, at the level of the anterior margin of the second premaxillary tooth. Posterior to this foramen, at the dorsal margin of the premaxilla main body, a shallow recess forms the narial fossa. The narial fossa marginates the posterior margin of the dorsal process of the premaxilla, reaching its deepest point at the lateral mid-length of the premaxillary body, around the level of the third premaxillary tooth. The shallow narial fossa is observed in both NHMD 164741 and NHMD 164758, although the exact anterior extend of this structure on the former is unknown. The narial fossa position, depth and shape differ from that of all *Plateosaurus* skulls described (i.e., AMNH FARB 6810; HMN XXIV; HMN MB.1927.19.1; MSF 11.4; MSF 16.1; MSF 1; SMNS 52968). The narial fossa of these specimens is marked by a ventral rim and is deeply depressed in the premaxilla. The condition observed in NHMD 164758 is closer to that of *U. tolentinoi* (UFSM 11069), *Mac. itaquii* (CAPPA/UFSM 0001a), *Mas. carinatus* (BP/1/4934), and *N. intlokoi* (BP/1/4779), as well as most early sauropodomorphs (such as *Ba. agudoensis* and *Bu. schultzi*).

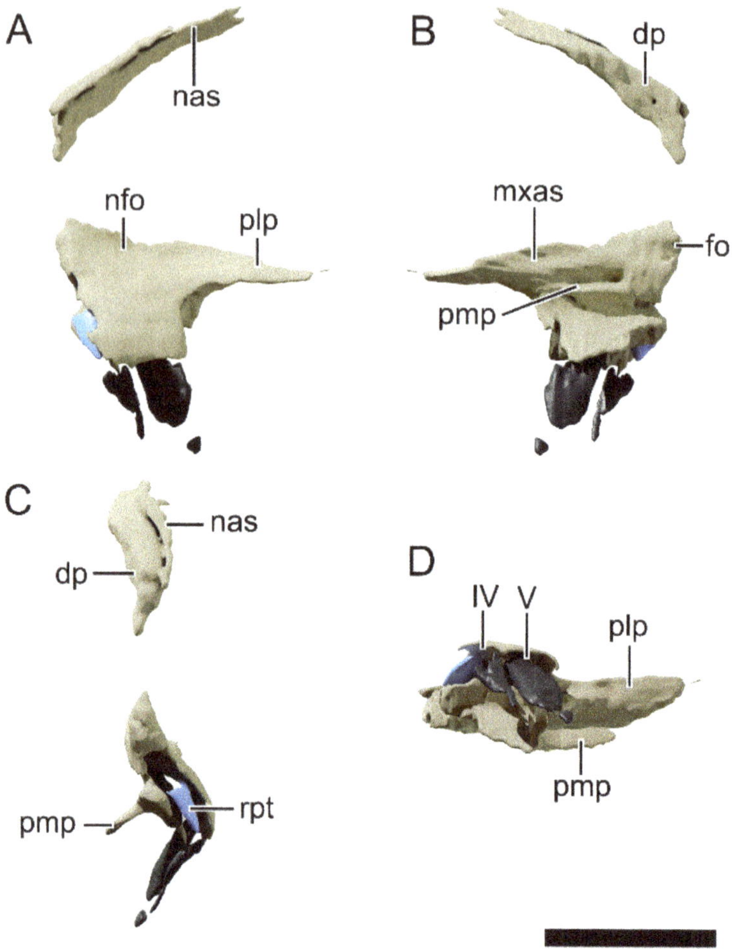

Figure 8. Digital reconstruction of the left premaxilla of NHMD 164741. (**A**) Lateral view. (**B**) Medial view. (**C**) Anterior view. (**D**) Ventral view. Abbreviations: dp, dorsal process; fo, foramen; mxas, articular surface for the maxilla; nas, articular surface for the nasal; nfo, narial fossa; plp, posterolateral process; pmp, posteromedial process; rpt, replacement tooth. IV and V represent the tooth position. Scale bar = 20 mm.

The medial surface of the premaxilla is almost straight at the contact with its counterpart, which occurs in the first half of the premaxillary main body. As in *Pl. trossingensis*, the teeth row forms an angle of 20° with the symphysis, resulting in a V-shaped dorsal and ventral profiles. The maxillary articular facet is delimited by a ventral sharp ridge that constitutes the posteroventral process and a smoother dorsal recess that forms the ventral margin of the posterolateral (=maxillary) process of the premaxilla. This later rests on the anterodorsal margin of the maxillary body and tapers to a point posteriorly until it contacts the rostroventral process of the nasal. Both NHMD 164741 and NHMD 164758 possess a round foramen at the dorsomedial surface of the premaxilla main body, at the base of the dorsal process of the premaxilla (Figures 8 and 9). This foramen is not observed on the disarticulated or exposed medial surface of the premaxillae of the plateosaurids *U. tolentinoi* (UFSM11069) [18] and *Pl. trossingensis* (AMNH FARB 6810 [52], MSF 16.1 [53] and MSF 15.8.935 [53]), nor the derived massospondylid *Mas. carinatus* (BP/1/5241) [16]. However, the saturnaliid sauropodomorph *Pam. berberenai* (ULBRA-PVT016) preserved a similar foramen at the medial surface of the premaxilla, although more posteriorly and ventrally located in the latter [49]. The dorsal (=nasal) process of the premaxilla slopes posteriorly at an angle of 61° with the main body of the premaxilla. This deflection is lower than in *Mac. itaquii* (45°), *U. tolentinoi* (40°), but similar to some *Plateosaurus* specimens, particularly MB.R.1937. The dorsal process encloses the anterior margin of the narial fenestra and extends posteriorly until the level of the posterior process of the premaxilla. This process contacts its counterpart medially, flattening dorsoventrally at the posterior end. This flattening results in a distal expansion of this process, as in *Plateosaurus*, *Mac. itaquii* and *Coloradisaurus brevis* [43] PVL 3967.

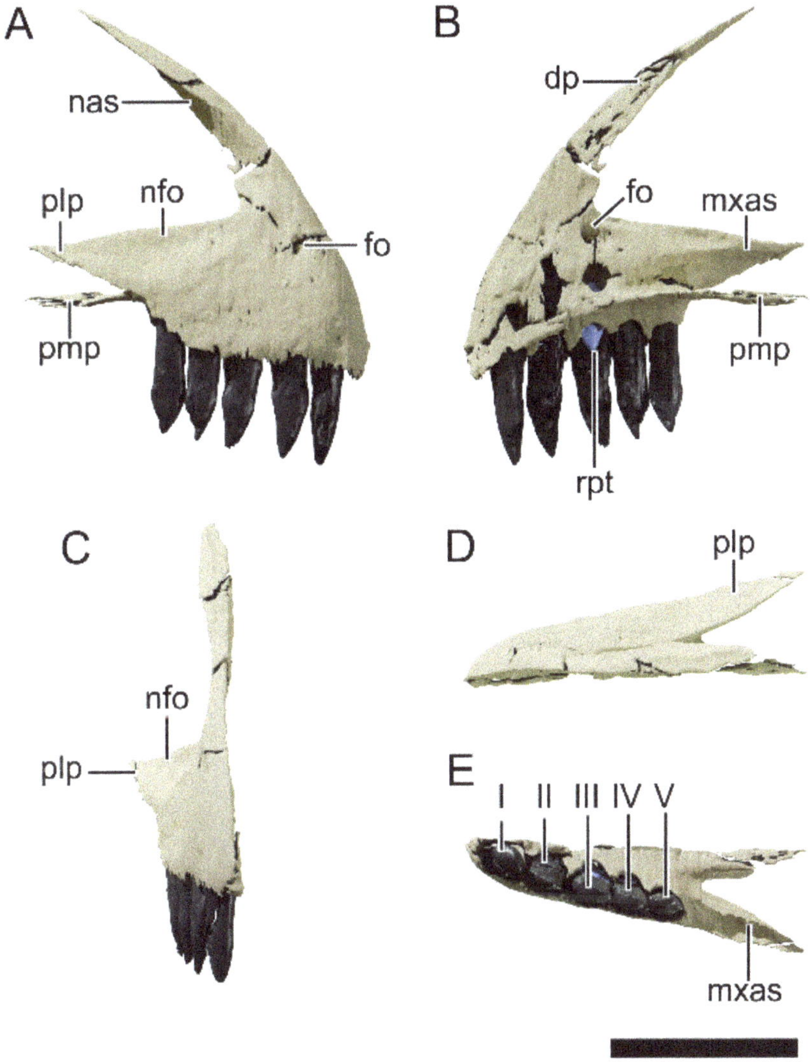

Figure 9. Digital reconstruction of the right premaxilla of NHMD 164758. (**A**) Lateral view. (**B**) Medial view. (**C**) Anterior view. (**D**) Dorsal view. (**E**) Ventral view. Abbreviations: dp, dorsal process; fo, foramen; mxas, articular surface for the maxilla; nas, articular surface for the nasal; nfo, narial fossa; plp, posterolateral process; pmp, posteromedial process; rpt, replacement tooth. I to V represent the tooth position. Scale bar = 20 mm.

4.8.3. Maxilla

The best-preserved maxillae are the left element in NHMD 164741 (Figure 10) and the right element in NHMD 164758 (Figure 11). This bone is triradiate in the lateral profile, with the horizontal main ramus divided by the vertical dorsal process of the maxilla. The main ramus is straight throughout its anteroposterior length, with both dorsal and ventral margins parallel to each other, only tapering at the posteriormost region. The dorsal process deflects slightly posteriorly and marginates the subtriangular antorbital fossa. A maximum of 23 alveoli in NHMD 164758 and 24 alveoli in NHMD 164741 are preserved, both specimens with four alveoli in the anterior segment of the main ramus. In other sauropodomorphs such as *Mac. itaquii*, *U. tolentinoi* (UFSM 11069), *Leyesaurus* (PVSJ 706) and *Mas. carinatus* (BP/1/4934) there are indeed four alveoli in this segment of the maxilla, but in *Plateosaurus* (such as AMNH FARB 6810, MSF 12.3, SMNS 13200) there are five alveoli anterior to the dorsal process of the maxilla.

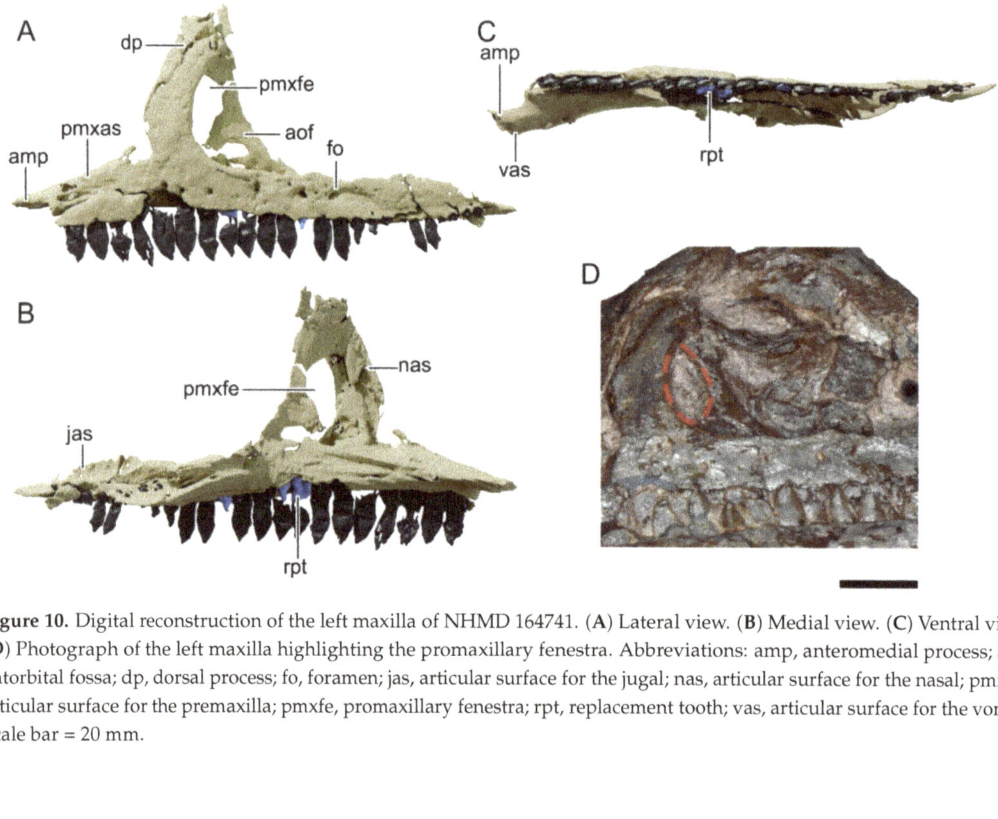

Figure 10. Digital reconstruction of the left maxilla of NHMD 164741. (**A**) Lateral view. (**B**) Medial view. (**C**) Ventral view. (**D**) Photograph of the left maxilla highlighting the promaxillary fenestra. Abbreviations: amp, anteromedial process; aof, antorbital fossa; dp, dorsal process; fo, foramen; jas, articular surface for the jugal; nas, articular surface for the nasal; pmxas, articular surface for the premaxilla; pmxfe, promaxillary fenestra; rpt, replacement tooth; vas, articular surface for the vomer. Scale bar = 20 mm.

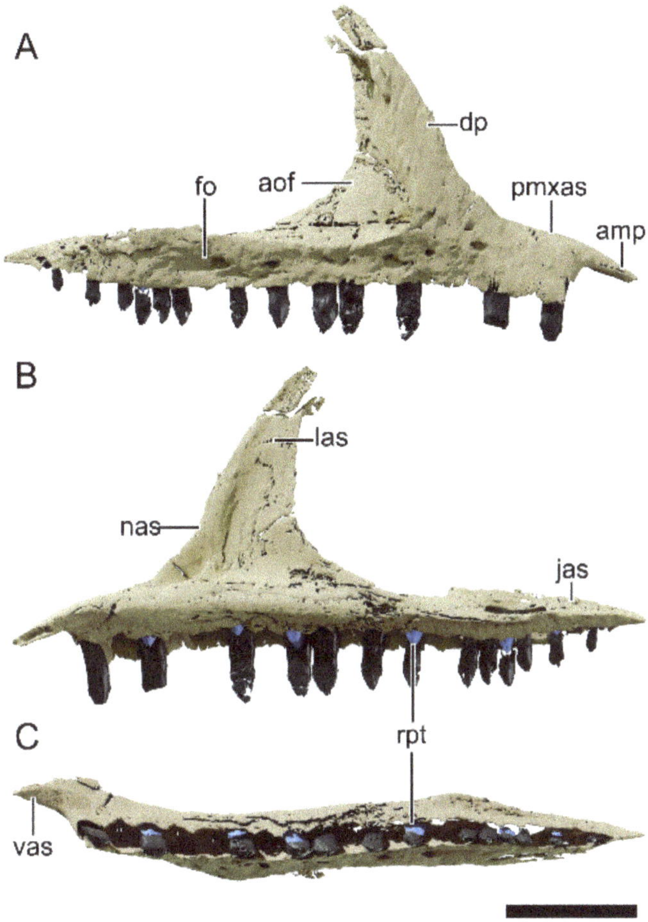

Figure 11. Digital reconstruction of the right maxilla of NHMD 164758. (**A**) Lateral view. (**B**) Medial view. (**C**) Ventral view. Abbreviations: amp, anteromedial process; aof, antorbital fossa; dp, dorsal process; fo, foramen; jas, articular surface for the jugal; las, articular surface for the lacrimal; nas, articular surface for the nasal; pmxas, articular surface for the premaxilla; rpt, replacement tooth; vas, articular surface for the vomer. Scale bar = 20 mm.

The anterior segment of the main ramus of the maxilla is anteroposteriorly shorter than the posterior segment and has no lateral neurovascular foramina. An anterior process of the maxilla extends medial to the premaxilla, tapering to a point anteriorly. This process is gently curved mediolaterally, with a sharp ridge at its dorsal margin, where it contacts laterally the maxillary process of the premaxilla. The anterior process of the maxilla contacts ventrally with the posteroventral process of the premaxilla and medially with the vomer, as in *Plateosaurus*.

The dorsal (=ascending) process of the maxilla tapers dorsally, with a slight dorsoposterior inclination. This process is almost perpendicular to the main ramus of the maxilla, differently from pre-Norian sauropodomorphs, such as non-Bagualosaurian (sensu [49]) and *Ba. agudoensis*. Anteriorly this process bounds the posterior margin of the external naris and laterally contacts the ventral (=maxillary) process of the nasal. The dorsal margin of the antorbital fossa converges to the dorsal tip of the maxillary dorsal process forming an apex, as in *Lufengosaurus huenei* [62] (IVPP V15), but differing from the posteriorly extended antorbital fossae of *Mac. itaquii*. Medially, the dorsal process of the maxilla forms the anterior margin of the antorbital fossa. In NHMD 164758 the antorbital fossa is fully closed, differing from unaysaurids (i.e., *Mac. itaquii* and *U. tolentinoi*, *sensu* [19,22]), whose antorbital fossae are perforated by a large promaxillary fenestra. In NHMD 164741 this region is fragmented on both left and right maxillae.

However, in the left antorbital fossa, the presence of the promaxillary fenestra (or a blind ridge, as in the right maxilla of *Mac. itaquii* CAPPA/UFSM 0001b, Müller, 2019) is not discarded, as the anteroventral margin of the antorbital fossa possesses an undamaged recess (Figure 10D). The lateral surface of the posterior process of the maxilla is perforated by foramina, the last of which being the largest and opening posteriorly. The dorsal and ventral margins of the posterior process are parallel throughout most of their length, tapering posteriorly with a slight dorsal deflection. Along this deflection, a medial groove is formed on the dorsal margin, where the jugal articulates with the maxilla.

4.8.4. Nasal

The nasals are dorsoventrally thin tetraradiate bones (Figures 3–7), anteroposteriorly longer than lateromedially wide. The nasal is relatively shorter in NHMD 164741 and NHMD 164758 than in *Plateosaurus*. In *Plateosaurus*, the nasal is longer than half the skull roof length, a distinctive feature for the genus [52,53,63]. The main body of the nasal is dorsally convex and overlaps laterally the apex of the maxillary dorsal process. At its posterior region, the nasal contacts the prefrontal laterally and the frontal ventrally. Anteriorly, the nasal radiates in two ventrally oriented processes, separated by a dorsal concavity of the nasal. This area encompasses the posterodorsal region of the narial fenestra. The anteromedial (=premaxillary) process of the nasal contacts the dorsal process of the premaxilla laterally. The lateroventral (=maxillary) process marginates the dorsal process of the maxilla as a ventrally oriented triangular blade. Ventrally, this process tapers to a point that finishes just before contacting the posteriormost tip of the posterolateral process. A point-contact between the distal part of the lateroventral process of the nasal and the posterolateral process of the premaxilla is observed in *Plateosaurus* but not in *Mac. itaquii* and *Mas. carinatus*.

4.8.5. Lacrimal

The best-preserved lacrimals are the left element in NHMD 164741 (Figure 12) and the right element in NHMD 164758 (Figure 13), although in the former the distal part of the maxillary process is broken, and in the latter, the lateral surface is weathered. The lacrimal bounds the posterior margin of the antorbital fenestra and the anterior margin of the orbit. The lacrimal is shaped like an inverted L, with a long anterodorsal (=maxillary) process extending anteriorly. This process is obscured dorsally by the nasal. The anterodorsal process of the lacrimal is hollow and subcircular in cross-section, with the dorsal margin tapering into a ridge. This ridge extends from the dorsomedial margin of the main shaft of the lacrimal until it bisects at the anteriormost region of the process. The ventral projection at the bifid junction is longer than the dorsal projection. The anterodorsal process of the lacrimal bends laterally at its distal part to articulate to the medial margin of the dorsal process of the maxilla. The overall shape of the anterodorsal process of the lacrimal differs from that of *Pl. trossingensis* (AMNH FARB 6810), in which the process is triangular in cross-section at the bifid region, with the dorsal projection of the process being longer than the ventral.

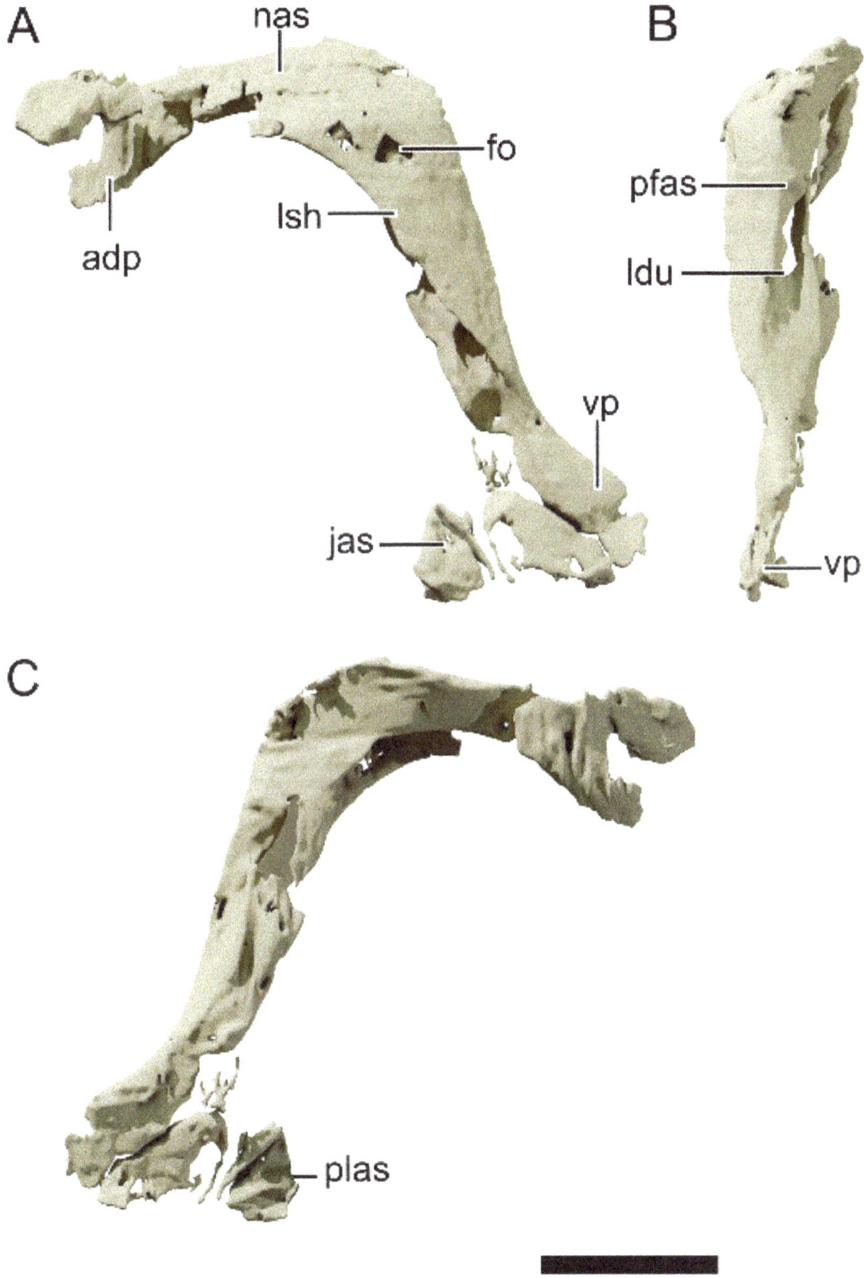

Figure 12. Digital reconstruction of the left lacrimal of NHMD 164741. (**A**) Lateral view. (**B**) Posterior view. (**C**) Medial view. Abbreviations: adp, anterodorsal process; fo, foramen; jas, articular surface for the jugal; ldu, lacrimal duct; lsh, lateral sheet of bone; nas, articular surface for the nasal; pfas, articular surface for the prefrontal; plas, articular surface for the palatine; vp, ventral process. Scale bar = 20 mm.

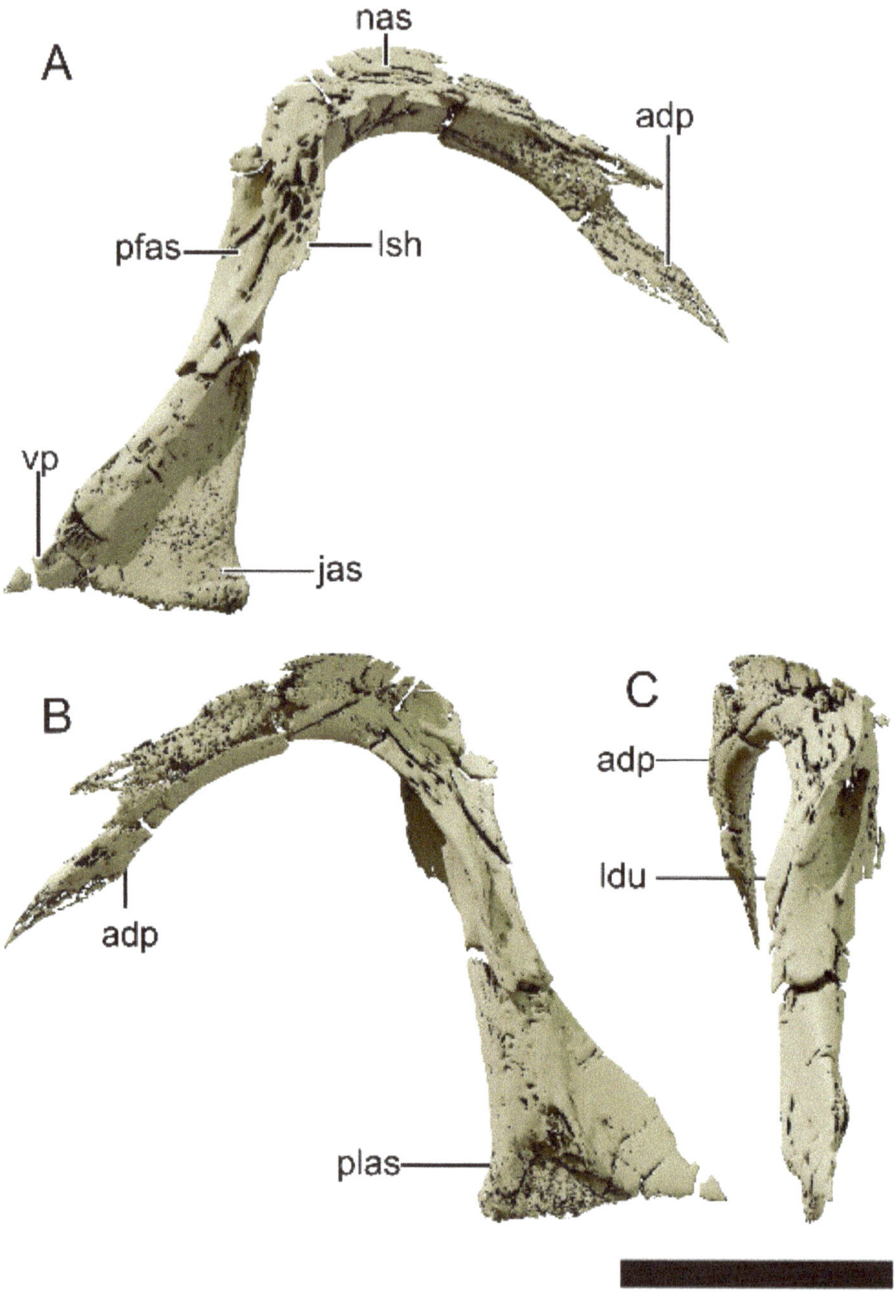

Figure 13. Digital reconstruction of the right lacrimal of NHMD 164758. (**A**) Lateral view. (**B**) Medial view. (**C**) Posterior view. Abbreviations: adp, anterodorsal process; fo, foramen; jas, articular surface for the jugal; ldu, lacrimal duct; lsh, lateral sheet of bone; nas, articular surface for the nasal; pfas, articular surface for the prefrontal; plas, articular surface for the palatine; vp, ventral process. Scale bar = 20 mm.

The main shaft of the lacrimal is inclined anteriorly at an angle of 50° to the long axis of the skull, contrasting with the sub-vertical lacrimal of *Mac. itaquii* and the 30° tilted lacrimal of *Plateosaurus*. The lateral flange of the shaft is well-developed in NHMD 164741, as in *Pl. trossingensis* and differing from the condition of *E. minor* and *Pl. gracilis*. The dorsal half of the anterior margin of this flange is concave in NHMD 164741. In most specimens of *Pl. trossingensis*, the flange is convex, except for MSF 15.4 and MSF 16.1, in which it is concave. However, this concavity may have been caused by diagenetic deformation in these specimens. Therefore, the condition in NHMD 164741 is similar to that of *N. intlokoi* and *Lu. huenei* but differs from that of *Mas. carinatus* [17].

The lateral surface of the lacrimal is perforated by a laterally opened foramen near the dorsal margin of the main shaft in NHMD 164741, as in *Lu. huenei* (IVPP V15) and *Mac. itaquii*, although on the latter the foramen is reduced in diameter. The posterior margin of the lacrimal is perforated by the ventral-facing lacrimal duct. Ventral to this duct, the lacrimal articulates with the ventral (=lacrimal) process of the prefrontal. The ventral process of the lacrimal projects anteriorly and posteriorly, forming a fin-like shelf, as in *Plateosaurus*. The anterior projection is overlapped laterally by the anterior process of the jugal and medially by the palatine.

4.8.6. Prefrontal

Both NHMD 164741 and NHMD 164758 preserved only fragments of the prefrontals (Figures 3–7). The prefrontal is a dorsoventrally thin bone, anteroposteriorly longer than wide. It forms the anterodorsal margin of the orbit, being slightly concave at its ventral margin, whereas straight at its dorsal margin. In lateral view, the bone is T-shaped, with its posterior process being longer than the anterior, and reaching over half the length of the orbit. This condition is also observed in most post-Carnian sauropodomorphs, but not in *Mac. itaquii*, in which the posterior process does not reach half the orbital length. Medially, the prefrontal is concave, forming a dorsal and a ventral shelf that project medially. An elongated and slender lacrimal process projects ventrally from the medioventral margin of the postorbital. This process marginates the posterior margin of the lacrimal beneath the lacrimal duct opening and participates in the anterior margin of the orbit, although not reaching its anteroventral corner.

The posterior process of the prefrontal is expanded in both NHMD 164741 and NHMD 164758, being anteroposteriorly longer than the anterior process of the prefrontal. However, this elongation does not restrict the participation of the frontal in the orbit, as seen in *Plateosaurus* [2]. In dorsal view, this process tapers distally as in *Plateosaurus* and contacts the anterior region of the frontal medially.

4.8.7. Postorbital

Only the left postorbitals are preserved on both NHMD 164741 (Figure 14) and NHMD 164758 (Figure 15). The postorbital is a triradiate, Y-shaped bone, forming the posterodorsal margin of the orbit, the anterior margin of the infratemporal fenestra and the anterolateral margin of the supratemporal fenestra. The anterodorsal process of the postorbital is oriented anteromedially and dorsally. It is lateromedially broader than dorsoventrally tall. The medial margin of the anterodorsal process is bifurcated to embrace the posterolateral process of the frontal, as in *Mac. itaquii* and *Plateosaurus*. In NHMD 164741, the anteroposterior length of the anterodorsal process of the postorbital is slightly longer than the anteroposterior length of the posterodorsal process of the postorbital. This process is not preserved in NHMD 164758. In *Plateosaurus*, the anteroposterior process of the postorbital is shorter than the other postorbital processes.

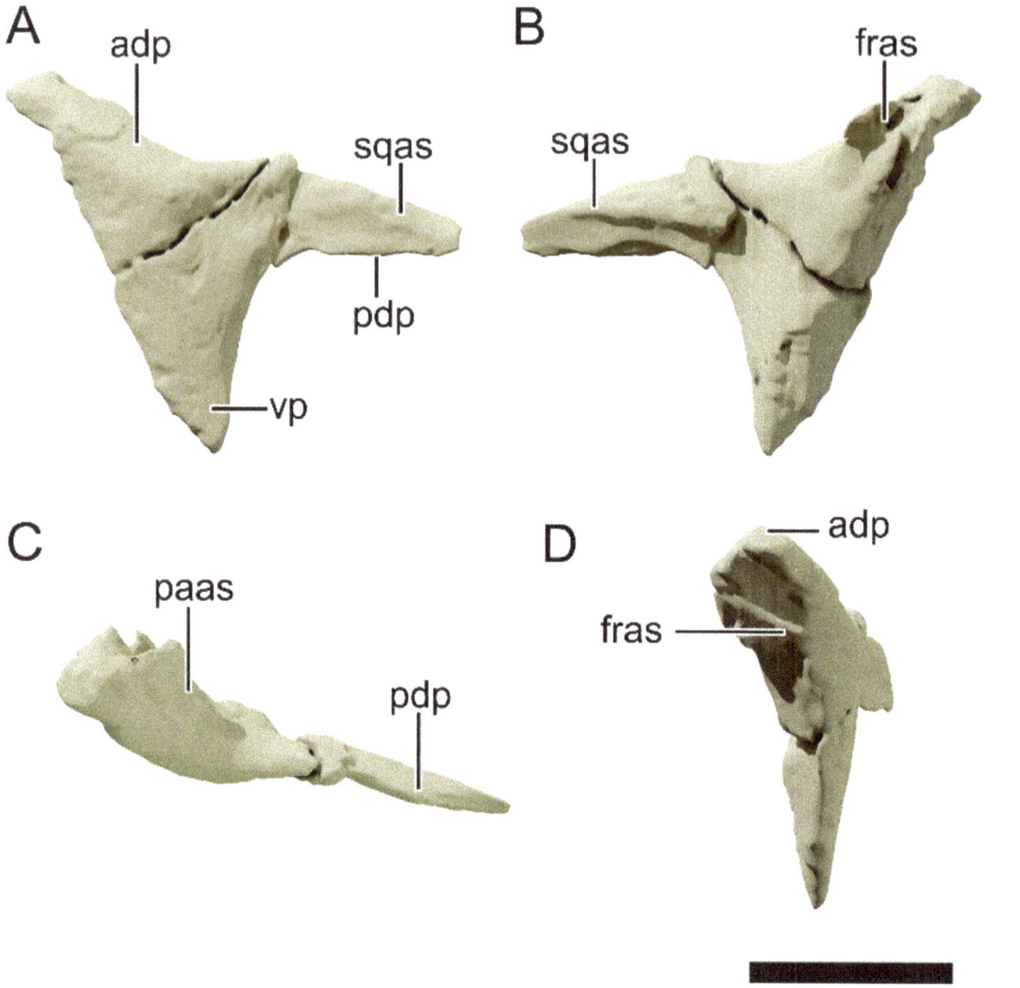

Figure 14. Digital reconstruction of the left postorbital of NHMD 164741. (**A**) Lateral view. (**B**) Medial view. (**C**) Dorsal view. (**D**) Anterior view. Abbreviations: adp, anterodorsal process; fras, articular surface for the frontal; paas, articular surface for the parietal; pdq, posterodorsal process; sqas, articular surface for the squamosal; vp, ventral process. Scale bar = 20 mm.

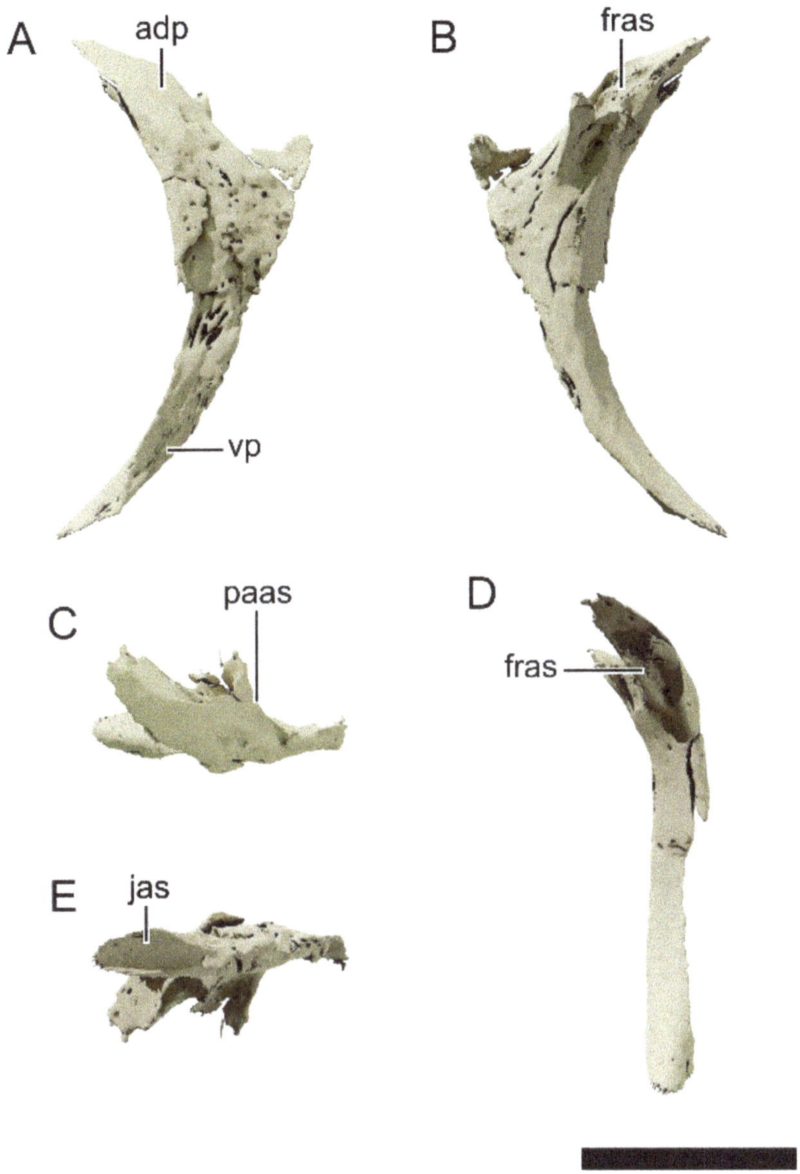

Figure 15. Digital reconstruction of the left postorbital of NHMD 164758. (**A**) Lateral view. (**B**) Medial view. (**C**) Dorsal view. (**D**) Ventral view. (**E**) Anterior view. Abbreviations: adp, anterodorsal process; fras, articular surface for the frontal; jas, articular surface for the jugal; paas, articular surface for the parietal; vp, ventral process. Scale bar = 20 mm.

The posterior process of the postorbital is only preserved in NHMD 164741. This process is lateromedially slender and dorsally convex. It articulates dorsally and medially to the anterior process of the squamosal, with the dorsolateral surface partially covered by the squamosal. The posterior process of the postorbital forms an angle to the anterior process of 149° in NHMD 164741 and 134° in NHMD 164758. In *Plateosaurus* these angles vary between 160° in AMNH FARB 6810 and MSF 11.4, and 110° in MSF 12.3. This latter is thought to be closer to the original condition, as the former were mediolaterally compressed.

The ventral process of the postorbital is broken in NHMD 164741 and best preserved in NHMD 164758 and is the longest process of the postorbital. This process is strongly concave anteriorly bounding the posterior surface of the orbit. It contacts the dorsal process of the jugal posteroventrally and excludes the jugal from the posterior margin of the orbit.

4.8.8. Squamosal

Only the left squamosal of NHMD 164741 is preserved (Figure 16). The squamosal is a tetraradiate bone that bound the dorsoposterior margin of the infratemporal fenestra and the dorsolateral margin of the supratemporal fenestra. The anterolateral (=postorbital) process of the squamosal sheets the posterior process of the postorbital and forms the posteromedial margin of the supratemporal fenestra. The anteromedial (=parietal) process is slightly laterally compressed, diverging from the anterolateral process at an angle of 45°. In *Pl. trossingensis* MSF 12.3 this divergence is 60° and possibly closer to the "in vivo" state [53]. This process marginates the posterior margin of the supratemporal fenestra laterally and contacts the posterior process of the parietal medially. The posterior process of the squamosal is 1.33 times longer than the anterior process, comprising over half the total dorsal length of the squamosal. In *Plateosaurus* and *Mac. itaquii*, the posterior process of the squamosal is shorter than the anterior process in all other described specimens (see Table 3). Ventrally, this process encapsulates the dorsalmost region of the quadrate head. This process is slightly ventrolaterally oriented, with a concave median margin that contacts part of the posterior process of the parietal and the distal surface of the paroccipital process of the otoccipital.

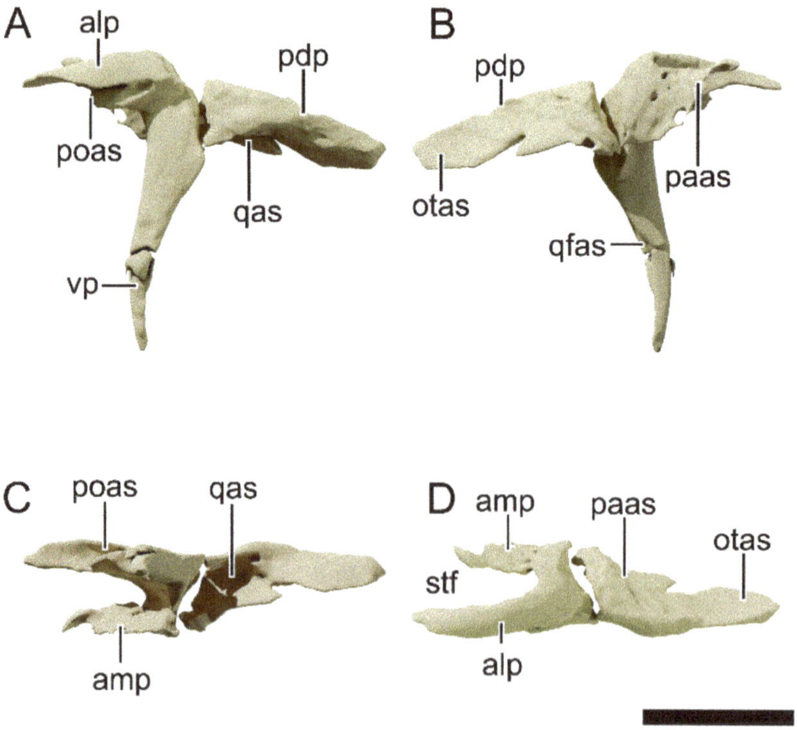

Figure 16. Digital reconstruction of the left squamosal of NHMD 164741. (**A**) Lateral view. (**B**) Medial view. (**C**) Ventral view. (**D**) Dorsal view. Abbreviations: alp, anterolateral process; amp, anteromedial process; otas, articular surface for the otoccipital; paas, articular surface for the parietal; pdt, posterodorsal process; poas, articular surface for the postorbital; qas, articular surface for the quadrate; qfas, articular surface for the quadrate flange; stf, supratemporal fenestra. Scale bar = 20 mm.

Table 3. Relative length of the posterior to the anterolateral processes of the squamosal across basal sauropodomorphs. Measurements were taken on the 3D models when available.

Squamosal—Posterior Process Length to Anterolateral Process Length		
Issi saaneq	NHMD 164741	1.31
Pl. trossingensis	AMNH FARB 6810	0.55
Pl. trossingensis	MSF 16.1	0.41
Pl. trossingensis	NAAG_00011238	0.48
Pl. trossingensis	MSF 12.3	0.53
Pl. trossingensis	MSF 15.4	0.83
Mac. itaquii	CAPPA/UFSM 0001b	0.77
Mas. carinatus	BP/1/5241	0.86
Bu. schultzi	CAPPA/UFSM 0035	0.88

The ventral (=quadrate) process is the longest of the squamosal processes, tapering distally and comprising the dorsoposterior margin of the infratemporal fenestra. It extends ventrally to about 60% of the infratemporal fenestra dorsoventral height. The distal end is deflected posteriorly as in *Plateosaurus*. The articular facet to the squamosal ramus of the quadratojugal is exposed laterally, meaning that it contacts the quadratojugal medially.

4.8.9. Jugal

The left jugal of NHMD 164758 is the best-preserved jugal element from both specimens (Figure 17), only missing the posteroventral process. This bone forms the ventral margin of the orbit and is concave dorsally and straight ventrally.

The anterior process of the jugal is five times anteroposteriorly longer than dorsoventrally tall, as in *Pl. trossingensis*, being relatively taller than in *Mac. itaquii* and *E. minor.* A dorsoventrally high jugal was found to a derived feature of *Pl. trossingensis* [2]. The anterior process of the jugal is laterally concave and divided in the mid-height by a longitudinal ridge. This process articulates to the maxilla beneath this ridge, at its anteroventral half, and medially to the lacrimal anteriorly and the ectopterygoid posterior to it. The posterodorsal (=postorbital) process of the jugal is deflected posteriorly, forming an angle of 137° to the anterior process. It contacts the ventral process of the postorbital anteriorly and bounds the anteroventral margin of the infratemporal fenestra. Medioventrally to this process sits the jugal fossa. The posterior process of the jugal is disarticulated from the main body in NHMD 164741. This process is slender and anteroposteriorly long and articulates with the quadratojugal lateroventrally.

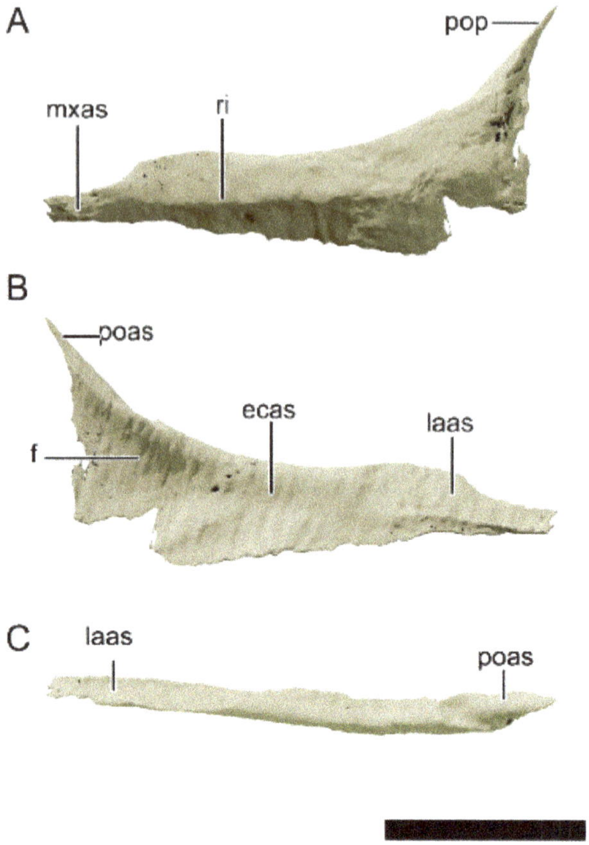

Figure 17. Digital reconstruction of the left jugal of NHMD 164758. (**A**) Lateral view. (**B**) Medial view. (**C**) Dorsal view. Abbreviations: ecas, articular surface for the ectopterygoid; f, fossa; laas, articular surface for the lacrimal; mxas, articular surface for the maxilla; poas, articular surface for the postorbital; pop, postorbital process; ri, ridge. Scale bar = 20 mm.

4.8.10. Quadratojugal

The left quadratojugal of NHMD 164741 is preserved (Figures 4 and 5), however the distal ends of the anteroventral (=jugal) and posterodorsal (=squamosal) processes are missing. The quadratojugal delimits the posteroventral corner of the infratemporal fenestra. The main body of the quadratojugal is posteriorly convex and slightly posteroventrally oriented. The anteroventral and posterodorsal processes are almost perpendicular, being separated by an 84° angle. In some specimens of *Pl. trossingensis*, this angle is less than 45° or subparallel, however, these acute angles are possibly the result of plastic deformation and were originally close to 70° [53,63]. The anteroventral process is laterally convex and articulates to the posteroventral process of the jugal both medially and ventrally, as observed by a ventral fossa in this process. The posterodorsal process is inclined medially to contact the lateral margin of the quadrate.

4.8.11. Quadrate

NHMD 164741 preserves the only complete quadrate (left quadrate, Figure 18), whereas both quadrates of NHMD 164758 have only the distal ends preserved (Figure 19). The main shaft of the quadrate is gently concave posteriorly and forms an angle of 153° to the quadrate head. The anterolateral and dorsal surfaces of the quadrate head are obscured by the ventral process of the squamosal. The head is anteroposteriorly expanded with a ridge at its lateral surface that continues ventrally to the lateral flange of the quadrate, which is poorly preserved in NHMD 164741. In anterior view, the quadrate is straight for most of its length, but slightly laterally oriented distally. The lateral medial condyle of the quadrate is ventrally positioned in relation to the lateral condyle, and its lateromedially inflated distally. The quadrate in NHMD 164741 is relatively tall when compared to the rostrum height. This ratio is also higher in NHMD 164741 than in other sauropodomorphs (Table 4). The medial (=pterygoid) flange of the quadrate is poorly preserved in NHMD 164741 but is in articulation to the posterolateral process of the pterygoid in the right element of NHMD 164758. This articulation occurs at the medial surface of this flange.

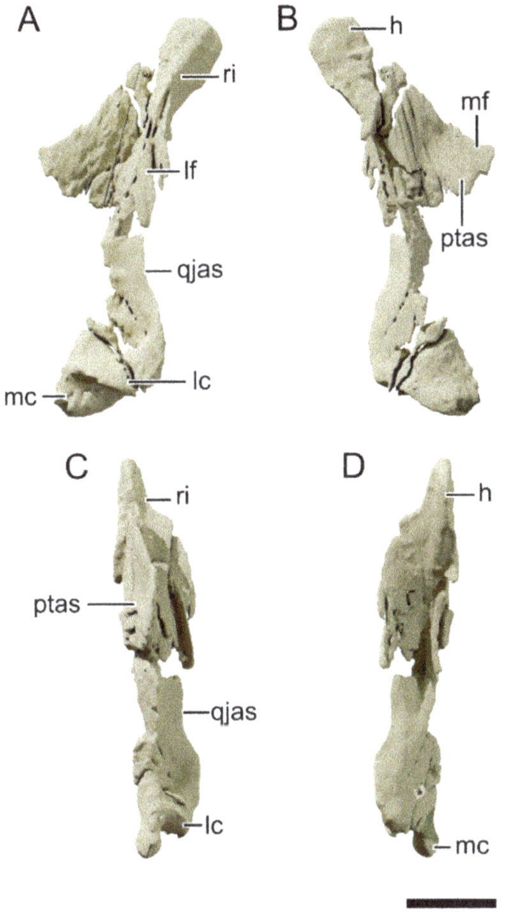

Figure 18. Digital reconstruction of the left quadrate of NHMD 164741. (**A**) Lateral view. (**B**) Medial view. (**C**) Anterior view. (**D**) Posterior view. Abbreviations: h, head; lc, lateral condyle; lf, lateral flange; mc, medial condyle; mf, medial flange; ptas, articular surface for the pterygoid; qjas, articular surface for the quadratojugal; ri, ridge. Scale bar = 20 mm.

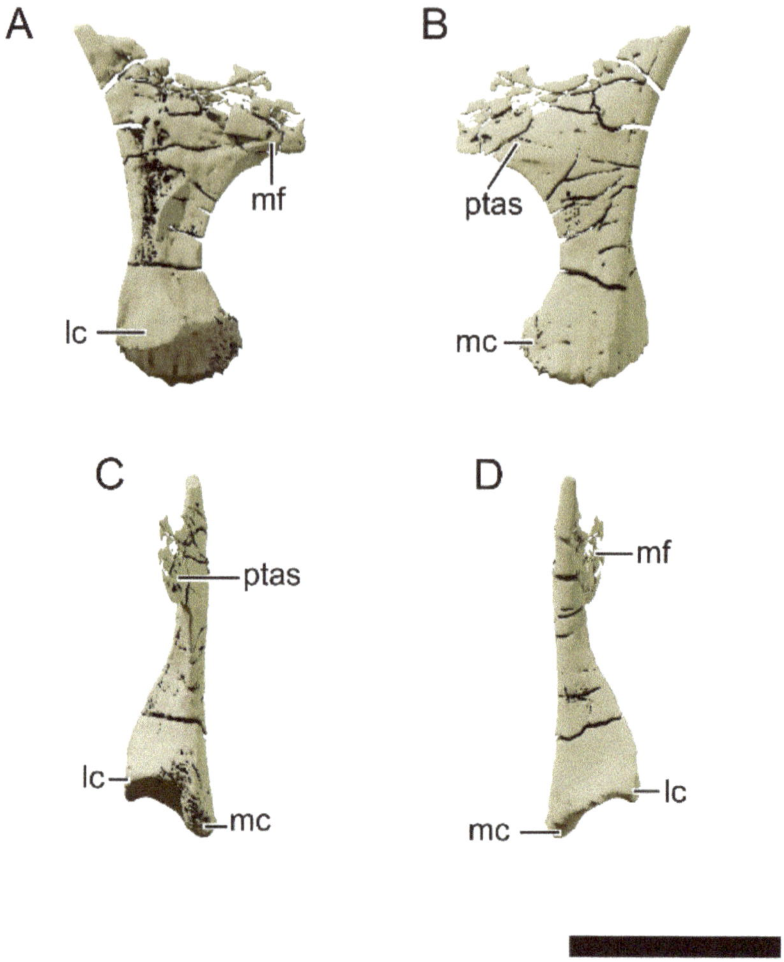

Figure 19. Digital reconstruction of the left quadrate of NHMD 164758. (**A**) Lateral view. (**B**) Medial view. (**C**) Anterior view. (**D**) Posterior view. Abbreviations: lc, lateral condyle; mc, medial condyle; mf, medial flange; ptas, articular surface for the pterygoid. Scale bar = 20 mm.

Table 4. Quadrate dorsoventral height to rostrum dorsoventral height ratio. The rostrum height is measured at the posterior margin of the external naris, from the ventral tip of the maxilla to the dorsal tip of the nasal. Measurements were taken on the 3D models when available.

Quadrate—Dorsoventral Height to Rostrum Dorsoventral Height		
Issi saaneq	NHMD 164741	1.57
Pl. trossingensis	AMNH FARB 6810	1.18
Pl. trossingensis	NAAG_00011238	1.30
Pl. trossingensis	MSF 11.4	1.27
Pl. trossingensis	MSF 15.4	1.13
Mac. itaquii	CAPPA/UFSM 0001a	1.28
Bu. schultzi	CAPPA/UFSM 0035	1.40
Mas. carinatus	BP/1/5241	1.12
N. intlokoi	BP/1/4779	1.11

4.8.12. Frontal

Both frontals of NHMD 164741 are displaced, with the left frontal preserving most of its total length (Figure 20), whereas in NHMD 16758 only the left frontal is preserved (Figure 21). The frontal delimits the orbit lateroventrally but is excluded from the supratemporal fenestra by the parietal. The dorsal margin of the frontal is slightly concave in lateral view. The anterodorsal margin of the frontal forms a depression at its mid-width, raising the lateral and medial margins of the bone. Lateral to this depression, an indentation encapsulates the posterior process of the prefrontal, as in *Plateosaurus*. In dorsal view, directly behind this indentation, the lateral margin of the frontal broadens, forming the posterolateral (=postorbital) process of the frontal. This process is well developed and extends further laterally in NHMD 164741, but not as much in NHMD 164758. The distal end of the process forms a groove for the insertion of the anterodorsal process of the postorbital. The medioventral margin of this process is bound by the parietal in NHMD 164758. The lateral half of the posterior margin of the posterolateral process is excavated forming the anterior margin of the supratemporal fossa, as in *Mac. itaquii* and *Plateosaurus* but absent in *Coloradisaurus* and *Mas. carinatus*.

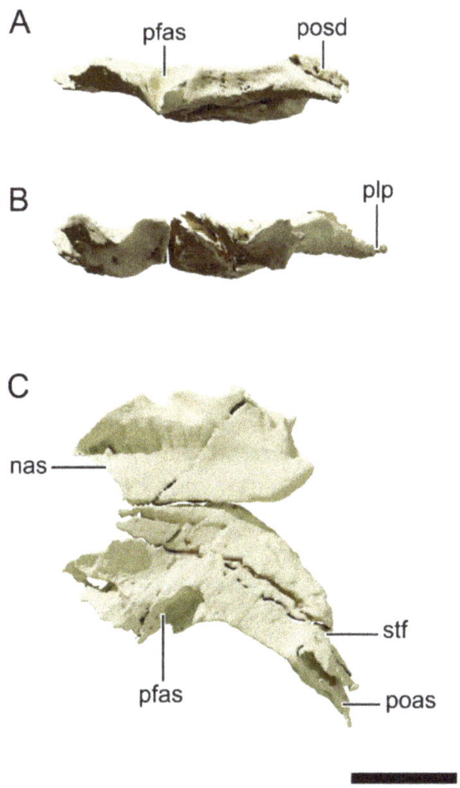

Figure 20. Digital reconstruction of the frontals of NHMD 164741. (**A**) Lateral view. (**B**) Anterior view. (**C**) Dorsal view. The left and right frontals were digitally articulated. Abbreviations: nas, articular surface for the nasal; pfas, articular surface for the prefrontal; poas, articular surface for the postorbital; plp, posterolateral process; stf, supratemporal fossa. Scale bar = 20 mm.

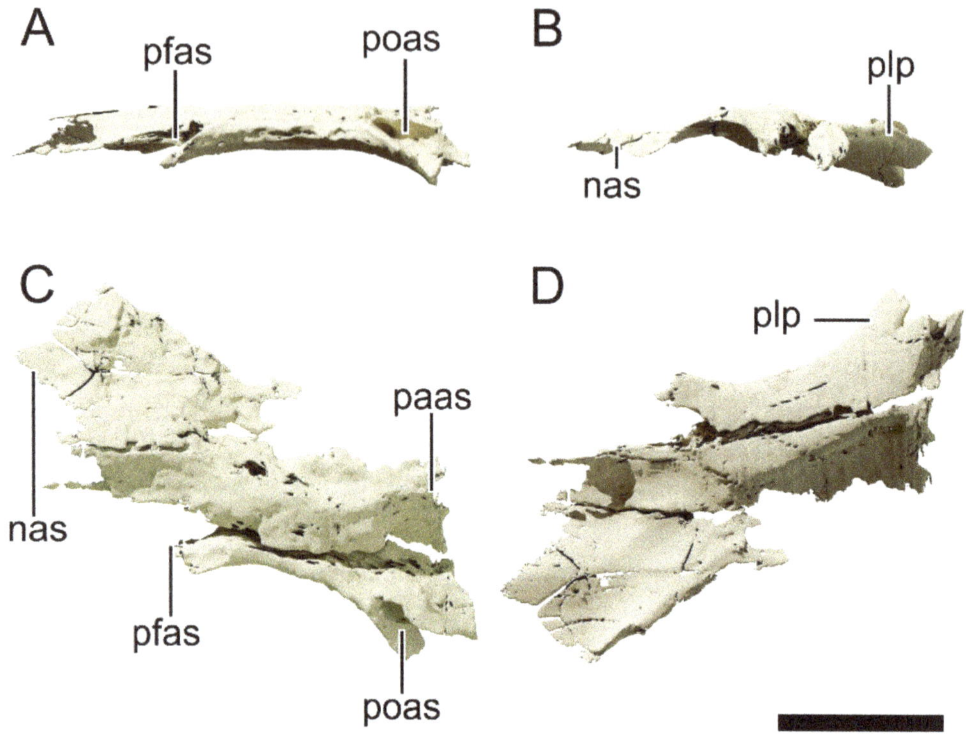

Figure 21. Digital reconstruction of the frontals of NHMD 164758. (**A**) Lateral view. (**B**) Anterior view. (**C**) Dorsal view. (**D**) Posterior view. The left and right frontals were digitally articulated. Abbreviations: nas, articular surface for the nasal; paas, articular surface for the parietal; pfas, articular surface for the prefrontal; poas, articular surface for the postorbital; plp, posterolateral process; stf, supratemporal fossa. Scale bar = 20 mm.

4.8.13. Parietal

Only the anterior part of the left parietal is preserved in NHMD 164758 (Figure 22) and the posterolateral process of the left parietal in NHMD 164741 (Figures 4 and 5). The anterolateral process of the parietal articulates to the mediodorsal process of the postorbital, excluding the frontal from the supratemporal fenestra. This process bounds the anteromedial margin of the supratemporal fenestra. The frontal-parietal suture is rugose and slightly elevated medially, as in *Pl. trossingensis*. The posterolateral process of the parietal is gently convex laterally, anteroposteriorly longer than dorsoventrally tall and mediolaterally flat. Its lateral surface contacts the anteromedial and the proximal half of the posterior processes of the squamosal, and contacts medially the lateral surface of the paroccipital process of the otoccipital. The distal end of this process slopes ventrally.

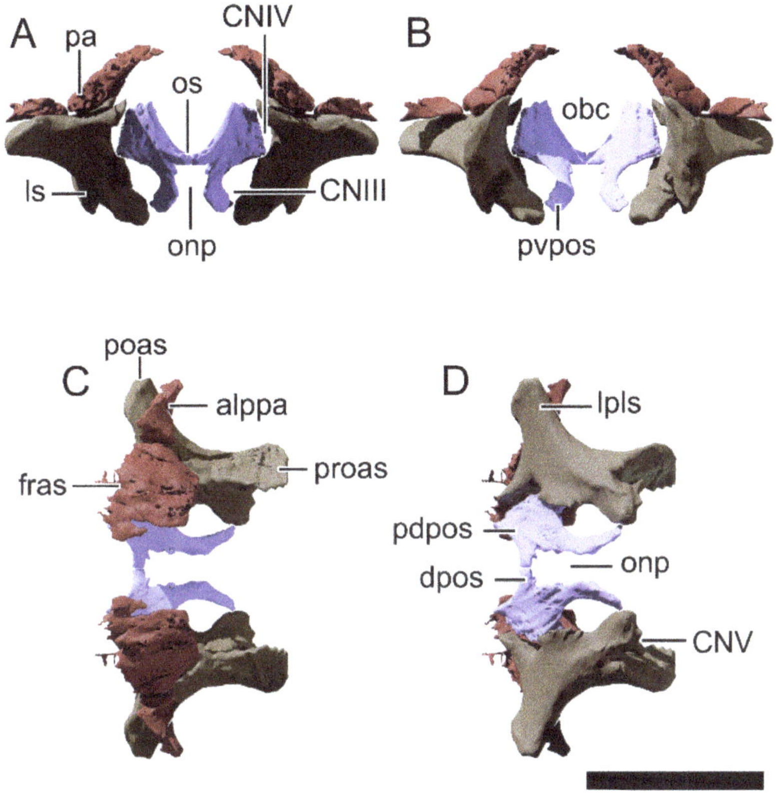

Figure 22. Digital reconstruction of the left upper braincase elements of NHMD 164758. (**A**) Anterior view. (**B**) Posterior view. (**C**) Dorsal view. (**D**) Ventral view. The elements were articulated and mirrored digitally for the reconstruction. Abbreviations: alppa, anterolateral process of the parietal; CN, cranial nerve; dpos, dorsal process of the orbitosphenoid; fras, articular surface for the frontal; lpls, lateral process of the laterosphenoid; ls, laterosphenoid; obc, olfactory bulb canal; onp, optic nerve passage; os, orbitosphenoid; pa, parietal; pdpos, posterodorsal process of the orbitosphenoid; poas, articular surface for the postorbital; proas, articular surface for the prootic; pvpos, posteroventral process of the orbitosphenoid. Scale bar = 20 mm.

4.8.14. Orbitosphenoid

Only the left orbitosphenoid of NHMD 164758 was preserved in the Greenland specimens, corresponding to a thin sheet of bone. When articulated in its original position (Figure 22), the orbitosphenoid forms the anterior wall of the braincase and contacts the laterosphenoid in three points. The orbitosphenoid contains three processes, one on the anterior portion (dorsal process) that contacts its counterpart medially, and two (posterodorsal and posteroventral processes) on the posterior portion, articulating with the laterosphenoid. The dorsal process forms the anteroventral margin of the olfactory bulb. The posteroventral process is the longest. It differs from the condition in *Mas. carinatus* (BP/1/5241) where the posterodorsal process is the longest. The distal half of that process is laterally deflected and its medial surface forms the dorsolateral margins of the optic nerve passage.

The contacts between the laterosphenoid and the orbitosphenoid were digitally reconstructed. These contacts would form two lateral foramina: a dorsal foramen that is smaller in diameter and a larger ventral foramen. This is another feature that distinguishes NHMD 164758 from *Mas. carinatus* (BP/1/5241), as the dorsal foramen is the largest in the latter.

The dorsal foramen would have formed the passage of cranial nerve IV (CNIV), whereas the ventral would have formed the passage of cranial nerve III (CNIII).

4.8.15. Laterosphenoid

The laterosphenoids of both NHMD 164741 and NHMD 164758 are disarticulated and displaced. When articulated in its original position in NHMD 164758 (Figure 22), the laterosphenoid contacts the orbitosphenoid anteriorly, the frontal anterodorsally, the parietal dorsally, and the postorbital laterally. However, it remains unclear if there is a ventral contact to the basisphenoid, as in *Pl. trossingensis*. The anterodorsal ramus of the laterosphenoid is a finger-like anterior projection that contacts the frontal distally and dorsally, and the orbitosphenoid medially. The lateral (=postorbital) process extends laterodorsally and has a distal inflated articular surface for the postorbital. This process is dorsoventrally robust and contacts the medial surface of the anterodorsal process of the postorbital. In *Pl. trossingensis* (AMNH FARB 6810), the postorbital process is proportionally much longer, slender and anterodorsally curving than in NHMD 164758 [52]. In this sense, the laterosphenoid of NHMD 164758 is closer to the condition observed in *Mas. carinatus* and *N. intlokoi*. The posterior region of the laterosphenoid articulates with the prootic and contains a deep notch that forms the anterior margin of the large trigeminal foramen (CNV), as in *Coloradisaurus* and *Plateosaurus*, differing from the gently concave condition in *N. intlokoi* and *Mas. carinatus*.

4.8.16. Otoccipital

The paroccipital process of NHMD 164741 (Figures 4 and 5) is lateromedially flattened and dorsoventrally expanded. Distally it contacts the posterior process of the squamosal. Anteriorly, at its base. the process expands to create the basioccipital articular surface. A deep, oval groove is present at the ventral surface of the proximal part of the paroccipital process, as in *Pl. trossingensis* (AMNH FARB 6810).

4.8.17. Basioccipital

Only a small fragment of the occipital condyle is preserved in NHMD 164741 (Figures 4 and 5), and part of the occipital condyle and the right basal tubera is preserved in NHMD 164758 (Figures 6 and 7). The occipital condyle is convex ventrally and concave dorsally at the foramen magnum exit. The basal tubera is separated from the occipital condyle by a deep lateral fossa. This condition is not as extreme as in *Pl. gracilis* (GPIT 18318a). Due to its disarticulated preservation, it is not possible to discern if the basioccipital is located dorsally to the basisphenoid.

4.8.18. Basisphenoid

The main body of the basisphenoid is missing on NHMD 164758 but preserved disarticulated in NHMD 164741 (Figure 23). Both specimens preserve the right basipterygoid process of the basisphenoid, and the parasphenoid process (Figure 24). As this element is poorly preserved in both specimens, so that the autapomorphic feature of a high interbasipterygoid septum with a median process of *Pl. trossingensis* cannot be assessed. The basipterygoid process is wrapped by the median "hook-like" process of the pterygoid. This feature was considered autapomorphic for *Pl. trossingensis* [10] but was also described for *Mas. carinatus* (BP/1/5241) [16], possibly being a variable trait among early sauropodomorphs [53]. In NHMD 164758, the basipterygoid process extends anteroventrally, as in the juvenile *Mas. carinatus* (BP/1/4376), whereas in NHMD 164741, the process appears to extend ventrally and only slightly anteriorly, as in the adult *Mas. carinatus* (BP/1/5241), *U. tolentinoi* (UFSM11069) and *Thecodontosaurus* (YPM 2192). In *Pl. trossingensis* (MSF 15.8.1043 and MSF 07.M) this process is posteroventrally oriented. The parasphenoid process is a long and slender anteriorly oriented process. In lateral view, it is smooth and does not feature the lateral deep groves observed in *Pl. trossingensis* (MSF 07.M, MSF 08.M, and MSF 15.8.1043). In lateral view, the ventral margin of the parasphenoid is straight and the dorsal margin is gently convex distally. The dorsal surface of the parasphenoid bears a deep groove, making the cross-section of this bone U-shaped. This condition is similar to that observed in *Mas. carinatus* (BP/1/5241) and appears to be so in *Pl. trossingensis* (MSF 15.8.1043).

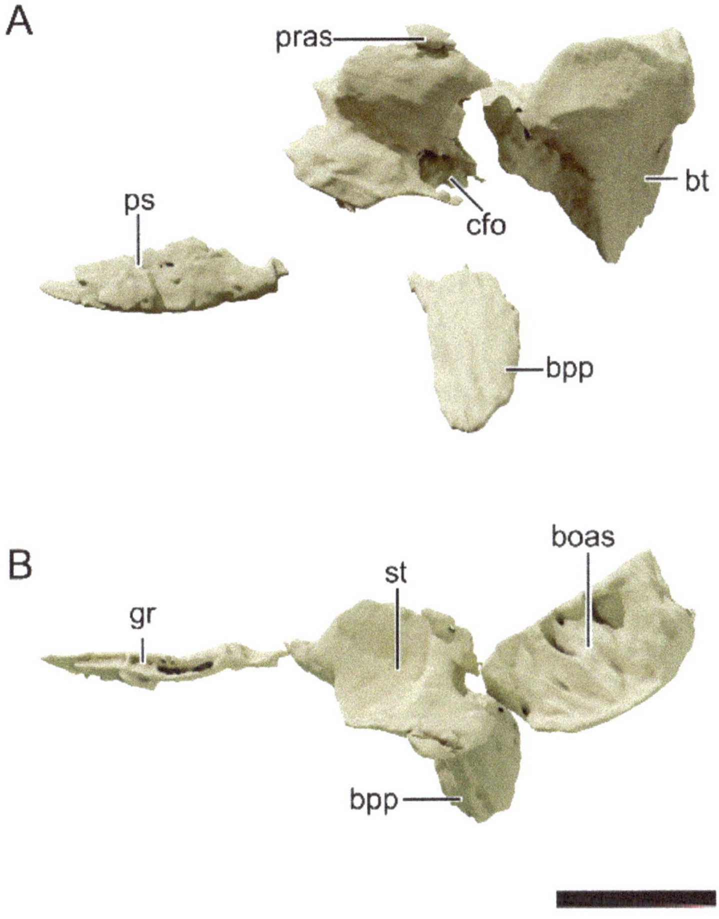

Figure 23. Digital reconstruction of the basisphenoid of NHMD 164741. (**A**) Lateral view. (**B**) Dorsal view. The elements were articulated digitally for the reconstruction. Abbreviations: boas, articular surface for the basioccipital; bpp, basipterygoid process; bt, basal tubera; cfo, carotid foramen; gr, groove; pras, articular surface for the prootic; ps, parasphenoid; st, sella turcica. Scale bar = 20 mm.

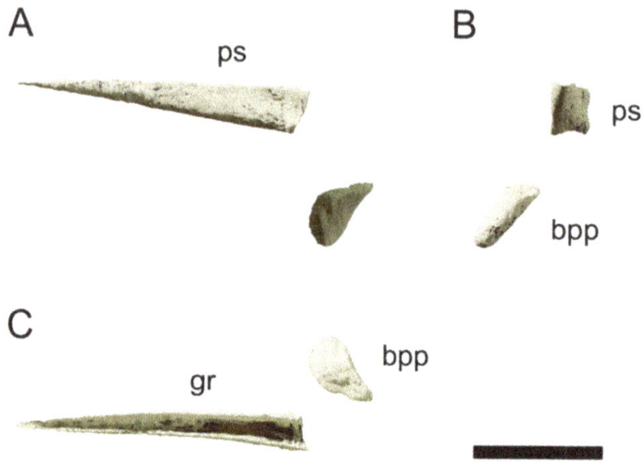

Figure 24. Digital reconstruction of the basisphenoid of NHMD 164758. (**A**) Lateral view. (**B**) Anterior view. (**C**) Dorsal view. The elements were articulated digitally for the reconstruction. Abbreviations: bpp, basipterygoid process; gr, groove; ps, parasphenoid. Scale bar = 20 mm.

4.8.19. Palate

The palate is best preserved in NHMD 164758 (Figures 25 and 26), with all composing elements partially articulated, almost complete and in anatomical position. NHMD 164741 preserves both pterygoids, the left ectopterygoid, fragments of both palatines and fragments of the left vomer. NHMD 164758 preserves both pterygoids, both ectopterygoids, both palatines and both vomers. The postpalatine fenestra is bound anteriorly by the palatine, laterally by the maxillae, medially by the pterygoids and posteriorly by the ectopterygoids. This fenestra is slightly anteroposteriorly longer than lateromedially wide (Figure 25).

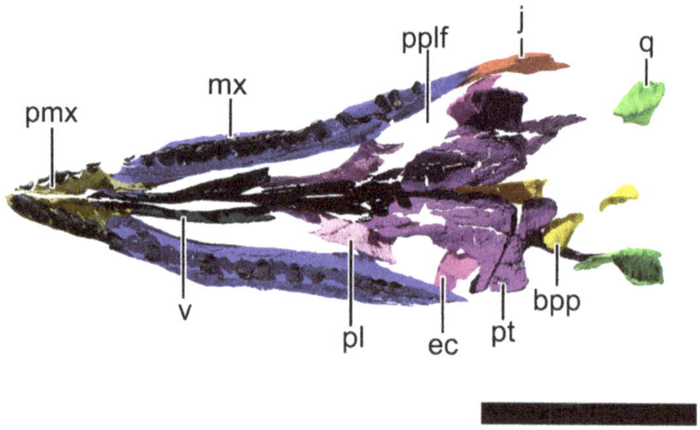

Figure 25. Digital reconstruction of the skull of NHMD 164758 in ventral view. The elements were articulated digitally for the reconstruction. Abbreviations: bpp, basipterygoid process; ec, ectopterygoid; j, jugal; mx, maxilla; pl, palatine; pmx, premaxilla; pplf, postpalatine fenestra; pt, pterygoid; q, quadrate; v, vomer. Scale bar = 50 mm.

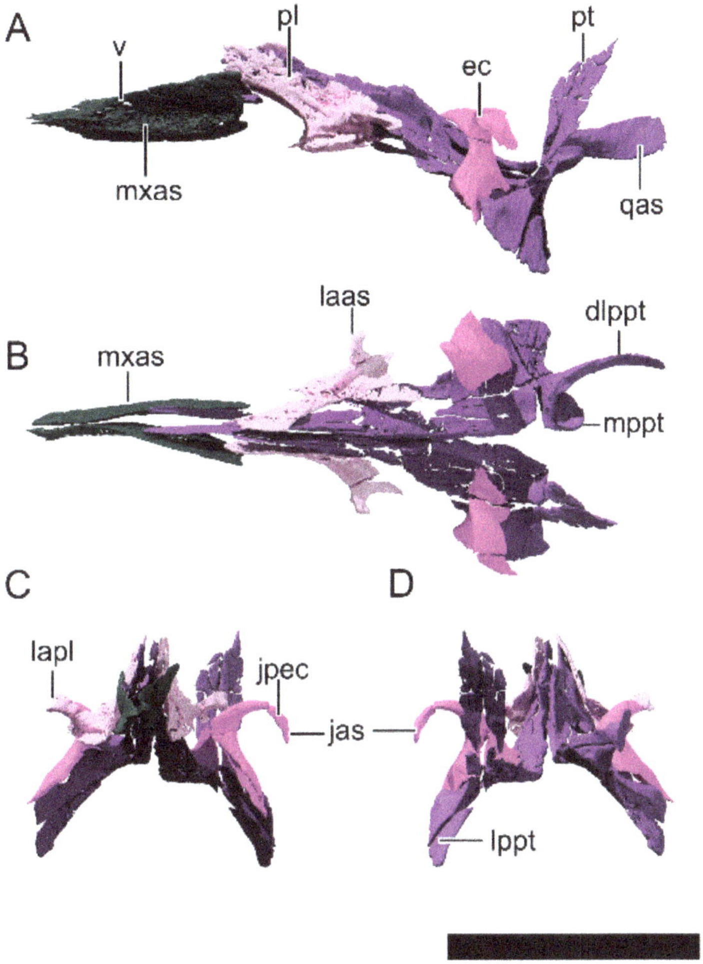

Figure 26. Digital reconstruction of the palate of NHMD 164758. (**A**) Left lateral view. (**B**) Dorsal view. (**C**) Anterior view. (**D**) Posterior view. The elements were articulated digitally for the reconstruction. Abbreviations: dlppt, dorsolateral process of the pterygoid; ec, ectopterygoid; ecf, ectopterygoid fossa; jas, articular surface for the jugal; jpec, jugal process of the ectopterygoid; laas, articular surface for the lacrimal; lapl, lateral process of the palatine; lppt, lateral process of the pterygoid; mppt, posteromedial process of the pterygoid; mxas, articular surface for the maxilla; pl, palatine; pt, pterygoid; qas, articular surface for the quadrate; v, vomer. Scale bar = 20 mm.

4.8.20. Pterygoid

None of the pterygoids are complete, but the best-preserved are the left element in NHMD 164741 (Figure 27) and the right one in NHMD 164758 (Figure 28). The pterygoid is the largest component of the palate. It is a tetraradiate bone that composes the posterolateral area of the palate. The anterior process of the pterygoid is the longest and is subdivided into two areas, a distal half that contacts the vomer laterally, and a proximal half that contacts the palatine dorsally. However, the distalmost part seems to be broken in the latter. The proximal half of the anterior process is lateroventrally expanded forming a dorsoventrally high lamina.

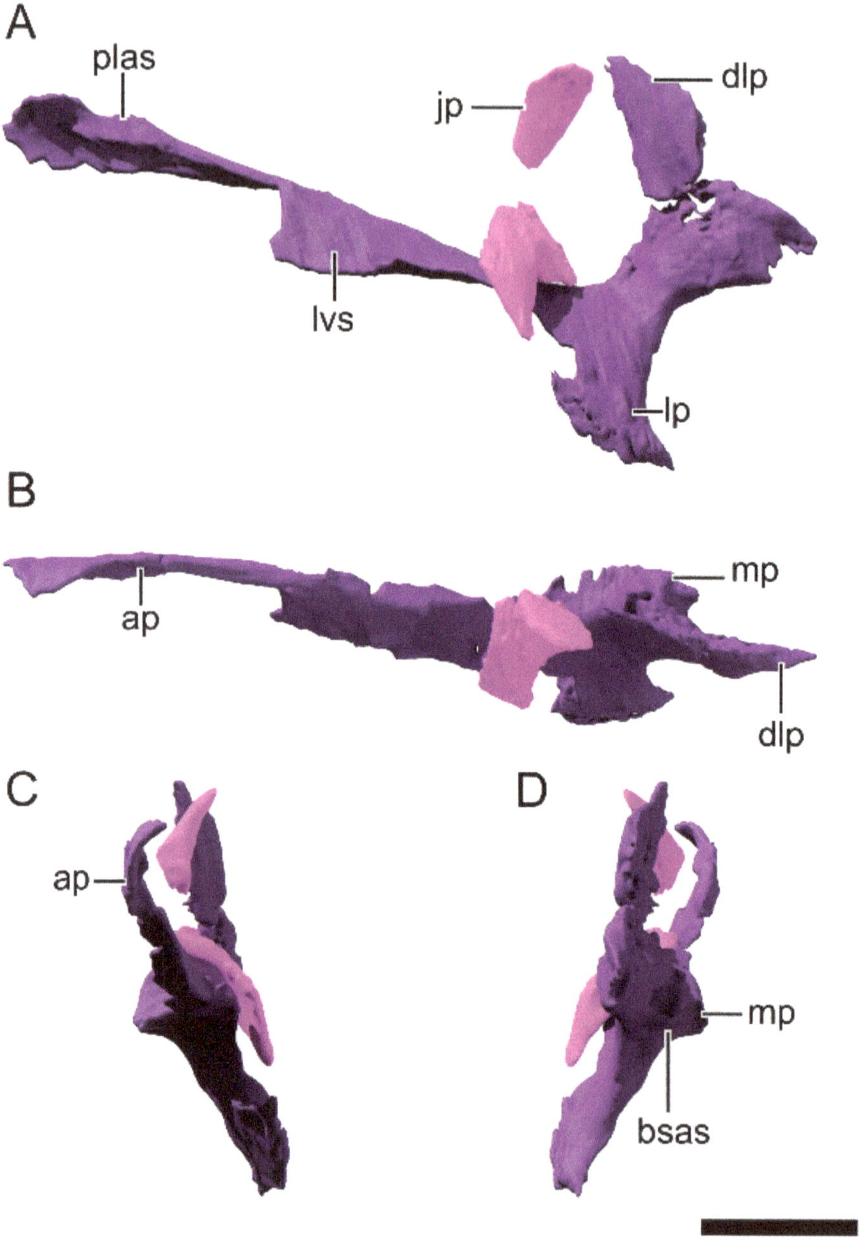

Figure 27. Digital reconstruction of the left pterygoid and ectopterygoid of NHMD 164741. (**A**) Lateral view. (**B**) Dorsal view. (**C**) Anterior view. (**D**) Posterior view. Abbreviations: ap, anterior process; bass, articular surface for the basisphenoid; dlp, dorsolateral process; ecf, ectopterygoid fossa; jp, jugal process; lp, lateral process; lvs, lateroventral sheet of bone; mp, medial process; plas, articular surface for the palatine. Scale bar = 20 mm.

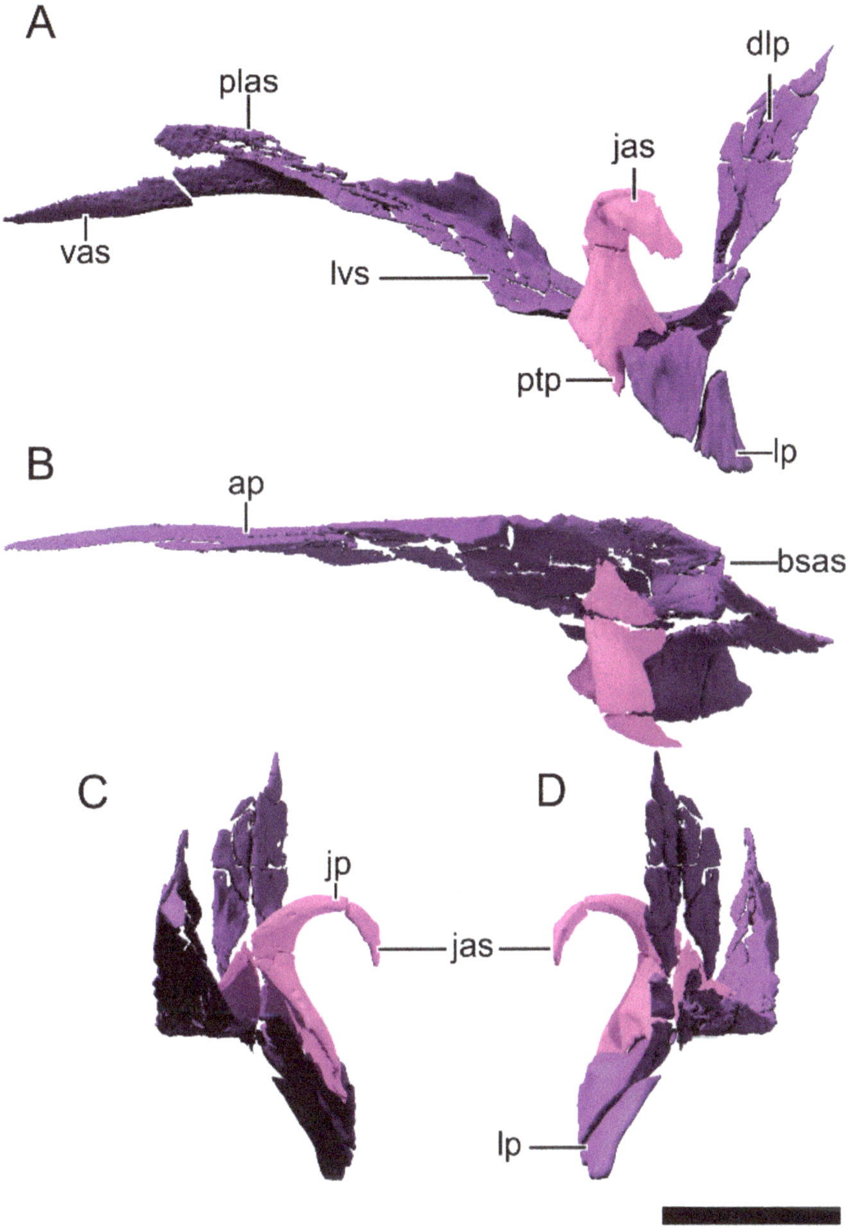

Figure 28. Digital reconstruction of the left pterygoid and ectopterygoid of NHMD 164758. (**A**) Lateral view. (**B**) Dorsal view. (**C**) Anterior view. (**D**) Posterior view. Abbreviations: ap, anterior process; bsas, articular surface for the basisphenoid; dlp, dorsolateral process; ecf, ectopterygoid fossa; jas, articular surface for the jugal; jp, jugal process; lp, lateral process; lvs, lateroventral sheet of bone; mp, medial process; plas, articular surface for the palatine. Scale bar = 20 mm.

The lateral process of the pterygoid is anteroposteriorly expanded and dorsoventrally flattened. This process is separated from the anterior process by an almost 90° angle in lateral view. The medial surface that separates both processes is highly dorsoventrally concave. The distal area of the lateral process is inflated both anteroposteriorly and dorsoventrally. This process contacts the ectopterygoid laterally and anteriorly.

The dorsolateral (=quadrate) process of the pterygoid is broken in both specimens. This process curves lateroposteriorly and contacts the pterygoid flange of the quadrate laterally. Medially at the proximal surface of this process, it contacts the lateral surface of the basipterygoid process. As mentioned before, the medial process of the pterygoid is hook-shaped and wraps the basipterygoid laterally.

4.8.21. Ectopterygoid

The left ectopterygoid of NHMD 164758 is the best preserved (Figure 28). This bone is divided into dorsal (=jugal) and ventral (=pterygoid) processes. The dorsal process is hook-shaped and expands dorsolaterally in the proximal half and ventrally in the distal half. The distal half contacts laterally the posteroventral half of the jugal. This contact forms the posterior margin of the postpalatine fenestra in dorsal view. The contact margin of the dorsal process of the ectopterygoid is similar to that of *Mac. itaquii*, contrasting with the anteriorly projected, T-shaped tip of *Pl. trossingensis* (AMNH FARB 6810).

The ventral process of the ectopterygoid is dorsoventrally elongated, expanding over the lateral surface of the pterygoid. Its distal tip is dorsally obscured by the distal expansion of the lateral process of the pterygoid.

4.8.22. Palatine

Both palatines are preserved in NHMD 164758 (Figure 29) as well as in NHMD 164741, but are fragmented in the latter. The palatine is anteroposteriorly elongated and lateromedially flattened. Its anterior process expands anterodorsally and slightly laterally. This process contacts the posterior margin of the vomer distally and the anterior process of the pterygoid medially. Ventral to the contact with the pterygoid, the palatine expands medially and posteriorly. This expansion forms a medial ridge and a small tubercle posteriorly. This tubercle, however, does not show the autapomorphic peg-like morphology of *Pl. trossingensis* [52,63]. The lateral process of the palatine forms the anterior and anteromedial margins of the postpalatine fenestra. This process contacts the ventral process of the lacrimal laterally. Ventral to this process, the palatine contacts the maxilla laterally. The posterior process of the palatine contacts the proximal half of the anterior process of the pterygoid dorsomedially.

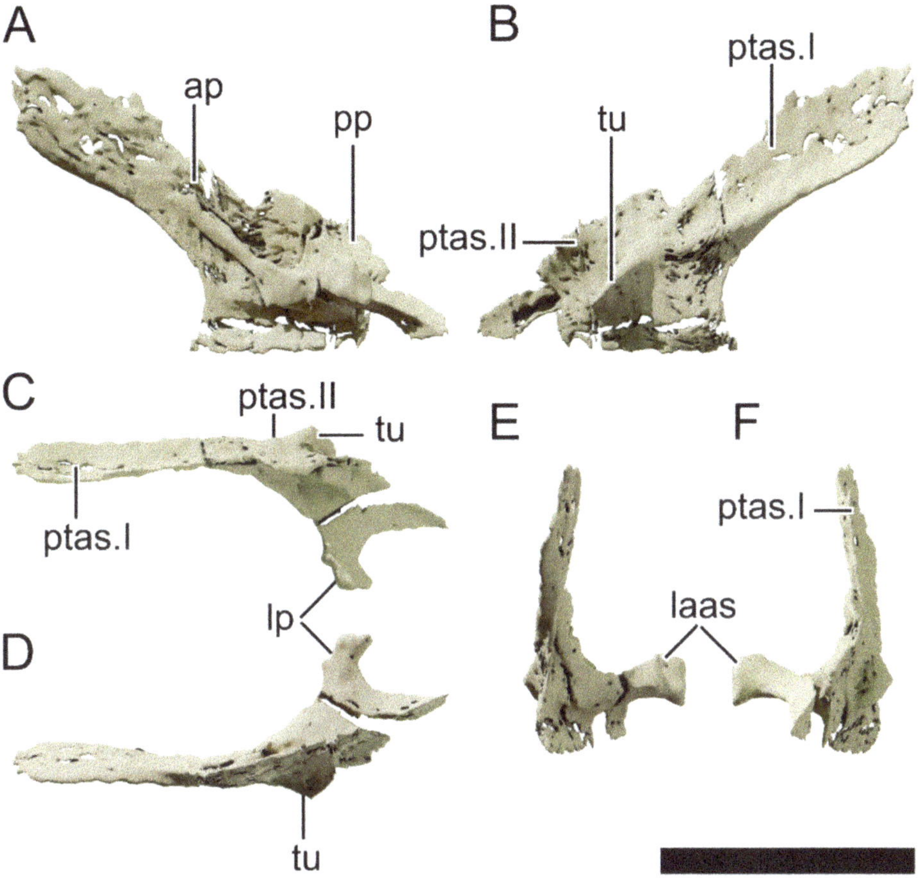

Figure 29. Digital reconstruction of the left palatine of NHMD 164758. (**A**) Lateral view. (**B**) Medial view. (**C**) Dorsal view. (**D**) Ventral view. (**E**) Anterior view. (**F**) Posterior view. Abbreviations: ap, anterior process; laas, articular surface for the lacrimal; lp, lateral process; pp, posterior process; ptas, articular surface for the pterygoid; tu, tubercle.

4.8.23. Vomer

The vomers are better preserved in NHMD 164758 and form the anterior region of the palate (Figures 25 and 26). This bone is anteroposteriorly elongated and triangular in lateral view, with a tapered anterior margin and a dorsoventrally tall posterior margin. The overall shape of the vomer in lateral view and the lack of foramina piecing this bone resemble the anatomy of the vomer in *Pl. trossingensis* (AMNH FARB 6810) and differs from the S-shaped vomer in *Mas. carinatus* (BP/1/5421). The anteroposterior length of the vomer is slightly over 0.25 of the total skull anteroposterior length. This ratio is around 0.18 in *Pl. trossingensis* (AMNH FARB 6810), 0.22 in *N. intlokoi* (BP/1/4779) and 0.31 in *Mas. carinatus* (BP/1/5241). The vomer is laterally convex in anterior view, having a ventral lateral expansion that articulates with the ventromedial surface of the maxilla. The medial surface of the posterior part of the vomer contacts the anterior process of the pterygoid over a shallow medial ridge in the vomer. The posterior surface contacts the anterior process of the palatine in an almost straight margin.

4.8.24. Dentary

NHMD 164741 preserves both dentaries but lacks the anteriormost region of both (Figure 30). NHMD 164758 preserves both complete dentaries partially in articulation (the right mandible is slightly ventrally deflected) (Figure 31). The dentary is the largest bone in the mandible. It articulates posteromedially with the splenial, dorsomedially with the coronoid, posteroventrally with the angular and posterodorsally with the surangular. Posteriorly, the dentary bounds the anterior margin of the mandibular fenestra. The dentary is slender and long, being over six times longer than tall, and over twice taller than wide (see Table 2 for measurements). In lateral view, the dorsal margin of the dentary is straight, and the ventral margin is slightly concave, as in *Pan. protos, Ba. agudoensis, Bu. Schultzi, U. tolentinoi, Mac. itaquii* and *Plateosaurus*, and different from the straight margin of early sauropodomorphs such as *Bu. Schultzi* and *Sat. tupiniquim*. The symphysis in NHMD 164758 is straight at the medial contact surface to its counterpart. Posterior to the symphysis, the dentary is laterally deflected. The lateral and medial surfaces are parallel for most of the dentary extent, tapering posteriorly to the last alveolus. The lateral surface of the anteriormost region of the left dentary in NHMD 164741 is missing, exposing three dentary tooth roots. In both NHMD 164741 and 164758, the dentary preserved 18 alveoli, but this number might be underestimated in NHMD 164741 due to the lack of the anteriormost region. However, its anteriormost medial margin includes the posterior part of the symphysis, meaning that it does not miss much of the anterior area. Therefore, NHMD 164741 could have a maximum of 20 dentary teeth, a feature seen in mature individuals [53].

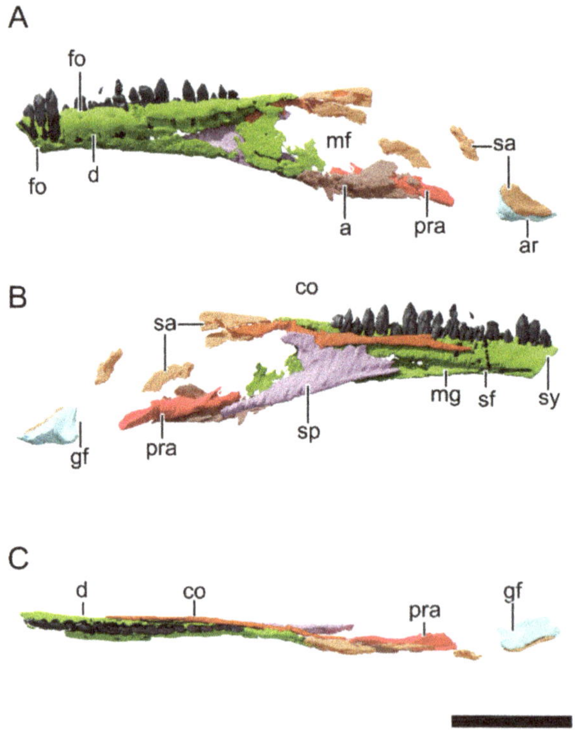

Figure 30. Digital reconstruction of the left mandible of NHMD 164741. (**A**) Lateral view. (**B**) Medial view. (**C**) Dorsal view. Abbreviations: a, angular; ar, articular; co, coronoid; d, dentary; fo, foramen; gf, glenoid fossa; mf, mandibular fenestra; mg, Meckelian groove; pra, prearticular; sa, surangular; sf, secondary fossa; sp, splenial; sy, symphysis. Scale bar = 50 mm.

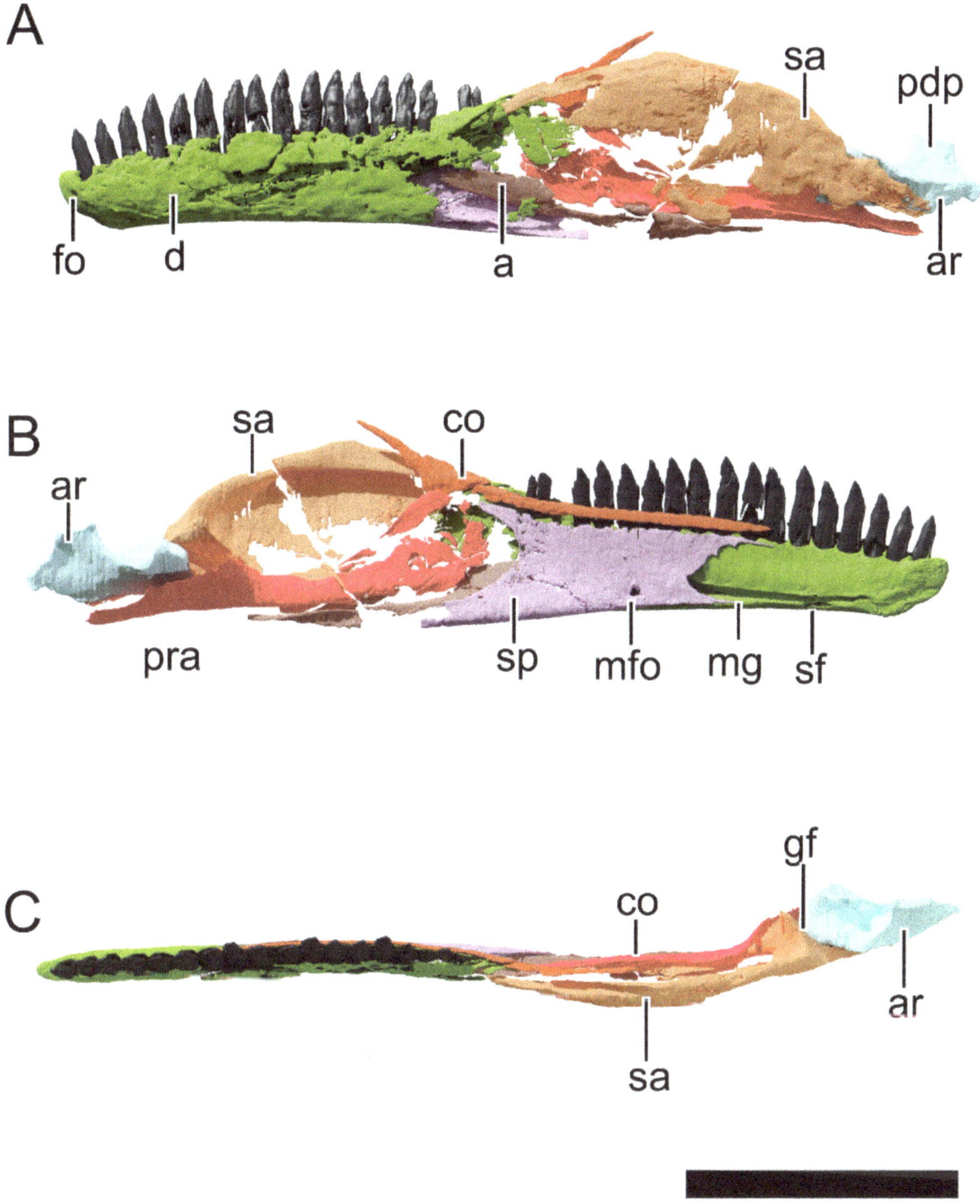

Figure 31. Digital reconstruction of the left mandible of NHMD 164758. (**A**) Lateral view. (**B**) Medial view. (**C**) Dorsal view. Abbreviations: a, angular; ar, articular; co, coronoid; d, dentary; fo, foramen; gf, glenoid fossa; mf, mandibular fenestra; mfo, mylohyoid foramen; mg, Meckelian groove; pdp, posterodorsal process; pra, prearticular; sa, surangular; sf, secondary fossa; sp, splenial; sy, symphysis. Scale bar = 50 mm.

A row of foramina is present at the first half of the dorsolateral surface of the dentary, just below the alveolar margin, as in pre-Norian sauropodomorphs, *U. tolentinoi*, *Mac. itaquii* and *Plateosaurus*. Different from the 9–12 foramina in *Pl. trossingensis* (AMNH FARB 6810, MSF 01 and MSF 16.1), both Greenland specimens only preserve 3–4 foramina.

In NHMD 164741, these foramina are large and open anterodorsally, whereas in NHMD 164758, they are smaller and open dorsally. Both dentaries of NHMD 164758 preserve a small anteroventral foramen at the lateral surface near the ventral corner of the dentary. This foramen was not reported for any other plateosaurid (*Mac. itaquii*, *U. tolentinoi* and *Plateosaurus*), but this might be due to the poorer preservation of this area in most specimens. In NHMD 164741 this feature cannot be assessed with confidence due to damage in this area. In *Pam. berberenai* (ULBRA-PVT016) this foramen is situated more dorsally. *Ba. agudoensis* (UFRGS-PV-1099-T in [39] Figure 2) seems to possess a simitar foramen as NHMD 164758.

The Meckelian groove extends longitudinally along the ventral part of the medial surface of the dentary. Anteriorly, this groove is constricted, but posteriorly it is dorsoventrally expanded and covered by the splenial medially. An anterior sheet of bone expands posteroventrally posterior to the symphysis, covering the anteriormost region of the Meckelian groove. This feature was observed in *Mac. itaquii* and *U. tolentinoi*, but is more developed in the latter. A deep, secondary fossa is present ventral to the anterior region of the Meckelian groove, as in *U. tolentinoi*, and unlike the single fossa of *Mac. itaquii* and *Plateosaurus*.

4.8.25. Splenial

NHMD 164758 preserves both complete splenials (Figure 31), and NHMD 164741 only the left element (Figure 30). The splenial is a rectangular sheet of bone that obscures the posteromedial surface of the dentary. Both the anterior and posterior margins bear two processes, one dorsally and one ventrally, making the anterior and posterior margins of the splenial concave in lateral view. The medial surface of the splenial is gently concave and contains a mylohyoid foramen ventrally at its anterior third. In *Bu. schultzi* (CAPPA/UFSM 0035), this foramen is located at the dorsoventral midpoint. In *Plateosaurus* this foramen is not visible in medial view. The lateral surface of the splenial contains a ventral sheet of bone that lies on the dorsal margin of the ventral region of the dentary. This sheet of bone continues posteriorly, forming the elongated posteroventral process. This process accommodates the anteroventral margin of the prearticular dorsally and the anteroventral margin of the angular ventrally. The dorsal margin of the splenial contacts the coronoid dorsally.

4.8.26. Intercoronoid/Coronoid

The intercoronoid is preserved on both mandibles of NHMD 164758 (Figure 31) and in the left mandible of NHMD 164741 (Figure 30). It is a slender and elongated bone that marginates the lingual surface of the dentary teeth. It extends anteriorly to the first third of the dentary. The coronoid is posterodorsally oriented and contacts the surangular laterally. The ventral margin of the coronoid is anteroposteriorly expanded and ventrally deflected, forming a triangular base as in *Pl. trossingensis* (AMNH FARB 6810). The coronoid is deflected dorsally in *N. intlokoi*.

4.8.27. Surangular

The left surangular in NHMD 164758 is the best-preserved surangular among the Greenland specimens (Figure 31). The bone composes most of the posterolateral surface of the mandible and forms the posterior margin of the mandibular fenestra. The surangular is sigmoid in lateral view, with the anterior half being more dorsoventrally expanded compared to the posterior half. The anterior (=dentary) process of the surangular is dorsally convex. It covers the posterodorsal corner of the dentary and is partially covered anteromedially by the coronoid. The anterior process is medioventrally expanded, forming a ventral groove that covers the dorsal corner of the adductor fossa. The lateral surface of the anterior process is slightly convex in anterior view and extends ventrally to contact the angular at its ventral margin. A medial expansion of the surangular is present posterior to its anterior process. This expansion forms a mediolaterally wide flange that articulates anteroventrally to the prearticular and posterodorsally forms the surangular contribution to the glenoid fossa. This contribution is marked laterally by a deep transverse groove, as in

Pam. berberenai, but is not as deep and marked in *Pl. trossingensis*. Posterior to the glenoid the surangular tapers dorsoventrally, marginating the lateral surface of the articular.

4.8.28. Angular

The left angular of NHMD 164741 is the best-preserved, although broken at its posterior half (Figure 30), whereas in NHMD 164758 both angulars are fragmented (Figure 31). It is a lateromedially flat bone with a medial concavity in anterior view. It contacts the posteroventral process of the dentary medially, the prearticular medially and the surangular dorsally. Anterodorsally it contributes to the posteroventral corner of the mandibular fenestra.

4.8.29. Prearticular

Both prearticulars are present in NHMD 164758 (Figure 31), but only the main shaft of the left prearticular of NHMD 164741 is preserved (Figure 30). Anteriorly, the prearticular is dorsoventrally expanded and lateromedially flat. This anterior region contacts the splenial anteriorly and the surangular dorsally and is sub-circular in lateral view. The main shaft is constricted dorsoventrally and concave laterally. Medially it bounds the ventral corner of the internal mandibular fenestra. The posteromedial process expands dorsally to contact the medial process of the surangular anteriorly and the medial surface of the articular laterally.

4.8.30. Articular

The anterior part of NHMD 164741 left articular is preserved, and both articulars of NHMD 164758 are preserved and in articulation (Figures 31 and 32). The anterior process of the articular forms the posterior half of the glenoid fossa dorsally. It is bound by a dorsolaterally oriented ridge and a well-developed medial pyramidal process, as in *Pl. trossingensis* (AMNH FARB 6810). In *Mac. itaquii*, this medial expansion is less developed. The lateral surface of the articular is concave in anterior view and contacts the posterior process of the surangular, whereas medially the articular is slightly convex and contacts the posterior process of the prearticular. The articular of NHMD 164758 has a peculiar dorsoposterior process, situated posterior to the glenoid fossa (Figure 32). This process is as tall as anteroposteriorly long and squared shaped in lateral view. This morphology differs from the dorsoventrally short, anteroposteriorly elongated and gently dorsally convex dorsoposterior process of the articular in *Mac. itaquii* (CAPPA/UFSM 0001b) [42] and the short (or even absent) and posteriorly convex dorsoposterior process of *Pl. trossingensis* (AMNH FARB 6810) [52]. This process is absent in *Bu. schultzi* (CAPPA/UFSM 0035) [41]. This process tapers lateromedially towards its dorsal end, being triangular in cross-section. The retroarticular process tapers posteriorly and ends posterior to this dorsal process.

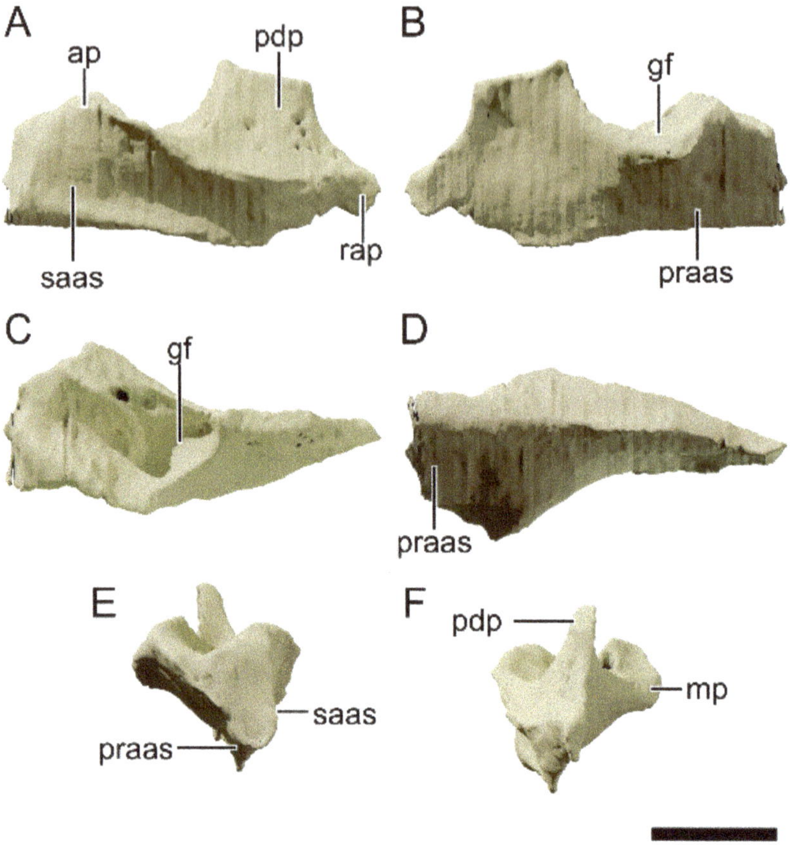

Figure 32. Digital reconstruction of the left articular of NHMD 164758. (**A**) Lateral view. (**B**) Medial view. (**C**) Dorsal view. (**D**) Ventral view. (**E**) Anterior view. (**F**) Posterior view. Abbreviations: ap, anterior process; gf, glenoid fossa; mp, medial process; pdp, posterodorsal process; praas, articular surface for the prearticular; rap, retroarticular process; saas, articular surface for the surangular. Scale bar = 10 mm.

4.8.31. Dentition

The premaxillae of NHMD 164758 preserve a total of five tooth alveoli each. Their roots comprise around two-thirds of the total tooth height on the most complete preserved teeth. The premaxillary tooth crowns are conical and slightly bent distally. In NHMD 164758, the first tooth (mesialmost tooth) of the premaxilla is the tallest in the series, differing from the condition in non-Bagualosaurian sauropodomorphs, and similar to that in *Plateosaurus*, *Ba. agudoensis*, and *Mas. carinatus*. A constriction separates the crown from the root. The mesial carina is convex and lacks denticles, as in most sauropodomorphs. The distal carina is concave and preserves coarse denticles set perpendicular to the tooth margin. These denticles are set apical to the basal third of the crown height and end just before reaching the apex. Both premaxillae contain replacement teeth erupting from the lingual side of the descended teeth. By the time the root of the replacement tooth starts to form, the crown and its denticles are fully developed, as observed in NHMD 164758 (Figure 33).

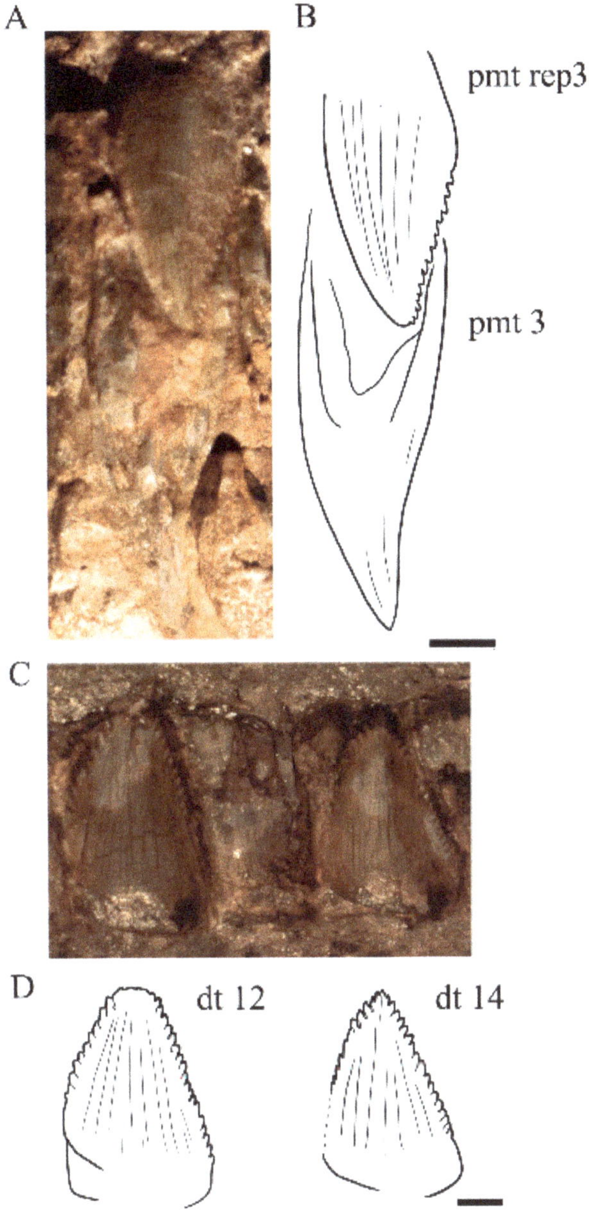

Figure 33. Photograph and schematic drawing of the teeth of NHMD 164741 and NHMD 164758. (**A**) Photograph of the left premaxillary teeth of NHMD 164758 in labial view. (**B**) Schematic drawing of the left premaxillary teeth of NHMD 164758 in labial view. (**C**) Photograph of the left dentary teeth 12 to 14 of NHMD 164741 in labial view. (**D**) Schematic drawing of the left dentary teeth 12 to 14 of NHMD 164741 in labial view. Abbreviations: dt, dentary tooth; pmt, premaxillary tooth. The numbers indicate the tooth position. Scale bar = 2 mm.

The mesialmost teeth (i.e., anterior left premaxilla, anterior right maxilla, anterior left and right dentaries) are missing in NHMD 164741. The left premaxilla preserves the two distalmost teeth in NHMD 164741. Both teeth are descended teeth with the replacement teeth already contained inside the alveoli. NHMD 164741 and NHMD 164758 preserve 24 and 23 maxillary alveoli on the left maxilla, respectively, and 18 dentary alveoli on the left dentary. Both on the maxilla and dentary, the tooth size decreases at the caudal-most part of the series.

The maxillary and dentary tooth crowns are leaf-shaped, with both labial and lingual surfaces slightly convex and the mesial and distal carinas covered with coarse denticles. The denticles are deflected apically and are occupy more area on the distal carina. As in the premaxillary teeth, the denticles start apical to the basal third of the crown (Figure 33C,D). The apex of the teeth lacks denticles. The tooth roots are slender and comprises most of the tooth height (average root height = 14.2 mm; average crown height = 8.6 mm in NHMD 164741). The labial surface of the root is slightly convex, whereas the lingual surface of the root is marked by a shallow groove, making the root overall B-shaped in cross-section (Figure 34).

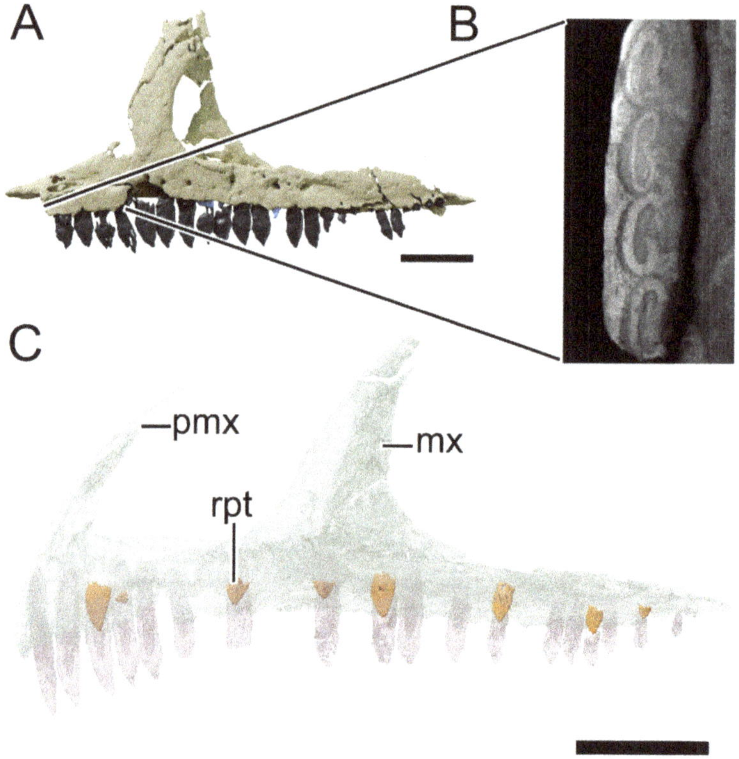

Figure 34. Digital reconstruction and CT-scan slide of the maxilla of NHMD 164741 and digital reconstruction of the right premaxilla and maxilla of NHMD 164758 in transparent display. (**A**) Digital reconstruction of the left maxilla of NHMD 164741 in lateral view. (**B**) Slice of the anterior teeth of the maxilla of NHMD 164741 showing the teeth roots in cross-section. (**C**) Mirrored reconstruction of the upper jaw of NHMD 164758. Abbreviations: mx, maxilla; pmx, premaxilla; rpt, replacement tooth. Scale bar = 50 mm.

4.8.32. Hyoid

Four long and slender bone fragments pertaining to the hyoid apparatus are preserved in NHMD 164758. All fragments pertain to the left ceratobranchial, with different diameters pertaining possibly to the position of these fragments when the bone was complete. One of these rod-like fragments is thicker and longer than the other three fragments (Figures 6 and 7). It is sigmoid in lateral view, with a convex dorsal margin at its anterior half and a concave dorsal margin at its posterior half. This bone has an anterior rounded and blunt edge and ends posteriorly in an expanded and lateromedially flattened process. The posterior process is laterally depressed. The three slender fragments are located behind the left articular and are equal in diameter. These elements are arched anteroposteriorly but only preserved in the mid-shaft region.

4.8.33. Sclerotic Ring

A total of 18 lateromedially thin and square-shaped plates are partially in articulation in NHMD 164758 left orbit (Figure 35). These plates form a circular arrangement, but the identification of positive and negative plates is not clear, nor is it possible to determine if the 18 plates constitute the whole sclerotic ring.

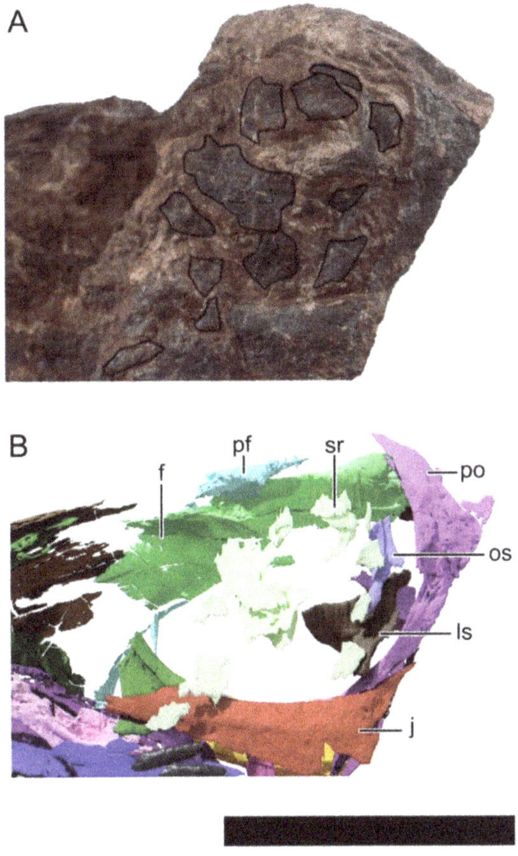

Figure 35. Photograph and digital reconstruction of the left orbital region of NHMD 164758. (**A**) Photograph of the left orbital region of NHMD 164758. (**B**) Digital reconstruction of the left orbital region of NHMD 164758. Scale bar = 50 mm.

4.9. Phylogenetic Analysis

The phylogenetic analysis (see Section 3.4) recovered a total of 400 MPTs of 1582 steps, with consistency index (CI) of 0.286 and retention index of 0.657. Their strict consensus tree with Bremer Support values is found below (Figure 36). The Greenland sauropodomorphs were nested together in our analysis as a sister clade of *Pl. trossingensis* plus *Pl. gracilis*. This clade is characterized by four unambiguous cranial synapomorphies: the postorbital rim of the orbit raised and projecting laterally (54, 1), shared with Massospondylidae; a central tubercle on the ventral surface of the palatine (90, 1), also shared with Massospondylidae; vomer anteroposterior length over 0.25 of the skull total length [92, (1)], shared with *Melanorosaurus*; and five or more premaxillary teeth (106, 1).

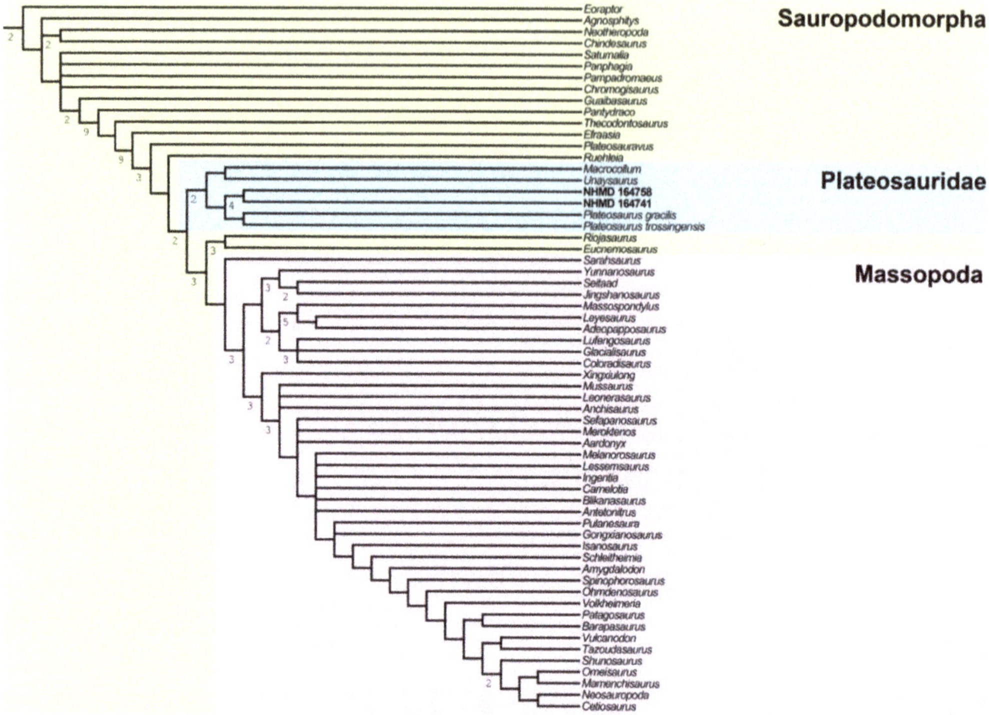

Figure 36. Strict consensus tree of the Greenland specimens. Strict consensus tree of Sauropodomorpha including the Greenland specimens as separate OTUs. Bremer support number (>1) are found under the nodes.

The specimens NHMD 164741 and NHMD 164758 form a clade supported by six unambiguous synapomorphies: weakly developed narial fossa (character 10, state 0), shared with *Mac. Itaquii*, and other more derived sauropodomorphs; small subnarial foramen (12, 1), shared with *Sarahsaurus* and early sauropodomorphs; anterior margin of the external naris anterior to the mid-length of the premaxilla [17, (0)], reversal of a plesiomorphic feature for Sauropodomorpha; anteroposterior length of the antorbital fossa less than the orbit (28, 1), shared with Massopoda (*sensu* [15]); antorbital fossa ending before the ventral process of the lacrimal (41, 1), shared with *Yunnanosaurus*; strongly curved jugal process of the ectopterygoid (86, 1), shared with *Pantydraco*, *Leyesaurus* and *Sarahsaurus*.

The genus *Plateosaurus* (i.e., *Pl. trossingensis* and *Pl. gracilis*) bears three unambiguous cranial synapomorphies: point contact of the posterolateral process of the premaxilla and the anteroventral process of the nasal (7, 1), shared with *E. minor* and *U. tolentinoi*; a depression behind the naris at the dorsal profile of the skull (19, 1), shared with Massospondylidae (*sensu* [15]); length of the posterior process of the prefrontal equal to that of the orbit (42, 1), shared with Massopoda.

The clade Plateosauridae is supported by two cranial and two post-cranial synapomorphies: basipterygoid processes and the parasphenoid rostrum below the level of the basioccipital condyle and the basal tuberae (80, 1), not visible in the Greenland sauropodomorphs and *Mac. itaquii*; strongly ventrally curved dentary symphyseal end (98, 1), also shared with Massospondylidae; ventrolateral twisting of the transverse axis of the distal end of the first phalanx of manual digit one relative to its proximal end of 60° (244, 2), not preserved on NHMD 164741, NHMD 164758 and *Pl. gracilis*; transverse width of the conjoined distal ischial expansions less than their sagittal depth (292, 1), only observed in *Mac. itaquii* and *Pl. trossingensis*.

In contrast to the analysis of Müller [19], *Mac. itaquii* forms a clade with *U. tolentinoi* at the base of Plateosauridae. This clade is supported by four ambiguous post-cranial synapomorphies: absence of ventral keels on the cervical vertebrae (132, 0); transverse width of the distal end of the humerus over 0.33 of the total length of the bone (217, 1); the first phalanx of manual digit one shorter than the first metacarpal (245, 1); femoral length between 200 and 399 mm (379, 1).

5. Discussion

5.1. *Arguments for a New Taxon and Comparisons*

The taxonomy of early sauropodomorphs, and plateosaurids specifically, is still not fully resolved and any attempt to erect a new taxon should be done carefully and be well-grounded. The Greenland sauropodomorphs were recovered from rocks of similar age to the central European *Plateosaurus* [28], therefore age separation provides no additional argument for the validity of a separate taxon. Both described skulls suffered some degree of taphonomic deformation, but this degree is not sufficient to explain the differences observed between the Greenland specimens and other sauropodomorphs.

As mentioned above, there are six unambiguous synapomorphies shared between NHMD 164741 and NHMD 164758 and three unambiguous apomorphies supporting *Plateosaurus* as the sister clade for the Greenland specimens. Although these features cannot be caused by taphonomic deformation alone, some of the apomorphies of *Issi saaneq* could indeed be attributed to intraspecific variation. The developmental plasticity of the closely related *Plateosaurus* was noted for specimens from Trossingen [64] and Frick [53]. This plasticity ranges from the total size and proportions of individuals to skull characters usually employed in phylogenetic studies. Phylogenetic characters such as the shape and size of the narial and antorbital fossae or the position of the external naris in the premaxilla found as apomorphic for *Issi saaneq* are highly variable among *Pl. trossingensis* specimens [53]. Therefore, these characters should be revised in future phylogenetic analyses. Nevertheless, some apomorphies of *Issi saaneq*, such as the weak development of the narial fossa and the strongly curved jugal ramus of the ectopterygoid, differ clearly from the condition observed in the described specimens of *Plateosaurus* [2,45,52,53,65]. Furthermore, the three above mentioned synapomorphies supporting the clade *Plateosaurus* are clearly different from the conditions observed in the Greenland specimens.

Issi saaneq possess a unique combination of traits that is not observed in other early sauropodomorphs. NHMD 164741 and NHMD 164758 share a foramen in the medial surface of the premaxilla, absent in *U. tolentinoi*, *Plateosaurus* and *Mas. carinatus*, and different from the more posteriorly located foramen in *Pam. berberenai*. The squamosal of NHMD 164741 preserves a uniquely elongated posterior process. This process is relatively longer anteroposteriorly than in other early sauropodomorphs (Figure 16 and see Table 3). NHMD 164741 also preserved a dorsoventrally elongated and mediolaterally slender quadrate that differs from this condition in all other closely related sauropodomorphs (Figure 18 and see Table 4). This makes the skull of NHMD 164741 dorsoventrally taller at its posterior half than that of other sauropodomorphs such as *Pl. trossingensis*, *Mac. itaquii*, *Bu. schultzi*, *N. intlokoi* and *Mas. carinatus*. NHMD 164758 preserves a unique morphology of the articular among early sauropodomorphs (Figure 32), not preserved in NHMD 164741. The dorsoposterior process of the articular is well-developed and dorsally tall, forming a squared blade in lateral view.

A combination of additional features further differentiates *Issi saaneq* gen. et sp. nov. from *Plateosaurus*. The nasals of NHMD 164741 and NHMD 164758 occupy 0.41 and 0.48, respectively, of the total skull roof length, whereas in *Plateosaurus* the nasal occupies over 0.5 of that length [52,53,63]. The lateral process of the laterosphenoid is elongated and arches laterally in *Pl. trossingensis* (AMNH FARB 6810) [52], whereas in NHMD 164758 this process is short and straight, as is the condition in other sauropodomorphs such as *Mac. itaquii*, *N. intlokoi* and *Mas. carinatus*. The autapomorphic feature of a ventrally located, central peg-like process of the palatine of *Pl. trossingensis* is absent in both NHMD 164741 and NHMD 164758. Only two of the autapomorphic features for *Pl. trossingensis* are present in *Issi saaneq*. The lateral sheet of bone in the lacrimal is present in NHMD 164741, although differing from that of *Pl. trossingensis* in having a foramen in the dorsolateral surface of this flange, similar to the lateral foramen of *Mac. itaquii*. In *Issi saaneq*, this flange is concave in its anterior margin, whereas it is convex in *Pl. trossingensis*. The dorsoventrally high jugal of *Pl. trossingensis* is visible in NHMD 164758 left jugal. This bone is morphologically similar to that of most *Pl. trossingensis* specimens. *Issi saaneq* also shares features with the Brazilian plateosaurids *Mac. itaquii* and *U. tolentinoi*, that are lacking in *Plateosaurus*. These features include a weakly developed narial fossa (observed on *Issi saaneq* and the Brazilian plateosaurids), a promaxillary fenestra (observed only in NHMD 164741 and the Brazilian plateosaurids), a foramen in the lateral surface of the lacrimal (observed in NHMD 164741 and *Mac. itaquii*) and a secondary fossa ventral to the Meckelian groove (observed in both *Issi saaneq* specimens and *U. tolentinoi*).

5.2. Arguments for a Single New Species

The specimens NHMD 164741 and NHMD 164758 are almost complete skulls of different sizes, varying only slightly in their morphology. Both were recovered from the same locality, meaning that both specimens existed coevally. Both specimens received the same scores for every coded character in the phylogenetic analysis, forming together the sister clade of *Plateosaurus*. However, due to incomplete preservation only one autapomorphy is preserved in both NHMD 164741 and NHMD 164758, the medial foramen of the premaxilla. The squamosal and dorsal half of the quadrates are missing in NHMD 164758 and the posterior half of the articular is missing in NHMD 164741, hampering the assessment of these autapomorphies. NHMD 164741 differs from NHMD 164758 in having a promaxillary fenestra in the antorbital fossa (Figures 10 and 11), the shape of the anterodorsal process of the lacrimal (which is distally broken in NHMD 164741, Figures 12 and 13) and in bulkier postorbitals (Figures 14 and 15) and frontals (Figures 20 and 21). The presence of a promaxillary fenestra was recovered as a synapomorphy for the Brazilian "unaysaurids" *Mac. itaquii* and *U. tolentinoi* [18,19,22]. Because this feature is plastic and can even be present in one element and absent in the other for the same individual [19], the presence of a promaxillary fenestra does not support a separation of the Greenland specimens in different species. The difference in the anterodorsal process of the lacrimal and the bulkier postorbital in NHMD 164741 could be attributed to ontogeny or even intraspecific variation, as is the case of the main body of the postorbital in *Pl. trossingensis* [53]. Due to the lack of features distinguishing both Greenland specimens, NHMD 164741 and NHMD 164758 are here regarded as belonging to the same new genus and species, *Issi saaneq* (Figures 37 and 38).

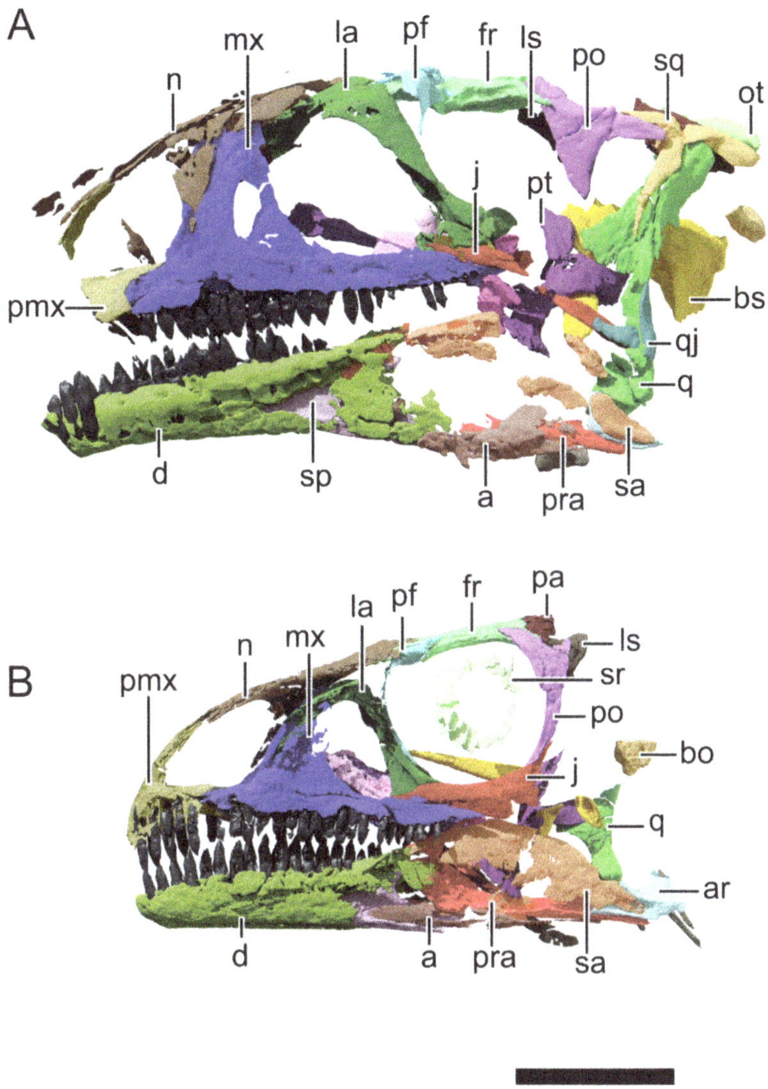

Figure 37. Digital reconstruction of the skulls NHMD 164741 and NHMD 164758 in articulation. Digital arrangement of preserved bone elements of NHMD 164741 (**A**) and NHMD 164758 (**B**) in left lateral view. Abbreviations: ar, articular; bo, basioccipital; bs, basisphenoid; d, dentary; fr, frontal; j, jugal; la, lacrimal; ls, laterosphenoid; mx, maxilla; n, nasal; ot, otoccipital; pf, prefrontal; pmx, premaxilla; po, postorbital; pra, prearticular; pt, pterygoid; q, quadrate; qj, quadratojugal; sa, surangular; sp, splenial; sq, squamosal; sr, sclerotic ring. Scale bar = 50 mm.

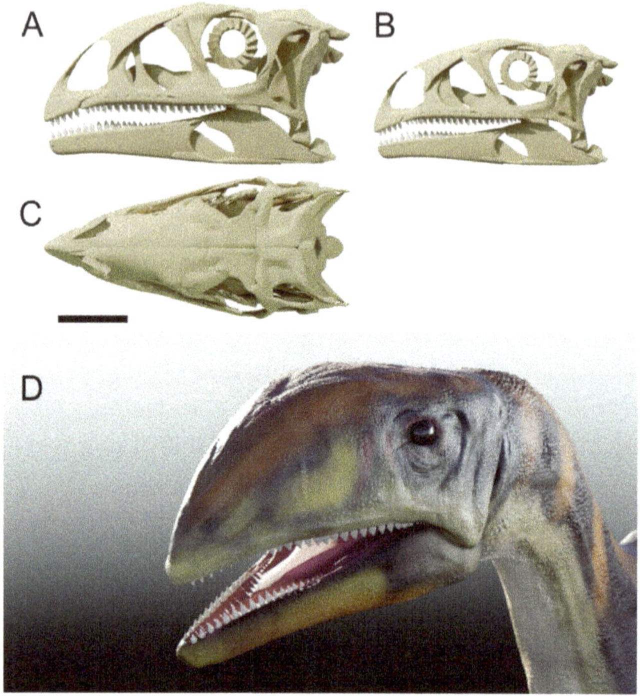

Figure 38. Digital interpretative reconstruction of the skulls NHMD 164741 and NHMD 164758 and living representation of *Issi saaneq*. (**A**) Digital interpretative reconstruction of the skull NHMD 164741 in left lateral view. (**B**) Digital interpretative reconstruction of the skull NHMD 164758 in left lateral view. (**C**) Digital interpretative reconstruction of skull NHMD 164741 in dorsal view. (**D**) Living representation of *Issi saaneq*. Scale bar = 50 mm.

5.3. Ontogeny

Although representing the same taxon, NHMD 164741 and NHMD 164758 differ in size and the skull proportions. Because size is not a good proxy for the ontogenetic stage in the closely related *Pl. trossingensis* [64,66], this may be also the case for the Greenland specimens. NHMD 164741 measures 1.45 times the anteroposterior length of that of NHMD 164758, even with the former lacking the anteriormost region of the skull (which would account for half the premaxilla main body length, possibly making the skull 15 to 20 mm longer). The orbit length in NHMD 164741 can only be estimated with the digital articulation of the preserved bones. This estimation ranges from 41 to 45.5 mm in length, meaning that the orbit length to skull length of NHMD 164741 would range from 0.17 to 0.20 (Figure 37A). NHMD 164758 preserved the orbit region still in articulation, with minor deformation (Figure 33). The orbit length measures 47.2 mm in this specimen, so that the orbit to skull length ratio (approximately 0.28) is higher than in NHMD 164741. A larger orbit is supposed to be a juvenile feature, as shown by the late-stage juvenile of *Pl. trossingensis* (MSF 12.3) [53]. The reduced dentary teeth count (<20 tooth position) is also a feature related to ontogeny in plateosaurids [53]. NHMD 164758 shows only 18 tooth positions in its dentaries, and NHMD 164741 preserves a minimum of 18 tooth positions. In the latter this number might be underestimated, as the anterior region of the dentary is missing, so that it could have reached at least 20 teeth when complete. The reduced gap between the first premaxillary tooth and the anterior tip of the premaxilla is observed in NHMD 164758, representing another juvenile feature for this specimen [53]. The orientation of the basipterygoid process of the basisphenoid can also be an indicative of ontogenetic stage in some sauropodomorphs, such as in *Mas. carinatus* [16]. The differences between the slightly anteriorly deflected basipterygoid process of NHMD 164758 and the sub-perpendicular basipterygoid process of NHMD 164741 may be related to their ontogenetic stages. All these discussed features position NHMD 164758 in an earlier stage of development than NHMD 164741, the former being interpreted as an early-stage juvenile and the latter as a possible late-stage juvenile or young adult.

5.4. Paleobiogeographic and Chronological Implications

Plateosaurid sauropodomorphs were found in Upper Triassic (Norian) strata of Brazil, Germany, France, Switzerland and Norway [19,22,52,53,63,67–71] and are now confidently reported for Greenland. *Issi saaneq* is the first sauropodomorph to reach paleolatitudes of over 40° N and is the first known non-*Plateosaurus* plateosaurid for Laurasia. The shared features of *Issi saaneq* and *Plateosaurus* were expected due to their contemporaneity during the mid-Norian (around 218-214 Mya, but possibly constrained to 217 Mya according to [28]) and paleogeographic proximity (the distance between localities is roughly estimated at 1000-2000 km). However, *Issi saaneq* shares additional features with *Mac. itaquii* and *U. tolentinoi*. The plateosaurid sauropodomorphs from Brazil were recovered in the Candelária Sequence at an estimated absolute age of 225.4 Mya (mid-Norian according to [72]). Due to their older age, the Brazilian plateosaurids may represent a key moment in the early evolution and dispersal of plateosaurids. However, the possible timing and evolution of plateosaurid sauropodomorphs can only be addressed with additional findings for this clade, either in North America or Africa. As of now, mid-Norian outcrops which could yield sauropodomorphs are rare in Africa, whereas Late Triassic dinosaurs are relatively well-known from North America [73]. The latter, however, lacks Triassic sauropodomorph dinosaurs until now [73]. The first basal sauropodomorphs from North America include *Anchisaurus polyzelus* [74] from the Portland Formation, USA and *Sarahsaurus aurifontanalis* [75] from the Kayenta Formation, USA; both Early Jurassic (Sinemurian-Pliensbachian) in age.

6. Conclusions

Two skulls of the new basal sauropodomorph (plateosaurid) dinosaur taxon *Issi saaneq* gen. nov. sp. nov. from the Late Triassic (Norian) of Jameson Land, central East Greenland are described based on data retrieved with μCT-scan image segmentation and photogrammetry. Both specimens, NHMD 164741 and NHMD 164758, were recovered from the uppermost Malmros Klint Formation in Greenland, and due to strong morphological similarities and no robust distinguishing features between them, are here regarded as a single taxon. The smaller NHMD 164758 represents an early-stage juvenile, due to the reduced gap between the first premaxillary teeth and the anterior margin of the premaxilla, large orbit, low number of teeth positions in the dentary and an anteriorly deflected basipterygoid process of the basisphenoid. The specimen NHMD 164741 represents either a late-stage juvenile or a young adult, due to a proportionally shorter orbit than NHMD 164758 and having a possible maximum of 20 dentary teeth positions. *Issi saaneq* differs from all other basal sauropodomorphs in four observed autapomorphies: (1) the presence of a small foramen at the medial surface of the premaxilla at the base of the lateral process of the premaxilla; (2) an anteroposteriorly elongated dorsoposterior process of the squamosal; (3) a quadrate relatively tall in comparison to the rostrum height; and (4) a well-developed, square-shaped in lateral view posterodorsal process of the articular.

Six ambiguous synapomorphies position *Issi saaneq* as the sister clade to *Plateosaurus* (*Pl. trossingensis* and *Pl. gracilis*). The Brazilian sauropodomorphs were recovered at the base of Plateosauridae and forming the sister clade to the clade containing *Issi saaneq* and *Plateosaurus*. *Issi saaneq* possesses a set of features thought to be exclusive of the Brazilian plateosaurids. *Issi saaneq* is the first sauropodomorph to reach the Northernmost parts of Laurasia and increases our understanding of the diversity of plateosaurids.

Supplementary Materials: The following is available online at https://www.mdpi.com/article/10.3390/d13110561/s1, Data S1: Data matrix 01.

Author Contributions: Conceptualization, V.B., O.M., O.W., J.M., L.B.C.; methodology, V.B., O.M., O.W.; software, V.B.; formal analysis, V.B., O.M., O.W., J.M., L.B.C.; investigation, V.B., O.M., O.W., J.M., L.B.C.; data curation, V.B.; writing—original draft preparation, V.B., O.M., O.W., J.M., L.B.C.; writing—review and editing, V.B., O.M., O.W., J.M., L.B.C.; visualization, V.B.; figures and 3D models, V.B.; supervision, O.M. and L.B.C. All authors have read and agreed to the published version of the manuscript.

Funding: This research benefited from the grant GeoBioTec-GeoBioSciences, GeoTechnologies and GeoEngineering NOVA [GeoBioCiências, GeoTecnologias e GeoEngenharias], grant UIDB/04035/2020 by the Fundação para a Ciência e Tecnologia. The geological work of LBC was supported by the Carlsberg Foundation, the Independent Research Fund Denmark, and Geocenter Møns Klint.

Institutional Review Board Statement: Not applicable.

Informed Consent Statement: Not applicable.

Data Availability Statement: All specimens in this study are housed at the Natural History Museum of Denmark (NHMD). All CT-scan data and 3D models of both specimens will be available in MorphoSource at: (ark:/87602/m4/393344).

Acknowledgments: We would like to thank the team that uncovered both specimens here described, William W. Amaral, William R. Downs, Stephen M. Gatesy, Neil H. Shubin and Niels Bonde and Farish Jenkins. We also would like to thank Marco Marzola, Filippo Rotatori and Alexandra Fernandes for the CT-scan of the specimens.

Conflicts of Interest: The authors declare no conflict of interest.

References

1. Meyer, H. Briefliche mitteilung an Prof. Bronn über *Plateosaurus engelhardti*. *Neues Jahrb. Mineral. Geogn. Geol. Petrefakten-Kunde* **1837**, *24*, 314–316.
2. Yates, A.M. The species taxonomy of the sauropodomorph dinosaurs from the Lowenstein Formation (Norian, Late Triassic) of Germany. *Palaeontology* **2003**, *46*, 317–337. [CrossRef]
3. Galton, P.M. Prosauropod Dinosaur *Plateosaurus* (=*Gresslyosaurus*) (*Saurischia: Sauropodomorpha*) from the Upper Triassic of Switzerland. *Geol. Palaeontol.* **1986**, *20*, 167–183.
4. Rütimeyer, L. Reptilien knochen aus dem Keuper von Liestal. *Verh. Schweiz. Nat. Ges.* **1856**, *41*, 62–64.
5. Huene, F.F. Über die Trias-Dinosaurier Europas. *Z. Dtsch. Geol. Ges.* **1905**, *57*, 345–349.
6. Huene, F.F. *Die Dinosaurier der Europäischen Triasformation mit Berücksichtigung der Aussereuropäischen Vorkommnisse*; G. Fischer: Jena, Germany, 1908; Volume 1.
7. Fraas, E. Die neuesten Dinosaurierfunde in der Schwäbischen Trias. *Naturwissenschaften* **1913**, *1*, 1097–1100. [CrossRef]
8. Yates, A.M. Solving a dinosaurian puzzle: The identity of *Aliwalia Rex* Galton. *Hist. Biol.* **2007**, *19*, 93–123. [CrossRef]
9. Yates, A.M. A revision of the problematic sauropodomorph dinosaurs from Manchester, Connecticut and the Status of *Anchisaurus* Marsh: The taxonomic status of Anchisaurus. *Palaeontology* **2010**, *53*, 739–752. [CrossRef]
10. Galton, P.M.; Kermack, D. The anatomy of *Pantydraco Caducus*, a very basal Sauropodomorph Dinosaur from the Rhaetian (Upper Triassic) of South Wales, UK. *Rev. Palóbiologie* **2010**, *29*, 341–404.
11. Apaldetti, C.; Martinez, R.N.; Alcober, O.A.; Pol, D. A new basal Sauropodomorph (*Dinosauria: Saurischia*) from Quebrada Del Barro formation (Marayes-El Carrizal Basin), Northwestern Argentina. *PLoS ONE* **2011**, *6*, e26964. [CrossRef]
12. Rauhut, O.W.M.; Holwerda, F.M.; Furrer, H. A derived Sauropodiform Dinosaur and other sauropodomorph material from the Late Triassic of Canton Schaffhausen, Switzerland. *Swiss J. Geosci.* **2020**, *113*, 8. [CrossRef]
13. Galton, P.M. Case 3560 *Plateosaurus Engelhardti* Meyer, 1837 (*Dinosauria, Sauropodomorpha*): Proposed replacement of unidentifiable name-bearing type by a neotype. *Bull. Zool. Nomencl.* **2012**, *69*, 203–212. [CrossRef]
14. ICZN. Opinion 2435 (Case 3560) *Plateosaurus* Meyer, 1837 (*Dinosauria, Sauropodomorpha*): New type species designated. *Bull. Zool. Nomencl.* **2019**, *76*, 144–145. [CrossRef]
15. Yates, A.M. The first complete skull of the triassic dinosaur *Melanorosaurus Haughton* (*Sauropodomorpha: Anchisauria*). *Spec. Pap. Palaeontol.* **2007**, *77*, 9–55.
16. Chapelle, K.E.J.; Choiniere, J.N. A revised cranial description of *Massospondylus Carinatus* owen (*Dinosauria: Sauropodomorpha*) based on computed tomographic scans and a review of cranial characters for basal Sauropodomorpha. *PeerJ* **2018**, *6*, e4224. [CrossRef] [PubMed]
17. Chapelle, K.E.J.; Barrett, P.M.; Botha, J.; Choiniere, J.N. *Ngwevu Intloko*: A new early sauropodomorph dinosaur from the lower jurassic elliot formation of South Africa and comments on cranial ontogeny in *Massospondylus carinatus*. *PeerJ* **2019**, *7*, e7240. [CrossRef]

18. McPhee, B.W.; Bittencourt, J.S.; Langer, M.C.; Apaldetti, C.; Da Rosa, Á.A.S. Reassessment of *Unaysaurus Tolentinoi* (Dinosauria: Sauropodomorpha) from the Late Triassic (Early Norian) of Brazil, with a consideration of the evidence for monophyly within non-sauropodan sauropodomorphs. *J. Syst. Palaeontol.* **2020**, *18*, 259–293. [CrossRef]
19. Müller, R.T. Craniomandibular osteology of *Macrocollum Itaquii* (*Dinosauria: Sauropodomorpha*) from the Late Triassic of Southern Brazil. *J. Syst. Palaeontol.* **2020**, *18*, 805–841. [CrossRef]
20. Leal, L.A.; Azevedo, S.A.K.; Kellner, A.W.A.; Rosa, Á.A.S. A new early dinosaur (Sauropodomorpha) from the Caturrita Formation (Late Triassic), Paraná Basin, Brazil. *Zootaxa* **2004**, *690*, 1–24. [CrossRef]
21. McPhee, B.W.; Choiniere, J.N.; Yates, A.M.; Viglietti, P.A. A second species of *Eucnemesaurus* Van Hoepen, 1920 (*Dinosauria, Sauropodomorpha*): New information on the diversity and evolution of the sauropodomorph fauna of South Africa's Lower Elliot Formation (Latest Triassic). *J. Vertebr. Paleontol.* **2015**, *35*, e980504. [CrossRef]
22. Müller, R.T.; Langer, M.C.; Dias-da-Silva, S. An exceptionally preserved association of complete dinosaur skeletons reveals the oldest long-necked sauropodomorphs. *Biol. Lett.* **2018**, *14*, 20180633. [CrossRef] [PubMed]
23. Novas, F.E.; Ezcurra, M.D.; Chatterjee, S.; Kutty, T.S. New dinosaur species from the upper Triassic upper Maleri and lower Dharmaram formations of Central India. *Earth Environ. Sci. Trans. R. Soc. Edinb.* **2011**, *101*, 333–349. [CrossRef]
24. Novas, F.E.; Ezcurra, M.D.; Chatterjee, S.; Kutty, T.S. *Late Triassic Continental Vertebrates and Depositional Environments of the Fleming Fjord Formation, Jameson Land, East. Greenland*; Jenkins, F.A., Shubin, N.H., Amaral, W.W., Gatesy, S.M., Schaff, C.R., Clemmensen, L.B., Downs, W.R., Davidson, A.R., Bonde, N., Osbaeck, F., Eds.; Meddelelser om Grønland Geoscience: Brenderup, Denmark, 1994; ISBN 978-87-601-4573-5.
25. Clemmensen, L.B.; Kent, D.W.; Mau, M.; Mateus, O.; Milàn, J. Triassic lithostratigraphy of the Jameson land basin (Central East Greenland), with emphasis on the New Fleming Fjord Group. *Bull. Geol. Soc. Den.* **2020**, *68*, 95–132. [CrossRef]
26. Marzola, M.; Mateus, O.; Milàn, J.; Clemmensen, L.B. A review of Palaeozoic and Mesozoic Tetrapods from Greenland. *Bull. Geol. Soc. Den.* **2018**, *66*, 21–46. [CrossRef]
27. Guarnieri, P.; Brethes, A.; Rasmussen, T.M. Geometry and kinematics of the Triassic rift basin in Jameson Land (East Greenland): Triassic rift basin East Greenland. *Tectonics* **2017**, *36*, 602–614. [CrossRef]
28. Kent, D.V.; Clemmensen, L.B. Northward dispersal of dinosaurs from Gondwana to Greenland at the Mid-Norian (215–212 Ma, Late Triassic) dip in atmospheric pCO_2. *Proc. Natl. Acad. Sci. USA* **2021**, *118*, e2020778118. [CrossRef] [PubMed]
29. Clemmensen, L.B.; Kent, D.V.; Jenkins, F.A. A late triassic lake system in East Greenland: Facies, depositional cycles and palaeoclimate. *Palaeogeogr. Palaeoclimatol. Palaeoecol.* **1998**, *140*, 135–159. [CrossRef]
30. Sellwood, B.W.; Valdes, P.J. Mesozoic climates: General circulation models and the rock record. *Sediment. Geol.* **2006**, *190*, 269–287. [CrossRef]
31. Agnolin, F.L.; Mateus, O.; Milàn, J.; Marzola, M.; Wings, O.; Adolfssen, J.S.; Clemmensen, L.B. *Ceratodus Tunuensis*, Sp. Nov., a new lungfish (Sarcopterygii, Dipnoi) from the Upper Triassic of Central East Greenland. *J. Vertebr. Paleontol.* **2018**, *38*, e1439834. [CrossRef]
32. Sulej, T.; Wolniewicz, A.; Bonde, N.; Błażejowski, B.; Niedźwiedzki, G.; Tałanda, M. New perspectives on the Late Triassic vertebrates of East Greenland: Preliminary results of a Polish—Danish palaeontological expedition. *Pol. Polar Res.* **2014**, *35*, 541–552. [CrossRef]
33. Clemmensen, L.B.; Milàn, J.; Adolfssen, J.S.; Estrup, E.J.; Frobøse, N.; Klein, N.; Mateus, O.; Wings, O. The vertebrate-bearing Late Triassic Fleming Fjord Formation of Central East Greenland revisited: Stratigraphy, palaeoclimate and new palaeontological data. *Geol. Soc. Lond. Spec. Publ.* **2016**, *434*, 31–47. [CrossRef]
34. Marzola, M.; Mateus, O.; Shubin, N.H.; Clemmensen, L.B. *Cyclotosaurus Naraserluki* Sp. Nov., a new Late Triassic Cyclotosaurid (*Amphibia, Temnospondyli*) from the Fleming Fjord Formation of the Jameson Land Basin (East Greenland). *J. Vertebr. Paleontol.* **2017**, *37*, e1303501. [CrossRef]
35. Niedźwiedzki, G.; Sulej, T. Theropod dinosaur fossils from the Gipsdalen and Fleming fjord formations (Carnian-Norian, Upper Triassic), East Greenland. In Proceedings of the 34th Nordic Geological Winter Meeting, Oslo, Norway, 8–10 January 2020; Nakrem, H.A., Hus, A.M., Eds.; Geological Society of Norway: Oslo, Norway, 2020; Volume 2020, p. 151.
36. Milàn, J.; Octávio, M.; Mau, M.; Rudra, A.; Sanei, H.; Clemmensen, L.B. A possible phytosaurian (*Archosauria, Pseudosuchia*) coprolite from the Late Triassic Fleming Fjord Group of Jameson Land, Central East Greenland. *Bull. Geol. Soc. Den.* **2021**, *69*, 71–80. [CrossRef]
37. Jenkins, F.A.; Gatesy, S.M.; Shubin, N.H.; Amaral, W.W. Haramiyids and Triassic mammalian evolution. *Nature* **1997**, *385*, 715–718. [CrossRef]
38. Sulej, T.; Krzesiński, G.; Tałanda, M.; Wolniewicz, A.S.; Błażejowski, B.; Bonde, N.; Gutowski, P.; Sienkiewicz, M.; Niedźwiedzki, G. The earliest-known mammaliaform fossil from Greenland sheds light on origin of mammals. *Proc. Natl. Acad. Sci. USA* **2020**, *117*, 26861–26867. [CrossRef] [PubMed]
39. Pretto, F.A.; Langer, M.C.; Schultz, C.L. A new dinosaur (*Saurischia: Sauropodomorpha*) from the Late Triassic of Brazil provides insights on the evolution of Sauropodomorph body plan. *Zool. J. Linn. Soc.* **2019**, *185*, 388–416. [CrossRef]
40. Cabreira, S.F.; Kellner, A.W.A.; Dias-da-Silva, S.; Roberto da Silva, L.; Bronzati, M.; de Almeida Marsola, J.C.; Müller, R.T.; de Souza Bittencourt, J.; Batista, B.J.; Raugust, T.; et al. A unique Late Triassic Dinosauromorph assemblage reveals dinosaur ancestral anatomy and Diet. *Curr. Biol.* **2016**, *26*, 3090–3095. [CrossRef] [PubMed]

41. Müller, R.T.; Langer, M.C.; Bronzati, M.; Pacheco, C.P.; Cabreira, S.F.; Dias-da-Silva, S. Early evolution of Sauropodomorphs: Anatomy and phylogenetic relationships of a remarkably well-preserved dinosaur from the Upper Triassic of Southern Brazil. *Zool. J. Linn. Soc.* **2018**, *184*, 1187–1248. [CrossRef]
42. Müller, R.T.; Ferreira, J.D.; Pretto, F.A.; Bronzati, M.; Kerber, L. The endocranial anatomy of *Buriolestes Schultzi* (*Dinosauria: Saurischia*) and the early evolution of brain tissues in sauropodomorph dinosaurs. *J. Anat.* **2021**, *238*, 809–827. [CrossRef]
43. Bonaparte, J.F. *Coloradia Brevis* n. g. et n. Sp. (*Saurischia, Prosauropoda*), a Plateosaurid dinosaur from the Los Colorados Formation, Upper Triassic of La Rioja, Argentina. *Ameghiniana* **1978**, *15*, 7.
44. Apaldetti, C.; Martinez, R.N.; Pol, D.; Souter, T. Redescription of the skull of *Coloradisaurus Brevis* (*Dinosauria, Sauropodomorpha*) from the Late Triassic Los Colorados Formation of the Ischigualasto-Villa Union Basin, Northwestern Argentina. *J. Vertebr. Paleontol.* **2014**, *34*, 1113–1132. [CrossRef]
45. Galton, P.M. Cranial anatomy of the prosauropod dinosaur *Plateosaurus* from the Knollenmergel (Middle Keuper, Upper Triassic) of Germany. II. All the cranial material and details of soft-part anatomy. *Geol. Palaeontol.* **1985**, *19*, 119–159.
46. Bronzati, M.; Rauhut, O.W.M. Braincase redescription of *Efraasia Minor*. Huene, 1908 (*Dinosauria: Sauropodomorpha*) from the Late Triassic of Germany, with comments on the evolution of the sauropodomorph braincase. *Zool. J. Linn. Soc.* **2018**, *182*, 173–224. [CrossRef]
47. Barrett, P.M.; Upchurch, P.; Xiao-Lin, W. Cranial osteology of *Lufengosaurus Huenei* young (*Dinosauria: Prosauropoda*) from the lower jurassic of Yunnan, People's Republic of China. *J. Vertebr. Paleontol.* **2005**, *25*, 806–822. [CrossRef]
48. Barrett, P.M. A new basal sauropodomorph dinosaur from the Upper Elliot Formation (Lower Jurassic) of South Africa. *J. Vertebr. Paleontol.* **2009**, *29*, 1032–1045. [CrossRef]
49. Langer, M.C.; McPhee, B.W.; Marsola, J.C.D.A.; Roberto-Da-Silva, L.; Cabreira, S.F. Anatomy of the dinosaur *Pampadromaeus Barberenai* (*Saurischia—Sauropodomorpha*) from the Late Triassic Santa Maria Formation of Southern Brazil. *PLoS ONE* **2019**, *14*, e0212543. [CrossRef]
50. Martinez, R.N.; Alcober, O.A. A Basal Sauropodomorph (Dinosauria: Saurischia) from the Ischigualasto Formation (Triassic, Carnian) and the early evolution of Sauropodomorpha. *PLoS ONE* **2009**, *4*, e4397. [CrossRef] [PubMed]
51. Galton, P.M. The Prosauropod Dinosaur Plateosaurus Meyer, 1837 (*Saurischia: Sauropodomorpha*; Upper Triassic). II. Notes on the referred species. *Rev. Paléobiologie* **2001**, *20*, 435–502.
52. Prieto-Márquez, A.; Norell, M.A. Redescription of a Nearly complete skull of *Plateosaurus* (*Dinosauria: Sauropodomorpha*) from the Late Triassic of Trossingen (Germany). *Am. Mus. Novit.* **2011**, *3727*, 1–58. [CrossRef]
53. Lallensack, J.N.; Teschner, E.; Pabst, B.; Sander, P.M. New skulls of the Basal Sauropodomorph *Plateosaurus Trossingensis* from Frick, Switzerland: Is there more than one species? *Acta Palaeontol. Pol.* **2021**, *66*, 1–28. [CrossRef]
54. Langer, M.C.; Abdala, F.; Richter, M.; Benton, M.J. A Sauropodomorph dinosaur from the Upper Triassic (Carnian) of Southern Brazil. *Académie Sci.* **1999**, *329*, 511–517.
55. Mallison, H.; Wings, O. Photogrammetry in paleontology—A practical guide. *J. Paleontol. Tech.* **2014**, *12*, 1–31.
56. Goloboff, P.A.; Catalano, S.A. TNT version 1.5, including a full implementation of phylogenetic morphometrics. *Cladistics* **2016**, *32*, 221–238. [CrossRef]
57. Owen, R. *Report on British Fossil Reptiles. Part. II*; Report of the British Association for the Advancement of Science; British Association for the Advancement of Science: London, UK, 1842.
58. Seeley, H.G. I. On the classification of the fossil animals commonly named Dinosauria. *Proc. R. Soc. Lond.* **1887**, *43*, 165–171. [CrossRef]
59. Huene, F.F. *Die Fossile Reptil-Ordnung Saurischia, ihre Entwicklung und Geschichte*; Verlag von Gebrüder Borntraeger: Leipzic, Germany, 1932.
60. Marsh, O.C. On the affinities and classification of the dinosaurian reptiles. *Am. J. Sci.* **1895**, *3*, 483–498. [CrossRef]
61. Clemmensen, L.B. Triassic lithostratigraphy of East Greenland between Scoresby Sund and Kejser Franz Josephs Fjord. *Grønlands Geol. Undersøgelse Bull.* **1980**, *139*, 56.
62. Young, C.-C. A Complete Osteology of Lufengosaurus Heunei Young (Gen. et Sp. Nov.) from Lufeng, Yunnan, China. *Geol. Surv. China* **1941**, *7*, 53.
63. Galton, P.M.; Upchurch, P. 12. Prosauropoda. In *The Dinosauria*, 2nd ed.; University of California Press: Berkeley, CA, USA, 2004; pp. 232–258.
64. Sander, P.M.; Klein, N. Developmental plasticity in the life history of a prosauropod dinosaur. *Science* **2005**, *310*, 1800–1802. [CrossRef] [PubMed]
65. Galton, P.M. Cranial anatomy of the prosauropod dinosaur *Plateosaurus* from the Knollenmergel (Middle Keuper, Upper Triassic) of Germany. I. Two complete skulls from Trossingen/Württ. with comments on the diet. *Geol. Palaeontol.* **1984**, *18*, 139–171.
66. Klein, N.; Sander, P.M. Bone histology and growth of the prosauropod dinosaur *Plateosaurus Engelhardti* von Meyer, 1837 from the Norian Bonebeds of Trossingen (Germany) and Frick (Switzerland). *Spec. Pap. Palaeontol.* **2007**, *77*, 169–206.
67. Sander, P.M. The norian *Plateosaurus* Bonebeds of Central Europe and Their yaphonomy. *Palaeogeogr. Palaeoclimatol. Palaeoecol.* **1992**, *93*, 255–299. [CrossRef]
68. Moser, M. *Plateosaurus engelhardti* Meyer, 1837 (*Dinosauria: Sauropodomorpha*) aus dem Feuerletten (Mittelkeuper; Obertrias) von Bayern. *Zitteliana* **2003**, *24*, 3–186.

69. Hurum, J.H.; Bergan, M.; Müller, R.; Nystuen, J.P.; Klein, N. A Late Triassic dinosaur bone, Offshore Norway. *Norwigian J. Geol.* **2006**, *86*, 117–123.
70. Mallison, H. The digital *Plateosaurus* I: Body mass, mass distribution and posture assessed using Cad and Cae on a digitally mounted complete skeleton. *Palaeontol. Electron.* **2010**, *13*, 26.
71. Nau, D.; Lallensack, J.N.; Bachmann, U.; Sander, P.M. Postcranial osteology of the first early-stage juvenile skeleton of *Plateosaurus Trossingensis* (Norian, Frick, Switzerland). *Acta Palaeontol. Pol.* **2020**, *65*, 679–708. [CrossRef]
72. Langer, M.C.; Ramezani, J.; Da Rosa, Á.A.S. U-Pb age constraints on dinosaur rise from South Brazil. *Gondwana Res.* **2018**, *57*, 133–140. [CrossRef]
73. Nesbitt, S.J.; Irmis, R.B.; Parker, W.G. A Critical re-evaluation of the Late Triassic dinosaur taxa of North America. *J. Syst. Palaeontol.* **2007**, *5*, 209–243. [CrossRef]
74. Marsh, O.C. Names of extinct reptiles. *Am. J. Sci.* **1885**, *29*, 169.
75. Rowe, T.B.; Sues, H.-D.; Reisz, R.R. Dispersal and diversity in the earliest North American Sauropodomorph dinosaurs, with a description of a New Taxon. *Proc. R. Soc. B* **2011**, *278*, 1044–1053. [CrossRef]

diversity

MDPI

Article

Seagrass Patch Complexity Affects Macroinfaunal Community Structure in Intertidal Areas: An In Situ Experiment Using Seagrass Mimics

Fernando G. Brun [1], José F. Cobo-Díaz [2], Vanessa González-Ortiz [1], José L. Varela [1], José Lucas Pérez-Lloréns [1] and Juan J. Vergara [1,*]

[1] Departamento de Biología, Facultad de Ciencias del Mar y Ambientales, Instituto Universitario de Investigación Marina (INMAR), CEI·MAR, Universidad de Cádiz, 11510 Puerto Real, Spain; fernando.brun@uca.es (F.G.B.); vgonzalezortizic@gmail.com (V.G.-O.); joseluis.varela@uca.es (J.L.V.); joselucas.perez@uca.es (J.L.P.-L.)
[2] Departamento de Higiene y Tecnología de los Alimentos, Universidad de León, 24071 León, Spain; josecobo1983@gmail.com
* Correspondence: juanjose.vergara@uca.es

Citation: Brun, F.G.; Cobo-Díaz, J.F.; González-Ortiz, V.; Varela, J.L.; Pérez-Lloréns, J.L.; Vergara, J.J. Seagrass Patch Complexity Affects Macroinfaunal Community Structure in Intertidal Areas: An In Situ Experiment Using Seagrass Mimics *Diversity* **2021**, *13*, 572. https://doi.org/10.3390/d13110572

Academic Editor: Michael Wink

Received: 30 September 2021
Accepted: 5 November 2021
Published: 9 November 2021

Publisher's Note: MDPI stays neutral with regard to jurisdictional claims in published maps and institutional affiliations.

Copyright: © 2021 by the authors. Licensee MDPI, Basel, Switzerland. This article is an open access article distributed under the terms and conditions of the Creative Commons Attribution (CC BY) license (https://creativecommons.org/licenses/by/4.0/).

Abstract: Seagrasses, as key ecosystem engineers in coastal ecosystems, contribute to enhancing diversity in comparison with nearby bare areas. It has been proved mainly for epifauna, but data on infauna are still scarce. The present study addresses how seagrass structural complexity (i.e., canopy properties) affects the diversity of infaunal organisms inhabiting those meadows. Canopy attributes were achieved using seagrass mimics, which were used to construct in situ vegetation patches with two contrasting canopy properties (i.e., shoot density and morphology) resembling the two seagrass species thriving in the inner Cadiz Bay: *Zostera noltei* and *Cymodocea nodosa*. After three months, bare nearby areas, two mimicked seagrass patches ('*Zostera*' and '*Cymodocea*'), and the surrounding natural populations of *Zostera noltei* were sampled in a spatially explicit way. Shifts in organism diets were also determined using ^{15}N and ^{13}C analyses in available food sources and main infaunal organisms, mixing models, and niche metrics (standard ellipse area). Seagrass-mimicked habitats increased the species richness (two-fold), organism abundance (three to four times), and functional diversity compared with bare nearby areas. The clam *Scrobicularia plana* (deposit/filter feeder) and the worm *Hediste diversicolor* (omnivore) were dominant in all of the samples (> 85%) and showed an opposite spatial distribution in the reconstructed patches: whilst *S. plana* accumulated in the outer-edge parts of the meadow, *H. diversicolor* abounded in the center. Changes in the isotopic signature of both species depending on the treatment suggest that this faunal distribution was associated with a shift in the diet of the organisms. Based on our results, we concluded that facilitation processes (e.g., reduction in predation and in bioturbation pressures) and changes in food availability (quality and quantity) mediated by seagrass canopies were the main driving forces structuring this community in an intertidal muddy area of low diversity.

Keywords: diversity; ecosystem engineers; facilitation; edge effect; seagrass; hydrodynamics; food availability; stable isotopes; mimics

1. Introduction

Seagrass ecologists broadly agree that anthropogenic activities are the main drivers of the current regression of seagrass ecosystems worldwide [1]. This brings into awareness that most of the crucial functions and services provided by these valuable ecosystems [2] may decline or even become lost. One of the most important functions of seagrasses is to act as ecosystem engineers [3], supporting higher biodiversity levels than bare areas [4–6]. The ongoing regression of seagrasses and/or associated diversity might impair the functioning of coastal ecosystems, with profound consequences for human welfare [7–9]. Deepening our understanding on the potential positive interactions between seagrasses and their associated communities is essential for successful conservation and restoration strategies [10,11].

Seagrasses constitute a reduced group of vascular plants that successfully colonized coastal areas through different morphological, physiological, and ecological adaptations allowing for overcoming most of the constraints imposed by the marine environment (e.g., salinity, tides, hydrodynamics, etc.). As a main outcome for such adaptations, most of the seagrass species show an adaptive convergence for morphological (e.g., very flexible aboveground structures and buried rhizomes and roots) and/or ecological traits (e.g., hydrophilic pollination and clonal growth) [12,13]. Seagrasses can occur either as large meadows or as scattered patches along the shore. However, regardless of the spatial pattern, the structural complexity of these habitats, both above and below the sediment surface, is higher than that in neighboring bare areas [14]. Seagrass canopy (i.e., aboveground structures) creates a boundary contrast with the surrounding unvegetated areas [15], which modifies the strength of both top-down and bottom-up processes [16]. Among the most widely studied biotic (i.e., top-down) mechanisms positively affecting the seagrass-inhabiting fauna are (1) the reduction in the strength of the predation intensity; (2) the creation of new habitats and niches, and (3) the reduction in the bioturbation activity of some organisms [5,6,17]. Within the abiotic (i.e., bottom-up), hydrodynamics is probably among the most important variables. In fact, the flexible aboveground biomass (AG) allows plants to reduce the drag force they support by bending in the same direction as the water flow [18,19], reducing the flow velocity throughout the canopy [20] and creating more favorable environmental conditions for fauna development [21–26].

Although the effect of increasing AG structural complexity (e.g., total biomass, shoot density, and length) on epifauna diversity has already been largely addressed [5,26–30], similar studies on infauna are scarce. Furthermore, the underlying mechanisms affecting infauna and epifauna are probably different. For instance, whereas a high AG complexity could increase epifaunal diversity either by widening the range of new colonizing areas or by reducing predation rates, such structural complexity could not produce the same outcome for infauna. High AG structural complexity can benefit infauna by (1) reducing predation rates [14], (2) increasing the settling probability of juveniles within the canopy [28,31], (3) improving the environmental conditions for instance for hydrodynamics [24,32], (4) modifying the food availability [33], and (5) excluding bioturbator species [17]. However, AG structural complexity may also produce negative effects on infauna since it may limit the water turnover within the canopy reducing food availability for organisms [33–37]. Furthermore, the dynamics of AG and belowground (BG) compartments in seagrasses are rather coupled [38,39]. Thus, large AG standing stocks are usually accompanied by high BG biomass values, which may affect negatively to infauna due to the reduction in the penetrability of the sediment [17,40]. Therefore, the predicted effects of increasing AG and BG structural complexity on infaunal organisms are not so straightforward as those described for epifauna and are less known so far.

Interactions between seagrass canopy and water flow may alter resource and particle fluxes from and towards the beds affecting seagrass meadows itself [20,41,42] as well as accompanying organisms [22,23,43–45]. That is, the quantity and quality (e.g., size) of edible particles is expected to change when water crosses through seagrass canopy, since both the reduction in velocity and the collision of suspended particles with shoots may produce a differential spatial arrangement of the suspended material and may also affect the resuspension of the particles already settled on the sediment surface. The resulting gradients of resource availability within the canopy might play a critical role in the spatial distribution patterns of benthic organisms inhabiting seagrass meadows, since macrofauna, especially filter feeders (such as bivalves), are highly selective for particle collection, filtration, and rejection [33,46–48].

The present study aims to explore two hypotheses: first, how changes in the seagrass meadow complexity (i.e., shoot density and morphometry) affect the macroinfaunal community structure (e.g., diversity) of soft-bottom intertidal communities. To answer this question, an in situ experiment was carried out using seagrass mimics to build artificial patches with different canopy properties (i.e., complexity) resembling two species (*Zostera noltei* and *Cymodocea nodosa*) thriving on the intertidal areas of Cádiz Bay. Both species have contrasting effects on canopy volumetric flow rate [20,32], which can be considered a proxy for food availability for filter feeders inhabiting seagrass patches [23]. The second hypothesis was to determine whether food availability (i.e., quantity and quality) promoted by changes in canopy properties may induce a diet shift in the macroinfaunal community and therefore contribute to explaining the expected changes in community structure. The stable isotope composition (i.e., ^{13}C and ^{15}N) in the collected macroinfaunal species and in the potential food sources, combined with mixing models and niche metrics (standard ellipse area), were used to determine changes in the diet of the organisms that could be ascribed to the alteration in the hydrodynamically driven food supply by the artificial seagrass patches.

2. Materials and Methods

2.1. Sampling Site

Los Toruños salt marsh is a sea arm at Cádiz Bay Natural Park (36°33′35.11″ N, 6°12′25.69″ W, Figure 1), where mono-specific as well as mixed meadows of the seagrasses *Zostera noltei* and *Cymodocea nodosa* occur in intertidal areas at mean depths between 1 and 1.5 m (relative to mean high water level, MHWL [49]). While mixed meadows usually thrive at the sandy mouth of this sea arm, the muddy edges of the channel are colonized by mono-specific *Z. noltei* stands either in a continuous or a patchy distribution [49]. During ebb and flow tides, there is a strong unidirectional flow parallel to the shoreline, reaching high flow velocities and turbulence levels (up to 75 cm s^{-1}) that completely mix the water body [50]. The sediment is muddy with a high organic matter content, which promotes anoxic conditions and low redox potentials at the sediment surface, with Eh values from −122 mV to −28.3 mV [51].

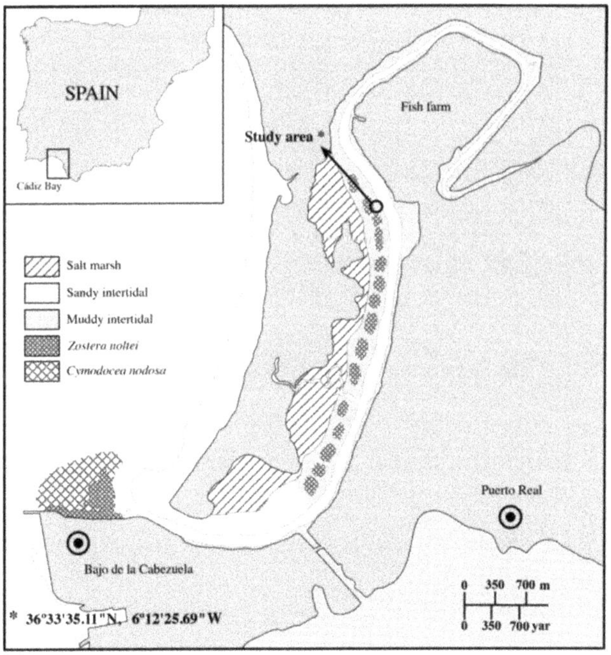

Figure 1. Location of the in situ experiments. Main characteristics (vegetation and type of sediment) and the situation of an aquaculture plant are indicated in the map. Asterisk indicates study area.

2.2. Mimics Design

Two types of artificial seagrass shoots (henceforward 'mimics') were designed to easily build artificial patches of *Zostera noltei* (ZNAP) and of *Cymodocea nodosa* (CNAP) with different canopy properties (i.e., shoot size and density) as well as to imitate the autogenic ecosystem engineer role played by above- and belowground structures [3]. Mimics were made with a silicon tube sealed at both ends (AG structure) attached to a wood stick (BG part) (Figure 2; Figure S1 in Supplementary Material [17,52]). Silicon tubes were rather flexible when interacting with water flow, while the central air chamber, resembling air-lacunae, warranted buoyancy at high tide allowing mimics to be held upright. During emersion, mimics lie on the sediment surface as seagrass leaves do. Wood sticks kept the structures anchored to the sediment while imitating the physical BG network to some extent. Mimic lengths and densities were within those recorded for *Z. noltei* and *C. nodosa* meadows in this area [23,53] (Table 1).

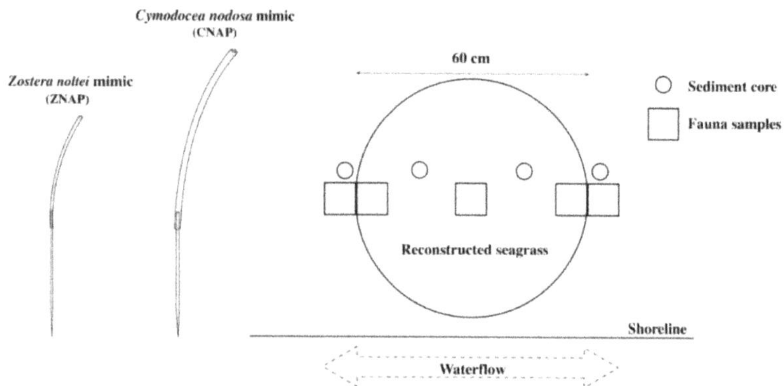

Figure 2. Scheme of the sampling procedure carried out. Fauna and sediment samples were collected following a spatial gradient parallel to the main flow direction. Mimics resembled the size and density of the two main species thriving in the area, *Zostera noltei* and *Cymodocea nodosa*. ZNAP, *Zostera noltei* artificial patches; CNAP, *Cymodocea nodosa* artificial patches.

Table 1. Canopy and morphometric properties of *Zostera noltei* natural populations (ZNNP), and artificial patches of *Z. noltei* (ZNAP) and *Cymodocea nodosa* (CNAP). Values are mean ± SE. The percentage of volume occupied by the belowground biomass of *Z. noltei* was calculated through the biometric data collected on the rhizome-root system and considering that these structures are situated within the first 7.5 cm of the sediment [54].

Variable	ZNNP	ZNAP	CNAP
Density (shoots m^{-2})	8594 ± 1.198	3500	1700
Above-ground biomass (g DW m^{-2})	110.7 ± 35.6	-	-
Leaf length (cm)	19.1 ± 0.92	20	35
Leaf width (cm)	0.12 ± 0.02	0.5	0.6
LAI ($m^2 \times m^{-2}$)	3.27 ± 0.34	11.0	11.2
Belowground biomass ($g \times DW \times m^{-2}$)	140.9 ± 22.8	-	-
Belowground length (cm)	-	30	30
Belowground width (cm)	-	0.32	0.44
Volume occupied by belowground structures (%)	15.7 ± 3.3	2.7	2.6
Meadow area (m^2)	-	≈ 0.30	≈ 0.30

2.3. In Situ Experimental Set-Up

The experiment took place in spring and lasted three months (from March to June) in an intertidal muddy-clay location where natural populations of *Zostera noltei* thrive.

This location was selected because of (1) the lack of direct anthropogenic disturbances (it was only accessible by boat), and (2) the low benthic macrofaunal diversity in bare neighboring areas [17]. Mimics were individually planted into the bare sediment to build *Zostera* (3500 mimics m^{-2}) and *Cymodocea* (1700 mimics m^{-2}) artificial patches (ZNAP and CNAP, respectively; 0.6 m in diameter each) with similar AG surface areas (i.e., leaf area index, LAI; Table 1). The belowground complexity, estimated as the percentage of sediment occupied, was similar for both types of patches (≈2.7% volume; Table 1). Six artificial patches (1 ZNAP and 1 CNAP × 3 replicates each) were randomly placed into the bare sediment (BS), leaving a minimum gap of 3 m between them and more than 5 m from the nearby natural seagrass populations.

2.4. Sampling Procedure

After 3 months, sediment samples were collected at low tide from ZNAPs, CNAPs, and BSs for infaunal studies. Furthermore, to compare the behavior of artificial patches with natural ones, surrounding *Zostera noltei* natural patches (ZNNP) of 10 m wide and up to 100 m long) were also sampled. A spatial explicit sampling was designed (Figure 2) to detect if there was a spatial gradient within the artificial patches. Thus, 5 samples (10 cm × 10 cm area, 25 cm depth) were collected per artificial patch along a transect parallel to the shoreline (i.e., main tidal flow direction): two samples at the outer-edges, two samples at the inner-edges, and one sample at the center of the patch (Figure 2). The sampling procedure for BS and ZNNP was different from that for the artificial patches. Previous sampling in the area showed that organism density and diversity levels were quite low in BS. For that reason in BS, five samples were randomly collected using a metal frame (16 cm × 16 cm area, 25 cm depth), while for ZNNP, five samples (10 cm × 10 cm area, 25 cm depth) were taken from the central part of the meadow. Since the highest proportion of the infauna inhabits within the uppermost 10 cm, this depth was considered large enough to ensure the collection of most of the organisms [55–57]. All of the samples were cleaned of muddy sediment in situ using a mesh bag (1 mm) and were transported refrigerated to the laboratory within two hours. All mimics were also collected, wrapped into plastic bags, and transported to the laboratory to measure epiphyte coverage and the presence of other settling organisms (e.g., egg masses, bryozoans, etc.).

Sediment organic matter (SOM) was also determined by measuring its concentration and its isotopic signature. Thus, additional sediment samples, close to the previously collected ones, were taken with a core (3 cm diameter, 5 cm depth; Figure 2). Cores were transported refrigerated to the laboratory. In addition, three independent water samples (1.5 L per sample) were collected from the main channel at approximately 1 m depth, kept in darkness, and refrigerated to measure particulate organic matter (POM) and its isotopic signatures. All of these samples were collected at the end of the experimental period.

2.5. Analytical Methods

Once in the laboratory, water samples were split in fine (<200 μm) and large (>200 μm) fractions with a mesh and then filtered at low vacuum in pre-combusted filters (Whatman GF/F, 0.7 μm) to retain all edible particles for infauna (e.g., plankton, feces, uneaten food particles from fish farm, etc.). Subsequently, filters were freeze-dried and sent for isotope analysis ($\delta^{13}C$ and $\delta^{15}N$) of POM.

Individual sediment samples were mixed and split. One portion was used for isotopic analysis of SOM. A sub-sample was acidified by adding 2N HCl drop-by-drop to remove carbonates (cessation of bubbling was used as a signal to stop the acidification) and subsequently freeze-dried, ground, and sent for $\delta^{13}C$ analysis, while for $\delta^{15}N$ analysis, non-acidified replicates were used [58]. The remaining portion of the sediment was oven-dried (60 °C until a constant weight) and, afterwards, burned in a muffle oven (525 °C until constant weight, ≈ 3 h) in pre-weighed ceramic cups. SOM was estimated as weight losses and expressed in %DW [59].

Ten mimics per artificial patch were randomly selected, and epiphytes were removed from the AG structures. The collected epiphytes were oven dried (60 °C) for 24 h and weighed. The data were scaled to AG area and expressed as gDW epiphytes cm^{-2} mimic. Five extra mimics from each artificial patch were cleaned of epiphytes, and the scraped material was subsequently freeze-dried, ground, and used for isotope analysis of the epiphyte community.

At a first glance, infauna samples revealed that the worm *Hediste diversicolor* (Müller, 1776) and the clam *Scrobicularia plana* (da Costa, 1778) were the dominant species (abundance >85 %). Accordingly, a minimum of five individuals of each species per sample were quickly selected once in the laboratory. These specimens were individually weighted, freeze-dried (only soft tissues in *S. plana*), ground (5 organisms per sample), and used for isotope analysis. Remaining fauna material was soaked in rose bengal (70% ethanol), identified to the species level, and weighed after drying (60 °C for 48–72 h). The species were also sorted out into functional groups based on the feeding type [44]. Seagrass material from ZNNP was split into AG and BG biomass, dried (48 h at 60 °C), and weighed (Table 2). Before drying, some shoots were collected from each sample, freeze-dried, ground, and used for isotope analysis.

Table 2. Effect of artificial patches on species feeding behavior and richness, percentage of *Scrobicularia plana* and *Hediste diversicolor*, epiphyte production, and sediment organic matter (SOM). BS, bare sediment; ZNAP, artificial patches of *Zostera noltei*; CNAP, artificial patches of *Cymodocea nodosa*; ZNNP, natural patches of *Z. noltei*. SF, suspension feeder; DF, deposit feeder; O, omnivore; C, carnivore. X, indicates species presence in the treatment. *, feeding behavior depends on species level. #, not determined. Data are expressed as mean ± SE.

	BS	ZNAP	CNAP	ZNNP
Number of species	4	8	8	7
Scrobicularia plana (%)	82.6 ± 3.3	63.0 ± 2.4	61.7 ± 1.8	86.9 ± 2.6
Hediste diversicolor (%)	6.5 ± 0.8	33.0 ± 1.6	31.7 ± 2.1	7.1 ± 0.7
Epiphyte (g DW × m^{-2})	-	505 ± 29	1563 ± 111	-
SOM (%)	9.2 ± 0.5	8.3 ± 0.3	8.4 ± 0.3	9.2 ± 0.4
Species and feeding behaviour				
Scrobicularia plana (SF/DF)	X	X	X	X
Cerastoderma edule (SF)	X	X	X	X
Venerupis rhomboides (SF)	X			
Hediste diversicolor (O)	X	X	X	X
Marphysa sanguinea (DF)		X	X	X
Bulla striata (C)				X
Venerupis sp. (SF)				X
Venerupis philippinarum (SF)		X		
Solen marginatus (SF)		X		
Diopatra neapolitana (C)		X	X	X
Xantho pilipes (C)			X	
Nephtys sp. (C/O) *			X	
Crustacean (SF) #		X	X	

2.6. Carbon and Nitrogen Composition and Isotopic Analysis

Samples were analyzed in an elemental analyzer coupled with an isotope ratio mass spectrometer (Europa Hydra IRMS coupled to a Carlo Erba NC250) for the determination of nitrogen and carbon content (% DW), and stable isotopes. Stable isotope ratios were converted to ‰ notation using Peedee Belamite (PDB) and air-N_2 as standards for C and N, respectively. The stable isotope ratio in the tissues of organisms (*Scrobicularia plana* and *Hediste diversicolor*) is directly related to its diet (e.g., potential food sources such as POM, SOM, epiphytes, and *Zostera noltei* leaves [60,61]). As the number of potential food sources exceeded the number of isotopes analyzed plus 1 (2 isotopes), a Bayesian mixing model (v4.0.3) (MixSIAR [62]) was applied to estimate the contribution of the different sources to diet, employing trophic enrichment factors previously used for *H. diversicolor* and *S. plana* ($\Delta^{13}C = 0.30 \pm 0.21$ and $\Delta^{15}N = 2.5 \pm 0.05$) [63]. Since epiphytes from artificial patches of *Z. noltei* and *C. nodosa* were only available under such treatments, these food sources were only utilized in ZNAP and CNAP treatments. The isotopic niche width for each treatment in both species was estimated using stable isotope Bayesian ellipses in R (SIBER [64]), which generates standard ellipse corrected areas (SEAc, containing 40% of the data) in a $\delta^{13}C$-$\delta^{15}N$ bi-plot space. SEAc overlap between treatments was calculated as the proportion of the non-overlapping area (total overlap area divided by the sum of the areas of two ellipses minus the total overlap area). SEAc overlap values $\geq$ 60% were considered biologically significant [65].

2.7. Statistics

A one-way ANOVA test was used to determine the effects of AG mimics with different canopy properties on infauna species richness and abundance. Therefore, we specifically tested the factors (1) bare sediment (BS), (2) *Zostera noltei* artificial patches (ZNAP), (3) *Cymodocea nodosa* artificial patches (CNAP), and (4) *Z. noltei* natural populations (ZNNP). This latter factor was used for comparison purposes with the artificial patches. A two-way ANOVA was also applied to check differences across spatial gradient and between both types of artificial patches. Therefore, in this second case, we specifically tested the factors (1) position within the patch, (2) type of artificial patches, and (3) interaction between factors. In those cases where significant differences were found, post hoc Tukey tests were accomplished. Differences in isotopic signatures ($\delta^{13}C$ and $\delta^{15}N$) between treatments were also checked with a one-way ANOVA analysis. Homoscedasticity and normality of the data were checked before conducting ANOVA tests, and data were log- or arcsine-transformed since heteroscedasticity was found in some of the variables. The data are shown as means $\pm$ standard error (SE). The significance level was set at 5% probability ($\alpha = 0.05$).

A multidimensional scaling MDS [66,67], ANOSIM [68], and permutational multivariate analyses of variance (PERMANOVA [69,70]) were applied to identify community similarities between treatments. After four-root transformation, a Bray–Curtis resemblance matrix was obtained from an abundance benthic data matrix. Additionally, the taxa contributing to dissimilarities observed were checked by the SIMPER analysis [71]. The multivariate analysis was developed using PRIMER (Plymouth Routines In Multivariate Ecological Research, 6.1.13 software [72]).

3. Results

3.1. Effects of Canopy Properties on Species Richness and Abundance

Overall, infaunal species richness, functional diversity (measured as increase in feeding types) and organism density were higher in artificial patches (ZNAP and CNAP) than in BS (Figure 3; Table 2). Particularly, species richness doubled that in BS and was similar to ZNNP. Such an increase was associated with a rise in functional diversity, where mainly carnivore polychaetes increased in number (Table 2). Infaunal abundance also increased 4 to 6 times ($F_{3,31} = 6.14$, $p < 0.001$) in comparison with BS but without significant differences between ZNAP and CNAP (Figure 3).

Differences in species composition were found among treatments. Even though the MDS plot was a relatively poor 2-D representation (stress = 0.18; 3-D, stress = 0.11) (Figure 4), the differences among the dominant species were significant (ANOSIM: R = 0.367, $p < 0.001$; PERMANOVA $p < 0.001$). Maximum differences were found between BS and artificial patches (ANOSIM: BS-ZNAP, R = 0.795, $p < 0.005$; BS-CNAP, R = 0.723, $p < 0.005$) with the latter displaying the highest values. In contrast, no significant differences were found between artificial patches (ZNAP-CNAP, R = −0.039, $p = 0.08$). The clam *Scrobicularia plana* was the dominant species in all of the sampled points and had the maximum contribution to similarity between treatments (SIMPER). The second-most abundant species was the worm *Hediste diversicolor* (Table 2). Although both species were found in all of the sampled plots, *H. diversicolor* was recorded only in one of the five samples from BS. Furthermore, although the total contribution of both species to the total infauna was quite constant in all of the sampled points (>85%), it changed drastically when artificial patches were included in the system (Table 2); whereas *H. diversicolor* represented less than 7% in untreated plots (BS and ZNNP), its abundance increased up to 32% in ZNAP and CNAP (Table 2).

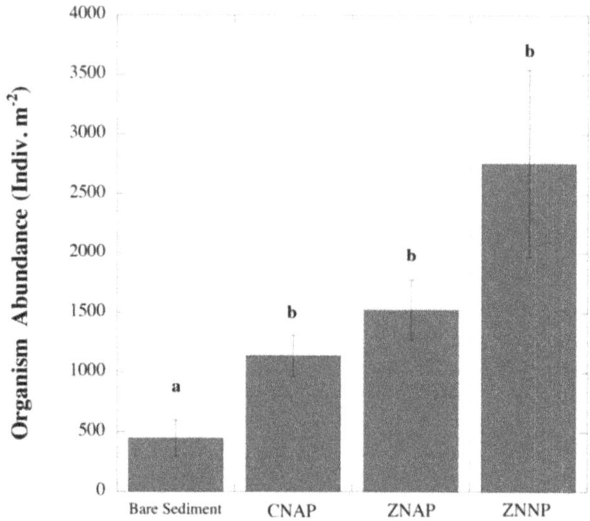

Figure 3. Organism density (indiv. m^{-2}) recorded in the different habitats: Bare sediment-BS; artificial patches of *Zostera noltei*-ZNAP, artificial patches of *Cymodocea nodosa*-CNAP, and natural patches of *Zostera noltei*-ZNNP. The inset letters indicate significant differences among treatments using the Tuckey test. Data are represented as mean ± SE.

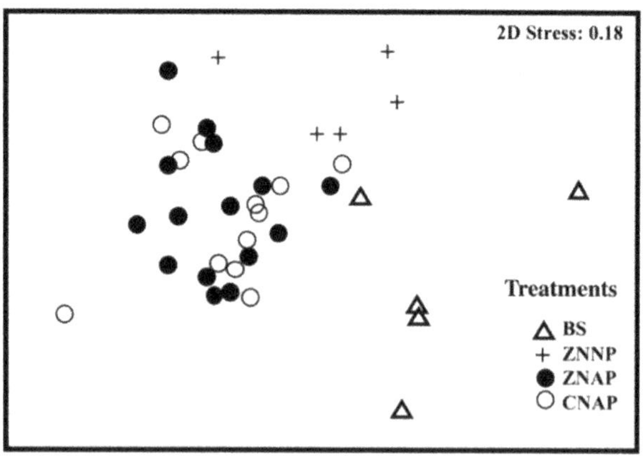

Figure 4. Multidimensional scaling plot of species composition associated with each habitat type. Bare sediment-BS; artificial patches of *Zostera noltei*-ZNAP, artificial patches of *Cymodocea nodosa*-CNAP, and natural patches of *Zostera noltei*-ZNNP.

3.2. Effects of Canopy Properties on Scrobicularia Plana and Hediste Diversicolor Abundance

Although the abundance of both species was higher in artificial patches than in BS (Figure 5A), significant differences were only found for *Hediste diversicolor,* since large within-patch variability was found for *Scrobicularia plana*. The weights of *S. plana* and

H. diversicolor individuals were not statistically different among treatments (BS, ZNAP, and CNAP), while the weight and size (data not shown) of *S. plana* were significantly lower in ZNNP (Figure 5B). The total biomass recorded within both types of artificial patches was similar for each species but significantly higher than that found either in BS or ZNNP (Figure 5C).

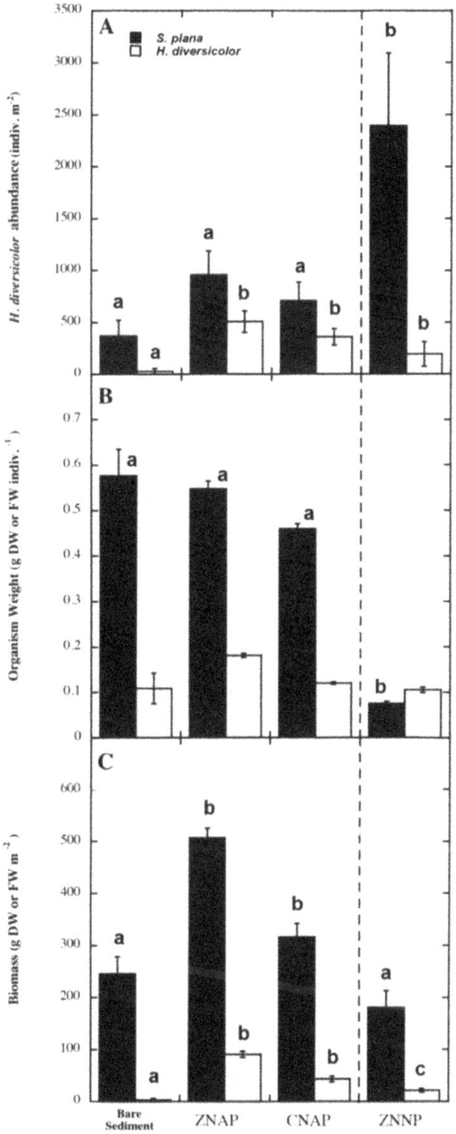

Figure 5. Density (**A**), organism individual weight (**B**), and total biomass (**C**) of the two main species recorded (>85%; *Scrobicularia plana* and *Hediste diversicolor*) in the collected samples from the different habitats: bare sediment-BS; artificial patches of *Zostera noltei*-ZNAP, artificial patches of *Cymodocea nodosa*-CNAP, and natural patches of *Zostera noltei*-ZNNP. The weight of *S. plana* included the shell and the soft tissues and is given in dry weight, while *H. diversicolor* is expressed as fresh weight. The inset letters indicate significant differences among treatments using the Tuckey test. Data are represented as mean ± SE.

3.3. *Spatial Explicit Gradients on S. plana and H. diversicolor Distribution within the Canopy*

A clear and significant species-specific spatial explicit gradient was recorded for both species regardless of the artificial patch type: whereas *S. plana* abundance increased centrifugally (Figure 6A), the pattern was centripetal for *H. diversicolor* (Figure 6B). This spatial gradient was accompanied, mainly in the case of *S. plana*, by changes in the individual body weight and size (Figure 6C,D): individuals of *S. plana* were, in general, smaller (weight and size) in the outer-edges than in the center of artificial patches (Figure 6C). As for individual abundance, total biomass for *S. plana* accumulated significantly at the periphery of the artificial patches. In addition, there were significant interactions between patch types (i.e., ZNAP or CNAP) and position (i.e., inner edge) in the total biomass of *S. plana* (Figure 6E). The total biomass for *H. diversicolor* was higher at the patch regardless of the patch type (Figure 6F), although only spatial explicit significant differences were found for ZNAP. In addition, lower biomass values were recorded at BS than at either ZNAP or CNAP for *H. diversicolor* (Figure 6F).

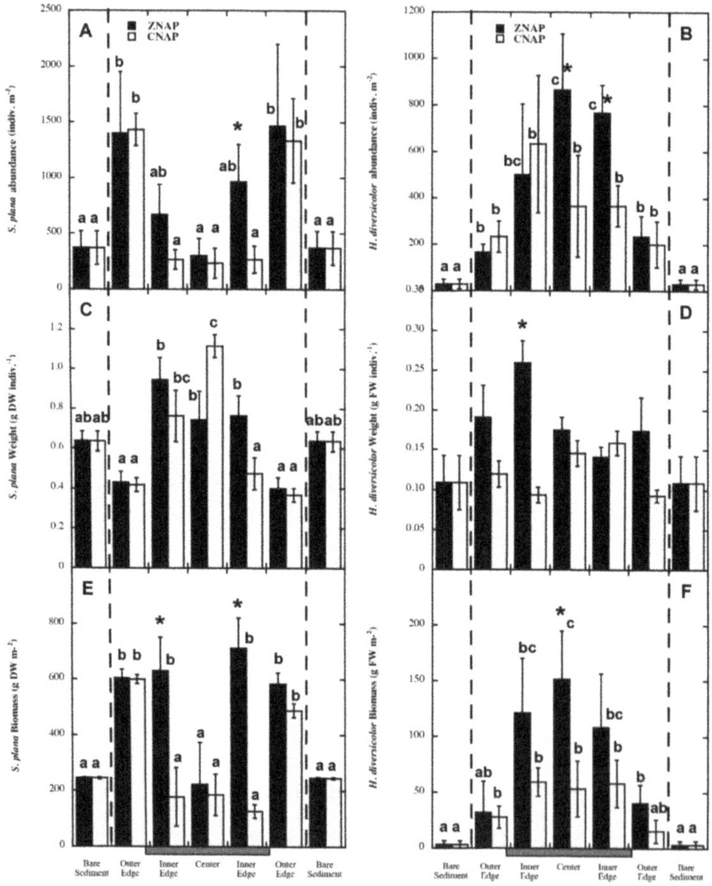

Figure 6. Recorded spatial gradients in (**A**,**B**) density, (**C**,**D**)organism individual weight, and (**E**,**F**) total biomass within the artificial seagrass patches for *Scrobicularia plana* (**A**,**C**,**E**) and *Hediste diversicolor* (**B**,**D**,**F**). The inset letters indicate significant differences between the different positions within each patch type (i.e., ZNAP or CNAP) using the Tuckey test. Inset asterisks indicate significant differences for the same position between the two patch types using the Tuckey test. The weight of *S. plana* included the shell and the soft tissues and is given in dry weight, while *H. diversicolor* is expressed as fresh weight. Data are represented as mean ± SE.

3.4. Organic Matter Content, Epiphytes, and Egg Masses

Overall, sediment organic matter content (SOM) was relatively high (Table 2). Although a minor decrease in SOM was recorded in ZNAP and CNAP in comparison with BS, differences were not statistically significant. Spatial gradients within the patches did not reveal any significant effect on the SOM content either (data not shown). Canopy properties affected epiphyte (mostly green algae) abundance, with a three-fold increase in CNAP compared with ZNAP at the end of the experiment (Table 2). The net production of epiphytes, estimated from biomass accumulation along the experiment (92 days) was 5.5 and 17 g DW $\times$ m^{-2} meadow d^{-1} for ZNAP and CNAP. Mimics also favored the anchorage of egg-laying. The most abundant and widely distributed egg masses in all artificial patches were from the cuttlefish *Sepia officinalis*, with a mean value of 52 $\pm$ 32 and 72 $\pm$ 42 eggs m^{-2} meadow for ZNAP and CNAP, respectively.

3.5. Deriving Diet Shifts from Stable Isotopes

The analysis of stable isotopes revealed that the different compartments studied could be easily distinguished of each other using the $\delta^{15}N$ and $\delta^{13}C$ signatures (Figure 7). In general, $\delta^{13}C$ values showed higher variability than $\delta^{15}N$ ones. *Zostera noltei* leaves were enriched in ^{13}C with δ values close to -12 ‰, while SOM and the gross fraction of the POM showed the lowest values (δ values from -20 to -26 ‰). Moreover, the lowest values for $\delta^{15}N$ were also found in SOM and POM (from 6.3 to 8.8 ‰). Interestingly, very large differences in $\delta^{13}C$ and, to some extent, in $\delta^{15}N$ were found between the fine and coarse fractions of POM (Figure 7). The ANOVA revealed nonsignificant differences among treatments (BS, ZNAP, and CNAP) for the $\delta^{13}C$ composition of SOM, *Scrobicularia plana*, and *Hediste diversicolor* (data not shown). Some significant differences among treatments were found for $\delta^{15}N$. For instance, $\delta^{15}N$ values for *S. plana* were significantly different between BS and CNAP ($F_{3,26} = 3.19$, $p < 0.05$), while $\delta^{15}N$ from SOM did not show any significant differences among treatments (data not shown). Moreover, $\delta^{15}N$ values of *H. diversicolor* from both types of artificial patches were significantly different from those of BS ($F_{3,32} = 4.73$, $p < 0.001$). The isotopic Bayesian mixing model (MixSIAR) showed that *S. plana* inhabiting artificial patches fed mainly on the fine fraction of the POM with a minor contribution of other sources to the diet. However, when *S. plana* inhabited natural populations of *Z. noltei* or bare sediment the contribution of all the food sources was more uniform (Table 3).

Table 3. Feasible contribution of the different food sources to the diet of *Scrobicularia plana* and *Hediste diversicolor* using the MixSIAR model in the different treatments. Five food sources were used: SOM, POM (<200 µm), POM (>200 µm), epiphytes (EPI), and *Zostera noltei* leaves (ZNL). SP, *S. plana*; HD, *H. diversicolor*; BS, bare sediment; ZNAP, artificial patches of *Zostera noltei*; CNAP, artificial patches of *Cymodocea nodosa*; and ZNNP, natural patches of *Z. noltei*. Values are presented as mean with their credible intervals at 5% and 97.5% (between parentheses).

Species-Site	ZNL	POM (<200 µm)	POM (>200 µm)	SOM	EPI
SP-BS	0.37 (0.24–0.47)	0.27 (0.07–0.46)	0.13 (0.03–0.30)	0.23 (0.05–0.47)	-
SP-ZNAP	0.25 (0.07–0.38)	0.53 (0.37–0.67)	0.08 (0.002–0.20)	0.07 (0.002–0.21)	0.07 (0.007–0.16)
SP-CNAP	0.20 (0.07–0.32)	0.61 (0.49–0.73)	0.05 (0.003–0.13)	0.11 (0.02–0.23)	0.05 (0.005–0.10)
SP-ZNNP	0.35 (0.18–0.50)	0.30 (0.04–0.55)	0.11 (0.01–0.34)	0.24 (0.03–0.46)	-
HD-BS	0.64 (0.46–0.82)	0.10 (0.004–0.31)	0.13 (0.01–0.31)	0.13 (0.01–0.37)	-
HD-ZNAP	0.42 (0.23–0.59)	0.15 (0.03–0.33)	0.09 (0.02–0.24)	0.08 (0.0–0.23)	0.26 (0.11–0.41)
HD-CNAP	0.44 (0.30–0.59)	0.13 (0.02–0.30)	0.08 (0.007–0.22)	0.12 (0.02–0.24)	0.22 (0.11–0.35)
HD-ZNNP	0.73 (0.60–0.84)	0.06 (0.001–0.23)	0.09 (0.01–0.23)	0.12 (0.01–0.26)	-

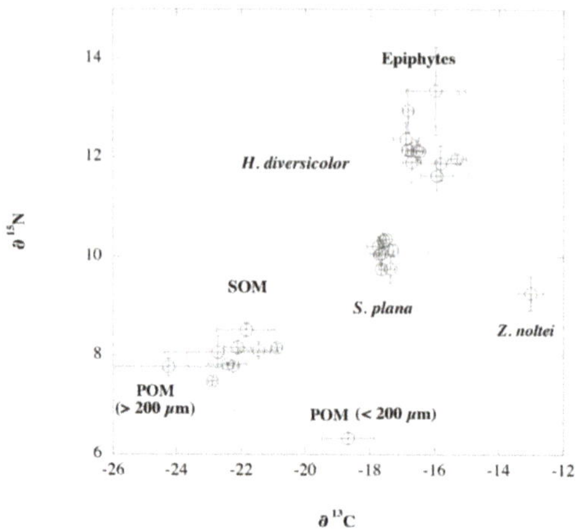

Figure 7. Isotopic signature ($\delta^{15}N$ and $\delta^{13}C$) of *Scrobicularia plana* and *Hediste diversicolor* and for the main likely food sources for these organisms collected in the different treatments (BS, ZNAP, CNAP, and ZNNP): Z. noltei, *Zostera noltei* leaves; Epiphytes, epiphytes attached to the mimics; SOM, sediment organic matter; and POM, fractionated particulate organic matter from the water column (i.e., higher or lower to 200 μm). Data are represented as mean ± SE.

Contrastingly, *H. diversicolor* seemed to feed from all available food sources in a quite variable proportion, although *Z. noltei* leaves were the source more consumed in all of the treatments (Table 3). However, the results derived from MixSIAR should be interpreted with caution, since a potential food source might be not considered as indicated by the model. The isotopic niche and overlap estimated from isotopic data by SIBER showed that, in both species, the largest niche width was observed in ZNAP. Otherwise, SIBER overlap was significant between ZNAP and CNAP for *H. diversicolor*. A considerable overlap was also estimated between ZNAP and CNAP, and between ZNAP and ZNNP for *S. plana* (Figure 8 A,B, and Table 4).

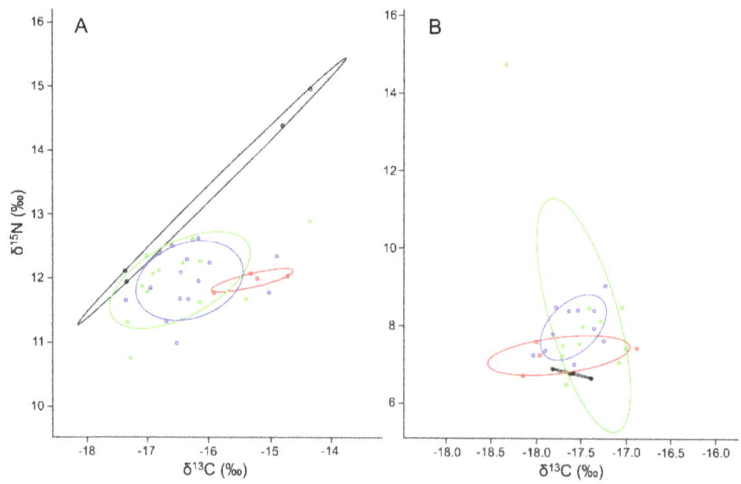

Figure 8. $\delta^{13}C$ and $\delta^{15}N$ bi-plots and standard ellipse corrected areas (SEAc, an ellipse that contains 40% of the data regardless of sample size). (**A**) *Hediste diversicolor* and (**B**) *Scrobicularia plana*. BS, bare sediment (black); ZNNP, *Zostera noltei* natural populations (red); ZNAP, *Z. noltei* artificial patches (green); and CNAP, *Cymodocea nodosa* artificial patches (blue).

Table 4. Trophic niche width (SEAc) and overlap (SEAc overlap in %) estimated by SIBER analysis (Stable Isotope Bayesian Ellipses in R) for Scrobicularia plana and Hediste diversicolor in the different experimental treatments. SEAc, corrected standard ellipse area. BS, bare sediment; ZNAP, artificial patches of Zostera noltei; CNAP, artificial patches of Cymodocea nodosa; and ZNNP, natural patches of Z. noltei.

Group	SEAc	SEA_c Overlap (%)
H. diversicolor		
BS (1)	0.76	1 vs. 2 (0.00)
		1 vs. 3 (0.00)
		1 vs. 4 (0.00)
ZNNP (2)	0.19	2 vs. 3 (0.18)
		2 vs. 4 (1.76)
ZNAP (3)	1.38	3 vs. 4 (62.4)
CNAP (4)	0.99	
S. plana		
BS (1)	0.02	1 vs. 2 (0.38)
		1 vs. 3 (0.25)
		1 vs. 4 (0.00)
ZNNP (2)	0.96	2 vs. 3 (13.0)
		2 vs. 4 (9.37)
ZNAP (3)	2.33	3 vs. 4 (17.3)
CNAP (4)	0.54	

4. Discussion

This study supports previous studies reporting positive effects of seagrasses (in this case, seagrass mimics) on diversity levels (i.e., species richness, abundance, and functional diversity). However, most of these studies were conducted in epifauna [5], while the present work was focused on infauna. Furthermore, the present results evidenced that, in the short term, the colonization of artificial seagrass patches followed a spatially explicit pattern and that the diet of the organisms shifted depending on their location (BS, ZNNP, and artificial patches), and even considering the position within the artificial meadows. It suggests that the alteration of the hydrodynamically driven food supply may be an important underlying factor contributing to structure seagrass communities.

4.1. Response of Macrobenthic Community

Our results agree with the general finding that seagrasses increase habitat complexity and ecological diversity (i.e., species richness and abundance) in harsh areas with low diversity [73–75] by fostering positive effects on fauna (e.g., facilitation) [17,26,30,76,77]. The presence of flexible mimics resembling the role played by wild seagrasses could facilitate the entrance of new species because of the amelioration of environmental conditions fostered by aboveground structures (mainly flow reduction), the reduction in predation rates, the presence of new food sources (epiphytes and preys), and the generation of gradients in food availability. In addition, belowground parts may also play an important role in diversity levels by reducing the effects of bioturbator species, as it has been demonstrated in the studied area [17]. However, belowground complexity (i.e., architectural attributes) was quite similar between both artificial patches and should affect the infauna in both treatments in a similar way.

Our results agreed with the general finding that the increase in habitat complexity associated with ecosystem engineers leads also to higher diversity levels in areas of low diversity. This enhancement of species richness and abundance in artificial patches was accompanied by both a shift in the abundance of some dominant species (mainly *Scrobicularia plana* and *Hediste diversicolor*) and by the increase in some functional groups. For instance, infaunal predators increased in artificial patches compared with BS. This could be the result of an increase in the abundance of prey, which also attracts predators ([78]

and references therein), and/or the role of the artificial canopy as a refuge for predators against their own [79].

The enhancement of species richness and organism abundance in artificial patches seemed to be mainly promoted by an active behavior of the organisms that moved from nearby areas towards the patches, since no significant differences in organism size or weight were found between artificial patches and bare sediment (Figure 5B). Field observations and laboratory experiments with *Scrobicularia plana* indicated that this species is capable of a substantial horizontal movement (<20cm) in response to some unfavorable conditions, but territoriality or overcrowding did not trigger such migration responses [55]. Therefore, the accumulation of *S. plana* in the periphery of the artificial patches suggests that this species actively migrated from nearby bare areas because of a more favorable habitat within the mimics. Such favorable growing conditions in the periphery are supported by previous studies in the area [33], where both the concentration of suspended edible particles and the food intake rate of cockles (i.e., an active filter feeding organisms) significantly increased in the leading edge of a reconstructed seagrass patch subjected to unidirectional flow.

For the two dominant species, *Scrobicularia plana* and *Hediste diversicolor*, the spatial explicit sampling design revealed that processes controlling their abundances differed. For instance, *S. plana* accumulated at the outer edges of the artificial patches (Figure 6A), where it is not expected that they would benefit from shelter against predators (mainly oystercatcher and other shorebirds [55,80]) siphon nipping (mainly by crabs or fishes [55]) or hydrodynamics [81] as it would occur within the patch. In addition, while there were not significant differences in the weight of the specimens between BS and both types of artificial patches, individuals of *S. plana* occurring within the mimics were significantly larger than those thriving in the outer edges (Figure 6C). This indicates that belowground structures did not hamper *S. plana* presence within the artificial patches.

4.2. Shifts of Diets for Macrobenthic Fauna

A likely explanation for the accumulation of *S. plana* at the outer patch edges could be related to a higher food availability at that point, as it was demonstrated for pipefishes and in artificial seagrass meadows in flume tank studies [17,82]. The interaction between canopy edges and flow results in a sudden decrease in velocity and a higher particle collision and loss of momentum of these particles, favoring the sedimentation of the suspended material [42,83–85]. It would increase the food availability at the patch periphery, stimulating the preferential occurrence of *S. plana* in this area (Figure 6A–C) as occurred for cockles in flume tank studies [17]. The measured differences in the isotopic signature and niche width in *S. plana* ($\delta^{15}N$) depending on its location (i.e., in BS or within artificial patches) is a clear sign of diet shifting, which agrees with this explanation. Moreover, the niche width (SEAc values in Table 4; Figure 8B) becomes wider within natural and artificial seagrass populations when compared with bare sediments. It may indicate that *S. plana* has access to more diverse food sources under such conditions as it was found in *S. plana* [86] and in other suspension-feeders [87], or by contrast, the higher density values recorded of *S. plana* may induce food competition [33], driving individuals to feed on other less preferred available food sources.

Abundance and total biomass of *H. diversicolor* increased centripetally in both types of artificial patches (Figure 6B,D), although only spatial significant differences were found for ZNAP. The effects on abundance might be partially attributed to an enhanced protection against predation, in comparison with BS, since this species is highly predated by fishes and shore birds [88,89]. However, a lower predation pressure cannot totally explain the preferential accumulation of *H. diversicolor* at the outer edge of the patches, where it does not obtain any protection against predation. In addition, differences between both types of artificial patches were also found, in spite of their similar above and belowground complexity (i.e., architectural attributes), which may render a similar predation effectiveness of predator species [90]. It is known that *H. diversicolor* is highly predated by different species and that it has a large plasticity in feeding behavior [61,91]. This feeding plasticity allows

H. diversicolor to occur in a variety of habitats [88]. The isotope composition (Figure 8; Tables 3 and 4) confirmed this point, showing that (1) *H. diversicolor* fed from all the potential sources studied and (2) the contribution of the different sources shifted depending on the habitat (mainly BS versus artificial patches—ZNAP and CNAP—and ZNNP). This might indicate that food availability (quantity and/or quality) depended highly on the type of habitat. It is important to note the large contribution of *Z. noltei* leaves to the diet of *H. diversicolor* (Table 3), which agrees with previous findings of a direct assimilation of plant detritus by this species [61] and with field observations where individuals of *H. diversicolor* were observed feeding directly over seagrass leaves [Brun, personal observations]. Therefore, changes in food availability could partially explain the distribution pattern of *H. diversicolor* within the artificial patches. However, our experimental set-up did not allow us to reach further conclusions; additional experimentation is required to determine the strength of different processes affecting its distribution within patches.

Mimics were heavily loaded by epiphytes and egg masses (i.e., *Sepia officinalis*) (Table 2). This serves as additional evidence of the role of seagrass meadows as physical structures increasing not only the diversity but also providing shelter and nursery grounds for species of commercial importance [92]. The net production of epiphytes along the experimental period was high and even higher than the total production of *Zostera noltei* populations in the area [49]. The daily net epiphyte production measured in CNAP was almost threefold higher than that of ZNAP and five times higher than the epiphyte production measured in natural populations of *Cymodocea nodosa* in the area (circa. 3.4 ± 1.1 g DW m^{-2} d^{-1} [93]). This result could be partly explained by the lack of aboveground biomass turnover and of chemical defenses in mimics in comparison with natural seagrass meadows [94,95], thus facilitating the epiphyte accumulation in mimics.

4.3. Ecological Significance

The presence of artificial seagrass patches resembling the autogenic ecosystem-engineering role of those plants positively affected species richness, organism abundance, and functional diversity of infauna. Such increases showed a distinctive component (both in canopy properties and macrofaunal species) and seemed to be the result of (1) changes in the hydrodynamically driven food supply to the organisms and (2) facilitation processes mediated by physical structures (reduction in predation pressure, amelioration of environmental conditions, reduction in bioturbators activity, etc.). In addition, the presence of egg masses and large epiphyte loads in the experimental plots highlight the importance of seagrass populations as "diversity and productivity boosters", since its presence allows for not only increases in the diversity levels of fauna but also enhancements in the presence of other primary producers that contribute in a significant manner to the community and ecosystem productivity.

Supplementary Materials: The following are available online at https://www.mdpi.com/article/10.3390/d13110572/s1, Figure S1: Images of the mimic set-up in Rio San Pedro.

Author Contributions: Conceptualization, F.G.B., J.L.P.-L. and J.J.V.; methodology, V.G.-O., J.F.C.-D., F.G.B., J.L.V. and J.J.V.; field experiments, F.G.B., J.F.C.-D., V.G.-O., J.L.P.-L. and J.J.V.; isotope analysis, J.L.V.; data curation, F.G.B., J.F.C.-D. and J.L.V.; writing—original draft preparation, F.G.B., J.L.V., J.L.P.-L. and J.J.V.; writing—review and editing, F.G.B., J.L.P.-L. and J.J.V.; supervision, F.G.B. and J.J.V.; project administration, J.L.P.-L. and F.G.B.; funding acquisition, F.G.B. and J.L.P.-L. All authors have read and agreed to the published version of the manuscript.

Funding: This work was supported by the Spanish Project PAVAROTTI (CTM2017-85365-R) from the Ministry of Science and Innovation and by the Junta de Andalucía Excellence Project PAMBIO (P08-RNM-03783).

Institutional Review Board Statement: Not applicable. Ethical review and approval were waived for this study due to neither humans nor protected animals being involved. The Cadiz Bay Natural Park approved the experiments as the responsible environmental agency for the site.

Informed Consent Statement: Not applicable.

Data Availability Statement: All data generated or analyzed during this study are included in this published article. Any additional information is available upon request.

Conflicts of Interest: The authors declare no conflict of interest.

References

1. Waycott, M.; Duarte, C.M.; Carruthers, T.J.B.; Orth, R.J.; Dennison, W.C.; Olyarnik, S.; Calladine, A.; Fourqurean, J.W.; Heck, K.L., Jr.; Hughes, A.R.; et al. Accelerating loss of seagrasses across the globe threatens coastal ecosystems. *Proc. Natl. Acad. Sci. USA* **2009**, *106*, 12377–12381. [CrossRef]
2. Vassallo, P.; Paoli, C.; Rovere, A.; Montefalcone, M.; Morri, C.; Bianchi, C.N. The value of the seagrass *Posidonia oceanica*: A natural capital assessment. *Mar. Pol. Bull.* **2013**, *75*, 157–167. [CrossRef]
3. Jones, C.G.; Lawton, J.H.; Shachak, M. Organisms as ecosystem engineers. *Oikos* **1994**, *69*, 373–386. [CrossRef]
4. Hemminga, M.A.; Duarte, C.M. *Seagrass Ecology*; Cambridge University Press: New York, NY, USA, 2000; 310p. [CrossRef]
5. Boström, C.; O'Brien, K.; Roos, C.; Ekebom, J. Environmental variables explaining structural and functional diversity of seagrass macrofauna in an archipelago landscape. *J. Exp. Mar. Biol. Ecol.* **2006**, *335*, 52–73. [CrossRef]
6. Larkum, A.W.D.; Orth, R.J.; Duarte, C.M. (Eds.) *Seagrasses: Biology, Ecology and Conservation*; Springer: Dordrecht, The Netherlands, 2006; 691p. [CrossRef]
7. Duffy, J.E. Biodiversity and the functioning of seagrass ecosystems. *Mar. Ecol. Prog. Ser.* **2006**, *311*, 233–250. [CrossRef]
8. Stachowicz, J.J.; Bruno, J.F.; Duffy, J.E. Understanding the effects of marine biodiversity on communities and ecosystems. *Annu. Rev. Ecol. Evol. Syst.* **2007**, *38*, 739–766. [CrossRef]
9. Telesca, L.; Belluscio, A.; Criscoli, A.; Ardizzone, G.; Apostolaki, E.T.; Fraschetti, S.; Gristina, M.; Knittweis, L.; Martin, C.S.; Pergent, G.; et al. Seagrass meadows (*Posidonia oceanica*) distribution and trajectories of change. *Sci. Rep.* **2015**, *5*, 12505. [CrossRef] [PubMed]
10. Thrush, S.F.; Dayton, P.K. Disturbance to marine benthic habitats by trawling and dredging: Implications for marine biodiversity. *Annu. Rev. Ecol. Syst.* **2002**, *33*, 449–473. [CrossRef]
11. Norkko, A.; Hewitt, J.E.; Thrush, S.F.; Funnell, G.A. Conditional outcomes of facilitation by a habitat-modifying subtidal bivalve. *Ecology* **2006**, *87*, 226–234. [CrossRef]
12. Vogel, S. *Life in Moving Fluids*; Willard Grant: Boston, MA, USA, 1981; 352p.
13. Kuo, J.; den Hartog, C. Seagrass morphology, anatomy, and ultrastructure. In *Seagrasses: Biology, Ecology and Conservation*; Larkum, A.W.D., Orth, R.J., Duarte, C.M., Eds.; Springer: Dordrecht, The Netherlands, 2006; pp. 51–87. [CrossRef]
14. Hovel, K.A.; Lipcius, R.N. Effects of seagrass habitat fragmentation on juvenile blue crab survival and abundance. *J. Exp. Mar. Biol. Ecol.* **2002**, *271*, 75–98. [CrossRef]
15. Holmquist, J.G. Permeability of patch boundaries to benthic invertebrates: Influences of boundary contrast, light level, and faunal density and mobility. *Oikos* **1998**, *81*, 558–566. [CrossRef]
16. Hovel, K.A.; Regan, H.M. Using an individual-based model to examine the roles of habitat fragmentation and behavior on predator–prey relationships in seagrass landscapes. *Landsc. Ecol.* **2008**, *23*, 75–89. [CrossRef]
17. González-Ortiz, V.; Egea, L.G.; Jiménez-Ramos, R.; Moreno-Marín, F.; Pérez-Lloréns, J.L.; Bouma, T.J.; Brun, F.G. Interactions between seagrass complexity, hydrodynamic flow and biomixing alter food availability for associated filter-feeding organisms. *PLoS ONE* **2014**, *9*, e104949. [CrossRef] [PubMed]
18. Verduin, J.J.; Backhaus, J.O. Dynamics of plant–flow interactions for the seagrass *Amphibolis antarctica*: Field observations and model simulations. *Estuar. Coast. Shelf Sci.* **2000**, *50*, 185–204. [CrossRef]
19. Bouma, T.J.; De Vries, M.B.; Low, E.; Peralta, G.; Tánczos, C.; Van de Koppel, J.; Herman, P.M.J. Trade-offs related to ecosystem engineering: A case study on stiffness of emerging macrophytes. *Ecology* **2005**, *86*, 2187–2199. [CrossRef]
20. Morris, E.P.; Peralta, G.; Brun, F.G.; van Duren, L.A.; Bouma, T.J.; Pérez-Lloréns, J.L. Interaction between hydrodynamics and seagrass canopy structure: Spatially explicit effects on ammonium uptake rates. *Limnol. Oceanogr.* **2008**, *53*, 1531–1539. [CrossRef]
21. Irlandi, E.A.; Peterson, C.H. Modification of animal habitat by large plants: Mechanisms by which seagrasses influence clam growth. *Oecologia* **1991**, *87*, 307–318. [CrossRef]
22. Peterson, B.J.; Heck, K.L., Jr. Positive interactions between suspension-feeding bivalves and seagrass - a facultative mutualism. *Mar. Ecol. Prog. Ser.* **2001**, *213*, 143–155. [CrossRef]
23. Brun, F.G.; Van Zetten, E.; Cacabelos, E.; Bouma, T.J. Role of two contrasting ecosystem engineers (*Zostera noltii* and *Cymodocea nodosa*) on the food intake rate of *Cerastoderma edule*. *Helgol. Mar. Res.* **2009**, *63*, 19–25. [CrossRef]
24. Tagliapietra, D.; Pessa, G.; Cornello, M.; Zitelli, A.; Magni, P. Temporal distribution of intertidal macrozoobenthic assemblages in a *Nanozostera noltii*-dominated area (Lagoon of Venice). *Mar. Environ. Res.* **2016**, *114*, 31–39. [CrossRef]
25. Jiménez-Ramos, R.; Egea, L.G.; Vergara, J.J.; Bouma, T.J.; Brun, F.G. The role of flow velocity combined with habitat complexity as a top–down regulator in seagrass meadows. *Oikos* **2019**, *128*, 64–76. [CrossRef]
26. Rodil, I.F.; Lohrer, A.M.; Attard, K.M.; Hewitt, J.E.; Thrush, S.F.; Norkko, A. Macrofauna communities across a seascape of seagrass meadows: Environmental drivers, biodiversity patterns and conservation implications. *Biodivers. Conserv.* **2021**, *30*, 3023–3043. [CrossRef]
27. Attrill, M.J.; Strong, J.A.; Rowden, A.A. Are macroinvertebrate communities influenced by seagrass structural complexity? *Ecography* **2000**, *23*, 114–121. [CrossRef]

28. Hovel, K.A.; Fonseca, M.S. Influence of seagrass landscape structure on the juvenile blue crab habitat-survival function. *Mar. Ecol. Prog. Ser.* **2005**, *300*, 170–191. [CrossRef]
29. Sirota, L.; Hovel, K.A. Simulated eelgrass *Zostera marina* structural complexity: Effects of shoot length, shoot density, and surface area on the epifaunal community of San Diego Bay, California, USA. *Mar. Ecol. Prog. Ser.* **2006**, *326*, 115–131. [CrossRef]
30. Gagnon, K.; Rinde, E.; Bengil, E.G.T.; Carugati, L.; Christianen, M.J.A.; Danovaro, R.; Gambi, C.; Govers, L.L.; Kipston, S.; Meysick, L.; et al. Facilitating foundation species: The potential for plant–bivalve interactions to improve habitat restoration success. *J. Appl. Ecol.* **2020**, *57*, 1161–1179. [CrossRef]
31. Barbier, P.; Meziane, T.; Forêt, M.; Tremblay, R.; Robert, R.; Olivier, F. Nursery function of coastal temperate benthic habitats: New insight from the bivalve recruitment perspective. *J. Sea Res.* **2017**, *121*, 11–23. [CrossRef]
32. Peralta, G.; Morris, E.P.; van Duren, L.A.; Bouma, T.J. Consequences of shoot density and stiffness for ecosystem engineering by benthic macrophytes in flow dominated areas: A hydrodynamic flume study. *Mar. Ecol. Prog. Ser.* **2008**, *368*, 103–115. [CrossRef]
33. González-Ortiz, V.; Alcazar, P.; Vergara, J.J.; Pérez-Lloréns, J.L.; Brun, F.G. Effects of two antagonistic ecosystem engineers on infaunal diversity. *Estuar. Coast. Shelf Sci.* **2014**, *139*, 20–26. [CrossRef]
34. Reusch, T.B.H.; Williams, S.L. Macrophyte canopy structure and the success of an invasive marine bivalve. *Oikos* **1999**, *84*, 398–416. [CrossRef]
35. Allen, B.J.; Williams, S.L. Native eelgrass *Zostera marina* controls growth and reproduction of an invasive mussel through food limitation. *Mar. Ecol. Prog. Ser.* **2003**, *254*, 57–67. [CrossRef]
36. Tsai, C.; Yang, S.; Trimble, A.C.; Ruesink, J.L. Interactions between two introduced species: *Zostera japonica* (dwarf eelgrass) facilitates itself and reduces condition of *Ruditapes philippinarum* (Manila clam) on intertidal flats. *Mar. Biol.* **2010**, *157*, 1929–1936. [CrossRef]
37. Carroll, J.M.; Peterson, B.J. Ecological trade-offs in seascape ecology: Bay scallop survival and growth across a seagrass seascape. *Landsc. Ecol.* **2013**, *28*, 1401–1413. [CrossRef]
38. Brun, F.G.; Cummaudo, F.; Olivé, I.; Vergara, J.J.; Pérez-Lloréns, J.L. Clonal extent, apical dominance and networking features in the phalanx angiosperm *Zostera noltii* Hornem. *Mar. Biol.* **2007**, *151*, 1917–1927. [CrossRef]
39. Brun, F.G.; Olivé, I.; Malta, E.-J.; Vergara, J.J.; Hernández, I.; Pérez-Lloréns, J.L. Increased vulnerability of *Zostera noltii* to stress caused by low light and elevated ammonium levels under phosphate deficiency. *Mar. Ecol. Prog. Ser.* **2008**, *365*, 67–75. [CrossRef]
40. Goshima, S.; Peterson, C. Both below- and aboveground shoalgrass structure influence whelk predation on hard clams. *Mar. Ecol. Prog. Ser.* **2012**, *451*, 75–92. [CrossRef]
41. Barrón, C.; Middelburg, J.J.; Duarte, C.M. Phytoplankton trapped within seagrass (*Posidonia oceanica*) sediments are a nitrogen source: An in situ isotope labelling experiment. *Limnol. Oceanogr.* **2006**, *51*, 1648–1653. [CrossRef]
42. Barcelona, A.; Oldham, C.; Colomer, J.; Garcia-Orellana, J.; Serra, T. Particle capture by seagrass canopies under an oscillatory flow. *Coast. Eng.* **2021**, *169*, 103972. [CrossRef]
43. Reusch, T.B.H.; Williams, S.L. Variable responses of native eelgrass *Zostera marina* to a non-indigenous bivalve *Musculista senhousia*. *Oecologia* **1998**, *113*, 428–441. [CrossRef] [PubMed]
44. Bouma, T.J.; Ortells, V.; Ysebaert, T. Comparing biodiversity effects among ecosystem engineers of contrasting strength: Macrofauna diversity in *Zostera noltii* and *Spartina anglica* vegetations. *Helgol. Mar. Res.* **2009**, *63*, 3–18. [CrossRef]
45. Giraldo-Ospina, A.; Ladah, L.B.; Hovel, K.A. Changes in within-canopy environmental conditions and the population structure of the speckled scallop associated to localized loss of above-ground seagrass cover. *J. Exp. Mar. Biol. Ecol.* **2021**, *534*, 151486. [CrossRef]
46. Self, R.F.L.; Jumars, P.A. Cross-phyletic patterns of particle selection by deposit feeders. *J. Mar. Res.* **1988**, *46*, 119–143. [CrossRef]
47. Defossez, J.-M.; Hawkins, A.J.S. Selective feeding in shellfish: Size-dependent rejection of large particles within pseudofaeces from *Mytilus edulis, Ruditapes philippinarum* and *Tapes decussatus*. *Mar. Biol.* **1997**, *129*, 139–147. [CrossRef]
48. Sobral, P.; Widdows, J. Effects of increasing current velocity, turbidity and particle-size selection on the feeding activity and scope for growth of *Ruditapes decussatus* from Ria Formosa, southern Portugal. *J. Exp. Mar. Biol. Ecol.* **2000**, *245*, 111–125. [CrossRef]
49. Brun, F.G.; Pérez-Lloréns, J.L.; Hernández, I.; Vergara, J.J. Patch distribution and within-patch dynamics of the seagrass *Zostera noltii* Hornem. in Los Toruños Salt-Marsh, Cádiz Bay Natural park, Spain. *Bot. Mar.* **2003**, *46*, 513–524. [CrossRef]
50. González-Gordillo, J.I.; Arias, A.M.; Rodríguez, A.; Drake, P. Recruitment patterns of decapod crustacean megalopae in a shallow inlet (SW Spain) related to life history strategies. *Estuar. Coast. Shelf Sci.* **2003**, *56*, 593–607. [CrossRef]
51. Ferrón, S.; Ortega, T.; Forja, J.M. Benthic fluxes in a tidal salt marsh creek affected by fish farm activities: Río San Pedro (Bay of Cádiz, SW Spain). *Mar. Chem.* **2009**, *113*, 50–62. [CrossRef]
52. Brun, F.G.; González-Ortiz, V.; Vergara, J.J.; Pérez-Lloréns, J.L. Unidad artificial flexible individual de angiosperma marina. Flexible artificial single seagrass unit. National Patent 201200489, 7 May 2012.
53. Brun, F.G.; Vergara, J.J.; Peralta, G.; García-Sánchez, M.P.; Hernández, I.; Pérez- Lloréns, J.L. Clonal building, simple growth rules and phylloclimate as key steps to develop functional-structural seagrass models. *Mar. Ecol. Prog. Ser.* **2006**, *323*, 133–148. [CrossRef]
54. Duarte, C.M.; Merino, M.; Agawin, N.S.; Uri, J.; Fortes, M.D.; Gallegos, M.E.; Marbá, N.; Hemminga, M.A. Root production and belowground seagrass biomass. *Mar. Ecol. Prog. Ser.* **1998**, *171*, 97–108. [CrossRef]
55. Hughes, R.N. Population Dynamics of the Bivalve *Scrobicularia plana* (Da Costa) on an Intertidal Mud-Flat in North Wales. *J. Anim. Ecol.* **1970**, *39*, 333–356. [CrossRef]

56. Jacobs, R.P.W.M.; Hegger, H.H.; Ras-Willems, A. Seasonal variations in the structure of a *Zostera* community on tidal flats in the SW Netherlands, with special reference to the benthic fauna. *Proc. Kon. Ned. Akad. Wet. Ser. C* **1983**, *86*, 347–375.
57. Boström, C.; Bonsdorff, E. Community structure and spatial variation of benthic invertebrates associated with *Zostera marina* (L.) beds in the northern Baltic Sea. *J. Sea Res.* **1997**, *37*, 153–166. [CrossRef]
58. Carabel, S.; Godínez-Domínguez, E.; Verísimo, P.; Fernández, L.; Freire, J. An assessment of sample processing methods for stable isotope analyses of marine food webs. *J. Exp. Mar. Biol. Ecol.* **2006**, *336*, 254–261. [CrossRef]
59. Skilleter, G.A.; Pryor, A.; Miller, S.; Cameron, B. Detecting the effects of physical disturbance on benthic assemblages in a subtropical estuary: A Beyond BACI approach. *J. Exp. Mar. Biol. Ecol.* **2006**, *338*, 271–287. [CrossRef]
60. Kanaya, G.; Takagi, S.; Nobata, E.; Kikuchi, E. Spatial dietary shift of macrozoobenthos in a brackish lagoon revealed by carbon and nitrogen stable isotope ratios. *Mar. Ecol. Prog. Ser.* **2007**, *345*, 117–127. [CrossRef]
61. Kanaya, G.; Takagi, S.; Kikuchi, E. Dietary contribution of the microphytobenthos to infaunal deposit feeders in an estuarine mudflat in Japan. *Mar. Biol.* **2008**, *155*, 543–553. [CrossRef]
62. Stock, B.C.; Semmens, B.X. MixSIAR GUI User Manual. Version 3.1. 2016. Available online: https://github.com/brianstock/MixSIAR (accessed on 7 September 2021).
63. Davoult, D.; Surget, G.; Stiger-Pouvreau, V.; Noisette, F.; Riera, P.; Stagnol, D.; Androuin, T.; Poupart, N. Multiple effects of a *Gracilaria vermiculophylla* invasion on estuarine mudflat functioning and diversity. *Mar. Environ. Res.* **2017**, *131*, 227–235. [CrossRef]
64. Jackson, A.L.; Inger, R.; Parnell, A.C.; Bearhop, S. Comparing isotopic niche widths among and within communities: SIBER—Stable Isotope Bayesian Ellipses in R. *J. Anim. Ecol.* **2011**, *80*, 595–602. [CrossRef]
65. Varela, J.L.; Spares, A.D.; Stokesbury, M.J. Feeding ecology of Atlantic bluefin tuna (*Thunnus thynnus*) in the Gulf of Saint Lawrence, Canada. *Mar. Environ. Res.* **2020**, *161*, 105087. [CrossRef]
66. Kruskal, J.B. Nonmetric multidimensional scaling: A numerical method. *Psychometrika* **1964**, *29*, 115–129. [CrossRef]
67. Shepard, R.N. The analysis of proximities: Multidimensional scaling with an unknown distance function, I. *Psychometrika* **1962**, *27*, 125–140. [CrossRef]
68. Clarke, K.R.; Green, R.H. Statistical design and analysis for a 'biological effects' study. *Mar. Ecol. Prog. Ser.* **1988**, *46*, 213–226. [CrossRef]
69. Anderson, M.J. A new method for non-parametric multivariate analysis of variance. *Aust. Ecol.* **2001**, *26*, 32–46. [CrossRef]
70. McArdle, B.H.; Anderson, M.J. Fitting multivariate models to community data: A comment on distance-based redundancy analysis. *Ecology* **2001**, *82*, 290–297. [CrossRef]
71. Clarke, K.R. Non-parametric multivariate analyses of changes in community structure. *Aust. J. Ecol.* **1993**, *18*, 117–143. [CrossRef]
72. Clarke, K.R.; Warwick, R.M. *Change in Marine Communities: An Approach to Statistical Analysis and Interpretation*, 2nd ed.; PRIMER-E Ltd: Plymouth, UK, 2001; 176p.
73. Crooks, J.A. Characterizing ecosystem-level consequences of biological invasions: The role of ecosystem engineers. *Oikos* **2002**, *97*, 153–166. [CrossRef]
74. Wright, J.P.; Jones, C.G. The concept of organisms as ecosystem engineers ten years on: Progress, limitations, and challenges. *Bioscience* **2006**, *56*, 203–209. [CrossRef]
75. Bouma, T.J.; Olenin, S.; Reise, K.; Ysebaert, T. Ecosystem engineering and biodiversity in coastal sediments: Posing hypotheses. *Helgol. Mar. Res.* **2009**, *63*, 95–106. [CrossRef]
76. Bertness, M.D.; Callaway, R. Positive interactions in communities. *Trends Ecol. Evol.* **1994**, *9*, 191–193. [CrossRef]
77. Bruno, J.F.; Stachowicz, J.J.; Bertness, M.D. Inclusion of facilitation into ecological theory. *Trends Ecol. Evol.* **2003**, *18*, 119–125. [CrossRef]
78. Heck, K.L.; Orth, R.J., Jr. Predation in seagrass beds. In *Seagrasses: Biology, Ecology and Conservation*; Larkum, A.W.D., Orth, R.J., Duarte, C.M., Eds.; Springer: Dordrecht, The Netherlands, 2006; pp. 537–550. [CrossRef]
79. Bowden, D.A.; Rowden, A.A.; Attrill, M.J. Effect of patch size and in-patch location on the infaunal macroinvertebrate assemblages of *Zostera marina* seagrass beds. *J. Exp. Mar. Biol. Ecol.* **2001**, *259*, 133–154. [CrossRef]
80. Pérez-Hurtado, A. Ecología Alimentaria de aves Limícolas Invernantes en la Bahía de Cádiz. Distribución y uso del Hábitat. Ph.D. Thesis, University of Seville, Seville, Spain, 1992. Available online: https://idus.us.es/handle/11441/79650 (accessed on 7 September 2021).
81. Meysick, L.; Ysebaert, T.; Jansson, A.; Montserrat, F.; Valanko, S.; Villnäs, A.; Boström, C.; Norkko, J.; Norkko, A. Context-dependent community facilitation in seagrass meadows along a hydrodynamic stress gradient. *J. Sea Res.* **2019**, *150*, 8–23. [CrossRef]
82. Macreadie, P.I.; Hindell, J.S.; Keough, M.J.; Jenkins, G.P.; Connolly, R.M. Resource distribution influences positive edge effects in a seagrass fish. *Ecology* **2010**, *91*, 2013–2021. [CrossRef]
83. Agawin, N.S.R.; Duarte, C.M. Evidence of direct particle trapping by a tropical seagrass meadow. *Estuaries* **2002**, *25*, 1205–1209. [CrossRef]
84. Hendriks, I.E.; Sintes, T.; Bouma, T.J.; Duarte, C.M. Experimental assessment and modeling evaluation of the effects of the seagrass *Posidonia oceanica* on flow and particle trapping. *Mar. Ecol. Prog. Ser.* **2008**, *356*, 163–173. [CrossRef]
85. Hendriks, I.E.; Bouma, T.J.; Morris, E.P.; Duarte, C.M. Effects of seagrasses and algae of the *Caulerpa* family on hydrodynamics and particle-trapping rates. *Mar. Biol.* **2010**, *157*, 473–481. [CrossRef]

86. Rossi, F.; Baeta, A.; Marques, J.C. Stable isotopes reveal habitat-related diet shifts in facultative deposit-feeders. *J. Sea Res.* **2015**, *95*, 172–179. [CrossRef]
87. Jones, A.G.; Dubois, S.F.; Desroy, N.; Fournier, J. Intertidal ecosystem engineer species promote benthic-pelagic coupling and diversify trophic pathways. *Mar. Ecol. Prog. Ser.* **2021**, *660*, 119–139. [CrossRef]
88. Arias, A.M.; Drake, P. Distribution and production of the polychaete *Nereis diversicolor* in a shallow coastal lagoon in the Bay of Cadiz (SW Spain). *Cah. Biol. Mar.* **1995**, *36*, 201–210. [CrossRef]
89. Rosa, S.; Granadeiro, J.P.; Vinagre, C.; França, S.; Cabral, H.N.; Palmeirim, J.M. Impact of predation on the polychaete *Hediste diversicolor* in estuarine intertidal flats. *Estuar. Coast. Shelf Sci.* **2008**, *78*, 655–664. [CrossRef]
90. Mattila, J.; Heck, K.L., Jr.; Millstein, E.; Miller, E.; Gustafsson, C.; Williams, S.; Byron, D. Increased habitat structure does not always provide increased refuge from predation. *Mar. Ecol. Prog. Ser.* **2008**, *361*, 15–20. [CrossRef]
91. Riisgård, H.U. Filter-feeding in the polychaete *Nereis diversicolor*: A review. *Neth. J. Aquat. Ecol.* **1994**, *28*, 453–458. [CrossRef]
92. Pile, A.J.; Lipcius, R.N.; Van Montfrans, J.; Orth, R.J. Density dependent settler-recruit-juvenile relationships in blue crabs. *Ecol. Monogr.* **1996**, *66*, 277–300. [CrossRef]
93. Martínez-Schönemann, A. Estimación de la Producción Primaria Neta de Epífitos en Praderas de Fanerógamas de la Bahía de Cádiz. Master's Thesis, Univesity of Cádiz, Cádiz, Spain, 2014; 30p.
94. Grignon-Dubois, M.; Rezzonico, B.; Alcoverro, T. Regional scale patterns in seagrass defences: Phenolic acid content in *Zostera noltii*. *Estuar. Coast. Shelf Sci.* **2012**, *114*, 18–22. [CrossRef]
95. Manck, L.; Quintana, E.; Suarez, R.; Brun, F.G.; Hernández, I.; Ortega, M.J.; Zubia, E. Profiling of phenolic natural products in the seagrass *Zostera noltei* by UPLC-MS. *Nat. Prod. Commun.* **2017**, *12*, 687–690. [CrossRef] [PubMed]

diversity

MDPI

Article

Predation Pressure of Invasive Marsh Frogs: A Threat to Native Amphibians?

Fabien Pille *, Laura Pinto and Mathieu Denoël

Laboratory of Ecology and Conservation of Amphibians (LECA), Freshwater and Oceanic Science Unit of Research (FOCUS), University of Liège, 4020 Liege, Belgium; laura.pinto.etu@outlook.fr (L.P.); Mathieu.Denoel@uliege.be (M.D.)

* Correspondence: fabien.pille@uliege.be

Abstract: Anurans have been introduced in many parts of the world and have often become invasive over large geographic areas. Although predation is involved in the declines of invaded amphibian populations, there is a lack of quantitative assessments evaluating the potential risk posed to native species. This is particularly true for *Pelophylax* water frogs, which have invaded large parts of western Europe, but no studies to date have examined their predation on other amphibians in their invaded range. Predation of native amphibians by marsh frogs (*Pelophylax ridibundus*) was assessed by stomach flushing once a month over four months in 21 ponds in southern France. Nine percent of stomachs contained amphibians. Seasonality was a major determinant of amphibian consumption. This effect was mediated by body size, with the largest invaders ingesting bigger natives, such as tree frogs. These results show that invasive marsh frogs represent a threat through their ability to forage on natives, particularly at the adult stage. The results also indicate that large numbers of native amphibians are predated. More broadly, the fact that predation was site- and time-specific highlights the need for repeated samplings across habitats and key periods for a clear understanding of the impact of invaders.

Keywords: amphibian decline; invasive alien species; predatory risk; size-selective predation; *Pelophylax ridibundus*; water frogs

Citation: Pille, F.; Pinto, L.; Denoël, M. Predation Pressure of Invasive Marsh Frogs: A Threat to Native Amphibians? *Diversity* **2021**, *13*, 595. https://doi.org/10.3390/d13110595

Academic Editor: Michael Wink

Received: 27 October 2021
Accepted: 16 November 2021
Published: 19 November 2021

Publisher's Note: MDPI stays neutral with regard to jurisdictional claims in published maps and institutional affiliations.

Copyright: © 2021 by the authors. Licensee MDPI, Basel, Switzerland. This article is an open access article distributed under the terms and conditions of the Creative Commons Attribution (CC BY) license (https://creativecommons.org/licenses/by/4.0/).

1. Introduction

Freshwater habitats often have to cope with invasive alien species [1]. Although invaders may disturb native organisms in various ways, trophic interactions appear to be one of the main impacts, especially through predation [2–4]. Indeed, the introduction of novel predators induces additional predation pressure to which native organisms may not respond effectively [5]. Furthermore, invasive alien predators may reach high densities because of their life-history traits (e.g., high reproduction rates), various associations with humans, or through the absence of natural predators and parasites [6]. As a result, native species may suffer from excessive predation pressure, which influences the dynamics of their populations [7]. Therefore, predation by invasive alien species is an often-reported cause of native species decline [3], especially in fully or partially aquatic species [8,9].

The detrimental impacts of invasive alien species are worrying, particularly with regard to the vulnerable and declining native class of amphibians [9,10]. Nowadays, 16% of amphibians listed in the International Union for Conservation of Nature (IUCN) red list are threatened by invasive alien species [11], and since the 1980s, an increasing number of studies have emphasized the negative impact of invasive alien predators on native amphibian populations [3,4]. Interestingly, anurans are one of the most reported invaders [4,12,13]. Due to their ease of translocation, high reproduction rates, and generalist diet, invasive anurans have become a challenging issue for conservation efforts [14]. Predator–prey interactions between invasive alien anurans and natives can impact many species of higher, equivalent, or lower trophic rank [4]. For instance, invasive cane toads (*Rhinella marina*) are known to poison native predators that feed upon them [15] but to rarely feed on native amphibians [16]. Other widely introduced invasive amphibians, such as the American bullfrog (*Lithobates catesbeianus*) and the African clawed frog (*Xenopus laevis*), consume native amphibians, including metamorphosed anurans [17–20], tadpoles [19,21], newts [19] and, more rarely, eggs [22].

Although the ecological impacts of the most widely introduced anurans are now documented, there has been little quantitative assessment of predation across space and time. However, biological invasions may have contrasting effects, depending on localities and time periods (i.e., depending on the phenology of invaders and natives), indicating the need for information regarding temporal patterns at a large number of sites [6]. Furthermore, the risk posed by predation by invasive alien anurans may also depend on the biological traits of the invaders, such as body size, which may influence prey selection due to gape-size limitations [23]. Therefore, this trait should be integrated for predation assessments.

Despite their high prevalence in some parts of the globe, the predatory pressure of some major amphibian invaders has not been studied to date [24]. For instance, in Europe, the animal trade has resulted in multiple invasions of *Pelophylax* water frogs (Ranidae). Recent increases in introductions and expansion in recent decades in western Europe have made these species the most widespread amphibian invaders on the continent [25]. In particular, the marsh frog (*P. ridibundus*) is considered native only east of the Rhine but is now present in large parts of France, Belgium, Luxembourg, and Italy [25–30]. Particularly in invaded areas with sister *Pelophylax* species present, introductions have often led to complex and cryptic invasions only identifiable through genetic assessment. As a result, the ecological consequences of water-frog invasions have remained understudied and underestimated. More recent studies have considered the replacement of native *Pelophylax* water frogs to be a severe threat due to hybridogenetic replacement [31,32]. However, the other risks of these *Pelophylax* invasions for native amphibians have still not been investigated. In their natural range, the diet of marsh frogs has been studied in freshwater habitats and indicates little or no predation on native amphibians. Most reported cases were cannibalism of metamorphosed individuals and tadpoles [33–38], whereas newts, tree frogs, and tadpoles were rarely detected in the diet [39–42]. Furthermore, a recent behavioral observation also confirmed the ability of marsh frogs to forage on tree frogs [43]. These observations suggest that marsh frogs may also feed on native amphibians in invaded areas. This implies the need for integrative studies to quantify the predation pressure exerted by water frog invaders in colonized areas. One such area is the Larzac plateau in southern France [26], historically devoid of *Pelophylax* frogs [44]. *P. ridibundus* has now colonized a large part of the plateau and is still expanding ([26]; M. Denoël, pers. obs.). Due to the high amphibian diversity value of Larzac [44–46], these invasions raise concerns about their effect on populations of native amphibians. This study system therefore provides an ideal model with which to determine the predation risk posed by invasive water frogs.

In this context, this study aimed to provide a quantitative assessment of the occurrence of amphibians in the diet of invasive marsh frogs in their invaded range (Larzac) in a replicated design across space and time. More specifically, our objectives were to highlight the species at risk of predation, the importance of seasonality, and the role of the body size of the invaders. Because *Pelophylax* are successful invaders, we hypothesized that they exert a predatory pressure on the different life stages of the native taxa, including adult amphibians that may be potential prey, given the large body size of the invaders. Furthermore, we expected predation pressure to reach its maximum during the breeding period of native amphibians, when they massively join ponds.

2. Materials and Methods

2.1. Study Area and Organisms

The study took place on the Larzac karst plateau in southern France (Hérault, France; area delineated from 43°48′ N to 43°54′ N and from to 3°21′ E to 3°33′ E; Supplementary Material Figure S1). The study area is mainly characterized by open, traditionally managed landscapes surrounded by forests. It hosts a large number and diversity of ponds, primarily used for watering cattle and sheep and inhabited by several species of amphibians [44,47]. Multiple surveys in 85ponds in the 1970s showed that Larzac was historically devoid of any species of *Pelophylax* ([44]; J. Gabrion, pers. comm.). Native *Pelophylax* species were restricted to lower-elevation sites outside the Larzac region [48]. This area was subsequently invaded by *Pelophylax* marsh frogs at the end of the 20th century and possibly in the early 2000s ([26]; M. Denoël, pers. obs.). Direct translocations occurred in Larzac, as testified by local inhabitants (personal communications to M. Denoël and F. Pille). One previous genetic study identified marsh frogs in the studied area as lineages of *Pelophylax ridibundus* [26]. This species is also named *Pelophylax fortis*, according to debated nomenclatural revisions [49], but the traditional taxonomic assignment (i.e., *P. ridibundus*) is used here. Two morphologically cryptic lineages (the Balkan marsh frog *P. kurtmuelleri* and the marsh frog *sensu stricto P. ridibundus*) were genetically identified in Larzac populations, both of non-indigenous origin (i.e., from eastern/southeastern Europe [26]). As they show closely related phylogenetic divergence and are morphologically cryptic, these two lineages were studied as *P. ridibundus sensu lato*. Although *Pelophylax perezi* populations were found in southern France [26,48], there were no populations in the surveyed area (C. Dufresnes, G. Mazepa and M. Pabijan, pers. comm.).

2.2. Sampling and Prey Identification

To gain a broad overview of the predation of invasive *Pelophylax* water frogs on amphibians, marsh frogs were sampled in 21 ponds (mean surface area $\pm$ SE = 158.94 $\pm$ 34.98 m^2) typically used to water cattle and surrounded by traditionally managed landscapes. Frogs were caught at dusk and early night manually or with dip nets during the active season from early April to the end of July 2019. Each pond was sampled monthly (i.e., four times over the whole study period). Frogs were kept individually in tanks and released within hours of capture. Frogs were identified individually using PIT-tags implanted in the back under the skin (Biolog-ID, 134.2 KHz; Agrident reader) to consider potential recaptures over time. PIT-tagging is an effective method for marking amphibians [50,51]. The snout-vent length (SVL) of each individual was measured from the tip of the snout to the end of the cloaca with a caliper. Only adult marsh frogs—that is, individuals with a minimum SVL of 50 mm—were included in the sampled population (i.e., minimum SVL of males exhibiting nuptial callosities and vocal sacs in this study). Stomach contents were collected by flushing frog stomachs, following the method described by Solé et al. [52], by gently injecting water into the stomach using a rounded, soft silicone canula through the mouth, causing prey items to be regurgitated. Water was injected into stomachs using a sprayer to maintain a slow, continuous flow. This technique does not affect the survival or behavior of amphibians [53]. No individuals were hurt during the study. All captures and manipulation followed ethical standards and were approved by the Direction Régionale de l'Environnement, de l'Aménagement et du Logement (Hérault). Stomach contents were stored individually in ethanol. A total of 1062 stomach contents (736 individuals) was sampled: 271 in April, 235 in May, 266 in June, and 290 in July (Supplementary Material Table S1). All manipulations were conducted at each study site to avoid displacement of individuals.

The occurrence of native Mediterranean tree frogs (*Hyla meridionalis*) and palmate newts (*Lissotriton helveticus*) was determined in order to account for their period of residency in water and to estimate the period during which they might be vulnerable to marsh frogs in the 21 studied ponds. This study focused on these two native species because they spend long periods in ponds during the sampling period and because they are potential prey present mainly at the water surface (tree frogs) and in the water (palmate newts). The number of adult marsh frogs was determined every month at each pond by visual counting (using 10 $\times$ 42 Swarowski binoculars from a distance to the pond and then walking slowly around the pond) and used as an index of abundance. Preliminary research showed that visual counting is proportional and close to the number of individuals present simultaneously at each site as estimated by capture-mark-recapture experiments ($R^2 = 0.73$, $p < 0.001$, $n = 14$, F. Pille & M. Denoël, unpublished data), therefore providing an estimate of potential pressure at each visit.

Amphibian prey items were identified under a stereoscopic microscope (Zeiss Stemi 2000). They were identified at the species level and developmental stage when their state of digestion allowed (i.e., post-metamorphic versus larval [54]). The SVL of amphibians preyed upon by marsh frogs was measured using a morphometric measurement tool on stomach-content pictures (TPSdig2 software; [55]). Amphibians ingested by marsh frogs were then classified according to three categories: metamorphosed anurans, tadpoles, and newts. Adults and larvae of newts were considered to belong to the same category because of their low occurrence in stomach contents and their similar habitat preferences.

2.3. Statistical Analyses

Using general linear mixed models (GLMM) with a binomial distribution, the potential effects of two variables were tested on the presence of each type of prey (metamorphosed anurans, tadpoles, newts; 0 = absence of prey, 1 = presence of prey) in the stomachs of marsh frogs: SVL of marsh frogs (continuous variable, in mm, ln transformed values) and time (month, from April to July). Because newts and tadpoles were found in all ponds and thus were available for predation by marsh frogs, GLMMs included the 21 ponds for these two groups of amphibians. However, according to visual detection, native metamorphosed anuran species predated by invasive marsh frogs occurred in only 15 ponds. Thus, GLMMs assessing predation occurrence on metamorphosed anurans were performed on this reduced dataset. Individuals and ponds were considered nested random factors. Akaike information criterium (AIC) and AIC weights were calculated to rank each candidate model [56]. All possible combinations of factors were tested using automated selections. The most parsimonious models with the lowest AIC were considered to be the best models (i.e., models with ΔAIC < 2 were considered equal models). Averaging of these models (ΔAIC < 2), which computes the weighted means of the parameter estimates, was performed. The 95% CI of estimates of average models were calculated to assess the magnitude of the effects. Differences in SVL between marsh frogs that preyed on native amphibians and those that did not were assessed by computing Cohen's d with confidence intervals (bootstraps, $n = 10{,}000$ repetitions). An effect was considered important when the 95% CI did not include zero [57]. The relationship between the SVL of preyed-upon amphibians and the SVL of marsh frogs (ln transformed values for both SVL to reach normal distributions) was assessed using a linear model. Confidence intervals of means presented in the results were calculated using bootstraps ($n = 10{,}000$ repetitions). All analyses were performed in R 3.6.1 using the lme4 and MuMIn packages [58,59].

Numbers of adult Mediterranean tree frogs and adult palmate newts that could have been consumed by marsh frogs during the whole study period were estimated from the stomach samples. The analysis focused on these two species because they were the only two ubiquitous native amphibians at the study sites. Two scenarios were tested: (1) a conservative scenario, which limits overestimation of predation pressure by considering previously estimated gastric-evacuation times for frogs [60] and which includes only fresh prey in calculation (i.e., consumed within the day), but that ignores the occurrence of more digested prey and may therefore underestimate predation rates; (2) a non-conservative scenario, for which estimates were computed based on the total number of consumed adults, whether recently ingested or in an advanced stage of digestion, but which can therefore overestimate predation, as some prey might have been ingested one day earlier. In both scenarios, estimates of predation pressure were computed for each month for which prey were found in the ponds by extrapolating the number of prey items observed in stomach contents to the maximum number of counted marsh frogs per visit during each month. These values were multiplied by the number of days per month to obtain the monthly estimated number of consumed individuals, and each estimation was divided by the number of ponds in which each species was observed (Mediterranean tree frogs: $n = 15$ ponds; palmate newts: $n = 21$ ponds).

3. Results

3.1. Consumed Amphibians

From all samples (April–July), amphibians were found in 96 out of 1062 predator stomachs (i.e., 9.04% of samples); 6.02% of the marsh frogs had preyed on anurans (metamorphosed anurans: 1.31%; tadpoles: 4.8%), 2.25% on newts (adults: 0.94%; larvae: 1.12%; undetermined: 0.28%), and 1.31% on undetermined amphibian remains. Marsh frogs consumed all native amphibian species known in Larzac Plateau either at the larval stage (anurans: *Bufo spinosus*, *Epidalea calamita*, *Pelobates cultripes*, *Alytes obstetricans*; caudates: *L. helveticus*, *Triturus marmoratus*) or the post-metamorphic stage (anurans: *H. meridionalis*, *Pelodytes punctatus*; caudates: *L. helveticus*; Figure 1; Supplementary Material Table S2). A single paedomorphic *L. helveticus* (i.e., a gilled adult) was identified in the stomach contents. Metamorphosed anurans, tadpoles, and newts represented, respectively, 8.46%, 73.54%, and 17.98% of the relative abundance of amphibians preyed upon by marsh frogs (0.02% non-determined). Cannibalism occurred in a very small proportion of captured marsh frogs and included predation on both juveniles (0.09%) and tadpoles (0.47%). Predation on amphibians occurred in 20 of the 21 sites.

Figure 1. Marsh frog (*Pelophylax ridibundus*) preying on an adult palmate newt (*Lissotriton helveticus*) in a pond of the Larzac Plateau. Photography by V. Renard.

3.2. Temporal Variations

Model selection on the different GLMMs showed that time (i.e., months) was a good predictor (proportion decreasing through months) of the occurrence of anurans in water-frog stomachs. This parameter was selected in the model with the lowest AIC for the metamorphosed anurans, but it did not have a strong effect according to the model averaging performed on the two best models ($\Delta AIC \leq 2$; Table 1). In tadpoles, it was selected in the two best models used in model averaging ($\Delta AIC \leq 2$; Table 1). Model averaging confirmed a negative effect of the time predictor in tadpoles (estimate = -0.68, 95% CI: -1.006 to -0.371; Figure 2, Supplementary Material Table S3) and suggested a similar but weak tendency in metamorphosed anurans (estimate = -0.34, 95% CI: -1.018 to 0.323; Figure 2,

Supplementary Material Table S3). This is explained by a decrease in the frequency of occurrence of metamorphosed anurans and tadpoles in the stomachs of marsh frogs between April and July (Figure 3). The effect of time was not strong for newt occurrence (Figure 2), being selected in only one of the three best models ($\Delta AIC \leq 2$).

Table 1. Comparison of models explaining the occurrence of amphibians in stomach contents. Models (all shown) are ranked from the best to the worst according to AIC. Explanatory parameters were time (i.e., months) and SVL (snout-vent length) of marsh frogs.

Variable	Rank	Time	SVL	AIC	ΔAIC	Weight
Metamorphosed anuran occurrence	1	−0.514	8.213	124.9	0.00	0.677
	2		8.752	126.4	1.50	0.323
	3	−0.759		138.6	13.70	0.001
	4			145.2	20.30	0.000
	Parameter importance	0.678	0.678			
Tadpole occurrence	1	−0.684		351.0	0.00	0.708
	2	−0.699	−0.441	352.8	1.80	0.292
	3			371.3	20.30	0.000
	4		0.664	372.8	21.80	0.000
	Parameter importance	>0.999	0.292			
Newt occurrence	1			131.1	0.00	0.449
	2		5.081	132.0	0.90	0.279
	3	−0.109		133.0	1.90	0.168
	4	−0.107	5.121	134.0	2.90	0.104
	Parameter importance	0.272	0.383			

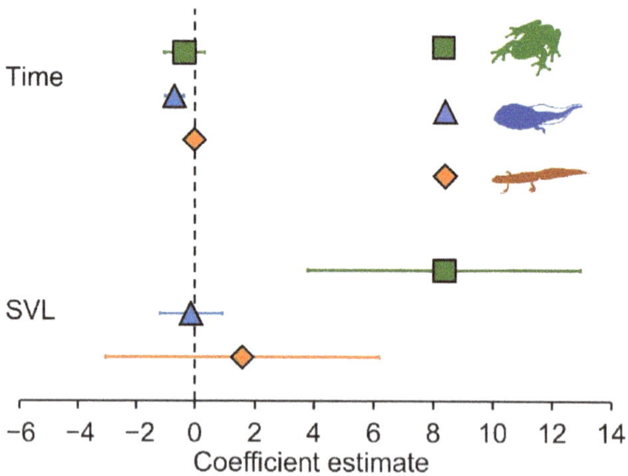

Figure 2. Effect of parameters in average GLMM models assessing amphibian occurrence in stomach contents of marsh frogs (estimates, 95% CI). Green lines and squares: metamorphosed anurans; blue lines and triangles: tadpoles; orange lines and diamonds: newts. Time: months (April to July); SVL (snout-vent length): body size of marsh frogs.

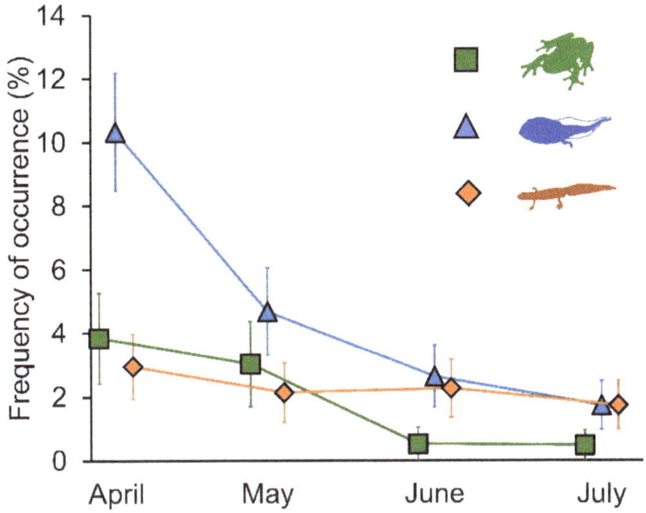

Figure 3. Monthly frequency of occurrence of amphibians in stomach contents of marsh frogs (n = 1062). Green lines and squares: metamorphosed anurans; blue lines and triangles: tadpoles; orange lines and diamonds: newts. Whiskers: SE.

3.3. Body Size

The effect of the body size of marsh frogs (SVL) on the consumption of metamorphosed anurans was retained in the two best models (ΔAIC $\leq$ 2; Table 1). Indeed, marsh frogs that consumed metamorphosed anurans were larger than those that did not prey on these organisms (Cohen's d = 1.14, 95% CI = 0.81 to 1.49; Figure 4). Model averaging confirmed this very strong effect (zero excluded from CI; estimate = 8.38, 95% CI: 3.802 to 12.971; Figures 2 and 4). SVL was selected in only one of the three best models (ΔAIC $\leq$ 2) used for model averaging in newts and did not show any effect, as highlighted by the confidence intervals crossing zero (estimate = 1.58, 95% CI: −3.029 to 6.192; Figure 2). For tadpole consumption, SVL was selected in one of the two best models (ΔAIC $\leq$ 2); model averaging did not confirm this effect (estimate = −0.12, 95% CI: −1.170 to 0.912; Table 1, Figure 2). Marsh frogs that consumed tadpoles and newts were larger than individuals that did not (respectively, Cohen's d = 0.29, 95% CI = 0.03 to 0.55 and Cohen's d = 0.50, 95% CI = 0.08 to 0.89; Figure 4). Linear regression showed a positive relationship between the SVL of preyed-upon amphibians (metamorphosed anurans, tadpoles, and newts combined) and the SVL of marsh frogs (R^2 = 0.35; slope estimate = 1.68, 95% CI: 1.13–2.22; Figure 5).

3.4. Predation Pressure

Among the 21 ponds that were surveyed for four months, Mediterranean tree frogs were observed in 15 ponds and palmate newts in all ponds. Across all ponds, the total numbers of adult marsh frogs observed by visual counting over the four-month period from April to July were 289, 445, 521, and 535, respectively (Supplementary Material Table S4). Following the conservative scenario (i.e., considering only fresh prey from the stomach contents of marsh frogs) and the time period of coexistence between marsh frogs and the native species (Supplementary Material Table S4), a minimum global consumption of 96 adult Mediterranean tree frogs and 496 adult palmate newts was estimated in the studied ponds over the study period. Per pond, this translates to, on average, estimates of 6.4 and 23.6 individuals, respectively. The non-conservative scenario (i.e., considering all consumed adults, whatever the state of digestion) provided higher estimations of predation rates, with a maximum global consumption of 412 adult Mediterranean tree frogs and 790 adult palmate newts. Predation pressure of invasive marsh frogs at the pond level was highest in

April for adult Mediterranean tree frogs, ranging from six in April to zero over the following months in the conservative scenario and 12.8, 10.8, 3.9, and 0 for April, May, June, and July, respectively, in the non-conservative scenario (Supplementary Material Figure S2). For palmate newts, the conservative scenario provided estimates of 15.2, 5.6, 2.8, and 0 for April, May, June, and July, respectively. The non-conservative scenario estimated predation pressure of 15.2, 5.6, 16.8, and 0 over the same period (Supplementary Material Table S4).

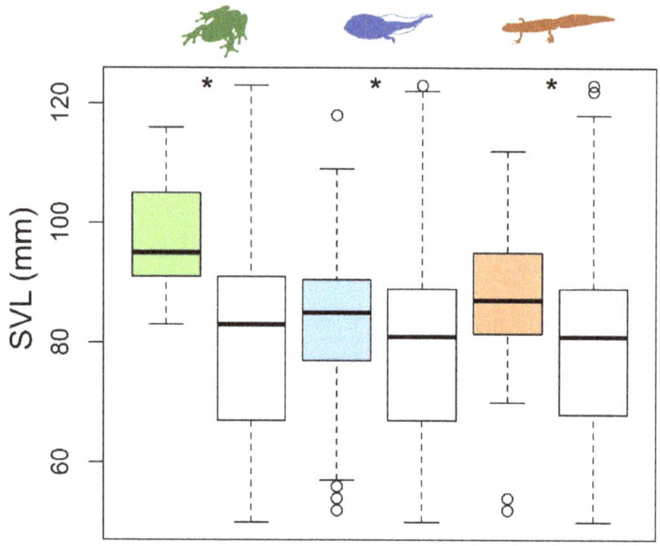

Figure 4. Size (snout-vent length, in mm) of marsh frogs that consumed amphibians (colored bars) and those that did not consume amphibians (clear bars). Green (**left**): metamorphosed anurans; blue (middle): tadpoles; orange (**right**): newts. Boxplots represent median (25–75%) ± 1.5 IQR. Open circles represent outliers. Asterisks indicate significant differences between groups (see text).

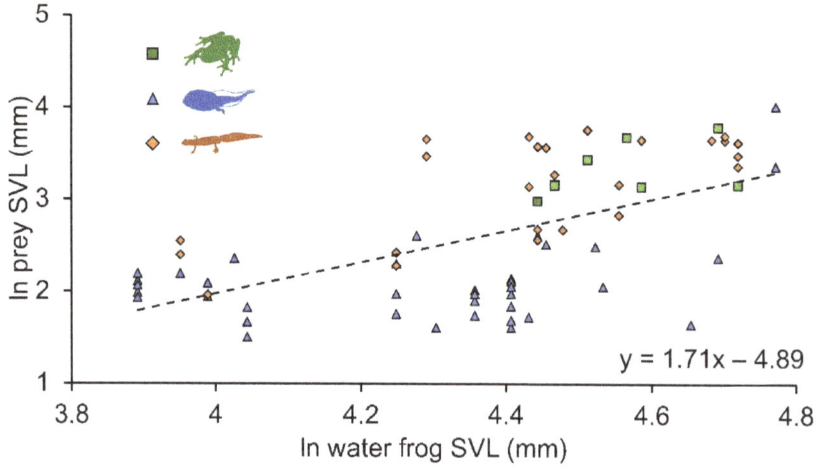

Figure 5. Size-selective predation: Size of consumed amphibians as a function of size of *Pelophylax* water frog. SVL: snout-vent length. Green squares: metamorphosed anurans; blue triangles: tadpoles; orange diamonds: newts. Dotted line represents the linear regression between the SVL of marsh frogs and the SVL of ingested amphibians (y = 1.71x − 4.89).

4. Discussion

The present study provides a quantitative assessment of the consumption of native amphibians by invasive marsh frogs in ponds and highlights the predation pressure on both Mediterranean tree frogs and newts. By encompassing multiple sites over time, these results shed new light on our understanding of the diet of the invaders and show that time and context can mask or exacerbate the observed effects, depending on sampling design. Moreover, these replicated analyses give better insight in terms of global estimates of amphibians predated during a complete active period at ponds. As *Pelophylax* water frogs are now invading most of western Europe [26–29], our findings have broad implications for the conservation of native amphibians.

4.1. Predation on Natives

Most anurans are generalist feeders, foraging on a wide variety of prey [61]. Within an invasion context, this life-history trait suggests that invasive frogs could predate on native amphibians [62]. In the present study, 9% of the invaders had consumed amphibians, suggesting opportunistic predation rather than trophic specialization. This predation rate appeared to be higher than in other invasive frogs, for which native amphibian consumption is typically low, such as in Cuban treefrogs [63], African clawed frogs [22], or cane toads [16]. Our results show predation rates similar to the ranges reported in invasive bullfrogs (4–32%) [18,19,21]. Despite having a larger gape width than marsh frogs, bullfrogs, which belong to the same family (Ranidae), share biological traits with them and are known to have strong impacts on native amphibian populations, mainly through predation [20]. On the other hand, our results show higher predation rates on native amphibians by *P. ridibundus* water frogs from the invaded range than previously shown from other stomach-content-based studies carried out in the native range (Table 2). Indeed, in most of these studies, no native amphibians were predated by native marsh frogs [33–38,64–68], whereas in a few others, frequency of predation on heterospecific amphibians ranked between 0.10 and 4.61% [39–42]. These differences might be due to differences in the sampled habitats and the study design. Indeed, our study focused on a large set of ponds in an area where native amphibians are widespread and diversified, whereas previous studies in the native range included lakes, rivers, and ponds where co-occurrence patterns were not shown.

In the present study, all eight native amphibian species were preyed upon by marsh frogs. Two of the species that were preyed upon have a high-risk conservation status; the spadefoot toad *P. cultripes* is considered high-priority at the regional and national scale [48,69], and *T. marmoratus* is also considered to be in decline according to the regional IUCN red list [48]. Although less threatened, even common species, such as palmate newts, were shown declining in large parts of their range, including in Larzac [70]. Therefore, predation pressure from invasive water frogs, such as marsh frogs, may act as a new stressor, possibly contributing to the current declines (see Section 4.6 for impact assessment).

Whether this predation rate is facilitated by some form of naïveté of native populations is unknown. Indeed, native amphibians that did not co-evolve with their new, exotic predators may lack anti-predator mechanisms [11]. In Larzac, and more broadly in southern France, there were indeed no *P. ridibundus* naturally present, but phylogenetically related species of *Pelophylax* exist in other parts of the range of the native species [26]. The naïveté is therefore possibly less marked than that relating to the continental translocation of the other species of invaders listed hereabove.

Table 2. Frequency of occurrence of amphibian consumption by marsh frogs according to previous diet studies and the present study (highlighted in bold). Studies that did not mention if consumed amphibians were linked to cannibalism or predation on heterospecifics were not included.

Country	Area	Taxa	Sample Size	Habitat	Number of Sites	Season	Predation Occurrence (%)		Consumed Natives	References
							Homospecifics	Heterospecifics		
Bulgaria	Native	*P. ridibundus*	375	Ponds, river	NA	Autumn	2.78–8.33	0.00		[37]
Bulgaria	Native	*P. ridibundus*	118	Ponds, river	NA	Spring, Autumn	1.69	0.00		[66]
Greece	Native	*P. ridibundus (kurtmuelleri)*	75	Ponds, river	2	Spring	2.99	0.00		[38]
Iran	Native	*P. ridibundus*	188	NA	7	Spring, Summer	10.08	0.00		[36]
Iran	Native	*P. ridibundus*	129	Ponds	19	Spring, Summer	0.00	1.63–4.61	Frogs	[41]
Romania	Native	*P. ridibundus*	364	Canal	1	Full year	0.27	0.00		[33]
Romania	Native	*P. ridibundus*	NA	Floodplains	2	Summer, Autumn, Winter	23.00	NA		[35]
Romania	Native	*P. ridibundus*	31	River	1	Spring	0.00	3.22	Newts	[39]
Romania	Native	*P. esculentus* complex	NA	NA	NA	Spring	0.00	1.36–4.10	Tadpoles, frogs	[40]
Romania	Native	*P. ridibundus*	100	Ponds	NA	Spring	0.00	0.00		[64]
Romania	Native	*P. esculentus* complex	74	Swamp	1	Spring, Summer	0.00	0.00		[67]
Romania	Native	*P. esculentus* complex	84	NA	1	Spring, Summer	0.00	0.00		[68]
Romania	Native	*P. ridibundus*	50	Ditch	1	Summer	2.00	NA		[71]
Russia	Native	*P. ridibundus*	101	River, lakes, ponds	NA	Spring, Summer, Autumn	0.1–2.2	0.1–2.2	Tadpoles	[42]
Serbia	Native	*P. esculentus* complex	270	Marsh	1	Spring, Summer, Autumn	NA	NA	Newts	[72]
Turkey	Native	*P. ridibundus*	82	Lakes	5	NA	0.00–10.00	0.00		[34]
Turkey	Native	*P. ridibundus*	53	NA	3	Summer	3.77	0.00		[65]
France	Invasive	*P. ridibundus*	1062	Ponds	21	Spring, Summer	0.47	8.57	Tadpoles, newts, frogs	This study

4.2. Predation on Larval and Metamorphosed Life Stages

Every life stage of the native amphibians in our study (except eggs) was impacted by predation by marsh frogs. Mediterranean tree frogs and palmate newts were the most-preyed-upon amphibians at the post-metamorphic adult stage. These predation events indicate that marsh frogs forage on organisms that have different microhabitat/space-use habits. Indeed, palmate newts are entirely aquatic and specifically benthic in ponds, except when moving up to breathe, whereas treefrogs spend most of their time at the water/air interface when present at ponds ([48]; M. Denoël & F. Pille, personal observation). Additionally, most native anurans were found in stomach contents exclusively at the larval stage. This also indicates predation on fully aquatic amphibians, which exploit different parts of the ponds, such as the bottom and the water column. Preliminary observations show that marsh frogs can ingest them from the water-air interface in the middle of ponds, but it is also possible that marsh frogs chase their prey during underwater movements as well. Detailed quantitative behavioral observations are needed to determine the predatory strategies of marsh frogs and therefore in which microhabitat they forage, particularly on amphibians.

4.3. Phenology of Predation

Time was the main determinant of native-anuran consumption by marsh frogs. This underlines the importance of conducting studies across a broad range of the active period of amphibians to assess predation by invasive alien species, as was done here. Indeed, by focusing on shorter surveys, predation by invasive frogs could have been overlooked, therefore leading to misleading conclusions regarding the conservation of impacted native species. In this study, predation on anurans (tadpoles and metamorphosed anurans) occurred mainly in April, suggesting a strong effect of the abundance of native species related to phenology. Because water frogs spend a great deal of time in ponds (i.e., in spring and summer), they can interfere with any crucial period in the phenology of native amphibians. For the Mediterranean tree frog, predation by marsh frogs occurred almost exclusively in April and included gravid females. This indicates that predation mostly occurred during the main period of reproductive activity at the study sites, when breeding adults reached high abundances in ponds. Furthermore, some other anuran species, such as *Bufo spinosus*, lay eggs at the very end of winter, resulting in high numbers of small tadpoles in ponds in April [48]. For *Pelodytes punctatus* and *Alytes obstetricans*, some tadpoles remain at the larval stage during winter and thus occur in ponds in early spring [48]. Therefore, as a result of their phenology, all these native amphibians are available as prey for marsh frogs at the beginning of spring. This is especially the case for tadpoles, which are highly abundant in ponds during spring and then metamorphose before summer.

Thus, patterns of phenology are more likely to be a reflection of the availability of native amphibians rather than a change in the behavior of marsh frogs.

By contrast, despite their higher numbers in early spring, newts can remain underwater in summer at the study sites [73], making them potential prey for marsh frogs during the overall survey period. However, although the present results fit well with the phenology of the preyed-upon amphibians, future research should investigate how additional factors may also explain the temporal variation of amphibian consumption.

4.4. Size-Selective Predation

Most trophic interactions are governed by size-selective predation (i.e., the fact that predators are gape-size-limited) [74]. In a recent review, Measey et al. [62] showed that the size of predatory frog species was a dominant predictor of anurophagy. Our results demonstrate that in marsh frogs, this pattern also occurs at the intraspecific level. This effect was particularly strong for metamorphosed anurans, which are large prey only consumed by the largest marsh frogs. The stomach contents of large marsh frogs included both small and large prey items. This is consistent with the general observation that larger predators, while able to catch large prey, continue to prey on smaller prey organisms [74]. In marsh frogs, this general trend is reflected by the fact that smaller individuals can prey upon tadpoles and small newts (e.g., larvae), whereas larger individuals consume a broader spectrum of amphibian prey ranging from adult native anurans and newts to anuran and newt larvae. In terms of conservation, predation on adult amphibians only by the largest marsh frogs shows that they are the most problematic for populations. This is reinforced by the fact that water frogs, such as marsh frogs (mean SVL: 80 mm; this study) and other invasive ranids (bullfrog mean SVL: 143 mm [21] and 110 mm [23]), can reach larger sizes than most natives (e.g., Mediterranean tree frogs SVL: 45 mm, this study).

4.5. Environmental Context

The structure of habitats influences the composition of amphibian species and their abundance in ponds [75]. Because frogs are generalist predators, this variability may influence diet composition. For instance, habitat influences the predation patterns of invasive bullfrogs [21]. In Larzac, ponds can be heterogeneous according to environmental parameters such as size, depth, vegetation composition, and cover. To test for the influence of these various parameters on the predation rates of invasive marsh frogs turns out to be a difficult task in a field study. Indeed, without knowing the precise availability of each type of prey following gradients of habitat heterogeneity, it is not possible to clearly identify site-specific strategies of frogs. However, by including a large number of replicated sites (21 ponds), this study encompasses habitat heterogeneity and provides insights about global predation patterns across pond networks, with predation on amphibians found in almost all studied ponds. Furthermore, despite possible differences in the composition of amphibian species across ponds, the most abundant and ubiquitous prey (i.e., *H. meridionalis* and *L. helveticus*) were present in the majority of ponds (15 and 21, respectively), making them available prey for marsh frogs. These two species were predated in at least four and eight ponds, respectively (i.e., in at least 26.6% and 38% of the studied ponds). However, more fine-scale study is needed to better understand how pond complexity could impact predatory strategies [21,38], though this would also require behavioral observations.

4.6. Predation Pressure and Conservation Concerns

Consumption by invasive predators is considered a direct impact on native species because it involves the death of individuals. Thus, predation could contribute to population decline by directly increasing mortality rates, leading to the removal of native species in the worst cases [76]. Although the present study did not measure native amphibian population dynamics and therefore potential decline patterns, the results show that metamorphosed native amphibians, especially Mediterranean tree frogs and palmate newts, were preyed upon by invasive marsh frogs in several surveys and ponds. An occurrence of 9% of heterospecific amphibians may seem low at first sight, but such a rate must be contextualized with the number of invasive frogs present in the ponds and the period in which prey and predators co-occur. As large numbers of marsh frogs were present in the studied ponds over a long period, predation pressure was estimated to reach hundreds of predated metamorphosed individuals over an active season in the studied ponds. This suggests that predation by invasive water frogs may be deleterious for native populations of Mediterranean tree frogs and may also affect those of palmate newts. Both scenarios, even the most conservative, which considers only fresh prey for estimations, give estimates of high numbers of consumed individuals. Because of the difference in data selection used for each scenario (i.e., fresh prey versus all prey), they may be considered "optimistic" and "pessimistic" points of view, respectively. The predations occurring during the breeding period is particularly critical because reproduction may be altered, and native females may be predated before laying their eggs. On the other hand, our results highlight a neglected long-term aspect of predation: that even if amphibians were found only in the stomach contents of some of the studied invaders, the fact that these invaders remain continuously in ponds for months increases their predation rate on a full-year basis. However, further studies focusing on densities of native newts and tree frogs across time are needed to precisely describe the impacts of invasive water frogs.

Our results show some predation pressure against the two targeted native amphibians 10–20 years after the start of the biological invasion in the study area (M. Denoël, personal observation). Considering this time lapse, one likely hypothesis is that predation pressures might have been higher at an earlier stage of the invasion because of possible prey naiveté or higher densities of available native amphibians. Unfortunately, there are currently no long-term quantitative datasets of population changes of native amphibians in association with the invasion of water frogs. Despite this, qualitative observations suggest that tree frogs were once more commonly seen in Larzac ponds than they are today. In some of these ponds, it has become exceptional to find tree frogs now that they have been invaded by marsh frogs (M. Denoël, personal observation). Future studies should therefore focus on such temporal patterns of changes in local communities. From this perspective, the present results provide indication for a plausible detrimental direct consumptive effect of water frogs, the global consequences of which remain to be determined. However, it is becoming difficult, as marsh frogs have already invaded most ponds in the area. Furthermore, predation from invasive herpetofauna may lead to various impacts on native species, such as changes in behavior, spatial ecology, or defense mechanisms [4]. These non-consumptive effects may have a strong influence [12,77] and should be considered in addition to the consumptive effects shown in this study.

5. Conclusions

Although *Pelophylax* water frogs were previously considered a threat mainly because of the risk of genetic pollution of related taxa (see e.g., [27]), the present study on marsh frogs shows another risk in pond environments: direct predation. Specifically, through a detailed account over time and across multiple water bodies, this study provides evidence for the overall ecological threat posed by invasive water frogs. By predating upon all local species, including both caudates and anurans at different life-history stages, marsh frogs interact negatively with all components of native amphibian communities. Because invasive alien *Pelophylax* water frogs are now well established in large geographic areas across western Europe, we argue that these invasive alien predators may have a negative impact on native amphibian populations similar to that found in other invasive taxa [78,79]. Because the global impact of invaders is likely complex [9,11], future fine-scale research is needed to understand the long-term consequences of invasions and how invaded communities change with time and pressures from invaders.

Supplementary Materials: The following are available online at https://www.mdpi.com/article/10.3390/d13110595/s1: Figure S1: Sampling area of invasive marsh frogs in Larzac. Background indicates relief with lowest elevation in dark. The insert shows France and three departments (Hérault, Gard, Aveyron). Detailed coordinates are not given for conservation purposes. Figure S2: Mean observed abundance of native Mediterranean tree frogs in the 15 ponds where they were observed (visual counting). Whiskers: 95% CI. Table S1: Sample size of studied marsh frogs per pond and per month. Table S2: Total number (N), number of occurrence (O), and frequency of occurrence (FO) of amphibian prey types in stomach of marsh frogs. ND: not determined. Table S3: Results of model averaging of the GLMM models showing factors influencing amphibian occurrence in stomach contents of marsh frogs. MA: metamorphosed anuran. Table S4: Estimated numbers of adult Mediterranean tree frogs (*Hyla meridionalis*) and adult palmate newts (*Lissotriton helveticus*) preyed by marsh frogs (MF) during the whole study period according to a conservative scenario (intact fresh prey items only) and to a non-conservative scenario (all consumed amphibians).

Author Contributions: Conceptualization, F.P. and M.D.; methodology, M.D.; validation, M.D.; formal analysis, F.P. and M.D.; investigation, F.P., M.D. and L.P.; resources, M.D.; data curation, F.P. and L.P.; writing—original draft preparation, F.P.; writing—review and editing, M.D.; visualization, F.P.; supervision, M.D.; project administration, M.D.; funding acquisition, M.D. All authors have read and agreed to the published version of the manuscript.

Funding: This research was supported by PDR grant number T.0070.19 of the Fonds de la Recherche Scientifique—FNRS.

Institutional Review Board Statement: Capture and manipulations were carried out under permit from Direction Régionale de l'Environnement, de l'Aménagement et du Logement (Hérault).

Informed Consent Statement: Not applicable.

Data Availability Statement: Data are available from the authors upon reasonable request.

Acknowledgments: We thank C. Duret, L. Lorrain, B. Lejeune, P. Levionnois, L. Seger, P. Padilla, A. Berna for their help in the field, V. Renard for permission to use her photograph taken during fieldwork, A. Lacombe for her help with statistical analysis, the landowners and municipalities for allowing access to their ponds, and the Institut Géographique National for providing maps through the INSPIRE program. M. Denoël is a Research Director of the Fonds de la Recherche Scientifique—FNRS.

Conflicts of Interest: The authors declare that they have no conflict of interest.

References

1. Strayer, D.L. Alien species in fresh waters: Ecological effects, interactions with other stressors, and prospects for the future. *Freshw. Biol.* **2010**, *55*, 152–174. [CrossRef]
2. Gallardo, B.; Clavero, M.; Sánchez, M.I.; Vilà, M. Global ecological impacts of invasive species in aquatic ecosystems. *Glob. Chang. Biol.* **2016**, *22*, 151–163. [CrossRef]
3. Kats, L.B.; Ferrer, R.P. Alien predators and amphibian declines: Review of two decades of science and the transition to conservation. *Divers. Distrib.* **2003**, *9*, 99–110. [CrossRef]
4. Kraus, F. Impacts from invasive reptiles and amphibians. *Annu. Rev. Ecol. Evol.* **2015**, *46*, 75–97. [CrossRef]
5. Sih, A.; Bolnick, D.I.; Luttbeg, B.; Orrock, J.L.; Peacor, S.D.; Pintor, L.M.; Preisser, E.; Rehage, S.; Vonesh, J.R. Predator-prey naïveté, antipredator behavior, and the ecology of predator invasions. *Oikos* **2010**, *119*, 610–621. [CrossRef]
6. Park, K. Assessment and management of invasive alien predators. *Ecol. Soc.* **2004**, *9*, 12. [CrossRef]
7. Salo, P.; Korpimäki, E.; Banks, P.B.; Nordström, M.; Dickman, C.R. Alien predators are more dangerous than native predators to prey populations. *Proc. R. Soc. B* **2007**, *274*, 1237–1243. [CrossRef]
8. Castaldelli, G.; Pluchinotta, A.; Milardi, M.; Lanzoni, M.; Giari, L.; Rossi, R.; Fano, E.A. Introduction of exotic fish species and decline of native species in the lower Po basin, north-eastern Italy. *Aquat. Conserv. Mar. Freshw. Ecosyst.* **2013**, *23*, 405–417. [CrossRef]
9. Falaschi, M.; Melotto, A.; Manenti, R.; Ficetola, G.F. Invasive species and amphibian conservation. *Herpetologica* **2020**, *76*, 216–227. [CrossRef]
10. Beebee, T.J.C.; Griffiths, R.A. The amphibian decline crisis: A watershed for conservation biology? *Biol. Conserv.* **2005**, *125*, 271–285. [CrossRef]
11. Nunes, A.L.; Fill, J.M.; Davies, S.J.; Louw, M.; Rebelo, A.D.; Thorp, C.J.; Vimercati, G.; Measey, J. A global meta-analysis of the ecological impacts of alien species on native amphibians. *Proc. R. Soc. B* **2019**, *286*, 20182528. [CrossRef]
12. Bucciarelli, G.M.; Blaustein, A.R.; Garcia, T.S.; Kats, L.B. Invasion complexities: The diverse impacts of nonnative species on amphibians. *Copeia* **2014**, *4*, 611–632. [CrossRef]
13. Lever, C. *Naturalized Reptiles and Amphibians of the World*; Oxford University Press: Oxford, UK, 2003.
14. Pitt, W.C.; Vice, D.S.; Pitzler, M.E. Challenges of invasive reptiles and amphibians. In *Proceedings of the 11th Wildlife Damage Management Conference*; Wildlife Damage Management: Fort Collins, CO, USA, 2005; pp. 112–119.
15. Shine, R. The ecological impact of invasive cane toads (*Bufo marinus*) in Australia. *Q. Rev. Biol.* **2010**, *85*, 253–291. [CrossRef]
16. Shine, R. A review of ecological interactions between native frogs and invasive cane toads in Australia. *Austral Ecol.* **2014**, *39*, 1–16. [CrossRef]
17. Bissattini, A.M.; Buono, V.; Vignoli, L. Disentangling the trophic interactions between American bullfrogs and native anurans: Complications resulting from post-metamorphic ontogenetic niche shifts. *Aquat. Conserv. Mar. Freshw. Ecosyst.* **2019**, *29*, 270–281. [CrossRef]
18. Boelter, R.A.; Kaefer, I.L.; Both, C.; Cechin, S. Invasive bullfrogs as predators in a Neotropical assemblage: What frog species do they eat? *Anim. Biol.* **2012**, *62*, 397–408. [CrossRef]
19. Jancowski, K.; Orchard, S. Stomach contents from invasive American bullfrogs *Rana catesbeiana* (=*Lithobates catesbeianus*) on southern Vancouver Island, British Columbia, Canada. *NeoBiota* **2013**, *16*, 17–37. [CrossRef]
20. Oda, F.H.; Guerra, V.; Grou, E.; de Lima, L.D.; Proença, H.C.; Gambale, P.G.; Takemoto, R.M.; Teixeira, C.P.; Campiao, K.M.; Ortega, J.C.G. Native anuran species as prey of invasive American bullfrog *Lithobates catesbeianus* in Brazil: A review with new predation records. *Amphib. Reptile Conserv.* **2019**, *13*, 217–226.
21. Silva, E.T.; Da Filho, O.P.R.; Feio, R.N. Predation of native anurans by invasive Bullfrogs in Southeastern Brazil: Spatial variation and effect of microhabitat use by prey. *S. Am. J. Herpetol.* **2011**, *6*, 1–10. [CrossRef]

22. Courant, J.; Vogt, S.; Marques, R.; Measey, J.; Secondi, J.; Rebelo, R.; De Villiers, A.; Ihlow, F.; De Busschere, C.; Backeljau, T. Are invasive populations characterized by a broader diet than native populations? *PeerJ* **2017**, *5*, e3250. [CrossRef]
23. Wu, A.Z.; Li, Y.; Wang, Y.; Adams, M.J. Diet of introduced bullfrogs *Rana catesbeiana*: Predation on and diet overlap with native frogs on Daishan Island, China. *J. Herpetol.* **2005**, *39*, 668–674. [CrossRef]
24. Measey, G.J.; Vimercati, G.; De Villiers, F.A.; Mokhatla, M.M.; Davies, S.J.; Thorp, C.J.; Rebelo, A.D.; Kumschick, S. A global assessment of alien amphibian impacts in a formal framework. *Divers. Distrib.* **2016**, *22*, 970–981. [CrossRef]
25. Dufresnes, C.; Leuenberger, J.; Amrhein, V.; Bühler, C.; Thiébaud, G.; Bohnenstengel, T.; Dubey, S. Invasion genetics of marsh frogs (*Pelophylax ridibundus sensu lato*) in Switzerland. *Biol. J. Linn. Soc.* **2018**, *123*, 402–410. [CrossRef]
26. Dufresnes, C.; Denoël, M.; Di Santo, L.; Dubey, S. Multiple uprising invasions of *Pelophylax* water frogs, potentially inducing a new hybridogenetic complex. *Sci. Rep.* **2017**, *7*, 6506. [CrossRef]
27. Dufresnes, C.; Di Santo, L.; Leuenberger, J.; Schuerch, J.; Mazepa, G.; Grandjean, N.; Canestrelli, D.; Perrin, N.; Dubey, S. Cryptic invasion of Italian pool frogs (*Pelophylax bergeri*) across Western Europe unraveled by multilocus phylogeography. *Biol. Invasions* **2017**, *19*, 1407–1420. [CrossRef]
28. Ficetola, G.F.; Scali, S. Invasive amphibians and reptiles in Italy. *Atti. VIII Congr. Naz. Soc. Herpetol. Ital.* **2010**, 335–340.
29. Holsbeek, G.; Mergeay, J.; Volckaert, F.A.M.; De Meester, L. Genetic detection of multiple exotic water frog species in Belgium illustrates the need for monitoring and immediate action. *Biol. Invasions* **2010**, *12*, 1459–1463. [CrossRef]
30. Pagano, A.; Crochet, P.A.; Graf, J.D.; Joly, P.; Lodé, T. Distribution and habitat use of water frog hybrid complexes in France. *Glob. Ecol. Biogeogr.* **2001**, *10*, 433–441. [CrossRef]
31. Dufresnes, C.; Dubey, S. Invasion genomics supports an old hybrid swarm of pool frogs in Western Europe. *Biol. Invasions* **2020**, *22*, 205–210. [CrossRef]
32. Vorburger, C.; Reyer, H. A genetic mechanism of species replacement in European waterfrogs? *Conserv. Genet.* **2003**, *4*, 141–155. [CrossRef]
33. Bogdan, H.V.; Covaciu-Marcov, S.D.; Cupsa, D.; Cicort-Lucaciu, S.I. Food composition of a *Pelophylax ridibundus* (Amphibia) population from a thermal habitat in Banat Region (Southwestern Romania). *Acta Zool. Bulg.* **2012**, *64*, 253–262.
34. Çiçek, K.; Mermer, A. Feeding biology of the marsh frog, *Rana ridibunda* Pallas 1771, (Anura, Ranidae) in Turkey's lake district. *North-West. J. Zool.* **2006**, *2*, 57–72.
35. Cogălniceanu, D.; Palmer, M.W.; Ciubuc, C. Feeding in anuran communities on islands in the Danube floodplain. *Amphib.-Reptil.* **2000**, *22*, 1–19. [CrossRef]
36. Fathinia, B.; Rastegar-Pouyani, N.; Darvishnia, H.; Shafaeipour, A.; Jaafari, G. On the trophic spectrum of *Pelophylax ridibundus* (Pallas, 1771) (Amphibia: Anura: Ranidae) in western Iran. *Zool. Middle East* **2016**, *62*, 247–254. [CrossRef]
37. Mollov, I. Sex based differences in the trophic niche of *Pelophylax ridibundus* (Pallas, 1771) (Amphibia: Anura) from Bulgaria. *Acta Zool. Bulg.* **2008**, *60*, 277–284.
38. Plitsi, P.; Koumaki, M.; Bei, V.; Pafilis, P.; Polymeni, R.M. Feeding ecology of the Balkan water frog (*Pelophylax kurtmuelleri*) in Greece with emphasis on habitat effect. *North-West. J. Zool.* **2016**, *12*, 292–298.
39. Balint, N.; Indrei, C.; Racula, I.; Ursut, A. On the diet of the *Pelophylax ridibundus* (Anura, Ranidae) in Ticleni, Romania. *South-West. J. Hortic. Biol. Environ.* **2010**, *1*, 57–66.
40. Nicoara, A.; Nicoara, M.; Bianchini, F. Diet composition during breeding period in populations of *Bufo viridis, Pelobates fuscus* and *Rana esculenta* complex from Ciric river's basin (Iasi, Romania). *An. Stiintifice Univ "Al.I. Cuza" Iasi Sect. Biol. Anim.* **2005**, *51*, 179–187.
41. Pesarakloo, A.; Rastegar-Pouyani, N.; Rastegar-Pouyani, E.; Najibzadeh, M.; Shakarami, J.; Kami, H.G.; Shayestehfar, A. Feeding biology and food composition in *Pelophylax ridibundus* (Pallas 1771) in the Iranian plateau. *Russ. J. Herpetol.* **2017**, *24*, 91–98. [CrossRef]
42. Ruchin, A.B.; Ryzhov, M.K. On the diet of the marsh frog (*Rana ridibunda*) in the Sura and Moksha watershed, Mordovia. *Adv. Amph. Res. Former Sov. Union* **2002**, *7*, 197–205.
43. Katsiyiannis, P.; Tzoras, E. First record of *Pelophylax kurtmuelleri* preying on *Hyla arborea* in Greece. *Parnass. Arch.* **2020**, *8*, 17–18.
44. Gabrion, J. La Néoténie Chez *Triturus helveticus* Raz. Etude Morphofonctionnelle de la Fonction Thyroidienne. Ph.D. Thesis, Université des Sciences et Techniques du Languedoc, Montpellier, France, 1976.
45. Denoël, M. Priority areas of intraspecific diversity: Larzac, a global hotspot for facultative paedomorphosis in amphibians. *Anim. Conserv.* **2007**, *10*, 110–116. [CrossRef]
46. Denoël, M.; Lehmann, A. Multi-scale effect of landscape processes and habitat quality on newt abundance: Implications for conservation. *Biol. Conserv.* **2006**, *130*, 495–504. [CrossRef]
47. Durand-Tullou, A. Un Milieu de Civilisation Traditionnelle. Le Causse de Blandas. Ph.D. Thesis, Faculté des Lettres et Sciences Humaines de Montpellier, Montpellier, France, 1959.
48. Geniez, P.; Cheylan, M. *Les Amphibiens et Reptiles du Languedoc-Roussillon et Régions Limitrophes*; Biotope: Mèze, France; Muséum National D'histoire Naturelle: Paris, France, 2012.
49. Dufresnes, C.; Mazepa, G. Hybridogenesis in Water Frogs. *eLS* **2020**, *1*, 718–726. [CrossRef]
50. Christy, M.T. The efficacy of using Passive Integrated Transponder (PIT) tags without anaesthetic in free-living frogs. *Aust. J. Zool.* **1996**, *30*, 139–142. [CrossRef]

51. Winandy, L.; Denoël, M. The use of visual and automatized behavioral markers to assess methodologies: A study case on PIT-tagging in the Alpine newt. *Behav. Res. Methods* **2011**, *43*, 568–576. [CrossRef]
52. Solé, M.; Beckmann, O.; Pelz, B.; Kwet, A.; Engels, W. Stomach-flushing for diet analysis in anurans: An improved protocol evaluated in a case study in Araucaria forests, southern Brazil. *Stud. Neotrop. Fauna Environ.* **2005**, *40*, 23–28. [CrossRef]
53. Joly, P. Le régime alimentaire des amphibiens: Méthodes d'étude. *Alytes* **1987**, *6*, 11–17.
54. Nöllert, A.; Nöllert, C. *Guide des Amphibiens d'Europe: Biologie, Identification, Répartition*; Delachaux et Niestlé: Paris, France, 2003.
55. Rohlf, F.J. *tpsDig2, Version 2.31. Stony Brook*; Department of Ecology and Evolution, State University of New York: New York, NY, USA, 2007.
56. Burnham, K.P.; Anderson, D.R. *Model Selection and Multimodel Inference: A Practical Information-Theoretic Approach*; Springer: New York, NY, USA, 2002.
57. Mazerolle, M.J. Improving data analysis in herpetology: Using Akaike's Information Criterion (AIC) to assess the strength of biological hypotheses. *Amphib.-Reptil.* **2006**, *27*, 169–180. [CrossRef]
58. Bates, D.; Maechler, M.; Bolker, B.; Walker, S. Fitting Linear Mixed-Effects Models Using lme4. *J. Stat. Softw.* **2015**, *67*, 1–48. [CrossRef]
59. Barton, K. MuMIn: Multi-Model Inference. R Package Version 1.43.17. Available online: https://CRAN.R-project.org/package=MuMIn (accessed on 18 September 2020).
60. Loman, J. Food, feeding rates and prey-size selection in juvenile and adult frogs, *Rana arvalis* Nilss. and *R. temporaria* L. *Ekol. Polska* **1979**, *27*, 581–601.
61. Wells, K.D. *The Ecology and Behavior of Amphibians*; The University of Chicago Press: Chicago, IL, USA, 2007.
62. Measey, G.J.; Vimercati, G.; De Villiers, F.A.; Mokhatla, M.M.; Davies, S.J.; Edwards, S.; Altwegg, R. Frog eat frog: Exploring variables influencing anurophagy. *PeerJ* **2015**, *3*, e1204. [CrossRef]
63. Glorioso, B.M.; Waddle, J.H.; Crockett, M.E.; Rice, K.G.; Percival, H.F. Diet of the invasive Cuban Treefrog (*Osteopilus septentrionalis*) in pine rockland and mangrove habitats in South Florida. *Caribb. J. Sci.* **2012**, *46*, 346–355. [CrossRef]
64. Balint, N.; Citrea, L.; Memetea, A.; Jurj, N.; Condure, N. Feeding ecology of the *Pelophylax ridibundus* (Anura, Ranidae) in Dobromir, Romania. *Biharean Biol.* **2008**, *2*, 27–37.
65. Çiçek, K.; Mermer, A. Food composition of the marsh frog, *Rana ridibunda* Pallas, 1771, in Thrace. *Turk. J. Zool.* **2007**, *31*, 83–90.
66. Mollov, I.; Boyadzhiev, P.; Donev, A. Trophic role of the marsh frog *Pelophylax ridibundus* (Pallas, 1771) (Amphibia, Anura) in the aquatic ecosystems. *Bulg. J. Agric. Sci.* **2010**, *16*, 298–306.
67. Sas, I.; Kovács, É.; Covaciu-Marcov, S.D.; Strugariu, A.; Covaci, R.; Ferenţi, S. Food habits of a pool frog *Pelophylax lessonae*–edible frog *Pelophylax kl. esculentus* population from North-Western Romania. *Biota* **2007**, *8*, 71–78.
68. Sas, I.; Covaciu-Marcov, S.D.; Strugariu, A.; David, A.; Ilea, C. Food habit of *Rana (Pelophylax) kl. esculenta* females in a new recorded E-system population from a forested habitat in north-western Romania. *Turk. J. Zool.* **2009**, *33*, 1–5.
69. UICN France; MNHN; SHF. *La Liste rouge des espèces menacées en France—Chapitre reptiles et amphibiens de France métropolitaine*; UICN: Paris, France, 2015. Available online: https://inpn.mnhn.fr/docs/LR_FCE/UICN-LR-Reptile-Fascicule-m5-1.pdf (accessed on 27 February 2019).
70. Denoël, M. Newt decline in Western Europe: Highlights from relative distribution changes within guilds. *Biodivers. Conserv.* **2012**, *21*, 2887–2898. [CrossRef]
71. Dimancea, N.; David, A.; Cupsa, D.; Cicort-Lucaciu, A.-S.; Indrei, C. The trophic spectrum analysis of a *Pelophylax ridibundus* population from Scăpău locality, Mehedinți county, SW of Romania. *Muz. Olten. Craiova Oltenia. Studii Comunicări. Ştiinţele Naturii* **2010**, *26*, 177–181.
72. Paunovic, A.; Bjelic-Cabrilo, O.; Simic, S. The diet of water frogs (*Pelophylax esculentus* 'complex') from the Petrovaradinski Rit marsh (Serbia). *Arch. Biol. Sci.* **2010**, *62*, 797–806. [CrossRef]
73. Denoël, M. Seasonal variation of morph ratio in facultatively paedomorphic populations of the palmate newt *Triturus helveticus*. *Acta Oecol.* **2006**, *29*, 165–170. [CrossRef]
74. Cohen, J.E.; Pimm, S.L.; Yodzis, P.; Saldana, J. Body sizes of animal predators and animal prey in food webs. *J. Anim. Ecol.* **1993**, *62*, 67–78. [CrossRef]
75. Duellman, W.E.; Trueb, L. *Biology of Amphibians*, 2nd ed.; Johns Hopkins University Press: Baltimore, MD, USA, 1994.
76. Kraus, F. *Alien Reptiles and Amphibians: A Scientific Compendium and Analysis*; Springer: New York, NY, USA, 2009.
77. Preisser, E.L.; Bolnick, D.I.; Benard, M.E. Scared to death? The effects of intimidation and consumption in predator prey interactions. *Ecology* **2005**, *86*, 501–509. [CrossRef]
78. Courant, J.; Secondi, J.; Vollette, J.; Herrel, A.; Thirion, J.M. Assessing the impacts of the invasive frog, *Xenopus laevis*, on amphibians in western France. *Amphib. -Reptil.* **2018**, *39*, 219–227. [CrossRef]
79. Manenti, R.; Falaschi, M.; Monache, D.D.; Marta, S.; Ficetola, G.F. Network-scale effects of invasive species on spatially-structured amphibian populations. *Ecography* **2020**, *43*, 119–127. [CrossRef]

diversity

MDPI

Article

Salinity Affects Freshwater Invertebrate Traits and Litter Decomposition

Manuela Abelho [1,2,*], Rui Ribeiro [2] and Matilde Moreira-Santos [2]

1 Polytechnic of Coimbra, Coimbra Agriculture School, Bencanta, 3045-601 Coimbra, Portugal
2 CFE-Centre for Functional Ecology—Science for People and the Planet, Department of Life Sciences, University of Coimbra, Calçada Martim de Freitas, 3000-456 Coimbra, Portugal; rui.ribeiro@zoo.uc.pt (R.R.); matilde.santos@zoo.uc.pt (M.M.-S.)
* Correspondence: abelho@esac.pt

Abstract: We evaluated the effect of seawater intrusion in coastal ecosystems on the freshwater invertebrate community and on leaf litter decomposition under realistic scenarios in six outdoor freshwater mesocosms containing fauna and flora, to which increasing volumes of seawater were added. The resulting salinity values were 0.28 (control, freshwater only), 2.0, 3.3, 5.5, 9.3, and 15.3 mS cm^{-1}. The effect of salinity was assessed for 65 days after seawater intrusion, by computing the deviation of values in each treatment in relation to the control. Our results show that seawater intrusion into freshwaters will affect the invertebrate communities and organic matter decomposition, with salinities of up to 3.3–5.5 mS cm^{-1} having opposite effects to salinities of more than 9.3 mS cm^{-1}. There was a net negative effect of the two highest salinities on mass loss and richness of the invertebrates associated with the decomposing leaves. Regarding the invertebrate communities of the mesocosms, there was a net negative effect of the intermediate salinity levels on abundance and richness. Invertebrate life cycle traits conferring resilience and resistance tended to increase with low and decrease with high salinity values, while avoidance traits showed an opposite trend, and these responses were more pronounced on the later stage community. These wave-like responses of the invertebrate species traits to increasing salinity suggest that the life-history and physiological adaptations most suitable to cope with osmotic stress will differ between low and high salinity levels.

Keywords: freshwater invertebrates; invertebrate traits; litter decomposition; mesocosms; primary salinization; seawater intrusion

Citation: Abelho, M.; Ribeiro, R.; Moreira-Santos, M. Salinity Affects Freshwater Invertebrate Traits and Litter Decomposition *Diversity* **2021**, *13*, 599. https://doi.org/10.3390/d13110599

Academic Editor: Michael Wink

Received: 25 October 2021
Accepted: 18 November 2021
Published: 21 November 2021

Publisher's Note: MDPI stays neutral with regard to jurisdictional claims in published maps and institutional affiliations.

Copyright: © 2021 by the authors. Licensee MDPI, Basel, Switzerland. This article is an open access article distributed under the terms and conditions of the Creative Commons Attribution (CC BY) license (https://creativecommons.org/licenses/by/4.0/).

1. Introduction

As a result of global warming, mean sea level increased 0.15–0.25 m between 1901 and 2018, with rates steadily increasing over time. The global mean sea level will most certainly continue to rise throughout the 21st century [1]. Even under a scenario of very low greenhouse gas (GHG) emissions, the likely global mean sea level rise by 2100 is predicted to be 0.28–0.55 m, and the increase may reach as much as 0.63–1.01 m [1]. Moreover, due to relative sea level rise, one in a century extreme sea level events are projected to occur at least annually at more than half of all tide gauge locations by 2100, increasing the frequency and severity of flooding and salinization of coastal areas (including freshwaters) along all low-lying coasts [1]. Additionally, groundwater salinization due to prolonged droughts that promote the inland retreat of the seawater/freshwater interface is also predicted to increase due to climate change [1,2]. Freshwater ecosystems provide essential services and carry out key processes, which depend on the integrity of populations and communities. How freshwater organisms and ecosystem processes will respond to the increasing salinization of coastal freshwaters, predicted by the Intergovernmental Panel on Climate Change [1], is a fundamental quest for an integrated biodiversity and conservation management of coastal ecosystems at risk.

Studies on a wide variety of lakes, rivers, and wetlands have shown salinity effects on freshwater organisms and on ecosystem structure and function [3–6]. Regarding freshwater animals, salinity may have lethal effects once a threshold is exceeded, or sub-lethal effects, reducing both organism and population fitness, and resulting in the decline of richness with increasing salinity [3,6]. Salinity also affects organisms' spatial distribution, trophic and other interactions, biochemical cycles, and leaf litter decomposition (e.g., [3,7–9]), thus altering ecosystem structure and function. Organic matter decomposition is one of the most important ecosystem processes and, because it results from the activity of an array of species in different trophic levels and taxonomic groups, provides a consolidated measure of ecosystem integrity [10]. Field and laboratory studies show a consistent trend of reduced organic matter processing in freshwaters with increasing salinity due to decreased enzymatic activity and lower fungal biomass on decaying leaves (e.g., [11–13]). The decrease of microbial colonization may also indirectly affect the detritivore freshwater invertebrates, which rely on decomposing (i.e., microbially colonized) plant litter as a food resource and contribute, while feeding, to litter processing [14]. Thus, salinity may decrease organic matter processing in a feedback process involving microorganisms and invertebrate detritivores, which can be further enhanced when it has direct effects on invertebrates, as discussed above.

Most studies have assessed the effect of dryland and secondary salinization [3,12], although seawater intrusion (primary salinization) already contributes significantly to coastal salinization [5]. Natural seawater varies in ionic composition and differs from salinized freshwater sources and dryland salinity [5], and ionic composition has been shown to significantly affect the results obtained in other salinization studies (e.g., [15,16]). Thus, the effects of primary salinization on organisms and on ecosystem processes requires special consideration.

Species traits are linked to a myriad of environmental conditions presenting different solutions to abiotic stress, such as the case of salinization [17]. Moreover, contrarily to taxa identity, which varies geographically, traits allow the comparison of species assemblages between waterbodies separated by large geographical distances [17], thus constituting a useful tool to assess the effect of freshwater salinization worldwide.

In the present study, we aimed to contribute to the knowledge on the effects of primary salinization on key drivers of organic matter decomposition in low-lying coastal ecosystems. We assessed the effects of seawater intrusion events of five different magnitudes (2.0, 3.3, 5.5, 9.3, and 15.3 mS cm^{-1}) on leaf litter decomposition and invertebrate communities, using dilutions of ocean water in 1500 L outdoor mesocosms under realistic conditions. We hypothesized that increasing salinity will decrease (H1) leaf litter decomposition and (H2) invertebrate abundance and richness. Regarding invertebrate traits, we also hypothesized that increasing salinity will (H3) increase the proportion of traits conferring resilience (short lifespan, ovoviviparous reproduction, and aerial respiration) and resistance (multivoltinism), due to the selection of taxa showing traits most suitable to cope with a given salinity, and (H4) decrease the proportion of traits conferring avoidance (amphibiotic life cycles and terrestrial reproduction), as taxa not able to cope with the osmotic stress disperse to other aquatic bodies. Finally, if the studied salinity levels do not cause acute toxic effects, we hypothesized that (H5) the proportion of resilience and resistance will increase while avoidance traits will decrease over time.

2. Materials and Methods

2.1. Outdoor Mesocosms

The present study was carried out in six lentic mesocosms during November 2013–January 2014 using the setup described in [18]. The mesocosms consisted of black rigid polyvinyl chloride containers with a capacity of 1500 L (1.5 m diameter, 1 m depth), buried into the ground to surface level at Escola Superior Agrária de Coimbra (Coimbra, 40°12′ N, 8°27′ W). The mesocosms were prepared in September 2013, using sediment (300 L; Lagoa dos Teixoeiros 40°18′ N, 8°46′ W) and 1000 L of freshwater, 600 L from a spring-fed stream (Olhos d'Água do Rio Anços 39°58′ N, 8°24′ W) and 400 L from a local well. Mesocosms were evenly inoculated with the macrophytes *Potamogeton* sp. and *Myriophyllum* sp. (8 individuals each per mesocosm; Olhos d'Água do Rio Anços), macroinvertebrates and zooplankton (collected at Olhos d'Água do Rio Anços, Lagoa dos Teixoeiros, Lagoa das Braças 40°14′ N, 8°48′ W, and Lagoa da Vela 40°16′ N, 8°47′ W), in addition to the phytoplankton, zooplankton, macroinvertebrates, and microbial organisms indirectly introduced via the sediment, water, and macrophytes. The mesocosms were left stabilizing for 7 weeks before seawater intrusion.

Seawater intrusion was simulated using Atlantic Ocean water (West coast of Central Portugal; 52.7 mS cm^{-1}). On 20 November 2013 (day 0), a water volume was removed from each mesocosm (between 32 and 280 L, filtered to return back organisms bigger than 50 µm), and the same seawater volume was added to the respective mesocosm (using a 1.65-fold dilution factor), resulting in the following five electrical conductivity values, hereafter used to designate the various treatments: 2.0, 3.3, 5.5, 9.3, and 15.3 mS cm^{-1}; a sixth mesocosm (freshwater only) was left intact to be used as the control treatment (0.28 mS cm^{-1}). Preliminary laboratory toxicity tests with a range of seawater dilutions revealed lethal effects on invertebrates close to 15 mS cm^{-1} (at higher than 12 mS cm^{-1} for the cladoceran *Daphnia magna* and the cnidarian *Hydra attenuata*, and at higher than 18 mS cm^{-1} for the insect *Chironomus riparius*), whereas [15] found seawater sublethal effects on the rotifer *Brachionus calyciflorus* and *Daphnia longispina* between 2 and 5 mS cm^{-1}.

Electrical conductivity, pH, and dissolved oxygen were measured in situ two times before (days -14 and -6) and four to six times after seawater intrusion (days 2, 7, 14, 21, 28, and 65) with field probes (Wissenschaftlich Technische Werkstätten, Weilheim, Germany: WTW conductivity440i, WTW pH330i, and WTW OXI 330i, respectively). Water temperature was recorded every 15 min with a Hobo Pendant Temp/Light logger UA-002-64 (Onset company, Cape Cod, MA, USA), except in treatment 3.3 mS cm^{-1}, where the equipment malfunctioned. Water samples (one per mesocosm) were collected on the same occasions (except for day -14), and a Hach DR/2000 photometer (Hach company, Loveland, CO, USA) was used to quantify the following nutrient concentrations (mgL^{-1}; three replicate measurements per water sample), according to Hach manual: ammonia ($N\text{-}NH_3$; method 8038), nitrate ($N\text{-}NO_3{}^{-}$; method 8192), orthophosphate ($PO_4{}^{3-}$; method 8048), and sulphate ($SO_4{}^{2-}$; method 8051). Monthly air temperature and precipitation values during the study period were obtained from a nearby meteorologic station (Escola Superior Agrária, Bencanta: 40°12′ N, 8°41′ W, altitude 17 m).

2.2. Leaf Litter Decomposition and Associated Invertebrates

The effect of seawater intrusion on organic matter decomposition was assessed using alder (*Alnus glutinosa* (L.) Gaertn.) leaves, a common riparian species in the study area. Alder is considered a key riparian plant species and a model litter widely used in decomposition experiments across Europe and elsewhere [19]. Five days before seawater intrusion (15 November 2013), groups of 5 air-dried leaves (1.15 g ± 0.09 standard deviation [SD]) were sprinkled with distilled water to avoid damage, tied by the petiole to form leaf packs, and nine leaf packs were introduced and let sunk naturally in each mesocosm. On the day of seawater intrusion (day 0: 20 November 2013), three replicates were removed to determine mass loss due to leaching and establish initial leaf mass values. Sampling was carried out after 16 (6 December 2013) and 63 days (22 January 2014) of immersion.

On each sampling occasion, three leaf packs were retrieved from each mesocosm, introduced individually in plastic bags, and transported in an ice-chest for immediate processing. In the laboratory, the leaves were gently washed with tap water over a 0.2 mm mesh to detach adhering debris and invertebrates, oven-dried (60 °C, 3 days), and weighed to the nearest 0.1 mg to determine oven-dry mass.

The oven-dry mass of the leaves used in the experiment was estimated by multiplying their air-dry mass by the humidity factor. Three extra leaf packs were weighed, oven-dried (60 °C, 3 days), and weighed again (±0.1 mg) to determine the humidity factor (oven-dry/air-dry mass; mean ± 1 SD: 0.857 ± 0.015). To account only for the effect of biological activity on mass loss, the oven-dry mass remaining on day 0 was used to determine the leaching factor (final/initial oven-dry mass; overall mean ± 1 SD: 0.915 ± 0.012). Oven-dry mass remaining at each sampling day was divided by the leaching factor to correct for mass loss due to leaching during the 5 days immersion period before seawater intrusion. Final values were expressed as the percentage loss of initial oven-dry mass after leaching of soluble compounds.

The invertebrates associated with the leaves were collected, sorted, stored in the cold, identified to order or family level [18], and counted under a binocular stereoscope within the next 24 h. Values were expressed as number of individuals (abundance) or number of taxa (richness) per leaf pack.

2.3. *Invertebrate Communities of the Mesocosms*

The data and methods are from [18]. Briefly, the invertebrates were sampled at 14 and 6 days before seawater intrusion to assess the similarity of the mesocosms, and six times after (2, 7, 14, 21, 28, and 65 days) to assess the effect of seawater intrusion. Sampling was carried out with a hand net (0.5 mm mesh) and consisted of three equidistant passages of the net across the diameter of the mesocosms at the macrophyte level. The invertebrates were carefully collected and stored in 50 mL plastic containers with 70% ethanol, identified to family level and counted. Values were expressed as number of individuals (abundance) or number of taxa (richness) per mesocosm.

2.4. *Invertebrate Life Cycle Traits*

Using the rationale of [17], five life cycle and one physiological trait categories from [20] were aggregated as combinations of traits providing resilience, resistance, and avoidance to osmotic stress. Life cycles shorter than one-year reduce time and energy to reach adult stage with earlier reproduction, ovoviviparity provides egg protection from environmental conditions, and aerial respiration allows coping with low dissolved oxygen concentrations, which may occur at high salinities [21] and were aggregated as combinations providing resilience to osmotic stress. Life cycles with more than one generation per year provide resistance as they increase the capacity to reach high population numbers and recover after disturbance. Finally, amphibiotic life cycles (taxa with non-aquatic life stages) and terrestrial reproduction provide independence from the aquatic environmental conditions and allow dispersion to other aquatic environments thus providing avoidance capacities.

The proportion of a trait category per treatment was calculated according to [20]. Each taxon was attributed an affinity score of 0 (no affinity) to 3 (strong affinity) for the trait category, and the frequency of that category in a taxon (Table S1 in Supplementary Materials) was calculated by dividing the score of the category by the sum of scores of all possible categories. Finally, the frequency of the taxon presenting the trait category (or the average frequency of the various categories in the cases of resilience and avoidance) was multiplied by the proportion of individuals in that taxon, and the resulting values summed for each sample.

2.5. *Data Analysis*

The experimental design includes only one mesocosm per treatment, i.e., no replicates, with measurements along time. Thus, initial abiotic and biotic conditions before seawater intrusion as well as the physicochemical parameters during the study period were compared with a repeated-measures procedure using the collection dates as the repeated-measures factor. To avoid violation of the sphericity assumption, the procedure was applied to groups of variables as a multivariate analysis of variance (MANOVA, [22]), as follows. Before seawater intrusion: one dependent variable was composed of dissolved oxygen, pH, temperature, and electrical conductivity and the other of invertebrate abundance and richness. During the experimental period, one dependent variable was composed of ammonia and sulphate, which increased with increasing salinity, and the other was composed of dissolved oxygen, pH, nitrate, and phosphate. Prior to analysis, normality was checked with the Shapiro–Wilk's test and homoscedasticity was checked with the Bartlett's test (normal distribution data) or the Levene's test (nonnormal distribution data). Nonnormal and/or heteroscedastic data was transformed using the square root transformation, $x' = \sqrt{x + 3/8}$ in the case of invertebrate abundance and richness or the logarithmic transformation, $x' = ln(x+1)$, in the case of the physicochemical data [22]. Given that the MANOVA is robust, operating well even with considerable heterogeneity of variances and deviations from normality [22], the procedure was still carried out even when normality or homoscedasticity were not achieved after transformation. Whenever a significant difference among treatments was found, post-hoc multiple comparisons were carried out with the Tukey test. All these procedures were performed with STATISTICA Version 10.0 (StatSoft, Inc., Tulsa, OK, USA) with the level of significance set at 0.05.

Regarding the biotic variables, rather than comparing the treatments, our aim was to assess the departure of the values in each treatment in relation to the control. Thus, the effect of salinity on litter decomposition and on the invertebrates was assessed by computing the average deviation and the 95% confidence limits (CL) of values in each treatment in relation to the values in the control, using all samples collected over the time period. For each response variable, the deviation of the value in each sample of a treatment in relation to the value in each sample of the control (2 sampling dates × 3 leaf packs for the decomposition experiment, and 6 sampling dates for the mesocosms' invertebrates) was computed as: (value in treatment-value in control)/value in control. We considered that there was a net effect of salinity in a treatment in relation to the control whenever the 95% CL were lower than the mean deviation, indicating that the treatment and the control samples came from different statistical populations.

3. Results

3.1. Abiotic Variables

Average air temperature in the study area ranged from 9.8 to 11.5 °C during the study period. Total precipitation increased along the time period, with 17.2 mm in November, 163.8 mm in December, and 204.6 mm in January. There were no significant differences among the mesocosms in the abiotic variables (electrical conductivity, dissolved oxygen, pH, and temperature) before seawater intrusion (MANOVA: $F_{0.05(1)10,10} = 1.90$, $p = 0.17$; Table S2 in Supplementary Materials); it was not possible to compare the mesocosms regarding nutrients because there was only one measurement before seawater intrusion (Table 1).

Table 1. Average values ± 1 SD (and range) of water physicochemical characteristics before (one to two measurements per mesocosm) and during the 65 days study period after seawater intrusion (four to six measurements per mesocosm).

Treatment	Electrical Conductivity (mS cm^{-1})	Dissolved Oxygen (mg L^{-1})	Temperature (°C)	pH	Nitrate (mg $N\text{-}NO_3$ L^{-1})	Ammonia (mg $N\text{-}NH_3$ L^{-1})	Orthophosphate (mg $PO_4{}^{3-}$ L^{-1})	Sulfate (mg $SO_4{}^{2-}$ L^{-1})
Before	n = 2	n = 2	n = 2	n = 2	n = 1	n = 1	n = 1	n = 1
All mesocosms	0.27 ± 0.01 (0.26–0.27)	10.9 ± 0.6 (9.6–11.9)	14.9 ± 2.4 (12.3–17.3)	8.0 ± 0.1 (7.8–8.2)	2.9 ± 0.9 (1.8–4.2)	0.19 ± 0.10 (0.07–0.34)	0.05 ± 0.03 (0.01–0.08)	10 ± 3 (7–16)
During	n = 6	n = 4	n = 5	n = 6	n = 6	n = 6	n = 6	n = 5
Control (0.28 mS cm^{-1})	0.26 ± 0.05 (0.18–0.29)	12.0 ± 1.7 (10.1–14.2)	8.0 ± 1.6 (6.0–10.0)	8.2 ± 0.6 (7.8–9.4)	4.3 ± 0.7 (2.0–6.9)	0.14 ± 0.12 (0.02–0.33)	0.07 ± 0.06 (0.01–0.17)	15 ± 12 (7–36)
2.0 mS cm^{-1}	1.86 ± 0.31 (1.26–2.05)	17.5 ± 2.5 (13.9–19.2)	8.1 ± 1.6 (6.1–10.0)	8.3 ± 0.5 (7.8–9.2)	3.8 ± 0.8 (3.3–4.8)	0.36 ± 0.36 (0.00–0.48)	0.09 ± 0.06 (0.05–0.16)	104 ± 54 (56–184)
3.3 mS cm^{-1}	3.05 ± 0.53 (2.10–3.34)	14.4 ± 0.5 (13.9–15.0)	no data	8.3 ± 0.4 (7.9–9.0)	3.6 ± 0.4 (3.0–4.0)	0.25 ± 0.20 (0.01–0.60)	0.08 ± 0.06 (0.04–0.15)	132 ± 57 (74–220)
5.5 mS cm^{-1}	5.10 ± 0.66 (3.85–5.52)	14.5 ± 1.1 (12.8–15.3)	8.5 ± 1.6 (6.6–10.3)	8.3 ± 0.4 (7.9–9.2)	3.7 ± 1.1 (2.2–5.2)	0.36 ± 0.33 (0.01–0.95)	0.09 ± 0.04 (0.05–0.17)	233 ± 170 (112–504)
9.3 mS cm^{-1}	8.58 ± 1.02 (6.61–9.22)	14.2 ± 1.8 (12.7–16.3)	8.3 ± 1.7 (6.3–10.1)	8.4 ± 0.4 (7.9–9.0)	4.1 ± 1.6 (2.9–6.9)	0.70 ± 0.70 (0.02–1.86)	0.08 ± 0.05 (0.03–0.15)	341 ± 136 (188–520)
15.3 mS cm^{-1}	14.01 ± 1.62 (11.00–15.12)	14.8 ± 0.9 (13.6–15.7)	8.2 ± 1.7 (6.1–10.0)	8.5 ± 0.4 (8.0–9.0)	4.3 ± 1.7 (2.0–6.9)	2.12 ± 0.93 (0.29–2.75)	0.11 ± 0.06 (0.08–0.21)	466 ± 241 (282–828)

During the 65 days following the seawater intrusion, dissolved oxygen, pH, nitrate and phosphate remained similar among the treatments (MANOVA: $F_{0.05(1)5,18} = 0.53$, $p = 0.75$; Table S2 in Supplementary Materials), but there was a significant effect of treatment on ammonia and sulphate (MANOVA: $F_{0.05(1)5,22} = 113.56$, $p < 0.0001$; Table S2 in Supplementary Materials), which increased with increasing salinity (Figure S1 in Supplementary Materials) and were only similar between treatments 2.0 and 3.3 mS cm^{-1} and between treatments 3.3 and 5.5 mS cm^{-1}. Electrical conductivity decreased in all treatments as precipitation increased, attaining minimum values by day 65, but the values in the treatments never overlapped (Table 1). Ammonia concentrations also decreased along time, while sulphate values increased (Figure S1 in Supplementary Materials), resulting in a signification interaction between treatment and time (Table S2 in Supplementary Materials).

3.2. *Leaf Litter Decomposition and Associated Invertebrates*

On average, leaf packs in treatments 3.3 and 5.5 mS cm^{-1} lost more mass than the control, with mass loss decreasing with increasing salinity afterwards (Figure 1a). However, salinity had a significant effect only in treatment 15.3 mS cm^{-1}, where mass loss showed a net decrease in relation to the control (Table 2).

Table 2. Average deviation ± 95% confidence limits of values in a treatment in relation to the control for the response variables of the leaf litter decomposition experiment and for the invertebrate communities in the mesocosms. Values in red represent a net (negative or positive) effect of a given salinity value on the response variable. Proportion of invertebrates with (a) short lifespan, ovoviviparous reproduction, and aerial respiration, (b) more than one generation per year, and (c) non-aquatic life stages and terrestrial reproduction.

Leaf Litter Decomposition					
All data (n = 6)	**2.0 mS cm^{-1}**	**3.3 mS cm^{-1}**	**5.5 mS cm^{-1}**	**9.3 mS cm^{-1}**	**15.3 mS cm^{-1}**
Mass loss	−0.21 ± 0.26	0.06 ± 0.34	0.15 ± 0.37	−0.59 ± 0.67	−0.78 ± 0.55
Invertebrate abundance	0.64 ± 0.80	0.14 ± 0.41	0.85 ± 1.28	0.33 ± 0.88	0.17 ± 0.38
Invertebrate richness	0.22 ± 0.55	−0.13 ± 0.32	−0.06 ± 0.31	−0.16 ± 0.13	−0.33 ± 0.21
(a) Resilience traits	−0.09 ± 0.18	−0.01 ± 0.29	0.33 ± 0.14	0.19 ± 0.26	0.22 ± 0.38
(b) Resistance traits	−0.34 ± 0.47	0.02 ± 0.57	0.74 ± 0.54	0.50 ± 0.80	0.69 ± 0.85
(c) Avoidance traits	0.39 ± 0.21	−0.19 ± 0.41	−0.27 ± 0.33	−0.23 ± 0.35	−0.40 ± 0.49
Day 63 (n = 3)	**2.0 mS cm^{-1}**	**3.3 mS cm^{-1}**	**5.5 mS cm^{-1}**	**9.3 mS cm^{-1}**	**15.3 mS cm^{-1}**
(a) Resilience traits	0.05 ± 0.16	0.18 ± 0.34	0.44 ± 0.02	0.40 ± 0.13	0.47 ± 0.11
(b) Resistance traits	0.04 ± 0.17	0.40 ± 0.60	1.17 ± 0.35	1.13 ± 0.49	1.17 ± 0.33
(c) Avoidance traits	0.56 ± 0.10	−0.40 ± 0.58	−0.54 ± 0.12	−0.46 ± 0.36	−0.67 ± 0.65
Invertebrate Communities in the Mesocosms					
All sampling days (n = 6)	**2.0 mS cm^{-1}**	**3.3 mS cm^{-1}**	**5.5 mS cm^{-1}**	**9.3 mS cm^{-1}**	**15.3 mS cm^{-1}**
Invertebrate abundance	−0.17 ± 0.35	−0.34 ± 0.24	−0.25 ± 0.20	0.01 ± 0.55	−0.15 ± 0.87
Invertebrate richness	−0.14 ± 0.26	−0.28 ± 0.20	−0.08 ± 0.24	−0.07 ± 0.26	0.07 ± 0.24
(a) Resilience traits	0.06 ± 0.04	0.13 ± 0.07	0.08 ± 0.08	0.04 ± 0.10	−0.09 ± 0.14
(b) Resistance traits	0.06 ± 0.07	0.16 ± 0.10	0.05 ± 0.14	0.05 ± 0.13	−0.14 ± 0.18
(c) Avoidance traits	0.01 ± 0.13	0.11 ± 0.19	−0.05 ± 0.12	0.02 ± 0.10	0.06 ± 0.23
Days 21–65 (n = 3)	**2.0 mS cm^{-1}**	**3.3 mS cm^{-1}**	**5.5 mS cm^{-1}**	**9.3 mS cm^{-1}**	**15.3 mS cm^{-1}**
(a) Resilience traits	0.08 ± 0.08	0.11 ± 0.12	0.12 ± 0.10	0.03 ± 0.22	−0.15 ± 0.07
(b) Resistance traits	0.06 ± 0.13	0.15 ± 0.14	0.10 ± 0.14	0.04 ± 0.26	−0.21 ± 0.14
(c) Avoidance traits	−0.12 ± 0.10	0.04 ± 0.16	−0.08 ± 0.08	0.02 ± 0.21	0.18 ± 0.33

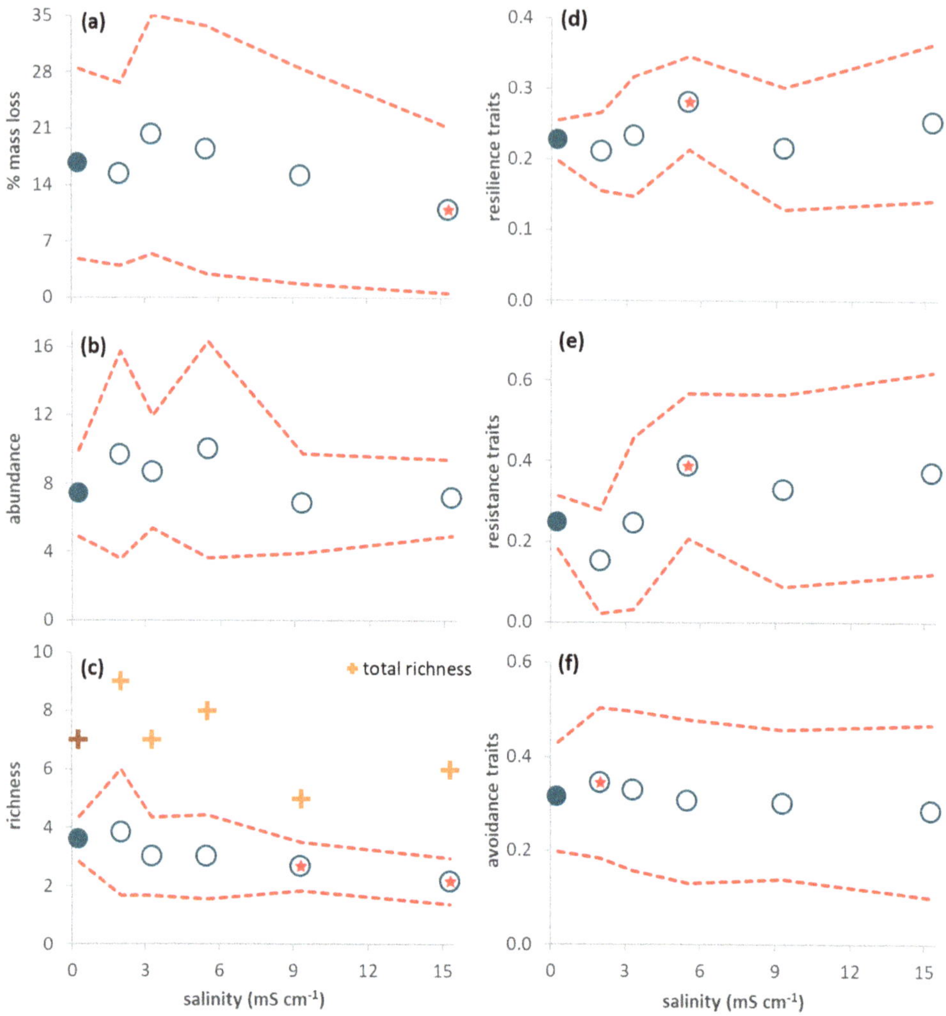

Figure 1. Response variables of the decomposition experiment during the 63-day study period (average per leaf pack ± 95% confidence limits (red dashed line); n = 6). (**a**) Leaf mass loss; (**b**) Invertebrate abundance; (**c**) Invertebrate richness (plus symbols refer to total (instead of average) richness: dark color for the control and light color for the treatments); (**d**) Proportion of resilience traits (short lifespan, ovoviviparous reproduction, and aerial respiration); (**e**) Proportion of resistance traits (multivoltinism); (**f**) Proportion of avoidance traits (non-aquatic life stages and terrestrial reproduction). A red star indicates a net average deviation of the values in a treatment (open circles) in relation to the values in the control (close circles) according to Table 2.

The abundance of invertebrates colonizing the leaves reached, at the most, 11–24 individuals per leaf pack (Figure 1b). Four taxa constituted 89% of all individuals: juveniles of Anisoptera (Corduliidae; 32%) and Zygoptera (Coenagrionidae; 19%), Chironomidae (24%), and Ostracoda (14%). The other taxa (Table S3 in Supplementary Materials) were rare and represented, at the most, 3% of all individuals. There were no leaf-shredding invertebrates; predators were the most abundant functional feeding group (56%), followed by gathering-collectors (23%), filtering-collectors (19%) and scrapers (3%). There was no significant effect of salinity on abundance (Table 2), although average invertebrate abundance was higher in salinities ranging from 2.0 to 5.5 mS cm^{-1} than in the control

(Figure 1b). Average richness decreased with increasing salinity (Figure 1c) and was significantly lower in treatments 9.3 and 15.3 mS cm^{-1} than in the control (Table 2).

The proportion of traits conferring resilience and resistance was significantly higher in treatment 5.5 mS cm^{-1} than in the control (Figure 1d,e; Table 2) while the proportion of traits conferring avoidance was significantly higher in treatment 2.0 mS cm^{-1} than in the control (Figure 1f; Table 2). When only data from day 63 were considered, resilience and resistance traits significantly increased from salinity level 5.5 mS cm^{-1} upwards (Figure 2a,b; Table 2) while avoidance traits (Figure 2c) significantly decreased for salinities higher than 5.5 mS cm^{-1} and increased in the lowest salinity treatment (2.0 mS cm^{-1}) in relation to the control (Table 2).

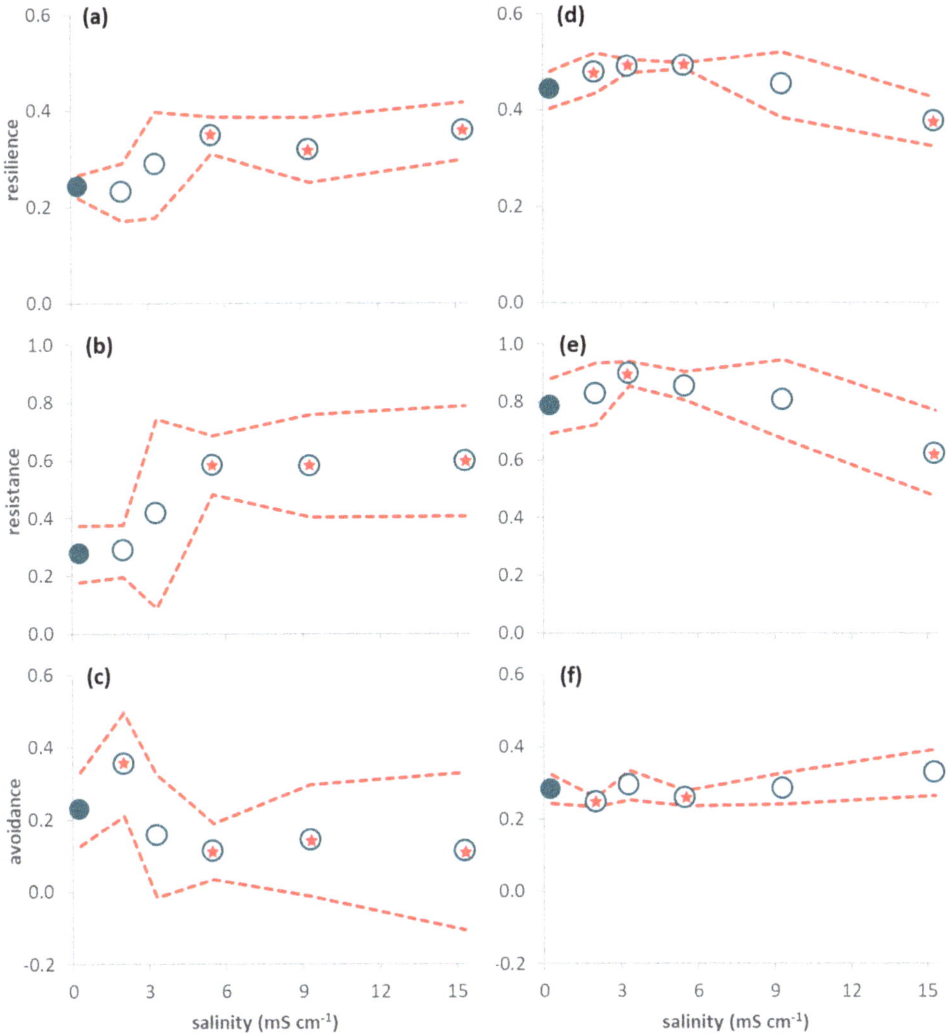

Figure 2. Proportion of traits (average per leaf pack (**a–c**) or average per sample (**d–f**) ± 95% confidence limits (red dashed line); n = 3) of invertebrates associated with decomposing leaf litter on day 63 (**a–c**) and sampled in the mesocosms during the period 21–65 days after seawater intrusion (**d–f**). (**a**,**d**) Resilience traits (short lifespan, ovoviviparous reproduction, and aerial respiration); (**b**,**e**) Resistance traits (multivoltinism); (**c**,**f**) Avoidance traits (non-aquatic life stages and terrestrial reproduction). A red star indicates a net average deviation of the values in a treatment (open circles) in relation to the values in the control (close circles) according to Table 2.

3.3. Invertebrate Coomunities in the Mesocosms

There were no significant differences among the mesocosms in invertebrate abundance and richness before seawater intrusion (MANOVA: $F_{0.05(1)5,6} = 0.92$, $p = 0.53$; Table S2 in Supplementary Materials); average abundance per mesocosm ranged from 15.0 ± 0.0 SD to 23.5 ± 3.5 SD, while average richness per mesocosm ranged from 3.0 ± 0.0 SD to 5.5 ± 0.7 SD (Table S4 in Supplementary Materials). Baetidae constituted 60% of all individuals, followed by Chironomidae (17%), Corduliidae and Corixidae (both with 7%).

After seawater intrusion, average abundance per sampling day ranged from 39 individuals in treatment 3.3 to 49 individuals in the control (Figure 3a). Baetidae constituted 61% of all individuals, followed by Chironomidae (14%), Corduliidae (13%) and Coenagrionidae (4%). The other taxa were rare and represented, at the most, 2% of all individuals (Table S4 in Supplementary Materials). Invertebrate abundance decreased with seawater intrusion (Figure 3a), with a net negative effect of salinity in treatments 3.3 and 5.5 mS cm^{-1} (Table 2). Average richness decreased with low salinity levels, and increased afterwards, with no significant deviations in relation to the control (Figure 3b; Table 2).

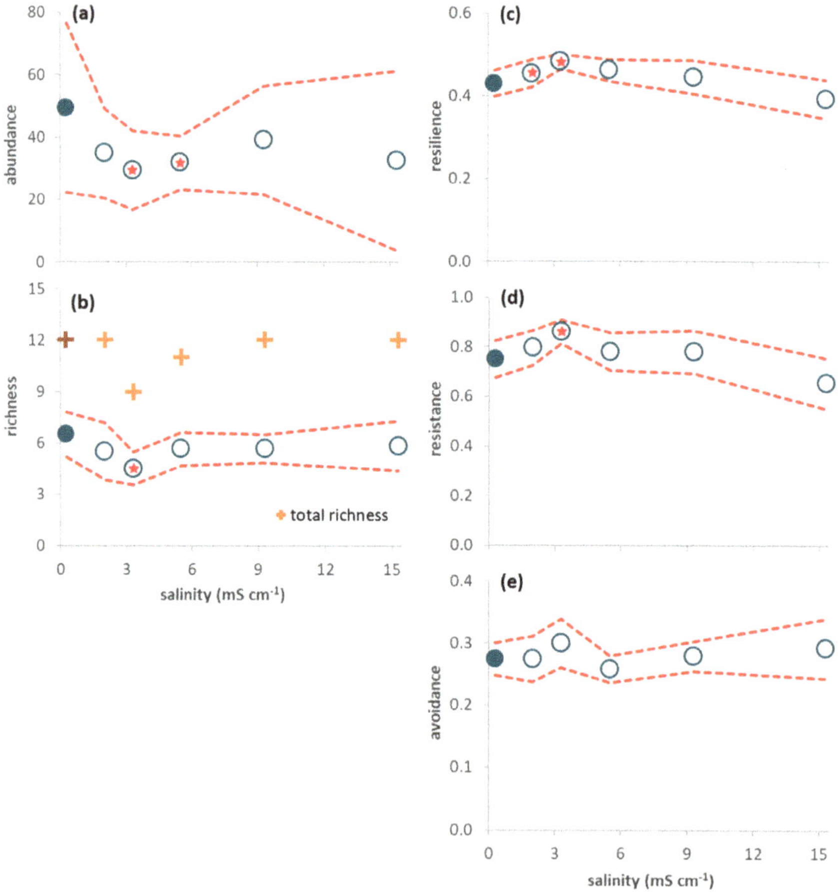

Figure 3. Response variables of the invertebrates sampled in the mesocosms during the 65-day period (average per sample ± 95% confidence limits (red dashed line); n = 6). (**a**) Invertebrate abundance; (**b**) Invertebrate richness (plus symbols refer to total (instead of average) richness: dark color for the control and light color for the treatments); (**c**) Proportion of resilience traits (short lifespan, ovoviviparous reproduction, and aerial respiration); (**d**) Proportion of resistance traits (multivoltinism); (**e**) Proportion of avoidance traits (non-aquatic life stages and terrestrial reproduction). A red star indicates a net average deviation of the values in a treatment (open circles) in relation to the values in the control (close circles) according to Table 2.

There was a net increase in the proportion of resilience (treatments 2.0 and 3.3 mS cm^{-1}) and resistance (treatment 3.3 mS cm^{-1}) traits in relation to the control (Figure 3c,d; Table 2), while avoidance traits showed no significant deviations in relation to the control (Figure 3e; Table 2). When only data from days 21–65 were considered (Figure 2d–f; Table 2), resilience traits showed a net increase in treatments 2.0 to 5.5 mS cm^{-1} and a net decrease in treatment 15.3 mS cm^{-1} (Figure 2d), resistance traits showed a net increase in treatment 3.3 mS cm^{-1} and a net decrease in treatment 15.3 mS cm^{-1} (Figure 2e), and avoidance traits showed a net decrease in treatments 2.0 and 5.5 mS cm^{-1} in relation to the control (Figure 2f; Table 2).

4. Discussion

4.1. *Leaf Litter Decomposition and Associated Invertebrates*

Mass loss estimation excluded the leaching period, which has been shown to be affected by salinity associated parameters [5,23] and there were no invertebrate shredders colonizing the leaves. We may thus assume that mass loss was mainly due to microbial degradation, and that the 61% lower mass loss in treatment 15.3 mS cm^{-1} was a result of decreased microbial activity on leaf litter. Increasing salinity has been systematically found to decrease autochthonous and allochthonous organic matter decomposition, microbial activity and/or fungal biomass associated with decomposing litter [11,12,23]. A significant decrease of mass loss of the North American species (family Sapindaceae) *Acer rubrum* leaf litter in mesocosms has been found with increasing salinity (0 to 13 psu) and mass loss correlated with bacterial diversity, with distinct bacterial communities associated to each of the four tested salinities [24]. Interestingly, bacterial diversity in the latter study increased with salinity during the first 18 days, suggesting that some small compensation by bacteria mitigated the effect of salinity, but the effect may be temporary since the pattern was reversed the longer the term. In the present study, mass loss was higher (although not significantly) at intermediate salinity (3.3 mS cm^{-1}) in relation to the control, suggesting that at low salinity levels, microbial activity may be enhanced by the presence of ions, such as Ca^{2+} and Mg^{2+} [11]. Overall, the results found here support our first hypothesis (H1) indicating a decrease in mass loss if salinity rises to 15.3 mS cm^{-1} with consequences for the aquatic detritivore food webs.

Richness of the invertebrates associated with decomposing leaves decreased with salinities equal to or higher than 9.3 mS cm^{-1}. However, most taxa were rare, and the response of richness to salinity may just be an artifact related to the choice of this habitat for reasons other than salinity. For instance, the invertebrates were dominated by juvenile Zygoptera and Anisoptera, predators that probably encountered in the leaf packs appropriate shelter during the earlier life stages. Since the invertebrates colonizing the leaves are only a subset of the source community, we thus considered mainly the invertebrate communities of the mesocosms for assessing the effects of salinity.

4.2. *Invertebrates in the Mesocosms*

It has been suggested that salinities as low as 1 g L^{-1} (<2 mS cm^{-1}) may already have toxic effects to invertebrates with sublethal effects occurring at even lower salinities, resulting in the reduction of diversity and abundance with increasing salinity [25]. In the present study, abundance and richness were decreased in all salinity treatments in relation to the control, although only significantly at intermediate salinity levels (3.3 and 5.5 mS cm^{-1}). This could be due to the low taxonomic resolution of our data set, not adequate to detect significant differences. A sharp decrease in taxonomic diversity may occur above conductivities of 5 to 20 mS cm^{-1} [21], but most studies showing a reduction in richness with increasing salinity were carried out along salinity gradients where the invertebrate communities were already adapted to the environmental conditions (e.g., [26]), and may not adequately represent the short-term effect of increasing salinity on the invertebrate assemblages. Additionally, [27] suggested that changes in invertebrate diversity will occur only at salinities $\geq$ 10 g L^{-1} (approximately 15.5 mS cm^{-1}). Thus, the salinities tested in the present study may have been too low to show a significant decrease in richness of the invertebrate community or, given the observations regarding the taxonomic groups (below), most probably the timelapse of 65 days was too short to show the disappearance of some taxa. Given the above results, our second hypothesis (H2) was supported only for low salinity levels.

Most of the taxa were rare, precluding conclusions on the effect of salinity on richness even when they were absent from the salinity treatments, as in the case of the dipterans Culicidae (absent in treatment 15.3 mS cm^{-1}) and the riffle bugs Veliidae (absent in treatments 9.3 and 15.3 mS cm^{-1}). The gastropod Physidae represented only 2.7% of all individuals associated with the leaves and 0.5% of the individuals in the mesocosms, but it was absent in both cases from salinities $\geq$ 9.3 mS cm^{-1}, conforming to the maximum salinity tolerance of 6.8 g L^{-1} determined by [28]. The performance of these gastropods has been found to be increased by salinity levels of up to 1 mS cm^{-1} and decrease at 3 (reproduction) to 5 mS cm^{-1} (growth), which would explain the disappearance of these gastropods at salinities $\geq$ 9.3 mS cm^{-1} [29]. Moreover, these scrapers could have been affected directly by salinity and indirectly by salinity-induced changes in their food source [25].

On the other hand, the ephemeropterans Baetidae were abundant. Total Baetidae (cf. genus *Cloeon* and/or *Centroptilum*) abundance was lower in all salinity treatments (88–151 individuals) than in the control (177 individuals), with the lowest abundance in treatment 15.3 mS cm^{-1}. Moreover, average Baetidae abundance in treatment 15.3 mS cm^{-1} decreased 86% from the first seven days (35 individuals per sample) to day 14 after seawater intrusion (average of five individuals per sample) and maintained these low values until day 65, with a significant deviation in relation to the control during that period (-0.84 ± 0.06 95% CL, n = 4). This pattern conforms to the salinity tolerance of 9.8 g L^{-1} and to LC_{50} values of 13–14 mS cm^{-1} determined in other studies [28,30,31] and suggests the disappearance of this taxon if salinity rises to 15.3 mS cm^{-1}.

Chironomids were also abundant; abundance was higher in the control than in the salinity treatments and followed a V-shaped curve with lowest values in treatment 5.5 mS cm^{-1}. In other studies, *Chironomus* sp. showed an inverted U-shape curve of abundance with highest survival at as much as 5 mS cm^{-1} and a steep decline at 10–15 mS cm^{-1} [32]. As chironomids are the largest family of aquatic insects [33] and their salinity tolerance varies widely among species (3.9 to 53.8 g L^{-1} [28]), it is not possible, from our results, to predict the outcomes of rising salinity levels on this taxonomic group.

Finally, the other abundant taxonomic groups were the odonatans Corduliidae and Coenagrionidae. Corduliidae showed a V-shaped response to salinity, with minimum at 3.3 mS cm^{-1}, whereas Coenagrionidae showed a U-shaped response with minimums at 2.0–9.3 mS cm^{-1}. Values in literature indicate salinity tolerances for Anisoptera varying from 5.9 to 14.8 mg L^{-1} [34], while Coenagrionidae have a salinity tolerance of 9.1 to 18 g L^{-1} [28]. However, given the obtained response, it is not possible to determine a maximum value above which changes in salinity will have effects on these taxonomic groups.

4.3. Invertebrate Traits

Contrarily to [21], who found a tendency for a decrease in the proportion of organisms with short lifespan and an increase in the proportion of multivoltine life cycles in rivers with increasing natural salinity, the present study found that the proportion of resilience and resistance traits were increased only at low salinity levels of up to 3.3 mS cm^{-1}, decreasing with increasing salinity afterwards. Several species' traits may concur for adapting to a particular environment and often they may respond in opposite directions to the same environmental stress through trade-offs [17]. For instance, while short lifespan may confer resilience by decreasing the amount of energy necessary to cope with osmotic stress [21], long development time may also confer resilience to high salinity [17]. At high salinity levels, the trade-offs with (other) traits conferring resilience and resistance may have surpassed the effects on the traits assessed in the present study, suggesting that the strategies for coping with osmotic stress differed in low and high salinity levels. Thus, our third hypothesis (H3) of increased resilience and resistance with salinity was only partially accepted, since the proportion of these traits decreased in the highest salinity treatment. We found no effects of salinity on avoidance, thus rejecting our fourth hypothesis (H4). Finally, our fifth hypothesis (H5) that resilience and resistance traits would increase while avoidance traits would decrease with time was only partially accepted. For the low salinity levels of up to 5.5 mS cm^{-1}, resilience and resistance increased while avoidance decreased with time after seawater intrusion. However, the opposite occurred at the highest tested salinity (15.3 mS cm^{-1}), once again suggesting different strategies to cope with low and high salinity levels.

5. Conclusions

In conclusion, our results show that seawater intrusion into freshwaters will affect both ecosystem structure and function, at least in respect to the invertebrate communities and to organic matter decomposition. This occurred even though the salinity levels in the treatments decreased along time due to precipitation events (Table 1, Figure S1 in Supplementary Materials), suggesting that the effects found in the present study may occur even at lower salinities than the ones tested in the experiment. During the 2 months after seawater intrusion, we obtained a wave-like response to increasing salinity for all variables. However, contrary to the hypothesis of [25], the hump at low salinity levels occurred only for mass loss, while for both abundance and richness of the invertebrate community, there was a depression at low salinity levels. In any case, the direction of the effect of salinity will depend on the magnitude of the intrusion events, with salinities of up to 3.3–5.5 mS cm^{-1} having opposite effects to salinities of more than 9.3 mS cm^{-1}. Moreover, there was also a wave-like response of the species traits to increasing salinity, suggesting that the life-history and physiological adaptations most suitable to cope with osmotic stress will differ between low and high salinity levels.

Supplementary Materials: The following are available online at https://www.mdpi.com/article/10.3390/d13110599/s1, Figure S1: Temporal variation of the physicochemical values in each treatment during the 65 days after seawater intrusion, Table S1: Invertebrate trait categories used to characterize resilience, resistance, and avoidance to osmotic stress, Table S2: Statistical results for the variables characterizing initial abiotic and biotic conditions of the mesocosms and of the physicochemical parameters during the 65 days after seawater intrusion, Table S3: Abundance of the invertebrates colonizing the leaf litter, Table S4: Abundance of the invertebrates sampled in the mesocosms.

Author Contributions: Conceptualization, R.R., M.M.-S. and M.A.; methodology, M.A., R.R. and M.M-S.; formal analysis, M.A.; investigation, M.A., R.R. and M.M-S.; data curation, M.A. and M.M.-S.; writing—original draft preparation, M.A.; writing—review and editing, M.A., R.R. and M.M-S.; project administration, M.M.-S.; funding acquisition, M.M.-S. All authors have read and agreed to the published version of the manuscript.

Funding: This research was funded by the projects Saltfree (PTDC/AAC-CLI/111706/2009-FCOMP-01-0124-FEDER-014016) and Saltfree II (POCI-01-0145-FEDER-031022), both co-funded by the Portuguese Foundation for Science and Technology (FCT) and the European Regional Development Fund (FEDER). It was also funded by the Centre for Functional Ecology Strategic Project (UID/BIA/04004/2020), within the PT2020 Partnership Agreement and Compete 2020. MM-S is a contracted researcher (IT057-18-7285, nr. 71) supported by FCT (nr. 1370).

Institutional Review Board Statement: Ethical review and approval were waived for this study because invertebrates do not require authorization.

Data Availability Statement: Most of the data presented in this study are available in the Supplementary Materials. Other data are available on request from the corresponding author.

Acknowledgments: We thank Cândida Shinn, Isabel Lopes and Cátia Venâncio for their help with the setup of the mesocosms and with the measurement of the physicochemical parameters, and Manuel Nunes for providing the meteorological data.

Conflicts of Interest: The authors declare no conflict of interest.

References

1. IPCC. Summary for Policymakers. In *Climate Change 2021: The Physical Science Basis*; Masson-Delmotte, V., Zhai, P., Pirani, A., Connors, S.L., Péan, C., Berger, S., Caud, N., Chen, Y., Goldfarb, L., Gomis, M.I., Eds.; Contribution of Working Group I to the Sixth Assessment Report of the Intergovernmental Panel on Climate Change; Cambridge University Press: Cambridge, UK, 2021; in press.
2. Werner, A.D.; Bakker, M.; Post, V.E.A.; Vandenbohede, B.; Lu, C.; Ataie-Ashtiani, A.; Simmons, C.T.; Barry, D.A. Seawater intrusion processes, investigation and management: Recent advances and future challenges. *Adv. Water Resour.* **2012**, *51*, 3–26. [CrossRef]
3. Cañedo-Argüelles, M.; Kefford, B.; Schäfer, R. Salt in freshwaters: Causes, effects and prospects-introduction to the theme issue. *Phil. Trans. R. Soc. B.* **2018**, *374*, 20180002. [CrossRef]
4. Iglesias, M.C.-A. A review of recent advances and future challenges in freshwater salinization. *Limnetica* **2020**, *39*, 185–211. [CrossRef]
5. Kaushal, S.S.; Likens, G.E.; Pace, M.L.; Reimer, J.E.; Maas, C.M.; Galella, J.G.; Utz, R.M.; Duan, S.; Kryger, J.R.; Yaculak, A.M.; et al. Freshwater salinization syndrome: From emerging global problem to managing risks. *Biogeochemistry* **2021**, *154*, 255–292. [CrossRef]
6. Venâncio, C.; Ribeiro, R.; Lopes, I. Seawater intrusion: An appraisal of taxa at most risk and safe salinity levels. *Biol. Rev.* **2021**. [CrossRef] [PubMed]
7. Chandramohan, G.; Arivoli, S.; Venkatesan, P. Effect of salinity on the predatory performance of *Diplonychus rusticus* (Fabricius). *J. Environ. Biol.* **2008**, *29*, 287–290. [PubMed]
8. Venâncio, C.; Anselmo, E.; Soares, A.; Lopes, I. Does increased salinity influence the competitive outcome of two producer species? *Environ. Sci. Pollut. Res.* **2017**, *24*, 5888–5897. [CrossRef]
9. Venâncio, C.; Ribeiro, R.; Lopes, I. Active emigration from climate change-caused seawater intrusion into freshwater habitats. *Environ. Pollut.* **2020**, *258*, 113805. [CrossRef]
10. Gessner, M.O.; Chauvet, E. A case for using litter breakdown to assess functional stream integrity. *Ecol. Appl.* **2002**, *12*, 498–510. [CrossRef]
11. Roache, M.C.; Bailey, P.C.; Boon, P.I. Effects of salinity on the decay of the freshwater macrophyte, *Triglochin procerum*. *Aquat. Bot.* **2006**, *84*, 45–52. [CrossRef]
12. Berger, E.; Frör, O.; Schäfer, R.B. Salinity impacts on river ecosystem processes: A critical mini-review. *Phil. Trans. R. Soc. B* **2018**, *374*, 20180010. [CrossRef]
13. Júnior, E.S.A.; Martínez, A.; Gonçalves, A.L.; Canhoto, C. Combined effects of freshwater salinization and leaf traits on litter decomposition. *Hydrobiologia* **2020**, *847*, 3427–3435. [CrossRef]
14. Graça, M.A.S. The Role of Invertebrates on Leaf Litter Decomposition in Streams—A Review. *Int. Rev. Hydrobiol.* **2001**, *86*, 383–393. [CrossRef]
15. Kefford, B.J.; Dalton, A.; Palmer, C.G.; Nugegoda, D. The salinity tolerance of eggs and hatchlings of selected aquatic macroinvertebrates in south-east Australia and South Africa. *Hydrobiologia* **2004**, *517*, 179–192. [CrossRef]
16. Venâncio, C.; Castro, B.B.; Ribeiro, R.; Antunes, S.C.; Abrantes, N.; Soares, A.M.V.M.; Lopes, I. Sensitivity of freshwater species under single and multigenerational exposure to seawater intrusion. *Phil. Trans. R. Soc. B* **2018**, *374*, 20180252. [CrossRef]
17. Verberk, W.C.E.P.; Siepel, H.; Esselink, H. Life-history strategies in freshwater macroinvertebrates. *Freshw. Biol.* **2008**, *53*, 1722–1738. [CrossRef]
18. Venâncio, C.A.R. Salinization Effects on Coastal Terrestrial and Freshwater Ecosystems. Ph.D. Thesis, Universidade de Aveiro, Aveiro, Portugal, 2017; pp. 139–166.
19. Pérez, J.; Basaguren, A.; López-Rojo, N.; Tonin, A.M.; Correa-Araneda, F.; Boyero, L. The Role of Key Plant Species on Litter Decomposition in Streams: Alder as Experimental Model. In *The Ecology of Plant Litter Decomposition in Stream Ecosystems*; Swan, C.M., Boyero, L., Canhoto, C., Eds.; Springer: Berlin, Germany, 2021; pp. 143–161. [CrossRef]
20. Tachet, H.; Richoux, P.; Bournaud, M.; Usseglio-Polatera, P. *Invertébrés D'eau Douce–Systématique, Biologie, Ecologie*; CNRS Éditions: Paris, France, 2000.
21. Gutiérrez-Cánovas, C.; Sánchez-Fernández, D.; Cañedo-Argüelles, M.; Millán, A.; Velasco, J.; Acosta, R.; Fortuño, P.; Otero, N.; Soler, A.; Bonada, N. Do all roads lead to Rome? Exploring community trajectories in response to anthropogenic salinization and dilution of rivers. *Phil. Trans. R. Soc. B* **2019**, *374*, 20180009. [CrossRef]
22. Zar, J.H. *Biostatistical Analysis*; Prentice Hall, Inc.: Upper Saddle River, NJ, USA, 1996; ISBN 0-13-086398-X.
23. Sauer, F.G.; Bundschuh, M.; Zubrod, J.P.; Schäfer, R.B.; Thompson, K.; Kefford, B.J. Effects of salinity on leaf breakdown: Dryland salinity versus salinity from a coalmine. *Aquat. Toxicol.* **2016**, *177*, 425–432. [CrossRef]
24. Werba, J.A.; Stucy, A.L.; Peralta, A.L.; McCoy, M.W. Effects of diversity and coalescence of species assemblages on ecosystem function at the margins of an environmental shift. *Peer J.* **2020**, *8*, e8608. [CrossRef]
25. Hart, B.T.; Bailey, P.; Edwards, R.; Hortle, K.; James, K.; McMahon, A.; Meredith, C.; Swadling, K. Effects of salinity on river, stream and wetland ecosystems in Victoria, Australia. *Water Res.* **1990**, *24*, 1103–1117. [CrossRef]
26. Piscart, C.; Moreteau, J.-C.; Beisel, J.-N. Biodiversity and structure of macroinvertebrate communities along a small permanent salinity gradient (Meurthe River, France). *Hydrobiologia* **2005**, *551*, 227–236. [CrossRef]

27. Herbert, E.R.; Boon, P.; Burgin, A.J.; Neubauer, S.C.; Franklin, R.B.; Ardón, M.; Hopfensperger, K.N.; Lamers, L.P.M.; Gell, P. A global perspective on wetland salinization: Ecological consequences of a growing threat to freshwater wetlands. *Ecosphere* **2015**, *6*, 206. [CrossRef]
28. Rutherford, J.C.; Kefford, B.J. Effects of Salinity on Stream Ecosystems: Improving Models for Macroinvertebrates. In *CSIRO Land and Water Technical Report 22/05*; CSIRO Land and Water: Canberra, Australia, 2005.
29. Kefford, B.J.; Nugegoda, D. No evidence for a critical salinity threshold for growth and reproduction in the freshwater snail *Physa acuta*. *Environ. Pollut.* **2005**, *134*, 377–383. [CrossRef]
30. Kefford, B.J.; Papas, P.J.; Nugegoda, D. Relative salinity tolerance of macroinvertebrates from the Barwon River, Victoria, Australia. *Mar. Freshw. Res.* **2003**, *54*, 755–765. [CrossRef]
31. Dunlop, J.E.; Horrigan, N.; McGregor, G.; Kefford, B.J.; Choy, S.; Prasad, R. Effect of spatial variation on salinity tolerance of macroinvertebrates in Eastern Australia and implications for ecosystem protection trigger values. *Environ. Pollut.* **2008**, *151*, 621–630. [CrossRef]
32. Hassell, K.L.; Kefford, B.J.; Nugegoda, D. Sub-lethal and chronic salinity tolerances of three freshwater insects: *Cloeon* sp. and *Centroptilum* sp. (Ephemeroptera: Baetidae) and *Chironomus* sp. (Diptera: Chironomidae). *J. Exp. Biol.* **2006**, *209*, 4024–4032. [CrossRef]
33. Hilsenhoff, W.L. Diversity and Classification of Insects and Collembola. In *Ecology and Classification of North American Freshwater Invertebrates*, 2nd ed.; Thorp, J.H., Covich, A.P., Eds.; Academic Press: Cambridge, MA, USA, 2001; pp. 661–731. [CrossRef]
34. Kefford, B.J.; Fields, E.J.; Clay, C.; Nugegoda, D. Salinity tolerance of riverine microinvertebrates from the southern Murray-Darling Basin. *Mar. Freshw. Res.* **2007**, *58*, 1019–1031. [CrossRef]

diversity

MDPI

Article

Integrative Descriptions of Two New *Mesobiotus* Species (Tardigrada, Eutardigrada, Macrobiotidae) from Vietnam †

Daniel Stec [1,2]

1 Institute of Systematics and Evolution of Animals, Polish Academy of Sciences, Sławkowska 17, 31-016 Kraków, Poland; daniel.stec@isez.pan.krakow.pl
2 Department of Invertebrate Evolution, Institute of Zoology and Biomedical Research, Faculty of Biology, Jagiellonian University, Gronostajowa 9, 30-387 Kraków, Poland

† urn:lsid:zoobank.org:act:6ABF8C3D-FDD1-4DE0-88C8-54F49E21EFB4 (*Mesobiotus imperialis* sp. nov.); urn:lsid:zoobank.org:act:26C5E830-9A84-4019-B3A4-301339FE3220 (*Mesobiotus marmoreus* sp. nov).

Abstract: To date, 34 tardigrade taxa have been recorded from Vietnam and this includes only two macrobiotid species belonging to the genus *Mesobiotus*. In this paper, two additional species of this genus, one of the *M. harmsworthi* group and one of the *M. furciger* group, are reported and described as new for science (*Mesobiotus imperialis* **sp. nov.**, *Mesobiotus marmoreus* **sp. nov.**). Both descriptions have an integrative character providing detailed morphological and morphometric data collected by phase contrast and scanning electron microscopy that are linked to genetic data. The latter constitute DNA sequences of molecular markers that are commonly used in tardigrade taxonomy. The genus phylogeny is also provided, elucidating the phylogenetic position of the newly discovered taxa.

Keywords: *Mesobiotus harmsworthi* group; *Mesobiotus furciger* group; morphogroup; new species; tardigrades; taxonomy; phylogeny

Citation: Stec, D. Integrative Descriptions of Two New *Mesobiotus* Species (Tardigrada, Eutardigrada, Macrobiotidae) from Vietnam. *Diversity* **2021**, *13*, 605. https://doi.org/10.3390/d13110605

Academic Editor: Michael Wink

Received: 2 November 2021
Accepted: 16 November 2021
Published: 21 November 2021

Publisher's Note: MDPI stays neutral with regard to jurisdictional claims in published maps and institutional affiliations.

Copyright: © 2021 by the author. Licensee MDPI, Basel, Switzerland. This article is an open access article distributed under the terms and conditions of the Creative Commons Attribution (CC BY) license (https://creativecommons.org/licenses/by/4.0/).

1. Introduction

Tardigrades, also known as water bears, are a phylum of microscopic animals whose body size usually does not exceed 1 mm. These organisms are ubiquitous as they are found in marine, freshwater and various limno-terrestrial habitats all over the world [1]. To date, more than 1300 species have been formally described and, interestingly, a great majority of them were found in mosses and lichens [2–4].

The genus *Mesobiotus* was founded five years ago by Vecchi et al. [5] based on morphological distinctions from other genera within Macrobiotidae. The composition was further supported by phylogenetic analyses confirming the newly proposed taxon to be monophyletic [5] and, as such, was also recovered in the recent phylogeny of the family Macrobiotidae [6]. Now, the genus comprises 71 nominal species that are grouped into two unformal complexes, namely the *Mesobiotus harmsworthi* group and the *Mesobiotus furciger* group [4,7,8]. Although this morphological clustering of *Mesobiotus* taxa helps other researchers in taxonomic studies devoted to these macrobiotids, it has been demonstrated that the grouping does not reflect the phylogenetic relationship within the genus [6,7,9]. Out of 34 species representing he currently known tardigrade fauna of Vietnam [10–16], only two belong to the genus *Mesobiotus*. The first one is *Mesobiotus harmsworthi* (Murray, 1907) [17], the type species for the genus, as well as the recently discovered *Mesobiotus datanlanicus* Stec, 2019 [15]. Notably, according to the recent redescription the occurrence of *M. harmsworthi* in Vietnam, it should be treated with great caution [9].

In the present study, two new *Mesobiotus* species are described by means of an integrative taxonomy approach. Both descriptions comprise detailed morphological and morphometric data collected under phase contrast and scanning electron microscopy (PCM and SEM, respectively). Furthermore, phenotypic data DNA sequences of molecular markers used as a standard in tardigrade taxonomy are provided for each analysed species. Finally, the phylogenetic tree presenting the position of both new taxa is also presented.

2. Material and Methods

2.1. Sample Processing

Two moss samples containing new species were collected in Huế and in the Marble Mountains, south of Đà Nẵng city (Vietnam). The samples were collected by Daniel Stec and Krzysztof Miler in August 2018 from tree bark and a stone walkway, respectively. The samples were examined for terrestrial tardigrades using standard methods (e.g., Stec et al. [18]). A total of 75 and 56 animals as well as 55 and 13 eggs of the two new species were extracted from both samples, respectively. In order to perform integrative taxonomic descriptions, the isolated animals and eggs were split into three groups for specific analyses: Morphological analysis with phase contrast light microscopy, morphological analysis with scanning electron microscopy and DNA sequencing (for details please see sections "Material examined" provided below for each description).

2.2. Microscopy and Imaging

Specimens for light microscopy were mounted on microscope slides in a small drop of Hoyer's medium and secured with a cover slip, following the protocol by Morek et al. [19]. Slides were then dried for five to seven days at 60 °C. Dried slides were sealed with a transparent nail polish and examined under an *Olympus BX53* light microscope with phase contrast (PCM), as well as with an *Olympus DP74* digital camera. Immediately after mounting the specimens in the medium, slides were checked under PCM for the presence of males and females in the studied population, as the spermatozoa in testis and *vas deferens* are visible only for several hours after mounting [20,21]. In order to obtain clean eggs for SEM, eggs were processed according to the protocol by Stec et al. [18]. In short, eggs were first subjected to a water/ethanol and an ethanol/acetone series, then to CO_2 critical-point drying and finally sputter-coated with a thin layer of gold. Specimens were examined under high vacuum in a *Versa 3D DualBeam* Scanning Electron Microscope at the ATOMIN facility of the Jagiellonian University, Kraków, Poland. All figures were assembled in *Corel Photo-Paint X6*. For structures that could not be satisfactorily focused on in a single photograph, a stack of 2–6 images were taken with an equidistance of ca. 0.2 μm and assembled manually into a single deep-focus image.

2.3. Morphometrics and Morphological Nomenclature

All measurements are given in micrometres (μm). Sample size was adjusted following recommendations by Stec et al. [22]. Structures were measured only if their orientation was suitable. Body length was measured from the anterior extremity to the end of the body, excluding the hind legs. The buccal apparatus and claws were classified according to Pilato and Binda [23] and Vecchi et al. [5], respectively. The terminology used to describe the oral cavity armature and the egg-shell morphology follows Michalczyk and Kaczmarek [24] and Kaczmarek and Michalczyk [25]. The macroplacoid length sequence is given according to Kaczmarek et al. [26] whereas morphological states of cuticular bars on legs follow Kiosya et al. [27]. The buccal tube length and the level of the stylet support insertion point were measured according to Pilato [28]. The *pt* index is the ratio of the length of a given structure to the length of the buccal tube expressed as a percentage (Pilato 1981). All other measurements and nomenclature follow Kaczmarek and Michalczyk [25]. The buccal tube width was measured as the external and internal diameter at the level of the stylet support insertion point. The lengths of the claw branches were measured from the base of the claw (i.e., excluding the lunula) to the top of the branch, including accessory points. The distance between egg processes was measured as the shortest distance between the base edges of the two closest processes. Morphometric data were handled using the "Parachela" ver. 1.8 template available from the Tardigrada Register [29] and are given in Supplementary Materials (SM.1 and SM.2). T-test comparisons of morphometric characters of one of the new species and *Mesobiotus philippinicus* Mapalo, Stec, Mirano-Boscos & Michalczyk, 2016 [30] were conducted using the statistical programming language R [31]. Since multiple testing inflates the Type I error rate, the Benjamini–Hochberg correction to the α-level was applied [32] independently to each of the three sets of *t*-tests (absolute and relative animal measurements as well as egg measurements). Results of the *t*-tests are given in Supplementary Materials (SM.3). The taxonomic keys for the genus *Mesobiotus* by Kaczmarek et al. [7] and Tumanov [8] were used to determine whether the isolated species had previously been described. The tardigrade taxonomy follows Stec et al. [6].

2.4. DNA Sequencing

The DNA was extracted from individual animals following a *Chelex*® *100* resin (*Bio-Rad*) extraction method by Casquet et al. [33] with modifications described in detail in Stec et al. [34]. Four DNA fragments differing in mutation rates were sequenced. Namely, the small ribosome subunit (18S rRNA, nDNA), the large ribosome subunit (28S rRNA, nDNA), the internal transcribed spacer (ITS-2, nDNA), and the cytochrome oxidase subunit I (COI, mtDNA). All fragments were amplified and sequenced according to the protocols described in Stec et al. [34]; primers are listed in Table 1. Sequencing products were read with the *ABI 3130xl* sequencer at the Molecular Ecology Lab, Institute of Environmental Sciences of the Jagiellonian University, Kraków, Poland. Sequences were processed in *BioEdit* ver. 7.2.5 [35] and submitted to GenBank. Prior to submission, all obtained COI sequences were translated into protein sequences in *MEGA7* version 7.0 [36] to check against pseudogenes.

Table 1. Primers with their original references used for amplification of the four DNA fragments sequenced in the study.

DNA Marker	Primer Name	Primer Direction	Primer Sequence (5′-3′)	Primer Source
18S rRNA	18S_Tar_Ff1	forward	AGGCGAAACCGCGAATGGCTC	[37]
	18S_Tar_Rr1	reverse	GCCGCAGGCTCCACTCCTGG	
28S rRNA	28S_Eutar_F	forward	ACCCGCTGAACTTAAGCATAT	[38]
	28SR0990	reverse	CCTTGGTCCGTGTTTCAAGAC	[39]
ITS-2	ITS2_Eutar_Ff	forward	CGTAACGTGAATTGCAGGAC	[40]
	ITS2_Eutar_Rr	reverse	TCCTCCGCTTATTGATATGC	
COI	LCO1490-JJ	forward	CHACWAAYCATAAAGATATYGG	[41]
	HCO2198-JJ	reverse	AWACTTCVGGRTGVCCAAARAATCA	

2.5. Phylogenetic Analysis and Genetic Comparisons

To establish phyletic positions of both new species, a phylogenetic tree was constructed using the dataset from Kaczmarek et al. [7] with the addition of sequences obtained in this study as well as sequences that were published to date (Table 2). DNA sequences of *Macrobiotus kamilae* Coughlan & Stec, 2019 [20] and *Macrobiotus hannae* Nowak & Stec, 2018 [42] were used as the outgroup. The sequences were aligned using the AUTO method (for COI and ITS-2) and the Q-INS-I method (for ribosomal markers: 18S rRNA and 28S rRNA) of MAFFT version 7 [43,44] and manually checked against non-conservative alignments in BioEdit. Then, the aligned sequences were trimmed to 1016 (18S rRNA), 811 (28S rRNA), 554 (ITS-2), and 658 (COI) bp and concatenated using SequenceMatrix [45]. Before partitioning, the concatenated alignment was divided into 6 data blocks constituting three separate blocks of ribosomal markers and three separate blocks of three codon positions in the COI dataset. Using PartitionFinder [46] under the Akaike Information Criterion (AIC), the best scheme of partitioning and substitution models were chosen for posterior phylogenetic analysis (SM.04). Bayesian inference (BI) marginal posterior probabilities were calculated for the concatenated (18S rRNA + 28S rRNA + ITS-2 + COI) dataset using MrBayes v3.2 [47]. Random starting trees were used, and the analysis was run for 10 million generations, sampling the Markov chain every 1000 generations. An average standard deviation of split frequencies of <0.01 was used as a guide to ensure the two independent analyses had converged. The program Tracer v1.6 [48] was then used to ensure Markov chains had reached stationarity, and to determine the correct 'burn-in' for the analysis, which was the first 10% of generations. The ESS values were greater than 200 and the consensus tree was obtained after summarising the resulting topologies and discarding the 'burn-in'. The consensus tree was viewed and visualised by FigTree v.1.4.3 available from http://tree.bio.ed.ac.uk/software/figtree (accessed on 10 August 2018). Uncorrected pairwise distances were calculated using *MEGA7* and are given in Supplementary Materials (SM.5).

Table 2. Sequences used for phylogenetic analysis and genetic comparisons (see Material and Methods section for details). Bold font indicates sequences obtained in this study.

Species	18S rRNA	28S rRNA	ITS-2	COI	Source
M. ethiopicus Stec & Kristensen, 2017 [49]	MF678793	MF678792	MN122776	MF678794	[15,49]
M. datanlanicus Stec, 2019 [15]	MK584659	MK584658	MK584657	MK578905	[15]
M. dilimanensis Itang et al., 2020 [50]	MN257048	MN257049	MN257050	MN257047	[50]
M. philippinicus Mapalo et al., 2016 [30]	KX129793	KX129794	KX129795	KX129796	[30]
M. insanis Mapalo et al., 2017 [51]	MF441488	MF441489	MF441490	MF441491	[51]
M. hilariae Vecchi et al., 2016 [5]	KT226070			KT226108	[5]
M. radiatus (Pilato et al., 1991) [52]	MH197153	MH197152	MH197267 MH197268	MH195147 MH195148	[53]
M. romani Roszkowska et al., 2018 [54]	MH197158	MH197151	MH197150	MH195149	[54]
M. harmsworthi (Murray, 1907) [17]	MH197146	MH197264	MH197154	MH195150 MH195151	[9]
M. occultatus Kaczmarek et al., 2018 [9]	MH197147		MH197155	MH195152	[9]
M. furciger group species NO	MH197148	MH197265	MH197156	MH195153	[9]
M. harmsworthi group species RU	MH197149	MH197266	MH197157	MH195154	[9]
M. furciger (Murray, 1907) [55]				JX865306 JX865308 JX865314	[56]
M. fiedleri Kaczmarek et al., 2020 [7]	MH681585	MH681693	MH681724	MH676056	[7]
"M. harmsworthi"				GU113140	Li and Xiao (unpublished)
M. anastasiae Tumanov, 2020 [8]	MT903468	MT903612	MT903470	MT904513	[8]
M. skoracki Kaczmarek et al., 2018 [9]		MW680636		MW656257	[57]
***M. imperialis*sp. nov.**	**OL257854** **OL257855**	**OL257866** **OL257867**		**OL311514** **OL311515**	**this study** **this study**
***M. marmoreus*sp. nov.**	**OL257856** **OL257857** **OL257858**	**OL257868** **OL257869** **OL257870**	**OL257861** **OL257862** **OL257863**	**OL311516** **OL311517** **OL311518**	**this study** **this study** **this study**
M. cf. *barabanovi*	MN310392	MN310388	MN310390	MN313170	[7]
Macrobiotus kamilae Coughlan & Stec, 2019 [20]	MK737070	MK737064	MK737067	MK737920 MK737921	[20]
Macrobiotus hannae Nowak & Stec, 2018 [42]	MH063922	MH063924	MH063923	MH057764	[42]

3. Results

3.1. Taxonomic Account of the New Species

Phylum: Tardigrada Doyère, 1840 [58].
Class: Eutardigrada Richters, 1926 [59].
Order: Parachela Schuster, Nelson, Grigarick & Christenberry, 1980 [60].
Superfamily: Macrobiotoidea Thulin, 1928 [61] (in [62]).
Family: Macrobiotidae Thulin, 1928 [61].
Genus: *Mesobiotus* Vecchi, Cesari, Bertolani, Jönsson, Rebecchi & Guidetti, 2016 [5].

3.2. Description of the New Species

***Mesobiotus imperialis* sp. nov.**
ZooBank: urn:lsid:zoobank.org:act:6ABF8C3D-FDD1-4DE0-88C8-54F49E21EFB4

(Tables 3 and 4, Figures 1–6).

Table 3. Measurements [in µm] and *pt* values of selected morphological structures of animals of *Mesobiotus imperialis* **sp. nov.**; specimens mounted in Hoyer's medium; N—number of specimen/structures measured, RANGE refers to the smallest and the largest structure among all measured specimens; SD—standard deviation.

CHARACTER	N	RANGE						Mean		SD		Holotype	
		µm			pt			µm	pt	µm	pt	µm	pt
Body length	20	313	–	539	*895*	–	*1219*	389	*1012*	51	*75*	436	*1053*
Buccal tube													
Buccal tube length	20	30.5	–	44.2		–		38.4	–	3.4	–	41.4	–
Stylet support insertion point	20	22.9	–	34.1	*75.0*	–	*77.3*	29.2	*76.0*	2.7	*0.8*	31.2	*75.4*
Buccal tube external width	20	5.2	–	7.6	*15.7*	–	*18.1*	6.5	*16.8*	0.6	*0.6*	7.0	*16.9*
Buccal tube internal width	20	4.0	–	5.7	*10.8*	–	*13.8*	4.9	*12.6*	0.5	*0.7*	5.7	*13.8*
Ventral lamina length	20	19.9	–	27.0	*56.2*	–	*65.2*	23.2	*60.4*	1.8	*2.2*	23.6	*57.0*
Placoid lengths													
Macroplacoid 1	20	3.8	–	6.0	*11.7*	–	*14.4*	5.1	*13.4*	0.6	*0.9*	5.6	*13.5*
Macroplacoid 2	20	2.9	–	4.5	*7.7*	–	*11.3*	3.7	*9.7*	0.5	*0.9*	4.2	*10.1*
Macroplacoid 3	20	3.1	–	5.8	*10.2*	–	*13.1*	4.5	*11.6*	0.6	*0.8*	4.7	*11.4*
Microplacoid	20	3.2	–	5.2	*8.3*	–	*12.7*	4.1	*10.7*	0.5	*1.1*	4.2	*10.1*
Macroplacoid row	20	11.9	–	19.4	*38.6*	–	*43.9*	16.0	*41.5*	1.8	*1.6*	17.9	*43.2*
Placoid row	20	16.0	–	25.7	*52.0*	–	*58.5*	21.3	*55.4*	2.4	*2.2*	23.8	*57.5*
Claw I heights													
External primary branch	20	7.5	–	10.6	*20.7*	–	*25.6*	8.7	*22.8*	0.8	*1.4*	9.6	*23.2*
External secondary branch	18	6.4	–	8.5	*17.2*	–	*22.3*	7.3	*19.0*	0.7	*1.5*	7.9	*19.1*
Internal primary branch	20	6.7	–	10.0	*19.7*	–	*22.6*	8.2	*21.2*	0.9	*0.9*	9.2	*22.2*
Internal secondary branch	18	5.1	–	8.0	*15.8*	–	*19.2*	6.7	*17.3*	0.7	*1.1*	7.4	*17.9*
Claw II heights													
External primary branch	20	7.3	–	11.1	*22.3*	–	*26.0*	9.2	*24.0*	0.9	*1.0*	10.3	*24.9*
External secondary branch	18	6.6	–	8.8	*17.6*	–	*22.9*	7.6	*19.9*	0.7	*1.5*	7.3	*17.6*
Internal primary branch	20	7.2	–	9.6	*19.2*	–	*23.6*	8.1	*21.2*	0.7	*1.2*	8.6	*20.8*
Internal secondary branch	13	6.1	–	8.0	*16.4*	–	*20.3*	7.1	*18.4*	0.6	*1.4*	7.1	*17.1*
Claw III heights													
External primary branch	20	7.8	–	11.1	*21.4*	–	*27.9*	9.3	*24.2*	0.9	*1.6*	10.4	*25.1*
External secondary branch	18	6.4	–	8.8	*17.2*	–	*23.1*	7.6	*19.8*	0.7	*1.6*	8.8	*21.3*
Internal primary branch	19	6.0	–	9.7	*16.2*	–	*23.3*	8.0	*20.9*	1.0	*1.6*	8.5	*20.5*
Internal secondary branch	15	5.9	–	8.1	*16.9*	–	*20.8*	7.1	*18.3*	0.7	*1.1*	8.0	*19.3*
Claw IV heights													
Anterior primary branch	20	8.0	–	11.2	*23.4*	–	*28.2*	9.7	*25.3*	0.9	*1.2*	10.8	*26.1*
Anterior secondary branch	18	6.5	–	8.8	*17.9*	–	*22.1*	7.8	*20.3*	0.7	*1.1*	8.5	*20.5*
Posterior primary branch	20	8.5	–	12.1	*25.0*	–	*30.1*	10.3	*27.0*	0.9	*1.2*	11.0	*26.6*
Posterior secondary branch	9	7.7	–	9.3	*20.2*	–	*22.1*	8.3	*21.2*	0.5	*0.7*	?	*?*

Table 4. Measurements [in µm] of the eggs of *Mesobiotus imperialis* **sp. nov.**; eggs mounted in Hoyer's medium; process base/height ratio is expressed as percentage; N—number of eggs/structures measured, RANGE refers to the smallest and the largest structure among all measured specimens; SD—standard deviation.

Character	N	RANGE			Mean	SD
Egg bare diameter	20	53.9	–	70.2	62.8	4.0
Egg full diameter	20	74.7	–	94.6	85.7	4.9
Process height	60	8.3	–	15.3	11.9	1.3
Process base width	60	6.9	–	12.5	10.0	1.1
Process base/height ratio	60	65%	–	116%	85%	11%
Inter-process distance	60	1.7	–	3.9	2.7	0.5
Number of processes on the egg circumference	20	15	–	18	16.2	0.8

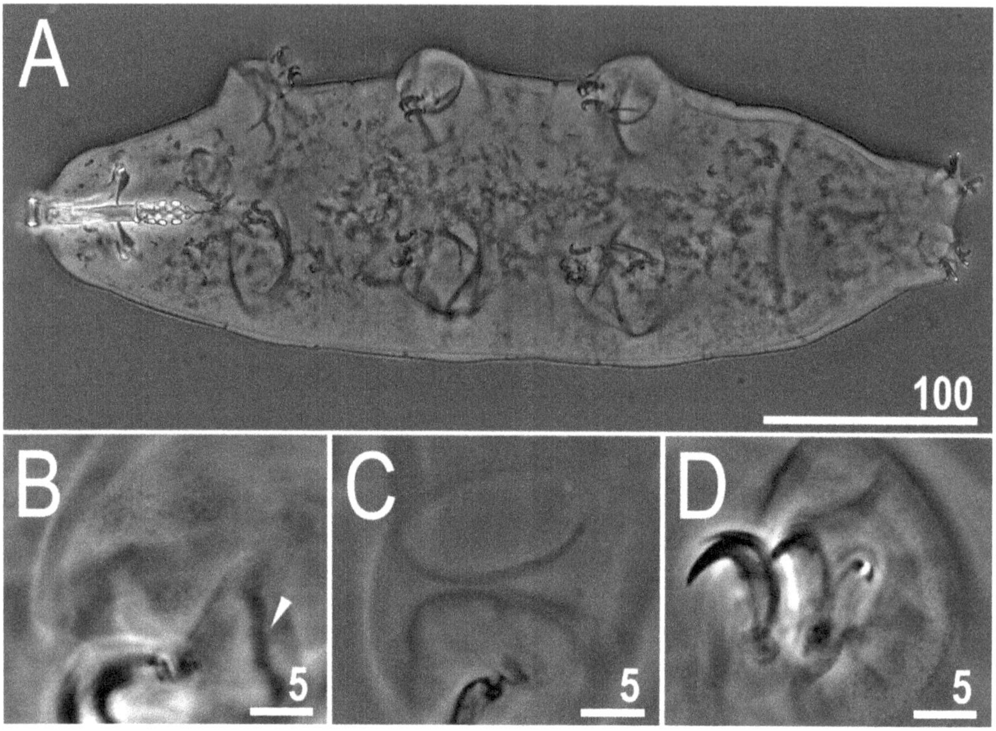

Figure 1. *Mesobiotus imperialis* **sp. nov.**—PCM image of habitus and leg's cuticle morphology: (**A**)—dorso-ventral projection (holotype); (**B**)—granulation on the external surface of leg II (holotype); (**C**)—a pulvinus-like cuticular bulge on the internal surface of leg III (paratype); (**D**)—granulation on dorsal and lateral surface of leg IV (paratype). Filled flat arrowhead indicates a single continuous cuticular bar above the claws. Scale bar in μm.

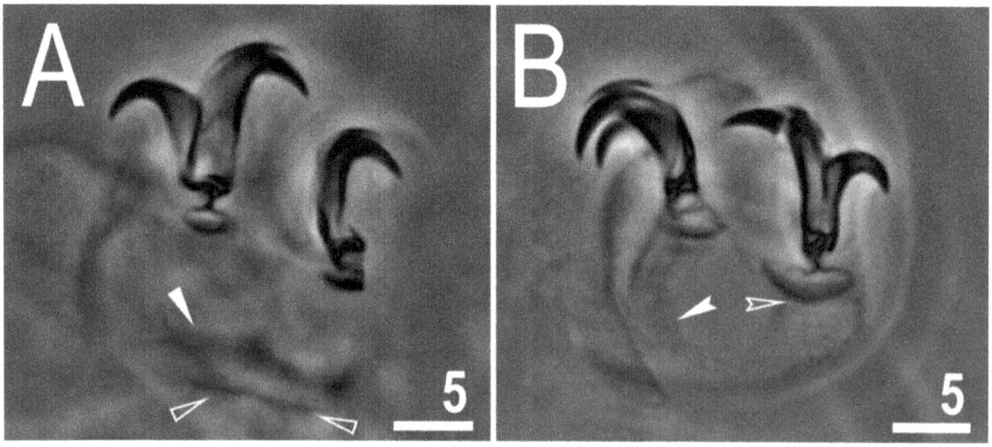

Figure 2. *Mesobiotus imperialis* **sp. nov.**—PCM images of claws: (**A**)—claw III with smooth lunulae (holotype); (**B**)—claw IV with smooth lunulae (paratype). Filled flat arrowhead indicates a single continuous cuticular bar above the claws, empty flat arrowheads indicate paired muscles attachments, filled indented arrowhead indicates horseshoe structure connecting the anterior and the posterior claw, empty indented arrowheads indicate faint dentation in lunula IV. Scale bars in μm.

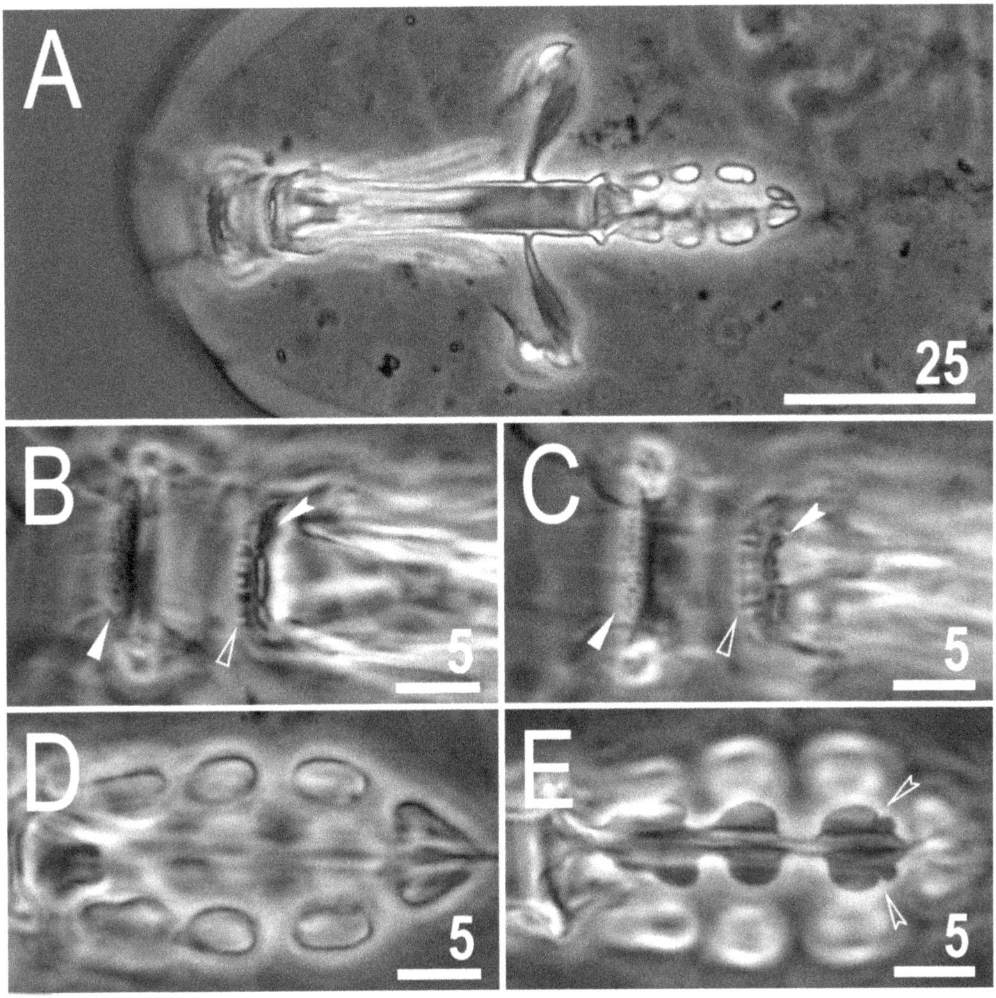

Figure 3. *Mesobiotus imperialis* **sp. nov.**—PCM images of the buccal apparatus: (**A**)—an entire buccal apparatus (paratype); (**B**,**C**)—the oral cavity armature, dorsal and ventral teeth, respectively (paratype); (**D**,**E**)—placoid morphology, dorsal and ventral placoids, respectively (holotype). Filled flat arrowheads indicate the first band of teeth, empty flat arrowheads indicate the second band of teeth, filled indented arrowheads indicate the third band of teeth, empty indented arrowheads indicate subterminal constrictions in the third macroplacoid. Scale bars in μm.

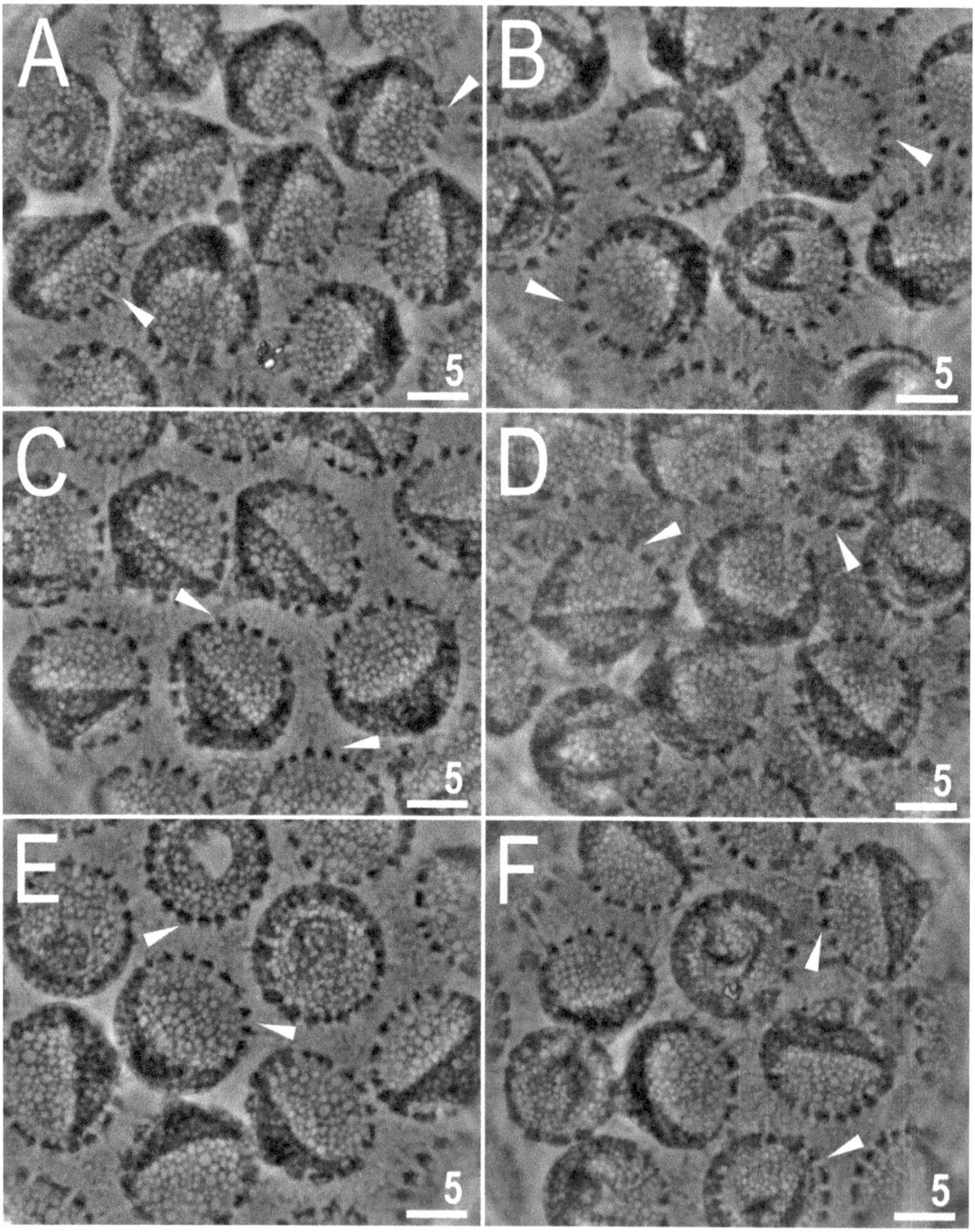

Figure 4. *Mesobiotus imperialis* **sp. nov.**—PCM images of the egg surface under ×1000 magnification. Arrowheads indicate crowns of strong thickenings around the process bases. Scale bars in μm.

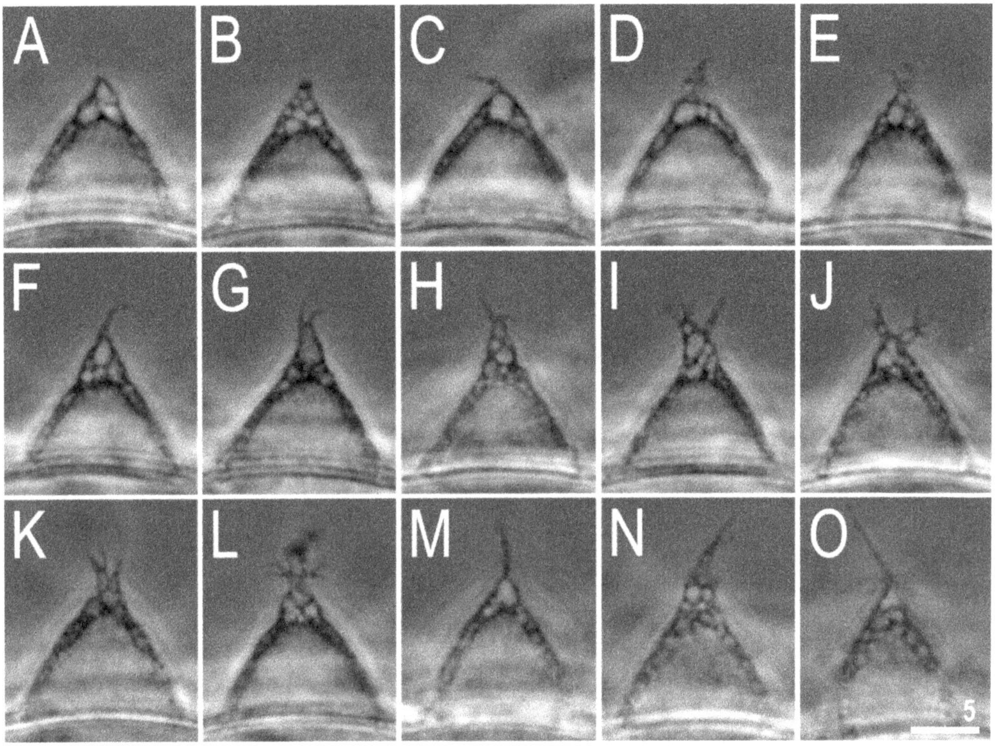

Figure 5. *Mesobiotus imperialis* **sp. nov.**—PCM images of the egg processes midsections under ×1000 magnification. Scale bar in μm.

3.2.1. Material Examined

In total, 73 animals, 50 eggs mounted on microscope slides in Hoyer's medium (some of the eggs were embryonated), 5 eggs fixed on a SEM stub (19.14) and 2 specimens were processed for DNA sequencing.

3.2.2. Type Locality

16°28′04″ N, 107°34′37″ E; 6 m asl: Vietnam, Huế, Imperial City, Kiến Trung Palace (Điện Kiến Trung), bark of a dying tree near a pat walk, coll. Daniel Stec and Krzysztof Miler, August 2018.

3.2.3. Etymology

The species is named after the place where it was discovered. Namely, it is Imperial City, a walled enclosure within the citadel of the city of Huế and the former imperial capital of Vietnam.

3.2.4. Type Depositories

The holotype with 6 paratypes (slide VN.061.03) and 47 paratypes (slides: VN.061.*, where the asterisk can be substituted by any of the following numbers: 01–02, 04–09) and 26 eggs (slides: VN.061.*: 12–15) are deposited at the Institute of Systematics and Evolution of Animals, Polish Academy of Sciences, Sławkowska 17, 31-016, Kraków, Poland;

Nineteen paratypes (slides: VN.061.*: 10–11), 24 eggs (slides: VN.061.*: 16–17) and SEM stub: 19.14 are deposited at the Institute of Zoology and Biomedical Research, Jagiellonian University, Gronostajowa 9, 30-387, Kraków, Poland.

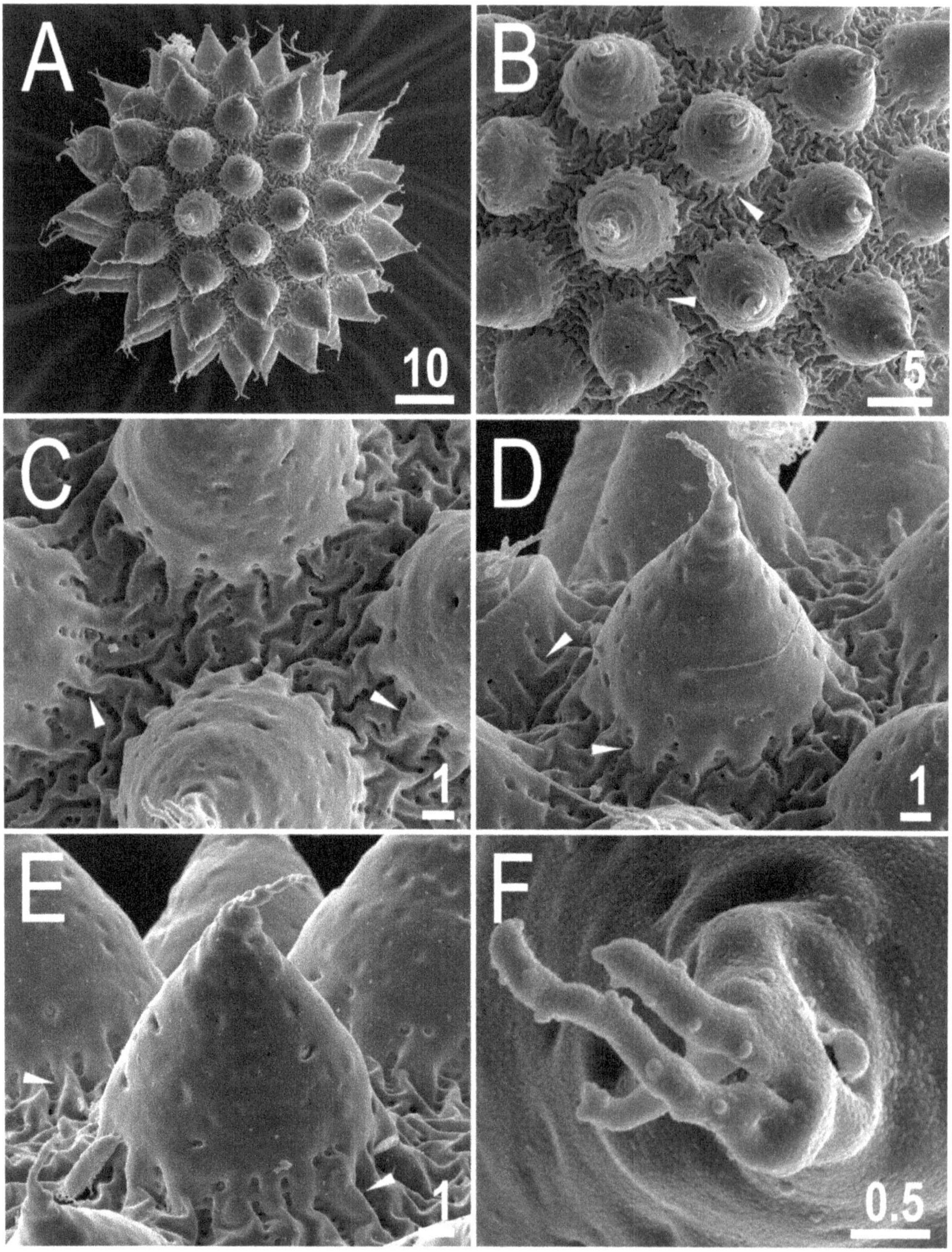

Figure 6. *Mesobiotus imperialis* **sp. nov.**—SEM images of eggs: (**A**)—entire view of the egg; (**B**,**C**)—details of the egg surface between processes; (**D**,**E**)—egg processes; (**F**)—top part of the processes divided into several flexible filaments covered with fine granulation. Arrowheads indicate strong thickenings around the process bases. Scale bars in μm.

3.2.5. Animals

The body is almost transparent in small specimens and whitish in adults; after fixation in Hoyer's medium, the body is transparent (Figure 1A). Eyes are present in alive animals and dissolved by Hoyer's medium in about 50% of all mounted specimens. The body cuticle is smooth, i.e., without pores or sculpturing. Fine granulation is present on the external surface of all legs I–III (Figure 1B) as well as on the lateral and dorsal surfaces of legs IV (Figure 1D). A cuticular bulge/fold, resembling a pulvinus, is present on the internal surface of legs I–III (Figure 1C). Claws of the *Mesobiotus* type were observed, with a peduncle connecting the claw to the lunula, a basal septum and well-developed accessory points situated in parallel to the primary branch (Figure 2A,B). Lunulae under claws I–III are smooth (Figure 2A) and those under claws IV are slightly dentate (Figure 2B; a character visible in about 50% of specimens mounted in Hoyer's medium). A single continuous cuticular bar and double muscle attachments are present above claws I–III (Figures 1B and 2A), whereas a horseshoe-shaped structure connects the anterior and posterior lunulae on claws IV (Figure 2B).

The mouth is antero-ventral. The Bucco-pharyngeal apparatus is of the *Macrobiotus* type, with the ventral lamina and ten small peribuccal lamellae (Figure 3A). The oral cavity armature is well developed and is composed of three bands of teeth (Figure 3B,C). The first band of teeth is composed of numerous small granules arranged in several discrete rows situated anteriorly in the oral cavity, just behind the bases of the peribuccal lamellae (Figure 3B,C). The second band of teeth is situated between the ring fold and the third band of teeth and is composed of ridges parallel to the main axis of the buccal tube that are larger than those in the first band (Figure 3B,C). The teeth of the third band are located within the posterior portion of the oral cavity, between the second band of teeth and the buccal tube opening (Figure 3B,C). The third band of teeth is discontinuous and divided into dorsal and ventral portions. Under PCM, dorsal teeth are visible as two lateral and one median transverse ridges/crests (Figure 3B) whereas ventral teeth consist of two lateral transverse ridges/crests, between which two to four (usually three) roundish and separated ventro-median teeth are present (Figure 3C). The pharyngeal bulb is ovoid (Figure 3A), with triangular apophyses, three rod-shaped macroplacoids and a large, elongated drop-shaped microplacoid placed close to the third macroplacoid (Figure 3D,E). The macroplacoid length sequence is $2 < 3 < 1$. The first macroplacoid is anteriorly narrowed and the third has a clearly defined sub-terminal constriction (Figure 3E). Measurements and statistics are presented in Table 3.

3.2.6. Eggs

Eggs are white, laid free, spherical in shape and equipped with conical processes (Figures 4–6). In PCM, the egg surface between processes seems to be rough with both dark and faintly light refracting dots (Figure 4), whereas in SEM, the surface is clearly wrinkled, with wrinkles radiating out from the process bases but not forming a connective network (Figure 6A–E). Small pores (up to 0.3 μm) are scattered across the inter-process surface with their lumen often being covered by a reticulate internal structure that seems to be a remnant of the reduced labyrinthine layer. The pores are clearly visible in SEM (Figure 6A–E), but under PCM, they are most probably seen as the mentioned faintly light refracting dots (Figure 4). The bases of egg processes are surrounded by crowns of strong thickenings that are evident in PCM as well as SEM (Figures 4 and 6A–E). The egg processes are evenly spaced, with a flexible upper portion often equipped with shorter flexible filaments (Figure 5). This flexible portion of the processes seems to be fragile and susceptible to fracture (Figure 5A–C). Often, in the upper portion of the egg processes, below the flexible part, a bubble-like structure is present and visible in the process midsection (Figure 5). The labyrinthine layer is visible under PCM as a reticulum in process walls, with varying mesh sizes uniformly distributed within the process walls (Figure 4). In SEM, the process walls are smooth with unevenly distributed depressions and faint tubercles and occasionally also pores often with closed lumen (Figure 6A–E). The top flexible portions of egg processes are irregularly covered with small granules that are visible only in SEM (Figure 6D,F). Measurements and statistics are presented in Table 4.

3.2.7. Reproduction

The examination of all individuals, freshly mounted in Hoyer's medium, under PCM did not reveal any testis or spermathecae filled with spermatozoa. Thus, it is most likely that the new species is parthenogenetic.

3.2.8. DNA sequences

The obtained sequences for three molecular markers analysed in this study were of good quality and were represented by single haplotypes. However, several attempts to amplify the ITS-2 marker for the new species failed, preventing me from obtaining these sequences for the new species.

The **18S rRNA** sequences (GenBank: OL257854-5), 1008 bp long.
The **28S rRNA** sequences (GenBank: OL257866-7), 774 bp long.
The **COI** sequences (GenBank: OL311514-5), 658 bp long.

3.3. Description of the New Species

Mesobiotus marmoreus **sp. nov.**
ZooBank: urn:lsid:zoobank.org:act:26C5E830-9A84-4019-B3A4-301339FE3220
(Tables 5 and 6, Figures 7–11).

Table 5. Measurements [in µm] and *pt* values of selected morphological structures of animals of *Mesobiotus marmoreus* **sp. nov.**; specimens mounted in Hoyer's medium; N—number of specimen/structures measured, RANGE refers to the smallest and the largest structure among all measured specimens; SD—standard deviation.

Character	N	RANGE						Mean		SD		Holotype	
		µm			*pt*			µm	*pt*	µm	*pt*	µm	*pt*
Body length	20	234	–	372	*883*	–	*1042*	320	*970*	32	*48*	308	*982*
Buccal tube													
Buccal tube length	20	26.5	–	37.1		–		33.0	–	2.7	–	31.3	–
Stylet support insertion point	20	20.2	–	28.7	*76.2*	–	*77.8*	25.5	*77.1*	2.2	*0.5*	24.0	*76.7*
Buccal tube external width	20	4.5	–	6.4	*16.2*	–	*18.4*	5.6	*17.0*	0.5	*0.6*	5.3	*16.9*
Buccal tube internal width	20	3.2	–	4.9	*12.0*	–	*14.6*	4.2	*12.7*	0.4	*0.6*	3.9	*12.5*
Ventral lamina length	20	15.9	–	22.5	*57.9*	–	*62.5*	19.9	*60.2*	1.6	*1.1*	19.2	*61.3*
Placoid lengths													
Macroplacoid 1	20	3.6	–	6.1	*12.9*	–	*16.4*	4.8	*14.6*	0.6	*0.9*	4.2	*13.4*
Macroplacoid 2	20	2.3	–	3.8	*8.5*	–	*10.8*	3.2	*9.6*	0.4	*0.6*	3.0	*9.6*
Macroplacoid 3	20	3.2	–	5.0	*10.6*	–	*13.6*	4.0	*12.0*	0.6	*1.1*	3.4	*10.9*
Microplacoid	20	3.0	–	4.7	*9.6*	–	*12.7*	3.6	*10.9*	0.4	*0.8*	3.5	*11.2*
Macroplacoid row	20	11.1	–	16.9	*38.9*	–	*45.6*	14.0	*42.4*	1.5	*1.6*	12.6	*40.3*
Placoid row	20	15.0	–	22.7	*53.7*	–	*61.2*	18.8	*56.7*	1.9	*1.8*	17.3	*55.3*
Claw I heights													
External primary branch	20	6.4	–	8.6	*19.5*	–	*25.4*	7.6	*23.0*	0.6	*1.5*	7.7	*24.6*
External secondary branch	16	5.1	–	7.6	*15.5*	–	*20.9*	6.3	*18.9*	0.6	*1.4*	6.2	*19.8*
Internal primary branch	20	6.2	–	8.4	*18.9*	–	*24.2*	7.3	*22.1*	0.6	*1.4*	6.8	*21.7*
Internal secondary branch	15	5.0	–	7.1	*14.3*	–	*19.6*	6.1	*18.0*	0.6	*1.3*	5.6	*17.9*
Claw II heights													
External primary branch	20	7.0	–	8.8	*21.0*	–	*26.6*	8.0	*24.3*	0.5	*1.5*	7.9	*25.2*
External secondary branch	18	5.3	–	7.6	*16.2*	–	*21.7*	6.5	*19.5*	0.5	*1.3*	6.8	*21.7*
Internal primary branch	20	6.2	–	8.9	*19.2*	–	*24.5*	7.3	*22.1*	0.6	*1.5*	7.1	*22.7*
Internal secondary branch	19	5.0	–	7.0	*16.3*	–	*20.1*	6.1	*18.6*	0.5	*1.2*	6.3	*20.1*
Claw III heights													
External primary branch	20	7.5	–	9.8	*21.5*	–	*28.3*	8.2	*24.9*	0.6	*1.7*	8.2	*26.2*
External secondary branch	15	5.8	–	7.6	*16.6*	–	*22.6*	6.6	*20.1*	0.6	*1.4*	6.8	*21.7*
Internal primary branch	20	6.4	–	8.8	*19.8*	–	*25.7*	7.5	*22.6*	0.6	*1.6*	7.0	*22.4*
Internal secondary branch	17	5.3	–	7.4	*16.2*	–	*20.9*	6.1	*18.7*	0.6	*1.3*	6.2	*19.8*

Table 5. *Cont.*

Character	N	RANGE						Mean		SD		Holotype	
		µm			*pt*			µm	*pt*	µm	*pt*	µm	*pt*
Claw IV heights													
Anterior primary branch	19	7.5	–	9.5	*21.5*	–	*29.1*	8.3	*25.1*	0.6	*1.8*	8.1	*25.9*
Anterior secondary branch	17	5.9	–	7.5	*17.5*	–	*22.5*	6.7	*20.5*	0.5	*1.2*	6.1	*19.5*
Posterior primary branch	18	8.0	–	10.1	*22.6*	–	*30.6*	9.0	*27.2*	0.6	*2.0*	8.5	*27.2*
Posterior secondary branch	12	6.2	–	7.9	*19.2*	–	*22.8*	7.2	*21.4*	0.5	*1.0*	7.0	*22.4*

Table 6. Measurements [in µm] of the eggs of *Mesobiotus marmoreus* **sp. nov.**; eggs mounted in Hoyer's medium; process base/height ratio is expressed as percentage; N—number of eggs/structures measured, RANGE refers to the smallest and the largest structure among all measured specimens; SD—standard deviation.

Character	N	RANGE			Mean	SD
Egg bare diameter	7	63.4	–	69.6	67.0	2.4
Egg full diameter	7	77.9	–	82.2	80.0	1.4
Process height	27	5.6	–	8.8	6.9	0.9
Process base width	27	3.4	–	6.5	5.0	0.7
Process base/height ratio	27	59%	–	94%	73%	10%
Inter-process distance	27	1.5	–	3.3	2.3	0.4
Number of processes on the egg circumference	7	26	–	30	28.3	1.5

3.3.1. Material Examined

In total, 53 animals, 9 eggs mounted on microscope slides in Hoyer's medium (some of the eggs were embryonated), 4 eggs fixed on an SEM stub (18.09) and 3 specimens were processed for DNA sequencing.

3.3.2. Type Locality

16°00′14″ N, 108°15′48″ E; 66 m asl: Vietnam, The Marble Mountains, south of Đà Nẵng, stone walkway, coll. Daniel Stec and Krzysztof Miler, August 2018.

3.3.3. Etymology

The species is named after the place where it was discovered, namely, The Marble Mountains, from Latin "marble" – "marmor".

3.3.4. Type Depositories

The holotype with 7 paratypes (slide VN.055.06) and 27 paratypes (slides: VN.055.*, where the asterisk can be substituted by any of the following numbers: 05, 07–08) and 7 eggs (slides: VN.055.*: 01–02) are deposited at the Institute of Systematics and Evolution of Animals, Polish Academy of Sciences, Sławkowska 17, 31-016, Kraków, Poland;

Eighteen paratypes (slides: VN.055.*: 09–10), two eggs (slides: VN.055.*: 03–04) and an SEM stub: 18.09 are deposited at the Institute of Zoology and Biomedical Research, Jagiellonian University, Gronostajowa 9, 30-387, Kraków, Poland.

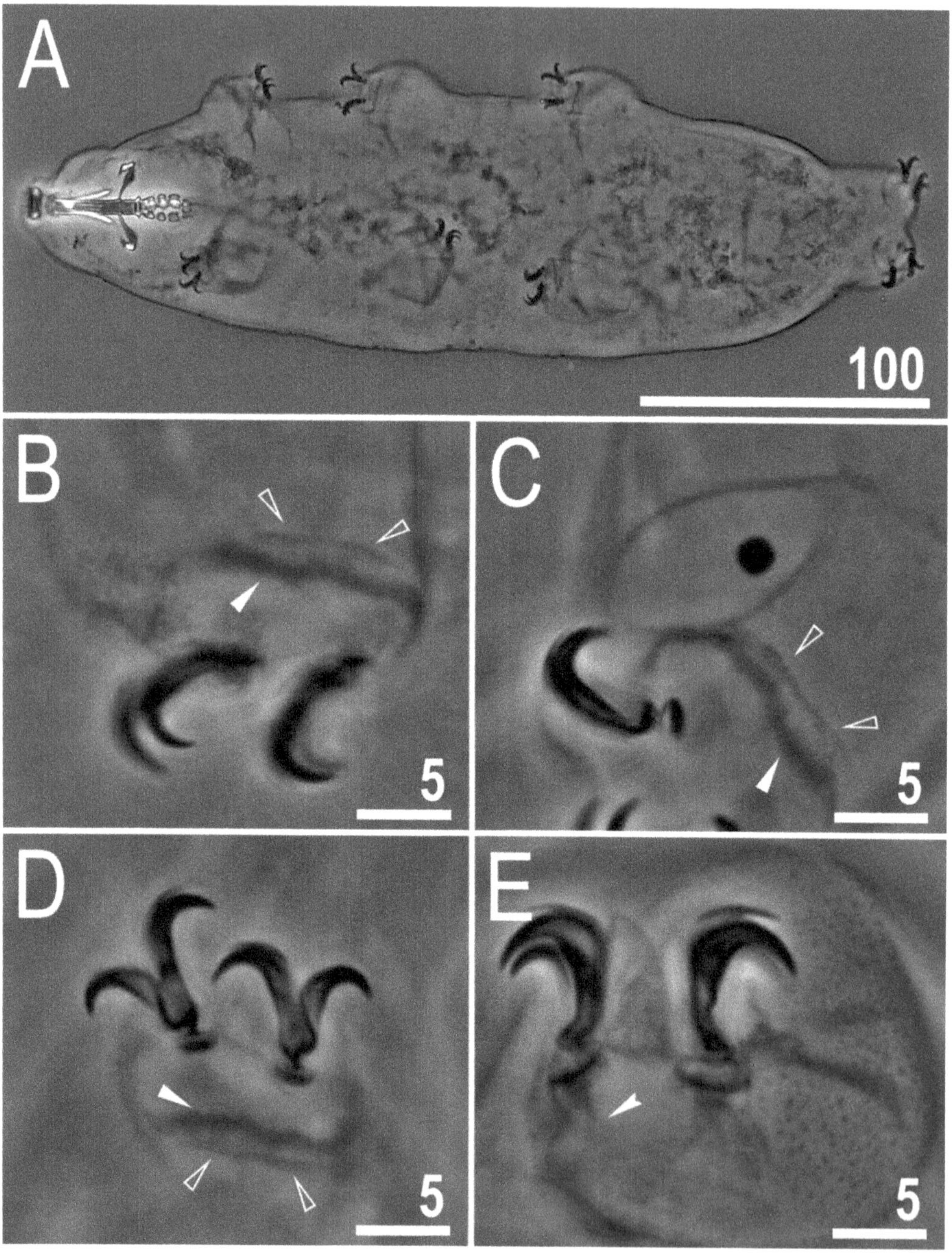

Figure 7. *Mesobiotus marmoreus* **sp. nov.**—PCM image of habitus and leg's cuticle morphology and claws: (**A**)—dorso-ventral projection (holotype); (**B**)—granulation on the external surface of leg III (holotype); (**C**)—a pulvinus-like cuticular bulge on the internal surface of leg II (paratype); (**D**)—claws I with smooth lunulae (holotype); (**E**)—granulation on dorsal and lateral surface and claws on leg IV (paratype). Filled flat arrowheads indicate a single continuous cuticular bar above the claws, empty flat arrowheads indicate paired muscles attachments, filled indented arrowhead indicates horseshoe structure connecting the anterior and the posterior claw. Scale bars in μm.

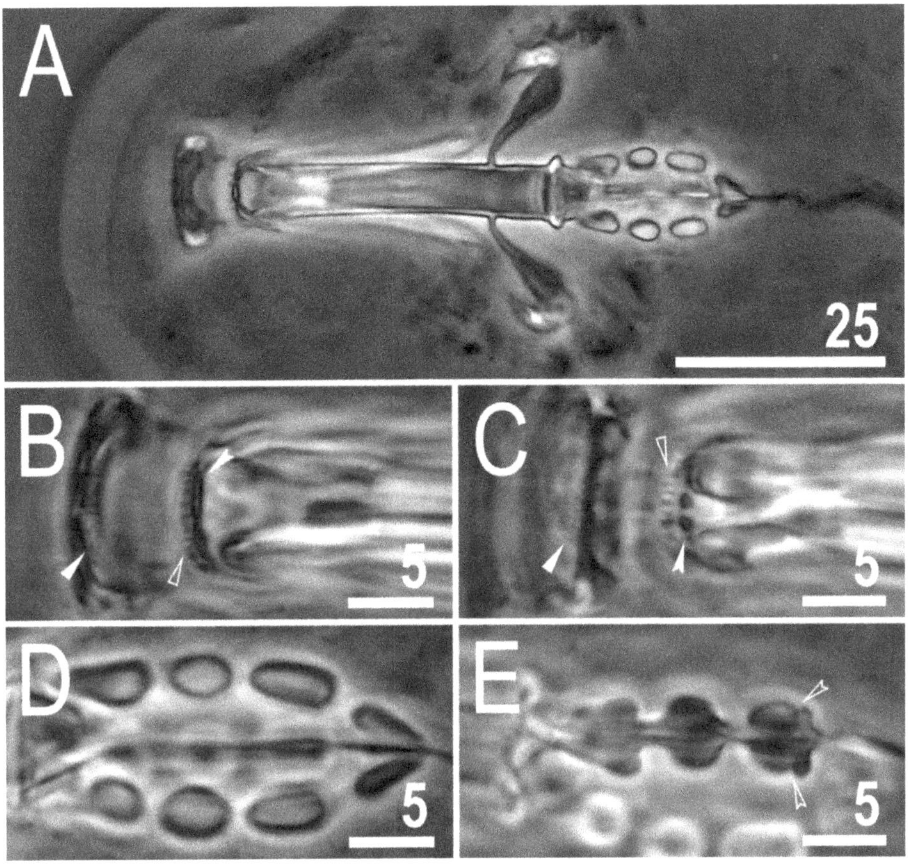

Figure 8. *Mesobiotus marmoreus* **sp. nov.**—PCM images of the buccal apparatus: (**A**)—an entire buccal apparatus (paratype); (**B**,**C**)—the oral cavity armature, dorsal and ventral teeth, respectively (paratype); (**D**,**E**)—placoid morphology, dorsal and ventral placoids, respectively (paratype). Filled flat arrowheads indicate the first band of tenth, empty flat arrowheads indicate the second band of teeth, filled indented arrowheads indicate the third band of teeth, empty indented arrowheads indicate subterminal constrictions in the third macroplacoid. Scale bars in μm.

3.3.5. Animals

The body is almost transparent in small specimens and whitish in adults; after fixation in Hoyer's medium, the body is transparent (Figure 7A). Eyes are absent in alive animals. The body cuticle is smooth, i.e., without pores or sculpturing. Granulation is present on the external surface of all legs I–III (Figure 7B) as well as on the lateral and dorsal surfaces of legs IV (Figure 1E). A cuticular bulge/fold, resembling a pulvinus, is present on the internal surface of legs I–III (Figure 7C). Claws are of the *Mesobiotus* type, with a peduncle connecting the claw to the lunula, a basal septum and well-developed accessory points situated in parallel to the primary branch (Figure 7D,E). Lunulae under all claws are smooth (Figure 7D,E). A single continuous cuticular bar and double muscle attachments are present above claws I–III (Figure 7B–D), whereas a horseshoe-shaped structure connects the anterior and posterior lunulae on claws IV (Figure 7E).

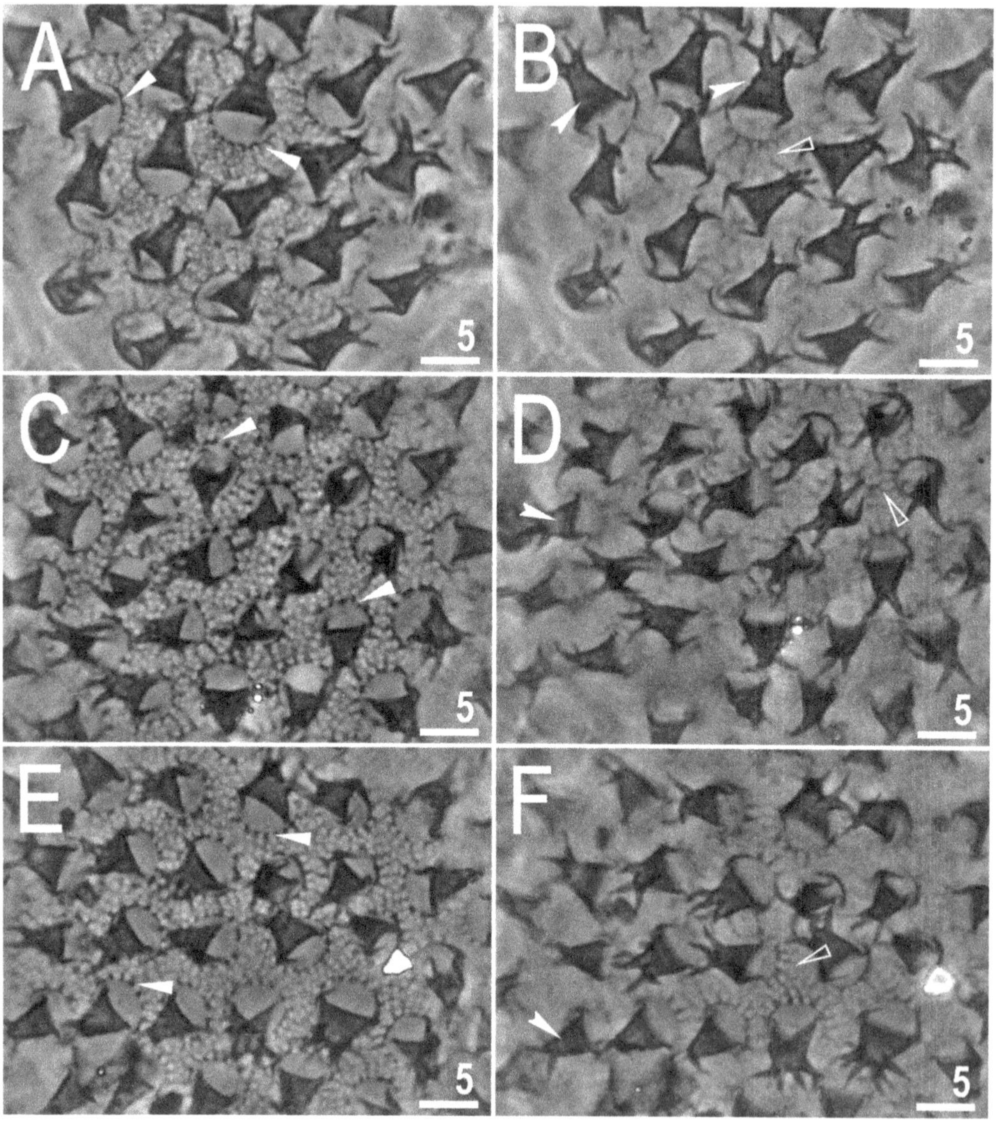

Figure 9. *Mesobiotus marmoreus* **sp. nov.**—PCM images of the egg surface under ×1000 magnification. Each row represents a different egg whereas columns represent different focus levels. Filled flat arrowheads indicate crowns of thickenings around the process bases, empty flat arrowheads indicate extending striae radiating from processes bases, filled indented arrowheads indicate faint thickenings and darkening in processes trunk that in SEM are visible as annulations. Scale bars in μm.

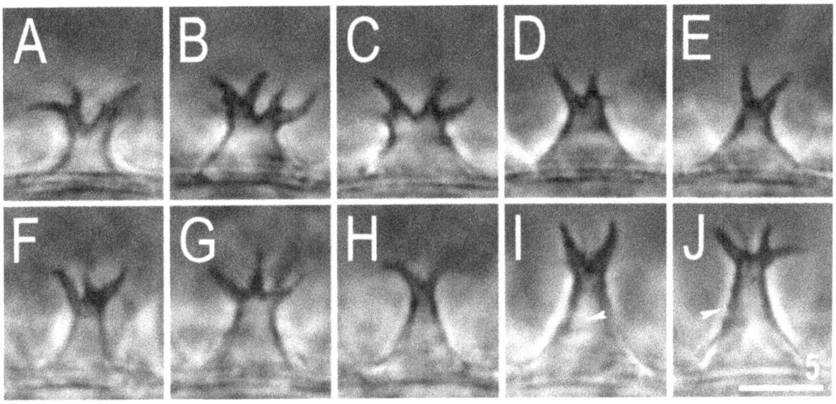

Figure 10. *Mesobiotus marmoreus* **sp. nov.**—PCM images of the egg processes midsections under ×1000 magnification. Filled indented arrowheads indicate faint thickenings and darkening that in SEM are visible as annulations (see Figure 11). Scale bar in μm.

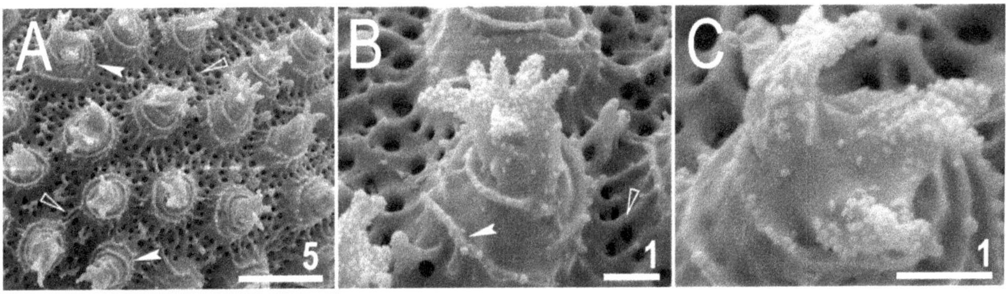

Figure 11. *Mesobiotus marmoreus* **sp. nov.**—SEM images of eggs: (**A**)—details of the egg surface; (**B**)—egg process; (**C**)—top part of the processes divided into several tentacular arms covered with fine granulation. Empty flat arrowheads indicate elevated bars of the reticulum that are visible as extending striae radiating from processes bases in PCM (see Figure 9), filled indented arrowheads indicate annulations present on the process trunks. Scale bars in μm.

The mouth is antero-ventral. The Bucco-pharyngeal apparatus is of the *Macrobiotus* type, with the ventral lamina and ten small peribuccal lamellae (Figure 8A). The oral cavity armature is well developed and composed of three bands of teeth (Figure 8B,C). The first band of teeth is composed of numerous small granules arranged in several discrete rows situated anteriorly in the oral cavity, just behind the bases of the peribuccal lamellae (Figure 8B,C). The second band of teeth is situated between the ring fold and the third band of teeth and is composed of ridges parallel to the main axis of the buccal tube that are larger than those in the first band (Figure 8B,C). The teeth of the third band are located within the posterior portion of the oral cavity, between the second band of teeth and the buccal tube opening (Figure 8B,C). The third band of teeth is discontinuous and divided intodorsal and ventral portions. Under PCM, dorsal teeth are visible as two lateral and one median transverse ridges/crests (Figure 8B) whereas ventral teeth consist of two lateral transverse ridges/crests between which two or three roundish and separated ventro-median teeth are present (Figure 8C). The pharyngeal bulb is ovoid (Figure 8A), with triangular apophyses, three rod-shaped macroplacoids and a large, elongated drop-shaped microplacoid placed close to the third macroplacoid (Figure 8D,E). The macroplacoid length sequence is 2 < 3 < 1. The first macroplacoid is anteriorly narrowed and the third has a clearly defined sub-terminal constriction (Figure 8E). Measurements and statistics are presented in Table 5.

3.3.6. Eggs

The eggs are spherical, whitish and laid freely, with processes in the shape of cones with multiple apices (Figures 9–11). In PCM, the egg surface is covered with a fine but clearly visible reticulum, typically with 2–5 rows of meshes between the neighbouring processes (Figure 9A). In SEM, the egg surface appears between porous and reticulated states, with pores (0.2–0.5 μm in diameter) similar in size to the width of mesh nodes and bars (0.1–0.6 μm; Figure 11A). In PCM, crowns of granular dark thickenings are present around the base of processes (Figure 9A,C,E), which extend into striae radiating from the process bases (Figure 9B,D,F). In SEM, these striae are also visible as elevated bars and nodes of the reticulum (Figure 11A,B). The egg processes exhibit one to three latitudinal annulations that are clearly visible only in SEM (Figure 11), whereas in PCM, they are only sometimes visible as faint, darkened lines in the process trunk (Figure 9B,D,F) or as faintly visible, thickening in the process midsection (Figure 10I,J). Under SEM, the annulations are seen as laminal rings with small granules present on their margins, giving the serrated impression (Figure 11). The process apex divided into multiple (typically 3–6), slender, varying in length, tentacular arms (Figures 9–11), which are covered by fine granulation, visible only in SEM (Figure 11). Measurements and statistics are presented in Table 6.

3.3.7. Reproduction

The examination of all individuals, freshly mounted in Hoyer's medium, under PCM has not revealed any testis or spermathecae filled with spermatozoa. Thus, it is most likely that the new species is parthenogenetic.

3.3.8. DNA Sequences

The obtained sequences for all four molecular markers analysed in this study were of good quality and were represented by single haplotypes.

The **18S rRNA** sequences (GenBank: OL257856-8), 1009 bp long.

The **28S rRNA** sequences (GenBank: OL257868-70), 799 bp long.

The **ITS-2** sequences (GenBank: OL257861-3), 405 bp long.

The **COI** sequences (GenBank: OL311516-8), 658 bp long;.

3.4. *Phylogenetic Position of the New Taxa*

The phylogenetic analysis of taxa belonging to the genus *Mesobiotus* did not indicate *M. harmsworthi* and *M. furciger* groups to be monophyletic (Figure 12). Species representing each of these groups are intermixed in the obtained tree (Figure 12). The analysis indicated *Mesobiotus imperialis* **sp. nov.** is closely related to *Mesobiotus philippinicus* (Figure 12). This is also obvious when inspecting the genetic distances that show a large amount of similarity between DNA sequences of nuclear markers (p-distance; 18S rRNA: 0.0%, 28S rRNA: 2.3%; SM.05). The same occurred in case of the COI dataset, where the lowest genetic distance out of all comparisons with other *Mesobiotus* taxa was 16.5% (p-distance; SM.05). In the tree, the closest relative of *Mesobiotus marmoreus* **sp. nov.** is *Mesobiotus dilimanensis* Itang, Stec, Mapalo Mirano-Bascos & Michalczyk, 2020 [50] (Figure 12). The genetic distances between these two species are also the lowest out of all conducted comparisons (p-distance; 18S rRNA: 0.1%, 28S rRNA: 1.5%, ITS-2: 9.9%, COI: 21.1%; SM.05).

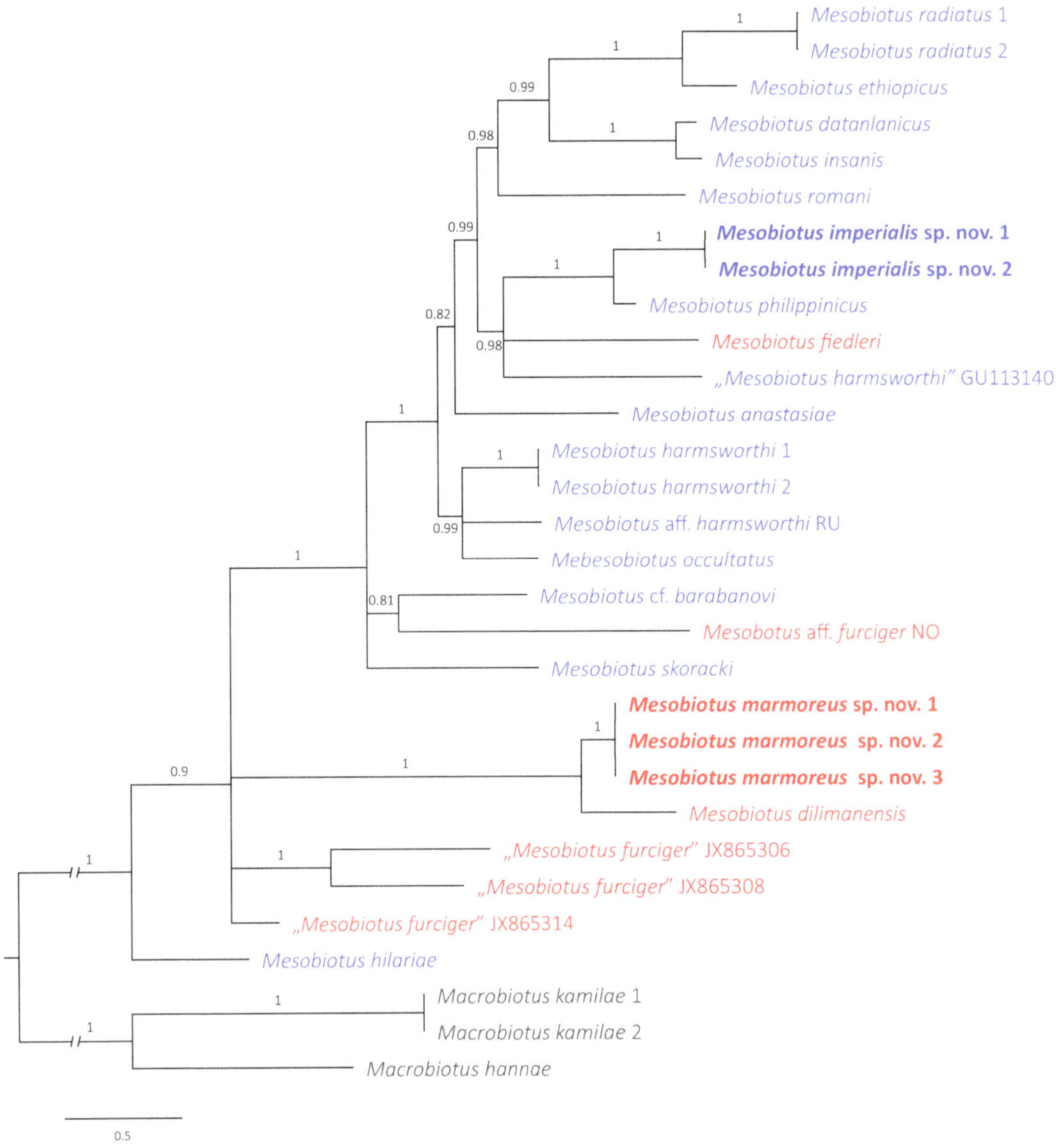

Figure 12. The Bayesian Inference (BI) phylogeny constructed from concatenated sequences (18S rRNA + 28S rRNA + ITS-2 + COI) of the genus *Mesobiotus*. Numbers at nodes indicate Bayesian posterior probability; nodes with values below 0.80 have been collapsed. Taxa newly sequenced in this study are marked with bolded font. Taxa of the *M. harmsworthi* and *M. furciger* complex are indicated by blue and red font, respectively. Outgroup is indicated by grey font. Quotation marks indicate misidentified *Mesobiotus* species or species with uncertain species identification. Scale bar represents substitutions per position.

4. Discussion

4.1. Differential Diagnosis of Mesobiotus imperialis sp. nov.

The new species belong to the informal *Mesobiotus harmsworthi* morphogroup as it exhibits rather large conical processes. After using the dichotomous key by Kaczmarek et al. [7] and Tumanov [8], the new species have been identified as *Mesobiotus philippinicus* known only from its type locality in Philippines [30]. Importantly, it should be also noted that both mentioned keys contain a mistake saying that *M. philippinicus* has the first band of teeth in the oral cavity armature not visible in light microscopy, which is not true [30]. However, despite the phenotypic match, the genetic data and phylogenetic analysis clearly indicate the Vietnamese population to be a distinct species. Closer comparison revealed minute morphological and morphometric differences based on which the new species is differentiated.

Mesobiotus imperialis **sp. nov.** differs from *M. philippinicus* by the presence of granulation on all legs that is visible in light microscopy (only granulation on leg IV faintly visible in some specimens of *M. philippinicus*), evidently more pronounced thickenings surrounding the bases of egg processes in the new species (crown of thickenings surrounding processes bases less pronounced in *M. philippinicus*), unevenly distributed depressions and faint tubercles in the egg processes walls (processes walls are smooth, without mentioned depressions and tubercles in *M. philippinicus*, with this character observable only in SEM), and having conical processes always stretched (egg processes covered with wrinkles forming a rose-like whorl in *M. philippinicus*; *remark*: Based on personal observations, this character is most probably an artefact caused by the culture environment and, importantly, *M. philippinicus* was described based on specimens form laboratory isoline). The morphometric comparisons of both populations revealed that ranges of measured characters greatly overlap. Therefore, statistical testing was involved to check for eventual differences between analysed species. T-test comparisons of morphometric characters revealed statistically significant differences between these two populations in almost all absolute and all relative claws measurements, with claws being larger in *M. philippinicus* ($p_{B\text{-}H} << 0.002$; SM.03). Out of the remaining animals' measurements, *pt* values for stylet support insertion point as well as ventral lamina length were also significantly different and larger in *M. philippinicus* and the new species, respectively ($p_{B\text{-}H} << 0.002$; SM.03). Moreover, there were also significant differences in egg measurements such as egg bare diameter, process height, process base–height ratio and inter-process distances ($p_{B\text{-}H} << 0.007$; SM.03). Nevertheless, as stated above, these latter differences in egg dimensions should be treated with great caution as they might be caused by culturing conditions.

4.2. Differential Diagnosis of *Mesobiotus marmoreus sp. nov.*

The new species belongs to the informal *Mesobiotus furciger* morphogroup as it exhibits rather small conical processes with branched apices. After using the dichotomous key by Kaczmarek et al. [7] and Tumanov [8], the new species could not be identified. By having reticulated egg surface between processes (at least visible as such in light microscopy) the new species is similar to the following taxa: *Mesobiotus creber* (Pilato & Lisi, 2009) [63], *M. dilimanensis*, *Mesobiotus divergens* (Binda, Pilato & Lisi, 2005) [64], *Mesobiotus kovalevi* (Tumanov, 2004) [65] and *Mesobiotus siamensis* (Tumanov, 2006) [66], but it differs specifically from the following:

Mesobiotus creber known only from the Seychelles Islands [63] by the presence of granulation on all legs (the granulation absent in *M. creber*); the medio-ventral tooth of the third band of teeth usually subdivided into three roundish teeth (only up to two roundish teeth present in *M. creber*); a more anteriorly positioned stylet support insertion point (*pt* = 76.2–77.8 in the new species vs. *pt* = 80.0–80.9 in *M. creber*); a more evident subdivisions of process apices that resemble tentacular arms (process apices subdivided into short, nodular terminal branches in *M. creber*); a larger egg bare diameter (63.4–69.6 μm in the new species vs. 52–60 μm in *M. creber*); a larger egg full diameter (77.9–82.2 μm in the new species vs. 59–66 μm in *M. creber*).

Mesobiotus dilimanensis, known only from the Philippines [50], by a different macroplacoid sequence (2 < 3 < 1 in the new species vs. 2 < 1 = 3 in *M. dilimanensis*); a more anteriorly positioned stylet support insertion point (*pt* = 76.2–77.8 in the new species vs. *pt* = 78.0–81.4 in *M. dilimanensis*); longer primary branches of external claws I (6.4–8.6 μm in the new species vs. 8.8–12.1 μm in *M. dilimanensis*); longer primary branches of external and internal claws II (7.0–8.8 and 6.2–8.9 μm, respectively in the new species vs. 10.0–12.9 and 9.2–12.0 μm, respectively in *M. dilimanensis*); longer primary branches of anterior and posterior claws IV (7.5–9.5 and 8.0–10.1 μm, respectively in the new species vs. 9.7–14.8 and 10.7–14.8 μm, respectively in *M. dilimanensis*); the presence of subdivisions in processes apices that resemble slender tentacular arms (process apices subdivided into multiple short, nodular, finger-like apices in *M. dilimanensis*); the presence of one to three latitudinal annulations on the processes trunks that are seen as laminal rings with small granules present on their margins giving the serrated impression (small globular tubercles present on the processes trunks in *M. dilimanensis*); a larger number of processes on the egg circumference (26–30 in the new species vs. 18–24 in *M. dilimanensis*).

Mesobiotus divergens, known only from New Zealand [64], by the presence of granulation on all legs (the granulation absent in *M. divergens*); the morphology of the stylet sheaths (typical in the new species vs. caudally thickened lateral portions of stylet sheaths in *M. divergens*); a relatively longer placoid row (*pt* = 53.7–61.2 in the new species vs. *pt* = 45.4–51.6 in *M. divergens*); a relatively larger microplacoid (*pt* = 9.6–12.7 in the new species vs. *pt* = 7.1–7.4 in *M. divergens*); a larger number of processes on the egg circumference (26–30 in the new species vs. 17 in *M. divergens*); a different point of division of the egg process apex (division closer to the process tip in the new species vs. division at half of the process height in *M. divergens*); the presence of subdivisions in process apices that resemble slender tentacular arms (processes subdivided into two or three stout branches that might be further subdivided into multiple, finger-like, nodular apices in *M. divergens*).

Mesobiotus kovalevi, known only from New Zealand [65], by the absence of eyes; the presence of granulation on all legs (the granulation absent in *M. kovalevi*); the presence of three bands of teeth in the oral cavity (the first and the second band of teeth absent or invisible in light microscopy in *M. kovalevi*); the presence of a medio ventral tooth of the third band of teeth subdivided into two or three roundish teeth (a single roundish medio-ventral tooth present in *M. kovalevi*); a different morphology of egg processes (in light microscopy stout processes with smooth trunks and apices divided into multiple slender, tentacular arms in the new species vs. elongated, conical processes only sometimes subdivided at the top with trunks covered with irregularly distributed minute spines in *M. kovalevi*); a smaller egg bare diameter (63.4–69.6 μm in the new species vs. 86–95 μm in *M. kovalevi*); shorter egg processes (5.6–8.8 μm in the new species vs. 12–17 μm in *M. kovalevi*), a slightly larger number of processes on the egg circumference (26–30 in the new species vs. up to 25 in *M. kovalevi*); a smaller meshes in the reticulum covering the egg surface between processes (mashes diameter 0.2–0.5 μm in the new species vs. nearly 1 μm in *M. kovalevi*).

Mesobiotus siamensis, known only from Thailand [66], by the presence of granulation on all legs (the granulation absent in *M. siamensis*); a more-developed first band of teeth in the oral cavity (always clearly visible in light microscopy in the new species vs. barely visible even in largest specimens of *M. siamensis*); the presence of a medio-ventral tooth of the third band of teeth subdivided into two or three roundish teeth (a medio-ventral tooth only almost broken into several granules in *M. siamensis*); a different morphology of lunulae IV (smooth in the new species vs. with undulated margins in *M. siamensis*); a different morphology of egg processes (in light microscopy stout processes with smooth trunks and apices divided into multiple slender, tentacular arms in the new species vs. bottle-shaped processes with an evidently elongated distal part that is subdivided at the top into short and pointed apices in *M. siamensis*); shorter processes (5.6–8.8 μm in the new species vs. 10.7–11.8 μm in *M. siamensis*); narrower process bases (3.4–6.5 μm in the new species vs. 7.4–10.0 μm in *M. siamensis*); a smaller egg bare diameter (63.4–69.6 μm in the new species vs. 70.3–77.7 μm in *M. siamensis*); a larger number of processes on the egg circumference (26–30 in the new species vs. up to 22 in *M. siamensis*).

4.3. Conclusions

Thanks to the integrative analysis of the two newly found *Mesobiotus* populations and their descriptions, as new to science, the number of Vietnamese tardigrade species was elevated to 36. The two new taxa presented herein have their closest relatives in Philippines as recovered by phylogenetic analysis also reflected in morphological similarities. This finding is not surprising when considering the geographic distance and the fact that both these regions belong to the generally speaking Oriental zoogeographic realm. Therefore, the more recent split of these evolutionary lineages should have been expected.

Supplementary Materials: The following are available online at https://www.mdpi.com/article/10.3390/d13110605/s1, SM.01. Raw morphometric data of *Mesobiotus imperialis* **sp. nov.** SM.02. Raw morphometric data of *Mesobiotus marmoreus* **sp. nov.** SM.03. Results of T-test comparisons. SM.04. Best-fit partitioning scheme and models suggested by PartitionFinder. SM.05. Uncorrected pairwise genetic distances.

Funding: The study was supported by the *Preludium* programme of the National Science Centre, Poland (grant no. 2018/31/N/NZ8/03096 to DS). During this study, DS was supported by the Foundation for Polish Science (FNP).

Institutional Review Board Statement: Not applicable.

Data Availability Statement: The author confirms that the data supporting the findings of this study are available within the article and its Supplementary Materials. The DNA sequences generated in this study are available in GenBank.

Acknowledgments: I am especially grateful to Krzysztof Miler for his magnificent help with samples collection.

Conflicts of Interest: The author declares no conflict of interest.

References

1. Nelson, D.R.; Guidetti, R.; Rebecchi, L. Phylum Tardigrada. In *Ecology and General Biology*, 4th ed.; Thorp, J., Rogers, D.C., Eds.; Thorp and Covich's Freshwater Invertebrates; Academic Press Inc.: Cambridge, MA, USA, 2015; Volume 1, pp. 347–380. [CrossRef]
2. Guidetti, R.G.; Bertolani, R.B. Tardigrade taxonomy: An updated check list of the taxa and a list of characters for their identification. *Zootaxa* **2005**, *845*, 1–46. [CrossRef]
3. Degma, P.; Guidetti, R. Notes to the current checklist of Tardigrada. *Zootaxa* **2007**, *1579*, 41–53. [CrossRef]
4. Degma, P.; Bertolani, R.; Guidetti, R. Actual Checklist of Tardigrada Species. Available online: https://doi.org/10.25431/11380_1178608 (accessed on 26 April 2021).
5. Vecchi, M.; Cesari, M.; Bertolani, R.; Jönsson, K.I.; Rebecchi, L.; Guidetti, R. Integrative systematic studies on tardigrades from Antarctica identify new genera and new species within Macrobiotoidea and Echiniscoidea. *Invertebr. Syst.* **2016**, *30*, 303–322. [CrossRef]
6. Stec, D.; Vecchi, M.; Calhim, S.; Michalczyk, Ł. New multilocus phylogeny reorganises the family Macrobiotidae (Eutardigrada) and unveils complex morphological evolution of the *Macrobiotus hufelandi* group. *Mol. Phylogenetics Evol.* **2020**, *160*, 106987. [CrossRef] [PubMed]
7. Kaczmarek, Ł.; Bartylak, T.; Stec, D.; Kulpa, A.; Kepel, M.; Kepel, A.; Roszkowska, M. Revisiting the genus Mesobiotus Vecchi et al. 2016 (Eutardigrada, Macrobiotidae)—remarks, updated dichotomous key and an integrative description of new species from Madagascar. *Zoöl. Anz. A J. Comp. Zoöl.* **2020**, *287*, 121–146. [CrossRef]
8. Tumanov, D.V. Integrative description of *Mesobiotus anastasiae* sp. nov. (Eutardigrada, Macrobiotoidea) and first record of Lobohalacarus (Chelicerata, Trombidiformes) from the Republic of South Africa. *Eur. J. Taxon.* **2020**, *726*, 102–131. [CrossRef]
9. Kaczmarek, Ł.; Zawierucha, K.; Buda, J.; Stec, D.; Gawlak, M.; Michalczyk, Ł.; Roszkowska, M. An integrative redescription of the nominal taxon for the *Mesobiotus harmsworthi* group (Tardigrada: Macrobiotidae) leads to descriptions of two new Mesobiotus species from Arctic. *PLoS ONE* **2018**, *13*, e0204756. [CrossRef]
10. Węglarska, B. Die Tardigraden Vietnams. *Acta Soc. Zool. Bohemoslov.* **1962**, *26*, 300–307.
11. Iharos, G. Einige Angaben zur Tardigradenfauna Vietnams. *Opusc. Zool. Bp.* **1969**, *9*, 273–277.
12. Pilato, G.; Binda, M.G. Isohypsibius barbarae, a new species of eutardigrade from Vietnam. *Bolletino Delle Sedute Dell'accademia Gioenia Di Sci. Nat. Catania* **2002**, *35*, 637–642.
13. Beasley, C.W.; Kaczmarek, Ł.; Michalczyk, Ł. Redescription of *Doryphoribius vietnamensis* (Iharos, 1969) comb. nov. on the basis of the holotype and additional material from China. *Acta Zool. Acad. Sci. Hung.* **2006**, *52*, 367–372. [CrossRef]
14. Tchesunov, A.V. Marine tardigrade *Halechiniscus jejuensis* Chang et Rho, 2002 (Arthrotardigrada: Halechiniscidae) found in Vietnam. *Invertzool* **2011**, *8*, 79–85. [CrossRef]
15. Stec, D. *Mesobiotus datanlanicus* sp. nov., a new tardigrade species (Macrobiotidae: *Mesobiotus harmsworthi* group) from Lâm Đồng Province in Vietnam. *Zootaxa* **2019**, *4679*, 164–180. [CrossRef] [PubMed]
16. Gąsiorek, P.; Vončina, K.; Nelson, D.R.; Michalczyk, Ł. The importance of being integrative: A remarkable case of synonymy in the genus Viridiscus (Heterotardigrada: Echiniscidae). *Zool. Lett.* **2021**, in press. [CrossRef]
17. Murray, J. XXV.—Arctic Tardigrada, collected by Wm. S. Bruce. *Trans. R. Soc. Edinb.* **1907**, *45*, 669–681. [CrossRef]

18. Stec, D.; Smolak, R.; Kaczmarek, Ł.; Michalczyk, Ł. An integrative description of *Macrobiotus paulinae* sp. nov. (Tardigrada: Eutardigrada: Macrobiotidae: Hufelandi group) from Kenya. *Zootaxa* **2015**, *4052*, 501–526. [CrossRef] [PubMed]
19. Morek, W.; Stec, D.; Gąsiorek, P.; Schill, R.O.; Kaczmarek, Ł.; Michalczyk, Ł. An experimental test of eutardigrade preparation methods for light microscopy. *Zoöl. J. Linn. Soc.* **2016**, *178*, 785–793. [CrossRef]
20. Coughlan, K.; Stec, D. Two new species of the *Macrobiotus hufelandi* complex (Tardigrada: Eutardigrada: Macrobiotidae) from Australia and India, with notes on their phylogenetic position. *Eur. J. Taxon.* **2019**, *573*, 1–38. [CrossRef]
21. Coughlan, K.; Michalczyk, Ł.; Stec, D. *Macrobiotus caelestis* sp. nov., a New Tardigrade Species (Macrobiotidae: Hufelandi Group) from the Tien Shan Mountains (Kyrgyzstan). *Ann. Zoöl.* **2019**, *69*, 499. [CrossRef]
22. Stec, D.; Gąsiorek, P.; Morek, W.; Kosztyła, P.; Zawierucha, K.; Michno, K.; Kaczmarek, Ł.; Prokop, Z.M.; Michalczyk, Ł. Estimating optimal sample size for tardigrade morphometry. *Zoöl. J. Linn. Soc.* **2016**, *178*, 776–784. [CrossRef]
23. Pilato, G.; Binda, M.G. Definition of families, subfamilies, genera and subgenera of the Eutardigrada, and keys to their identification. *Zootaxa* **2010**, *2404*, 1–54. [CrossRef]
24. Michalczyk, Ł.; Kaczmarek, Ł. A description of the new tardigrade *Macrobiotus reinhardti* (Eutardigrada: Macrobiotidae, harmsworthi group) with some remarks on the oral cavity armature within the genus Macrobiotus Schultze. *Zootaxa* **2003**, *331*, 1–24. [CrossRef]
25. Kaczmarek, Ł.; Michalczyk, Ł. The *Macrobiotus hufelandi* group (Tardigrada) revisited. *Zootaxa* **2017**, *4363*, 101–123. [CrossRef]
26. Kaczmarek, Ł.; Cytan, J.; Zawierucha, K.; Diduszko, D.; Michalczyk, Ł. Tardigrades from Peru (South America), with descriptions of three new species of Parachela. *Zootaxa* **2014**, *3790*, 357–379. [CrossRef]
27. Kiosya, Y.; Pogwizd, J.; Matsko, Y.; Vecchi, M.; Stec, D. Phylogenetic position of two Macrobiotus species with a revisional note on Macrobiotus sottilei Pilato, Kiosya, Lisi & Sabella, 2012 (Tardigrada: Eutardigrada: Macrobiotidae). *Zootaxa* **2021**, *4933*, 113–135. [CrossRef]
28. Pilato, G. Analisi di nuovi caratteri nello studio degli Eutardigradi. *Animalia* **1981**, *8*, 51–57.
29. Michalczyk, Ł.; Kaczmarek, Ł. The Tardigrada Register: A comprehensive online data repository for tardigrade taxonomy. *J. Limnol.* **2013**, *72*, e22. [CrossRef]
30. Mapalo, M.A.; Stec, D.; Mirano-Bascos, D.; Michalczyk, Ł. *Mesobiotus philippinicus* sp. nov., the first limnoterrestrial tardigrade from the Philippines. *Zootaxa* **2016**, *4126*, 411–426. [CrossRef]
31. R Core Team. *R: A Language and Environment for Statistical Computing*; R Foundation for Statistical Computing: Vienna, Austria, 2021; Available online: http://www.R-project.org/ (accessed on 26 April 2021).
32. Benjamini, Y.; Hochberg, Y. Controlling the False Discovery Rate: A Practical and Powerful Approach to Multiple Testing. *J. R. Stat. Soc. Ser. B Methodol.* **1995**, *57*, 289–300. [CrossRef]
33. Casquet, J.; Thébaud, C.; Gillespie, R.G. Chelex without boiling, a rapid and easy technique to obtain stable amplifiable DNA from small amounts of ethanol-stored spiders. *Mol. Ecol. Resour.* **2011**, *12*, 136–141. [CrossRef] [PubMed]
34. Stec, D.; Kristensen, R.M.; Michalczyk, Ł. An integrative description of *Minibiotus ioculator* sp. nov. from the Republic of South Africa with notes on Minibiotus pentannulatus Londoño et al., 2017 (Tardigrada: Macrobiotidae). *Zoöl. Anz. A J. Comp. Zoöl.* **2020**, *286*, 117–134. [CrossRef]
35. Hall, T.A. BioEdit: A user-friendly biological sequence alignment editor and analysis program for Windows 95/98/NT. *Nucleic Acids Symp. Ser.* **1999**, *41*, 95–98.
36. Kumar, S.; Stecher, G.; Tamura, K. MEGA7: Molecular Evolutionary Genetics Analysis Version 7.0 for Bigger Datasets. *Mol. Biol. Evol.* **2016**, *33*, 1870–1874. [CrossRef]
37. Stec, D.; Zawierucha, K.; Michalczyk, Ł. An integrative description of Ramazzottius subanomalus (Biserov, 1985 (Tardigrada) from Poland. *Zootaxa* **2017**, *4300*, 403–420. [CrossRef]
38. Gąsiorek, P.; Stec, D.; Zawierucha, K.; Kristensen, R.M.; Michalczyk, Ł. Revision of Testechiniscus Kristensen, 1987 (Heterotardigrada: Echiniscidae) refutes the polar-temperate distribution of the genus. *Zootaxa* **2018**, *4472*, 261–297. [CrossRef]
39. Mironov, S.V.; Dabert, J.; Dabert, M. A new feather mite species of the genus Proctophyllodes Robin, 1877 (Astigmata: Proctophyllodidae) from the Long-tailed Tit Aegithalos caudatus (Passeriformes: Aegithalidae)—morphological description with DNA barcode data. *Zootaxa* **2012**, *3253*, 54–61. [CrossRef]
40. Stec, D.; Morek, W.; Gąsiorek, P.; Michalczyk, Ł. Unmasking hidden species diversity within the Ramazzottius oberhaeuseri complex, with an integrative redescription of the nominal species for the family Ramazzottiidae (Tardigrada: Eutardigrada: Parachela). *Syst. Biodivers.* **2018**, *16*, 357–376. [CrossRef]
41. Astrin, J.J.; Stüben, P.E. Phylogeny in cryptic weevils: Molecules, morphology and new genera of western Palaearctic Cryptorhynchinae (Coleoptera:Curculionidae). *Invertebr. Syst.* **2008**, *22*, 503–522. [CrossRef]
42. Nowak, B. An integrative description of *Macrobiotus hannae* sp. nov. (Tardigrada: Eutardigrada: Macrobiotidae: Hufelandi group) from Poland. *Turk. J. Zoöl.* **2018**, *42*, 269–286. [CrossRef]
43. Katoh, K.; Misawa, K.; Kuma, K.; Miyata, T. MAFFT: A novel method for rapid multiple sequence alignment based on fast Fourier transform. *Nucleic Acids Res.* **2002**, *30*, 3059–3066. [CrossRef]
44. Katoh, K.; Toh, H. Recent developments in the MAFFT multiple sequence alignment program. *Briefings Bioinform.* **2008**, *9*, 286–298. [CrossRef]
45. Vaidya, G.; Lohman, D.J.; Meier, R. SequenceMatrix: Concatenation software for the fast assembly of multi-gene datasets with character set and codon information. *Cladistics* **2011**, *27*, 171–180. [CrossRef]

46. Lanfear, R.; Frandsen, P.B.; Wright, A.M.; Senfeld, T.; Calcott, B. PartitionFinder 2: New Methods for Selecting Partitioned Models of Evolution for Molecular and Morphological Phylogenetic Analyses. *Mol. Biol. Evol.* **2017**, *34*, 772–773. [CrossRef]
47. Ronquist, F.; Huelsenbeck, J.P. MrBayes 3: Bayesian phylogenetic inference under mixed models. *Bioinformatics* **2003**, *19*, 1572–1574. [CrossRef] [PubMed]
48. Rambaut, A.; Drummond, A.J. Tracer v1.6. 2014. Available online: http://beast.bio.ed.ac.uk/Tracer (accessed on 8 September 2021).
49. Stec, D.; Kristensen, R.M. An integrative description of *Mesobiotus ethiopicus* sp. nov. (Tardigrada: Eutardigrada: Parachela: Macrobiotidae: Harmsworthi group) from the northern Afrotropic region. *Turk. J. Zoöl.* **2017**, *41*, 800–811. [CrossRef]
50. Itang, L.A.M.; Stec, D.; Mapalo, M.A.; Mirano-Bascos, D.; Michalczyk, Ł. An integrative description of Mesobiotus dilimanensis, a new tardigrade species from the Philippines (Eutardigrada: Macrobiotidae: Furciger group). *Raffles Bull. Zool.* **2020**, *68*, 19–31.
51. Mapalo, M.; Stec, D.; Mirano-Bascos, D.; Michalczyk, Ł. An integrative description of a limnoterrestrial tardigrade from the Philippines, *Mesobiotus insanis*, new species (Eutardigrada: Macrobiotidae: Harmsworthi group). *Raffles Bull. Zool.* **2017**, *65*, 440–454.
52. Pilato, G.; Binda, M.G.; Catanzaro, R. Remarks on some tardigrades of the African fauna with the description of three new species of Macrobiotus Schultze 1834. *Trop. Zoöl.* **1991**, *4*, 167–178. [CrossRef]
53. Stec, D.; Roszkowska, M.; Kaczmarek, Ł.; Michalczyk, Ł. An integrative description of a population of *Mesobiotus radiatus* (Pilato, Binda & Catanzaro, 1991) from Kenya. *Turk. J. Zoöl.* **2018**, *42*, 523–540. [CrossRef]
54. Roszkowska, M.; Stec, D.; Gawlak, M.; Kaczmarek, Ł. An integrative description of a new tardigrade species *Mesobiotus romani* sp. nov. (Macrobiotidae: Harmsworthi group) from the Ecuadorian Pacific coast. *Zootaxa* **2018**, *4450*, 550–564. [CrossRef]
55. Murray, J. Encystment of Tardigrada. *Trans. R. Soc. Edinb.* **1908**, *45*, 837–854. [CrossRef]
56. Czechowski, P.; Sands, C.J.; Adams, B.J.; D'Haese, C.A.; Gibson, J.A.E.; McInnes, S.J.; Stevens, M.I. Antarctic Tardigrada: A first step in understanding molecular operational taxonomic units (MOTUs) and biogeography of cryptic meiofauna. *Invertebr. Syst.* **2012**, *26*, 526–538. [CrossRef]
57. Kayastha, P.; Roszkowska, M.; Mioduchowska, M.; Gawlak, M.; Kaczmarek, Ł. Integrative Descriptions of Two New Tardigrade Species along with the New Record of *Mesobiotus skorackii* Kaczmarek et al., 2018 from Canada. *Diversity* **2021**, *13*, 394. [CrossRef]
58. Doyère, P.L.N. Memoire sur les Tardigrades. *Ann. Des Sci. Nat.* **1840**, *14*, 269–362.
59. Richters, F. Tardigrada. In *Handbuch der Zoologie*; Kükenthal, W., Krumbach, T., Eds.; Walter de Gruyter & Co.: Berlin/Heidelberg, Germany, 1926; Volume 3, pp. 58–61.
60. Schuster, R.O.; Nelson, D.R.; Grigarick, A.A.; Christenberry, D. Systematic Criteria of the Eutardigrada. *Trans. Am. Microsc. Soc.* **1980**, *99*, 284–303. [CrossRef]
61. Thulin, G. Über die phylogenie und das system der tardigraden. *Hereditas* **2010**, *11*, 207–266. [CrossRef]
62. Marley, N.; McInnes, S.J.; Sands, C. Phylum Tardigrada: A re-evaluation of the Parachela. *Zootaxa* **2011**, *2819*, 51–64. [CrossRef]
63. Pilato, G.; Lisi, O.P.V. Tardigrades of the Seychelles Islands, with the description of three new species. *Zootaxa* **2009**, *2124*, 1–20. [CrossRef]
64. Binda, M.G.; Pilato, G.; Lisi, O.P.V. Remarks on Macrobiotus furciger Murray, 1906 and description of three new species of the furciger group (Eutardigrada, Macrobiotidae). *Zootaxa* **2005**, *1075*, 55–68. [CrossRef]
65. Tumanov, D.V. *Macrobiotus kovalevi*, a new species of Tardigrada from New Zealand (Eutadigrada, Macrobiotidae). *Zootaxa* **2004**, *406*, 1–8. [CrossRef]
66. Tumanov, D.V. *Macrobiotus siamensis* sp. n. (Eutardigrada, Macrobiotidae) from Thailand (Asia). *Zootaxa* **2006**, *1202*, 53–59. [CrossRef]

MDPI
St. Alban-Anlage 66
4052 Basel
Switzerland
www.mdpi.com

Diversity Editorial Office
E-mail: diversity@mdpi.com
www.mdpi.com/journal/diversity

Disclaimer/Publisher's Note: The statements, opinions and data contained in all publications are solely those of the individual author(s) and contributor(s) and not of MDPI and/or the editor(s). MDPI and/or the editor(s) disclaim responsibility for any injury to people or property resulting from any ideas, methods, instructions or products referred to in the content.

www.ingramcontent.com/pod-product-compliance
Lightning Source LLC
Chambersburg PA
CBHW061504220926
51850CB00004B/165
* 9 7 8 3 0 3 6 5 9 9 5 5 7 *